T0135346

Springer Handbooks of Computational Statistics

Series editors
James E. Gentle
Wolfgang K. Härdle
Yuichi Mori

More information about this series at http://www.springer.com/series/7286

Wolfgang Karl Härdle • Henry Horng-Shing Lu •
Xiaotong Shen
Editors

Handbook of Big Data Analytics

Editors
Wolfgang Karl Härdle
Ladislaus von Bortkiewicz Chair of Statistics
C.A.S.E. Center for Applied Statistics & Economics
Humboldt-Universität zu Berlin
Berlin, Germany

Henry Horng-Shing Lu
Institute of Statistics
National Chiao Tung University
Hsinchu, Taiwan

Xiaotong Shen
School of Statistics
University of Minnesota
Minneapolis, USA

ISSN 2197-9790 ISSN 2197-9804 (electronic)
Springer Handbooks of Computational Statistics
ISBN 978-3-030-13238-5 ISBN 978-3-319-18284-1 (eBook)
https://doi.org/10.1007/978-3-319-18284-1

© Springer International Publishing AG, part of Springer Nature 2018
Softcover re-print of the Hardcover 1st edition 2018
This work is subject to copyright. All rights are reserved by the Publisher, whether the whole or part of the material is concerned, specifically the rights of translation, reprinting, reuse of illustrations, recitation, broadcasting, reproduction on microfilms or in any other physical way, and transmission or information storage and retrieval, electronic adaptation, computer software, or by similar or dissimilar methodology now known or hereafter developed.
The use of general descriptive names, registered names, trademarks, service marks, etc. in this publication does not imply, even in the absence of a specific statement, that such names are exempt from the relevant protective laws and regulations and therefore free for general use.
The publisher, the authors and the editors are safe to assume that the advice and information in this book are believed to be true and accurate at the date of publication. Neither the publisher nor the authors or the editors give a warranty, express or implied, with respect to the material contained herein or for any errors or omissions that may have been made. The publisher remains neutral with regard to jurisdictional claims in published maps and institutional affiliations.

Printed on acid-free paper

This Springer imprint is published by the registered company Springer International Publishing AG part of Springer Nature.
The registered company address is: Gewerbestrasse 11, 6330 Cham, Switzerland

Preface

A tremendous growth of high-throughput techniques leads to huge data collections that are accumulating in an exponential speed with high volume, velocity, and variety. This creates the challenges of big data analytics for statistical science. These challenges demand creative innovation of statistical methods and smart computational Quantlets, macros, and programs to capture the often genuinely sparse informational content of huge unstructured data. Particularly, the development of analytic methodologies must take into account the co-design of hardware and software to handle the massive data corpus so that the veracity can be achieved and the inherent value can be revealed.

This volume of the *Handbook of Computational Statistics* series collects twenty-one chapters to provide advice and guidance to the fast developments of big data analytics. It covers a wide spectrum of methodologies and applications that provide a general overview of this new exciting field of computational statistics. The chapters present topics related to mathematics and statistics for high-dimensional problems, nonlinear structures, sufficient dimension reduction, spatiotemporal dependence, functional data analysis, graphic modeling, variational Bayes, compressive sensing, density functional theory, and supervised and semi-supervised learning. The applications include business intelligence, finance, image analysis, compress sensing, climate changes, text mining, neuroscience, and data visualization in very large dimensions. Many of the methods that we present are reproducible in R or MATLAB or Python language. Details of the Quantlets are found at http://www.quantlet.com/.

We would like to acknowledge the dedicated work of all the contributing authors, reviewers, and members in the editorial office of Springer, including Alice Blanck, Frank Holzwarth, Jessica Fäcks, and the related members. Finally, we also thank the great support of our families and friends for this long journey of editing process.

Berlin, Germany	Wolfgang Karl Härdle
Hsinchu, Taiwan	Henry Horng-Shing Lu
Minneapolis, USA	Xiaotong Shen
July 4, 2017	

Contents

Part I Overview

1 **Statistics, Statisticians, and the Internet of Things** 3
John M. Jordan and Dennis K. J. Lin

2 **Cognitive Data Analysis for Big Data** 23
Jing Shyr, Jane Chu, and Mike Woods

Part II Methodology

3 **Statistical Leveraging Methods in Big Data** 51
Xinlian Zhang, Rui Xie, and Ping Ma

4 **Scattered Data and Aggregated Inference** 75
Xiaoming Huo, Cheng Huang, and Xuelei Sherry Ni

5 **Nonparametric Methods for Big Data Analytics** 103
Hao Helen Zhang

6 **Finding Patterns in Time Series** 125
James E. Gentle and Seunghye J. Wilson

7 **Variational Bayes for Hierarchical Mixture Models** 151
Muting Wan, James G. Booth, and Martin T. Wells

8 **Hypothesis Testing for High-Dimensional Data** 203
Wei Biao Wu, Zhipeng Lou, and Yuefeng Han

9 **High-Dimensional Classification** 225
Hui Zou

10 **Analysis of High-Dimensional Regression Models Using Orthogonal Greedy Algorithms** 263
Hsiang-Ling Hsu, Ching-Kang Ing, and Tze Leung Lai

11 Semi-supervised Smoothing for Large Data Problems 285
Mark Vere Culp, Kenneth Joseph Ryan, and George Michailidis

12 Inverse Modeling: A Strategy to Cope with Non-linearity 301
Qian Lin, Yang Li, and Jun S. Liu

13 Sufficient Dimension Reduction for Tensor Data 325
Yiwen Liu, Xin Xing, and Wenxuan Zhong

14 Compressive Sensing and Sparse Coding 339
Kevin Chen and H. T. Kung

15 Bridging Density Functional Theory and Big Data Analytics with Applications 351
Chien-Chang Chen, Hung-Hui Juan, Meng-Yuan Tsai, and Henry Horng-Shing Lu

Part III Software

16 Q3-D3-LSA: D3.js and Generalized Vector Space Models for Statistical Computing 377
Lukas Borke and Wolfgang K. Härdle

17 A Tutorial on `Libra`: R Package for the Linearized Bregman Algorithm in High-Dimensional Statistics 425
Jiechao Xiong, Feng Ruan, and Yuan Yao

Part IV Application

18 Functional Data Analysis for Big Data: A Case Study on California Temperature Trends 457
Pantelis Zenon Hadjipantelis and Hans-Georg Müller

19 Bayesian Spatiotemporal Modeling for Detecting Neuronal Activation via Functional Magnetic Resonance Imaging 485
Martin Bezener, Lynn E. Eberly, John Hughes, Galin Jones, and Donald R. Musgrove

20 Construction of Tight Frames on Graphs and Application to Denoising 503
Franziska Göbel, Gilles Blanchard, and Ulrike von Luxburg

21 Beta-Boosted Ensemble for Big Credit Scoring Data 523
Maciej Zięba and Wolfgang Karl Härdle

Part I
Overview

Chapter 1
Statistics, Statisticians, and the Internet of Things

John M. Jordan and Dennis K. J. Lin

Abstract Within the overall rubric of big data, one emerging subset holds particular promise, peril, and attraction. Machine-generated traffic from sensors, data logs, and the like, transmitted using Internet practices and principles, is being referred to as the "Internet of Things" (IoT). Understanding, handing, and analyzing this type of data will stretch existing tools and techniques, thus providing a proving ground for other disciplines to adopt and adapt new methods and concepts. In particular, new tools will be needed to analyze data in motion rather than data at rest, and there are consequences of having constant or near-constant readings from the ground-truth phenomenon as opposed to numbers at a remove from their origin. Both machine learning and traditional statistical approaches will coevolve rapidly given the economic forces, national security implications, and wide public benefit of this new area of investigation. At the same time, data practitioners will be exposed to the possibility of privacy breaches, accidents causing bodily harm, and other concrete consequences of getting things wrong in theory and/or practice. We contend that the physical instantiation of data practice in the IoT means that statisticians and other practitioners may well be seeing the origins of a post-big data era insofar as the traditional abstractions of numbers from ground truth are attenuated and in some cases erased entirely.

Keywords Machine traffic · Internet of Things · Sensors · Machine learning · Statistical approaches to big data

---

J. M. Jordan
Department of Supply Chain & Information Systems, Smeal College of Business, Pennsylvania State University, University Park, PA, USA
e-mail: jmj13@psu.edu

D. K. J. Lin (✉)
Department of Statistics, Eberly College of Science, Pennsylvania State University, University Park, PA, USA
e-mail: dkl5@psu.edu

© Springer International Publishing AG, part of Springer Nature 2018

W. K. Härdle et al. (eds.), *Handbook of Big Data Analytics*, Springer Handbooks of Computational Statistics, https://doi.org/10.1007/978-3-319-18284-1_1

1.1 Introduction

Even though it lacks a precise definition, the notion of an "Internet of Things" refers generally to networks of sensors, actuators, and machines that communicate over the Internet and related networks. (Some years ago, the number of inanimate objects on the Internet surpassed the number of human beings with connections.) In this chapter, we will first elaborate on the components of the IoT and discuss its data components. The place of statistics in this new world follows, and then we raise some real-world issues such as skills shortages, privacy protection, and so on, before concluding.

1.1.1 The Internet of Things

The notion of an Internet of Things is at once both old and new. From the earliest days of the World Wide Web, devices (often cameras) were connected so people could see the view out a window, traffic or ski conditions, a coffee pot at the University of Cambridge, or a Coke machine at Carnegie Mellon University. The more recent excitement dates to 2010 or thereabouts and builds on a number of developments: many new Internet Protocol (IP) addresses have become available, the prices of sensors are dropping, new data and data-processing models are emerging to handle the scale of billions of device "chirps," and wireless bandwidth is getting more and more available.

1.1.2 What Is Big Data in an Internet of Things?

Why do sensors and connected devices matter for the study of statistics? If one considers the definition of a robot—an electromechanical device that can digitally sense and think, then act upon the physical environment—those same actions characterize large-scale Internet of Things systems: they are essentially meta-robots. The GE Industrial Internet model discussed below includes sensors on all manner of industrial infrastructure, a data analytics platform, and humans to make presumably better decisions based on the massive numbers from the first domain crunched by algorithms and computational resources in the second. Thus, the Internet of Things becomes, in some of its incarnations, an offshoot of statistical process control, six-sigma, and other established industrial methodologies.

Unlike those processes that operated inside industrial facilities, however, the Internet of Things includes sensors attached to or otherwise monitoring individual people in public. Google Glass, a head-mounted smartphone, generated significant controversy before it was pulled from distribution in 2015. This reaction was a noteworthy step in the adoption of Internet of Things systems: both technical details and cultural norms need to be worked out. Motion sensors, cameras, facial recognition, and voice recording and synthesis are very different activities on a factory floor compared to a city sidewalk.

Thus, the Internet of Things is both an extension of existing practices and the initial stage in the analysis of an ever-more instrumented public sphere. The IoT (1) generates substantially large bodies of data (2) in incompatible formats (3) sometimes attached to personal identity. Statisticians now need to think about new physical-world safety issues and privacy implications in addition to generating new kinds of quantitative tools that can scale to billions of data points per hour, across hundreds of competing platforms and conventions. The magnitude of the task cannot be overstated.

1.1.3 Building Blocks[1]

The current sensor landscape can be understood more clearly by contrasting it to the old state of affairs. Most important, sensor networks mimicked analog communications: radios couldn't display still pictures (or broadcast them), turntables couldn't record video, and newspapers could not facilitate two- or multi-way dialog in real time. For centuries, sensors in increasing precision and sophistication were invented to augment human senses: thermometers, telescopes, microscopes, ear trumpets, hearing aids, etc. With the nineteenth-century advances in electro-optics and electromechanical devices, new sensors could be developed to extend the human senses into different parts of the spectrum (including infrared, radio frequencies, measurement of vibration, underwater acoustics, etc.).

Where they were available, electromechanical sensors and later sensor networks

- Stood alone
- Measured one and only one thing
- Cost a lot to develop and implement
- Had inflexible architectures: they did not adapt well to changing circumstances

Sensors traditionally stood alone because networking them together was expensive and difficult. Given the lack of shared technical standards, in order to build a network of offshore data buoys for example, the interconnection techniques and protocols would be uniquely engineered to a particular domain, in this case, saltwater, heavy waves, known portions of the magnetic spectrum, and so on. Another agency seeking to connect sensors of a different sort (such as surveillance cameras) would have to start from scratch, as would a third agency monitoring road traffic.

[1]This section relies heavily on John M. Jordan, *Information, Technology, and Innovation* (Hoboken: John Wiley, 2012), ch. 23.

In part because of their mechanical componentry, sensors rarely measured across multiple yardsticks. Oven thermometers measured only oven temperature, and displayed the information locally, if at all (given that perhaps a majority of sensor traffic informs systems rather than persons, the oven temperature might only drive the thermostat rather than a human-readable display). Electric meters only counted watt-hours in aggregate. In contrast, today a consumer Global Positioning Satellite (GPS) unit or smartphone will tell location, altitude, compass heading, and temperature, along with providing weather radio.

Electromechanical sensors were not usually mass produced, with the exception of common items such as thermometers. Because supply and demand were both limited, particularly for specialized designs, the combination of monopoly supply and small order quantities kept prices high.

1.1.4 Ubiquity

Changes in each of these facets combine to help create today's emerging sensor networks, which are growing in scope and capability every year. The many examples of sensor capability accessible to (or surveilling) the everyday citizen illustrate the limits of the former regime: today there are more sensors recording more data to be accessed by more end points. Furthermore, the traffic increasingly originates and transits exclusively in the digital domain.

- Computers, which sense their own temperature, location, user patterns, number of printer pages generated, etc.
- Thermostats, which are networked within buildings and now remotely controlled and readable.
- Telephones, the wireless variety of which can be understood as beacons, bar-code scanners, pattern matchers (the Shazam application names songs from a brief audio sample), and network nodes.
- Motor and other industrial controllers: many cars no longer have mechanical throttle linkages, so people step on a sensor every day without thinking as they drive by wire. Automated tire-pressure monitoring is also standard on many new cars. Airbags rely on a sophisticated system of accelerometers and high-speed actuators to deploy the proper reaction for collision involving a small child versus a lamp strapped into the front passenger seat.
- Vehicles: the OBD II diagnostics module, the toll pass, satellite devices on heavy trucks, and theft recovery services such as Lojack, not to mention the inevitable mobile phone, make vehicle tracking both powerful and relatively painless.
- Surveillance cameras (of which there are over 10,000 in Chicago alone, and more than 500,000 in London).[2]

[2]Brian Palmer, "Big Apple is Watching You," Slate, May 3, 2010, http://www.slate.com/id/2252729/, accessed 29 March 2018.

- Most hotel door handles and many minibars are instrumented and generate electronic records of people's and vodka bottles' comings and goings.
- Sensors, whether embedded in animals (RFID chips in both household pets and race horses) or gardens (the EasyBloom plant moisture sensor connects to a computer via USB and costs only $50), or affixed to pharmaceutical packaging.

Note the migration from heavily capital-intensive or national-security applications down-market. A company called Vitality has developed a pill-bottle monitoring system: if the cap is not removed when medicine is due, an audible alert is triggered, or a text message could be sent.[3]

A relatively innovative industrial deployment of vibration sensors illustrates the state of the traditional field. In 2006, BP instrumented an oil tanker with "motes," which integrated a processor, solid-state memory, a radio, and an input/output board on a single 2" square chip. Each mote could receive vibration data from up to ten accelerometers, which were mounted on pumps and motors in the ship's engine room. The goal was to determine if vibration data could predict mechanical failure, thus turning estimates—a motor teardown every 2000 h, to take a hypothetical example—into concrete evidence of an impending need for service.

The motes had a decided advantage over traditional sensor deployments in that they operated over wireless spectrum. While this introduced engineering challenges arising from the steel environment as well as the need for batteries and associated issues (such as lithium's being a hazardous material), the motes and their associated sensors were much more flexible and cost-effective to implement compared to hard-wired solutions. The motes also communicate with each other in a mesh topology: each mote looks for nearby motes, which then serve as repeaters en route to the data's ultimate destination. Mesh networks are usually dynamic: if a mote fails, signal is routed to other nearby devices, making the system fault tolerant in a harsh environment. Finally, the motes could perform signal processing on the chip, reducing the volume of data that had to be transmitted to the computer where analysis and predictive modeling was conducted. This blurring of the lines between sensing, processing, and networking elements is occurring in many other domains as well.[4]

All told, there are dozens of billions of items that can connect and combine in new ways. The Internet has become a common ground for many of these devices, enabling multiple sensor feeds—traffic camera, temperature, weather map, social media reports, for example—to combine into more useful, and usable, applications, hence the intuitive appeal of "the Internet of Things." As we saw earlier, network effects and positive feedback loops mean that considerable momentum can develop as more and more instances converge on shared standards. While we will not

[3]Ben Coxworth, "Ordinary pill bottle has clever electronic cap," New Atlas, May 5, 2017, https://newatlas.com/pillsy-smart-pill-bottle/49393/, accessed 29 March 2018.

[4]Tom Kevan, "Shipboard Machine Monitoring for Predictive Maintenance," *Sensors Mag*, February 1, 2006. http://www.sensorsmag.com/sensors-mag/shipboard-machine-monitoring-predictive-maintenance-715?print=1

discuss them in detail here, it can be helpful to think of three categories of sensor interaction:

- Sensor to people: the thermostat at the ski house tells the occupants that the furnace is broken the day before they arrive, or a dashboard light alerting the driver that the tire pressure on their car is low.
- Sensor to sensor: the rain sensor in the automobile windshield alerts the antilock brakes of wet road conditions and the need for different traction-control algorithms.
- Sensor to computer/aggregator: dozens of cell phones on a freeway can serve as beacons for a traffic-notification site, at much lower cost than helicopters or "smart highways."

An "Internet of Things" is an attractive phrase that at once both conveys expansive possibility and glosses over substantial technical challenges. Given 20+ years of experience with the World Wide Web, people have long experience with hyperlinks, reliable inter-network connections, search engines to navigate documents, and Wi-Fi access everywhere from McDonalds to over the mid-Atlantic in flight. None of these essential pieces of scaffolding has an analog in the Internet of Things, however: garage-door openers and moisture sensors aren't able to read; naming, numbering, and navigation conventions do not yet exist; low-power networking standards are still unsettled; and radio-frequency issues remain problematic. In short, as we will see, "the Internet" may not be the best metaphor for the coming stage of device-to-device communications, whatever its potential utility.

Given that "the Internet" as most people experience it is global, searchable, and anchored by content or, increasingly, social connections, the "Internet of Things" will in many ways be precisely the opposite. Having smartphone access to my house's thermostat is a private transaction, highly localized and preferably NOT searchable by anyone else. While sensors will generate volumes of data that are impossible for most humans to comprehend, that data is not content of the sort that Google indexed as the foundation of its advertising-driven business. Thus, while an "Internet of Things" may feel like a transition from a known world to a new one, the actual benefits of networked devices separate from people will probably be more foreign than being able to say "I can connect to my appliances remotely."

1.1.5 Consumer Applications

The notion of networked sensors and actuators can usefully be subdivided into industrial, military/security, or business-to-business versus consumer categories. Let us consider the latter first. Using the smartphone or a web browser, it is already possible to remotely control and/or monitor a number of household items:

- Slow cooker
- Garage-door opener

- Blood-pressure cuff
- Exercise tracker (by mileage, heart rate, elevation gain, etc.)
- Bathroom scale
- Thermostat
- Home security system
- Smoke detector
- Television
- Refrigerator

These devices fall into some readily identifiable categories: personal health and fitness, household security and operations, and entertainment. While the data logging of body weight, blood pressure, and caloric expenditures would seem to be highly relevant to overall physical wellness, few physicians, personal trainers, or health insurance companies have built business processes to manage the collection, security, or analysis of these measurements. Privacy, liability, information overload, and, perhaps most centrally, outcome-predicting algorithms have yet to be developed or codified. If I send a signal to my physician indicating a physical abnormality, she could bear legal liability if her practice does not act on the signal and I subsequently suffer a medical event that could have been predicted or prevented.

People are gradually becoming more aware of the digital “bread crumbs” our devices leave behind. Progressive Insurance’s Snapshot campaign has had good response to a sensor that tracks driving behavior as the basis for rate-setting: drivers who drive frequently, or brake especially hard, or drive a lot at night, or whatever could be judged worse risks and be charged higher rates. Daytime or infrequent drivers, those with a light pedal, or people who religiously buckle seat belts might get better rates. This example, however, illustrates some of the drawbacks of networked sensors: few sensors can account for all potentially causal factors. Snapshot doesn’t know how many people are in the car (a major accident factor for teenage drivers), if the radio is playing, if the driver is texting, or when alcohol might be impairing the driver’s judgment. Geographic factors are delicate: some intersections have high rates of fraudulent claims, but the history of racial redlining is also still a sensitive topic, so data that might be sufficiently predictive (postal codes traversed) might not be used out of fear it could be abused.

The “smart car” applications excepted, most of the personal Internet of Things use cases are to date essentially remote controls or intuitively useful data collection plays. One notable exception lies in pattern-recognition engines that are grouped under the heading of “augmented reality.” Whether on a smartphone/tablet or through special headsets such as Google Glass, a person can see both the physical world and an information overlay. This could be a real-time translation of a road sign in a foreign country, a direction-finding aid, or a tourist application: look through the device at the Eiffel Tower and see how tall it is, when it was built, how long the queue is to go to the top, or any other information that could be attached to the structure, attraction, or venue.

While there is value to the consumer in such innovations, these connected devices will not drive the data volumes, expenditures, or changes in everyday life that will emerge from industrial, military, civic, and business implementations.

1.1.6 The Internets of [Infrastructure] Things

Because so few of us see behind the scenes to understand how public water mains, jet engines, industrial gases, or even nuclear deterrence work, there is less intuitive ground to be captured by the people working on large-scale sensor networking. Yet these are the kinds of situations where networked instrumentation will find its broadest application, so it is important to dig into these domains.

In many cases, sensors are in place to make people (or automated systems) aware of exceptions: is the ranch gate open or closed? Is there a fire, or just an overheated wok? Is the pipeline leaking? Has anyone climbed the fence and entered a secure area? In many cases, a sensor could be in place for years and never note a condition that requires action. As the prices of sensors and their deployment drop, however, more and more of them can be deployed in this manner, if the risks to be detected are high enough. Thus, one of the big questions in security—in Bruce Schneier's insight, not "Does the security measure work?" but "Are the gains in security worth the costs?"—gets difficult to answer: the costs of IP-based sensor networks are dropping rapidly, making cost-benefit-risk calculations a matter of moving targets.

In some ways, the Internet of Things business-to-business vision is a replay of the RFID wave of the mid-aughts. Late in 2003, Walmart mandated that all suppliers would use radio-frequency tags on their incoming pallets (and sometimes cases) beginning with the top 100 suppliers, heavyweight consumer packaged goods companies like Unilever, Procter & Gamble, Gillette, Nabisco, and Johnson & Johnson. The payback to Walmart was obvious: supply chain transparency. Rather than manually counting pallets in a warehouse or on a truck, radio-powered scanners could quickly determine inventory levels without workers having to get line-of-sight reads on every bar code. While the 2008 recession contributed to the scaled-back expectations, so too did two powerful forces: business logic and physics.

To take the latter first, RFID turned out to be substantially easier in labs than in warehouses. RF coverage was rarely strong and uniform, particularly in retrofitted facilities. Electromagnetic noise—in the form of everything from microwave ovens to portable phones to forklift-guidance systems—made reader accuracy an issue. Warehouses involve lots of metal surfaces, some large and flat (bay doors and ramps), others heavy and in motion (forklifts and carts): all of these reflect radio signals, often problematically. Finally, the actual product being tagged changes radio performance: aluminum cans of soda, plastic bottles of water, and cases of tissue paper each introduce different performance effects. Given the speed of assembly lines and warehouse operations, any slowdowns or errors introduced by a new tracking system could be a showstopper.

The business logic issue played out away from the shop floor. Retail and consumer packaged goods profit margins can be very thin, and the cost of the RFID tagging systems for manufacturers that had negotiated challenging pricing schedules with Walmart was protested far and wide. The business case for total supply chain transparency was stronger for the end seller than for the suppliers, manufacturers, and truckers required to implement it for Walmart's benefit. Given that the systems delivered little value to the companies implementing them, and given that the technology didn't work as advertised, the quiet recalibration of the project was inevitable.

RFID is still around. It is a great solution to fraud detection, and everything from sports memorabilia to dogs to ski lift tickets can be easily tested for authenticity. These are high-value items, some of them scanned no more than once or twice in a lifetime rather than thousands of times per hour, as on an assembly line. Database performance, industry-wide naming and sharing protocols, and multiparty security practices are much less of an issue.

While it's useful to recall the wave of hype for RFID circa 2005, the Internet of Things will be many things. The sensors, to take only one example, will be incredibly varied, as a rapidly growing online repository makes clear (see http://devices.wolfram.com/).[5] Laboratory instruments are shifting to shared networking protocols rather than proprietary ones. This means it's quicker to set up or reconfigure an experimental process, not that the lab tech can see the viscometer or Geiger counter from her smart phone or that the lab will "put the device on the Internet" like a webcam.

Every one of the billions of smartphones on the planet is regularly charged by its human operator, carries a powerful suite of sensors—accelerometer, temperature sensor, still and video cameras/bar-code readers, microphone, GPS receiver—and operates on multiple radio frequencies: Bluetooth, several cellular, and Wi-Fi. There are ample possibilities for crowdsourcing news coverage, fugitive hunting, global climate research (already, amateur birders help show differences in species' habitat choices), and more using this one platform.

Going forward, we will see more instrumentation of infrastructure, whether bridges, the power grid, water mains, dams, railroad tracks, or even sidewalks. While states and other authorities will gain visibility into security threats, potential outages, maintenance requirements, or usage patterns, it's already becoming clear that there will be multiple paths by which to come to the same insight. The state of Oregon was trying to enhance the experience of bicyclists, particularly commuters. While traffic counters for cars are well established, bicycle data is harder to gather. Rather than instrumenting bike paths and roadways, or paying a third party to do so, Oregon bought aggregated user data from Strava, a fitness-tracking smartphone app. While not every rider, particularly commuters, tracks his mileage, enough do that the bike-lane planners could see cyclist speeds and traffic volumes by time of day, identify choke points, and map previously untracked behaviors.

[5] Wolfram Connected Devices Project, (http://devices.wolfram.com/), accessed 29 March 2018.

Strava was careful to anonymize user data, and in this instance, cyclists were the beneficiaries. Furthermore, cyclists compete on Strava and have joined with the expectation that their accomplishments can show up on leader boards. In many other scenarios, however, the Internet of Things' ability to "map previously untracked behaviors" will be problematic, for reasons we will discuss later. To provide merely one example, when homes are equipped with so-called smart electrical meters, it turns out that individual appliances and devices have unique "fingerprints" such that outside analysis can reveal when the toaster, washing machine, or hair dryer was turned on and off.[6] Multiply this capability across toll passes, smartphones, facial recognition, and other tools, and the privacy threat becomes significant.

1.1.7 Industrial Scenarios

GE announced its Industrial Internet initiative in 2013. The goal is to instrument more and more of the company's capital goods—jet engines are old news, but also locomotives, turbines, undersea drilling rigs, MRI machines, and other products—with the goal of improving power consumption and reliability for existing units and to improve the design of future products. Given how big the company's footprint is in these industrial markets, 1% improvements turn out to yield multibillion-dollar opportunities. Of course, instrumenting the devices, while not trivial, is only the beginning: operational data must be analyzed, often using completely new statistical techniques, and then people must make decisions and put them into effect.

The other striking advantage of the GE approach is financial focus: 1% savings in a variety of industrial process areas yields legitimately huge cost savings opportunities. This approach has the simultaneous merits of being tangible, bounded, and motivational. Just 1% savings in aviation fuel over 15 years would generate more than $30 billion, for example. To realize this promise, however, GE needs to invent new ways of networking, storage, and data analysis. As Bill Ruh, the company's vice president of global software services, stated, "Our current jet aircraft engines produce one terabyte of data per flight. ... On average an airline is doing anywhere from five to ten flights a day, so that's 5–10 terabytes per plane, so when you're talking about 20,000 planes in the air you're talking about an enormous amount of data per day."[7] Using different yardsticks, Ruh framed the scale in terms of variables: 50 million of them, from 10 million sensors.

To get there, the GE vision is notably realistic about the many connected investments that must precede the harvesting of these benefits.

[6] Ariel Bleicher, "Privacy on the Smart Grid," IEEE Spectrum, 5 October 2010, http://spectrum.ieee.org/energy/the-smarter-grid/privacy-on-the-smart-grid, accessed 29 March 2018.

[7] Danny Palmer, "The future is here today: How GE is using the Internet of Things, big data and robotics to power its business," Computing 12 March 2015, http://www.computing.co.uk/ctg/feature/2399216/the-future-is-here-today-how-ge-is-using-the-internet-of-things-big-data-and-robotics-to-power-its-business, accessed 29 March 2018.

1. The technology doesn't exist yet. Sensors, instrumentation, and user interfaces need to be made more physically robust, usable by a global workforce, and standardized to the appropriate degree.
2. Information security has to protect assets that don't yet exist, containing value that has yet to be measured, from threats that have yet to materialize.
3. Data literacy and related capabilities need to be cultivated in a global workforce that already has many skills shortfalls, language and cultural barriers, and competing educational agendas. Traditional engineering disciplines, computer science, and statistics will merge into new configurations.[8]

1.2 What Kinds of Statistics Are Needed for Big IoT Data?

The statistical community is beginning to engage with machine learning and computer science professionals on the issue of so-called big data. Challenges abound: data validation at petabyte scale; messy, emergent, and dynamic underlying phenomena that resist conventional hypothesis testing; and the need for programming expertise for computational heavy lifting. Most importantly, techniques are needed to deal with flowing data as opposed to static data sets insofar as the phenomena instrumented in the IoT can be life-critical: ICU monitoring, the power grid, fire alarms, and so on. There is no time for waiting for summarized, normalized data because the consequences of normal lags between reading and analysis can be tragic.

1.2.1 Coping with Complexity

In the Internet of Things, we encounter what might be called "big^2 data": all the challenges of single-domain big data remain, but become more difficult given the addition of cross-boundary complexity. For example, astronomers or biostatisticians must master massive data volumes of relatively homogeneous data. In the Internet of Things, it is as if a geneticist also had to understand data on particle physics or failure modes of carbon fiber.

Consider the example of a military vehicle instrumented to determine transmission failure to facilitate predictive maintenance. The sensors cannot give away any operational information that could be used by an adversary, so radio silencing and data encryption are essential, complicating the data acquisition process. Then comes the integration of vast quantities of multiple types of data: weather (including tem-

[8]Peter C. Evans and Marco Annunziata, *Industrial Internet: Pushing the Boundaries of Minds and Machines*, 26 November 2012, p. 4, http://www.ge.com/docs/chapters/Industrial_Internet.pdf, p. 4., accessed 29 March 2018.

perature, humidity, sand/dust, mud, and so on); social network information (think of a classified Twitter feed on conditions and operational updates from the bottom of the organization up); vibration and other mechanical measurements; dashboard indicators such as speedometer, gearshift, engine temperature, and tachometer; text-heavy maintenance logs, possibly including handwriting recognition; and surveillance data (such as satellite imagery).

Moving across domains introduces multiple scales, some quantitative (temperature) and others not (maintenance records using terms such as "rough," "bumpy," and "intermittent" that could be synonymous or distinct). How is an X change in a driveshaft harmonic resonance to correlate with sandy conditions across 10,000 different vehicles driven by 50,000 different drivers? What constitutes a control or null variable? The nature of noise in such a complex body of data requires new methods of extraction, compression, smoothing, and error correction.

1.2.2 Privacy

Because the Internet of Things can follow real people in physical space (whether through drones, cameras, cell phone GPS, or other means), privacy and physical safety become more than theoretical concerns. Hacking into one's bank account is serious but rarely physically dangerous; having stop lights or engine throttles compromised is another matter entirely, as the world saw in the summer of 2015 when an unmodified Jeep was remotely controlled and run off the road.[9] Given the large number of related, cross-domain variables, what are the unintended consequences of optimization?

De-anonymization has been shown to grow easier with large, sparse data sets.[10] Given the increase in the scale and diversity of readings or measurements attached to an individual, it is theoretically logical that the more sparse data points attach to an individual, the simpler the task of personal identification becomes (something as basic as taxi fare data, which intuitively feels anonymous, can create a privacy breach at scale: http://research.neustar.biz/2014/09/15/riding-with-the-stars-passenger-privacy-in-the-nyc-taxicab-dataset/).[11] In addition, the nature of IoT measurements might not feel as personally risky at the time of data creation: logging into a financial institution heightens one's sense of awareness, whereas walking down the street, being logged by cameras and GPS, might feel more carefree than it

[9] Andy Greenberg, "Hackers Remotely Kill a Jeep on the Highway - With Me in It," Wired, 21 July 2015, http://www.wired.com/2015/07/hackers-remotely-kill-jeep-highway/, accessed 29 March 2018.

[10] Arvind Narayanan and Vitaly Shmatikov, "Robust De-anonymization of Large Sparse Datasets," no date, https://www.cs.utexas.edu/~shmat/shmat_oak08netflix.pdf, accessed 29 March 2018.

[11] Anthony Tockar, "Riding with the Stars: Passenger Privacy in the NYC Taxicab Dataset," neustar Research, 15 September 2014, accessed 29 March 2018 [Note: "neustar" is lower-case in the corporate branding].

perhaps should. The cheapness of computer data storage (as measured by something called Kreider's law) combines with the ubiquity of daily digital life to create massive data stores recording people's preferences, medications, travels, and social contacts. The statistician who analyzes and combines such data involving flesh-and-blood people in real space bears a degree of responsibility for their privacy and security. (Researchers at Carnegie Melon University successfully connected facial recognition software to algorithms predicting the subjects' social security numbers.[12]) Might the profession need a new code of ethics akin to the Hippocratic oath? Can the statistician be value-neutral? Is there danger of data "malpractice"?

1.2.3 Traditional Statistics Versus the IoT

Traditional statistical thinking holds that large samples are better than small ones, while some machine learning advocates assert that very large samples render hypotheses unnecessary.[13] At this intersection, the so-called the death of p-value is claimed.[14] However, fundamental statistical thinking with regard to significance, for example, still applies (although the theories may not be straightforwardly applied in very large data sets). Big data on its own cannot replace scientific/statistical thinking. Thus, a wishlist for needed statistical methodologies should have the following properties:

- High-impact problems

  Refining existing methodologies is fine, but more efforts should focus on working high-impact problems, especially those problems from other disciplines. Statisticians seem to keep missing opportunities: examples range from genetics to data mining. We believe that statisticians should seek out high-impact problems, instead of waiting for other disciplines to formulate the problems into statistical frames. Collaboration across many disciplines will be necessary, if unfamiliar, behavior. This leads to the next item.
- Provide structure for poorly defined problems

  A skilled statistician is typically most comfortable and capable when dealing with well-defined problems. Instead, statisticians should develop some methodologies for poorly defined problems and help devise a strategy of attack. There are many opportunities for statistical applications, but most of them are not in the "standard" statistics frame—it will take some intelligent persons to

[12]Deborah Braconnier, "Facial recognition software could reveal your social security number," Phys.org, 2 August 2011, https://phys.org/news/2011-08-facial-recognition-software-reveal-social.html, accessed 29 March 2018.

[13]Chris Anderson, "The End of Theory: The Data Deluge Makes the Scientific Method Obsolete," Wired, 23 June 2008, https://www.wired.com/2008/06/pb-theory/, accessed 29 March 2018.

[14]Tom Siegfried, "P value ban: small step for a journal, giant leap for science," ScienceNews, 17 March 2015, https://www.sciencenews.org/blog/context/p-value-ban-small-step-journal-giant-leap-science, accessed 29 March 2018.

formulate these problems into statistics-friendly problems (then to be solved by statisticians). Statisticians can devote more efforts to be such intelligent persons.

- Develop new theories

  Most fundamental statistical theories based upon iid (independently identically distributed) for one fixed population (such ascentral limit theorem, or law of large number) may need to be modified to be appropriately applied to big data world. Many (non-statisticians) believe that big data leads to "the death of p-value." The logic behind this is that when the sample size n becomes really large, all p-values will be significant—regardless how little the practical significance is. This is indeed a good example of misunderstanding the fundamentals. One good example is about "small n and large p" where the sparsity property is assumed. First, when there are many exploratory variables, some will be classified as active variables (whether or not this is true!). Even worse, after the model is built (mainly based on the sparsity property), the residuals may highly correlate with some remaining variables—this contradicts the assumption for all fundamental theorems that "error is independent with all exploratory variables." New measurement is needed for independence in this case.

1.2.4 A View of the Future of Statistics in an IoT World

Having those wishlist items in mind, what kinds of statistics are needed for big data? For an initial approximation, here are some very initial thoughts under consideration.

- Statistics and plots for (many) descriptive statistics. If conventional statistics are to be used for big data, and it is very likely there will be too many of them because of the heterogeneity of the data, what is the best way to extract important information from these statistics? For example, how to summarize thousands of correlations? How about thousands of p-values? ANOVAs? Regression models? Histograms? etc. Advanced methods to obtain "sufficient statistics" (whatever it means in a particular context: astrophysics and biochemistry will have different needs, for example) from those many conventional statistics are needed.
- Coping with heterogeneity. Numbers related to such sensor outputs as check-engine lights, motion detectors, and flow meters can be extremely large, of unknown quality, and difficult to align with more conventional measurement systems.
- Low-dimension behavior. Whatever method is feasible for big data (the main concern being the computational costs), the reduction in resolution as it is converted to low-dimension resolution (especially 2D graphs) is always important to keep in mind.
- As we have mentioned, analyzing real-time measurements that are derived from actual ground truth demands stream-based techniques that exceed standard practice in most statistical disciplines.

- Norm or Extreme. Depending on the problem, we could be interested in either norm or extreme, or both. Basic methods for both feature extraction (mainly for extremes) and pattern recognition (mainly for norm) are needed.
- Methods for new types/structures of data. A simple example would be "How to build up a regression model, when both inputs and outputs are network variables?" Most existing statistical methodologies are limited to numbers (univariate or multivariate), but there is some recent work for functional data or text data. How to extract the basic information (descriptive statistics) or even analysis (inferential statistics) of these new types of data are highly demanding. This includes network data, symbolic data, fingerprints data, 2D or 3D image data, just to name a few. There is more that can be done, if we are willing to open our minds.
- Prediction vs estimation. One difference between computer science and statistics methods has to do with the general goal—while CS people focus more on prediction, statisticians focus more on estimation (or statistical inference). Take Artificial Neural Networks (ANN) as an example: the method can fit almost anything, but what does it mean? ANN is thus popularly used in data mining, but has received relatively low attention from statisticians. For big data, it is clear that prediction is probably more feasible in most cases. **Note:** in some very fundamental cases, we believe that statistical inference remains important, always bearing in mind the essential research question at hand.

1.3 Big Data in the Real World

Moving statistical and analytical techniques from academic and laboratory settings into the physical world sensed and measured by the IoT introduces new challenges. Not surprisingly, organizational and technical matters are emerging, and even the limits of human cognition must be appreciated and accounted for.

1.3.1 Skills

Here's a quiz: ask someone in the IT shop how many of his of her colleagues are qualified to work in Hive, Pig, Cassandra, MongoDb, or Hadoop. These are some of the tools that are emerging from the front-runners in big data, web-scale companies including Google (that needs to index the entire Internet), Facebook (manage a billion users), Amazon (construct and run the world's biggest online merchant), or Yahoo (figure out what social media is conveying at the macro scale). Outside this small industry, big data skills are rare; then consider how few people understand both data skills and the intricacies of industrial and other behind-the-scenes processes, many of them life critical (e.g., the power grid or hospital ICU sensor networks).

1.3.2 Politics

Control over information is frequently thought to bring power within an organization. Big data, however, is heterogeneous, is multifaceted, and can bring performance metrics where they had not previously operated. If a large retailer, hypothetically speaking, traced its customers' purchase behavior first to social media expressions and then to advertising channel, how will the various budget-holders respond? Uncertainty as to ad spend efficacy is as old as advertising, but tracing ad channels to purchase activity might bring light where perhaps it is not wanted. Information sharing across organizational boundaries ("how are you going to use this data?") can also be unpopular. Once it becomes widely understood how one's data "bread crumbs" can be manipulated, will consumers/citizens demand stricter regulation?

1.3.3 Technique

Given that relational databases have been around for about 35 years, a substantial body of theory and practice makes these environments predictable. Big data, by contrast, is just being invented, but already there are some important differences between the two: Most enterprise data is generated by or about humans and organizations: SKUs are bought by people, bills are paid by people, health care is provided to people, and so on. At some level, many human activities can be understood at human scale. Big data, particularly social media, can come from people too, but in more and more cases, it comes from machines: server logs, point of sale scanner data, security sensors, and GPS traces. Given that these new types of IoT data don't readily fit into relational structures and can get massively large in terms of storage, it's nontrivial to figure out what questions to ask of these data types.

When data is loaded into relational systems, it must fit predefined categories that ensure that what gets put into a system makes sense when it is pulled out. This process implies that the system is defined at the outset for what the designers expect to be queried: the questions are known, more or less, before the data is entered in a highly structured manner. In big data practice, meanwhile, data is stored in as complete a form as possible, close to its original state. As little as possible is thrown out so queries can evolve and not be constrained by the preconceptions of the system. Thus, these systems can look highly random to traditional database experts. It's important to stress that big data will not replace relational databases in most scenarios; it's a matter of now having more tools to choose from for a given task.

1.3.4 Traditional Databases

Traditional databases are designed for a concrete scenario, then populated with examples (customers, products, facilities, or whatever), usually one per row: the questions and answers one can ask are to some degree predetermined. Big data can be harvested in its original form and format, and then analyzed as the questions emerge. This open-ended flexibility can of course be both a blessing and a curse.

Traditional databases measured the world in numbers and letters that had to be predicted: zip codes were 5 or 10 digits, SKU formats were company specific, or mortgage payments were of predictable amounts. Big data can accommodate Facebook "likes," instances of the "check engine" light illuminating, cellphone location mapping, and many other types of information.

Traditional databases are limited by the computing horsepower available: to ask harder questions often means buying more hardware. Big data tools can scale up much more gracefully and cost-effectively, so decision-makers must become accustomed to asking questions they could not contemplate previously. To judge advertising effectiveness, one cable operator analyzed every channel-surfing click of every remote across every household in its territory, for example: not long ago, such an investigation would have been completely impractical.

1.3.5 Cognition

What does it mean to think at large scales? How do we learn to ask questions of the transmission of every car on the road in a metropolitan area, of the smartphone of every customer of a large retail chain, or of every overnight parcel in a massive distribution center? How can more and more people learn to think probabilistically rather than anecdotally?

The mantra that "correlation doesn't imply causation" is widely chanted yet frequently ignored; it takes logical reasoning beyond statistical relationships to test what's really going on. Unless the data team can grasp the basic relationships of how a given business works, the potential for complex numerical processing to generate false conclusions is ever present. Numbers do not speak for themselves; it takes a human to tell stories, but as Daniel Kahneman and others have shown, our stories often embed mental traps. Spreadsheets remain ubiquitous in the modern enterprise, but numbers at the scale of Google, Facebook, or Amazon must be conveyed in other ways. Sonification—turning numbers into a range of audible tones—and visualization show a lot of promise as alternative pathways to the brain, bypassing mere and non-intuitive numerals. In the meantime, the pioneers are both seeing the trail ahead and taking some arrows in the back for their troubles. But the faster people, and especially statisticians, begin to break the stereotype that "big data is what we've always done, just with more records or fields," the faster the breakthrough questions, insights, and solutions will redefine practice.

1.4 Conclusion

There's an important point to be made up front: whether it originates in a financial system, public health record-keeping, or sensors on electrical generators, *big data is not necessarily complete, or accurate, or true.* Asking the right questions is in some cases learned through experience, or made possible by better theory, or a matter of luck. But in many instances, by the time investigators figure out what they should be measuring in complex systems, it's too late to instrument the "before" state to compare to the "after." Signal and noise can be problematic categories as well: one person's noise can be a goldmine for someone else. Context is everything. Value is in the eye of the beholder, not the person crunching the numbers. However, this is rarely the case. Big data is big, often because it is automatically collected. Thus, in many cases, it may not contain much information relative to noise. This is sometimes called a DRIP—Data Rich, Information Poor—environment. The IoT is particularly prone to these issues, given both (a) notable failure and error rates of the sensors (vs the machines they sense) and (b) the rarity of certain kinds of failures: frequencies of 1 in 10,000,000 leave many readings of normal status as their own type of noise. In any event, the point here is that bigger does not necessarily mean better when it comes to data.

Accordingly, big data skills cannot be purely a matter of computer science, statistics, or other processes. Instead, the backstory behind the creation of any given data point, category, or artifact can be critically important and more complex given the nature of the environments being sensed. While the same algorithm or statistical transformation might be indicated in a bioscience, a water main, and a financial scenario, knowing the math is rarely sufficient. Having the industry background to know where variance is "normal," for instance, comes only from a holistic understanding of the process under the microscope. As we move into unprecedented data volumes (outside the Large Hadron Collider perhaps), understanding the ground truth of the data being collected and the methods of its collection, automated and remote though they may be, will pose a significant challenge.

Beyond the level of the device, data processing is being faced with new challenges—in both scope and kind—as agencies, companies, and NGOs (to name but three interested parties) try to figure out how to handle billions of cellphone chirps, remote-control clicks, or GPS traces. What information can and should be collected? By what entity? With what safeguards? For how long? At what level of aggregation, anonymization, and detail? With devices and people opting in or opting out? Who is allowed to see what data at what stage in the analysis life cycle? For a time, both Google (in its corporate lobby) and Dogpile (on the web) displayed real-time searches, which were entertaining, revealing, and on the whole discouraging: porn constituted a huge percentage of the volume. Will ski-lift webcams go the same way in the name of privacy?

Once information is collected, the statistical and computer science disciplines are challenged to find patterns that are not coincidence, predictions that can be validated, and insights available in no other way. Numbers rarely speak for themselves, and the context for Internet of Things data is often difficult to obtain or manage given the wide variety of data types in play. The more inclusive the model, however, the more noise is introduced and must be managed. And the scale of this information is nearly impossible to fathom: according to IBM Chief Scientist Jeff Jonas, mobile devices in the United States alone generated 600 billion geo-tagged transactions every day—as of 2010.[15] Finally, the discipline of statistics is being forced to analyze these vast bodies of data in near real time—and sometimes within seconds—given how many sensors have implications for human safety and well-being.

In addition to the basic design criteria, the privacy issues cannot be ignored. Here, the history of Google Glass might be instructive: whatever the benefits that accrue to the user, the rights of those being scanned, identified, recorded, or searched matter in ways that Google has yet to acknowledge. Magnify Glass to the city or nation-state level (recall that England has an estimated 6 million video cameras, but nobody knows exactly how many[16]), as the NSA revelations appear to do, and it's clear that technological capability has far outrun the formal and informal rules that govern social life in civil society.

In sum, data from the Internet of Things will challenge both the technical capabilities and the cultural codes of practice of the data community: unlike other categories of big data, people's faces, physical movements, and public infrastructures define much of their identity and well-being. The analytics of these things becomes something akin to medicine in the gravity of its consequences: perhaps the numbers attached to the IoT should be referred to a "serious data" rather than merely being another category of "big."

[15]Marshall Kirkpatrick, "Meet the Firehose Seven Thousand Times Bigger than Twitter's," Readwrite 18 November 2010, http://readwrite.com/2010/11/18/meet_the_firehose_seven_thousand_times_bigger_than#awesm =~oIpBFuWjKFAKf9, accessed 29 March 2018.

[16]David Barrett, "One surveillance camera for every 11 people in Britain, says CCTV survey," The Telegraph, 10 July 2013, https://www.telegraph.co.uk/technology/10172298/One-surveillance-camera-for-every-11-people-in-Britain-says-CCTV-survey.html, accessed 29 March 2018.

Chapter 2
Cognitive Data Analysis for Big Data

Jing Shyr, Jane Chu, and Mike Woods

Abstract Cognitive data analysis (CDA) automates and adds cognitive processes to data analysis so that the business user or data analyst can gain insights from advanced analytics. CDA is especially important in the age of big data, where the data is so complex, and includes both structured and unstructured data, that it is impossible to manually examine all possible combinations. As a cognitive computing system, CDA does not simply take over the entire process. Instead, CDA interacts with the user and learns from the interactions. This chapter reviews IBM Corporation's (IBM SPSS Modeler CRISP-DM guide, 2011) Cross Industry Standard Process for Data Mining (CRISP-DM) as a precursor of CDA. Then, continuing to develop the ideas set forth in Shyr and Spisic's ("Automated data analysis for Big Data." WIREs Comp Stats 6: 359–366, 2014), this chapter defines a new three-stage CDA process. Each stage (Data Preparation, Automated Modeling, and Application of Results) is discussed in detail. The Data Preparation stage alleviates or eliminates the data preparation burden from the user by including smart technologies such as natural language query and metadata discovery. This stage prepares the data for specific and appropriate analyses in the Automated Modeling stage, which performs descriptive as well as predictive analytics and presents the user with starting points and recommendations for exploration. Finally, the Application of Results stage considers the user's purpose, which may be to directly gain insights for smarter decisions and better business outcomes or to deploy the predictive models in an operational system.

Keywords Business intelligence · Cognitive data analysis · Data lineage · Data quality · Entity analytics · Metadata discovery · Natural language query · Social media analytics

J. Shyr (✉) · J. Chu · M. Woods
IBM Business Analytics, Chicago, IL, USA
e-mail: jshyr@us.ibm.com; jchu@us.ibm.com; mwoods@us.ibm.com

© Springer International Publishing AG, part of Springer Nature 2018

W. K. Härdle et al. (eds.), *Handbook of Big Data Analytics*, Springer Handbooks of Computational Statistics, https://doi.org/10.1007/978-3-319-18284-1_2

2.1 Introduction

The combination of big data and little time, or simply the lack of data science expertise in an organization, motivates the need for cognitive data analytic processes. Cognitive data analysis (CDA) refers to research and applications that seek to fill this need.

This chapter begins by defining CDA and discussing its role in the era of big data and then provides a framework for understanding the stages of CDA. Subsequent sections explore each stage in depth.

2.1.1 Big Data

Much of the data science industry describes big data in three dimensions: volume (scale of data), velocity (speed of streaming data), and variety (different forms of data). IBM Corporation (2014) adds another dimension: veracity (uncertainty of data). These four V's of big data are summarized in Fig. 2.1.

Sources of big data include social media (e.g., Facebook, Twitter, and YouTube), streaming data (e.g., web log tracking, sensors, and smart meters), collections of customer activities (e.g., transactions and demographics), and publicly available data (e.g., the US government's data.gov web site).

Big data includes structured data and unstructured data. Traditionally, organizations analyzed structured data, like that in relational databases. But it is widely held that 80% or more of organizational data now comes in unstructured forms, such

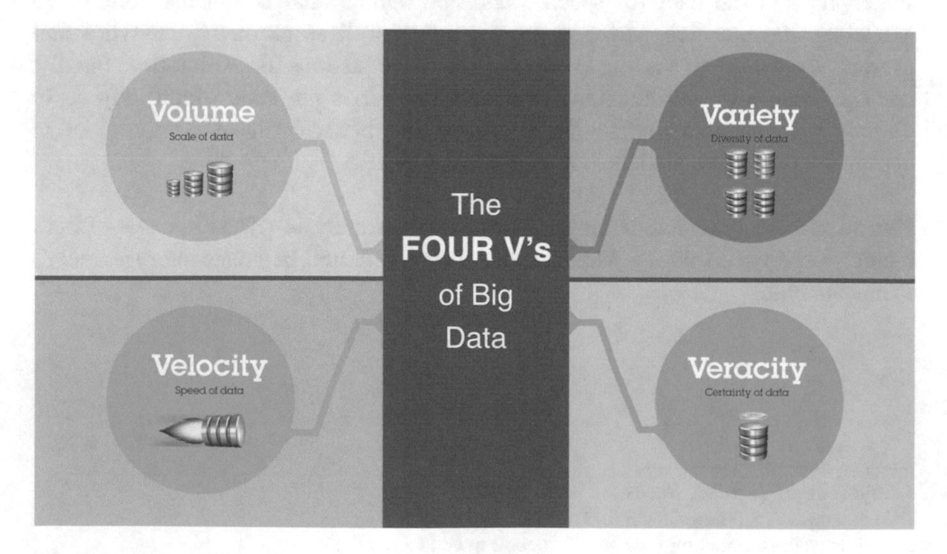

Fig. 2.1 The four V's of big data

as videos, images, symbols, and natural language—e.g., see Barrenechea (2013) or Maney (2014). Big data thus requires a new computing model for businesses to process and make sense of it and to enhance and extend the expertise of humans.

Clearly, unstructured data requires more effort to be wrangled into a form suitable for analysis, but both types of data present challenges for CDA. For example, with structured data it is easy to retrieve a customer's data given his or her customer ID. But it may not be so easy to assess the quality of the customer's data if it includes hundreds or thousands of transactions, along with demographics and survey responses. Numerous decisions need to be made when preparing the data for any analysis: Should the data be aggregated? Which fields are best for a segmentation analysis? How should invalid or missing values be identified or handled?

We argue that CDA needs to adequately handle structured data before it can claim to handle unstructured data. Both types are considered in this chapter, but we focus on structured data.

2.1.2 *Defining Cognitive Data Analysis*

CDA is a type of cognitive computing, which aims to develop in computers a capacity to learn and interact naturally with people, and to extend what either a human or machine could accomplish independently. Hurwitz et al. (2015) state that in a cognitive computing system, humans and computers work together to gain insights from data. In the realm of data science, these systems help humans to manage the complexity of big data and to harvest insights from such data.

CDA can be viewed as consolidating and automating the phases of the Cross-Industry Standard Process for Data Mining (CRISP-DM), which is a methodology and process model of data mining. Figure 2.2, from IBM Corporation (2011), shows a summary of the CRISP-DM process.

As shown in Fig. 2.2, CRISP-DM has six phases, beginning with *Business Understanding* and *Data Understanding* and proceeding to *Data Preparation*, *Modeling*, *Evaluation*, and *Deployment*. The arrows indicate important dependencies between phases, but these should be taken as guidelines. In practice, there may be dependencies and re-iteration among all phases.

One characteristic of CRISP-DM, which is not apparent in Fig. 2.2, is that the process is largely manual. The analyst needs to request or specify the various operations on the data. Furthermore, these operations involve many routine activities, including data cleaning, inspection of univariate statistics, and inspection of bivariate relationships. Moving from these initial activities to statistical modeling (regression analysis, time series analysis, neural networks, etc.), the analyst is faced with a complex array of decisions (what type of model to fit, which modeling criteria to use, how to assess model fit, etc.). Especially in the context of big data, this process is inefficient. The analyst cannot manually examine all possible

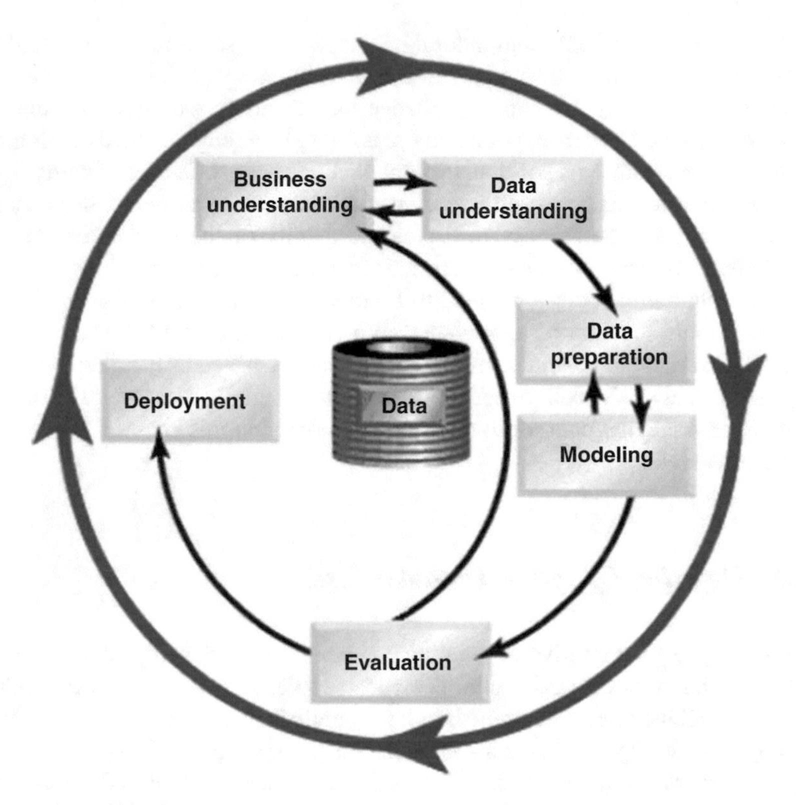

Fig. 2.2 The CRISP-DM process

combinations of data, and the choice of the best model can be daunting. Analytic results thus may be invalid or incomplete, and decisions based on such results may be substandard.

The journey toward an automated data analytic system with cognitive capabilities is described in Shyr and Spisic's (2014) "Automated Data Analysis for Big Data." This system automates routine data analytic tasks in a Data Analysis Preparation phase and then automates data exploration and modeling in a Relationship Discovery phase. But the system does not completely automate decision making or make obsolete any human involvement in the process. In contrast, the ideal is for the machine to handle technical issues so that the human can focus on understanding the results for better decision making. The system achieves this ideal by introducing cognitive processes that interact with the human analyst and learn from these interactions.

CDA, as defined in the current chapter, continues the journey by further exploration and development of ideas in adding automation and cognition to the system—including natural language query, data type detection, starting points, and system recommendations. CDA also directly considers the application of the results, whether for gaining insights about the data, or for sharing or deployment.

Finally, it should be stated that CDA is neither an all-or-nothing nor a one-size-fits-all enterprise. Any CDA system needs to give careful attention to the intended user base. A system geared toward more novice users automates more of the data preparation and analytic choices and puts more effort into clear and simple presentation of the analytic results to guide users to the best insights. In contrast, a system intended for more knowledgeable users allows or requires human intervention at key points and perhaps provides a more detailed presentation of the results for users to find extra insights. Similarly, a system might be developed for general purpose usage, or it might be domain specific.

2.1.3 Stages of CDA

CDA begins with a Data Preparation stage that combines CRISP-DM's *Data Understanding* and *Data Preparation* phases. The Data Preparation stage includes these features:

- Natural language query
- Data integration
- Metadata discovery
- Data quality verification
- Data type detection
- Data lineage

CRISP-DM's *Data Understanding* phase includes collecting data, describing data, exploring data, and verifying data quality. CDA does not specifically call for an initial assessment of the data separate from the final assessment but instead (potentially) automates the entire preparation of the data under the Data Preparation stage. Exploring the data, like modeling it, is then done in CDA's Automated Modeling stage.

After Data Preparation, CDA proceeds to an Automated Modeling stage. This stage not only discovers and fits an appropriate model but automates the evaluation of that model. Thus, Automated Modeling encompasses CRISP-DM's *Modeling* and *Evaluation* phases and includes designed interactions with the user (e.g., system recommendations). In sum, Automated Modeling includes these features:

- Descriptive analytics
- Predictive analytics
- Starting points
- System recommendations

The final stage of CDA, Application of Results, includes but is not limited to CRISP-DM's *Deployment* phase:

- Gaining insights
- Sharing and collaborating
- Deployment

Table 2.1 The relationship between CRISP-DM and CDA

CRISP-DM phase	CDA stage
Business Understanding	
Data Understanding	Data preparation
Data Preparation	Data preparation
Modeling	Automated modeling
Evaluation	Automated modeling
Deployment	Application of results

Table 2.1 captures the relationships between the CRISP-DM phases and the CDA stages.

Note that CRISP-DM's *Business Understanding* phase seems to be outside of the CDA system. This phase comprises determining the business objectives, assessing the situation (resources, risks, etc.), determining the data mining goals, and producing a project plan. In fact, this phase is inherent in the human interaction element of CDA and can/should be spread through all CDA stages, depending on the implementation.

The CDA user experience is also highly interactive, with human–machine interaction involved in the decision making at key points. Figure 2.3 depicts the CDA system. Although CDA automates much of the workflow, the human analyst is an essential part of the system.

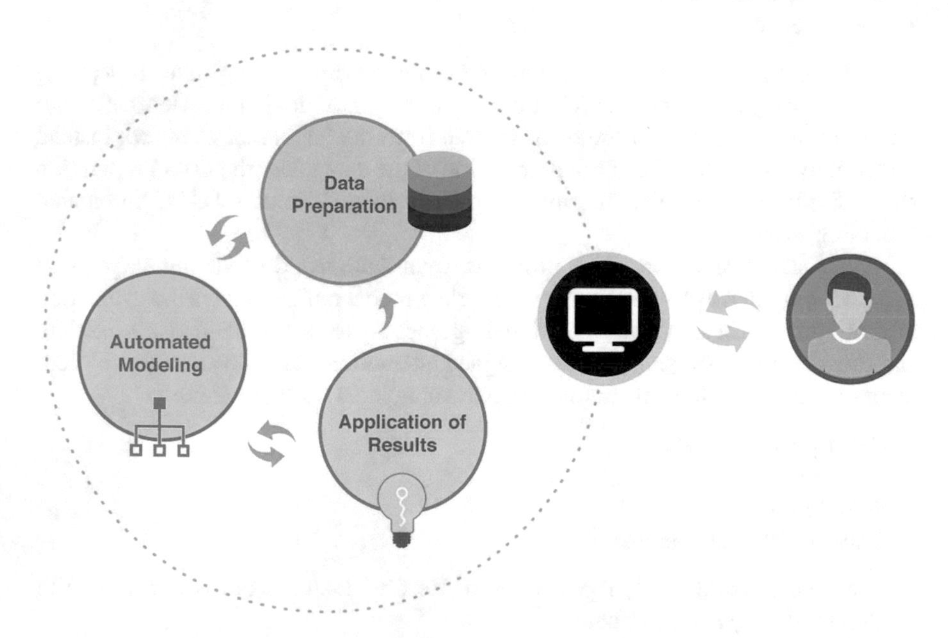

Fig. 2.3 The CDA system

The rest of this chapter is organized as follows. Section 2.2 discusses Data Preparation, giving in-depth descriptions of natural language query, data integration, metadata discovery, data quality verification, data type detection, and data lineage. Section 2.3 discusses Automated Modeling, covering the main categories of analytics (descriptive and predictive), as well as the main approaches to presenting analytic results to users in a CDA system (starting points versus system recommendations). Section 2.4 covers Application of Results, including summaries of gaining insights, sharing and collaborating, and deployment. Section 2.5 presents a use case involving a real-world data set and how it might be handled by an ideal CDA system. Finally, Sect. 2.6 ends the chapter with a reiteration of the CDA workflow and the place of the human analyst in it.

2.2 Data Preparation

For data scientists, it is common knowledge that data preparation is the most time-consuming part of data analysis. Lohr (2014) noted that data scientists claim to spend 50–80% of their time collecting and preparing data.

Data preparation was traditionally done through an ETL (Extract, Transform, Load) process, which facilitates data movement and transformation. ETL requires business experts to specify the rules regarding which data sources are extracted, how they are transformed, and where the resulting data is loaded, before the process can be scheduled to perform data movement jobs on a regular basis.

In the CDA system, some smart technologies and methods are applied to the different steps in data preparation, such that business users and data analysts can manage it without help from IT and business experts.

We separate these steps based on whether the data is ready for analysis. If there is no data or there are multiple data sources, then data needs to be found or merged/joined. These efforts are referred to as the "data acquisition" and "data integration" steps, respectively. Then, additional data preparation steps—such as metadata discovery, data quality verification, data type detection, and data cleaning and transformation—can be applied before analysis. Note that these steps are not a linear process but are interdependent.

For the data acquisition and integration steps, smart technologies include, but are not limited to:

- Using natural language processing to suggest data sets to answer the user's questions.
- Implementing entity analytics to detect related entities across disparate collections of data sources.
- Applying text mining to social media data to extract sentiment and identify author demographics.

For other steps, the CDA system can apply semantic data analysis to enhance metadata discovery, provide identification of data quality issues and novice-oriented detection of data types, and so forth. These steps run automatically, but the user can control the process and direction.

These smart technologies and methods are described in more detail next.

2.2.1 Natural Language Query

Current data discovery practices often assume that specific data sources have been collected and loaded into data warehouses, operational data stores, or data marts. Then, data preparation is applied to these data sources to answer the user's business questions. Because only the user understands the business questions, they would need to know which data sources to collect and what data preparation processes to conduct.

The CDA system employs natural language processing to decipher the user's business questions and connect the user to other possible data sources. Moreover, CDA can search both internal and external data sources to suggest the most relevant sources. The benefits of this additional functionality include an increased likelihood of getting better insights and more complete answers than might be obtained based only on the data at hand.

For example, the HR staff in a pharmaceutical company wants to understand the reasons for voluntary attrition among employees. The staff has access to a base data set containing tens of thousands of records for company employees in the last 10 years. The CDA system detects that this is an employee data set via semantic data analysis. Furthermore, the system uses natural language processing to understand the user's business question and suggests using the existing "Attrition" field as the target/outcome field. The data set also contains a "Salary" field, and the CDA system automatically searches for related data sources. Knowing from experience that "Salary" is likely to be an important predictor/input field of "Attrition," the system searches for data sources containing salary averages by industry, position, and geography.

Sometimes, a user forgets which data sets have some desired information, or the user simply has some business questions without data sets handy. The user can type in questions, and the CDA system's natural language query can be used to find and suggest relevant data sources.

For example, if an HR staff member cannot find the employee data set, he or she might enter "human resources" or "employee attrition" into a type-in interface, and the CDA system would proceed by finding relevant data sources, such as the employee attrition and salary averages discussed above.

Another example is a company that wants to forecast sales for its products. In this case, the system might query weather and social media data to assess how temperature and customers' sentiments about products impact sales.

2.2.2 Data Integration

Any data set, whether from the user or automatically found, presents an opportunity to apply transformations appropriate to the user's purpose. And if there are multiple data sets, all of them need to be integrated into one before analysis commences. The CDA system can provide relevant advice to, and possibly perform these actions for, the user.

For example, having found a data source of salary averages by different categories, the CDA system can join the new data to the base file—e.g., by adding a field containing national average salary conditional on the employee's position and experience. This kind of join goes beyond traditional data discovery tools that join by the same key field in all data sets. The CDA system can also automatically compute the difference between each employee's observed salary and the corresponding national average. Subsequent predictive analytics can assess both fields (the observed salary and the difference). A clear benefit of this approach is that it assesses more relationships than would be available using the base file only. It would be a valuable insight if our example discovered that the difference is a better predictor of attrition than the observed salary.

As mentioned above, these data sources might come from various structured and unstructured formats. Before integration, all sources need to be converted to a structured format. We describe below two smart technologies: social media analytics for handling unstructured social media data and entity analytics for detecting like and related entities from different data sources.

Social media data—such as Twitter, Facebook, blogs, message boards, and reviews—is an increasingly attractive source of information for public relationships, marketing, sales, and brand/product management for many companies. But its unstructured format makes it a challenge to use.

Social media analytics uses natural language processing to extract meaning from social media data. It applies a two-level approach. First, document-level analysis—with concept detection, sentiment detection, and author features extraction—is applied. Next, collection-level analysis—with evolving topics, consolidation of author profiles, influence detection, and virality modeling—is applied. The resulting data is a structured data set, which can be integrated with other data sets. Similarly, if the social media data is in image or sound format, the CDA system can apply appropriate tools to convert it to structured formats.

Another challenge is to integrate diverse company-wide data sources, say customer, product, and order, all of which may be managed by different divisions. One critical data preparation activity involves recognizing when multiple references to the same entity, across different data sets or within the same data set, are indeed the same entity. For example, it is important to know the difference between three loan applications done by three different people versus one person applying to all three loans. Entity analytics (Sokol and Jonas 2012) in the CDA system includes entity resolution (whether entities are the same, despite efforts to hide them) and entity relationships (how the resolved entities are related to each other, e.g., by the same birthday or address). In addition, entity analytics incrementally relates new data to previous data, updates resolved entities, and remembers these relationships. This process can improve data quality, which in turn may provide better analytic solutions.

2.2.3 Metadata Discovery

Traditional statistical tools have offered some basic metadata discovery, such as storage type (e.g., numeric, string, date), measurement level (continuous, nominal, ordinal), and field role (target, input, record ID). These characteristics are based on summary statistics with some or all of the data.

CDA employs semantic data analysis to enhance this basic metadata discovery. We describe some principles below. More details can be found in Rais-Ghasem et al. (2013) and Shyr and Spisic (2014).

Semantic data analysis uses natural language processing to map lexical clues to both common concepts (time, geography, etc.) and business concepts (revenue, sales channel, etc.) and can be trained to be sensitive to language-specific patterns. For example, semantic data analysis can identify fields named "Employee Number," "Employee No," or "Employ_Num" to be record identifiers, while "Number of Employees," "No. Employees," or "Num_Employees" are treated as count fields.

Metadata discovery can also build or detect hierarchies in the data. For example, it checks if unique one-to-many relationships exist in the values of fields to detect hierarchies. If semantic analysis finds that a country consists of states, and each state consists of different cities, then country-state-city is detected as a hierarchy. Similarly, it will build a hierarchy of year-month-day given a date field with value 02/28/2015, and include derived fields "Year," "Month," "Day." These derived fields are useful in the Automated Modeling stage.

Metadata discovery can further detect whether each time field is regular (equally spaced in time) or irregular (e.g., transactional data). If a time field is regular, it will further detect the time interval (e.g., days) and its related parameters (e.g., the beginning of week and the number of days per week). If a time field is irregular, it will suggest the most plausible time interval for aggregation to transform irregular transactional data to regular time series data. The basic idea is to select the smallest time interval in which most time bins have at least one record. These smart methods are important in determining data types. Details about the time interval detection method are described in the USPTO patent application by Chu and Zhong (2015).

Semantic data analysis also differentiates between numeric categorical and continuous fields. For example, a field named "Zip Code" may contain large integer values that would usually indicate a numeric field with a continuous measurement level. Matching the field name against a known concept reveals that the measurement level should be set to nominal.

These discovered characteristics are applied to narrow down the applicable algorithms used in the Automated Modeling stage. For example, the presence of one or more record ID fields greatly assists in automatically detecting the domain. A field named "Employee ID" with a different value for each record almost certainly indicates an employee data set, and the system might infer that the data set is from the HR domain. If a field named "Attrition" is detected and suggested as a target field because it can help to answer a user question such as "what influences employees to quit?," then the system might fit a logistic regression model. The system might also focus on finding predictors that are potentially mutable (e.g., salary or satisfaction ratings) rather than on those that are not (e.g., demographics)—assuming that the idea is to act on the results and try to retain some employees.

The CDA system could benefit further from these metadata discovery rules if they were used to group similar fields. For example, several field groups can be created for a data set, such as demographics, survey responses, and user metrics. Different processes can leverage this grouping information for specific tasks.

- A target detection process might ignore field groups unlikely to contain targets (e.g., demographics) and focus on groups likely to contain targets (e.g., user metrics).
- An automated segmentation process might restrict cluster features to groups likely to yield insights (e.g., survey responses).
- The system might provide a summary of records based on groups with relevant information (e.g., demographics).

The CDA system should also automatically detect field roles or problematic data values. At the automated extreme, these detected field roles could be taken as truth, problematic data values could be automatically corrected, and the system could proceed with the analysis. But if the detected roles or values are not handled properly, the user experience will suffer. Hence the CDA system should get users' input before decisions are committed. For example, the system might allow the user to review and adjust each field, including:

- Setting target fields
- Specifying missing values
- Selecting outliers handling
- Changing or specifying a hierarchy

2.2.4 Data Quality Verification

For any data set, the CDA system identifies data quality problems at several levels: the entire data set, each field/column, each record/row, or each cell in the table.

At the cell level, the system verifies whether each cell contains a missing/null or outlying value and confirms that a value in a cell is compliant to the domain of the column (e.g., text in a US zip code field).

At the field level, the system checks completeness by counting the number of missing values in each field, checks uniqueness by counting the number of repeated values, and identifies constant or nearly constant fields.

At the record level, the system finds records that violate automatic or manual rules (e.g., values in "Revenue" should be larger than those in "Profit") and detects duplicated records by leveraging entity analytics.

The CDA system might aggregate all the findings in a unified quality score for the entire data set and for each column. It can further categorize the score as low/medium/high quality and recommend actions for improving data quality. These recommendations can be delivered automatically or (ideally) interactively so that the user understands the issues and accepts them before the data are transformed.

Furthermore, more complex, domain-specific, data quality assessments might be performed. For example, consider a survey data set about customers' responses to 100 questions with a common Likert scale (e.g., Strongly Agree, Agree, Neutral, Disagree, Strongly Disagree). For such a long survey, the reliability of the responses could be assessed. Assuming the order of the questions/fields in the data corresponds to the order in which the questions were administered in the survey, the sequence might be split to do assessments. For example, the CDA system splits the sequence into quarters, computes the variance of responses for each question, and averages them for each quarter. Suppose the average variance for the last quarter is significantly lower than the preceding three. This might suggest customers were tired and not paying attention for the final questions. In this case, the questions/fields in the fourth quarter might be flagged as low quality or actively excluded. Or the CDA system might compare the response variances of the first 75 questions and the last 25 questions for each customer. If the former is much higher than the latter for some customers, it might suggest that the responses from these customers are not trustworthy and that the records should be excluded from the data set.

2.2.5 Data Type Detection

This section focuses on detecting data types for structured data sets. As mentioned previously, unstructured data is converted to a structured format before analysis is performed. The most commonly encountered structured data set types for business users are cross-sectional (characteristics about customers, products, etc.) and time series (metrics measured over time). The importance of data type detection is that certain analytic techniques are suitable for only specific data types, so the number of possible analytic techniques to consider will be smaller in the Automated Modeling stage. For example, regression models and tree models are used for cross-sectional data, and exponential smoothing and autoregressive integrated moving average (ARIMA) models are used for time series data, while neural network models might be useful for both.

However, there are many more types than these two. For example, panel/longitudinal data combines cross-sectional and time series, so none of analytics mentioned above are suitable. Instead, generalized linear mixed models or generalized estimating equation models should be considered.

Even when a data set contains one or more timestamp fields, it is not necessarily time series data. For example, if the timestamps are taken irregularly, the data might be transactional. In this case, traditional time series analysis might not be appropriate without aggregation, or possibly other algorithms should be applied depending on the intent/goal. In fact, a data set with time or duration information involving an event (e.g., death from a certain disease, broken pipe, crime committed again after a criminal is released from prison) might be about lifetime data, with the intent being to find survival rate past a certain time or what factors impact the survival time of an event. Lifetime analysis (also called survival analysis or reliability analysis) should be used in this condition.

In addition to temporal/time fields, spatial information is often embedded in business data. Spatial fields may be a predetermined set of location points/areas, or they may be irregular. As with time, the appropriate analytics will differ depending on the intent/goal.

If the data set has a hierarchical structure, then there are additional challenges. For example, the data includes hierarchies in both time (year-month-day) and geospatial (country-state-city) fields, and the user wants to forecast future monthly sales for all levels in the geospatial hierarchy. We can apply analysis to monthly time series at the city level to obtain forecasts and then aggregate from the bottom up to get forecasts for state and country levels. However, such forecasting results might not be optimal, and other algorithms are available to obtain better results.

In short, different types of data pose challenges to the CDA system but also offer a potentially rich source of insights from multiple perspectives.

2.2.6 Data Lineage

According to Techopedia (2018), data lineage is defined as a kind of data life cycle that includes the data's origins and where it moves over time. This term can also describe what happens to data as it goes through diverse processes.

In the age of big data, with challenges of massive scale, unstructured data types, uncertainty, and complex platforms, data lineage is more important than ever. The CDA system should store all data preparation steps/recipes, including user inputs, for reuse on the same data sets or extended data sets with additional records.

As a cognitive computing system, CDA will continuously learn from users, thereby improving the existing methods, or adding new methods of data preparation. Such learning will make data lineage more challenging, especially for users who are unaware of changes to the default settings of the methods mentioned previously. Ideally, the CDA system should offer user-friendly access to such changes.

2.3 Automated Modeling

Automated Modeling is the automated application of various types of analytics to the prepared data, to find patterns and uncover relationships. This stage is especially important for big data, where manual examination without guidance is unlikely to be efficient or complete.

Current data discovery tools use interactive user controls in this stage, including filtering, sorting, pivoting, linking, grouping, and user-defined calculation. These tools enable users to explore patterns manually. However, this approach, while important, is not sustainable in the age of big data. Especially when there are many fields, it is not feasible to manually explore all possible combinations of analytics. In such conditions, users tend to explore their own hypotheses, often using small samples, and will likely miss important patterns and relationships.

In the CDA system, suitable analytic techniques are applied to the data automatically. This automated approach tends to generate many results. The CDA system needs to score and rank these results and present the most important to the user. Moreover, the typical user is a business user or data analyst who might not have well-formed business questions to answer or hypotheses to test. The system should guide such users through starting points and recommendations. Starting points can be expressed as questions that lead the user to data exploration, while recommendations can present relevant results for further exploration.

The two main types of analytics applied in this stage are descriptive analytics and predictive analytics. Descriptive analytics applies business intelligence (BI) and basic statistical methods, while predictive analytics applies advanced statistical methods and data mining or machine learning algorithms. Current data discovery tools often focus on one type of analytics, or sometimes only on a few techniques (e.g., using decision tree models to build all predictive models). The ideal CDA system comprises both types of analytics, and multiple techniques within each, to give users a more complete and accurate picture of the data.

The next two sections discuss descriptive and predictive analytics in the context of CDA, after which starting points and recommendations are discussed in more detail.

2.3.1 Descriptive Analytics

Descriptive analytics uses BI and basic statistical techniques to summarize the data. BI and statistical analysis provide patterns and relationships from different angles and should complement each other.

Davenport and Harris (2007) define BI as "a set of technologies and processes that use data to understand and analyze business performance." Basically, BI includes reports, dashboards, and various types of queries by applying some filters on data before computing descriptive statistics (e.g., sums, averages, counts, percentages, min, max, etc.).

Statistical analysis, on the other hand, computes univariate statistics, such as the mean and variance, for each field. Bivariate statistics may also be computed to uncover the association between each pair of fields. In a CDA system, univariate and bivariate statistics will likely be computed and used in the metadata discovery process. Thus, there is no need to compute them again—though the computations will need to be updated if the user has adjusted the metadata, fixed data quality issues, or applied data transformations.

A big challenge to CDA is consolidating results from BI and statistical analysis, which tend to use different terminologies. For example, metrics and dimensions in BI correspond to continuous and categorical fields, respectively, in statistical analysis. Also, BI does not have the concept of a field role, while statistical analysis largely depends on each field's role (target, input, record identifier, etc.). BI metrics are often treated as targets in statistical analysis.

The method proposed by Shyr et al. (2013) bridges the gap between BI and statistical analysis. Basically, it produces a series of aggregated tabular reports that illustrate the important metric-dimension relationships by using some statistical exploratory and modeling techniques mentioned in this and subsequent sections of this chapter. These dimensions are referred to as "key drivers" for the specific metric/target and provide useful insights. Additional information on bivariate statistics and aggregated tabular reports can be found in Shyr and Spisic (2014) and the references therein.

2.3.2 Predictive Analytics

Predictive analytics builds models by using advanced statistical analysis and data mining or machine learning algorithms. One of the main purposes of predictive models is to extrapolate the model to generate predictions for new data. However, it is also possible to extract useful insights regarding the relationship between the target field and a set of input fields. In general, statistical algorithms are used to build a model based on some theory about how things work (relationships in the data), while machine learning algorithms focus more on prediction (generating accurate predictions). The CDA system needs to understand the user's intent to employ the right algorithms.

There are two types of modeling algorithms, depending on whether a target field is needed.

- Supervised learning algorithms require a target and include algorithms such as regression, decision trees, neural networks, and time series analysis.
- Unsupervised learning algorithms, in contrast, do not designate a target. These algorithms include cluster analysis and association rules.

As mentioned previously, not all algorithms would be applicable for all data. The CDA system will know which algorithms should be run for each specific type of data.

There are conditions when the same data set can be identified with multiple data types, so that seemingly disparate algorithms may be appropriate. For example, an employee data set might include fields "Starting Date" (for an employee's date of hire) and "Ending Date" (for the employee's date of leaving). Other fields might include "Salary," "Job Category," and "Attrition." If the business question asks what factors influence an employee to quit, then a decision tree algorithm is appropriate. In this case, the data set is treated as cross-sectional. However, if user's intent is to know the expected duration of time until attrition for an employee, then a Cox regression model is appropriate. In this case, the data set is treated as lifetime data.

Many business users and data analysts are confused by the idea that predictive analytics is not limited to the time domain. Prediction can be used for all data types to score new data. To avoid confusion, "forecast" is used for time series data and "predict" for other types of data.

Finally, there is often an endless amount of hyper-parameter choices for each algorithm. Parameters may include kernel type, pruning strategy, learning rate, number of trees in a forest, and so forth. As with the exploration of relationships, it is impossible for a user to try every combination of parameters for each algorithm. The CDA system should apply some additional algorithms, such as Bayesian optimization (Thornton et al. 2013) or model-based genetic algorithm (Ansotegui et al. 2015) to tune these hyper-parameters automatically.

Algorithm selection itself is not new, but the new thing is that the CDA system is cognitive and can learn from the user's past behavior to fine-tune algorithm selection for the future.

2.3.3 Starting Points

There are many ways to display the results of the data preparation steps to users. One approach uses system-generated questions. For example, if metadata discovery detects fields "Salary" and "Attrition" with measurement level continuous and nominal, and field role input and target, respectively, then relevant questions can be constructed and presented. Depending on the system's understanding of the user's intent, the system might ask "What are the values of Salary by Attrition?" or "What factors influence Attrition?"

Alternatively, the user might have specific business questions or hypotheses in mind. The CDA system should provide an interface in which the user can enter their questions. These questions can provide additional starting points for users to explore their data.

For the system-generated questions, the CDA system relies on declarative rules from semantic analysis to make recommendations on how to best analyze the data.

First, the CDA system selects all eligible fields, with the categorical ones further filtered with hierarchies from the semantic model or the user's input. The result is a list of valid fields or combinations of fields for the various analytic techniques mentioned above.

Second, the CDA system scores and ranks each item in the list based on metadata information, such as the fields' concept types, the data characteristics, and the association between fields. The system should also provide an interface that allows users to search or guide this process. If a user enters one or more field names, then the system might emphasize techniques that involve those fields and adjust the ranking accordingly. Similarly, if the user enters a question, the system might handle it with natural language processing and consider only relevant techniques.

Third, the CDA system creates a set of responses in which the wording reflects the underlying analytic techniques. The set should be constructed such that it demonstrates accuracy (the results are consistent with the user's intent) and variability (multiple similar results from different types of analytics are not presented at the same time).

2.3.4 System Recommendations

In addition to producing a set of starting points to guide the user in exploring the data, the CDA system can also provide automatic system recommendations. In this sense, the CDA system is both automated and guided. When the user is inspecting a result, the system can recommend other important results, thereby engaging the user in drill-down on key analytics or navigation to other analytics. These recommendations should be tied to the user's current context. For example, if the user is inspecting a predictive model with target field "Attrition," then recommendations might include generating attrition predictions on some new data. Or they might include exploring relationships among the predictor fields—e.g., whether "Salary" values differ across combinations of "Job Category" and "Education Level."

The analysis of any data set, especially big data, will produce many results from various analytics. These results should be ranked and the system should only recommend the most important and relevant results. Like starting points, a scoring and ranking process is needed for recommendations. For example, a system might derive interestingness scores based on statistics that characterize each analytic technique (e.g., predictive strength and unusualness of nodes for a tree model). The choice of which statistics to use for this purpose requires careful thought. For example, in a big data context where nearly any hypothesis test will be statistically significant, statistics other than the conventional p-value should be considered. One possibility is to rank the test results based on the effect size, which should not vary much with the number of records. For additional information on interestingness indices and effect size, see Shyr and Spisic (2014).

Again, CDA can learn the user's interactions to improve the results of both starting points and recommendations over time.

2.4 Application of Results

In addition to Data Preparation and Automated Modeling, it is important to consider how the user will apply the results. Often a user will have multiple purposes, but for convenience we consider three categories of application:

- The user wants to directly gain insights about the business problem.
- The user is interested in sharing insights, and possibly collaborating, with colleagues.
- The user wants to formally deploy the obtained model(s) in a larger process.

2.4.1 Gaining Insights

Perhaps the most common use case is when the user wants to directly gain insights about the business problem at hand. This section considers how to guide the user to the best insights, and how to present those insights so the user understands them.

Big data is so large and complex that it is difficult to analyze efficiently with traditional tools, which require many data passes to execute multiple analyses prior to exploration. Furthermore, data scientists with the necessary analysis skills are a scarce resource. Hence, for the business user or data analyst who may not be savvy in data science skills, it is important to find the right balance of automated versus guided analytics.

An automated system will do most or all the analyst's work and present important results first. In contrast, a guided analytics system is a conversation between the user and the system that guides the user through the data discovery journey.

Presentation style is important too. Traditional statistical software, such as SPSS and SAS, focused on numeric output in tables. Over time, as statistical software sought to become more user-friendly, more emphasis was put on visual output. Visualizations became especially important with the advent of big data, where a large amount of information often needs to be summarized in a single visualization.

Although visualizations are powerful means of communication, they do not always lead the user to the most significant results. The addition of natural language interpretation, either separately or in conjunction with visualizations, alleviates this problem. Ideally, natural language interpretations will follow the principles of plain language, so that the reader can "easily find what they need, understand what they find, and use that information" (International Plain Language Federation, 2018). Techniques for achieving this goal include using short sentences and common words.

Using plain language for analytic output interpretation presents many challenges. Analytics are highly technical, and terms will not always have common equivalents. A term such as "mean" can be replaced by its everyday equivalent "average," but there is no common equivalent to "standard deviation." Attempts to create a new, easily understood equivalent (e.g., "regular difference") can be unsatisfying at best.

In cases like this, one should consider whether the analytic is really needed by most users. If it is, then presenting it with the technical name, along with a hidden but accessible definition, may be a reasonable solution. If most users do not need it, then showing it in some other container (e.g., an under-the-hood container, where the user can find statistical details) may be best. Such a container also allows experts to assess the methods and detailed results of the CDA system.

The combined presentation of visual and text insights is a form of storytelling. Visual and text insights can be coupled and flat, or coupled and interactive. The latter can be presented as dynamic visual insights, for example, by modifying the visualization depending on the text selected (and being read) by the user. Alternatively, visual and text insights can be presented hierarchically in a progressive disclosure of insights. Under-the-hood insights, mentioned above, also should be available for more analytic-savvy users.

Finally, the CDA system should also allow user interactions such as the ability to "favorite" specific results or views, as well as the selection and assembling of the output in dashboards or reports.

2.4.2 Sharing and Collaborating

Another important area of application involves allowing users to share data sets or findings (as visualizations, dashboards, reports, etc.). The CDA system should include relevant options for sharing output on the cloud, either a public or a private cloud. In addition, there should be a capacity to post findings directly to social media (e.g., Facebook or Twitter).

Collaboration between users on the same project is also important. This can range from allowing other users to view one's project and add comments to allowing multiple users to work on the same project in real time. This is especially important in an analytic context where different users will provide different opinions about the results. Just as different analytic models give multiple perspectives of the same data set, a CDA system will learn from its interactions with multiple humans who provide their own perspectives on the data, analytics, and interpretation of the results.

2.4.3 Deployment

After predictive models have been built and validated, they can be deployed by embedding them into an operational system to make better business decisions. This extends predictive analytics to prescriptive analytics, where business rules and optimization tell the user what is likely to happen or what they should do. Prescriptive analytics can recommend alternative actions or decisions given a set of objectives, requirements, and constraints.

For example, suppose the marketing department of an insurance company uses past campaign data to gain insight into how historical revenue has been related to campaign and customer profile. They may also build a predictive model to understand the future and predict how campaign and other factors influence revenue. Then they may apply the model to a new set of customers to advise possible options for assigning campaigns to new customers to maximize expected revenue.

Prescriptive analytics can be complex. It is unlikely that business users or data analysts can directly specify their business problems as optimization models. And getting a good specification of the problem to solve may require iterations with the user. The CDA system should express a prescriptive problem based on the user's data set(s) and, in plain language, review interactively suggested actions/decisions, refine the specification of the problem, and compare multiple scenarios easily.

Prescriptive analytics can recommend one or more possible actions to guide the user to a desired outcome based on a predictive model. It is important that the selected predictors/inputs in the predictive model are attributes that the user can act upon, and the CDA system should focus on selecting such predictors if deployment is important.

For example, a department store has weekly sales data for the past 10 years and uses it to forecast future sales. In addition, the store has set the sales value to meet for each week of the next quarter and wants to know what actions it can take to achieve the planned sales values. The CDA system discovers that temperature and advertisement expenditure would influence sales. Because advertisement expenditure, but not temperature, is under the store's control, the time series model built by the system should include advertisement expenditure as a predictor to make the results actionable.

2.5 Use Case

This section presents a use case involving real-world data and how it might be handled by an ideal CDA system. For our use case, suppose that an analyst wants to analyze crime trends in Chicago. After launching the ideal system, the analyst is presented with a set of possible actions (e.g., analyze popular data sets, review previous interactions with the system, etc.). The system also offers a prompt, "What do you want to analyze?" The analyst enters "Chicago crimes."

The CDA system proceeds by searching the Internet, as well as all connected databases, for any data sets relevant to Chicago crimes. Then the system presents the analyst with a list of found data sets. The most relevant data set contains data about Chicago crimes from 2001 to the present. This public data set is available from the City of Chicago (2011). Table 2.2 presents a sample of five records. The analyst selects this data set.

Table 2.2 Five records from the Chicago Crimes Data Set

ID	Case number	Date	Block	IUCR	Primary type	Description	Location description	Arrest	Domestic
8519160	HV194208	1/1/2012 0:00	047XX N HARDING AVE	1562	SEX OFFENSE	AGG CRIMINAL SEXUAL ABUSE	RESIDENCE	0	0
9467940	HX120966	1/1/2012 0:00	015XX S KEELER AVE	840	THEFT	FINANCIAL ID THEFT: OVER $300	RESIDENCE	0	0
8904856	HV578917	1/1/2012 0:00	002XX E CHESTNUT ST	840	THEFT	FINANCIAL ID THEFT: OVER $300	RESIDENCE	0	0
8511440	HV188411	1/1/2012 0:00	030XX W 64TH ST	840	THEFT	FINANCIAL ID THEFT: OVER $300	RESIDENCE	0	0
9836248	HX484599	1/1/2012 0:00	001XX W GARFIELD BLVD	1130	DECEPTIVE PRACTICE	FRAUD OR CONFIDENCE GAME	CURRENCY EXCHANGE	0	0

Beat	District	Ward	Community area	FBI code	X Coordinate	Y Coordinate	Year	Updated on	Latitude	Longitude	Location
1723	17	39	14	17	1149166	1931138	2012	3/21/2012 8:38	41.9669718	−87.72692152	(41.96697176, −87.726921519)
1012	10	24	29	6	1148673	1892260	2012	1/21/2014 9:16	41.8602963	−87.72974025	(41.860296321, −87.729740248)
1833	18	42	8	6	1178595	1906435	2012	12/3/2012 17:07	41.8985635	−87.61947207	(41.898563507, −87.619472067)
823	8	15	66	6	1157194	1862082	2012	3/9/2012 16:01	41.7773154	−87.69927908	(41.777315422, −87.699279082)
225		3	37	11			2012	10/29/2014 12:37			

The system continues by using metadata discovery on the selected data set and discovers that it contains temporal and spatial fields. The system further detects that the time fields are irregular (not equally spaced in time), and it determines that this is a transactional data set. Based on these and other discoveries, the system enumerates various statistical methods that might be applied. For example, time series analysis could be used to assess trends, or spatial-temporal analysis could be used to discover spatial-temporal patterns, such as associations between crimes and spatial or temporal intervals.

For each possible statistical method, the system considers the types of data sets that might be joined with the crimes data to allow or enhance the analysis. The system also performs another search for these secondary data sets. For the time series analysis, weather data might be useful for enhancing the time series model with weather-related predictors. For the temporal-spatial analysis, a map data set containing details about the neighborhoods and streets, which can be inferred from the crime location fields in the crimes data set, would be needed for the associations algorithm.

Next, the system presents the analyst a set of possible analyses, where time series analysis is given a more prominent position than spatial-temporal analysis based on the analyst's past interactions. The system also presents information about recommended secondary data sets, giving the analyst full information and choice. The analyst selects the time series analysis to discover trends and chooses to join weather data as a source of possible predictors of crime.

The system now has two data sets to be integrated: the primary crimes data set and the secondary weather data set. The system next determines and suggests that the best time interval for aggregation of the crimes data set, to transform the irregular transactional data to regular time series data, is by date. That is, the system recommends aggregating the crimes data set to create a new data set containing daily crime counts.

The system informs the analyst of its findings and recommendations and waits for the analyst's consent. The system tells the analyst that the data needs to be aggregated into a regular time series data set to perform time series analysis. By default, the system will treat total crime count per day as the time series but allows the analyst to select other fields to be used instead or in addition. For example, based on the system's natural language processing, it has discovered that there is a "Primary Type" field giving the category of each crime event. These categories could be aggregated to get category counts for particular crime types. The analyst selects the option of aggregating by date, and treating the total crime count per day as the time series field, to get a general view of overall crime trends.

Next, the system extracts a subset of weather data corresponding to the time range in the crimes data and aggregates the weather data by date before joining the crimes and weather data sets by date. Table 2.3 presents a sample of 5 days from the new joined data set. For brevity, only two weather fields are presented in the table.

The system also automatically performs data quality checks and saves a history of all data checks and transformations. Now, with the data acquisition and preparation

Table 2.3 Five records from the Aggregated Chicago Crimes and Weather Data Set

Date	N crimes	Mean temperature F	Mean humidity
1/1/2012	1387	37	69
1/2/2012	699	23	66
1/3/2012	805	19	56
1/4/2012	799	34	59
1/5/2012	869	38	62

done—as with human analysts—most of the hard work is done, and the system proceeds to Automated Modeling.

The system computes basic descriptive analytics for all fields (e.g., the mean and variance for continuous fields). But more directly applicable to the analyst's request, the system fits various time series models, treating daily crime count as the time series. For example, the system uses exponential smoothing to create a traditional time series model containing only the time series field, daily crime count, where each day's crime count is predicted by the crime counts of the immediately preceding days. The system also fits a temporal causal model, treating various weather fields (e.g., temperature and humidity) as predictors of daily crime count. The system performs related analyses too and generates graphs to assist the analyst in understanding the trend, seasonality, and other properties of the time series.

The system presents these analytic results to the analyst in a meaningfully ordered interface—e.g., beginning with a line chart showing the daily crime count time series itself and proceeding to various time series models of the data. As the analyst interacts with the results, the system makes context-dependent recommendations for further examination. When the analyst views the exponential smoothing results, the system recommends a more sophisticated model, such as the temporal causal model with weather-related predictors. The system presents all analytic results with appropriate visual and written insights added, such as pointing out the trend and seasonal pattern in the crime count time series.

The system also offers various post-analysis applications, including options to share the results on the cloud or to deploy one of the models for forecasting. The analyst selects the option to deploy the traditional time series model for forecasting. The system automatically saves the model and provides an interface for the analyst to request a forecast. The analyst proceeds by deploying the selected time series model to forecast daily crime counts for the next 7 days.

Note that this entire sequence was initiated with two words from the analyst, namely "Chicago crimes." Throughout the CDA stages, the system checked in with the analyst regarding key decisions. Clearly, the entire process could be automated, and all decisions made using heuristics or other default rules. But with minimal inputs by the analyst, the system has provided a custom analysis geared to the analyst's needs.

2.6 Conclusion

In conclusion, we have seen that the CDA workflow includes Data Preparation, Automated Modeling, and Application of Results and furthermore that the human analyst is an integral part of CDA's operation.

Data preparation has traditionally been the most time-consuming part of data analysis for humans, and in CDA it is also the most intensive stage computationally. By incorporating natural language query and smart metadata discovery, CDA relieves much of the data preparation burden from the human analyst but in doing so necessarily adds to the backend complexity of CDA systems.

Similarly, Automated Modeling in CDA, being automated and exhaustive, also reduces the error-prone burden of manually sifting through large amounts of data looking for insights. And by the judicious presentation of discovered patterns as starting points and system recommendations, CDA offers the added benefit of guiding the human analyst to the *best* insights.

Finally, the Application of Results in CDA can be greatly enhanced by incorporating ideas from information and visualization design in the presentation of insights, as well as investing in sharing, collaboration, and deployment technologies.

Contrary to the idea that CDA automates all decision making, we have seen that CDA is in fact very human centered, both in its heavy reliance on human–machine interaction and in its goal of reducing human workload.

References

Ansotegui C, Malitsky Y, Samulowitz H, Sellmann M, Tierney K (2015) Model-based genetic algorithms for algorithm configuration. In: Proceedings of the twenty-fourth international joint conference on artificial intelligence. Association of for the Advancement of Artificial Intelligence Press, Palo Alto, CA, pp 733–739

Barrenechea M (2013) Big data: big hype? Forbes, February 4:2013. http://www.forbes.com/sites/ciocentral/2013/02/04/big-data-big-hype/

Chu J, Zhong WC (2015) Automatic time interval metadata determination for business intelligence and predictive analytics. US Patent Application 14/884,468. 15 Oct 2015

City of Chicago (2011) Crimes—2001 to present. City of Chicago Data Portal. https://data.cityofchicago.org/Public-Safety/Crimes-2001-to-present/ijzp-q8t2. Accessed 5 Apr 2018

Davenport TH, Harris JG (2007) Competing on analytics: the new science of winning. Harvard Business School Press, Boston, MA

Hurwitz JS, Kaufman M, Bowles A (2015) Cognitive computing and big data analytics. Wiley, Indianapolis, IN

IBM Corporation (2011) IBM SPSS Modeler CRISP-DM guide. IBM Corporation, Armonk, NY

IBM Corporation (2014) The four V's of big data. IBM Big Data & Analytics Hub. http://www.ibmbigdatahub.com/infographic/four-vs-big-data

International Plain Language Federation (2018). Plain Language definition. http://www.iplfederation.org/. Accessed 5 Apr 2018

Lohr S (2014) For big-data scientists, 'Janitor Work' is key hurdle to insights. New York Times, August 17 2014. http://www.nytimes.com/2014/08/18/technology/for-big-data-scientists-hurdle-to-insights-is-janitor-work.html?_r=0

Maney K (2014) 'Big Data' will change how you play, see the doctor, even eat. Newsweek, July 24 2014. http://www.newsweek.com/2014/08/01/big-data-big-data-companies-260864.html

Rais-Ghasem M, Grosset R, Petitclerc M, Wei Q (2013) Towards semantic data analysis. IBM Canada, Ltd., Ottawa, Ontario

Shyr J, Spisic D (2014) Automated data analysis for Big Data. WIREs Comp Stats 6:359–366

Shyr J, Spisic D, Chu J, Han S, Zhang XY (2013). Relationship discovery in business analytics. In: JSM Proceedings, Social Statistics Section. Alexandria, VA: American Statistical Association. pp 5146–5158

Sokol L, Jonas J (2012) Using entity analytics to greatly increase the accuracy of your models quickly and easily. IBM Corporation, Armonk, NY

Techopedia (2018). Data lineage. https://www.techopedia.com/definition/28040/data-lineage. Accessed 5 Apr 2018

Thornton C, Hutter F, Hoos HH, Leyton-Brown K (2013) Auto-WEKA: combined selection and hyperparameter optimization of classification algorithm. In: Proceedings of the 19th ACM SIGKDD international conference on knowledge discovery and data mining. New York, NY: Association for Computing Machinery. 847–855

Part II
Methodology

Chapter 3
Statistical Leveraging Methods in Big Data

Xinlian Zhang, Rui Xie, and Ping Ma

Abstract With the advance in science and technologies in the past decade, big data becomes ubiquitous in all fields. The exponential growth of big data significantly outpaces the increase of storage and computational capacity of high performance computers. The challenge in analyzing big data calls for innovative analytical and computational methods that make better use of currently available computing power. An emerging powerful family of methods for effectively analyzing big data is called statistical leveraging. In these methods, one first takes a random subsample from the original full sample, then uses the subsample as a surrogate for any computation and estimation of interest. The key to success of statistical leveraging methods is to construct a data-adaptive sampling probability distribution, which gives preference to those data points that are influential to model fitting and statistical inference. In this chapter, we review the recent development of statistical leveraging methods. In particular, we focus on various algorithms for constructing subsampling probability distribution, and a coherent theoretical framework for investigating their estimation property and computing complexity. Simulation studies and real data examples are presented to demonstrate applications of the methodology.

Keywords Randomized algorithm · Leverage scores · Subsampling · Least squares · Linear regression

3.1 Background

With the advance in science and technologies in the past decade, big data has become ubiquitous in all fields. The extraordinary amount of big data provides unprecedented opportunities for data-driven knowledge discovery and decision making. However, the task of analyzing big data itself becomes a significant

X. Zhang · R. Xie · P. Ma (✉)
Department of Statistics, University of Georgia, Athens, GA, USA
e-mail: xinlian.zhang25@uga.edu; ruixie@uga.edu; pingma@uga.edu

© Springer International Publishing AG, part of Springer Nature 2018
W. K. Härdle et al. (eds.), *Handbook of Big Data Analytics*, Springer Handbooks of Computational Statistics, https://doi.org/10.1007/978-3-319-18284-1_3

challenge. Key features of big data, including large volume, vast variety and high velocity, all contribute to the challenge of the analysis. Among these, the large volume problem is of great importance. On one hand, the number of predictors for big data may be ultra-large, and this is encountered frequently in genetics and signal processing study. The ultra-high dimension of predictors is referred to as the curse of dimensionality. One of the most pressing needs and continuous efforts in alleviating the curse of dimensionality is to develop new techniques and tools to achieve dimension reduction and variable selection with good properties (Bhlmann and van de Geer 2011; Friedman et al. 2001). On the other hand, we often encounter cases in which sample size is ultra-large. When the sample size reaches a certain scale, although it is considered as preferable in the classical regime of statistical theory, the computational costs of many statistical methods become too expensive to carry out in practice. The topic of this chapter focuses on analyzing big data with ultra-large sample sizes.

A Computer Engineering Solution A computer engineering solution to the big data problem is to build more powerful computing facilities. Indeed, in the past decade, high performance computing platforms such as supercomputers and cloud computing have been developed rapidly. However, none of these technologies by themselves can fully solve this big data problem. The supercomputers are precious computing resources and cannot be allocated to everyone. As for cloud computing, it does possess the advantage of large storage capacities and relatively cheap accessibility. However, problems arise when transferring big data over limited Internet uplink bandwidth. Not to mention that the transferring process also raises new privacy and security concerns. More importantly, the exponential growth of the volume of big data significantly outpaces the increase of the storage and computational capacity of high performance computers.

Computational Capacity Constrained Statistical Methods Given fixed computational capacity, analytical and computational methods need to be adapted to this constraint. One straightforward approach is *divide-and-conquer*. In this approach, one divides the large dataset into small and manageable pieces and performs statistical analysis on each of the small pieces. These results from small pieces are then combined together to provide a final result for the full sample. One notable feature of this procedure is the significant reduction in computing time in a distributed computing environment. However, divide-and-conquer method has its own limitations. On one hand, the efficiency of divide-and-conquer methods still relies on the parallel computing environment, which is not available at all times; on the other hand, it is challenging to develop a universal scheme for combining the results from smaller pieces to form a final estimate with good statistical properties. See Agarwal and Duchi (2011), Chen and Xie (2014), Duchi et al. (2012), Zhang et al. (2013) for applications of this approach.

Fundamentally novel analytical and computational methods are still much needed to harness the power and capture the information of these big data. Another emerging family of methods to tackle the super-large sample problem is the family of *statistical leveraging methods* (Drineas et al. 2006, 2010; Ma et al. 2013, 2014; Ma and Sun 2015; Mahoney 2011). The key idea is to draw a manageable small subsample from the full sample, and perform statistical analysis on this subsample. For the rest of this chapter, we focus on the statistical leveraging methods under a linear model setup. Consider the following linear model

$$y_i = \mathbf{x}_i^T \boldsymbol{\beta}_0 + \epsilon_i, \qquad i = 1, \ldots, n \tag{3.1}$$

where y_i is the response, $\mathbf{x}_i$ is the p-dimensional *fixed* predictor, $\boldsymbol{\beta}_0$ is the $p \times 1$ coefficient vector, and the noise term is $\epsilon_i \overset{\text{i.i.d.}}{\sim} N(0, \sigma^2)$. To emphasize, we are dealing with big data in cases where sample size n is ultra large, and $n \gg p$.

Written in vector-matrix format, the linear model in (3.1) becomes

$$\mathbf{y} = \mathbf{X}\boldsymbol{\beta}_0 + \boldsymbol{\epsilon}, \tag{3.2}$$

where $\mathbf{y}$ is the $n \times 1$ response vector, $\mathbf{X}$ is the $n \times p$ *fixed* predictor or design matrix, $\boldsymbol{\beta}_0$ is the $p \times 1$ coefficient vector, and the noise vector is $\boldsymbol{\epsilon} \sim N(\mathbf{0}, \sigma^2 I)$. In this case, the unknown coefficient $\boldsymbol{\beta}_0$ can be estimated through a least squares (LS) procedure, i.e.,

$$\hat{\boldsymbol{\beta}}_{\text{LS}} = \text{argmin}_{\boldsymbol{\beta}} ||\mathbf{y} - \mathbf{X}\boldsymbol{\beta}||^2,$$

where $||\cdot||$ represents the Euclidean norm on $\mathbb{R}^n$. If the predictor matrix $\mathbf{X}$ is of full column rank, the LS estimator $\hat{\boldsymbol{\beta}}_{\text{LS}}$ can be expressed as

$$\hat{\boldsymbol{\beta}}_{\text{LS}} = (\mathbf{X}^T\mathbf{X})^{-1}\mathbf{X}^T\mathbf{y}. \tag{3.3}$$

The general statistical leveraging method in linear model (Drineas et al. 2012, 2006; Mahoney 2011) is given below.

As a by-product, Step 1 of Algorithm 1 provides a random "sketch" of the full sample. Thus visualization of the subsamples obtained enables a surrogate visualization of the full data, which is one of the unique features of subsample methods.

Successful application of statistical leveraging methods relies on both effective design of subsampling probabilities, through which influential data points are sampled with higher probabilities, and an appropriate way to model the subsampled data. So we will review a family of statistical algorithms that employ different subsampling probabilities as well as different modeling approaches for the subsampled data.

Algorithm 1: Statistical leveraging in linear model

---

1 **Step 1(subsampling)**: Calculate the sampling probability $\{\pi_i\}_{i=1}^n$ based on leveraging-related methods with the full sample $\{y_i, \mathbf{x}_i\}_{i=1}^n$, use $\{\pi_i\}_{i=1}^n$ to take a random subsample of size $r > p$ and denote the subsample as $\{y_i^*, \mathbf{x}_i^*\}_{i=1}^r$.

2 **Step 2(model-fitting)**: Fit the linear model to the subsample $\{y_i^*, \mathbf{x}_i^*\}_{i=1}^r$ via a weighted LS with weights $1/\pi_i$, and return the estimator $\tilde{\boldsymbol{\beta}}$.

For the rest of this chapter, we first describe the motivation and detailed layout for statistical leveraging methods. Then corresponding asymptotic properties for these estimators are demonstrated. Furthermore, synthetic and real-world data examples are analyzed. Finally, we conclude this chapter with a discussion of some open questions in this area.

3.2 Leveraging Approximation for Least Squares Estimator

In this chapter, the idea of statistical leveraging is explained with a focus on the linear model. We mainly tackle the challenge of designing subsampling probability distribution, which is the core of statistical leveraging sampling. Various leveraging sampling procedures are discussed from different perspectives. A short summary of all discussed approaches for modeling the subsampled data concludes this section.

3.2.1 Leveraging for Least Squares Approximation

For linear model in (3.2), the LS estimator is $\hat{\boldsymbol{\beta}}_{\text{LS}} = (\mathbf{X}^T\mathbf{X})^{-1}\mathbf{X}^T\mathbf{y}$. The predicted response vector can be written as $\hat{\mathbf{y}} = \mathbf{H}\mathbf{y}$, where $\mathbf{H} = \mathbf{X}(\mathbf{X}^T\mathbf{X})^{-1}\mathbf{X}^T$ is the named as hat matrix for its purpose of getting $\hat{\mathbf{y}}$. The ith diagonal element of $\mathbf{H}$, $h_{ii} = \mathbf{x}_i^T(\mathbf{X}^T\mathbf{X})^{-1}\mathbf{x}_i$, where $\mathbf{x}_i^T$ is the ith row of $\mathbf{X}$, is the *statistical leverage* of ith observation. The concept of "statistical leverage" is historically originated from regression diagnostics, where the statistical leverage is used in connection with analyses aiming at quantifying the extent of how influential an observation is for model prediction (Chatterjee and Hadi 1986; Hoaglin and Welsch 1978; Velleman and Welsch 1981). If rank($\mathbf{H}$) $=$ rank($\mathbf{X}$) $= p$ (assuming that $n \gg p$), then we have trace($\mathbf{H}$) $= p$, i.e. $\sum_{i=1}^{n} h_{ii} = p$. A widely recommended rule of thumb in practice for "large" leverage score is $h_{ii} > 2p/n$ (Hoaglin and Welsch 1978). Also note that $\text{Var}(e_i) = \text{Var}(\hat{y}_i - y_i) = (1 - h_{ii})\sigma^2$. Thus if the ith observation is a "high leverage point," then the value of y_i has a large impact on the predicted value $\hat{y}_i$ and the corresponding residual has a small variance. To put it another way, the fitted regression line tends to pass closer to the high leverage data points.

As an illustration, we provide a toy example in Fig. 3.1. For $i = 1, \ldots, 10{,}000$, we simulate $y_i = -1 + x_i + \epsilon_i$, where x_i is generated from t-distribution with 2 degree of freedom and $\epsilon_i \overset{\text{i.i.d.}}{\sim} N(0, 4)$. The left panel of Fig. 3.1 displays the scatterplot for the full sample and associated fitted LS regression line. The right panel displays the scatterplot for a subsample of size 20 drawn using sampling probabilities constructed from leverage scores (h_{ii}), i.e.

$$\pi_i = \frac{h_{ii}}{\sum_{i=1}^{n} h_{ii}} = \frac{h_{ii}}{p}, \quad i = 1, \ldots, n. \tag{3.4}$$

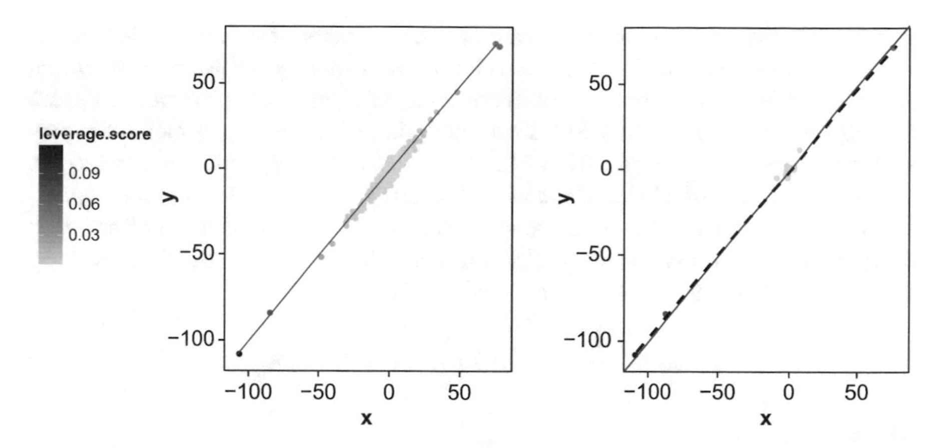

Fig. 3.1 Illustration of motivation of statistical leveraging. The left panel displays the scatterplot of full sample and the corresponding fitted LS regression line. In the right panel, only 20 points were chosen using sampling probabilities constructed from leverage scores (h_{ii}) are plotted. The solid line is the fitted LS regression line with full dataset and the dashed line is the fitted weighted LS regression line from 20 sampled points. In both panels, the color of points corresponds to the leverage score, the higher the leverage score, the darker the color

The solid line is the fitted LS regression line with full sample and dashed line is the fitted weighted LS regression (as described in Step 2 of Algorithm 1) line with 20 subsampled data points. As shown in the left panel of Fig. 3.1, there are several points on the upper right and lower left corner with relatively higher leverage scores, and they are close to the fitted regression line of the full sample. The right panel of Fig. 3.1 implies that fitted weighted LS regression line from the 20 subsampled points (dashed line) is good enough to recover the LS regression line using the full sample.

It is worth noting that one cannot take the subsample by simply using 20 points with highest leverage scores in a deterministic way. The reason is that once the observations are ordered with respect to corresponding leverage scores, then the joint distribution of certain part of the data, e.g. the top 20 observations, will be different from the distribution of original data. Estimators constructed from these deterministic subsamples are biased compared to both the true parameter and the LS estimators from the full sample (Coles et al. 2001). However, if the subsamples are collected in a random fashion, with the probabilities formed by normalizing leverage scores of all observations, the estimates based on the random samples are still guaranteed to be asymptotically unbiased to the true parameter as well as the LS estimators of the full sample. See Sect. 3.3 for more details.

Figure 3.1 also shows how the statistical leveraging methods provide a way of visualizing large datasets.

Following Algorithm 1, the statistical leveraging methods involve computing the exact or approximating statistical leverage scores of the observations, constructing subsample probabilities according to those scores as in Eq. (3.4), sampling data points randomly, and estimating the weighted LS estimator from subsampled data as the final estimate for the full data. We refer to this sampling procedure as *leverage-based sampling* and denote the estimator from this particular procedure as *basic leveraging estimator (BLEV)*. Statistical analysis of the leveraging method is provided in Drineas et al. (2006, 2010), Ma et al. (2013), Raskutti and Mahoney (2014).

As an extension of BLEV, Ma et al. (2014, 2013) proposed *shrinked leveraging-based sampling*, which takes subsample using a convex combination of an exact or approximate leverage scores distribution and the uniform distribution, thereby obtaining the benefits of both. For example, we let

$$\pi_i = \lambda \frac{h_{ii}}{p} + (1-\lambda)\frac{1}{n}, \quad i = 1, \ldots, n, \tag{3.5}$$

where $0 < \lambda < 1$.

We refer to the estimator resulting from sampling data points using (3.5), and estimating the weighted LS estimation on the sampled data points as *shrinked leveraging estimator (SLEV)*. In particular, we use the notation SLEV(λ) to differentiate various levels of shrinkage whenever needed.

3.2.2 A Matrix Approximation Perspective

Most of the modern statistical models are based on matrix calculation, so in a sense when applied to big data, the challenges they face can be solved through using easy-to-compute matrices to approximate the original input matrix, e.g., a low-rank matrix approximation. For example, one might randomly sample a small number of rows from an input matrix and use those rows to construct a low-rank approximation to the original matrix. In this way, it is not hard to construct "worst-case" inputs for which *uniform* random sampling performs very poorly (Drineas et al. 2006; Mahoney 2011). Motivated by this idea, a substantial amount of efforts have been devoted to developing improved algorithms for matrix-based problems that construct the random sample in a *nonuniform* and data-dependent approach (Mahoney 2011), such as least-squares approximation (Drineas et al. 2006, 2010), least absolute deviations regression (Clarkson et al. 2013; Meng and Mahoney 2013), and low-rank matrix approximation (Clarkson and Woodruff 2013; Mahoney and Drineas 2009).

In essence, these procedures are composed of the following steps. The first step is to compute exact or approximate (Drineas et al. 2012; Clarkson et al. 2013), statistical leverage scores of the design matrix. The second step is to use those scores to form a discrete probability distribution with which to sample columns and/or rows randomly from the input data. Finally, the solution of the subproblem is used as an approximation to the solution of the original problem. Thus, the leveraging methods can be considered as a special case of this approach. A detailed discussion of this approach can be found in the recent review monograph on randomized algorithms for matrices and matrix-based data problems (Mahoney 2011).

3.2.3 The Computation of Leveraging Scores

The key for calculating the leveraging scores lies in the hat matrix **H**, which can be expressed as $\mathbf{H} = \mathbf{U}\mathbf{U}^T$, where **U** is a matrix with its columns formed by any orthogonal basis for the column space of **X**, e.g., the Q matrix from a QR decomposition or the matrix of left singular vectors from the thin singular value decomposition (SVD). Thus, the leverage score of the ith observation, i.e. the ith diagonal element of **H**, h_{ii}, can also be expressed as

$$h_{ii} = ||\mathbf{u}_i||^2, \tag{3.6}$$

where $\mathbf{u}_i$ is the ith row of **U**. Using Eq. (3.6), the leverage scores h_{ii}, for $i = 1, 2, \ldots, n$ can be obtained. In practice, as a surrogate, $\tilde{h}_{ii}$ are computed as the approximated leverage score in some cases.

The theoretical and practical characterization of the computational cost of leveraging algorithms is of great importance. The running time of the leveraging algorithms depends on both the time to construct the sampling probabilities, $\{\pi_i\}_{i=1}^n$, and the time to solve the optimization problem using the subsample. For uniform sampling, the computational cost of subsampling is negligible and the computational cost depends on the size of the subsample. For statistical leveraging methods, the running time is dominated by computation of the exact or approximating leverage scores. A naïve implementation involves computing a matrix **U**, through, e.g., QR decomposition or SVD, spanning the column space of **X** and then reading off the Euclidean norms of rows of **U** to obtain the exact leverage scores. This procedure takes $O(np^2)$ time, which is at the same order with solving the original problem exactly (Golub and Van Loan 1996). Fortunately, there are available algorithms, e.g., Drineas et al. (2012), that compute relative-error approximations to leverage scores of **X** in roughly $O(np \log p)$ time. See Drineas et al. (2006) and Mahoney (2011) for more detailed algorithms as well as their empirical applications. These implementations demonstrate that, for matrices as small as several thousand by several hundred, leverage-based algorithms can be competitive in terms of running time with the computation of QR decomposition or the SVD with packages like LAPACK. See Avron et al. (2010), Meng et al. (2014) for more details on this topic. In the next part of this section, another innovative procedure will provide the potential for reducing the computing time to the order of $O(np)$.

3.2.4 An Innovative Proposal: Predictor-Length Method

Ma et al. (2016) introduced an algorithm that allows a significant reduction of computational cost. In the simple case of $p = 1$, and no intercept, we have that $h_{ii} = x_i^2$, i.e. the larger the absolute value of an observation, the higher the leverage score. The idea for our new algorithm is to extend this idea to the case of $p > 1$ by using simply Euclidean norm of each observation to approximate leverage scores and indicate the importance of the observation. That is, we define sampling probabilities

$$\pi_i = \frac{\|\mathbf{x}_i\|}{\sum_{i=1}^{n} \|\mathbf{x}_i\|}, \quad i = 1, \ldots, n. \tag{3.7}$$

If the step 1 of Algorithm 1 is carried out using the sampling probabilities in Eq. (3.7), then we refer to the corresponding estimator as the *predictor-length estimator (PL)*. It is very important to note that the computational cost for this procedure is only $O(np)$, i.e. we only need to go through each observation and calculate its Euclidean norm.

For illustration, in Figs. 3.2 and 3.3, we compare the subsampling probabilities in BLEV, SLEV(0.1), and PL using predictors generated from normal distribution and t-distribution. Compared to normal distribution, t-distribution is known to have heavier tail. So it is conceivable that observations generated from normal distribution tend to have more homogeneous subsampling probabilities, i.e. the circles in Fig. 3.2 are of similar sizes, whereas observations generated from t-distribution tend to have heterogeneous subsampling probabilities, i.e. high probabilities will be assigned to only a few data points, represented in Fig. 3.3 as that a few relatively huge circles on the upper right corner and down left corner. From Fig. 3.3, we clearly observe that the subsampling probabilities used to construct BLEV are much more dispersive than SLEV(0.1) especially in the case of t. It is interesting to note that the probabilities of observations for PL roughly lie in between that of BLEV and SLEV(0.1).

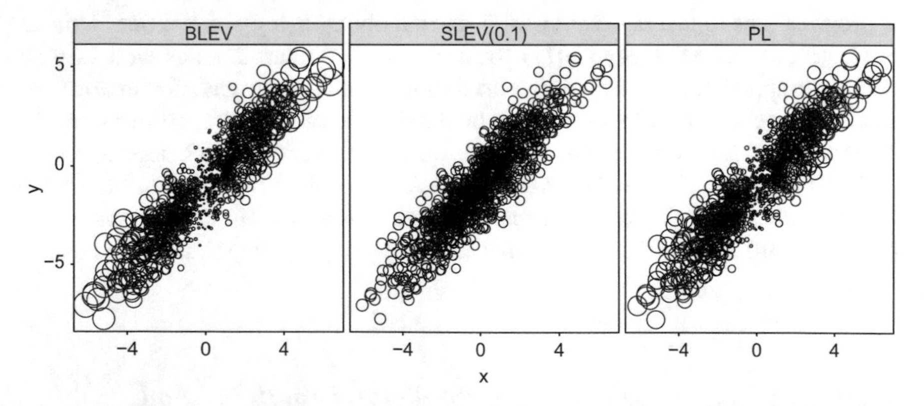

Fig. 3.2 Illustration of different subsampling probabilities with predictor generated from normal distribution. For $i = 1, \ldots, 1000$, $y_i = -1 + x_i + \epsilon_i$, where x_i is generated i.i.d. from $N(0, 4)$ and $\epsilon_i \sim N(0, 1)$. Each circle represents one observation, and the area of each circle is proportional to its *subsampling probability* under each scheme. The area of point with maximum probability in all three probability distributions are set to 100 times that of the point with minimum probability

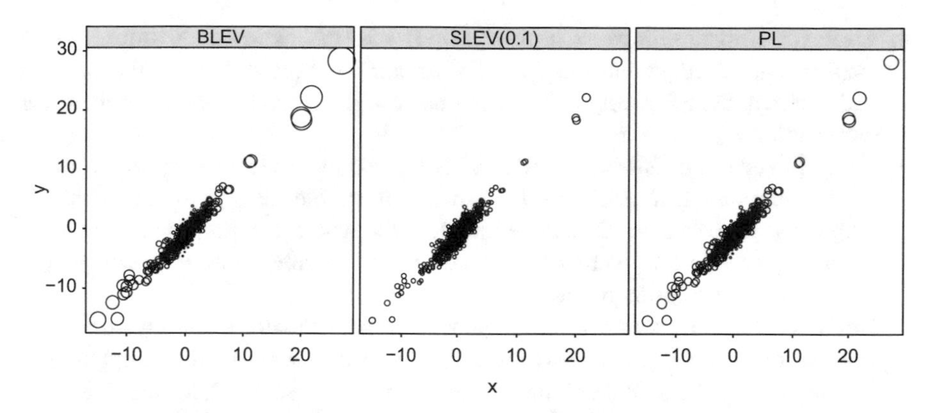

Fig. 3.3 Illustration of different subsampling probabilities with predictor generated from *t*-distribution. For $i = 1, \ldots, 1000$, $y_i = -1 + x_i + \epsilon_i$, where x_i is generated i.i.d. from *t*-distribution with *df*=2 and $\epsilon_i \overset{\text{i.i.d.}}{\sim} N(0, 1)$. Each circle represents one observation, and the area of each circle is proportional to its *subsampling probability* under each scheme. The area of point with maximum probability in all three probability distributions are set to 100 times that of the point with minimum probability

3.2.5 *More on Modeling*

As mentioned in Step 2 of Algorithm 1, after subsampling, the basic framework of statistical leveraging requires to *rescale, i.e. weight* subsamples appropriately using the same probability distribution as used for subsampling. The purpose for this *weighted LS* step is essentially to construct unbiased estimator of coefficient vector for the linear regression analysis (Drineas et al. 2012, 2006; Mahoney 2011). However, *unweighted leveraging estimators*, in which the modeling of sampled data is carried out by plain least square, is also considered in the literature. That means, in Step 2 of Algorithm 1, instead of solving weighted LS problem, we solve for an *unweighted LS estimator*. Ma et al. (2013) first proposed several versions of unweighted methods that suggest potential improvement over BLEV. The asymptotic properties of unweighted estimators will be discussed later.

3.2.6 *Statistical Leveraging Algorithms in the Literature: A Summary*

Based on the different combinations of subsampling and modeling strategies we list several versions of the statistical leveraging algorithms that are of particular interest in practice.

- **Uniform Subsampling Estimator (UNIF)** is the estimator resulting from *uniform subsampling* and *weighted LS estimation*. Note that when the weights are uniform, then the weighted LS estimator is the same as the unweighted LS estimator.
- **Basic Leveraging Estimator (BLEV)** is the estimator resulting from *leverage-based sampling* and *weighted LS estimation* on the sampled data, which is originally proposed in Drineas et al. (2006), where the empirical statistical leverage scores of X were used to construct the subsample and weight the subsample optimization problem.
- **Shrinked Leveraging Estimator (SLEV)** is the estimator resulting from sampling using probabilities in (3.5) and *weighted LS estimation* on the sampled data. The motivation for SLEV will be further elaborated in Sect. 3.3. Similar ideas are also proposed in the works of importance sampling (Hesterberg 1995).
- **Unweighted Leveraging Estimator (LEVUNW)** is the estimator resulting from *leverage-based sampling* and *unweighted LS estimation* on the sampled data. The sampling and reweighing steps in this procedure are done according to different distributions, so the results for the bias and variance of this estimator might differ from the previous ones.
- **Predictor Length Estimator (PL)** is the estimator resulting from sampling using probabilities in (3.7) and *weighted LS estimation* on the sampled data.

3.3 Statistical Properties of Leveraging Estimator

In this section, we provide an analytic framework for evaluating the statistical properties of statistical leveraging. We examine the results for bias and variance of leveraging estimator discussed in previous sections.

The challenges for analyzing the bias and variance of leveraging estimator come from two parts. One is the two layers of randomness in the estimation, randomness in the linear model and from the random subsampling; the other is that the estimation relies on random sampling through a nonlinear function of the inverse of random sampling matrix. A Taylor series analysis is used to overcome the challenges so that the leveraging estimator can be approximated as a linear combination of random sampling matrices.

3.3.1 Weighted Leveraging Estimator

We start with bias and variance analysis of leveraging estimator $\tilde{\boldsymbol{\beta}}$ in Algorithm 1. The estimator can be written as

$$\tilde{\boldsymbol{\beta}} = (\mathbf{X}^T\mathbf{W}\mathbf{X})^{-1}\mathbf{X}^T\mathbf{W}\mathbf{y}, \tag{3.8}$$

where $\mathbf{W}$ is an $n \times n$ diagonal matrix. We can treat $\tilde{\boldsymbol{\beta}}$ as a function of $\mathbf{w} = (w_1, w_2, \ldots, w_n)^T$, the diagonal entries of $\mathbf{W}$, denoted as $\tilde{\boldsymbol{\beta}}(\mathbf{w})$. Randomly sampling with replacement makes $\mathbf{w} = (w_1, w_2, \ldots, w_n)^T$ have a scaled multinomial distribution,

$$\mathbf{Pr}\left[w_1 = \frac{k_1}{r\pi_1}, w_2 = \frac{k_2}{r\pi_2}, \ldots, w_n = \frac{k_n}{r\pi_n}\right] = \frac{r!}{k_1!k_2!\ldots,k_n!}\pi_1^{k_1}\pi_2^{k_2}\cdots\pi_n^{k_n},$$

with mean $E[\mathbf{w}] = \mathbf{1}$. To analyze the statistical properties of $\tilde{\boldsymbol{\beta}}(\mathbf{w})$, Taylor series expansion is performed around the vector $\mathbf{w}_0$, which is set to be the all-ones vector, i.e., $\mathbf{w}_0 = \mathbf{1}$. As a result, $\tilde{\boldsymbol{\beta}}(\mathbf{w})$ can be expanded around the full sample ordinary LS estimator $\hat{\boldsymbol{\beta}}_{\mathrm{LS}}$, as we have $\tilde{\boldsymbol{\beta}}(\mathbf{1}) = \hat{\boldsymbol{\beta}}_{\mathrm{LS}}$. Then we come up with the following lemma, the proof of which can be found in Ma et al. (2013).

Lemma 1 *Let $\tilde{\boldsymbol{\beta}}$ be the output of the Algorithm 1, obtained by solving the weighted LS problem of (3.8), where $\mathbf{w}$ denotes the probabilities used to perform the sampling and reweighting. Then, a Taylor expansion of $\tilde{\boldsymbol{\beta}}$ around the point $\mathbf{w}_0 = \mathbf{1}$ yields*

$$\tilde{\boldsymbol{\beta}} = \hat{\boldsymbol{\beta}}_{LS} + (\mathbf{X}^T\mathbf{X})^{-1}\mathbf{X}^T Diag\left\{\hat{\mathbf{e}}\right\}(\mathbf{w} - \mathbf{1}) + R_W, \tag{3.9}$$

where $\hat{\mathbf{e}} = \mathbf{y} - \mathbf{X}\hat{\boldsymbol{\beta}}_{\mathrm{LS}}$ is the LS residual vector, and where R_W is the Taylor expansion remainder.

Given Lemma 1, we establish the expression for conditional and unconditional expectations and variances for the weighted sampling estimators in the following Lemma 2.

Conditioned on the data $\mathbf{y}$, the expectation and variance are provided by the first two expressions in Lemma 2; and the last two expressions in Lemma 2 give similar results, except that they are not conditioned on the data $\mathbf{y}$.

Lemma 2 *The conditional expectation and conditional variance for the algorithmic leveraging procedure Algorithm 1, i.e., when the subproblem solved is a weighted LS problem, are given by:*

$$\mathbf{E}_{\mathbf{w}}\left[\tilde{\boldsymbol{\beta}}|\mathbf{y}\right] = \hat{\boldsymbol{\beta}}_{LS} + \mathbf{E}_{\mathbf{w}}\left[R_W\right]; \tag{3.10}$$

$$\mathbf{Var}_{\mathbf{w}}\left[\tilde{\boldsymbol{\beta}}|\mathbf{y}\right] = (\mathbf{X}^T\mathbf{X})^{-1}\mathbf{X}^T\left[Diag\left\{\hat{\mathbf{e}}\right\}Diag\left\{\frac{1}{r\pi}\right\}Diag\left\{\hat{\mathbf{e}}\right\}\right]\mathbf{X}(\mathbf{X}^T\mathbf{X})^{-1} + \mathbf{Var}_{\mathbf{w}}\left[R_W\right], \tag{3.11}$$

where $\mathbf{W}$ specifies the probability distribution used in the sampling and rescaling steps. The unconditional expectation and unconditional variance for the algorithmic leveraging procedure Algorithm 1 are given by:

$$\mathbf{E}\left[\tilde{\boldsymbol{\beta}}\right] = \boldsymbol{\beta}_0; \tag{3.12}$$

$$\mathbf{Var}\left[\tilde{\boldsymbol{\beta}}\right] = \sigma^2(\mathbf{X}^T\mathbf{X})^{-1} + \frac{\sigma^2}{r}(\mathbf{X}^T\mathbf{X})^{-1}\mathbf{X}^T Diag\left\{\frac{(1-h_{ii})^2}{\pi_i}\right\}\mathbf{X}(\mathbf{X}^T\mathbf{X})^{-1} + \mathbf{Var}\left[R_W\right]. \tag{3.13}$$

The estimator $\tilde{\boldsymbol{\beta}}$, conditioning on the observed data $\mathbf{y}$, is approximately unbiased to the full sample LS estimator $\hat{\boldsymbol{\beta}}_{\mathrm{LS}}$, when the linear approximation is valid, i.e., when the $\mathbf{E}[R_W]$ is negligible; and the estimator $\tilde{\boldsymbol{\beta}}$ is unbiased relative to the true β_0 of the parameter vector β.

For the first term of conditional variance of Eq. (3.11) and the second term of unconditional variance of Eq. (3.13), both of them are inversely proportional to the subsample size r; and both contain a sandwich-type expression, the middle of which involves the leverage scores interacting with the sampling probabilities.

Based on Lemma 2, the conditional and unconditional expectation and variance for the BLEV, PLNLEV, and UNIF procedures can be derived, see Ma et al. (2013). We will briefly discuss the relative merits of each procedure.

The result of Eq. (3.10) shows that, given a particular data set $(\mathbf{X}, \mathbf{y})$, the leveraging estimators (BLEV and PLNLEV) can approximate well $\hat{\boldsymbol{\beta}}_{\mathrm{LS}}$. From the statistical inference perspective, the unconditional expectation result of Eq. (3.12) shows that the leveraging estimators can infer well $\boldsymbol{\beta}_0$.

For the BLEV procedure, the conditional variance and the unconditional variance depend on the size of the $n \times p$ matrix $\mathbf{X}$ and the number of samples r as p/r. If one chooses $p \ll r \ll n$, the variance size-scale can be controlled to be very small. The sandwich-type expression containing the leverage scores $1/h_{ii}$, suggests that the variance could be arbitrarily large due to small leverage scores. This disadvantage of BLEV motivates the SLEV procedure. In SLEV, the sampling/rescaling probabilities approximate the h_{ii} but are bounded from below, therefore preventing the arbitrarily large inflation of the variance. For the UNIF procedure, since the variance size-scale is large, e.g., compared to the p/r from BLEV, these variance expressions will be large unless r is nearly equal to n. Moreover, the sandwich-type expression in the UNIF procedure depends on the leverage scores in a way that is not inflated to arbitrarily large values by very small leverage scores.

3.3.2 *Unweighted Leveraging Estimator*

In this section, we consider the unweighted leveraging estimator, which is different from the weighted estimators, in that the sampling and reweighting are done according to different distributions. That is, modifying Algorithm 1, no weights are used for least squares. Similarly, we examine the bias and variance of the unweighted leveraging estimator $\tilde{\boldsymbol{\beta}}_{\mathrm{LEVUNW}}$. The Taylor series expansion is performed to get the following lemma, the proof of which may be found in Ma et al. (2013).

Lemma 3 *Let $\tilde{\boldsymbol{\beta}}_{LEVUNW}$ be the output of the modified Algorithm 1, obtained by solving the unweighted LS problem of (3.3), where the random sampling is performed with probabilities proportional to the empirical leverage scores. Then, a Taylor expansion of $\tilde{\boldsymbol{\beta}}_{LEVUNW}$ around the point $\mathbf{w}_0 = r\boldsymbol{\pi}$ yields*

$$\tilde{\boldsymbol{\beta}}_{LEVUNW} = \hat{\boldsymbol{\beta}}_{WLS} + (\mathbf{X}^T \mathbf{W}_0 \mathbf{X})^{-1} \mathbf{X}^T Diag\{\hat{\mathbf{e}}_w\} (\mathbf{w} - r\boldsymbol{\pi}) + R_{LEVUNW}, \quad (3.14)$$

where $\hat{\mathbf{e}}_w = \mathbf{y} - \mathbf{X}\hat{\boldsymbol{\beta}}_{WLS}$ *is the LS residual vector,* $\hat{\boldsymbol{\beta}}_{WLS} = (\mathbf{X}^T\mathbf{W}_0\mathbf{X})^{-1}\mathbf{X}\mathbf{W}_0\mathbf{y}$ *is the full sample weighted LS estimator,* $\mathbf{W}_0 = Diag\{r\boldsymbol{\pi}\} = Diag\{rh_{ii}/p\}$*, and* R_{LEVUNW} *is the Taylor expansion remainder.*

Even though Lemma 3 is similar to Lemma 1, the point about which the Taylor expansion is calculated, and the factors that left multiply the linear term, are different for the LEVUNW than they were for the weighted leveraging estimators due to the fact that the sampling and reweighting are performed according to different distributions.

Then the following Lemma 4, providing the expectations and variances of the LEVUNW, both conditioned and unconditioned on the data $\mathbf{y}$, can be established given the Lemma 3.

Lemma 4 *The conditional expectation and conditional variance for the LEVUNW procedure are given by:*

$$\begin{aligned}\mathbf{E}_{\mathbf{w}}\left[\tilde{\boldsymbol{\beta}}_{LEVUNW}|\mathbf{y}\right] &= \hat{\boldsymbol{\beta}}_{WLS} + \mathbf{E}_{\mathbf{w}}\left[R_{LEVUNW}\right];\\ \mathbf{Var}_{\mathbf{w}}\left[\tilde{\boldsymbol{\beta}}_{LEVUNW}|\mathbf{y}\right] &= (\mathbf{X}^T\mathbf{W}_0\mathbf{X})^{-1}\mathbf{X}^T Diag\{\hat{e}_w\}\,\mathbf{W}_0 Diag\{\hat{e}_w\}\,\mathbf{X}(\mathbf{X}^T\mathbf{W}_0\mathbf{X})^{-1}\\ &\quad + \mathbf{Var}_{\mathbf{w}}\left[R_{LEVUNW}\right].\end{aligned}$$

where $\mathbf{W}_0 = Diag\{r\boldsymbol{\pi}\}$*, and where* $\hat{\boldsymbol{\beta}}_{\mathrm{WLS}} = (\mathbf{X}^T\mathbf{W}_0\mathbf{X})^{-1}\mathbf{X}\mathbf{W}_0\mathbf{y}$ *is the full sample weighted LS estimator. The unconditional expectation and unconditional variance for the LEVUNW procedure are given by:*

$$\begin{aligned}\mathbf{E}\left[\tilde{\boldsymbol{\beta}}_{LEVUNW}\right] &= \boldsymbol{\beta}_0;\\ \mathbf{Var}\left[\tilde{\boldsymbol{\beta}}_{LEVUNW}\right] &= \sigma^2(\mathbf{X}^T\mathbf{W}_0\mathbf{X})^{-1}\mathbf{X}^T\mathbf{W}_0^2\mathbf{X}(\mathbf{X}^T\mathbf{W}_0\mathbf{X})^{-1}\\ &\quad + \sigma^2(\mathbf{X}^T\mathbf{W}_0\mathbf{X})^{-1}\mathbf{X}^T Diag\{I - P_{\mathbf{X},\mathbf{W}_0}\}\,\mathbf{W}_0 Diag\{I - P_{\mathbf{X},\mathbf{W}_0}\}\\ &\quad \times\mathbf{X}(\mathbf{X}^T\mathbf{W}_0\mathbf{X})^{-1} + \mathbf{Var}\left[R_{LEVUNW}\right]\end{aligned} \tag{3.15}$$

where $P_{\mathbf{X},\mathbf{W}_0} = \mathbf{X}(\mathbf{X}^T\mathbf{W}_0\mathbf{X})^{-1}\mathbf{X}^T\mathbf{W}_0$.

The estimator $\tilde{\boldsymbol{\beta}}_{\mathrm{LEVUNW}}$, conditioning on the observed data $\mathbf{y}$, is approximately unbiased to the full sample *weighted* LS estimator $\hat{\boldsymbol{\beta}}_{\mathrm{WLS}}$, when $\mathbf{E}_{\mathbf{w}}\left[R_{\mathrm{LEVUNW}}\right]$ is negligible; and the estimator $\tilde{\boldsymbol{\beta}}_{\mathrm{LEVUNW}}$ is unbiased relative to the "true" value $\boldsymbol{\beta}_0$ of the parameter vector $\boldsymbol{\beta}$.

Note that the unconditional variance in Eq. (3.15) is the same as the variance of uniform random sampling, since the leverage scores are all the same. The solutions to the weighted and unweighted LS problems are identical, since the problem being solved, when reweighting with respect to the uniform distribution, is not changed. Moreover, the variance is not inflated by small leverage scores. The conditional variance expression is also a sandwich-type expression. The center of the conditional variance, $\mathbf{W}_0 = Diag\{rh_{ii}/n\}$, is not inflated by very small leverage scores.

3.4 Simulation Study

In this section, we use some synthetic datasets to illustrate the efficiency of various leveraging estimators.

One hundred replicated datasets of sample size $n = 100{,}000$ were generated from

$$y_i = \mathbf{x}_i^T \boldsymbol{\beta}_0 + \epsilon_i,$$

where coefficient $\boldsymbol{\beta}_0$ was set to be $(\mathbf{1}_{10}, \mathbf{0.2}_{30}, \mathbf{1}_{10})^T$, and $\epsilon_i \overset{\text{i.i.d.}}{\sim} N(0, 3)$, the predictor $\mathbf{x}_i$ was generated from three different distributions: multivariate normal distribution (denoted as Normal here and after), multivariate t-distribution with $df = 2$ (denoted as T2 here and after), and multivariate Cauchy distribution (denoted as Cauchy here and after). Compared to normal distribution, t-distribution has a heavy tail. Tail of Cauchy distribution is even heavier compared to that of the t-distribution. For the multivariate normal distribution, the mean vector was set to be $\mathbf{1}_{50}$ and the covariance matrix to be Σ, the (i, j)th element of which was $\Sigma_{i,j} = 3 \times (0.6)^{|i-j|}$. For the multivariate t-distribution, we set the non-centrality parameter as $\mathbf{1}_{50}$, the covariance matrix as Σ and the degree of freedom as 2. For the multivariate Cauchy distribution, we used $\mathbf{1}_{50}$ for position vector and Σ defined above for dispersion matrix.

3.4.1 *UNIF* and *BLEV*

We applied the leveraging methods with different subsample sizes, $r = 2p, \ldots, 10p$, to each of 100 datasets, and calculated squared bias and variance of the leveraging estimators to the true parameters $\boldsymbol{\beta}_0$. In Fig. 3.4, we plotted the variance and squared bias of $\tilde{\boldsymbol{\beta}}_{\text{BLEV}}$ and $\tilde{\boldsymbol{\beta}}_{\text{UNIF}}$ for three multivariate distributions.

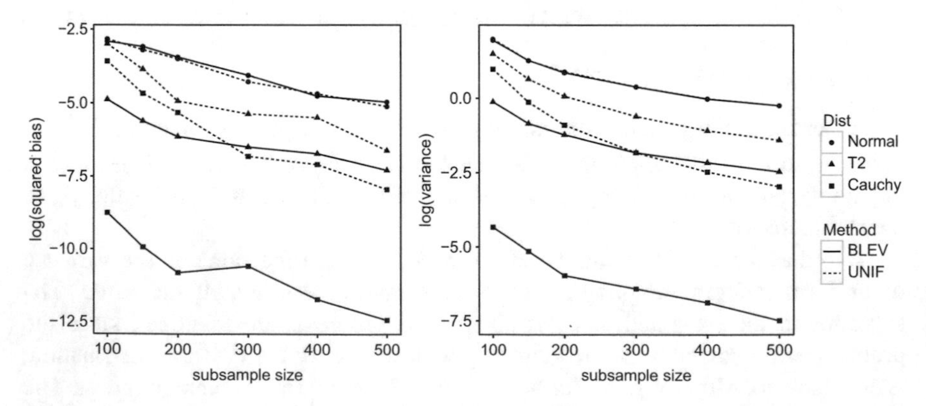

Fig. 3.4 Comparison of variances and squared biases of $\tilde{\boldsymbol{\beta}}_{\text{BLEV}}$ and $\tilde{\boldsymbol{\beta}}_{\text{UNIF}}$ in three distributions. In the graph, "Normal" stands for multivariate normal distribution, "T2" stands for multivariate t-distribution with degree of freedom 2, and "Cauchy" stands for the multivariate Cauchy distribution

Several features are worth noting about Fig. 3.4. First, in general the magnitude of bias is small compared to that of variance, corroborating our theoretical results on unbiasedness of estimators of leveraging methods in Sect. 1. Second, when the predictor vectors $\mathbf{x}_i$ were generated from normal distribution, the bias of BLEV and UNIF are close to each other, which is expected since we know from Fig. 3.2 that the leverage scores are very homogeneous. Same observation exists for the variance. In contrast, in case of T2 and Cauchy, both bias and variance of BLEV estimators are substantially smaller than bias and variance of UNIF estimators correspondingly.

3.4.2 *BLEV* and *LEVUNW*

Next, we turn to the comparison between BLEV and LEVUNW in Fig. 3.5. As we mentioned before, the difference between these two methods is from modeling approach. BLEV was computed using weighted least squares, whereas LEVUNW was computed by unweighted LS. Moreover, both $\tilde{\boldsymbol{\beta}}_{\text{BLEV}}$ and $\tilde{\boldsymbol{\beta}}_{\text{LEVUNW}}$ are unbiased estimator for the unknown coefficient $\boldsymbol{\beta}_0$. As shown in Fig. 3.5, the biases are in general small for both estimators; but when predictors were generated from T2 and Cauchy, LEVUNW consistently outperforms BLEV at all subsample sizes.

3.4.3 *BLEV* and *SLEV*

In Fig. 3.6, we compare the performance of BLEV and SLEV. In SLEV, the subsampling and weighting steps are performed with respect to a combination of the subsampling probability distribution of BLEV and UNIF. As shown, the SLEV(0.9) performs uniformly better than BLEV and SLEV(0.1); and BLEV is

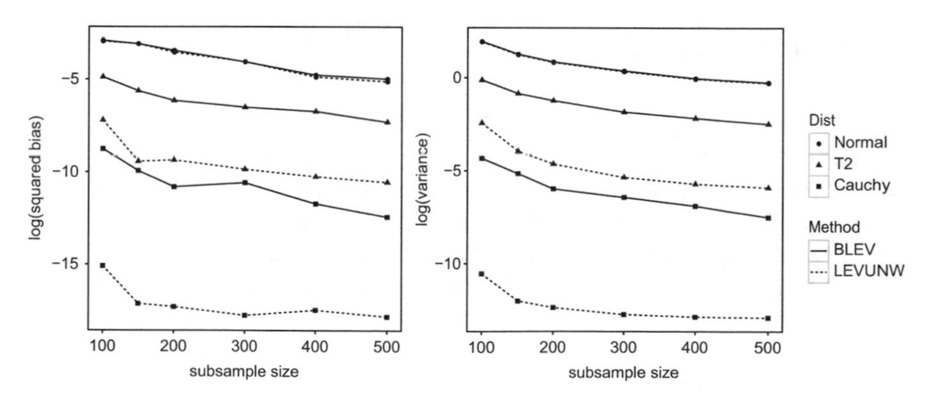

Fig. 3.5 Comparison of squared biases and variances of $\tilde{\boldsymbol{\beta}}_{\text{BLEV}}$ and $\tilde{\boldsymbol{\beta}}_{\text{LEVUNW}}$ in three distributions

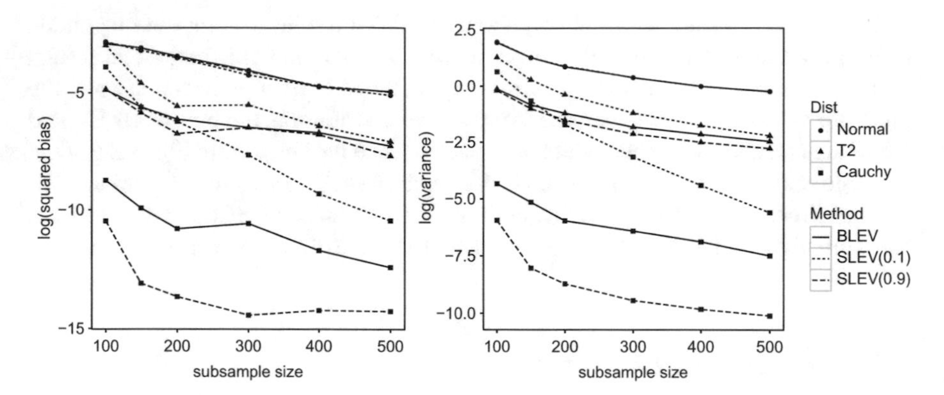

Fig. 3.6 Comparison of squared biases and variances of $\widetilde{\boldsymbol{\beta}}_{\text{BLEV}}$ and $\widetilde{\boldsymbol{\beta}}_{\text{SLEV}}$ in three distributions. In the graph, SLEV(0.1) corresponds to choosing $\pi_i = \lambda \frac{h_{ii}}{p} + (1-\lambda)\frac{1}{n}$, where $\lambda = 0.1$, and SLEV(0.9) corresponds to $\lambda = 0.9$

better than SLEV(0.1) in terms of both squared bias and variance. By construction, it is easy to understand that SLEV(0.9) and SLEV(0.1) enjoy unbiasedness, in the same way that UNIF and BLEV do. In Fig. 3.6, the squared biases are uniformly smaller than the variances for all estimators at all subsample sizes. Note that for SLEV $\pi_i \geq (1-\lambda)/n$, and the equality holds when $h_{ii} = 0$. Thus, the introduction of λ with uniform distribution in $\{\pi\}_{i=1}^n$ helps bring up extremely small probabilities and suppress extremely large probabilities correspondingly. Thus SLEV avoids the potential disadvantage of BLEV, i.e. extremely large variance due to extremely small probabilities and the unnecessary oversampling in BLEV due to extremely large probabilities. As shown in the graph, also suggested by Ma et al. (2013), as a rule of thumb, choosing $\lambda = 0.9$ strikes a balance between needing more samples and avoiding variance inflation.

3.4.4 *BLEV* and *PL*

In Fig. 3.7, we consider comparing BLEV and PL. As shown, the squared bias and variance for PL and BLEV are very close to each other in Normal distribution. In T2 distribution, as subsample size increases, we notice some slight advantage of PL over BLEV in both squared bias and variance. The superiority of PL over BLEV is most appealing in Cauchy distribution and as subsample size increases, the advantage of PL in terms of both bias and variance gets more significant.

3.4.5 *SLEV* and *PL*

Lastly, in Fig. 3.8 we compare SLEV(0.9) and PL. The squared bias and variance of SLEV(0.9) are slightly smaller than those of PL at all subsample sizes in T2

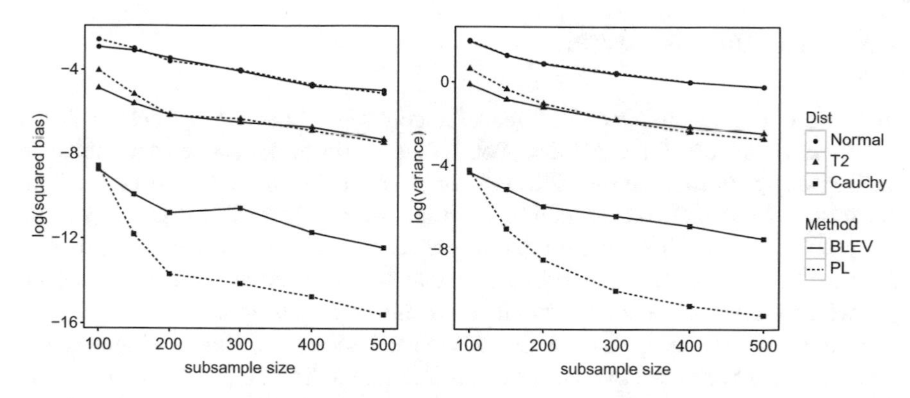

Fig. 3.7 Comparison of squared biases and variances of $\tilde{\boldsymbol{\beta}}_{\mathrm{PL}}$ and $\tilde{\boldsymbol{\beta}}_{\mathrm{BLEV}}$ in three distributions

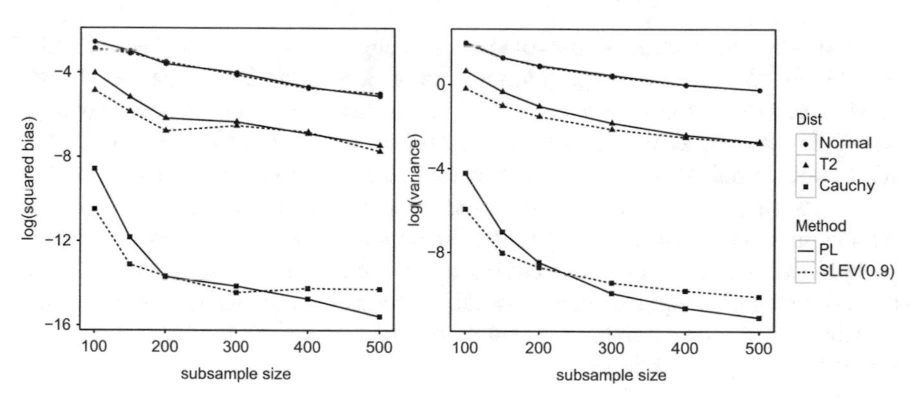

Fig. 3.8 Comparison of squared biases and variances of $\tilde{\boldsymbol{\beta}}_{\mathrm{SLEV}(0.9)}$ and $\tilde{\boldsymbol{\beta}}_{\mathrm{PL}}$ in three distributions

distribution. PL has smaller squared bias and variances compared to SLEV(0.9) at subsample sizes greater than 400. Considering that PL reduces computational time to $O(np)$, which makes PL especially attractive in extraordinarily large datasets.

Our simulation study shows that the statistical leveraging methods perform well in relatively large sample size data sets. Overall, compared to other methods, PL performs reasonably well in estimation in most cases being evaluated and it requires significantly less computing time than other estimators. In practice, we recommend starting analysis with constructing the PL subsampling probabilities, using the probability distribution to draw random samples from the full sample, and performing scatterplots on the sampled data to get a general idea of the dataset distribution. If the distribution of explanatory variables are more close to normal distribution, then we suggest also try BLEV, SLEV, LEVUNW using exact or approximating leverage scores; if the distribution of explanatory variables is close to t or Cauchy distribution, then we refer to the PL estimator.

3.5 Real Data Analysis

In this section, we analyze the "YearPredictionMSD" dataset (Lichman 2013), which is a subset of the Million Song Dataset (http://labrosa.ee.columbia.edu/millionsong/). In this dataset, 515,345 songs are included, and most of them are western, commercial tracks ranging from the year 1922 to 2011, peaking in the early 2000s. For each song, multiple segments are extracted and each segment is described by a 12-dimensional timbre feature vector. Then average, variance, and pairwise covariance are taken over all "segments" for each song.

In this analysis, the goal of analysis is to use 12 timbre feature averages and 78 timbre feature variance/covariances as predictors, totaling 90 predictors, to predict the year of release (response). We chose a linear model in (3.2) to accomplish this goal. Considering the large sample size, we opt to using the statistical leveraging methods.

First, we took random samples of size 100 using sampling probabilities of BLEV and PL to visualize the large dataset. Figure 3.9 shows the scatterplot for the subsample of one predictor using two different sampling probability distributions. In Fig. 3.9, we can see that there are several points with extra dark color standing out from the rest, indicating that the data distribution might be closer to t or Cauchy than to Normal distribution and that statistical leveraging methods will perform better than uniform sampling according to the simulation study. Also, these two sampling probability distributions are very similar to each other, suggesting that PL might be a good surrogate or approximation for BLEV. Since the computation of PL is more scalable, it is an ideal method for exploratory data analysis before other leveraging methods are applied to the data.

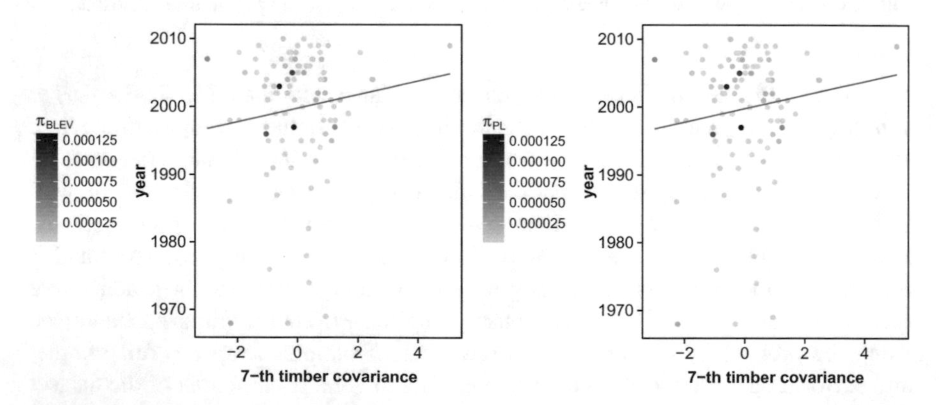

Fig. 3.9 Scatterplots of a random subsample of size 100 from "YearPredictionMSD" dataset (Lichman 2013). In this example, we scale each predictor before calculating the subsampling probabilities. The left panel displays the subsample drawn using subsampling probability distribution in BLEV in (3.4). The right panel displays the subsample drawn using subsampling probability distribution in PL in (3.7). Color of the points corresponds to the sampling probabilities: the darker the color, the higher the probability

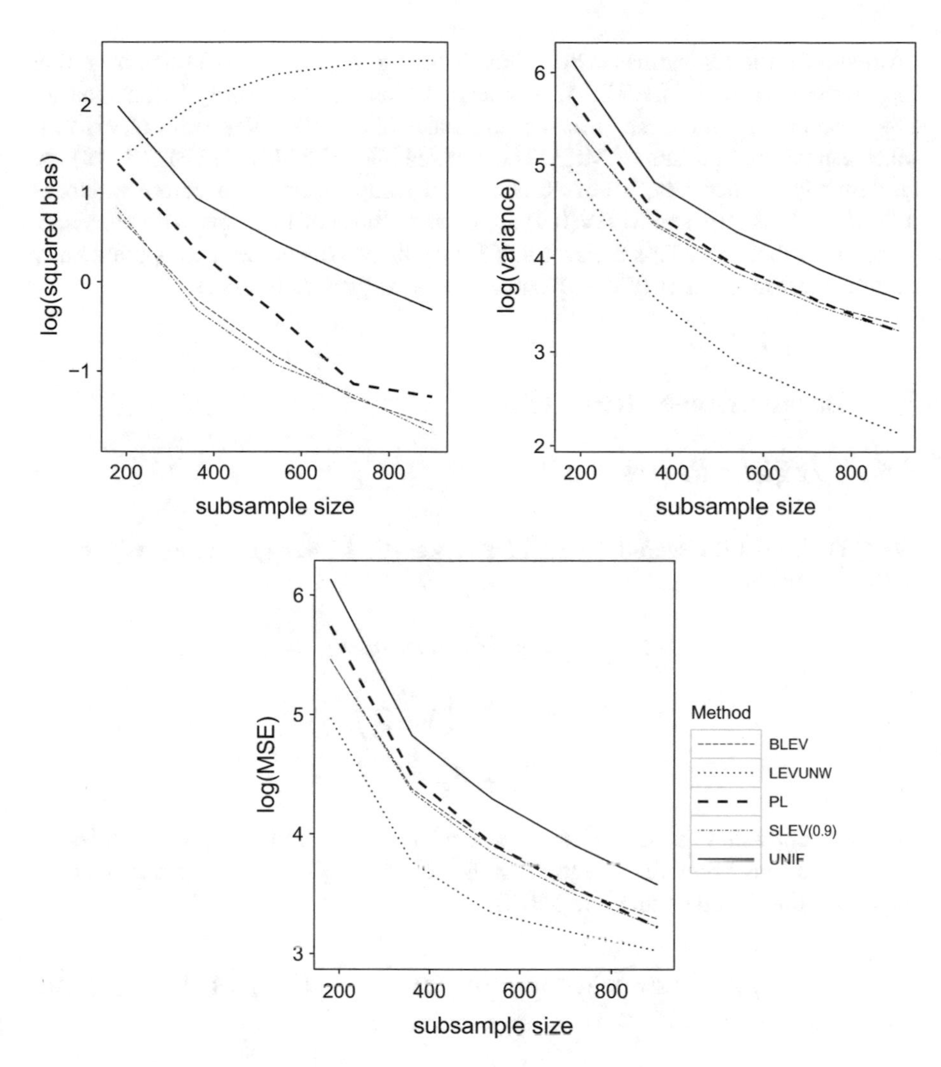

Fig. 3.10 The squared bias, variance and MSE (with respect to the LS estimator of whole dataset) of different statistical leveraging algorithms using "YearPredictionMSD" dataset

Next, we applied various reviewed statistical leveraging methods with different subsample sizes to the data. We scaled the predictors and centered the response prior to analysis. Each method is repeated 100 times and the bias, variance and mean squared error (MSE) to the full sample LS estimator are plotted in Fig. 3.10.

Consistent with results in the simulation study, the variance is larger compared to bias for all statistical leveraging estimators at all subsample sizes. As shown in Fig. 3.10, LEVUNW has the smallest MSE and variance, but the largest bias. As reviewed in Sect. 3.3 about the asymptotic properties of LEVUNW, we discerned that it is an unbiased estimator for the underlying true parameter $\boldsymbol{\beta}_0$ but a biased estimator for the LS estimator $\hat{\boldsymbol{\beta}}_{\mathrm{LS}}$. Thus in the figure, as subsample size is getting larger, the bias of the LEVUNW estimator becomes significantly larger than all other estimators. But since variance dominates bias, LEVUNW still outperforms other estimators in terms of MSE. The squared bias of BLEV and SLEV(0.9) are consistently smaller than that of PL at each subsample size; however, the variances and MSEs of BLEV and SLEV(0.9) are close to those of PL, especially at sample sizes larger than 400. This means that PL may be considered as a computationally practical surrogate for BLEV and SLEV(0.9), as suggested in Fig. 3.9.

3.6 Beyond Linear Regression

3.6.1 Logistic Regression

Wang et al. (2017) generalized the idea of statistical leveraging to logistic model defined as below:

$$
\begin{aligned}
y_i &\sim \mathrm{Binomial}(n_i, p_i) \\
\mathrm{logit}(p_i) &= \log\left(\frac{p_i}{1-p_i}\right) \\
&= \mathbf{x}_i^T \boldsymbol{\beta}_0.
\end{aligned}
$$

Logistic regression is one of the most widely used and typical examples of generalized linear models. The regression coefficients $\boldsymbol{\beta}_0$ are usually estimated by maximum likelihood estimation (MLE), i.e.

$$\hat{\boldsymbol{\beta}}_{\mathrm{MLE}} = \max_{\boldsymbol{\beta}} \sum_{i=1}^{n} [y_i \log p_i(\boldsymbol{\beta}) + (1-y_i)\log\{1-p_i(\boldsymbol{\beta})\}], \tag{3.16}$$

where $p_i(\boldsymbol{\beta}) = \exp(\mathbf{x}_i^T\boldsymbol{\beta})/\{1+\exp(\mathbf{x}_i^T\boldsymbol{\beta})\}$.

Unlike linear regression with normally distributed residuals as stated in (3.1), there exists no closed-form expression for (3.16). So an iterative procedure must be used instead; for example, Newton's method. As shown, Newton's method for numerically solving (3.16) corresponds to an iterative weighted least square (IWLS) problem (McCullagh and Nelder 1989). However, the generalization of BLEV is not trivial. The weights in each iterated WLS involve $\hat{\boldsymbol{\beta}}_{\mathrm{MLE}}$. Consequently, to construct the leverage scores one has to obtain $\hat{\boldsymbol{\beta}}_{\mathrm{MLE}}$ first. The mutual dependence of subsampling probabilities and $\hat{\boldsymbol{\beta}}_{\mathrm{MLE}}$ under GLM settings poses a dilemma for the intended generalization.

To tackle this problem, Wang et al. (2017) proposed a two-step procedure. First, we draw a subsample of size r_1 using uniform subsampling probabilities and obtain an estimate of coefficient values, denoted as $\tilde{\boldsymbol{\beta}}^{(1)}$. Second, with the $\tilde{\boldsymbol{\beta}}^{(1)}$, the weight matrix or variance-covariance matrix can be estimated and the subsampling probabilities for BLEV can be further constructed. Finally, we obtain another subsample of size r_2 to get the final estimator $\tilde{\boldsymbol{\beta}}$.

The computation time for $\tilde{\boldsymbol{\beta}}^{(1)}$ in the first step is $O(m_1 r_1 p^2)$ where m_1 is the number of iterations of IWLS in the first step; the computation time for construction of the subsampling probabilities can be as low as $O(np)$; the computation time for $\tilde{\boldsymbol{\beta}}$ in the second step is $O(m_2 r_2 p^2)$ where m_2 is the number of iterations of IWLS in the second step. So the overall time complexity is $O(np+m_1 r_1 p^2+m r_2 p^2)$. Considering that p is fixed and m_1, m_2, r_1, and r_2 are all much smaller than n, the time complexity of whole procedure stays at the order of $O(np)$.

Another remark about the two-step procedure concerns the balance between r_1 and r_2. On one hand, in order to get a reliable estimate $\tilde{\boldsymbol{\beta}}^{(1)}$, r_1 should not be too small; on the other hand, the efficiency of the two-step algorithm decreases if r_1 grows larger compared to r_2. In practice, we find that the algorithm works well when the ratio $r_1/r_1 + r_2$ is between 0.2 and 0.4 and this is the rule of thumb recommended by Wang et al. (2017).

3.6.2 *Time Series Analysis*

Although we have reviewed so far on the setting of independent and identically distributed data, the natural extension of statistical leveraging methods can be made to handle the dependent data settings, e.g. time series data. Time series in big data framework are widely available in different areas, e.g. sensors data, which is the most widespread and is a new type of time series data. With storage costs coming down significantly, there are significant efforts on analyzing these big time series data (including instrument-generated data, climatic data, and other types of sensor data). However, analyzing the big time series data has new challenge due to the computational cost. Autoregressive and moving average (ARMA) model has been extensively used for modeling time series. But the traditional ARMA model is facing the limit from the computational perspective on analyzing big time series data. The leveraging theory and method thus have been proposed for fitting ARMA model (Xie et al. 2017). A distinguished feature of the novel leveraging method is that instead of sampling individual data points, we subsample blocks of time series so that the time dependence can be well estimated. Such leveraging subsampling approach has a significant challenge related to stationarity. In time series analysis, it is necessary to assume that at least some features of the underlying probability are sustained over a time period. This leads to the assumptions of different types of stationarity. However, a block of time series in stationarity does not necessarily imply the stationarity of the whole time series. Thus novel statistical methods are needed.

In the context of ARMA model, Xie et al. (2017) propose a novel sequential leveraging subsampling method for non-explosive AR(p) series. The sequential leveraging subsampling method can adapt to the availability of computing resources. When only single (or a few) computing processor but a large memory is available, we design a sequential leveraging method starting with one single data point. The idea is that we sample a single data point base on leverage-probability, and then expand the single data point to its neighborhood to form a block of time series sequentially. The key is that as long as the single block is long enough, all the features of the full sample are captured in the single block time series. When there are a large number of computing processors, each of which has moderate memory, we sample several points and perform the leveraging sequential sampling on each of them so that we have a snapshot of the whole time series.

3.7 Discussion and Conclusion

When analyzing big data with large sample size, one faces significant computational challenge, i.e., the high computational cost renders many conventional statistics methods inapplicable in big data. There is an extensive literature in the computer science community on efficient storage and computation of the big data, such as parallel computing algorithms using GPUs, etc. However, very few of them overcome the computational challenge from statistical perspective, e.g. the bias and variance of the big data estimation. In this chapter, we reviewed the statistical leveraging methods for analyzing big data. The idea of statistical leveraging is very simple, i.e., to take a random subsample, on which all subsequent computation steps are performed. Sampling is one of the most common tools in statisticians' toolkit and has a great potential to overcome the big data challenge. The key to success of statistical leveraging methods relies on the effective construction of the sampling probability distribution, based on which influential data points are sampled. Moreover, we also presented some preliminary ideas about extending the statistical leveraging to GLM and time series model. But obviously the power of the leveraging methods has not been fully exploited in this chapter. The performance of leveraging methods is waiting to be examined on more complicated problems, such as penalized regression.

Acknowledgements This work was funded in part by NSF DMS-1440037(1222718), NSF DMS-1438957(1055815), NSF DMS-1440038(1228288), NIH R01GM122080, NIH R01GM113242.

References

Agarwal A, Duchi JC (2011) Distributed delayed stochastic optimization. In: Advances in neural information processing systems, pp 873–881

Avron H, Maymounkov P, Toledo S (2010) Blendenpik: supercharging LAPACK's least-squares solver. SIAM J Sci Comput 32:1217–1236

Bhlmann P, van de Geer S (2011) Statistics for high-dimensional data: methods, theory and applications, 1st edn. Springer, Berlin

Chatterjee S, Hadi AS (1986) Influential observations, high leverage points, and outliers in linear regression. Stat Sci 1(3):379–393

Chen X, Xie M (2014) A split-and-conquer approach for analysis of extraordinarily large data. Stat Sin 24:1655–1684

Clarkson KL, Woodruff DP (2013) Low rank approximation and regression in input sparsity time. In: Proceedings of the forty-fifth annual ACM symposium on theory of computing. ACM, New York, pp 81–90

Clarkson KL, Drineas P, Magdon-Ismail M, Mahoney MW, Meng X, Woodruff DP (2013) The Fast Cauchy Transform and faster robust linear regression. In: Proceedings of the twenty-fourth annual ACM-SIAM symposium on discrete algorithms. Society for Industrial and Applied Mathematics, Philadelphia, pp 466–477

Coles S, Bawa J, Trenner L, Dorazio P (2001) An introduction to statistical modeling of extreme values, vol 208. Springer, Berlin

Drineas P, Mahoney MW, Muthukrishnan S (2006) Sampling algorithms for ℓ_2 regression and applications. In: Proceedings of the 17th annual ACM-SIAM symposium on discrete algorithms, pp 1127–1136

Drineas P, Mahoney MW, Muthukrishnan S, Sarlós T (2010) Faster least squares approximation. Numer Math 117(2):219–249

Drineas P, Magdon-Ismail M, Mahoney MW, Woodruff DP (2012) Fast approximation of matrix coherence and statistical leverage. J Mach Learn Res 13:3475–3506

Duchi JC, Agarwal A, Wainwright MJ (2012) Dual averaging for distributed optimization: convergence analysis and network scaling. IEEE Trans Autom Control 57(3):592–606

Friedman J, Hastie T, Tibshirani R (2001) The elements of statistical learning, vol 1. Springer series in statistics. Springer, Berlin

Golub GH, Van Loan CF (1996) Matrix computations. Johns Hopkins University Press, Baltimore

Hesterberg T (1995) Weighted average importance sampling and defensive mixture distributions. Technometrics 37(2):185–194

Hoaglin DC, Welsch RE (1978) The hat matrix in regression and ANOVA. Am Stat 32(1):17–22

Lichman M (2013) UCI machine learning repository

Ma P, Sun X (2015) Leveraging for big data regression. Wiley Interdiscip Rev Comput Stat 7(1):70–76

Ma P, Mahoney MW, Yu B (2014) A statistical perspective on algorithmic leveraging. In: Proceedings of the 31st international conference on machine learning (ICML-14), pp 91–99

Ma P, Mahoney MW, Yu B (2015) A statistical perspective on algorithmic leveraging. J Mach Learn Res 16:861–911

Ma P, Zhang X, Ma J, Mahoney MW, Yu B, Xing X (2016) Optimal subsampling methods for large sample linear regression. Technical report, Department of Statistics, University of Georgia

Mahoney MW (2011) Randomized algorithms for matrices and data. Foundations and trends in machine learning. NOW Publishers, Boston. Also available at: arXiv:1104.5557

Mahoney MW, Drineas P (2009) CUR matrix decompositions for improved data analysis. Proc Natl Acad Sci 106(3):697–702

McCullagh P, Nelder JA (1989) Generalized linear models, vol 37. CRC, Boca Raton

Meng X, Mahoney MW (2013) Low-distortion subspace embeddings in input-sparsity time and applications to robust linear regression. In: Proceedings of the forty-fifth annual ACM symposium on theory of computing. ACM, New York, pp 91–100

Meng X, Saunders MA, Mahoney MW (2014) LSRN: a parallel iterative solver for strongly over-or underdetermined systems. SIAM J Sci Comput 36(2):C95–C118

Raskutti G, Mahoney MW (2016) A statistical perspective on randomized sketching for ordinary least-squares. J Mach Learn Res 17(214):1–31

Velleman PF, Welsch ER (1981) Efficient computing of regression diagnostics. Am Stat 35(4): 234–242

Wang H, Zhu R, Ma P (2017) Optimal subsampling for large sample logistic regression. J Am Stat Assoc (in press)

Xie R, Sriram TN, Ma P (2017) Sequential leveraging sampling method for streaming time series data. Technical report, Department of Statistics University of Georgia

Zhang Y, Duchi JC, Wainwright MJ (2013) Divide and conquer kernel ridge regression: a distributed algorithm with minimax optimal rates. CoRR. abs/1305.5029

Chapter 4
Scattered Data and Aggregated Inference

Xiaoming Huo, Cheng Huang, and Xuelei Sherry Ni

Abstract Scattered Data and Aggregated Inference (SDAI) represents a class of problems where data cannot be at a centralized location, while modeling and inference is pursued. Distributed statistical inference is a technique to tackle a type of the above problem, and has recently attracted enormous attention. Many existing work focus on the averaging estimator, e.g., Zhang et al. (2013) and many others. In this chapter, we propose a one-step approach to enhance a simple-averaging based distributed estimator. We derive the corresponding asymptotic properties of the newly proposed estimator. We find that the proposed one-step estimator enjoys the same asymptotic properties as the centralized estimator. The proposed one-step approach merely requires one additional round of communication in relative to the averaging estimator; so the extra communication burden is insignificant. In finite-sample cases, numerical examples show that the proposed estimator outperforms the simple averaging estimator with a large margin in terms of the mean squared errors. A potential application of the one-step approach is that one can use multiple machines to speed up large-scale statistical inference with little compromise in the quality of estimators. The proposed method becomes more valuable when data can only be available at distributed machines with limited communication bandwidth. We discuss other types of SDAI problems at the end.

Keywords Large-scale statistical inference · Distributed statistical inference · One-step method · M-estimators · Communication-efficient algorithms

---

X. Huo (✉) · C. Huang
Georgia Institute of Technology, School of Industrial and Systems Engineering, Atlanta, GA, USA
e-mail: huo@gatech.edu; c.huang@gatech.edu

X. S. Ni
Kennesaw State University, Department of Statistics and Analytical Sciences, Kennesaw, GA, USA
e-mail: sni@kennesaw.edu

© Springer International Publishing AG, part of Springer Nature 2018

W. K. Härdle et al. (eds.), *Handbook of Big Data Analytics*, Springer Handbooks of Computational Statistics, https://doi.org/10.1007/978-3-319-18284-1_4

4.1 Introduction

In many important contemporary applications, data are often partitioned across multiple servers. For example, a search engine company may have data coming from a large number of locations, and each location collects tera-bytes of data per day (Corbett et al. 2012). On a different setting, high volume of data (like videos) have to be stored distributively, instead of on a centralized server (Mitra et al. 2011). Given the modern “data deluge,” it is often the case that centralized methods are no longer possible to implement. It has also been notified by various researchers (e.g., Jaggi et al. 2014) that the speed of local processors can be thousands time faster than the rate of data transmission in a modern network. Consequently it is evidently advantageous to develop communication-efficient method, instead of transmitting data to a central location and then apply a global estimator.

We name the above scenario a *scattered data with aggregated inference* (SDAI) situation. The associated techniques can play an increasingly significant role in a modern society, such as:

- A major web search engine has to save its data in a range of platforms, storage units, and even geographical locations; a company-wide policy may require knowledge on these distributed data.
- An international organization may have information stored all over the world, while between different world offices, there are limited communication and data sharing bandwidth.
- A major supply chain company (or a superstore giant) has collected tremendous amounts of information at many different locations, and it can be costly and unrealistic to transfer them to a common storage unit.
- A government may have enormous quantities of information saved across different agencies and locations; sharing these data requires substantial political and/or administrative struggle.
- In public health surveillance, the Centers for Disease Control and Prevention (CDC) have a tremendous volume of potentially useful data across VA hospitals and city/county/state agencies; creating a warning system with few false alarms is a challenge.
- Hospital across the nation (or even worldwide) has enormous amount of health and/or disease related data. They all want to build some predictive models, however sharing the data is largely deterred due to privacy, legal, and proprietary concerns.

All the above illustrate possible scenarios of scattered data and aggregated inference, in which data were stored locally, with limited amount of information transferred to other nodes.

In statistical inference, estimators are introduced to infer some important hidden quantities. In ultimate generality, a statistical estimator of a parameter $\theta \in \Theta$ is a measurable function of the data, taking values in the parameter space Θ. Many statistical inference problems could be solved by finding the maximum likelihood estimators (MLE), or more generally, M-estimators. In either case, the task is to minimize an objective function, which is the average of a criterion function over the entire data, which is typically denoted by $S = \{X_1, X_2, \ldots, X_N\}$, where N is called the sample size. Here we choose a capitalized N to distinguish from a lower n that will be used later. Traditional centralized setting requires access to entire data set S simultaneously. However, due to the explosion of data size, it may be infeasible to store all the data in a single machine like we did during past several decades. Distributed (sometimes, it is called *parallel*) statistical inference would be an indispensable approach for solving these large-scale problems.

At a high level, there are at least two types of distributed inference problems. In the first type, each observation X_i is completely observed at one location; and different observations (i.e., X_i and X_j for $i \neq j$) may be stored at different locations. This chapter will focus to this type of problems. In the second type, it is possible that for the same sample X_i, different parts are available at different locations, and they are *not* available in a centralized fashion. The latter has been studied in the literature (see Gamal and Lai (2015) and the references therein). This paper will not study the second type.

For distributed inference in the first type of the aforementioned setting, data are split into several subsets and each subset is assigned to a processor. This paper will focus on the M-estimator framework, in which an estimator is obtained by solving a distributed optimization problem. The objective in the distributed optimization problem may come from an M-estimator framework (or more particularly from the maximum likelihood principle), empirical risk minimization, and/or penalized version of the above. Due to the type 1 setting, we can see that the objective functions in the corresponding optimization problem are separable; in particular, the global objective function is a summation of functions such that each of them only depends on data reside on one machine. The exploration in this paper will base on this fact. As mentioned earlier, a distributed inference algorithm should be communication-efficient because of high communication cost between different machines or privacy concerns (such as sensitive personal information or financial data). It is worth noting that even if the data could be handled by a single machine, distributed inference would still be beneficial for reducing computing time.

Our work has been inspired by recent progress in distributed optimization. We review some noticeable progress in numerical approaches and their associated theoretical analysis. Plenty of research work has been done in distributed algorithms for large-scale optimization problems during recent years. Boyd et al. (2011) suggest to use Alternating Direction Method of Multipliers (ADMM) to solve distributed optimization problems in statistics and machine learning. Using a trick of *consistency* (or sometimes called *consensus*) constraints on local variables and a global variable, ADMM can be utilized to solve a distributed version of the Lasso problem (Tibshirani 1996; Chen et al. 1998). ADMM has also been adopted in solving distributed logistic regression problems, and many more. ADMM is feasible for a wide range of problems, but it requires iterative communication between local machines and the center. In comparison, we will propose a method that only requires two iterations. Zinkevich et al. (2010) propose a parallelized stochastic gradient descent method for empirical risk minimization and prove its convergence. The established contractive mappings technique seems to be a powerful method to quantify the speed of convergence of the derived estimator to its limit. Shamir et al. (2014) present the Distributed Approximate Newton-type Method (DANE) for distributed statistical optimization problems. Their method firstly averages the local gradients then follows by averaging all local estimators in each iteration until convergence. They prove that this method enjoys linear convergence rate for quadratic objectives. For non-quadratic objectives, it has been showed that the value of the objective function has a geometric convergence rate. Jaggi et al. (2014) propose a communication-efficient method for distributed optimization in machine learning, which uses local computation with randomized dual coordinate descent in a primal–dual setting. They also prove the geometric convergence rate of their method. The above works focus on the properties of numerical solutions to the corresponding optimization problems. Nearly all of them require more than two rounds of communication. Due to different emphasis, they did not study the statistical asymptotic properties (such as convergence in probability, asymptotic normality, Fisher information bound) of the resulting estimators. Our paper will address these issues.

Now we switch the gear to statistical inference. Distributed inference has been studied in many existing works, and various proposals have been made in different settings. To the best of our knowledge, the distributed one-step estimator has *not* been studied in any of these existing works. We review a couple of state-of-the-art approaches in the literature. Our method builds on a closely related recent line of work of Zhang et al. (2013), which presents a straightforward approach to solve large-scale statistical optimization problem, where the local empirical risk minimizers are simply averaged. They showed that this averaged estimator achieves mean squared errors that decays as $O(N^{-1} + (N/k)^{-2})$, where N stands for the total number of samples and k stands for the total number of machines. They also showed that the mean squared errors could be even reduced to $O(N^{-1} + (N/k)^{-3})$ with one more bootstrapping sub-sampling step. Obviously, there exists efficiency loss in their method since the centralized estimator could achieve means squared error $O(N^{-1})$. Liu and Ihler (2014) propose an inspiring two-step approach: firstly find local maximum likelihood estimators, then subsequently combine them by minimizing the total Kullback–Leibler divergence (KL-divergence). They prove the exactness of their estimator as the global MLE for the full exponential family. They also estimate the mean squared errors of the proposed estimator for a curved exponential family. Due to the adoption of the KL-divergence, the effectiveness of this approach heavily depends on the parametric form of the underlying model. Chen and Xie (2014) propose a split-and-conquer approach for a penalized regression problem (in particular, a model with the canonical exponential distribution) and show that it enjoys the same oracle property as the method that uses the entire data set in a single machine. Their approach is based on a majority voting, followed by a weighted average of local estimators, which somewhat resembles a one-step estimator however is different. In addition, their theoretical results require $k \le O(N^{\frac{1}{5}})$, where k is the number of machines and N is again the total number of samples. This is going to be different from our needed condition for theoretical guarantees. Their work considers a high-dimensional however sparse parameter vector, which is not considered in this paper. Rosenblatt and Nadler (2014) analyze the error of averaging estimator in distributed statistical learning under two scenarios. The number of machines is fixed in the first one and the number of machines grows in the same order with the number of samples per machine. They present the asymptotically exact expression for estimator errors in both scenarios and show that the error grows linearly with the number of machines in the latter case. Their work does not consider the one-step updating that will be studied in this paper. Although it seems that their work proves the asymptotic optimality of the simple averaging, our simulations will demonstrate the additional one-step updating can improve over the simple averaging, at least in some interesting finite sample cases. Battey et al. (2015) study the distributed parameter estimation method for penalized regression and establish the oracle asymptotic property of an averaging estimator. They also discuss hypotheses testing, which is not covered in this paper. Precise upper bounds on the errors of their proposed estimator have been developed. We benefited from reading the technical proofs of their paper; however unlike our method, their method is restricted to linear regression problems with penalty and requires the number of machine $k = o(\sqrt{N})$. Lee et al. (2015) devise a one-shot approach, which averages "debiased" lasso estimators, to distributed sparse regression in the high-dimensional setting. They show that their approach converges at the same order of rate as Lasso when the data set is not split across too many machines.

It is worth noting that nearly all existing distributed estimators are *averaging* estimators. The idea of applying one additional updating, which correspondingly requires one additional round of communication, has not been explicitly proposed. We may notice some precursor of this strategy. For example, in Shamir et al. (2014), an approximate Newton direction was estimated at the central location, and then broadcasted to local machines. Another occurrence is that in Lee et al. (2015), some intermediate quantities are estimated in a centralized fashion, and then distributed to local machines. None of them explicitly described what we will propose.

In the theory on maximum likelihood estimators (MLE) and M-estimators, there is a one-step method, which could make a consistent estimator as efficient as MLE or M-estimators with a single Newton–Raphson iteration. Here, efficiency stands for the relative efficiency converges to 1. See van der Vaart (2000) for more details. There have been numerous papers utilizing this method. See Bickel (1975), Fan and Chen (1999), and Zou and Li (2008). A one-step estimator enjoys the same asymptotic properties as the MLE or M-estimators as long as the initial estimators are $\sqrt{n}$-consistent. A $\sqrt{n}$-consistent estimator is much easier to find than the MLE or an M-estimator. For instance, the simple averaging estimator [e.g., the one proposed by Zhang et al. (2013)] is good enough as a starting point for a one-step estimator.

In this paper, we propose a one-step estimator for distributed statistical inference. The proposed estimator is built on the well-analyzed simple averaging estimator. We show that the proposed one-step estimator enjoys the same asymptotic properties (including convergence and asymptotic normality) as the centralized estimator, which would utilize the entire data. Given the amount of knowledge we had on the distributed estimators, the above result may not be surprising. However, when we derive an upper bound for the error of the proposed one-step estimator, we found that we can achieve a slightly better one than those in the existing literature. We also perform a detailed evaluation of our one-step method, comparing with the simple averaging method and a centralized method using synthetic data. The numerical experiment is much more encouraging than the theory predicts: in nearly all cases, the one-step estimator outperformed the simple averaging one with a clear margin. We also observe that the one-step estimator achieves the comparable performance as the global estimator at a much faster rate than the simple averaging estimator. Our work may indicate that in practice, it is better to apply a one-step distributed estimator than a simple-average one.

This paper is organized as follows. Section 4.2 describes details of our problem setting and two methods—the simple averaging method and the proposed one-step method. In Sect. 4.3, we study the asymptotic properties of the one-step estimator in the M-estimator framework and analyze the upper bound of its estimation error. Section 4.4 provides some numerical examples of distributed statistical inference with synthetic data. A discussion on statistical inference for scattered data where aggregated inference is necessary is furnished in Sect. 4.5. This section presents other techniques that have been developed by other researchers. In surveying the current landscape of SDAI, we discuss some other types of SDAI problems in Sect. 4.6, though this chapter does not tackle these problems. We conclude in Sect. 4.7. Due to space limitation, all detailed proofs are relegated to an online document (Huang and Huo 2015).

4.2 Problem Formulation

This section is organized as follows. In Sect. 4.2.1, we describe some notations that will be used throughout the paper. In Sect. 4.2.2, the M-estimator framework is reviewed. The simple averaging estimator will be presented in Sect. 4.2.3. The proposed one-step distributed estimator will be defined in Sect. 4.2.4.

4.2.1 Notations

In this subsection, we will introduce some notations that will be used in this paper. Let $\{m(x;\theta) : \theta \in \Theta \subset \mathbb{R}^d\}$ denote a collection of criterion functions, which should have continuous second derivative. Consider a data set S consisting of $N = nk$ samples, which are drawn i.i.d. from $p(x)$ (for simplicity, we assume that the sample size N is a multiple of k). This data set is divided evenly at random and stored in

k machines. Let S_i denote the subset of data assigned to machine i, $i = 1, \ldots, k$, which is a collection of n samples drawn i.i.d. from $p(x)$. Note that any two subsets in those S_i's are not overlapping.

For each $i \in \{1, \ldots, k\}$, let the local empirical criterion function that is based on the local data set on machine i and the corresponding maximizer be denoted by

$$M_i(\theta) = \frac{1}{|S_i|} \sum_{x \in S_i} m(x; \theta) \text{ and } \theta_i = \arg\max_{\theta \in \Theta} M_i(\theta). \tag{4.1}$$

Let the global empirical criterion function be denoted by

$$M(\theta) = \frac{1}{k} \sum_{i=1}^{k} M_i(\theta). \tag{4.2}$$

And let the population criterion function and its maximizer be denoted by

$$M_0(\theta) = \int_{\mathscr{X}} m(x; \theta) p(x) dx \text{ and } \theta_0 = \arg\max_{\theta \in \Theta} M_0(\theta), \tag{4.3}$$

where $\mathscr{X}$ is the sample space. Note that θ_0 is the parameter of interest. The gradient and Hessian of $m(x; \theta)$ with respect to θ are denoted by

$$\dot{m}(x; \theta) = \frac{\partial m(x; \theta)}{\partial \theta}, \ddot{m}(x; \theta) = \frac{\partial^2 m(x; \theta)}{\partial \theta \, \partial \theta^T}. \tag{4.4}$$

We also let the gradient and Hessian of the local empirical criterion function be denoted by

$$\dot{M}_i(\theta) = \frac{\partial M_i(\theta)}{\partial \theta} = \frac{1}{|S_i|} \sum_{x \in S_i} \frac{\partial m(x; \theta)}{\partial \theta}, \ddot{M}_i(\theta) = \frac{\partial^2 M_i(x; \theta)}{\partial \theta \, \partial \theta^T} = \frac{1}{|S_i|} \sum_{x \in S_i} \frac{\partial^2 m(x; \theta)}{\partial \theta \, \partial \theta^T}, \tag{4.5}$$

where $i \in \{1, 2, \ldots, k\}$, and let the gradient and Hessian of global empirical criterion function be denoted by

$$\dot{M}(\theta) = \frac{\partial M(\theta)}{\partial \theta}, \ddot{M}(\theta) = \frac{\partial^2 M(\theta)}{\partial \theta \, \partial \theta^T}. \tag{4.6}$$

Similarly, let the gradient and Hessian of population criterion function be denoted by

$$\dot{M}_0(\theta) = \frac{\partial M_0(\theta)}{\partial \theta}, \ddot{M}_0(\theta) = \frac{\partial^2 M_0(\theta)}{\partial \theta \, \partial \theta^T}. \tag{4.7}$$

The vector norm $\|\cdot\|$ for $a \in \mathbb{R}^d$ that we use in this paper is the usual Euclidean norm $\|a\| = (\sum_{j=1}^d a_j^2)^{\frac{1}{2}}$. And we also use $|||\cdot|||$ to denote a norm for matrix $A \in \mathbb{R}^{d\times d}$, which is defined as its maximal singular value, i.e., we have

$$|||A||| = \sup_{u:u\in R^d, \|u\|\leq 1} \|Au\|.$$

The aforementioned matrix norm will be the major matrix norm that is used throughout the paper. The only exception is that we will also use the Frobenius norm in Huang and Huo (2015). And the Euclidean norm is the only vector norm that we use throughout this paper.

4.2.2 Review on M-Estimators

In this paper, we will study the distributed scheme for large-scale statistical inference. To make our conclusions more general, we consider M-estimators, which could be regarded as a generalization of the Maximum Likelihood Estimators (MLE). The M-estimator $\hat{\theta}$ could be obtained by maximizing the empirical criterion function, which means

$$\hat{\theta} = \arg\max_{\theta\in\Theta} M(\theta) = \arg\max_{\theta\in\Theta} \frac{1}{|S|}\sum_{x\in S} m(x;\theta).$$

Note that, when the criterion function is the log likelihood function, i.e., $m(x;\theta) = \log f(x;\theta)$, the M-estimator is exactly the MLE. Let us recall that $M_0(\theta) = \int_{\mathscr{X}} m(x;\theta)p(x)dx$ is the population criterion function and $\theta_0 = \arg\max_{\theta\in\Theta} M_0(\theta)$ is the maximizer of population criterion function. It is known that $\hat{\theta}$ is a consistent estimator for θ_0, i.e., $\hat{\theta} - \theta_0 \xrightarrow{P} 0$. See Chapter 5 of van der Vaart (2000).

4.2.3 Simple Averaging Estimator

Let us recall that $M_i(\theta)$ is the local empirical criterion function on machine i,

$$M_i(\theta) = \frac{1}{|S_i|}\sum_{x\in S_i} m(x;\theta).$$

And, θ_i is the local M-estimator on machine i,

$$\theta_i = \arg\max_{\theta\in\Theta} M_i(\theta).$$

Then as mentioned in Zhang et al. (2013), the simplest and most intuitive method is to take average of all local M-estimators. Let $\theta^{(0)}$ denote the average of these local M-estimators, we have

$$\theta^{(0)} = \frac{1}{k}\sum_{i=1}^{k}\theta_i, \tag{4.8}$$

which is referred to as the simple averaging estimator in the rest of this paper.

4.2.4 One-Step Estimator

Under the problem setting above, starting from the simple averaging estimator $\theta^{(0)}$, we can obtain the one-step estimator $\theta^{(1)}$ by performing a single Newton–Raphson update, i.e.,

$$\theta^{(1)} = \theta^{(0)} - [\ddot{M}(\theta^{(0)})]^{-1}[\dot{M}(\theta^{(0)})], \tag{4.9}$$

where $M(\theta) = \frac{1}{k}\sum_{i=1}^{k} M_i(\theta)$ is the global empirical criterion function, $\dot{M}(\theta)$ and $\ddot{M}(\theta)$ are the gradient and Hessian of $M(\theta)$, respectively. Throughout this paper, the dimension d of the parameter space Θ that was introduced at the beginning of Sect. 4.2.1 is assumed to be at most moderate. Consequently, the Hessian matrix $\ddot{M}(\theta)$, which should be $d \times d$, is not considered to be large. The whole process to compute one-step estimator can be summarized as follows.

1. For each $i \in \{1, 2, \ldots, k\}$, machine i compute the local M-estimator with its local data set,

$$\theta_i = \arg\max_{\theta\in\Theta} M_i(\theta) = \arg\max_{\theta\in\Theta} \frac{1}{|S_i|}\sum_{x\in S_i} m(x;\theta).$$

2. All local M-estimators are averaged to obtain simple averaging estimator,

$$\theta^{(0)} = \frac{1}{k}\sum_{i=1}^{k}\theta_i.$$

   Then $\theta^{(0)}$ is sent back to each local machine.
3. For each $i \in \{1, 2, \ldots, k\}$, machine i compute the gradient and the Hessian matrix of its local empirical criterion function $M_i(\theta)$ at $\theta = \theta^{(0)}$. Then send $\dot{M}_i(\theta^{(0)})$ and $\ddot{M}_i(\theta^{(0)})$ to the central machine.

4. Upon receiving all gradients and Hessian matrices, the central machine computes the gradient and the Hessian matrix of $M(\theta)$ by averaging all local information,

$$\dot{M}(\theta^{(0)}) = \frac{1}{k}\sum_{i=1}^{k} \dot{M}_i(\theta^{(0)}),\ \ddot{M}(\theta^{(0)}) = \frac{1}{k}\sum_{i=1}^{k} \ddot{M}_i(\theta^{(0)}).$$

   Then the central machine would perform a Newton–Raphson iteration to obtain a one-step estimator,

$$\theta^{(1)} = \theta^{(0)} - [\ddot{M}(\theta^{(0)})]^{-1}[\dot{M}(\theta^{(0)})].$$

Note that $\theta^{(1)}$ is not necessarily the maximizer of empirical criterion function $M(\theta)$ but it shares the same asymptotic properties with the corresponding global maximizer (M-estimator) under some mild conditions, i.e., we will show

$$\theta^{(1)} \xrightarrow{P} \theta_0,\ \sqrt{N}(\theta^{(1)} - \theta_0) \xrightarrow{d} \mathbf{N}(0, \Sigma), \text{as} N \to \infty,$$

where the covariance matrix Σ will be specified later.

The one-step estimator has advantage over simple averaging estimator in terms of estimation error. In Zhang et al. (2013), it is showed both theoretically and empirically that the MSE of simple averaging estimator $\theta^{(0)}$ grows significantly with the number of machines k when the total number of samples N is fixed. More precisely, there exists some constant $C_1, C_2 > 0$ such that

$$\mathbb{E}[\|\theta^{(0)} - \theta_0\|^2] \le \frac{C_1}{N} + \frac{C_2 k^2}{N^2} + O(kN^{-2}) + O(k^3 N^{-3}).$$

Fortunately, one-step method $\theta^{(1)}$ could achieve a lower upper bound of MSE with only one additional step. We will show the following in Sect. 4.3:

$$\mathbb{E}[\|\theta^{(1)} - \theta_0\|^2] \le \frac{C_1}{N} + O(N^{-2}) + O(k^4 N^{-4}).$$

4.3 Main Results

First of all, some assumptions will be introduced in Sect. 4.3.1. After that, we will study the asymptotic properties of one-step estimator in Sect. 4.3.2, i.e., convergence, asymptotic normality, and mean squared errors (MSE). In Sect. 4.3.3, we will consider the one-step estimator under the presence of information loss.

4.3.1 Assumptions

Throughout this paper, we impose some regularity conditions on the criterion function $m(x; \theta)$, the local empirical criterion function $M_i(\theta)$, and the population criterion function $M_0(\theta)$. We use the similar assumptions in Zhang et al. (2013). Those conditions are also standard in classical statistical analysis of M-estimators (cf. van der Vaart 2000).

The first assumption restricts the parameter space to be compact, which is reasonable and not rigid in practice. One reason is that the possible parameters lie in a finite scope for most cases. Another justification is that the largest number that computers could cope with is always limited.

Assumption 1 (Parameter Space) *The parameter space $\Theta \in \mathbb{R}^d$ is a compact convex set. And let $D \triangleq \max_{\theta,\theta' \in \Theta} \|\theta - \theta'\|$ denote the diameter of Θ.*

We also assume that $m(x; \theta)$ is concave with respect to θ and $M_0(\theta)$ has some curvature around the unique optimal point θ_0, which is a standard assumption for any method requires consistency. The constant λ below could depend on the sample size—this is a delicate issue and (due to space) is not addressed in the present paper.

Assumption 2 (Invertibility) *The Hessian of the population criterion function $M_0(\theta)$ at θ_0 is a nonsingular matrix, which means $\ddot{M}(\theta_0)$ is negative definite and there exists some $\lambda > 0$ such that $\sup_{u \in \mathbb{R}^d : \|u\| < 1} u^t \ddot{M}(\theta_0) u \leq -\lambda$.*

In addition, we require the criterion function $m(x; \theta)$ to be smooth enough, at least in the neighborhood of the optimal point θ_0, $B_\delta = \{\theta \in \Theta : \|\theta - \theta_0\| \leq \delta\}$. So, we impose some regularity conditions on the first and second derivatives of $m(x; \theta)$. We assume the gradient of $m(x; \theta)$ is bounded in moment and the difference between $\ddot{m}(x; \theta)$ and $\ddot{M}_0(\theta)$ is also bounded in moment. Moreover, we assume that $\ddot{m}(x; \theta)$ has Lipschitz continuity in B_δ.

Assumption 3 (Smoothness) *There exist some constants G and H such that*

$$\mathbb{E}[\|\dot{m}(X; \theta)\|^8] \leq G^8 \text{ and } \mathbb{E}\left[\left|\left|\left| \ddot{m}(X; \theta) - \ddot{M}_0(\theta) \right|\right|\right|^8 \right] \leq H^8, \forall \theta \in B_\delta.$$

For any $x \in \mathscr{X}$, the Hessian matrix $\ddot{m}(x; \theta)$ is $L(x)$-Lipschitz continuous,

$$\left|\left|\left| \ddot{m}(x; \theta) - \ddot{m}(x; \theta') \right|\right|\right| \leq L(x) \|\theta - \theta'\|, \forall \theta, \theta' \in B_\delta,$$

where $L(x)$ satisfies

$$\mathbb{E}[L(X)^8] \leq L^8 \text{ and } \mathbb{E}[(L(X) - \mathbb{E}[L(X)])^8] \leq L^8,$$

for some finite constant $L > 0$.

By Theorem 8.1 in Chapter XIII of Lang (1993), $m(x;\theta)$ enjoys interchangeability between the differentiation on θ and the integration on x, which means the following two equations hold:

$$\dot{M}_0(\theta) = \frac{\partial}{\partial\theta}\int_{\mathcal{X}} m(x;\theta)p(x)dx = \int_{\mathcal{X}} \frac{\partial m(x;\theta)}{\partial\theta}p(x)dx = \int_{\mathcal{X}} \dot{m}(x;\theta)p(x)dx,$$

and

$$\ddot{M}_0(\theta) = \frac{\partial^2}{\partial\theta^t\partial\theta}\int_{\mathcal{X}} m(x;\theta)p(x)dx = \int_{\mathcal{X}} \frac{\partial^2 m(x;\theta)}{\partial\theta^t\partial\theta}p(x)dx = \int_{\mathcal{X}} \ddot{m}(x;\theta)p(x)dx.$$

4.3.2 Asymptotic Properties and Mean Squared Errors (MSE) Bounds

Our main result is that the one-step estimator enjoys the oracle asymptotic properties and has the mean squared errors of $O(N^{-1})$ under some mild conditions.

Theorem 4 *Let* $\Sigma = \ddot{M}_0(\theta_0)^{-1}\mathbb{E}[\dot{m}(x;\theta_0)\dot{m}(x;\theta_0)^t]\ddot{M}_0(\theta_0)^{-1}$, *where the expectation is taken with respect to* $p(x)$. *Under Assumptions 1–3, when the number of machines* k *satisfies* $k = O(\sqrt{N})$, $\theta^{(1)}$ *is consistent and asymptotically normal, i.e., we have*

$$\theta^{(1)} - \theta_0 \xrightarrow{P} 0 \text{ and } \sqrt{N}(\theta^{(1)} - \theta_0) \xrightarrow{d} \mathbf{N}(0, \Sigma) \text{as} N \to \infty.$$

See an online document (Huang and Huo 2015) for a proof. The above theorem indicates that the one-step estimator is asymptotically equivalent to the centralized M-estimator.

It is worth noting that the condition $\|\sqrt{N}(\theta^{(0)} - \theta_0)\| = O_P(1)$ suffices for our proof to Theorem 4. Let $\tilde{\theta}^{(0)}$ denote another starting point for the one-step update, then the following estimator

$$\tilde{\theta}^{(1)} = \tilde{\theta}^{(0)} - \ddot{M}(\tilde{\theta}^{(0)})^{-1}\dot{M}(\tilde{\theta}^{(0)})$$

also enjoys the same asymptotic properties as $\theta^{(1)}$ does (and the centralized M-estimator $\hat{\theta}$) as long as $\sqrt{N}(\tilde{\theta}^{(0)} - \theta_0)$ is bounded in probability. Therefore, we can replace $\theta^{(0)}$ with any estimator $\tilde{\theta}^{(0)}$ that satisfies

$$\|\sqrt{N}(\tilde{\theta}^{(0)} - \theta_0)\| = O_P(1).$$

We also derive an upper bound for the MSE of the one-step estimator $\theta^{(1)}$.

Theorem 5 *Under Assumptions 1–3, the mean squared errors of the one-step estimator* $\theta^{(1)}$ *is bounded by*

$$\mathbb{E}[\|\theta^{(1)} - \theta_0\|^2] \leq \frac{2Tr[\Sigma]}{N} + O(N^{-2}) + O(k^4N^{-4}).$$

When the number of machines k *satisfies* $k = O(\sqrt{N})$, *we have*

$$\mathbb{E}[\|\theta^{(1)} - \theta_0\|^2] \leq \frac{2Tr[\Sigma]}{N} + O(N^{-2}).$$

See an online document (Huang and Huo 2015) for a proof.

In particular, when we choose the criterion function to be the log likelihood function, $m(x; \theta) = \log f(x; \theta)$, the one-step estimator has the same asymptotic properties as the maximum likelihood estimator (MLE) holds, which is described below.

Corollary 1 *If* $m(x; \theta) = \log f(x; \theta)$ *and* $k = O(\sqrt{N})$, *one-step estimator* $\theta^{(1)}$ *is a consistent and asymptotic efficient estimator of* θ_0,

$$\theta^{(1)} - \theta_0 \xrightarrow{P} 0 \text{ and } \sqrt{N}(\theta^{(1)} - \theta_0) \xrightarrow{d} \mathbf{N}(0, I(\theta_0)^{-1}), \text{ as } N \to \infty,$$

where $I(\theta_0)$ *is the Fisher's information at* $\theta = \theta_0$. *And the mean squared errors of* $\theta^{(1)}$ *is bounded as follows:*

$$\mathbb{E}[\|\theta^{(1)} - \theta_0\|^2] \leq \frac{2Tr[I^{-1}(\theta_0)]}{N} + O(N^{-2}) + O(k^4N^{-4}).$$

A proof follows immediately from Theorems 4, 5 and the definition of the Fisher's information.

4.3.3 Under the Presence of Communication Failure

In practice, it is possible that the information (local estimator, local gradient and local Hessian) from a local machine *cannot* be received by the central machine due to various causes (for instance, a network problem or a hardware crash). We assume that the communication failure on each local machine occurs independently.

We now derive a distributed estimator under the scenario with possible information loss. We will also present the corresponding theoretical results. We use $a_i \in \{0, 1\}, i = 1, \ldots, k$, to denote the status of local machines: when machine i successfully sends all its local information to the central machine, we have $a_i = 1$; when machine i fails, we have $a_i = 0$. The corresponding simple averaging estimator is computed as

$$\theta^{(0)} = \frac{\sum_{i=1}^{k} a_i \theta_i}{\sum_{i=1}^{k} a_i}.$$

And one-step estimator is as follows:

$$\theta^{(1)} = \theta^{(0)} - \left[\sum_{i=1}^{k} a_i \ddot{M}_i(\theta^{(0)})\right]^{-1} \left[\sum_{i=1}^{k} a_i \dot{M}_i(\theta^{(0)})\right].$$

Corollary 2 *Suppose r is the probability (or rate) that a local machine fails to send its information to the central machine. When $n = N/k \to \infty$, $k \to \infty$ and $k = O(\sqrt{N})$, the one-step estimator is asymptotically normal:*

$$\sqrt{(1-r)N}(\theta^{(1)} - \theta_0) \xrightarrow{d} \mathbf{N}(0, \Sigma).$$

And more precisely, unless all machines fail, we have

$$\mathbb{E}[\|\theta^{(1)} - \theta_0\|^2] \le \frac{2Tr[\Sigma]}{N(1-r)} + \frac{6Tr[\Sigma]}{Nk(1-r)^2} + O(N^{-2}(1-r)^{-2}) + O(k^2 N^{-2}).$$

See an online document (Huang and Huo 2015) for a proof. Note that the probability that all machines fail is r^k, which is negligible when r is small and k is large.

4.4 Numerical Examples

In this section, we will discuss the results of simulation studies. We compare the performance of the simple averaging estimator $\theta^{(0)}$ and the one-step estimator $\theta^{(1)}$, as well as the centralized M-estimator $\hat{\theta}$, which maximizes the global empirical criterion function $M(\theta)$ when the entire data are available centrally. Besides, we will also study the resampled averaging estimator, which is proposed by Zhang et al. (2013). The main idea of a resampled averaging estimator is to resample $\lfloor sn \rfloor$ observations from each local machine to obtain another averaging estimator $\theta_1^{(0)}$. Then the resampled averaging estimator can be constructed as follows:

$$\theta_{\text{re}}^{(0)} = \frac{\theta^{(0)} - s\theta_1^{(0)}}{1-s}.$$

In our numerical examples, the resampling ratio s is chosen to be $s = 0.1$ based on the past empirical studies. We shall implement these estimators for logistic regression (Sect. 4.4.1), Beta distribution (Sect. 4.4.2), and Gaussian Distribution (Sect. 4.4.4). We will also study the parameter estimation for Beta distribution with occurrence of communication failures (Sect. 4.4.3), in which some local machines could fail to send their local information to the central machine.

4.4.1 Logistic Regression

In this example, we simulate the data from the following logistic regression model:

$$y \sim \text{Bernoulli}(p), \text{where} p = \frac{\exp(x^t\theta)}{1+\exp(x^t\theta)} = \frac{\exp\left(\sum_{j=1}^d x_j\theta_j\right)}{1+\exp\left(\sum_{j=1}^d x_j\theta_j\right)}. \tag{4.10}$$

In this model, $y \in \{0, 1\}$ is a binary response, $x \in R^d$ is a continuous predictor, and $\theta \in R^d$ is the parameter of interest.

In each single experiment, we choose a fixed vector θ with each entry $\theta_j, j = 1, \ldots, d$, drawn from Unif$(-1, 1)$ independently. Entry $x_j, j = 1, \ldots, d$ of $x \in R^d$ is sampled from Unif$(-1, 1)$ independently. All x_j's are independent with parameters θ_j's too. After generating the parameter θ and the predictor x, we can compute the value of probability p and generate y according to (4.10). We fix the number of observed samples $N = 2^{17} = 131{,}072$ in each experiment, but vary the number of machines k. The target is to estimate θ with different number of parallel splits k of the data. The experiment is repeated for $K = 50$ times to obtain a reliable average error. And the criterion function is the log-likelihood function,

$$m(x, y; \theta) = yx^t\theta - \log(1 + \exp(x^t\theta)).$$

The goal of each experiment is to estimate parameter θ_0 maximizing the population criterion function

$$M_0(\theta) = \mathbb{E}_{x,y}[m(x, y; \theta)] = \mathbb{E}_{x,y}[yx^t\theta - \log(1 + \exp(x^t\theta))].$$

In this particular case, θ_0 is exactly the same with the true parameter.

In each experiment, we split the data into $k = 2, 4, 8, 16, 32, 64, 128$ non-overlapping subsets of size $n = N/k$. We compute a local estimator θ_i from each subset. And the simple averaging estimator is obtained by taking average on all local estimators: $\theta^{(0)} = \frac{1}{k}\sum_{i=1}^k \theta_i$. Then the one-step estimator $\theta^{(1)}$ could be computed by applying a Newton–Raphson update to $\theta^{(0)}$, i.e., Eq. (4.9).

The dimension is chosen to be $d = 20$ and $d = 100$, which could help us understand the performance of those estimators in both low and high dimensional cases. In Fig. 4.1, we plot the mean squared errors of each estimator versus the number of machines k. As we expect, the mean squared errors of the simple averaging estimator grows rapidly with the number of machines. But, the mean squared errors of the one-step estimator remains the same with the mean squared errors of the oracle estimator when the number of machines k is not very large. Even when the $k = 128$ and the dimension of predictors $d = 100$, the performance of the

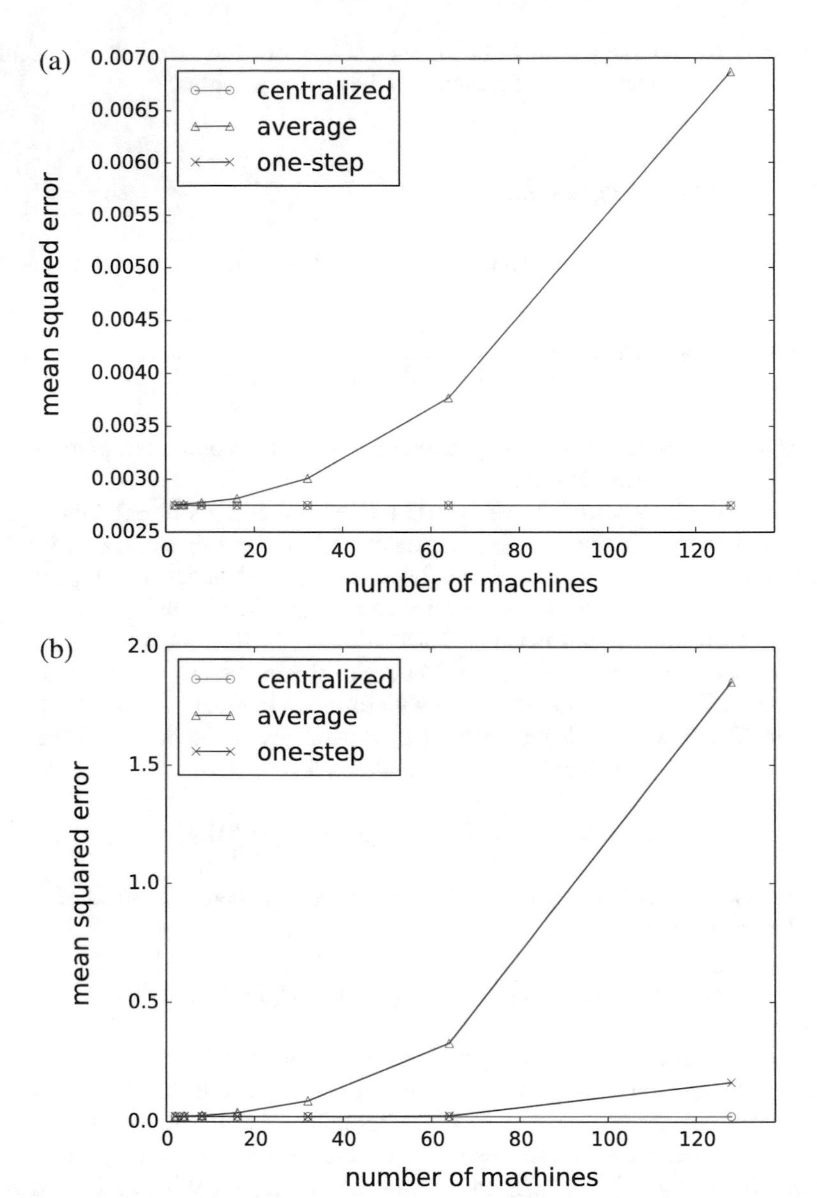

Fig. 4.1 Logistic Regression: the mean squared errors $\|\hat{\theta} - \theta_0\|^2$ versus number of machines, with 50 simulations. The "average" is $\theta^{(0)}$ and the "one-step" is $\theta^{(1)}$. The "centralized" denotes the oracle estimator with entire data. (**a**) $d = 20$. (**b**) $d = 100$

one-step estimator is significantly better than the simple averaging estimator. As we can easily find out from Fig. 4.1, the mean squared errors of the simple averaging estimator is about 10 times of that of the one-step estimator when $k = 128$ and $d = 100$. Detailed values of the mean squared errors are listed in Tables 4.1 and 4.2.

Table 4.1 Logistic regression ($d = 20$): detailed values of squared error $\|\hat{\theta} - \theta_0\|^2$

No. of machines	2	4	8	16	32	64	128
Simple avg ($\times 10^{-4}$)	28.036 (7.982)	28.066 (7.989)	28.247 (8.145)	28.865 (8.443)	30.587 (9.812)	38.478 (14.247)	69.898 (27.655)
One-step ($\times 10^{-4}$)	28.038 (7.996)	28.038 (7.996)	28.038 (7.996)	28.038 (7.996)	28.038 (7.996)	28.035 (7.998)	28.039 (8.017)
Centralized ($\times 10^{-4}$)	28.038 (7.996)						

In each cell, the first number is the mean squared errors in $K = 50$ experiments and the number in the brackets is the standard deviation of the mean squared errors

Table 4.2 Logistic regression ($d = 100$): detailed values of squared error $\|\hat{\theta} - \theta_0\|^2$

No. of machines	2	4	8	16	32	64	128
Simple avg ($\times 10^{-3}$)	23.066 (4.299)	23.818 (4.789)	26.907 (6.461)	38.484 (10.692)	87.896 (22.782)	322.274 (67.489)	1796.147 (324.274)
One-step ($\times 10^{-3}$)	22.787 (4.062)	22.784 (4.060)	22.772 (4.048)	22.725 (3.998)	22.612 (3.835)	24.589 (4.651)	151.440 (43.745)
Centralized ($\times 10^{-3}$)	22.787 (4.063)						

In each cell, the first number is the mean squared errors in $K = 50$ experiments and the number in the brackets is the standard deviation of squared error

From the tables, we can easily figure out that the standard deviation of the errors of the one-step estimator is significantly smaller than that of simple averaging, especially when the number of machines k is large, which means the one-step estimator is more stable.

4.4.2 Beta Distribution

In this example, we use data simulated from the Beta distribution $\text{Beta}(\alpha, \beta)$, whose p.d.f. is as follows:

$$f(x; \alpha, \beta) = \frac{\Gamma(\alpha + \beta)}{\Gamma(\alpha)\Gamma(\beta)} x^{\alpha-1}(1-x)^{\beta-1}.$$

In each experiment, we generate the value of parameter as $\alpha \sim \text{Unif}(1, 3)$ and $\beta \sim \text{Unif}(1, 3)$, independently. Once (α, β) is determined, we can simulate samples from the above density. In order to examine the performance of the two distributed methods when k is extremely large, we choose to use a data set with relatively small size $N = 2^{13} = 8192$ and let the number of machines vary in a larger range $k = 2, 4, 8, \ldots, 256$. And the objective is to estimate parameter (α, β) from the observed data. The experiment is again repeated for $K = 50$ times. The criterion function is $m(x; \theta) = \log f(x; \alpha, \beta)$, which implies that the centralized estimator is the MLE.

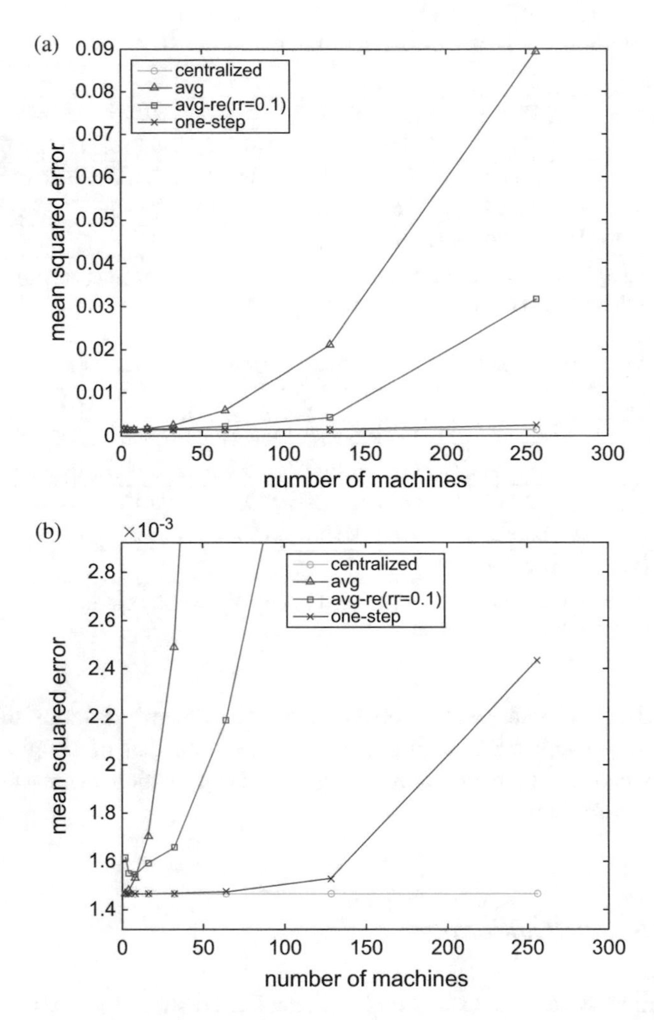

Fig. 4.2 Beta Distribution: the error $\|\theta - \theta_0\|^2$ versus the number of machines, with 50 simulations, where θ_0 is the true parameter. The "avg" is $\theta^{(0)}$, the "avg-re" is $\theta_{\text{re}}^{(0)}$ with resampling ratio $s = 10\%$ and the "one-step" is $\theta^{(1)}$. The "centralized" denotes maximum likelihood estimator with the entire data. (**a**) Overview. (**b**) Detailed view

Figure 4.2 and Table 4.3 show that the one-step estimator has almost the same performance with the centralized estimator in terms of MSE and standard deviation when the number of machines $k \leq \sqrt{N}$ (i.e., when $k \leq 64$). However, the one-step estimator performs worse than the centralized estimator when $k > \sqrt{N}$ (i.e., when $k = 128$ or 256), which confirms the necessity of condition $k = O(\sqrt{N})$ in Theorem 4. In addition, we can easily find out that both the simple averaging estimator and the resampled averaging estimator are worse than the proposed one-step estimator regardless of the value of k.

Table 4.3 Beta distribution: detailed values of the squared error $\|\hat{\theta} - \theta_0\|^2$

Number of machines	Simple avg ($\times 10^{-3}$)	Resampled avg ($\times 10^{-3}$)	One-step ($\times 10^{-3}$)	Centralized ($\times 10^{-3}$)
2	1.466 (1.936)	1.616 (2.150)	1.466 (1.943)	
4	1.480 (1.907)	1.552 (2.272)	1.466 (1.943)	
8	1.530 (1.861)	1.545 (2.177)	1.466 (1.943)	
16	1.704 (1.876)	1.594 (2.239)	1.466 (1.946)	1.466 (1.943)
32	2.488 (2.628)	1.656 (2.411)	1.468 (1.953)	
64	5.948 (5.019)	2.184 (3.529)	1.474 (1.994)	
128	21.002 (11.899)	4.221 (7.198)	1.529 (2.199)	
256	89.450 (35.928)	31.574 (36.518)	2.435 (3.384)	

In each cell, the first number is the mean squared errors with $K = 50$ experiments and the number in the brackets is the standard deviation of the mean squared errors

4.4.3 Beta Distribution with Possibility of Losing Information

Now, we would like to compare the performance of the simple averaging estimator and our one-step estimator under a more practical scenario, in which each single local machine could fail to send its information to the central machine. We assume those failures would occur independently with probability $r = 0.05$. The simulation settings are similar to the previous example in Sect. 4.4.2. However, we will generate $N = 409{,}600$ samples from the Beta distribution $\text{Beta}(\alpha, \beta)$, where α and β are chosen from $\text{Unif}(1, 3)$, independently. And the goal of this experiment is to estimate parameter (α, β). In each experiment, we let the number of machines vary: $k = 8, 16, 32, 64, 128, 256, 512$. We also compare the performance of the centralized estimator with entire data and the centralized estimator with $(1 - r) \times 100\% = 95\%$ of entire data. This experiment is repeated for $K = 50$ times (Table 4.4 and Fig. 4.3).

Table 4.4 Beta distribution with possibility of losing information: detailed values of the squared error $\|\hat{\theta} - \theta_0\|^2$

Number of machines	Simple avg ($\times 10^{-5}$)	One-step ($\times 10^{-5}$)	Centralized ($\times 10^{-5}$)	Centralized (95%) ($\times 10^{-5}$)
8	4.98 (10.76)	4.95 (10.62)		
16	4.82 (7.61)	4.75 (7.40)		
32	4.85 (9.65)	4.72 (9.31)		
64	4.51 (7.89)	4.10 (7.04)	4.07 (6.91)	4.76 (9.81)
128	5.25 (9.16)	4.48 (7.77)		
256	7.57 (12.26)	4.52 (7.70)		
512	16.51 (20.15)	5.24 (8.02)		

In each cell, the first number is the mean squared errors in $K = 50$ experiments and the number in the brackets is the standard deviation of the squared error

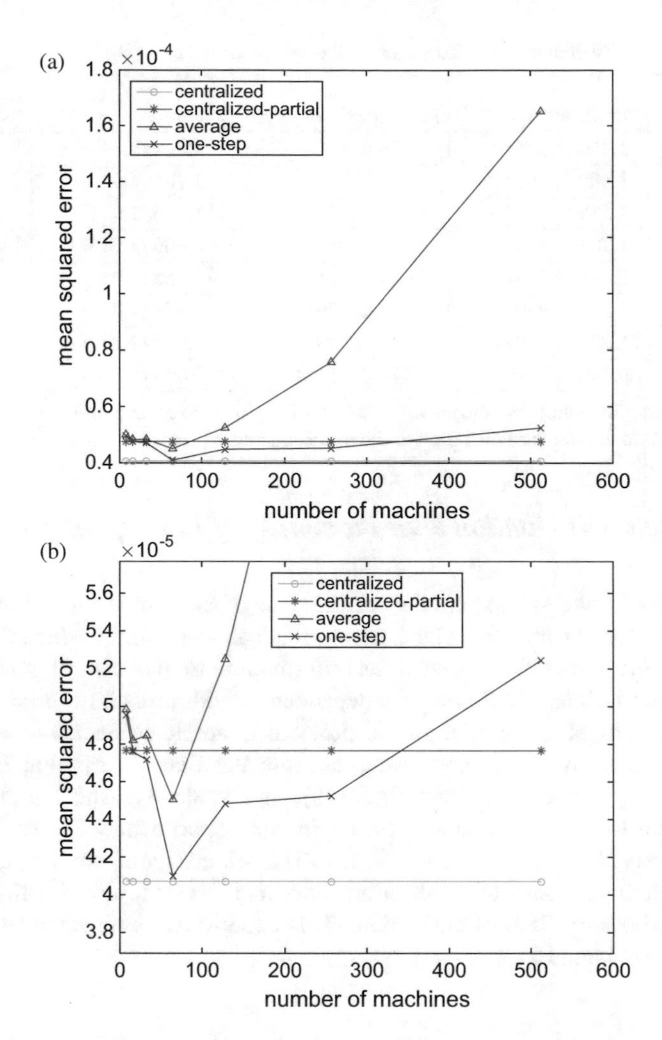

Fig. 4.3 Beta Distribution with Possibility of Losing Information: The error $\|\theta - \theta_0\|^2$ versus the number of machines, with 50 simulations, where θ_0 is the true parameter. The "average" is $\theta^{(0)}$ and the "one-step" is $\theta^{(1)}$. The "centralized" denotes the maximum likelihood estimator with the entire data. And the "centralized-partial" denotes the maximum likelihood estimator with $(1-r)\times 100\% = 95\%$ of data. (**a**) Overview. (**b**) Detailed view

In Fig. 4.3a, we plot the MSE of each estimator against the number of machines. As expected, the MSE of the simple averaging estimator grows significantly with the number of machines while the other three remain nearly the same. We can easily find out that the performance of the simple averaging estimator is far worse than the others, especially when the number of machines is large (for instance, when $k = 256 \text{or} 512$). If we take a closer look at the other three estimators from Fig. 4.3b, we will find that the performance of the one-step estimator is volatile but always remains in a reasonable range. And as expected, the error of the one-step estimator converges to the error of the oracle estimator with partial data when the number of machines k is large.

4.4.4 Gaussian Distribution with Unknown Mean and Variance

In this part, we will compare the performance of the simple averaging estimator, the resampled averaging estimator, and the one-step estimator when fixing the number of machines $k = \sqrt{N}$ and letting the value of N increase. We draw N samples from $N(\mu, \sigma^2)$, where $\mu \sim \text{Unif}(-2, 2)$ and $\sigma^2 \sim \text{Unif}(0.25, 9)$, independently. We let N vary in $\{4^3, \ldots, 4^9\}$ and repeat the experiment for $K = 50$ times for each N. We choose the criterion function to be the log-likelihood function

$$m(x; \mu, \sigma^2) = -\frac{(x - \mu)^2}{2\sigma^2} - \frac{1}{2}\log(2\pi) - \frac{1}{2}\log\sigma^2.$$

Figure 4.4 and Table 4.5 show that the one-step estimator is asymptotically efficient while the simple averaging estimator is absolutely not. It is worth noting that the resampled averaging estimator is not asymptotically efficient though it is better than the simple averaging estimator. When the number of samples N is relatively small, the one-step estimator is worse than the centralized estimator.

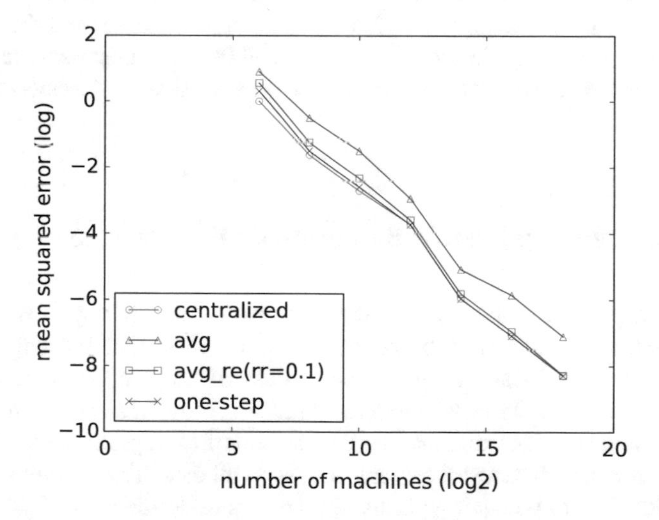

Fig. 4.4 Gaussian Distribution with Unknown Mean and Variance: the log error ($\log \|\theta - \theta_0\|^2$) versus the log number of machines ($\log_2 k$), with 50 repeated experiments for each N, where θ_0 is the true parameter. The "avg," "avg-re," and "one-step" denote $\theta^{(0)}$, $\theta_{\text{re}}^{(0)}$ with resampling ratio $s = 10\%$ and $\theta^{(1)}$, respectively. The "centralized" denotes the maximum likelihood estimator with the entire data. The sample size is fixed to be $N = k^2$

Table 4.5 Gaussian distribution with unknown mean and variance: detailed values of squared error $\|\hat{\theta} - \theta_0\|^2$

No. of machines	No. of samples	Simple avg	Resampled avg	One-step	Centralized
8	64	3.022104	2.153958	1.694668	1.388959
		(4.385627)	(3.458645)	(2.882794)	(2.424813)
16	256	0.739784	0.392389	0.318765	0.286175
		(1.209734)	(0.739390)	(0.621990)	(0.566140)
32	1024	0.118766	0.041050	0.034494	0.032563
		(0.151695)	(0.053808)	(0.046586)	(0.045779)
64	4096	0.026839	0.016519	0.014255	0.014414
		(0.046612)	(0.030837)	(0.029258)	(0.030533)
128	16,384	0.010996	0.004542	0.004329	0.004357
		(0.019823)	(0.009089)	(0.009453)	(0.009315)
256	65,536	0.002909	0.001158	0.001105	0.001099
		(0.005785)	(0.002733)	(0.002779)	(0.002754)
512	262,144	0.000843	0.000461	0.000376	0.000376
		(0.001426)	(0.000744)	(0.000596)	(0.000595)

In each cell, the first number is the mean squared errors in $K = 50$ experiments and the number in the brackets is the standard deviation of squared error

When the number of samples N grows large, the differences between the one-step estimator and the centralized estimator become minimal in terms of both mean squared errors and standard deviation. However, the error of the simple averaging estimator is significantly larger than both the one-step estimator and the centralized estimator. When the sample size $N = 4^9 \approx 250{,}000$, the mean squared errors of the simple averaging estimator is more than twice of that of the one-step and the centralized estimator.

4.5 Discussion on Distributed Statistical Inference

The M-estimator is a fundamental and high-impact methodology in statistics. The classic M-estimator theory is based on the assumption that the entire data are available at a central location, and can be processed/computed without considering communication issues. In many modern estimation problems arising in contemporary sciences and engineering, the classical notion of asymptotic optimality suffers from a significant deficiency: it requires access to all data. The asymptotic property when the data has to be dealt with distributively is under-developed. In this paper, we close this gap by considering a distributed one-step estimator.

Our one-step estimator is built on the existing *averaging* estimator. In a nutshell, after obtaining an averaging estimator, this initial estimate is broadcasted back to all local machines, to facilitate their computation of the gradients and Hessians of their objective functions. By doing so, the data do *not* need to be transmitted to the central machine. The central machine then collects the locally estimated gradients and Hessians, to produce a global estimate of the overall gradient and the overall Hessian. Consequently, a one-step update of the initial estimator is implemented. Just like the one-step approach has improved the estimator in the classical (non-distributed) setting, we found that the one-step approach can improve the performance of an estimator under the distributed setting, both theoretically and numerically.

Besides the works that have been cited earlier, there are many other results in the relevant literature, however they may not be directly technically linked to what's been done here. We discuss their influence and insights in the next few paragraphs.

An interesting split-and-merge Bayesian approach for variable selection under linear models is proposed in Song and Liang (2015). The method firstly split the ultrahigh dimensional data set into a number of lower dimensional subsets and select relevant variables from each of the subsets, and then aggregate the variables selected from each subset and eventually select relevant variables from the aggregated data set. Under mild conditions, the authors show that the proposed approach is consistent, i.e., the underlying true model will be selected in probability 1 as the sample size becomes large. This work differs from all the other approaches that we discussed in this paper: it splits the variables, while all the other approaches that we referenced (including ours) split the data according to observations. This paper certainly is in line with our research, however takes a very distinct angle.

An interesting piece of work that combines distributed statistical inference and information theory in communication is presented in Zhang et al. (2013). Their current results need to rely on some special model settings: a uniform location family $\mathscr{U} = \{P_\theta, \theta \in [-1, 1]\}$, where P_θ denotes the uniform distribution on the interval $[\theta - 1, \theta + 1]$, or Gaussian location families $N_d([-1, 1]^d) = \{N(\theta, \sigma^2 I_{d\times d}) \mid \theta \in \Theta = [-1, 1]^d\}$. It will be interesting to see whether or not more general results are feasible.

Neiswanger et al. (2013) propose an asymptotically exact, embarrassingly parallel MCMC method by approximating each sub-posterior with Gaussian density, Gaussian kernel or weighted Gaussian kernel. They prove the asymptotic correctness of their estimators and bound rate of convergence. Our paper does not consider the MCMC framework. The analytical tools that they used in proving their theorems are of interest.

Wang et al. (2014) propose a distributed variable selection algorithm, which accepts a variable if more than half of the machines select that variable. They give upper bounds for the success probability and mean squared errors (MSE) of the estimator. This work bears similarity with Song and Liang (2015) and Chen and Xie (2014), however with somewhat different emphases.

Kleiner et al. (2014) propose a scalable bootstrap (named 'bag of little bootstraps' (BLB)) for massive data to assess the quality of estimators. They also demonstrate its favorable statistical performance through both theoretical analysis and simulation studies. A comparison with this work will be interesting, however not included here.

Zhao et al. (2014) consider a partially linear framework for massive heterogeneous data and propose an aggregation type estimator for the commonality parameter that possesses the minimax optimal bound and asymptotic distribution when the number of sub-populations does not grow too fast.

A recent work Arjevani and Shamir (2015) shed interesting new light on the distributed inference problem. The authors studied the fundamental limits to communication-efficient distributed methods for convex learning and optimization, under different assumptions on the information available to individual machines, and the types of functions considered. The current problem formulation is more numerical than statistical properties. Their idea may lead to interesting counterparts in statistical inference.

Various researchers have studied communication-efficient algorithms for statistical estimation (e.g., see the papers Dekel et al. (2012), Balcan et al. (2012), Wainwright (2014), McDonald et al. (2010) and the references therein). They were not discussed in detail here, because they are pretty much discussed/compared in other references of this paper.

There is now a rich and well-developed body of theory for bounding and/or computing the minimax risk for various statistical estimation problems, e.g., see Yang and Barron (1999) and the references therein. In several cited references, researchers have started to derive the optimal minimax rate for estimators under the distributed inference setting. This will be an exciting future research direction.

Other numerical approaches have been proposed and studied. Paper Bradley et al. (2011) propose a parallel coordinate descent algorithm for minimizing L1-regularized losses. They presented a comprehensive empirical study of their methods for Lasso and sparse logistic regression. It's been reported that their approach outperforms other published solvers on a range of large problems, proving to be one of the most scalable algorithms for L1. Due to space, we cannot cite all the relevant works in this line of research, instead we just single out the above as a potential starting point for literature search.

4.6 Other Problems

There are many other learning problems, beyond regression. We review a few possible directions of research along this line, and describe potential future work.

- *EM algorithm.* Nowak (2003) proposed a distributed expectation-maximization (EM) algorithm for density estimation and clustering in sensor networks. Though the studied problem is technically different from ours, it provides an inspiring historic perspective: distributed inference has been studied more than 10 years ago.
- *Distributed principal component analysis.* Besides estimation, other distributed statistical technique may be of interest, such as the distributed principal component analysis (Balcan et al. 2014).

- *Distributed support vector machines.* Utilizing ADMM (Boyd et al. 2011), consensus-based distributed support vector machines were introduced (Forero et al. 2010).
- *Distributed nonnegative matrix factorization.* Nonnegative Matrix Factorization (NMF), as a data analysis technique, has been successfully used in many different areas. Historically, Lee and Seung (1999) applied NMF to extract parts-based features of human faces and semantic features of text. Since then, NMF has been used for extracting features from many kinds of sources, such as hyperspectral images (Pauca et al. 2006), speeches (Schmidt et al. 2007), and music (Fevotte et al. 2009). NMF has also been applied to other topics such as text mining (Xu et al. 2003), clustering (Ding et al. 2005), air emission control (Paatero and Tapper 1994), and blind source separation (Cichocki et al. 2009). The Separability condition (Donoho and Stodden 2003) for the exact recovery in NMF has played an important role in the recent literature. See recent algorithmic and theoretical developments in Mizutani (2014), Gillis and Luce (2014). A new condition was established in Huang et al. (2014). Distributed NMF and similar uniqueness conditions will be an interesting future research topic.
- *Sparse nonparametric models.* The simplest high-dimensional structured nonparametric model is the sparse additive model (Ravikumar et al. 2009; Fan 2012). Through the sieve methods, it can be formulated as a high-dimensional linear model with group structure (Yuan and Lin 2006). This idea can be applied to more complicated nonparametric models, via the sieve approximations and the "kernel tricks." Distributed inference techniques will be developed under this contact.
- *Mixture models for Big Data.* A stylized feature of Big Data is that they are often comprised of many heterogeneous subgroups (Fan et al. 2014) and are often modeled by a mixture model (Fan et al. 2014; Städler et al. 2010). An important application of Big Data is for precision treatments and personalized services, which hinge on identifications of the features of the mixed sub-populations. This involves optimizing a nonconvex objective function and requires intensive computation. This along with distributed storage of data sets makes it very compelling for the development of distributed inference.

Though various distributed algorithms have been introduced in most of the aforementioned scenarios, the corresponding knowledge on the statistical performance of them is at least underdeveloped. It will be interesting to extend the theoretical analysis from regression with low-dimensional (sparse) underlying structure to the above cases. Note that in the central estimation setting, most of the above approaches have corresponding supporting statistical theory. For example, it is known that the centralized support vector machines achieve statistical consistency (in the sense of approaching to the Bayes classifier) at an optimal rate, if the underlying distribution satisfies certain properties. These potential research topics demonstrate the broadness of SDAI problems.

4.7 Conclusion

Scattered Data and Aggregated Inference (SDAI) is a class of problems where data cannot be at a centralized location, while aggregated modeling and inference is desirable and/or necessary. Distributed statistical inference is a technique to tackle a type of the above problem, and this chapter describes one of them. It is noticed that many existing work focus on the averaging estimator, e.g., Zhang et al. (2013) together with many others. In this chapter, we propose a one-step approach to enhance a simple-averaging based distributed estimator. The corresponding asymptotic properties of the newly proposed estimator are derived. We found that the proposed one-step estimator enjoys the same asymptotic properties as the centralized estimator. The proposed one-step approach merely requires one additional round of communication in relative to the averaging estimator; so the extra communication burden is insignificant. In finite sample cases, numerical examples show that the proposed estimator outperforms the simple averaging estimator with a large margin in terms of the mean squared errors. The proposed method becomes more valuable when data can only be available at distributed machines with limited communication bandwidth. Other distinct types of SDAI problems are discussed at the end of this chapter. We hope our efforts will attract more future work on this topic.

References

Arjevani Y, Shamir O (2015) Communication complexity of distributed convex learning and optimization. Technical report. http://arxiv.org/abs/1506.01900. Accessed 28 Oct 2015

Balcan M-F, Blum A, Fine S, Mansour Y (2012) Distributed learning, communication complexity and privacy. https://arxiv.org/abs/1204.3514. Accessed 25 May 2012

Balcan M-F, Kanchanapally V, Liang Y, Woodruff D (2014) Improved distributed principal component analysis. Technical report. http://arxiv.org/abs/1408.5823. Accessed 23 Dec 2014

Battey H, Fan J, Liu H, Lu J, Zhu Z (2015) Distributed estimation and inference with statistical guarantees. https://arxiv.org/abs/1509.05457. Accessed 17 Sept 2015

Bickel PJ (1975) One-step Huber estimates in the linear model. J Am Stat Assoc 70(350):428–434

Boyd S, Parikh N, Chu E, Peleato B, Eckstein J (2011) Distributed optimization and statistical learning via the alternating direction method of multipliers. Found Trends Mach Learn 3(1):1–122

Bradley JK, Kyrola A, Bickson D, Guestrin C (2011) Parallel coordinate descent for L1-regularized loss minimization. In Proceedings of 28th international conference on Machine Learning. https://arxiv.org/abs/1105.5379. Accessed 26 May 2011

Chen X, Xie M-g (2014) A split-and-conquer approach for analysis of extraordinarily large data. Stat Sin 24:1655–1684

Chen S, Donoho DL, Saunders MA (1998) Atomic decomposition by basis pursuit. SIAM J Sci Comput 20(1):33–61

Cichocki A, Amari S-I, Zdunek R, Phan AH (2009) Non-negative matrix and tensor factorizations: applications to exploratory multi-way data analysis and blind source separation. Wiley-Blackwell, Hoboken

Corbett JC, Dean J, Epstein M, Fikes A, Frost C, Furman JJ, Ghemawat S, Gubarev A, Heiser C, Hochschild P et al. (2012) Spanner: Googles globally distributed database. In: Proceedings of the USENIX symposium on operating systems design and implementation
Dekel O, Gilad-Bachrach R, Shamir O, Xiao L (2012) Optimal distributed online prediction using mini-batches. J Mach Learn Res 13:165–202
Ding C, He X, Simon HD (2005) On the equivalence of nonnegative matrix factorization and spectral clustering. In: SIAM international conference on data mining, pp 606–610
Donoho D, Stodden V (2003) When does non-negative matrix factorization give a correct decomposition into parts? In: Advances in neural information processing systems. Stanford University, Stanford
El Gamal M, Lai L (2015) Are Slepian-Wolf rates necessary for distributed parameter estimation? Technical report. http://arxiv.org/abs/1508.02765. Accessed 10 Nov 2015
Fan J, Chen J (1999) One-step local quasi-likelihood estimation. J R Stat Soc Ser B Stat Methodol 61(4):927–943
Fan J, Feng Y, Song R (2012) Nonparametric independence screening in sparse ultra-high-dimensional additive models. J Am Stat Assoc 106:544–557
Fan J, Han F, Liu H (2014) Challenges of big data analysis. Natl Sci Rev 1:293–314
Fevotte C, Bertin N, Durrieu JL (2009) Nonnegative matrix factorization with the Itakura-Saito divergence: with application to music analysis. Neural Comput 21(3):793–830
Forero PA, Cano A, Giannakis GB (2010) Consensus-based distributed support vector machines. J Mach Learn Res 11:1663–1707
Gillis N, Luce R (2014) Robust near-separable nonnegative matrix factorization using linear optimization. J Mach Learn Res 15:1249–1280
Huang C, Huo X (2015) A distributed one-step estimator. Technical report. http://arxiv.org/abs/1511.01443. Accessed 10 Nov 2015
Huang K, Sidiropoulos ND, Swami A (2014) Non-negative matrix factorization revisited: uniqueness and algorithm for symmetric decomposition. IEEE Trans Signal Process 62(1):211–224
Jaggi M, Smith V, Takác M, Terhorst J, Krishnan S, Hofmann T, Jordan MI (2014) Communication-efficient distributed dual coordinate ascent. In: Advances in neural information processing systems, pp 3068–3076
Kleiner A, Talwalkar A, Sarkar P, Jordan MI (2014) A scalable bootstrap for massive data. J R Stat Soc Ser B Stat Methodol 76(4):795–816
Lang S (1993) Real and functional analysis, vol 142. Springer Science & Business Media, Berlin
Lee DD, Seung HS (1999) Learning the parts of objects by nonnegative matrix factorization. Nature 401:788–791
Lee JD, Sun Y, Liu Q, Taylor JE (2015) Communication-efficient sparse regression: a one-shot approach. arXiv preprint arXiv:1503.04337
Liu Q, Ihler AT (2014) Distributed estimation, information loss and exponential families. In: Advances in neural information processing systems, pp 1098–1106
McDonald R, Hall K, Mann G (2010) Distributed training strategies for the structured perceptron. In: North American chapter of the Association for Computational Linguistics (NAACL)
Mitra S, Agrawal M, Yadav A, Carlsson N, Eager D, Mahanti A (2011) Characterizing web-based video sharing workloads. ACM Trans Web 5(2):8
Mizutani T (2014) Ellipsoidal rounding for nonnegative matrix factorization under noisy separability. J Mach Learn Res 15:1011–1039
Neiswanger W, Wang C, Xing E (2013) Asymptotically exact, embarrassingly parallel MCMC. arXiv preprint arXiv:1311.4780
Nowak RD (2003) Distributed EM algorithms for density estimation and clustering in sensor networks. IEEE Trans Signal Process 51(8):2245–2253
Paatero P, Tapper U (1994) Positive matrix factorization: a nonnegative factor model with optimal utilization of error estimates of data values. Environmetrics 5(2):111–126
Pauca VP, Piper J, Plemmons RJ (2006) Nonnegative matrix factorization for spectral data analysis. Linear Algebra Appl 401(1):29–47

Ravikumar P, Lafferty J, Liu H, Wasserman L (2009) Sparse additive models. J R Stat Soc Ser B Stat Methodol 71(5):1009–1030
Rosenblatt J, Nadler B (2014) On the optimality of averaging in distributed statistical learning. arXiv preprint arXiv:1407.2724
Schmidt MN, Larson J, Hsiao FT (2007) Wind noise reduction using non-negative sparse coding. In: Machine learning for signal processing, IEEE workshop, pp 431–436
Shamir O, Srebro N, Zhang T (2014) Communication-efficient distributed optimization using an approximate Newton-type method. In: Proceedings of the 31st international conference on machine learning, pp 1000–1008
Song Q, Liang F (2015) A split-and-merge Bayesian variable selection approach for ultrahigh dimensional regression. J R Stat Soc B 77(Part 5):947–972
Städler N, Bühlmann P, Van De Geer S (2010) ℓ_1-Penalization for mixture regression models. Test 19(2):209–256
Tibshirani R (1996) Regression shrinkage and selection via the Lasso. J R Stat Soc Ser B 58(1):267–288
van der Vaart AW (2000) Asymptotic statistics. Cambridge series in statistical and probabilistic mathematics. Cambridge University Press, Cambridge
Wainwright M (2014) Constrained forms of statistical minimax: computation, communication, and privacy. In: Proceedings of international congress of mathematicians
Wang X, Peng P, Dunson DB (2014) Median selection subset aggregation for parallel inference. In: Advances in neural information processing systems, pp 2195–2203
Xu W, Liu X, Gong Y (2003) Document clustering based on non-negative matrix factorization. In: The 26th annual international ACM SIGIR conference on research and development in information retrieval, pp 267–273
Yang Y, Barron A (1999) Information-theoretic determination of minimax rates of convergence. Ann Stat 27(5):1564–1599
Yuan M, Lin Y (2006) Model selection and estimation in regression with grouped variables. J R Stat Soc Ser B Stat Methodol 68(1):49–67
Zhang Y, Duchi JC, Wainwright MJ (2013) Communication-efficient algorithms for statistical optimization. J Mach Learn Res 14:3321–3363
Zhang Y, Duchi JC, Jordan MI, Wainwright MJ (2013) Information-theoretic lower bounds for distributed statistical estimation with communication constraints. Technical report, UC Berkeley. Presented at the NIPS Conference 2013
Zhao T, Cheng G, Liu H (2014) A partially linear framework for massive heterogeneous data. arXiv preprint arXiv:1410.8570
Zinkevich M, Weimer M, Li L, Smola AJ (2010) Parallelized stochastic gradient descent. In: Advances in neural information processing systems, pp 2595–2603
Zou H, Li R (2008) One-step sparse estimates in nonconcave penalized likelihood models. Ann Stat 36(4):1509–1533

Chapter 5
Nonparametric Methods for Big Data Analytics

Hao Helen Zhang

Abstract Nonparametric methods provide more flexible tools than parametric methods for modeling complex systems and discovering nonlinear patterns hidden in data. Traditional nonparametric methods are challenged by modern high dimensional data due to the curse of dimensionality. Over the past two decades, there have been rapid advances in nonparametrics to accommodate analysis of large-scale and high dimensional data. A variety of cutting-edge nonparametric methodologies, scalable algorithms, and the state-of-the-art computational tools have been designed for model estimation, variable selection, statistical inferences for high dimensional regression, and classification problems. This chapter provides an overview of recent advances on nonparametrics in big data analytics.

Keywords Sparsity · Smoothing · Nonparametric estimation · Regularization · GAM · COSSO

5.1 Introduction

Nonparametric methods play a fundamental role in statistics and machine learning, due to their high flexibility and ability to discover nonlinear and complex relationship between variables. There is a very rich literature on nonparametrics and smoothing methods in statistics. Classical smoothing techniques include kernel estimators (Nadaraya 1964; Altman 1990; Tsybakov 2009), local weighted polynomial regression (Cleveland 1979; Fan and Gijbels 1996), regression splines and smoothing splines (Kimeldorf and Wahba 1971; de Boor 1978; Wahba 1990; Green and Silverman 1994; Stone et al. 1994; Mammen and van de Geer 1997; Gu 2002), and wavelets (Mallet 2008; Donoho and Johnstone 1994). When handling high dimensional data, nonparametric methods face more challenges than linear

H. H. Zhang (✉)
Department of Mathematics, ENR2 S323, University of Arizona, Tucson, AZ, USA
e-mail: hzhang@math.arizona.edu

© Springer International Publishing AG, part of Springer Nature 2018

W. K. Härdle et al. (eds.), *Handbook of Big Data Analytics*, Springer Handbooks of Computational Statistics, https://doi.org/10.1007/978-3-319-18284-1_5

methods, due to the curse of dimensionality and the intrinsic infinite-dimensional nature of multivariate functions. Many structured models have been developed to overcome the above difficulties and facilitate model estimation and inferences, including additive models (Friedman and Stuetzle 1981; Buja et al. 1989; Hastie and Tibshirani 1990; Fan and Jiang 2005), smoothing spline analysis of variance (SS-ANOVA; Wahba 1990; Wahba et al. 1995; Gu 2002) models, multivariate adaptive regression splines (MARS; Friedman 1991), and Friedman and Silverman (1989). The main idea of these works is to decompose a multivariate function into the sum of a series of functional components such as main-effect and interaction-effect terms, truncate the series by ignoring high-order terms, and estimate the truncated model by regularization approaches.

Driven by the need of analyzing high and ultra-high dimensional data, more and more commonly collected in modern sciences and industry, there has been a surge of research interest in developing statistical and machine learning methods which can effectively reduce data dimension without information loss, automatically conduct variable selection, and produce sparse models and interpretable results. Over the past two decades, a large body of penalization methods have been developed to simultaneously conduct variable selection and model estimation in linear models, such as the nonnegative garrote (Breiman 1995), the LASSO (Tibshirani 1996; Efron et al. 2004; Zhao and Yu 2006), the SCAD (Fan and Li 2001, 2004), the Dantzig selector (Candes and Tao 2007), the elastic net (Zou and Hastie 2005; Zou and Zhang 2009), the adaptive LASSO (Zou 2006; Zhang and Lu 2007), and the minimax concave penalty (MCP; Zhang 2010). By effectively removing noise variables from the model, these methods can greatly improve prediction accuracy and enhance model interpretability. A variety of state-of-the-art computational algorithms and tools have been designed to facilitate their implementation for high dimensional data. When the data dimension is ultra high, it is useful to pre-screen variables and reduce dimensionality to moderate or low levels before a refined variable selection procedure is applied. Sure screening methods have been developed to achieve this, including the SIS and iterative SIS (Fan and Lv 2008), forward regression screening (Wang 2009), screening for classification problems (FAIR; Fan and Fan 2008), and interaction screening (Hao and Zhang 2014). All these works mentioned above mainly focus on linear models.

Motivated by the success of penalization methods for linear models, there are rapid advances in multivariate nonparametric methods to achieve smooth and sparse function estimation for high dimensional regression and classification problems. Recent works include the basis pursuit (Chen et al. 1999; Zhang et al. 2004), the COmponent Selection and Smoothing Operator (COSSO; Lin and Zhang 2006; Zhang and Lin 2006), the adaptive COSSO (Storlie et al. 2011), the rodeo (Lafferty and Wasserman 2008), the sparse additive models (SpAM; Ravikumar et al. 2009), and the linear and nonlinear discover method (LAND; Zhang et al. 2011). This chapter reviews these works for both nonparametric regression and classification.

Assume the observations are $\{(\mathbf{x}_i, y_i), i = 1, \ldots, n\}$, where $\mathbf{x}_i$'s and y_i's are realizations of a vector of predictors $\mathbf{X} = (X_1, \ldots, X_p) \in \mathcal{X} \subset \mathbb{R}^p$ and the response Y, respectively. Without loss of generality, we assume $\mathcal{X} = [0, 1]^p$ throughout the chapter. The dimension p can be fixed, or increase with the sample size n, or much larger than the sample size with $p \gg n$. The output $Y \in \mathcal{Y} \subset \mathbb{R}$ is a random variable, which takes a continuous value (leading to a regression problem) or a categorical value from a discrete set (leading to a classification problem). Here $\mathbb{R}$ denotes the real line, and $\mathbb{R}^p$ denotes the p-dimensional real coordinate space. The random vector $(\mathbf{X}, Y)$ jointly follows a distribution $\Pr(\mathbf{X}, Y)$. The goal of supervised learning is to learn a function $f : \mathcal{X} \rightarrow \mathcal{Y}$ from the data to characterize the relationship between $\mathbf{X}$ and Y, predict future values of the response based on new predictors, and make optimal decisions.

5.2 Classical Methods for Nonparametric Regression

5.2.1 *Additive Models*

Additive models assume that

$$Y_i = b + \sum_{j=1}^{p} f_j(X_{ij}) + \epsilon_i, \quad i = 1, \ldots, n, \tag{5.1}$$

where b is the intercept term, f_j's are unspecified smooth functions of X_j, and ϵ_i's are i.i.d. random variables with mean zero and finite variance. To make b and f_j's identifiable, it is common to center f_j's over the samples by assuming

$$\sum_{i=1}^{n} f_j(x_{ij}) = 0, \quad j = 1, \ldots, p.$$

The f_j's are usually estimated by smoothing techniques such as cubic splines, local polynomial regression, and kernel smoothers. For example, if f_j's have the second-order continuous derivatives, they can be estimated by minimizing a penalized least squares

$$\min_{b, f_1, \ldots, f_p} \sum_{i=1}^{n} \left\{ y_i - b - \sum_{j=1}^{p} f_j(x_{ij}) \right\}^2 + \sum_{j=1}^{p} \lambda_j \int_0^1 [f_j''(x_j)]^2 dx_j, \tag{5.2}$$

where $\int [f_j''(x_j)]^2 dx_j$ is a roughness penalty imposed on functions to encourage a smooth fit, and $\lambda_j \geq 0$ are smoothing parameters. The minimization in (5.2) takes place in the function space $\{f_1, \ldots, f_p : \int_0^1 [f_j''(x)]^2 dx < \infty, j = 1, \ldots, p\}$. Each estimated function component in the solution of (5.2) is a cubic smoothing spline. The values of λ_j's are chosen adaptively from the data to assure optimal results. Commonly used tuning criteria include cross validation, generalized cross validation (GCV), and various types of information criteria. It is known that GCV has computational advantages over CV (Wahba 1990). The optimization problem in (5.2) is usually solved by a coordinate descent Gauss–Seidel procedure called the backfitting algorithm (Breiman and Friedman 1985; Buja et al. 1989; Hastie and Tibshirani 1990; Opsomer and Ruppert 1998).

Besides smoothing splines, additive models can be alternatively estimated via regression splines. For regression splines, each component f_j is represented in terms of a finite number of pre-selected basis functions

$$f_j(x_j) = \sum_{k=1}^{m} \beta_{jk}\phi_k(x_j), \quad j = 1, \ldots, p,$$

where ϕ_k's are basis functions like polynomial basis or B-splines, β_{jk}'s are coefficients of basis functions, and m is the number of basis functions. Define $\boldsymbol{\beta}_j = (\beta_{j1}, \ldots, \beta_{jm})^T, j = 1, \ldots, p$ and $\boldsymbol{\beta} = (\boldsymbol{\beta}_1^T, \ldots, \boldsymbol{\beta}_p^T)^T$. The parameter vector $\boldsymbol{\beta}$ is estimated by the least squares

$$\min_{b,\boldsymbol{\beta}} \sum_{i=1}^{n} \left\{ y_i - b - \sum_{j=1}^{p} \sum_{k=1}^{m} \beta_{jk}\phi_k(x_{ij}) \right\}^2, \tag{5.3}$$

In practice, B-splines are commonly used basis functions due to their local support property and high stability for large-scale spline interpolation (de Boor 1978). The basis dimension m is a smoothing parameter that controls the degree of smoothness of the fitted model, and its value needs to be properly chosen to balance the trade-off between model fit and function smoothness. In practice, the model fit is not very sensitive to the exact value of m as long as it is sufficiently large to provide adequate flexibility.

Nonparametric inferences for additive models are challenging due to the curse of dimensionality and complexity of the backfitting algorithm. Theoretical properties of backfitting estimators, such as oracle property and rates of convergence, are studied in Linton (1997), Fan et al. (1998), Mammen et al. (1999), Wand (1999), and Opsomer (2000). For example, Fan et al. (1998) show that any additive component in (5.11) can be estimated as well as if the rest of the components were known, and this is the so-called oracle property. Traditionally, variable selection for additive models is conducted by testing the null hypotheses: $f_j(x_j) = 0$ vs $f_j(x_j) \neq 0$ for $j = 1, \ldots, p$. Variables with large p-values are regarded as "not significant" or "not important." Fan and Jiang (2005) develop generalized likelihood ratio (GLR) tests and their bias-corrected versions for testing nonparametric hypotheses in additive models. Under the null models, the GLR test statistics are known to follow asymptotically rescaled chi-squared distributions, with the scaling constants and the degrees of freedom independent of the nuisance parameters, which is known as Wilks phenomenon (Fan and Jiang 2005). The GLR tests are also asymptotically optimal in terms of rates of convergence for nonparametric hypothesis.

5.2.2 *Generalized Additive Models (GAM)*

When the response Y is binary or count, generalized additive models (GAM; Hastie and Tibshirani 1990) is a popular and useful tool for nonparametric regression and classification. Given $\mathbf{X} = \mathbf{x}$, the conditional distribution of Y is assumed to be from the exponential family with density

$$\exp\left\{\frac{y\theta(\mathbf{x}) - h(\theta(\mathbf{x}))}{\eta} + c(y, \eta)\right\}, \tag{5.4}$$

where $h(\cdot)$ and $c(\cdot)$ are known functions, h is twice continuously differentiable and bounded away from zero, η is the *dispersion* parameter, and θ is the canonical parameter. It is easy to show that, for (5.4), $\mathrm{E}(Y|\mathbf{X}) = \mu(\mathbf{X}) = h'(\theta(\mathbf{X}))$ and $\mathrm{Var}(Y|\mathbf{X}) = h''(\theta(\mathbf{X}))\eta$. The exponential family includes normal, binomial, Poisson, and gamma distributions, as special cases.

Define the link function as $g = (h')^{-1}$. Generalized additive models (GAM) assume

$$g(\mu(\mathbf{X})) = \theta(\mathbf{X}) = b + \sum_{j=1}^{p} f_j(X_j). \tag{5.5}$$

The estimators of f_j's are obtained by minimizing a penalized negative log-likelihood

$$\min_f \sum_{i=1}^{n} l(y_i, f(\mathbf{x}_i)) + \sum_{j=1}^{p} \lambda_j \int [f_j''(x_j)]^2 dx_j, \tag{5.6}$$

where $l(y, f)$ is the log likelihood function. The optimization problem (5.6) is typically solved by minimizing the penalized iteratively re-weighted least squares (penalized IRLS). There are three R packages for fitting GAMS, and they are *gam* written by Trevor Hastie, *gss* written by Chong Gu, and *mgcv* written by Simon Wood. Bayesian confidence intervals of the true function components can be constructed using the estimated GAM estimators (Wahba 1983; Wood 2006).

5.2.3 *Smoothing Spline ANOVA (SS-ANOVA)*

Smoothing spline analysis of variance (SS-ANOVA; Wahba 1990) models are a general family of smoothing methods for multivariate nonparametric function estimation. In the functional ANOVA setup, any multivariate function $f(X_1, \ldots, X_p)$

can be decomposed into the sum of functional components

$$f(\mathbf{X}) = b + \sum_{j=1}^{p} f_j(X_j) + \sum_{j<k}^{p} f_{jk}(X_j, X_k) + \cdots, \tag{5.7}$$

where b is a constant, f_j's are main-effect terms, f_{jk}'s are two-way interactions, and so on. The identifiability of the terms in (5.7) is assured by imposing side conditions through averaging operators. In SS-ANOVA, assume $f_j \in \mathcal{H}_j$, which is the space of functions of X_j over $[0, 1]$ and satisfies $\mathcal{H}_j = \{1\} \oplus \bar{\mathcal{H}}_j$. Then the tensor product space of $\mathcal{H}_j$'s is

$$\bigotimes_{j=1}^{p} \mathcal{H}_j = \{1\} \oplus_{j=1}^{p} \bar{\mathcal{H}}_j \oplus_{j<k} [\bar{\mathcal{H}}_j \otimes \bar{\mathcal{H}}_k] \oplus \cdots. \tag{5.8}$$

In (5.8), the space $\otimes_{j=1}^{p} \mathcal{H}_j$ is decomposed into a direct sum of orthogonal subspaces such as main effect subspaces $\bar{\mathcal{H}}_j$'s, two-way interaction subspaces $\bar{\mathcal{H}}_j \otimes \bar{\mathcal{H}}_k$, and higher-order interaction subspaces. Correspondingly, each functional component in (5.7) lies in one corresponding subspace of (5.8), e.g., $f_j \in \bar{\mathcal{H}}_j, f_{jk} \in \bar{\mathcal{H}}_j \otimes \bar{\mathcal{H}}_k$, and so on.

To facilitate model estimation and interpretation, higher-order terms are often truncated in (5.7) and (5.8). For example, when only main effects are retained in the decomposition, then (5.7) reduces to the additive model

$$f(\mathbf{X}) = b + \sum_{j=1}^{p} f_j(X_j), \quad \forall f_j \in \bar{\mathcal{H}}_j \tag{5.9}$$

If main and pairwise interaction effects are kept in (5.7), we obtain the two-way interaction model

$$f(\mathbf{X}) = b + \sum_{j=1}^{p} f_j(X_j) + \sum_{j<k}^{p} f_{jk}(X_j, X_k), \quad \forall\ f_j \in \bar{\mathcal{H}}_j,\ f_{jk} \in \bar{\mathcal{H}}_j \otimes \bar{\mathcal{H}}_k.$$

A convenient way of estimating f is through the reproducing kernel Hilbert space framework, as shown in Wahba (1990). Assume each $\mathcal{H}_j$ is an RKHS. For example, $\mathcal{H}_j$ is the second-order Sobolev space on $[0, 1]$, $\mathcal{S}^2[0, 1] = \{g : g, g'$ are absolutely continuous, $\int_0^1 [g''(t)]^2 < \infty\}$. When endowed with the inner product

$$\langle g, h \rangle = \int_0^1 g(t)dt \int_0^1 h(t)dt + \int_0^1 g'(t)dt \int_0^1 h'(t)dt$$
$$+ \int_0^1 g''(t)h''(t)dt, \forall g, h \in \mathcal{S}^2[0, 1],$$

the space $\mathcal{S}^2[0,1] = \{1\} \oplus \bar{\mathcal{S}}^2[0,1]$ is an RKHS associated with the reproducing kernel

$$\mathcal{K}(s,t) = 1 + k_1(s)k_1(t) + k_2(s)k_2(t) - k_4(|s-t|),$$

where $k_1(s) = s - 0.5$, $k_2(s) = [k_1(s)^2 - 1/12]/2$ and $k_4(s) = [k_1^4(s) - k_1^2(s)/2 + 7/240]/24$.

Denote the truncated space by $\mathcal{H}$ and its norm by $\|\cdot\|_{\mathcal{H}}$. For the additive model (5.9), the estimation of $f \in \mathcal{H}$ is done by solving a regularization problem

$$\min_{f\in\mathcal{H}} \frac{1}{n}\sum_{i=1}^{n}\{y_i - f(\mathbf{x}_i)\}^2 + \tau_0 \sum_{j=1}^{p} \theta_j^{-1}\|P^j f\|_{\mathcal{H}}^2, \tag{5.10}$$

where $\tau_0 > 0$ and $\theta_j \geq 0$ are smoothing parameters, P^j is the orthogonal projection of f onto $\mathcal{H}_j$, and $\|\cdot\|_{\mathcal{H}}^2$ is a roughness penalty to control complexity of f_j. If we assume $f_j \in \mathcal{S}^2[0,1]$, then $\||P^j f\|_{\mathcal{H}_j}^2 = \int [f_j''(t)]^2 dt$. Wahba (1990) and Gu (2002) provide more details on the RKHS theory and implementation. The smoothing parameter τ_0 is confounded with θ_j's, but it is usually included to facilitate the computation. The tuning parameters need to be chosen properly to balance the trade-off between the data fit and function complexity. Computational algorithms for solving (5.10) are given in Wahba and Wendelberger (1980). One common procedure of selecting θ_j's is generalized cross validation (GCV; Craven and Wahba 1979; Wahba 1985).

5.3 High Dimensional Additive Models

In nonparametric regression, it is assumed that

$$Y_i = f(\mathbf{X}) + \epsilon_i, \quad i = 1, \ldots, n, \tag{5.11}$$

where the error terms ϵ_i's have mean zero and finite variance, and f is an unspecified multivariate function of $\mathbf{X}$. For high dimensional data, it is possible that Y depends on $\mathbf{X}$ only through its subset, i.e., not all X_j's have effects on Y. For example, in the additive model with $f(\mathbf{X}) = b + \sum_{j=1}^{p} f_j(X_j)$, some f_j's are zero functions, i.e.,

$$f(\mathbf{X}) = b + \sum_{f_j \neq 0} f_j(X_j) + \sum_{f_j \equiv 0} 0(X_j),$$

implying that only those predictors with nonzero f_j's contribute to the prediction of Y. In this note, we regard a function g as a zero function on $[0,1]$, $g \equiv 0$, if and only if $g(t) = 0, \forall t \in [0,1]$. Define $\mathcal{A} = \{X_j : f_j \neq 0, j = 1, \ldots, p\}$, which is the set of *important* variables. The process of identifying $\mathcal{A}$ is called the problem of variable selection, or subset selection, or feature selection.

Traditional methods of variable selection for nonparametric regression are either based on hypothesis tests or stepwise approaches. For example, the MARS procedure (Friedman 1991) and other related approaches (Stone et al. 1997) build the model f by adding and deleting basis functions in a stepwise fashion. These methods may suffer instability due to the discrete nature of the selection process (Breiman and Spector 1992). Furthermore, their theoretical properties are not clearly understood and difficult to investigate. Recently, a variety of modern selection techniques have been developed for additive models, such as COSSO, adaptive COSSO, and SpAM. These methods perform variable selection by applying a sparsity penalty functional to function components to achieve smooth and sparse estimation simultaneously. These methods are computationally more stable than stepwise methods, and they also enjoy better theoretical properties.

Given any variable selection method, let $\widehat{\mathcal{A}}$ be the set of variables selected by the method. A model selection procedure is *selection consistent* if it can identify the true set of important variables $\mathcal{A}$ correctly with probability going to 1 as the sample size n increases to infinity, i.e.,

$$\Pr(\widehat{\mathcal{A}} = A) \longrightarrow 1, \quad \text{as} \quad n \rightarrow \infty.$$

For linear models, a variable selection procedure has oracle properties if it can asymptotically select the correct subset of predictors with probability tending to one, and also estimate nonzero parameters as efficiently as would be possible if the true model structure were known (Fan and Li 2001). The SCAD and the adaptive LASSO methods have oracle procedures for variable selection in linear models. For additive models, Storlie et al. (2011) propose the concept of *nonparametric oracle*. A nonparametric estimator $\hat{f}$ has the nonparametric oracle *(np)-oracle* property if, as the sample size goes to infinity, $\hat{f}$ converges to the true regression function f at the optimal rate and $\hat{f}_j \equiv 0, \forall j \notin A$ with probability tending to one. The adaptive COSSO is nonparametric oracle as shown by Storlie et al. (2011).

5.3.1 COSSO Method

For the additive model, $f(\mathbf{X}) = b + \sum_{j=1}^{p} f_j(X_j)$, the problem of variable selection is equivalent to identifying nonzero function components f_j's in (5.7). The component smoothing and selection operator (COSSO; Lin and Zhang 2006; Zhang and Lin 2006) imposes a soft-thresholding functional to shrink function components to exactly zero function and therefore achieve sparsity. In particular, the COSSO procedure for regression solves

$$\min_{f \in \mathcal{H}} \frac{1}{n} \sum_{i=1}^{n} \{y_i - f(\mathbf{x}_i)\}^2 + \lambda \sum_{j=1}^{p} \|P^j f\|_{\mathcal{H}}, \tag{5.12}$$

where P^j is the projection of f onto $\mathcal{H}_j$ and λ is the smoothing parameter. The penalty term $\sum_{j=1}^{p} \|P^j f\|_{\mathcal{H}}$ is a sum of RKHS norms, and it is a convex functional designed to encourage both smoothness and sparsity of the solution in function estimation. The existence of the solution to (5.12) is established in Theorem 1 of Lin and Zhang (2006).

Theorem 1 (Lin and Zhang 2006) *Let $\mathcal{H}$ be an RKHS of functions over an input space $\mathcal{X}$, with a decomposition $\mathcal{H} = \{1\} \oplus_{j=1}^{p} \bar{\mathcal{H}}_j$. Then there exists a minimizer of* (5.12) *in $\mathcal{H}$.*

Furthermore, the minimizer of (5.12) has a finite-dimensional representation.

Lemma 1 (Lin and Zhang 2006) *Let $\hat{f} = \hat{b} + \sum_{j=1}^{p} \hat{f}_j$ be a minimizer of* (5.12) *in $\mathcal{H} = \{1\} \oplus_{j=1}^{p} \bar{\mathcal{H}}_j$, with $\hat{f}_j \in \bar{\mathcal{H}}_j$. Then $\hat{f}_j \in span\{\mathcal{K}_j(\mathbf{x}_i, \cdot), i = 1, \ldots, n\}$ for $j = 1, \ldots, p$, where $\mathcal{K}_j(\cdot, \cdot)$ is the reproducing kernel of $\bar{\mathcal{H}}_j$.*

Asymptotic properties of the COSSO estimators are studied by Lin and Zhang (2006). For the additive model with $f_j \in \bar{\mathcal{S}}^2[0, 1]$, if the error terms ϵ_i's are independent $N(0, \sigma^2)$ noise, then the COSSO estimator can achieve the optimal rate of convergence $n^{-2/5}$ if the parameter λ converges to zero at a proper rate. The selection consistency property of the COSSO is also studied in Lin and Zhang (2006). In the special case of a tensor product design with periodic functions, the COSSO can select the correct model structure with probability tending to one as the sample size goes to infinity. When f is truly a linear function of X_j's, then the COSSO estimator reduces to the LASSO estimator.

To compute the COSSO, we consider the following optimization problem

$$\min_{f \in \mathcal{H}, \boldsymbol{\theta} \geq 0} \frac{1}{n} \sum_{i=1}^{n} \{y_i - f(\mathbf{x}_i)\}^2 + \lambda_0 \sum_{j=1}^{p} \theta_j^{-1} \|P^j f\|_{\mathcal{H}}^2 + \lambda_1 \sum_{j=1}^{p} \theta_j, \tag{5.13}$$

where $\boldsymbol{\theta} = (\theta_1, \ldots, \theta_p)^T$ are nonnegative scale parameters, $\lambda_0 > 0$ is a constant, and λ_1 is the smoothing parameter. In Lemma 2 of Lin and Zhang (2006), it is shown that solving (5.12) and solving (5.13) are equivalent. In (5.13), the penalty term $\sum_{j=1}^{p} \theta_j$ shrinks θ_j's toward zero exactly and hence produces sparse solutions, which results in zero function components in the COSSO estimate. Based on the finite representer theorem of the smoothing spline, for any fixed θ_j's, the COSSO solution of (5.13) has the form

$$f(\mathbf{x}) = b + \sum_{i=1}^{n} c_i \left[\sum_{j=1}^{p} \theta_j \mathcal{K}_j(x_{ij}, x_j) \right],$$

where $\mathcal{K}_j$ is the reproducing kernel of $\bar{\mathcal{H}}_j$ for $j = 1, \ldots, p$. Therefore, for any fixed λ_0 and λ_1, the COSSO solution f is fully characterized by $b, \boldsymbol{\theta}$, and $\mathbf{c} = (c_1, \ldots, c_n)^T$. To solve (5.13), one can alternatively update $(b, \mathbf{c}^T)^T$ by fixing $\boldsymbol{\theta}$

at their current values, and update $\boldsymbol{\theta}$ by fixing $(b, \mathbf{c}^T)^T$ at their current values, and repeat this until convergence. Furthermore, Zhang and Lin (2006) propose the following one-step algorithm, which is efficient and provides very good solutions in practice.

One-Step Algorithm

Step 1. Initialization: Fix $\theta_j = 1$ for $j = 1, \ldots, p$, and solve for $(b, \mathbf{c}^T)^T$ in (5.13).
Step 2. Fixing $(b, \mathbf{c}^T)^T$ at their current values, solve for $\boldsymbol{\theta}$ in (5.13).
Step 3. Fixing $\boldsymbol{\theta}$ at their current values, solve for $(b, \mathbf{c}^T)^T$ in (5.13).

At step 2, the optimization problem amounts to solving a traditional smoothing spline, and step 3 amounts to solving the nonnegative garrote. Existing algorithms can then be adopted for optimization. An *R* package *COSSO* is developed to implement the one-step algorithm, available from Comprehensive *R* Archive Network (CRAN) at http://CRAN.R-project.org/package=cosso. To speed up computation in the context of Big data, the package implements a parsimonious basis method to reduce the number of parameters.

Example 1 Consider an additive model with $p = 60$, and the true regression function is

$$f(\mathbf{x}) = g_1(x_1) + g_2(x_2) + g_3(x_3) + g_4(x_4) + 1.5g_1(x_5) + 1.5g_2(x_6) + 1.5g_3(x_7) \\ +1.5g_4(x_8) + 2g_1(x_9) + 2g_2(x_{10}) + 2g_3(x_{11}) + 2g_4(x_{12}),$$

where $g_1(t) = t, g_2(t) = (2t-1)^2, g_3(t) = \frac{\sin(2\pi t)}{2\sin(2\pi t)}$, and $g_4(t) = 0.1\sin(2\pi t) + 0.2\cos(2\pi t) + 0.3\sin^2(2\pi t) + 0.4\cos^3(2\pi t) + 0.5\sin^3(2\pi t)$. The first 12 variables are important, and the remaining 48 are uninformative. The sample size $n = 500$. The variance of noise $\sigma^2 = 0.52$ and the signal-to-noise ratio is 3:1. Consider two types of covariance structure for $\mathbf{X}$: (1) compound symmetry with $\text{corr}(X_j, X_k) = t^2/(1+t^2)$ for any $j \neq k$; (2) AR(1) with $\text{corr}(X_j, X_k) = \rho^{|j-k|}$ for any $j \neq k$. We compare performance of the COSSO tuned with GCV, the COSSO tuned with fivefold cross validation, and the MARS. The prediction accuracy of each method is measured by the integrated squared error ISE $= E_{\mathbf{X}}\{\hat{f}(\mathbf{X}) - f(\mathbf{X})\}^2$, which is computed by Monte Carlo integration using 10,000 data points. The MARS is implemented in R with the function "mars" in the "mda" package. The simulation is repeated 100 times, and Table 5.1 reports the average ISE and model sizes of the three methods under two covariance structure settings. Standard errors are given in parentheses. Table 5.1 shows that the COSSO (5CV) produces the smallest ISE and the most parsimonious model among the three methods. The average model size of COSSO (5CV) is actually very close to the true model size 12 in all the settings.

The COSSO can also be used to select important main effects in generalized additive models (Zhang and Lin 2006), select nonparametric two-way interaction effects (Zhang and Lin 2006), and select important variables for nonparametric hazard regression models in survival data analysis (Leng and Zhang 2007). For

Table 5.1 Model estimation and variable selection results for Example 1

		Compound symmetry		AR(1)	
		$t = 0$	$t = 1$	$\rho = 0.5$	$\rho = -0.5$
ISE	COSSO (GCV)	201 (4)	178 (5)	199 (6)	183 (5)
	COSSO (5CV)	144 (4)	162 (5)	153 (4)	149 (5)
	MARS	353 (7)	302 (7)	286 (6)	280 (5)
Model size	COSSO (GCV)	18.0 (4.1)	18.0 (4.1)	19.0 (5.1)	18.0 (4.3)
	COSSO (5CV)	12.0 (0.2)	11.7 (1.4)	12.1 (1.4)	11.9 (1.0)
	MARS	35.2 (2.3)	36.1 (2.1)	35.2 (2.5)	35.9 (2.4)

variable selection in high dimensional classification, Zhang (2006) generalizes the COSSO for variable selection in support vector machines (SVMs). Very recently, Zhu et al. (2014) extend the COSSO to select important functional principal components (FPCs) in the context of functional additive regression.

5.3.2 *Adaptive COSSO*

In the COSSO estimation (5.12), all the functional components are equally penalized. To improve the COSSO, Storlie et al. (2011) propose to penalize different components differently, such that important components are less penalized and unimportant components are more heavily penalized. This leads to the procedure of the adaptive COSSO (ACOSSO). In particular, the adaptive COSSO employs a weighted penalty functional and solves

$$\min_{f \in \mathcal{H}} \frac{1}{n} \sum_{i=1}^{n} \{y_i - f(\mathbf{x}_i)\}^2 + \lambda \sum_{j=1}^{p} w_j \|P^j f\|_{\mathcal{H}}, \tag{5.14}$$

where w_j's are positive weights associated with each functional component. The choices of w_j's are crucial to guarantee desired theoretical and empirical properties of the adaptive COSSO estimator. Storlie et al. (2011) suggest to construct the weights as

$$w_j = \|P^j \tilde{f}\|_{L_2}^{-\gamma}, \quad j = 1, \ldots, p$$

where $\tilde{f}$ is some initial estimator of f, $\|P^j\tilde{f}\|_{L_2}$ is the L_2 norm of $P^j\tilde{f}$, and $\gamma > 0$ is a pre-specified constant. In practice, the initial estimator $\tilde{f}$ can be either the traditional smoothing spline solution or the COSSO solution. A nonparametric estimator $\hat{f}$ has the nonparametric-oracle property if $\|\hat{f} - f\|_n \longrightarrow 0$ at the optimal rate, and $\hat{f}_j \equiv 0$ for all $j \notin \mathcal{A}$ with probability tending to one, as the sample size increases to infinity. Here $\|f\|_n^2 = \frac{1}{n}\sum_{i=1}^{n}\{f(\mathbf{x}_i)\}^2$ is the squared norm of f evaluated at the design points.

The adaptive COSSO estimator has the nonparametric oracle (np-oracle) property when the weights are chosen properly, as shown in Corollary 2 of Storlie et al. (2011).

Corollary 2 (Nonparametric Oracle Property) *Assume that the input* $\mathbf{X}$ *follows a tensor product design. Let* $f \in \mathcal{H}$ *with* $\mathcal{H} = \{1\} \oplus \bar{\mathcal{S}}^2_{per,1} \oplus \cdots \oplus \bar{\mathcal{S}}^2_{per,p}$, *where* $\mathcal{S}^2_{per,j} = \{1\} \oplus \bar{\mathcal{S}}^2_{per,j}$ *is the second-order Sobolev space of periodic functions of* X_j *defined on* $[0, 1]$. *Assume the error terms* ϵ_i*'s are independent, mean zero, and uniformly sub-Gaussian. Define the weights,* $w_{j,n} = \|P^j\tilde{f}\|_{L_2}^{-\gamma}$, *where* $\tilde{f}$ *is given by the traditional smoothing spline with* $\tau_0 \sim n^{-4/5}$, *and* $\gamma > 3/4$. *If* $\lambda_n \sim n^{-4/5}$, *then the adaptive COSSO estimator has the np-oracle property.*

When the true model is linear, the adaptive COSSO reduces to the adaptive LASSO. The tuning parameter can be chosen by cross validation or BIC. In practice, when the number of parameters grows with the sample size, a modified BIC taking into account the dimension can be used. The computation of the adaptive COSSO is similar to COSSO, by iteratively fitting a traditional smoothing spline and solving a non-negative garrote problem until convergence. In practice, $\gamma = 2$ is often used.

Example 2 Consider an additive model with $p = 10$ and the true regression function is

$$f(\mathbf{x}) = 5g_1(x_1) + 3g_2(x_2) + 4g_3(x_3) + 6g_4(x_4),$$

with $\epsilon \sim N(0, 3.03)$ and the signal-to-noise ratio 3:1. Assume $\mathbf{X}$ is uniform in $[0, 1]^{10}$. We compare performance of adaptive COSSO (ACOSSO), COSSO, MARS, GAM, and the oracle. Four versions of ACOSSO are considered: ACOSSO-5CV-T, ACOSSO-5CV-C, ACOSSO-BIC-T, and ACOSSO-BIC-C, where (-T) and (-C) stand for using the traditional smoothing spline solution and the COSSO solution, respectively, as the initial estimator. Table 5.2 summarizes performance of all the methods over 100 simulations. Besides the average ISE, model size, we also report Type-I error (false negative rate) and power (true positive rate) for each method. Standard errors are given in parentheses. From Table 5.2, it is observed that

Table 5.2 Model estimation and variable selection results for Example 2

	ISE	Type-I error	Power	Model size
ACOSSO-5CV-T	1.204 (0.042)	0.252 (0.034)	0.972 (0.008)	5.4 (0.21)
ACOSSO-5CV-C	1.186 (0.048)	0.117 (0.017)	0.978 (0.007)	4.6 (0.11)
ACOSSO-BIC-T	1.257 (0.048)	0.032 (0.008)	0.912 (0.012)	3.8 (0.08)
ACOSSO-BIC-C	1.246 (0.064)	0.018 (0.006)	0.908 (0.014)	3.7 (0.07)
COSSO	1.523 (0.058)	0.095 (0.023)	0.935 (0.012)	4.3 (0.15)
MARS	2.057 (0.064)	0.050 (0.010)	0.848 (0.013)	3.7 (0.08)
GAM	1.743 (0.053)	0.197 (0.019)	0.805 (0.011)	4.4 (0.13)
ORACLE	1.160 (0.034)	0.000 (0.000)	1.000 (0.000)	4.0 (0.00)

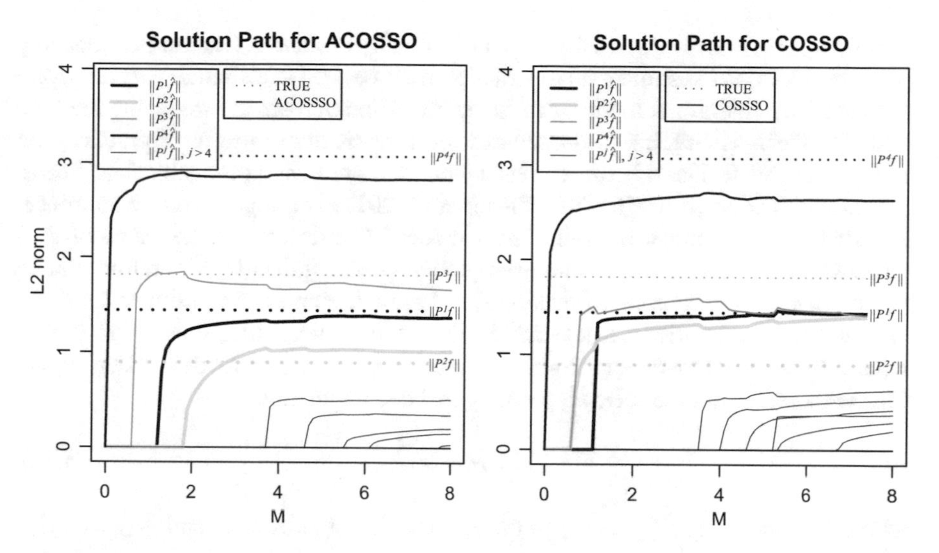

Fig. 5.1 Solution paths of ACOSSO and COSSO on a realization for Example 2

the ACOSSO procedures tend to work better than the COSSO by showing smaller ISE, lower Type-I error, and higher power.

Figure 5.1 plots the solution paths of ACOSSO and COSSO as the tuning parameter M changes. Here M is a tuning parameter, which has a one-to-one correspondence relationship with λ_1. Each plot shows how the magnitude of each estimated components $\|P^j\hat{f}\|_{L^2}$ changes as M increases for $j = 1, \ldots, 10$, in one simulation. Dashed lines represent the true functional component norms $\|P^j f\|_{L^2}$ for all j. It is observed that the ACOSSO solutions are generally closer to the truth than the COSSO solution.

5.3.3 *Linear and Nonlinear Discover (LAND)*

Partially linear models are more flexible than linear or nonparametric models by allowing some predictors to have linear effects while others to have nonlinear effects. The underlying true regression model $f(\mathbf{x})$ is additive and has the following structure

$$f(\mathbf{x}) = b + \sum_{j \in I_L} x_j \beta_j + \sum_{j \in I_N} f_j(x_j) + \sum_{j \in I_O} 0(x_j), \tag{5.15}$$

where b is the intercept, $0(\cdot)$ is zero function which maps any input value to zero, and I_L, I_N, I_O are the index sets for nonzero linear effects, nonzero nonlinear effects, and null effects, respectively. Denote the total index set by $I = \{1, \ldots, p\}$, then

$I = I_L \cup I_N \cup I_O$ and the three subgroups are mutually exclusive. One central question to model (5.15) is to estimate its true model structure, i.e., to estimate I_L, I_N, I_O from the data. Two traditional methods for structure estimation are the screening method and hypothesis testing based procedures. Both of them are useful in practice but not guaranteed to identify the correct model structure asymptotically. The linear and nonlinear discoverer (LAND; Zhang et al. 2011) is a regularization approach to structure selection for partially linear models. Constructed in the frame of SS-ANOVA, the LAND method can asymptotically distinguish linear and nonlinear terms, identify uninformative covariates, and provide a consistent estimate for f.

Consider the additive model (5.9) in the SS-ANOVA setup. Assume $f_j \in \bar{\mathcal{H}}_j$, where $\bar{\mathcal{H}}_j$ is the second-order Sobolev space of functions of X_j defined on [0, 1]. Furthermore, $\bar{\mathcal{H}}_j$ has the following orthogonal decomposition

$$\bar{\mathcal{H}}_j = \bar{\mathcal{H}}_{0j} \oplus \bar{\mathcal{H}}_{1j}, \tag{5.16}$$

where $\bar{\mathcal{H}}_{0j} = \{f_j : f_j''(t) \equiv 0\}$ is the linear contrast subspace, and $\bar{\mathcal{H}}_{1j} = \{f_j : \int_0^1 f_j(t)dt = 0, \int_0^1 f_j'(t)dt = 0, f_j'' \in \mathcal{L}_2[0,1]\}$ is the nonlinear contrast subspace. Both $\bar{\mathcal{H}}_{0j}$ and $\bar{\mathcal{H}}_{1j}$ are RKHS and associated with kernel $\mathcal{K}_{0j}(\cdot,\cdot)$ and $\mathcal{K}_{1j}(\cdot,\cdot)$, respectively. This leads to an orthogonal decomposition of $\mathcal{H}$

$$\mathcal{H} = \{1\} \bigoplus_{j=1}^{p} \bar{\mathcal{H}}_j = \{1\} \bigoplus_{j=1}^{p} \bar{\mathcal{H}}_{0j} \bigoplus_{j=1}^{p} \bar{\mathcal{H}}_{1j} = \{1\} \bigoplus \bar{\mathcal{H}}_0 \bigoplus \bar{\mathcal{H}}_1, \tag{5.17}$$

where $\bar{\mathcal{H}}_0 = \oplus_{j=1}^p \bar{\mathcal{H}}_{0j}$ and $\bar{\mathcal{H}}_1 = \oplus_{j=1}^p \bar{\mathcal{H}}_{1j}$. In this way, any function $f_j \in \bar{\mathcal{H}}_j$ can be expressed as

$$f_j(x_j) = \beta_j \left(x_j - \frac{1}{2} \right) + f_{1j}(x_j), \tag{5.18}$$

where the term $f_{0j}(x_j) = \beta_j(x_j - \frac{1}{2}) \in \bar{\mathcal{H}}_{0j}$ is the "linear" component of f_j, and $f_{1j}(x_j) \in \bar{\mathcal{H}}_{1j}$ is the "nonlinear" component of f_j. We say that X_j is a linear predictor if $\beta_j \neq 0$ and $f_{1j} \equiv 0$, and X_j is a nonlinear predictor if $f_{1j}(x_j)$ is not zero function. Therefore, we define

Linear index set:	$I_L = \{j = 1, \ldots, p : \beta_j \neq 0, f_{1j} \equiv 0\}$,
Nonlinear index set:	$I_N = \{j = 1, \ldots, p : f_{1j} \neq 0\}$,
Null index set:	$I_O = \{j = 1, \ldots, p : \beta_j = 0, f_{1j} \equiv 0\}$.

Then the nonlinear index set I_N can be expressed as $I_N = I_{PN} \cup I_{LN}$, where $I_{PN} = \{\beta_j = 0, f_{1j} \neq 0\}$ consists of purely nonlinear predictors, and $I_{LN} = \{\beta_j \neq 0, f_{1j} \neq 0\}$ consists of predictors whose linear and nonlinear terms are both nonzero.

Assume $f(\mathbf{x}) = b + \sum_{j=1}^{p} f_j(x_j)$, where f_j has the form (5.18). The LAND procedure solves

$$\min_{f \in \mathcal{H}} \frac{1}{n} \sum_{i=1}^{n} \{y_i - f(\mathbf{x}_i)\}^2 + \lambda_1 \sum_{j=1}^{p} w_{0j} \|P^{0j} f\|_{\bar{\mathcal{H}}_0} + \lambda_2 \sum_{j=1}^{p} w_{1j} \|P^{1j} f\|_{\bar{\mathcal{H}}_1}, \quad (5.19)$$

where P^{0j} and P^{1j} are the projection operators, respectively, from $\mathcal{H}$ to $\bar{\mathcal{H}}_{0j}$ and $\bar{\mathcal{H}}_{1j}$. For each predictor X_j, two types of regularization are employed in (5.19): $\|P_{0j} f\|_{\bar{\mathcal{H}}_0} = |\beta_j|$ is the L_1 penalty on linear coefficient, and $\|P^{1j} f\|_{\bar{\mathcal{H}}_1} = \{\int_0^1 [f_{1j}''(t)]^2 dt\}^{1/2}$ is the RKHS norm of f_j in $\bar{\mathcal{H}}_{1j}$. Two tuning parameters (λ_1, λ_2) control shrinkage on linear and nonlinear terms, respectively. The weights w_{0j} and w_{1j} are adaptively chosen from data such that unimportant components are penalized more and important components are penalized less. Assume $\tilde{f}$ is an initial and consistent estimator of $f \in \mathcal{H}$, then the weights can be constructed as:

$$w_{0j} = \frac{1}{|\widetilde{\beta}_j|^{\alpha}}, \quad w_{1j} = \frac{1}{\|\widetilde{f}_{1j}\|_2^{\gamma}}, \quad \text{for } j = 1, \ldots, p, \quad (5.20)$$

where $\tilde{\beta}_j, \tilde{f}_{1j}$ are from the decomposition of $\tilde{f}$ based on (5.18), $\|\cdot\|_2$ represents the L_2 norm, and $\alpha > 0$ and $\gamma > 0$ are pre-specified constants. A natural choice of $\tilde{f}$ is the standard SS-ANOVA solution to (5.10). For the LAND estimator $\hat{f}$, its model structure is defined as

$$\widehat{I}_L = \{j = 1, \ldots, p : \widehat{\beta}_j \neq 0, \widehat{f}_{1j} \equiv 0\}, \widehat{I}_N = \{j = 1, \ldots, p : \widehat{f}_{1j} \neq 0\}, \widehat{I}_O = I \backslash \{\widehat{I}_L \cup \widehat{I}_N\}.$$

To compute the LAND solution, we consider the following optimization problem

$$\min_{\boldsymbol{\theta} \geq \mathbf{0}, f \in \mathcal{H}} \frac{1}{n} \sum_{i=1}^{n} \{y_i - f(\mathbf{x}_i)\}^2 + \lambda_1 \sum_{j=1}^{p} w_{0j} \|P^{0j} f\|_{\bar{\mathcal{H}}_0}$$

$$+ \tau_0 \sum_{j=1}^{p} \theta_j^{-1} w_{1j} \|P^{1j} f\|_{\bar{\mathcal{H}}_1}^2 + \tau_1 \sum_{j=1}^{p} w_{1j} \theta_j,$$

$$\text{subject to } \theta_j \geq 0, \; j = 1, \ldots, p, \quad (5.21)$$

where $\boldsymbol{\theta} = (\theta_1, \ldots, \theta_p)^T$, $\tau_0 > 0$ is a constant, and (λ_1, τ_1) are tuning parameters. It can be shown that there is one-to-one correspondence between the solutions of (5.19) (for all (λ_1, λ_2)) and those of (5.21) (for all (λ_1, τ_1)). When $\boldsymbol{\theta}$ is fixed, solving (5.21) amounts to fitting a partial spline model, so its solution has a finite representation

$$\hat{f}(\mathbf{x}) = \hat{b} + \sum_{j=1}^{p} \hat{\beta}_j k_1(x_j) + \sum_{i=1}^{n} \hat{c}_i \left[\sum_{j=1}^{p} \hat{\theta}_j w_{1j}^{-1} \mathcal{K}_{1j}(x_{ij}, x_j) \right].$$

We iteratively update $\mathbf{c} = (c_1, \ldots, c_n)^T$ and $(\boldsymbol{\beta}^T, \boldsymbol{\theta}^T)^T$, until convergence.

Asymptotic properties of the LAND estimator are studied in Zhang et al. (2011). When (λ_1, λ_2) are chosen properly, the LAND is consistent in both structure selection and model estimation. In particular, the convergence rates of the LAND estimator is $n^{-2/5}$, which is optimal for nonparametric regression. Furthermore, under the tensor product design, the LAND can identify the correct model structure asymptotically, i.e., $\widehat{I}_L \to I_L, \widehat{I}_N \to I_N, \widehat{I}_O \to I_O$ with probability tending to one as $n \to \infty$. In other words, the LAND procedure can distinguish linear and nonlinear predictors as the sample size goes to infinity.

5.3.4 Adaptive Group LASSO

Consider a partition of [0, 1] by the knots $0 = \xi_0 < \xi_1 < \cdots < \xi_K < \xi_{K+1} = 1$ into K subintervals, $I_{Kt} = [\xi_t, \xi_{t+1}), t = 0, \ldots, K-1$ and $I_{KK} = [\xi_K, \xi_{K+1}]$, where the positive integer $K \equiv K_n = n^\nu$ with $0 < \nu < 0.5$ such that $\max_{1 \le k \le K+1} |\xi_k - \xi_{k-1}| = O(n^{-\nu})$. Furthermore, let $W_n^l[0, 1]$ be the space of polynomial splines of degree $l \ge 1$ consisting of functions s satisfying: (1) s is a polynomial degree of l on I_{Kt}, for $1 \le t \le K$; (2) for $l \ge 2$ and $0 \le l' \le l-2$, s is l' times continuously differentiable on [0, 1]. Define $m_n = K_n + l$. Then there exists a normalized B-spline basis $\{\phi_k, 1 \le k \le m_n\}$ for $W_n^l[0, 1]$ (Schumaker 1981).

For the additive model (5.9), assume $f_j \in W_n^l[0, 1]$ for each j. Then it can be expressed as

$$f_j(x_j) = \sum_{k=1}^{m_n} \beta_{jk}\phi_k(x_j), \quad 1 \le j \le p. \tag{5.22}$$

Define $\boldsymbol{\beta}_{nj} = (\beta_{j1}, \ldots, \beta_{jm_n})^T$ and $\boldsymbol{\beta}_n = (\boldsymbol{\beta}_{n1}^T, \ldots, \boldsymbol{\beta}_{np}^T)^T$. In (5.22), each functional component f_j is fully determined by a group of coefficients $\boldsymbol{\beta}_{nj}$, and therefore the problem of selecting nonzero components f_j's amounts to identify nonzero coefficient groups $\boldsymbol{\beta}_j$'s. Huang et al. (2010) propose to apply the adaptive group lasso penalty on the coefficients to select functional components

$$\min_{\boldsymbol{\beta}_n} \sum_{i=1}^{n} \left\{ y_i - b - \sum_{j=1}^{p}\sum_{k=1}^{m_n} \beta_{jk}\phi_k(x_{ij}) \right\}^2 + \lambda_n \sum_{j=1}^{p} w_j \|\boldsymbol{\beta}_{nj}\|_2, \tag{5.23}$$

$$\text{subject to } \sum_{i=1}^{n}\sum_{k=1}^{m_n} \beta_{jk}\phi_k(x_{ij}) = 0, \quad j = 1, \ldots, p,$$

where $\|\boldsymbol{\beta}_{nj}\|_2$ is the L_2 norm of $\boldsymbol{\beta}_{nj} \in \mathbb{R}^{m_n}$, the weights $\mathbf{w} = (w_1, \ldots, w_p)^T \ge 0$ are given constants, and $\lambda_n > 0$ is a tuning parameter. To assure identifiability of the parameters, the optimization (5.23) is subject to the centering constraints over the samples. Define the centralized response vector as $\mathbf{y}^c = (y_1 - \bar{y}, \ldots, y_n - \bar{y})^T$, where $\bar{y} = \sum_{i=1}^n y_i/n$. To get rid of the constraints, define $\bar{\phi}_{jk} = \frac{1}{n}\sum_{i=1}^n \phi_k(x_{ij})$ and $\psi_k(x_{ij}) = \phi_k(x_{ij}) - \bar{\phi}_{jk}$. Let $Z_{ij} = (\psi_1(x_{ij}), \ldots, \psi_{m_n}(x_{ij}))^T$, which consists of (center) values of basis functions evaluated at x_{ij}. Let $\mathbf{Z}_j = (Z_{1j}, \ldots, Z_{nj})^T$ and $\mathbf{Z} = (\mathbf{Z}_1, \ldots, \mathbf{Z}_p)$. Then (5.23) is converted to the following unconstrained minimization problem

$$\min_{\boldsymbol{\beta}_n} \|\mathbf{y}^c - \mathbf{Z}\boldsymbol{\beta}_n\|_2^2 + \lambda_n \sum_{j=1}^p w_j \|\boldsymbol{\beta}_{nj}\|_2 \tag{5.24}$$

Huang et al. (2010) suggest to construct the weights by

$$w_j = \|\tilde{\boldsymbol{\beta}}_{nj}\|_2^{-1}, \quad j = 1, \ldots, p,$$

where $\tilde{\boldsymbol{\beta}}_n$ is obtained by solving (5.24) with $w_1 = \cdots = w_p = 1$, i.e., the group lasso estimator. The parameter λ can be selected by BIC (Schwarz 1978) or EBIC (Chen and Chen 2008).

The selection and estimation properties of the adaptive group LASSO estimator are studied in Huang et al. (2010). Under certain regularity conditions, if the group lasso estimator is used as the initial estimator, then the adaptive group lasso estimator is shown to select nonzero components correctly with probability approaching one, as the sample size increases to infinity, and also achieve the optimal rate of convergence for nonparametric estimation of additive models. The results hold even when p is larger than n.

5.3.5 *Sparse Additive Models (SpAM)*

In the additive model (5.9), assume $\mathbf{y}$ is centered, so the intercept can be omitted. Ravikumar et al. (2009) propose to impose the L_2 norm penalty on nonparametric components for variable selection. In particular, assume $f_j \in \mathcal{H}_j$, a Hilbert space of measurable functions $f_j(x_j)$ such that $\mathbb{E}(f_j(X_j)) = 0$, $\mathbb{E}(f_j^2(X_j)) < \infty$, with inner product $\langle f_j, f_j' \rangle = \mathbb{E}\left(f_j(X_j)f_j'(X_j)\right)$. The SpAM estimation procedure solves

$$\min_{f_j \in \mathcal{H}_j} \mathbb{E}\left(Y - \sum_{j=1}^p f_j(X_j)\right)^2 \tag{5.25}$$

$$\text{subject to } \sum_{j=1}^p \sqrt{\mathbb{E}(f_j^2(X_j))} \le L,$$

$$\mathbb{E}(f_j) = 0, \quad j = 1, \ldots, p,$$

where L is the tuning parameter. The penalty function above shares similar spirits as the COSSO, and it can encourage smoothness and sparsity in the estimated functional components.

The optimization problem (5.25) is convex and therefore computationally feasible for high dimensional data analysis. The SpAM backfitting algorithm can be used to solve (5.25). Theoretical properties of the SpAM, including its risk consistency and model selection consistency, are studied in Ravikumar et al. (2009). Define the risk function of f by $R(f) = \mathbb{E}(Y - f(\mathbf{X}))^2$. An estimator $\hat{f}_n$ is said to be *persistent* relative to a class of functions $\mathcal{M}_n$ is

$$R(\hat{f}_n) - R(f_n^*) \longrightarrow_p 0,$$

where $f_n^* = \arg\min_{f \in \mathcal{M}_n} R(f)$ is the predictive oracle. The SpAM is persistent relative to the class of additive models when the tuning parameter L is chosen properly, under some technical assumptions. The SpAM is also persistent, i.e., $P(\hat{A}_n = A) \longrightarrow 1$ as $n \to \infty$ when the regularization parameter has a proper rate.

5.3.6 Sparsity-Smoothness Penalty

Consider the additive model (5.9). Without loss of generality, assume $\mathbf{y}$ is centered, so the intercept can be omitted. For each $j = 1, \ldots, p$, define $\|f_j\|_n^2 = \frac{1}{n}\sum_{i=1}^n f_j^2(x_{ij})$ and the smoothness measure $I^2(f_j) = \int_0^1 \left[f_j''(t)\right]^2 dt$. In order to achieve sparse and smooth function estimation, Meier et al. (2009) propose the following sparsity-smooth penalty function

$$J_{\lambda_1,\lambda_2}(f_j) = \lambda_1 \sqrt{\|f_j\|_n^2 + \lambda_2 I^2(f_j)},$$

where $\lambda_1, \lambda_2 \geq 0$ are tuning parameters to control the degree of penalty on functional components. The estimator is given by solving

$$\min_{f \in \mathcal{H}} \frac{1}{n} \sum_{i=1}^{n} \{y_i - f(\mathbf{x}_i)\}^2 + \sum_{j=1}^{p} J_{\lambda_1,\lambda_2}(f_j). \tag{5.26}$$

The solution of (5.26), $\hat{f}_j$'s, are natural cubic splines with knots at x_{ij}, $i = 1, \ldots, n$, so each function component has a finite-representer expression. Each f_j can be expressed by a set of cubic B-spline basis, $f_j(x_j) = \sum_{k=1}^{m} \beta_{jk}\phi_k(x_j)$ as in (5.22), where ϕ_k are B-spline basis. One typical choice is that $m_n - 4 \asymp \sqrt{n}$. For each $j = 1, \ldots p$, let $B_{ij} = (\phi_1(x_{ij}), \ldots, \phi_{m_n}(x_{ij}))^T$, which consists of the values of basis functions evaluated at x_{ij}. Let $\mathbf{B}_j = (B_{1j}, \ldots, B_{nj})^T$ and $\mathbf{B} = (\mathbf{B}_1, \ldots, \mathbf{B}_p)$. Then the optimization problem (5.26) is equivalent to

$$\arg\min_{\boldsymbol{\beta}} \|\mathbf{y} - \mathbf{B}\boldsymbol{\beta}\|_n^2 + \lambda_2 \sum_{j=1}^{p} \sqrt{\frac{1}{n}\boldsymbol{\beta}_j^T \mathbf{B}_j^T \mathbf{B}_j \boldsymbol{\beta}_j + \lambda_2 \boldsymbol{\beta}_j^T \Omega_j \boldsymbol{\beta}_j}, \tag{5.27}$$

where Ω_j is an $m_n \times m_n$ matrix with its kl entry equal to $\int_0^1 \phi_k''(x)\phi_l''(x)dx$, with $1 \leq k, l \leq m_n$. The optimization problem (5.27) can be seen as a group lasso problem, so efficient coordinate-wise algorithms can be used to obtain the solution. Theoretical properties of the penalized estimator are studied in Meier et al. (2009). In particular, an oracle inequality is derived for the penalized estimator under the compatibility condition.

5.4 Nonparametric Independence Screening (NIS)

In Big data analytics, it is possible that the number of predictors p grows at a much faster rate than the sample size n, e.g., $\log(p) = O(n^\alpha)$ with $0 < \alpha < 1$. For such ultra-high dimensional data, it is useful to first reduce data dimensionality to a moderate scale using a pre-screening procedure, before a refined variable selection is applied. This is the idea of variable *screening*. A variable screening method is said to be *screening-consistent* if

$$\Pr(\widehat{\mathcal{A}} \subset A) \longrightarrow 1, \quad \text{as} \quad n \longrightarrow \infty,$$

where A is the set of important variables for the true model f, and $\hat{\mathcal{A}}$ is the model selected by the procedure based on the data. In the context of linear models, a variety of sure screening methods have been developed, including Fan and Lv (2008), Wang (2009), Fan and Fan (2008), Hao and Zhang (2014).

Fan et al. (2011) extend the concept to nonparametric independence screening (NIS) by ranking the importance of predictors by the goodness fit of marginal models. Assume the random sample $(\mathbf{x}_i, y_i), i =, \ldots, n$ follow the nonparametric regression model

$$Y = f(\mathbf{X}) + \epsilon, \quad E(\epsilon) = 0. \tag{5.28}$$

Consider the marginal nonparametric regressions problem,

$$\min_{f_j \in L_2(P)} E\left\{Y - f_j(X_j)\right\}^2, \quad j = 1, \ldots, p, \tag{5.29}$$

where P is the joint distribution of $(\mathbf{X}, Y)$ and $L_2(P)$ is the class of square integrable functions under the measure P. The minimizer of (5.29) is given by $f_j = E(Y|X_j)$. Furthermore, assume $f_j \in W_n^l[0, 1]$, the space of polynomial splines of degree $l \geq 1$

defined in Sect. 5.3.4, then it can be expressed as a linear combination of a set of B-spline basis as (5.22). Then the estimator of f_j is given by

$$\hat{f}_j = \arg \min_{f_j \in W_n^l[0,1]} \frac{1}{n} \sum_{i=1}^{n} \{y_i - f_j(x_{ij})\}^2 .$$

The nonparametric independent screening (NIS; Fan et al. 2011) selects a set of variables by ranking the magnitude of marginal estimators $\hat{f}_j$'s,

$$\widehat{\mathcal{M}}_{\nu_n} = \{j = 1, \ldots, p : \|\hat{f}_j\|_n^2 \geq \nu_n\},$$

where $\|\hat{f}_j\|_n^2 = \frac{1}{n} \sum_{i=1}^n \hat{f}_j^2(x_{ij})$ and ν_n is pre-specified thresholding value. The NIS procedure can reduce the data dimensionality from p to $|\widehat{\mathcal{M}}_n|$, which is typically much smaller than p. Under some technical conditions, Fan et al. (2011) show that the SIS has the sure screening property, i.e., $P(A \subset \widehat{\mathcal{M}}_{\nu_n}) \longrightarrow 1$, when ν_n is selected properly. This sure screening result holds even if p grows at an exponential rate of the sample size n. Furthermore, the false selection rate converges to zero exponentially fast. To further reduce the false positive rate and increase stability, an iterative NIS (INIS) is proposed and studied by Fan et al. (2011).

References

Altman NS (1990) Kernel smoothing of data with correlated errors. J Am Stat Assoc 85:749–759
Breiman L (1995) Better subset selection using the non-negative garrote. Technometrics 37:373–384
Breiman L, Friedman JH (1985) Estimating optimal transformations for multiple regression and correlations (with discussion). J Am Stat Assoc 80:580–619
Breiman L, Spector P (1992) Subset selection and evaluation in regression: the X-random case. Int Stat Rev 60:291–319
Buja A, Hastie TJ, Tibshirani RJ (1989) Linear smoothers and additive models. Ann Stat 17:453–555
Candes E, Tao T (2007) The Dantzig selector: statistical estimation when p is much larger than n. Ann Stat 35:2313–2351
Chen J, Chen Z (2008) Extended Bayesian information criteria for model selection with large model space. Biometrika 95:759–771
Chen S, Donoho DL, Saunders MA (1999) Atomic decomposition by basis pursuit. SIAM J Sci Comput 20(1):33–61
Cleveland W (1979) Robust locally weighted fitting and smoothing scatterplots. J Am Stat Assoc 74:829–836
Craven P, Wahba G (1979) Smoothing noise data with spline functions: estimating the correct degree of smoothing by the method of generalized cross validation. Numer Math 31:377–403
de Boor C (1978) A practical guide to splines. Springer, New York
Donoho D, Johnstone I (1994) Ideal spatial adaptation by wavelet shrinkage. Biometrika 81:425–455
Efron B, Hastie T, Johnstone I, Tibshirani R (2004) Least angle regression. Ann Stat 32:407–451

Fan J, Fan Y (2008) High-dimensional classification using features annealed independence rules. Ann Stat 36:2605–2637
Fan J, Gijbels I (1996) Local polynomial modeling and its applications. Chapman and Hall, Boca Raton
Fan J, Jiang J (2005) Nonparametric inference for additive models. J Am Stat Assoc 100:890–907
Fan J, Li R (2001) Variable selection via nonconcave penalized likelihood and its oracle property. J Am Stat Assoc 96:1348–1360
Fan J, Li R (2004). New estimation and model selection procedures for semiparametric modeling in longitudinal data analysis. J Am Stat Assoc 96:1348–1360
Fan J, Lv J (2008) Sure independence screening for ultrahigh dimensional feature space. J R Stat Soc B 70:849–911
Fan J, Härdle W, Mammen E (1998) Direct estimation of additive and linear components for high-dimensional Data. Ann Stat 26:943–971
Fan J, Feng Y, Song R (2011) Nonparametric independence screening in sparse ultra-high-dimensional additive models. J Am Stat Assoc 106:544–557
Friedman JH (1991) Multivariate adaptive regression splines (invited paper). Ann Stat 19:1–141
Friedman JH, Silverman BW (1989) Flexible parsimonious smoothing and additive modeling. Technometrics 31:3–39
Friedman JH, Stuetzle W (1981) Projection pursuit regression. J Am Stat Assoc 76:817–823
Green P, Silverman B (1994) Nonparametric regression and generalized linear models: a roughness penalty approach. Chapman & Hall, Boca Raton
Gu C (2002) Smoothing spline ANOVA models. Springer, Berlin
Hao N, Zhang HH (2014) Interaction screening for ultra-high dimensional data. J Am Stat Assoc 109:1285–1301
Hastie TJ, Tibshirani RJ (1990) Generalized additive models. Chapman and Hall, Boca Raton
Huang J, Horovitz J, Wei F (2010) Variable selection in nonparametric additive models. Ann Stat 38:2282–2313
Kimeldorf G, Wahba G (1971) Some results on Tchebycheffian spline functions. J Math Anal Appl 33:82–95
Lafferty J, Wasserman L (2008) RODEO: sparse, greedy nonparametric regression. Ann Stat 36:28–63
Leng C, Zhang HH (2007) Nonparametric model selection in hazard regression. J Nonparametric Stat 18:417–429
Lin Y, Zhang HH (2006) Component selection and smoothing in multivariate nonparametric regression. Ann Stat 34:2272–2297
Linton OB (1997) Efficient estimation of additive nonparametric regression models. Biometrika 84:469–473
Mallet S (2008) A wavelet tour of signal processing: the sparse way. Elsevier, Burlington, MA
Mammen E, van de Geer S (1997) Locally adaptive regression splines. Ann Stat 25:387–413
Mammen E, Linton O, Nielsen J (1999) The existence and asymptotic properties of a backfitting projection algorithm under weak conditions. Ann Stat 27:1443–1490
Meier L, Van De Geer S, Buhlmann P (2009) High-dimensional additive modeling. Ann Stat 37:3779–3821
Nadaraya E (1964) On estimating regression. Theory Probab Appl 9:141–142
Opsomer JD (2000) Asymptotic properties of backfitting estimators. J Multivar Anal 73:166–179
Opsomer JD, Ruppert D (1998) A fully automated bandwidth selection method for fitting additive models. J Am Stat Assoc 93:605–619
Ravikumar P, Liu H, Lafferty J, Wasserman L (2009) Sparse additive models. J R Stat Soc Ser B 71:1009–1030
Schumaker L (1981) Spline functions: basic theory. Cambridge mathematical library. Cambridge University Press, Cambridge
Schwarz G (1978) Estimating the dimension of a model. Ann Stat 6:461–464
Stone C, Buja A, Hastie T (1994) The use of polynomial splines and their tensor-products in multivariate function estimation. Ann Stat 22:118–184

Stone C, Hansen M, Kooperberg C, Truong Y (1997) Polynomial splines and their tensor products in extended linear modeling. Ann Stat 25:1371–1425
Storlie C, Bondell H, Reich B, Zhang HH (2011) The adaptive COSSO for nonparametric surface estimation and model selection. Stat Sin 21:679–705
Tibshirani R (1996) Regression shrinkage and selection via the Lasso. J R Stat Soc Ser B 58:147–169
Tsybakov AB (2009) Introduction to Nonparametric Estimation. Springer, New York
Wahba G (1983) Bayesian "confidence intervals" for the cross-validated smoothing spline. J R Stat Soc Ser B 45:133–150
Wahba G (1985) A comparison of GCV and GML for choosing the smoothing parameter in the generalized spline smoothing problems. Ann Stat 13:1378–1402
Wahba G (1990) Spline models for observational data. In: SIAM CBMS-NSF regional conference series in applied mathematics, vol 59
Wahba G, Wendelberger J (1980) Some new mathematical methods for variational objective analysis using splines and cross-validation. Mon Weather Rev 108:1122–1145
Wahba G, Wang Y, Gu C, Klein R, Klein B (1995) Smoothing spline ANOVA for exponential families, with application to the Wisconsin epidemiological study of diabetic retinopathy. Ann Stat 23:1865–1895
Wand MP (1999) A central limit theorem for local polynomial backfitting estimators. J Multivar Anal 70:57–65
Wang H (2009) Forward regression for ultra-high dimensional variable screening. J Am Stat Assoc 104:1512–1524
Wood S (2006) Generalized additive models: an introduction with R. CRC Press, Boca Raton
Zhang HH (2006) Variable selection for support vector machines via smoothing spline ANOVA. Stat Sin 16:659–674
Zhang C-H (2010) Nearly unbiased variable selection under minimax concave penalty. Ann Stat 38:894–942
Zhang HH, Lin Y (2006) Component selection and smoothing for nonparametric regression in exponential families. Stat Sin 16:1021–1042
Zhang HH, Lu W (2007) Adaptive-LASSO for Cox's proportional hazard model. Biometrika 94:691–703
Zhang HH, Wahba G, Lin Y, Voelker M, Ferris M, Klein R, Klein B (2004) Variable selection and model building via likelihood basis pursuit. J Am Stat Assoc 99:659–672
Zhang HH, Cheng G, Liu Y (2011) Linear or nonlinear? Automatic structure discovery for partially linear models. J Am Stat Assoc 106:1099–1112
Zhao P, Yu B (2006) On model selection of lasso. J Mach Learn Res 7:2541–2563
Zhu H, Yao F, Zhang HH (2014) Structured functional additive regression in reproducing kernel Hilbert spaces. J R Stat Soc B 76:581–603
Zou H (2006) The adaptive lasso and its oracle properties. J Am Stat Assoc 101:1418–1429
Zou H, Hastie T (2005) Regularization and variable selection via the elastic net. J R Stat Soc B 67:301–320
Zou H, Zhang HH (2009) On the adaptive elastic-net with a diverging number of parameters. Ann Stat 37:1733–1751

Chapter 6
Finding Patterns in Time Series

James E. Gentle and Seunghye J. Wilson

Abstract Large datasets are often time series data, and such datasets present challenging problems that arise from the passage of time reflected in the datasets. A problem of current interest is clustering and classification of multiple time series. When various time series are fitted to models, the different time series can be grouped into clusters based on the fitted models. If there are different identifiable classes of time series, the fitted models can be used to classify new time series.

For massive time series datasets, any assumption of stationarity is not likely to be met. Any useful time series model that extends over a lengthy time period must either be very weak, that is, a model in which the signal-to-noise ratio is relatively small, or else must be very complex with many parameters. Hence, a common approach to model building in time series is to break the series into separate regimes and to identify an adequate local model within each regime. In this case, the problem of clustering or classification can be addressed by use of sequential patterns of the models for the separate regimes.

In this chapter, we discuss methods for identifying changepoints in a univariate time series. We will emphasize a technique called alternate trends smoothing.

After identification of changepoints, we briefly discuss the problem of defining patterns. The objectives of defining and identifying patterns are twofold: to cluster and/or to classify sets of time series, and to predict future values or trends in a time series.

Keywords Data smoothing · Changepoints · Clustering · Classification · Alternating trand smoothing · Pattern recognition · Bounding lines · Time series

---

J. E. Gentle (✉) · S. J. Wilson
George Mason University, Fairfax, VA, USA
e-mail: jgentle@gmu.edu; swilso16@masonlive.gmu.edu

© Springer International Publishing AG, part of Springer Nature 2018

W. K. Härdle et al. (eds.), *Handbook of Big Data Analytics*, Springer Handbooks of Computational Statistics, https://doi.org/10.1007/978-3-319-18284-1_6

6.1 Introduction

Many really large datasets are time series, and such datasets present unique problems that arise from the passage of time reflected in the datasets. A problem of current interest is clustering and classification of multiple time series; see, for example, Lin and Li (2009), Fu (2011), Zhou et al. (2012), and He et al. (2014). When various time series are fitted to models, the different time series can be grouped into clusters based on the fitted models. If there are different identifiable classes of time series, the fitted models can be used to classify new time series.

For massive time series datasets, any assumption of stationarity is not likely to be met. It is generally futile to attempt to model large time series using traditional parametric models.

In all statistical models, we seek to identify some random variable with zero autocorrelations whose realizations are components of the observable variables. The model is then composed of two parts, a systematic component plus a random component.

The problem in modeling time series is identification of any such random variable in a model over a long time period, or even in a short time period when the data are massive.

Any useful time series model that extends over a lengthy time period must either be very weak, that is, a model in which the signal-to-noise ratio is relatively small, or else must be very complex with many parameters.

A common approach to model building in time series is to break the series into separate regimes and to identify an adequate local model within each regime. In this case, the problem of clustering or classification can be addressed by use of sequential patterns of the models for the separate regimes.

6.1.1 Regime Descriptors: Local Models

Within a particular time regime the time series data exhibit some degree of commonality that is captured in a simple model. The model may specify certain static characteristics such as average value (mean, median, and so on) or scale (variance, range, and so on). The model may also specify certain time-dependent characteristics such as trends or autocorrelations. The model within any regime may be very specific and may fit most of the observations within that regime, or it may be very general with many observations lying at some distance from their fitted values. For the data analyst, the choice of an appropriate model presents the standard tradeoff between a smooth fit and an "over" fit.

The beginning and ending points of each regime are important components of the model. Independent of the actual beginning and ending points, the length of the regime is also an important characteristic.

The model may be formulated in various ways. For purposes of clustering and classification of time series, it may be desirable for the model to be one of a small pre-chosen set of models. It may also be desirable that the individual characteristics be specified as one a particular small set. For example, if the model specifies location, that is some average value within the regime, we may use categorical labels to specify ranked levels of location; "a" may denote small, "b" may denote somewhat larger average values, and so on. These relative values are constant within a given regime, but the set of possible categories depends on the values within other regimes in the time series.

Specifying a model in place of the full dataset allows for significant data reduction. Substitution of the individual values within a regime by the sufficient statistical descriptors is an important form of data reduction in time series.

6.1.2 Changepoints

Once we accept that different models (or models with different fitted parameters) are needed in different regimes, the main problem now becomes identification of the individual regimes; that is, identification of the *changepoints* separating regimes.

The complexity of this problem depends to a large extent on the "smoothness" of our individual models; if the models are linear in time, then changepoints are easier to identify than if the models are nonlinear in time or if they involve features other than time, such as autoregressive models.

The two change points that determine the extent of a regime together with the sufficient statistical descriptors describing the regime may be an adequate reduction of the full set of time series data within the regime.

6.1.3 Patterns

Once regimes within a time series are identified, the patterns of interest now become the sequences—or subsequences—of local models for the regimes.

Between any two changepoints, we have a local model, say $\mu_i(t)$. A particular sequence of local models, $\mu_i(t), \mu_{i+1}(t), \ldots, \mu_{i+r}(t)$, defines a *pattern*. We will often denote a pattern in the form P_{ri}, where $P_{ri} = (\mu_i(t), \mu_{i+1}(t), \ldots, \mu_{i+r}(t))$. While the model is a *function* of time together with descriptions of other model components of the temporal relationships, such as the probability distribution of a random "error" component, we may represent each $\mu_i(t)$ as a vector whose elements quantify all relevant aspects of the model.

6.1.4 Clustering, Classification, and Prediction

There is considerable interest currently in learning in time series data. "Learning" generally means clustering and/or classification of time series. This is one of the main motivations of our work in pattern recognition within time series.

Forecasting or prediction is also an important motivation for time series analysis, whether we use simple trend analysis, ARMA-type models, or other techniques of analysis.

Prediction in time series, of course, is often based on unfounded hopes. We view the prediction problem as a simple classification problem, in which statistical learning is used to develop a classifier based on patterns. The response could be another local model within the class of local models used between changepoints, or the response could be some other type of object, such as a simple binary "up" or "down." The length of time over which the prediction is made must, of course, be considered in the classification problem.

6.1.5 Measures of Similarity/Dissimilarity

Clustering or classification is often based on some metric for measuring dissimilarity of elements in a set. For clustering and classification of time series or subsequences of time series based on patterns, we need a metric $\rho(P_{ri}, P_{sj})$, where P_{ri} is a pattern consisting of a sequence $\mu_i, \mu_{i+1}, \ldots, \mu_{i+r}$ and P_{sj} is a pattern consisting of a sequence over s regimes beginning at the jth one.

6.1.6 Outline

In the following we discuss methods for identifying changepoints in a univariate time series. In massive datasets a major challenge is always that of overfitting. With so much data, very complex models can be developed, but model complexity does not necessarily result in better understanding or in more accurate predictions.

We will generally consider linear models, either simple constant models or simple linear trends. By restricting our attention to models that are linear in time, we avoid some kinds of overfitting. In smoothing time series using a sequence of linear models, "overfitting" is the identification of spurious changepoints.

Our main concern will be on the identification of changepoints, and we will emphasize a technique called *alternate trends smoothing*.

After identification of changepoints, we briefly discuss the problem of defining patterns. The objectives of defining and identifying patterns are twofold: to cluster and/or to classify sets of time series, and to predict future values or trends in a time series.

Although we do not emphasize any specific area of application, some of our work has been motivated by analysis of financial time series, so we occasionally refer to financial time series data, in particular, to series of stock prices or of rates of return.

6.2 Data Reduction and Changepoints

Analysis of massive data sets, whether they are time series or not, often begins with some form of data reduction. This usually involves computation of summary statistics that measure central tendencies and other summary statistics that measure spread. These two characteristics of a dataset are probably the most important ones for a stationary population in which a single simple model is adequate.

Even assuming a single model for all data, just concentrating on summary measures can miss important information contained in some significant individual observations. These significant observations are often the extreme or outlying points in the dataset. One simple method of analyzing a time series is just to assume a single constant model and to identify the extreme points, say the 10% outliers, with respect to that model. These outliers may carry a significant amount of information contained in the full dataset. The set of outliers may be further reduced. Fink and Gandhi (2011), for example, described a method for successively identifying extreme points in a time series for the purpose of data reduction. The extreme points alone provide a useful summary of the entire time series.

Another type of significant point in a time series is one that corresponds to a change in some basic characteristic of the time series. A changepoint may or may not be an extreme point. Changepoints can also be used for data reduction because they carry the most significant information, at least from one perspective.

In a time series, a changepoint is a point in time at which some property of interest changes. A changepoint, therefore, has meaning only in the context of a model. The model for the observable data may be some strong parametric model, such as an ARMA model, or it may be some weak parametric model, such as constant median and nothing more. In the former case, a changepoint would be a point in time at which one of the parameters changes its value. (Here, we are assuming ARMA models with constant parameters.) In the latter case, a changepoint would be any point at which the median changes. A changepoint may also be a point in time at which the class of appropriate model changes. Perhaps an ARMA model is adequate up to a certain point and then beyond that the constant variance assumption becomes entirely untenable.

From one perspective, the problem of identification of changepoints can be viewed as just a part of a process of model building. This, of course, is not a well-posed problem without further restrictions, such as use of some pre-selected class of models and specification of criteria for ranking models.

In the following, we will focus on identification of changepoints in simple piecewise linear models of an observable random variable. We do not assume finite moments, so we will refer to the parameter of central tendency as the "median," and the parameter of variability as the "scale." We will also focus most of our study on univariate time series, although we will consider some extensions to multivariate series.

There is a vast literature on identification of changepoints, but we do not attempt any kind of general review; rather we discuss some of the methods that have proven useful for the identification of patterns.

6.2.1 Piecewise Constant Models

The simplest model for time series with changepoints is one in which each regime is modeled by a constant. The constant is some average value of the data over that regime. For our purposes, the nature of that "average" is not relevant; however, because of possibly heavy tails in the frequency distributions and asymmetry of the data, we often think of that average as a median.

There are various approaches to modeling time series with median values that change over time. The first step in any event is to determine the breakpoints. Sometimes, when the data-generating process is indeed a piecewise constant model, the breakpoints may be quite apparent, as in Fig. 6.1.

In other cases, we may choose to approximate the data with a piecewise constant model, as in Fig. 6.2, even though it is fairly obvious that the underlying data-generating process is not piecewise constant or even piecewise linear.

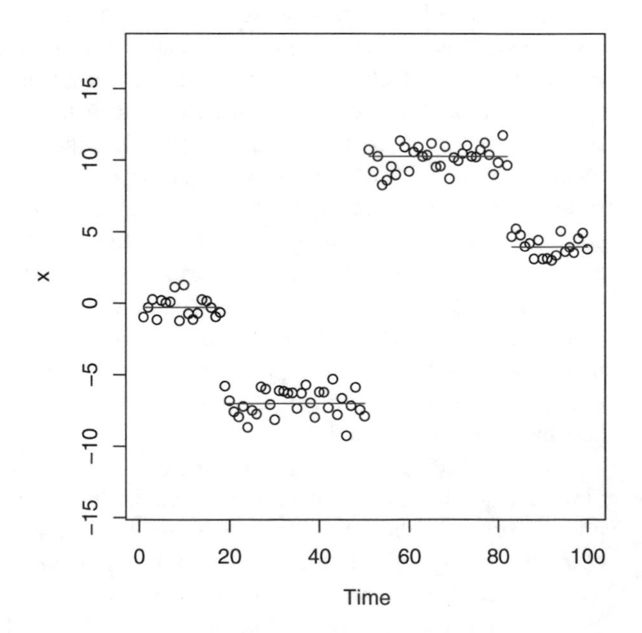

Fig. 6.1 Time series following a piecewise constant model

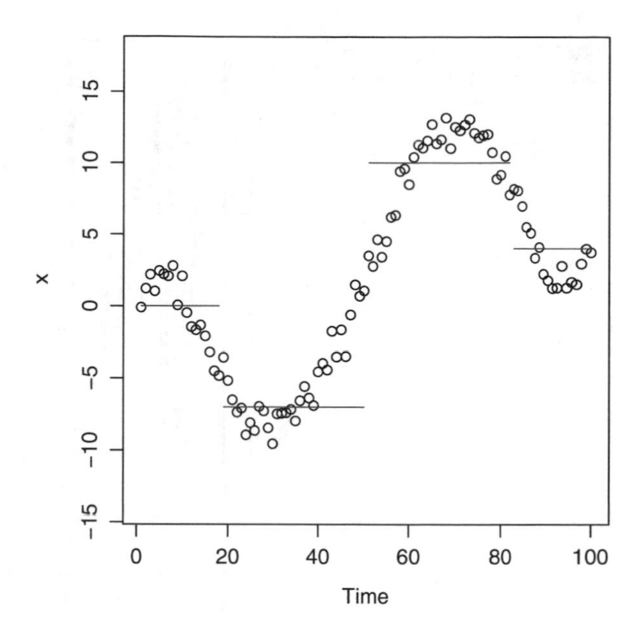

Fig. 6.2 Time series approximated by a piecewise constant model

There are various straightforward ways of determining the values of the approximating constants. A simple batch process is to use sample quantiles of the data that correspond to some parametric model, such as a normal distribution.

6.2.2 Models with Changing Scales

Piecewise constant models, such as the data in Fig. 6.1 seem to follow, or other simple models for changing location may not to be of much interest, but there is a type of derived financial data that exhibit similar behavior. It is rates of return. The standard way of defining the rate of return for stock prices or stock indexes from time t to time $t + 1$ is $\log X_{t+1} - \log X_t$, where X_{t+1} and X_t are the prices at the respective times.

A stylized property of rates of return is volatility clustering. Figure 6.3 is an illustration of this property for a small monthly sequence of the S&P 500 Index over a period from January, 2010, through November, 2015. (This short time series was just chosen arbitrarily. More data and data over other time spans may illustrate this better; but here, our emphasis is on a simple exposition. See Gentle and Härdle (2012) for more complete discussions of volatility clustering and other properties of financial time series.)

Volatility clustering is an example of changes in a time series in which the location (mean or median) may be relatively unchanged, but the scales change from one regime to another. The changepoints in this case are points in the time series where the scales change. The derive time series, that is, the volatility time series can be approximated with a piecewise constant model.

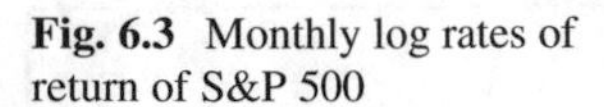

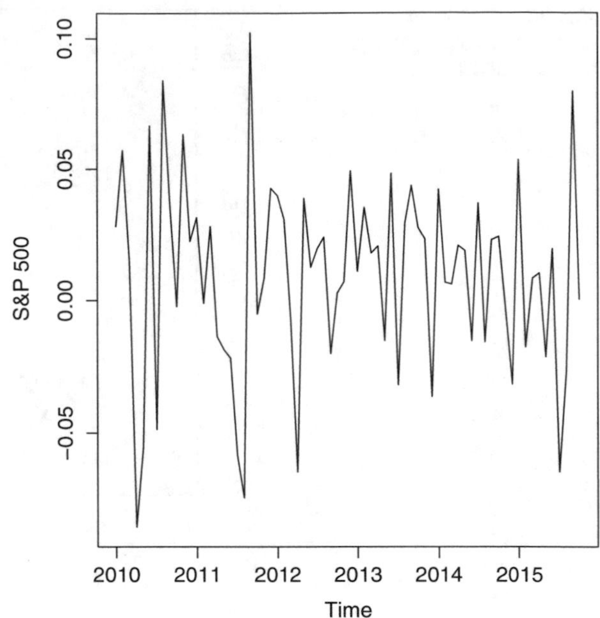

Fig. 6.3 Monthly log rates of return of S&P 500

In order to identify changes in scale or volatility, we must have some measure of volatility. It may not be obvious how to measure volatility in a time series, and this is especially true if the volatility is changing. A simple measure, called "statistical volatility" by economists, is just the sample standard deviation, which of course ignores any autocorrelations. To illustrate, however, we compute the statistical volatilities over the apparent separate regimes of the log returns shown in Fig. 6.3. This type of analysis results in a piecewise constant time series shown in Fig. 6.4 of the type we discussed in Sect. 6.2.1. There are various methods for detecting changepoints for scales, but we will not discuss them here.

6.2.3 Trends

The main interest in patterns in time series most often focuses on changes in trends. This is particularly true in financial time series, see, for example, Bao and Yang (2008) and Badhiye et al. (2015) for methods that focus solely on trends.

More interesting simple linear models in time series are those that exhibit a "trend" either increasing or decreasing. Changepoints are the points at which the trend changes.

Identification of changepoints is one of the central aspects of technical analysis of financial data, and is the main feature of the so-called point and figure charts that have been used for many years (Dorsey 2007). Point and figure charts are good for identification of changepoints and the amount of change within an up or a down trend.

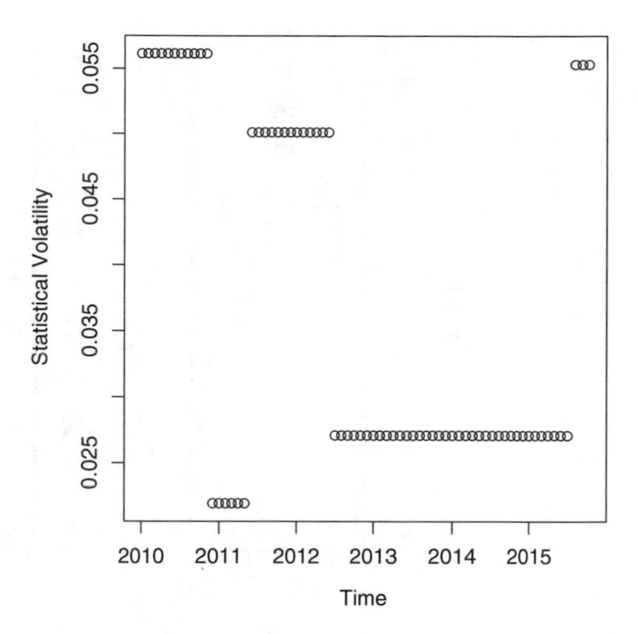

Fig. 6.4 Monthly statistical volatility of log returns of S&P 500

Some interesting patterns are easily seen in a point and figure chart. For example, a pattern that many technical analysts believe carries strong predictive powers is the "head-and-shoulders" pattern. Figure 6.5 shows the stock price for Intel Corporation (NASDAQ:INTC), and on the right side, a modified point and figure chart. (The modifications, suggested in Gentle (2012), among other things involve the definition of threshold change.) The head-and-shoulders pattern is clearly visible in both the graph of the raw prices on the left and the trend chart on the right. (This is a very strong head-and-shoulders pattern; most head-and-shoulders patterns that technical analysts would identify are not this clear.)

Notice in the trend chart in Fig. 6.5 that the time axis is transformed into an axis whose values represent only the ordered changepoints. One of the major deficiencies of point and figure charts and trend charts is that information about the length of time between changepoints is not preserved.

A very effective smoothing method is use of piecewise linear models. Piecewise linear fits are generalizations of the piecewise constant models, with the addition of a slope term. There are many variations on this type of fit, including the criterion for fitting (ordinary least squares is most common) and restrictions such as continuity (in which case the piecewise linear fit is a first degree spline). New variations on the basic criteria and restrictions are suggested often; see, for example, Zhou et al. (2012).

Some breaks between trends are more interesting than others, depending on the extent to which the trend changes. Within a regime in which a single trend is dominant, shorter trends of different direction or of different magnitude may occur. This raises the issue of additional regimes, possibly leading to overfitting or of leaving a regime in which many points deviate significantly from the fitted model.

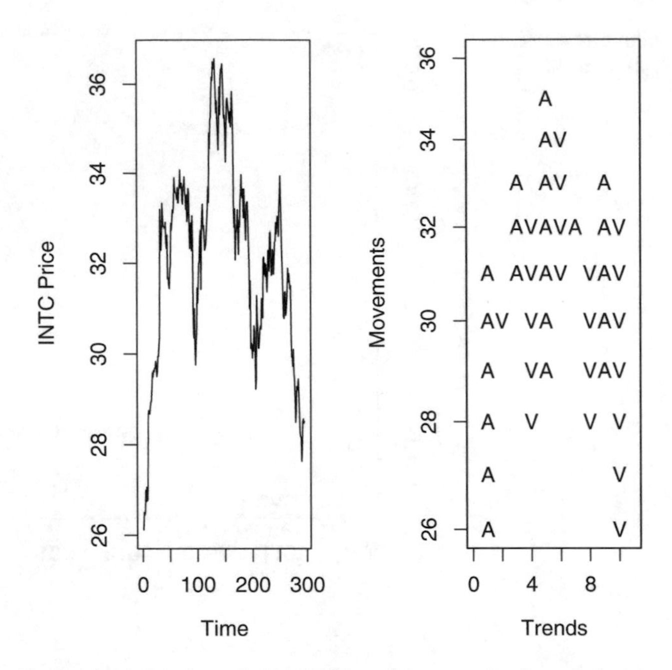

Fig. 6.5 Intel price, 2014-6-1, through 2015-7-31, and the associated trend chart

6.3 Model Building

One of the main objectives of building a model for a time series is to reduce the amount of data by use of an approximate representation of the dataset. While the main objectives of the standard models for analysis of time series, such as ARIMA and GARCH extensions in the time domain and Fourier series or wavelets in the frequency domain, may be to understand the data-generating process better, such models also provide an approximation or smoothing of the data and thereby achieve significant data reduction. Several approximations based on simple piecewise models, each with its three-letter-acronym, have been proposed. These obviously depend on identification of changepoints prior to or in conjunction with the modeling within the individual regimes. Representation of the sequence of models then becomes an important issue. While a model is usually represented as a parameterized equation, a common method of simplifying the representation further is to define a set of models, often of a common form, but each instantiated with fixed values of all parameters, and then to associate a symbol with each instantiation of each model. Two methods following this approach are symbolic aggregate approximation (SAX), see Lin et al. (2007), and nonparametric symbolic approximate representation (NSAR), see He et al. (2014). Fu (2011) provides a general review of various methods of smoothing time series.

Because the identification of changepoints, that is, the identification of regimes, is intimately tied to the identification and fitting of models within the individual regimes, it is not possible to separate those two steps. Usually, a form of the model is chosen and then regimes are chosen based on the goodness of fits of potential models of that form. Often, especially in the analysis of stock prices, there is no model within the regimes other than simple increasing or decreasing trends. Bao and Yang (2008) and Gentle (2012), for example, described methods for determining changepoints between increasing and decreasing price trends.

6.3.1 Batch Methods

For fitting piecewise constant models, there are various straightforward ways of determining the values of the approximating constants. If all of the data are available as we mentioned above, a simple batch process is to use sample quantiles of the data that correspond to some parametric model, such as a normal distribution, and then just identify regimes as those subsequences clustering around the sample quantiles.

Another simple batch approach is to fit a single model of the specified form, and then to identify subsequences based on points that are outliers with respect to a fitted model. This process is repeated recursively on subsequences, beginning with the full sequence.

Given a single linear trend over some regime, Fu et al. (2008) defined measures for "perceptually important points," which would be candidate changepoints. The perceptually important points are ones that deviate most (by some definition) from a trendline.

6.3.2 Online Methods

Batch methods, such as ones that base local models on sample quantiles of the whole time series, or those that recursively identify subsequences with local models, have limited applicability. In most applications of time series analysis, new data are continually being acquired, and so an online method is preferable to a batch method.

An online method accesses the data one observation at a time and can retain only a predetermined amount of data to use in subsequent computations.

6.4 Model Building: Alternating Trends Smoothing

A method of identifying changepoints in a time series based on alternating up and down linear trends, called alternating trends smoothing, or ATS, is given in Algorithm 1. It depends on a smoothing parameter, h, which specifies the step size within which to look for changepoints.

Algorithm 1: Alternating trends smoothing (h)

1. Set $d = 1$ (changepoint counter)
2. While (more data in first time step)
   (a) for $i = 1, 2, \ldots, m$, where $m = h$ if h additional data available or else m is last data item:
   input x_i;
   (b) set $b_d = 1$; $c_d = x_1$
   (c) determine $j_+, j_-, x_{j_+}, x_{j_-}$ such that
   $x_{j_+} = \max x_1, \ldots, x_h$ and $x_{j_-} = \min x_1, \ldots, x_h$
   (d) set $s = (x_k - x_i)/(k - i)$ and $r = \text{sign}(s)$
   (e) while $r = 0$, continue inputting more data; stop with error at end of data
3. Set $j = i$ (index of last datum in previous step); and set $d = d + 1$
4. While (more data)
   (a) for $i = j + 1, j + 2, \ldots, j + m$, where $m = h$ if h additional data available or else $j + m$ is last data item:
   input x_i;
      i. while ($\text{sign}(s) = r$)
         A. set $k = \min(i + h, n)$
         B. if ($k = i$) break
         C. set $s = (x_k - x_j)/(k - j)$
         D. set $j = k$
      ii. determine j_+ such that rx_{j_+} is the maximum of $rx_{j+1}, \ldots, rx_{j+m}$
      iii. set $b_d = j_+$; and set $c_d = x_{j_+}$
      iv. set $d = d + 1$; set $j = j_+$; and set $r = -r$
   (b) set $b_d = j_+$; and set $c_d = x_{j_+}$

The output of this algorithm applied to a time series $x_1, x_2, \ldots$ is

$$(b_1, c_1), (b_2, c_2), \ldots,$$

where $b_1 = 1$, $c_1 = x_1$, $b_2 = t^{(2)}$, and $c_1 = x_{t^{(2)}}$, where $t^{(2)}$ is the time at which the first trend changes sign.

Between two breakpoints the trend is represented by the slope of the time series values at the two points divided by the time between the two points, and the smoothed time series is the piecewise linear trendlines that connect the values at the changepoints. The method is effective for finding interesting patterns. For example, the head-and-shoulders pattern in the Intel stock price, shown in Fig. 6.5, is very apparent in the ATS representation of the time series shown in Fig. 6.6. A step size of 30 was used in that fit.

The output of the ATS algorithm applied to the INTC data in Fig. 6.6 is

$$(1, 26.11), (69, 34.07), (97, 29.75), (132, 36.57), (206, 29.23), (251, 33.94), (290, 27.63).$$

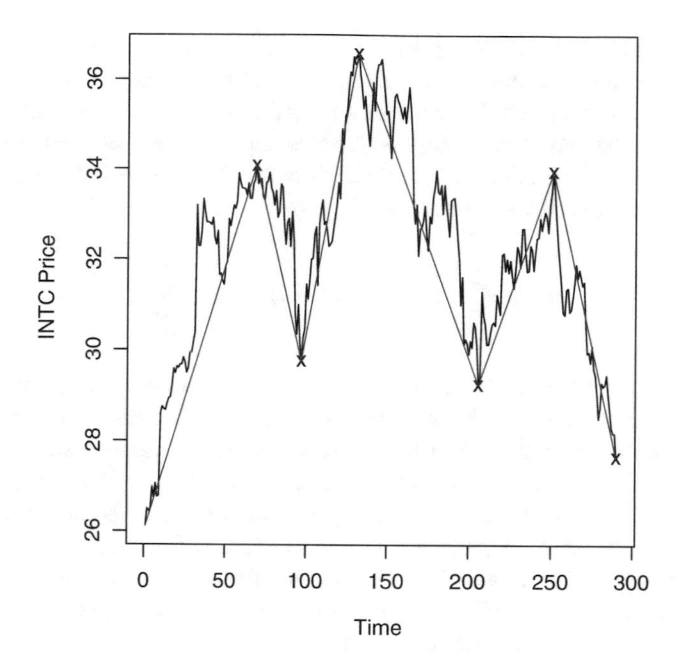

Fig. 6.6 A head-and-shoulders pattern in ATS

Thus, the 290 raw data points are summarized in the seven pairs of numbers representing the changepoints and their values.

While the ATS fit emphasizes only the signs of the trends, the actual slopes are very easily computed from the values at the changepoints.

6.4.1 The Tuning Parameter

The tuning parameter h in Algorithm 1 is a "step size." The process of identifying the next changepoint begins with the datum one step size beyond the current changepoint. Larger values of h tend to increase the distances between changepoints, but the actual distance between changepoints can be smaller than h; in fact, the distance between changepoints can be as small as 1 time unit.

Although in the standard implementation, the identification of trends in ATS is based on individual points, the aggregate behavior tends to dominate, especially after the first trend. In identifying the changepoint at the end of the first regime, however, the dependence of the trend on the point one step size beyond can lead to misidentification of the trend. This is because ATS works by identifying changepoints based on changing trends, and in the first trend there is no previous trend for comparison. This can result in a trend determined by points x_1 and x_h differing completely from the apparent trend in the points $x_1, \ldots, x_{h-1}$. For this reason, a simple modification of the ATS algorithm in the first step is to use some other criterion for determining the trend. One simple approach is a least-squares fit of a line through x_1 and that comes close to the points $x_2, \ldots, x_h$. Because of the constraint that the line goes through x_1, the least-squares criterion might be weighted by the leverages of the other points. If the time-spacing is assumed to be equal over the points $x_1, \ldots, x_h$, the least-squares slope is

$$\arg\min_s \sum_{i=2}^{h} (x_i - si)^2/i = (h^2 - h)/\sum_{i=2}^{h} x_i. \tag{6.1}$$

If all of the first h observations follow the same trend, the modification has no effect.

This modification can also be at each step, and it often results in what appears visually to be a better fit. Nevertheless, it is not always easy to pick a good rule for determining the direction of a trend. Because of the way the algorithm looks backwards after detecting a change in the sign of the trend, the modification does not have as much effect in subsequent regimes after the first one.

If the length of the time series is known in advance, a step size equal to about one tenth of the total length seems to work reasonably well. Even so, it is often worthwhile to experiment with different step sizes.

Figure 6.7 shows ATS applied to the daily closing price of the stock of International Business Machines Corporation (NYSE:IBM) from January 1, 1970, through December 31, 2014. There are $n = 11{,}355$ points. Over this full period, a step size of $h = 1136$ (a step size of $n/10$) was used. This resulted in the alternating trend lines shown as solid red line segments.

There are only five changepoints identified in the IBM daily closing prices, and the changes are as likely to occur in the area of "low action" (the earlier times) as in the areas of higher volatility. This is because ATS operates in an online fashion; when processing the data in the earlier time regimes, it is not known that the trends will become more interesting. Only one changepoint from the observation at the time index of 5971 through the end of the series is identified. A smaller step size

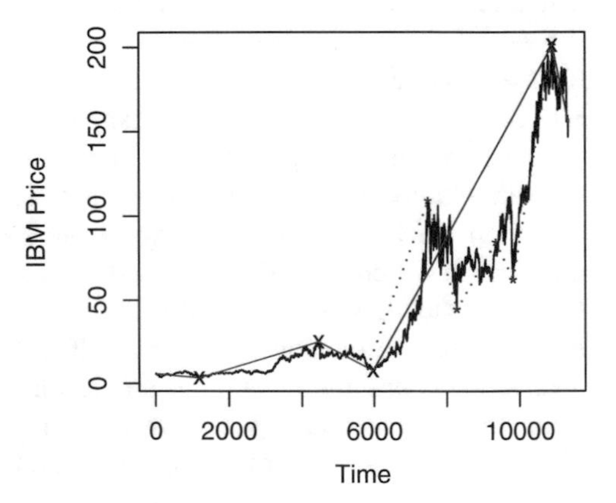

Fig. 6.7 ATS with different stepsizes over different regions

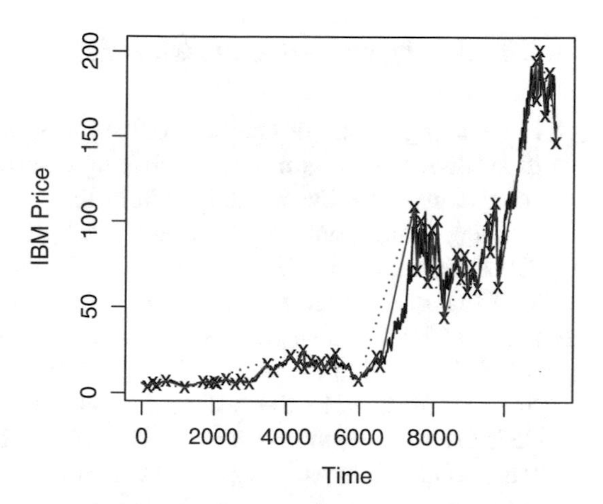

Fig. 6.8 ATS applied twice

may be appropriate. Alternating trend segments were then determined for these data, beginning with January 1, 1993, (a time index of 5815 in the original series), and using a stepsize of 554. These are shown as dashed blue line segments in Fig. 6.7. The ATS fit resulting from those two different stepsizes better captures the pattern of the time series.

Alternating trends smoothing can be applied recursively. Once the original data are smoothed, ATS can be applied to the changepoints determined in the original smoothing. This is illustrated in Fig. 6.8 using the same IBM daily closing price data as before.

First, a step size of 100 was used. This resulted in 53 changepoints. These are shown in Fig. 6.8, and the trend lines connecting them are shown as solid red line segments. Next, ATS was applied to the changepoints (55 points in all, including the first and last observations). A stepsize of five was used for this smoothing. This resulted in nine changepoints. The alternating trend lines connecting them are shown as dashed blue line segments in Fig. 6.8. The original set of 54 trendline segments may be considered too noisy to be a good overall model. That model may be considered to be overfit. The subsequent fit of the changepoints is of course much smoother than the fit of the original data.

This repeated ATS fitting is iterative data reduction. The first fit reduced the data to 55 points (including the two endpoints). These points may contain sufficient information to summarize the original 11,355 points. Carrying the reduction further by applying ATS to the changepoints, we reduce the data to 11 points (again, including the two endpoints), and this may be sufficient for our purposes.

Making transformations to a time series before applying ATS results in different changepoints being applied. For example, using a log transformation of the IBM price data in Fig. 6.7 would result in different changepoints, and even using only one stepsize for the whole time series, those changepoints on the log data may be more meaningful.

6.4.2 Modifications and Extensions

An alternating trends fit can be modified in several ways. One very simple way is to subdivide the regimes by identifying changepoints within any regime based on deviations from the trendline within that regime. This type of procedure for identifying changepoints has been suggested previously in a more general setting.

Consider again the IBM data in Fig. 6.7, with the ATS fit using a stepsize of 1136. (This is the solid red line in the figure.) Over the region from a time of 5971 to a time of 10,901, a single model was fit. This model is the line segment from the point $(5971, 7.63)$ to the point $(10901, 200.98)$. The most deviate point within this regime, as measured by the vertical residuals is at time point 9818 where the actual price is 61.90, while the point on the trendline is 158.51.

This point can be considered to be a changepoint, as shown in Fig. 6.9. In this case the trends on either side of the changepoint are both positive; that is, they are not alternating trends.

Continuing to consider changepoints within regimes identified with a single trend, in this case we would likely identify a changepoint at about time 7500, which would correspond to the maximum deviation from the single trendline between the original changepoint at time 5971 and the newly identified changepoint at time 9818. In this case, while the single trend is positive, the first new trend would be positive and the second new trend would be negative.

While in many applications in finance, only the sign of a trend is of primary interest, occasionally, the magnitude may also be of interest, and a changepoint might be identified as a point at which the slope of the trend changes significantly. The basic ideas of ATS can be adapted to this more general definition of changepoints; however, some of the simplicity of the computations of ATS would be lost.

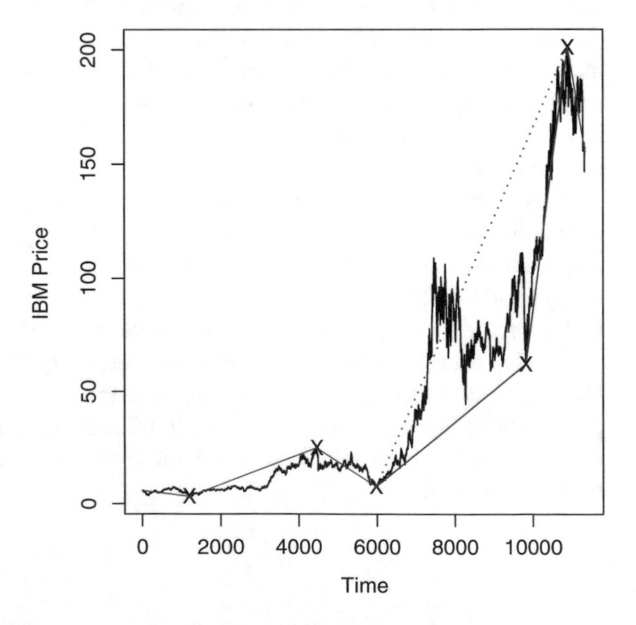

Fig. 6.9 A linear trend broken into two trends at the most extreme point

This approach would also require an additional tuning parameter to quantify "significance" of change in trend, when the sign of the trend may not change.

Another possible improvement to the basic ATS algorithm is to allow the stepsize to be adjusted within the computations. The alterations to the stepsizes could be based on the number of changepoints or on goodness-of-fit within a regime, and in either of these general approaches, there are several possible ways of doing it. The modification shown in Fig. 6.9 could be performed routinely as a postprocessing step in any ATS fit. That modification of course would require an additional tuning parameter to be used in deciding whether or not to break up an existing regime.

In very noisy data goodness-of-fit measures can often be misleading. This is because single or a few outliers can cause the measure to indicate an overall lack of fit. (Note that the basic ATS fitting, although individual points are used in determining changepoints, the method is generally resistant to outliers.) As in most data analysis, outliers often must be treated in ad hoc ways. This is because they often contain completely new information.

6.5 Bounding Lines

In statistical modeling it is common practice to associate "confidence bounds" with a fitted model. These are usually based on some underlying probability distribution, and they can take various forms depending on the model, which includes the relevant probability distributions.

In our objective of finding patterns in the data, we have not assumed any specific probability model. In other applications of trend analysis, it is common to identify bounding lines within a given regime that generally are in the same directions as the trend over that region, but which are either above all the points in the regime (a "resistance line") or below all the points (a "support line"). Such bounding lines do not depend on any probability model.

As we have emphasized, finding the changepoints is the paramount problem. Once the regimes are identified, however, it may be of interest to identify bounding lines for those regimes.

For a given time series $\{x_t : t = 1, \ldots, n\}$, we seek a lower bounding line $x = a + bt$ that satisfies the optimization problem

$$\begin{aligned} &\min_{a,b} \textstyle\sum_{t=1}^{n} \rho(x_t - a + bt) \\ &\text{s.t. } x_t \geq a + bt, \quad \text{for } t = 1, \ldots, n, \end{aligned} \tag{6.2}$$

where $\rho(\cdot)$ is a nonnegative function such that for a vector $v = (v_1, \ldots, v_n)$, $\|v\| = \sum_{t=1}^{n} \rho(v_t)$ is a norm. An upper bounding line is defined the same way, except that the inequality in the constraint is reversed.

In general, this is a hard optimization problem, but for the L_1 norm, that is, when $\rho(\cdot) = |\cdot|$, it is straightforward; and a method is given in Algorithm 2. The method depends on the fact that the L_1 norm satisfies the triangular inequality with equality: that is, $\|y + z\|_1 = \|y\|_1 + \|z\|_1$. The method also depends on the fact that the data are equally spaced along the time axis.

Algorithm 2: L_1 lower bounding line: $x_t = \tilde{a} + \tilde{b}t$

1. Fit the data $x_t = a + bt$ by a minimum L_1 criterion to obtain parameters a_* and b_*.
2. Determine the position of the minimum residual, k, and adjust a_*:
   $a_* \leftarrow x_k - b_* k$.
3. If $k - 1 = n - k$, then set $\tilde{b} = b_*$ and $\tilde{a} = a_*$, and stop.
4. Else if $k \leq n/2$,
   (a) rotate the line $x = a_* + b_* t$ clockwise about the point (k, x_k) until for some point (i, x_i), $x_i = a_* + b_* i$.
   (b) set $\tilde{b} = (x_i - x_k)/(i - k)$ and $\tilde{a} = x_k - \tilde{b}k$, and stop.
5. Else if $k > n/2$,
   (a) rotate the line $x = a_* + b_* t$ counterclockwise about the point (k, x_k) until for some point (j, x_j), $x_j = a_* + b_* j$.
   (b) set $\tilde{b} = (x_k - x_j)/(k - j)$ and $\tilde{a} = x_k - \tilde{b}k$, and stop.

Theorem 1 *Algorithm 2 yields a solution to optimization problem (6.2), when* $\rho(\cdot) = |\cdot|$.

Proof Let a_* and b_* be such that $\sum_{t=1}^{n} |x_t - a_* - b_* t| = \min_{a,b} \sum_{t=1}^{n} |x_t - a - bt|$. Let k be such that

$$x_k = \arg\min_{x_i}(x_i - a_* - b_* i).$$

For the optimal values of $\tilde{a}$ and $\tilde{b}$, we must have $x_k \geq \tilde{a} + \tilde{b}k$.

There are three cases to consider: $k = (n + 1)/2$ (corresponding to step 3 in Algorithm 2), $k \leq n/2$ (corresponding to step 4 in the algorithm), and $k > n/2$ (corresponding to step 5 in the algorithm). We will first consider the case $k \leq n/2$. The case $k > n/2$ follows the same argument. Also following that argument, it will be seen that the solution given for case $k = (n + 1)/2$ is optimal.

If $k \leq n/2$, consider the line $x = x_k - b_* k + b_* t$. (The intercept here is what is given in step 2 of the algorithm.) This line goes through (k, x_k), and also satisfies the constraint $x_t \geq x_k - b_* k + b_* t$, for $t = 1, \ldots, n$. The residuals with respect to this line are $x_t - x_k + b_* k - b_* t$ and all residuals are positive. Any point t for $t < k$ is balanced by a point $\tilde{t} > k$, and there are an additional $n - 2k$ points $x_{\tilde{t}}$ with $\tilde{t} > k$. If any point $x_{\tilde{t}}$ with $\tilde{t} > k$ lies on the line, that is, $x_{\tilde{t}} = x_k - b_* k + b_* \tilde{t}$, then this line satisfies the optimization problem (6.2) because any change in either the intercept $x_k - b_* k$ or the slope b_* would either violate the constraints or would change the residuals in a way that would increase the norm of the residuals. In this case the solution is as given at the end of step 4, because $\tilde{b} = b_*$.

The step now is to rotate the line in a clockwise direction, which results in an increase in any residuals indexed by t for $t < k$ and a decrease of the same amount in the same number of residuals $x_{\tilde{t}}$ with $\tilde{t} > k$. (This number may be 0.) It is the decrease in the residuals of the additional points indexed by $\tilde{t} > k$ that allows for a possible reduction in the residual norm (and there is a positive number of such points).

Consider the line $x = x_k - b_* k + (b_* + \delta_b)t$ such that δ_b is the minimum value such that there is a point $x_{\tilde{t}}$ with $\tilde{t} > k$ such that $x_{\tilde{t}} = x_k - b_* k + (b_* + \delta_b)\tilde{t}$. This line satisfies the constraints and by the argument above is optimal. Hence, the solution is as given at the end of step 4, because $\tilde{b} = b_* + \delta_b$.

Now consider the case $k = (n + 1)/2$. (In this case, n is an odd integer, and $k - 1 = n - k$ as in Algorithm 2.) Following the same argument as above, we cannot change the intercept or the slope because doing so would either violate the constraints or would change the residuals in a way that would increase the norm of the residuals. Hence, we have $\tilde{b} = b_*$ and $\tilde{a} = x_k - b_* k$ as given in the algorithm is a solution to optimization problem (6.2) when $\rho(\cdot) = |\cdot|$. ∎

One possible concern in this method is that the L_1 fit may be nonunique. This does not change any of the above arguments about an optimal solution to optimization problem (6.2) when $\rho(\cdot) = |\cdot|$. It is possible that the solution to this optimization problem is nonunique, and that is the case independently of whether or not the initial fit in Algorithm 2 is nonunique.

The discussion above was for lower bounding lines under an L_1 criterion. Upper bounding lines are determined in the same way following a reversal of the signs on the residuals.

Bounding lines can easily be drawn over any region of a univariate time series. They may be more meaningful if separate ones are drawn over separate regimes of a time series, as in Fig. 6.10, where separate bounding lines are shown for the six regimes corresponding to alternating trends, that were shown in Fig. 6.6.

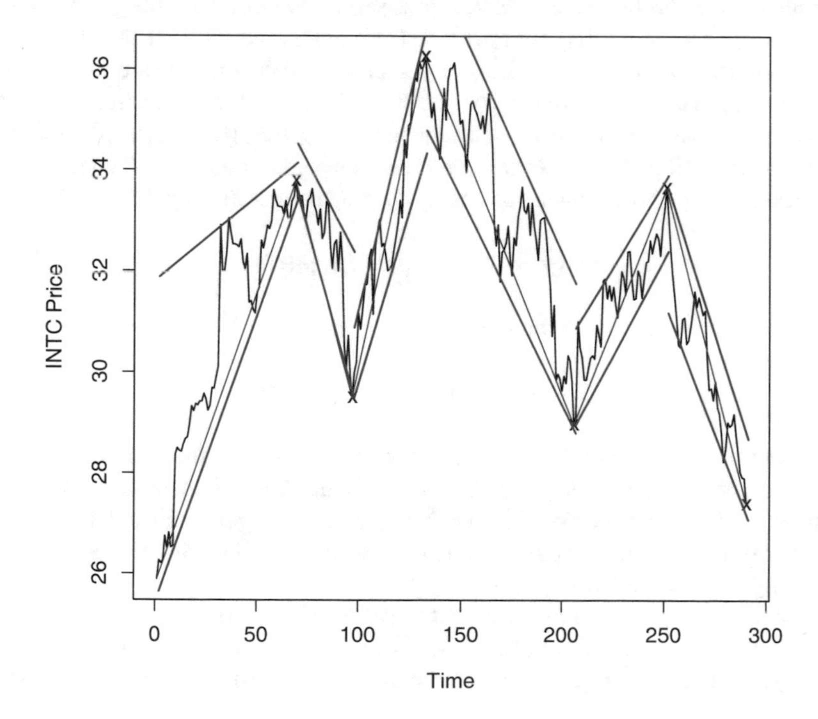

Fig. 6.10 Bounding lines in six regimes

6.6 Patterns

One of our motivations for fitting time series is for clustering and classification of time series based on similarities of trends and patterns. Usually, in large time series datasets a single model does not fit well, so our approach has been to identify a sequence of local models, $\mu_1(t), \mu_2(t), \ldots, \mu_k(t)$, in k regimes.

The models $\mu_i(t)$ may be of various forms, and they may contain various levels of information. For example, in piecewise constant modeling, the form of the model is

$$\mu_i(t) = c_i \mathrm{I}_{[t_{i1}, t_{i2}]}(t), \tag{6.3}$$

where $\mathrm{I}_S(t)$ is the indicator function: $\mathrm{I}_S(t) = 1$ if $t \in S$, and $\mathrm{I}_S(t) = 0$ otherwise. (This formulation allows the global model to be written as $\sum_i \mu_i(t)$.) The form of the model in ATS is

$$\mu_i(t) = (a_i + s_i t)\mathrm{I}_{[t_{i1}, t_{i2}]}(t), \tag{6.4}$$

where in the notation of Algorithm 1, $a_i = c_i$, $s_i = (c_{i+1} - c_i)/(b_{i+1} - b_i)$, $t_{i1} = c_i$ and $t_{i2} = c_{i+1}$, and of course a global is just the sum of these local models.

For comparing different time series for clustering or classification, we may focus on patterns of models, $\mu_i(t), \mu_{i+1}(t), \ldots, \mu_{i+r}(t)$, on r successive regimes, not necessarily beginning at the start of the time series and not necessarily extending over the full extent of the time series. We compare the pattern $P_{ri} = (\mu_i(t), \mu_{i+1}(t), \ldots, \mu_{i+r}(t))$ with patterns from other time series. The obvious basis for comparison would be a metric, or a measure with some of the properties of a metric, applied to the patterns; that is, we define a a metric function $\rho(P_{ri}, P_{sj})$, where P_{ri} is a pattern consisting of a sequence $\mu_i(t), \mu_{i+1}(t), \ldots, \mu_{i+r}(t)$ and P_{sj} is a pattern consisting of a sequence over s regimes beginning at the jth one.

Because the patterns depend on fitted models, the fact that a pattern

$$(\mu_i(t), \quad \mu_{i+1}(t), \quad \mu_{i+2}(t))$$

in one time series is exactly the same as a pattern

$$(\mu_j(t), \quad \mu_{j+1}(t), \quad \mu_{j+2}(t))$$

in another time series does not mean that the actual values in the two time series are the same over those regimes or even that the values have some strong association, such as positive correlation, with each other. This is generally not inconsistent with our objectives in seeking patterns in time series or in using those patterns in clustering and classification.

In this section, we discuss some of the issues in clustering and classification of time series, once a sequence of regimes is identified. An important consideration in analyzing multiple time series is registration of the different time series; that is, shifting and scaling the time series so that the regimes in the separate time series can be compared. We also briefly indicate possible approaches for further data reduction. Once these issues have been addressed, the problems of clustering and classification are similar to those in other areas of application. Examples and the details of the use of the clustering and classification methods are discussed by Wilson (2016).

6.6.1 Time Scaling and Junk

Prior to comparing two time series or the patterns in the two series, we generally need to do some registration of the data. Usually, only subsequences of time series are to be compared, and so the beginning and ending points in the two series must be identified. The actual time associated with the subsequences may be different, and the subsequences may be of different lengths. There are various methods of registering two subsequences. The most common methods are variants of "dynamic time warping" (DTW), which is a technique that has been around for many years. There are several software libraries for performing DTW.

In the overall task of identifying and comparing patterns in time series, the registration step, whether by DTW or some other method, can be performed first or later in the process. Our preference generally is to identify breakpoints prior to registration.

Similarity of patterns is not an absolute or essential condition. Similarity, or dissimilarity, depends on our definition of similarity, which in turn depends on our purposes. We may wish to consider two patterns to be similar even in the time intervals of the piecewise models do not match. We also may wish to ignore some models within a pattern, especially models of brief duration.

In Fig. 6.11 we see three patterns that for some purposes we would wish to consider to be similar to each other. The times as well as the actual values are rather different, however. Among the three time series, the times are both shifted (the starting time of course is almost always arbitrary) and scaled. The unit of time is not always entirely arbitrary. It depends on our ability to sample at different frequencies, and the sampling rate is not always adjustable. The unit of time may also be important in an entirely different way. It is a well-recognized property of markets that frequency of trading (or frequency of recording data) results in different market structures (see, for example, Gentle and Härdle 2012).

Figure 6.11 also illustrates another problem for comparing patterns. The blip in the time series on the right side results in two additional model terms in the fitted time series. We would actually like to compare the patterns

$$(\mu_1, \mu_2, \mu_3, \mu_4)$$

and

$$(\nu_1, \mathcal{S}(\nu_2, \nu_3, \nu_4), \nu_5, \nu_6),$$

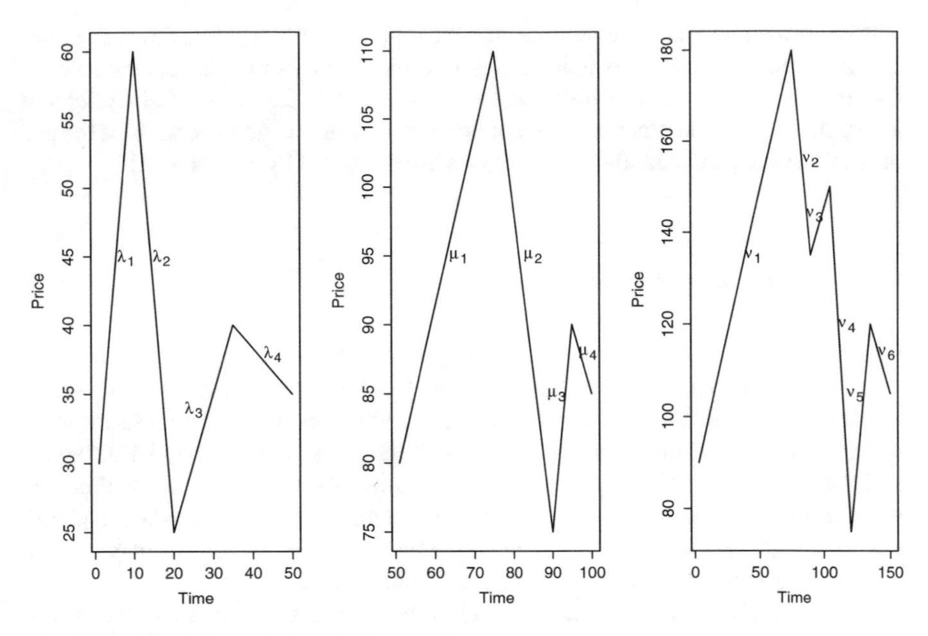

Fig. 6.11 Three patterns that are similar

where the specific correspondences are $\mu_1 \approx \nu_1$, $\mu_3 \approx \nu_5$, $\mu_4 \approx \nu_6$, and $\mu_2 \approx \mathcal{S}(\nu_2, \nu_3, \nu_4)$, where $\mathcal{S}(\nu_2, \nu_3, \nu_4)$ is some smooth of the three models ν_2, ν_3, and ν_4.

The plot on the right side of Fig. 6.11 compared with the other two plots illustrates the intimate connection between smoothing or model fitting and pattern recognition. A smoother fit of the time series shown on the right side would have resulted in just four models (three changepoints), in which the second model would be some smooth of ν_2, ν_3, and ν_4, that is, $\mathcal{S}(\nu_2, \nu_3, \nu_4)$.

The extra blip in the time series on the right side of Fig. 6.11 is "junk" at a lower level of resolution.

6.6.2 *Further Data Reduction: Symbolic Representation*

While the individual components of a pattern P_{ri} may contain various details of the models, in some cases some details can be suppressed while salient features of the pattern, that is, the *sequence*, are retained. For example, for a sequence of constant models such as in Eq. (6.3), the most important features of the sequence may be a sequence of indicators whether the c_i were small, mid-size, or large; that is, the pattern is a sequence of the form *abcde*. . ., where each of the *a*s, *b*s, and so on, are just the values `s`, `m`, and `l`, indicating "small," "medium," and "large." For example, the pattern `sllmls` would indicate a sequence of six piecewise constant models of the form of Eq. (6.3), in which c_1 is (relatively) small, c_2 and c_2 are large, and so on.

This representation, of course, does not include information about the t_{i1}s or t_{i2}s or even the exact values of the c_is, but the *patterns* of small, medium, and large may be information for clustering or classifying time series.

This further step of data reduction of forming categories of models and associating each category with a single symbol can be very useful in data mining of time series, and has been used in various ways for some time. One widely used symbolic approximation for time series is SAX, which is based on a modification of a sequence of piecewise constant models (called PAA), see Lin et al. (2007). Another type of symbolic approximation of a time series, called NSAR for nonparametric symbolic approximate representation, was described by He et al. (2014). This method is based on preliminary wavelet transformations and so enjoys the multi-resolution properties of wavelets.

The transformation of models with quantitative parameters to models of categorical symbols requires some a priori definition of the symbols, possibly a complete listing or catalogue of the symbols, or at least some formula for defining new symbols. In batch processing of the data, the range of possible models can be determined before the transformation of quantitative models to categorical models.

In ATS, each model is characterized by four real numbers. A sequence of r models is characterized by $2r + 2$ real numbers. To reduce the data further, the real numbers are binned into ordered groups. These ordered bins can be associated with a unique set of symbols.

The replacement of models with continuous numeric parametrizations by symbolic representations results in loss of data. A linear model in a given regime may be transformed into a model that carries only the information that a particular coefficient is large, relative to the same coefficient in other regimes. The symbolic approximations may even lose information concerning the time of the changepoints.

6.6.3 Symbolic Trend Patterns (STP)

The symbolic approximation of SAX is based on a type of piecewise constant modeling called "piecewise aggregate approximation" ("PAA") as described by Lin et al. (2007). The same idea of SAX can be applied to the models in ATS, as described by Gentle (2012) who called it "symbolic trend patterns," or "STP." These symbols consist of pairs of symbols or syllables. They are formed by selection of a consonant

$$\mathrm{J, K, L, M, N}$$

that represents duration of an upward trend, or of a consonant

$$\mathrm{P, Q, R, S, T}$$

that represents duration of an downward trend, and selection of a vowel

A, E, I, O, U

that represents magnitude of a trend. In many cases, however, the vowels could represent the magnitude of the change in value instead of the magnitude of a trend, that is the slope of the segment between the changepoints.

In each case, the individual letters as listed above represent increasing magnitudes. Thus, "P" represents a downward trend of short duration, and "A" represents a trend (up or down) of very small magnitude.

If these symbols are defined and assigned in a batch fashion, they can represent quantiles (or approximate quantiles) of the fully observed time series.

For example, the ATS fit of the INTC data in the head-and-shoulders pattern shown in Fig. 6.6 could be represented by the STP symbolic approximation

LO, PE, KO, RO, JE, QI

where the vowel is used to represent magnitude of change, rather than rate of change. The ATS fit of the IBM price data shown by the solid red lines in Fig. 6.7 could be represented by the symbolic approximation

PA, LE, QE, NU

where again the vowel represents magnitude of change.

Of course, because the trends in ATS alternate, if a single direction is given, then there would be no need for different symbols to be used to designate up and down moves.

6.6.4 Patterns in Bounding Lines

Following any kind of data reduction, there are enhanced opportunities for identifying patterns. Trends are a simple form of data reduction that offer various methods of pattern identification. Likewise, the bounding lines discussed in Sect. 6.5 may be used to develop patterns. The bounding lines have the same kinds of characteristics as the trend lines of Sect. 6.4; they have slopes and duration. When bounding lines are determined in regions determined by the ATS their slopes will generally (but not necessarily) have the same sign as the slopes of the trends.

Another interesting characteristic of bounding lines is their relationship to each other; in particular, whether they seem to be converging or diverging. (By their definition, they can never cross within the region for which they are defined.) In Fig. 6.10, for example, we see that the bounding lines in the leftmost regime seem to be converging, while those in the second regime from the left seem to be diverging. Technical security analysts sometimes attach meaning to such patterns.

6.6.5 Clustering and Classification of Time Series

In clustering and classification of data we need appropriate measures of similarity or dissimilarity. The most useful measures of dissimilarity are metrics, because of their uniqueness and the ordering they induce, and the most useful metrics on $\mathbb{R}^d$ are those induced by a norm. For a given class of patterns, whether defined on a finite set of symbols or on $\mathbb{R}^d$, there is a wide choice of possible metrics. For metrics induced by norms the equivalence relation among any set of norms yields an equivalence of metrics. This equivalence carries over to metrics on a set of ordered bins or symbols (see Gentle 2012).

The problem of clustering or classification on a set of time series is essentially the problem of clustering or classification on a set of patterns. Despite the equivalence of metrics, on a given class of patterns, different metrics can lead to different clusters or different classifiers.

The most challenging problem in clustering and classification of time series arises from the time scaling and "junk" models that constitute a pattern. The three similar time series shown in Fig. 6.11, for example, may be associated with the STP approximations

JO, PU, ME, RA

MO, RU, KE, PA

MO, PI, JA, RI, KE, PA

The question here is how to recognize the similarity of the patterns. These patterns exhibit different time scalings and in one case include a superfluous model. An approach to the problem at this point is the same approach that was used from the start: further discretization; that is, further data reduction, with its concomitant loss of information.

One way of dealing with the time scale is a further discretization; instead of ten different values, we may just use two, up or down. The first two patterns are now the same:

$$+O, -U, +E, -A$$

A model with both short duration and small change in magnitude is a candidate for a superfluous modes; that is, one that can be smoothed away by combinations with nearby models. Applying this approach to the smoothed time series on the right side of Fig. 6.11 would result in the second through fourth models being combined into a single model, which would be represented as RU or –U.

This approach to the problem involves combinations and adjustments of any or all of the models in a set of patterns, and so is obviously not entirely satisfactory. For clustering and classification, of course, we do not need for the patterns to be exactly alike, so another approach would be based on use of appropriate metric.

A metric that weights differences in direction of a trend much more heavily than differences in length of two trends in the same direction would achieve some of the same effect as considering the duration to be a binary variable.

Classification of time series is closely related to the standard problem of prediction or forecasting in time series. For a given pattern, the predicted value is merely the predicted class of the pattern.

References

Badhiye SS, Hatwar KS, Chatur PN (2015) Trend based approach for time series representation. Int J Comput Appl 113:10–13

Bao D, Yang Z (2008) Intelligent stock trading system by turning point confirming and probabilistic reasoning. Expert Syst Appl 34:620–627

Dorsey TJ (2007) Point and figure charting: the essential application for forecasting and tracking market prices, 3rd edn. Wiley, Hoboken

Fink E, Gandhi HS (2011) Compression of time series by extracting major extrema. J Exp Theor Artif Intell 23:255–270

Fu T (2011) A review on time series data mining. Eng Appl Artif Intell 24:164–181

Fu T-C, Chung KF-L, Luk RWP, man Ng C (2008) Representing financial time series based on data point importance. Eng Appl Artif Intell 21:277–300. https://doi.org/10.1016/j.engappai.2007.04.009

Gentle JE (2012) Mining for patterns in financial time series. In: JSM 2012 proceedings. American Statistical Association, Alexandria, pp 2978–2988

Gentle JE, Härdle WK (2012) Modeling asset prices. In: Handbook of computational finance. Springer, Heidelberg, pp 15–33

He X, Shao C, Xiong Y (2014) A non-parametric symbolic approximate representation for long time series. Pattern Anal Appl 15. https://doi.org/10.1007/s10044-014-0395-5

Lin J, Li Y (2009) Finding structural similarity in time series data using bag-of-patterns representation. In: Statistical and scientific database management, pp 461–477. https://doi.org/10.1007/978-3-642-02279-1_33

Lin J, Keogh E, Wei L, Lonardi S (2007) Experiencing SAX: a novel symbolic representation of time series. Data Min Knowl Discov 15:107–144. https://doi.org/10.1007/s10618-007-0064-z

Wilson SJ (2016) Statistical learning in financial time series data. Dissertation, George Mason University

Zhou J, Ye G, Yu D (2012) A new method for piecewise linear representation of time series data. Phys. Procedia 25:1097–1103

Chapter 7
Variational Bayes for Hierarchical Mixture Models

Muting Wan, James G. Booth, and Martin T. Wells

Abstract In recent years, sparse classification problems have emerged in many fields of study. Finite mixture models have been developed to facilitate Bayesian inference where parameter sparsity is substantial. Classification with finite mixture models is based on the posterior expectation of latent indicator variables. These quantities are typically estimated using the expectation-maximization (EM) algorithm in an empirical Bayes approach or Markov chain Monte Carlo (MCMC) in a fully Bayesian approach. MCMC is limited in applicability where high-dimensional data are involved because its sampling-based nature leads to slow computations and hard-to-monitor convergence. In this chapter, we investigate the feasibility and performance of variational Bayes (VB) approximation in a fully Bayesian framework. We apply the VB approach to fully Bayesian versions of several finite mixture models that have been proposed in bioinformatics, and find that it achieves desirable speed and accuracy in sparse classification with finite mixture models for high-dimensional data.

Keywords Bayesian inference · Generalized linear mixed models · Large p small n problems · Linear mixed models · Markov chain Monte Carlo · Statistical bioinformatics

M. Wan
New York Life Insurance Company, New York, NY, USA

J. G. Booth
Department of Biological Statistics and Computational Biology, Cornell University, Ithaca, NY, USA
e-mail: jim.booth@cornell.edu

M. T. Wells (✉)
Department of Statistical Science, Cornell University, Ithaca, NY, USA
e-mail: mtw1@cornell.edu

© Springer International Publishing AG, part of Springer Nature 2018

W. K. Härdle et al. (eds.), *Handbook of Big Data Analytics*, Springer Handbooks of Computational Statistics, https://doi.org/10.1007/978-3-319-18284-1_7

7.1 Introduction

Variational Bayes (VB) methods in statistics arose from the family of variational approximation methods (Jaakkola 2000) for performing approximate Bayesian inference for graphical models with latent variables (Bishop 1999; Attias 2000; Beal 2003). Since then, VB has been promoted and employed in several fields of modern applications, such as signal processing (Smídl and Quinn 2005; Blei and Jordan 2006; Tzikas et al. 2008), political science (Grimmer 2011), bioinformatics (Logsdon et al. 2010; Li and Sillanpää 2012), and in medical research such as brain imaging (Friston et al. 2011; Goldsmith et al. 2011). In the early years, VB was applied to Gaussian mixture models, factor analysis, principal component analysis, hidden Markov models, and their mixtures, for model learning (Bishop 1999; Ghahramani and Beal 2000; Bishop et al. 2002; Beal 2003) and model selection (Corduneanu and Bishop 2001; Teschendorff et al. 2005; McGrory and Titterington 2007). Since then, VB has also been used for learning nonlinear latent variable models (Honkela and Valpola 2005; Salter-Townshend and Murphy 2009), conducting functional regression analysis (Goldsmith et al. 2011), dealing with missing data in regression (Faes et al. 2011), and fitting location-scale models that contain elaborate distributional forms (Wand et al. 2011). Recently, Logsdon et al. (2010) apply VB based on the model proposed by Zhang et al. (2005) and show that their VB solution outperforms single-marker testing in QTL analysis. Li et al. (2011) utilize VB as an alternative to Markov chain Monte Carlo (MCMC) for hierarchical shrinkage-based regression models in QTL mapping with epistasis.

VB has been promoted as a fast deterministic method of approximating marginal posterior distributions and therefore as an alternative to MCMC methods for Bayesian inference (Beal 2003; Bishop 2006; Ormerod and Wand 2010). Posterior means computed from VB approximated marginal posterior density have been observed to be extremely accurate. For example, Bishop (2006) illustrates the VB solution of a simple hierarchical Gaussian model. Comparison with Gelman et al. (2003), where the true marginal posteriors for the same model is provided, shows that the VB solution in Bishop (2006) recovers the posterior mean exactly. However, it has also been observed that VB often underestimates posterior variance. This property is explained in Bishop (2006) to be due to the form of Kullback–Leibler divergence employed in the VB theory. In the setting where n observations follow a Gaussian distribution with large known variance and a zero-mean Gaussian-distributed mean whose precision is assigned a Gamma prior, Rue et al. (2009) use a simple latent Gaussian model to illustrate that VB may underestimate posterior variance of the precision parameter by a ratio of $O(n)$. Ormerod (2011) proposes a grid-based method (GBVA) that corrects the variance in VB approximated marginal posterior densities for the same model.

Finite mixture models have been widely used for model-based classification (McLachlan and Peel 2004; Zhang et al. 2005). Mixtures of distributions provide flexibility in capturing complex data distributions that cannot be well described by a single standard distribution. Finite mixture models, most commonly finite Gaussian mixture models, facilitate classification where each component distribution characterizes a class of "similar" data points.

Many problems in bioinformatics concern identifying sparse non-null features in a high-dimensional noisy null features space. This naturally leads to sparse classification with a finite mixture of "null" and "non-null" component distributions and latent indicator variables of component membership. The two-groups model in Efron (2008) is a finite mixture model with an implicit latent indicator variable for detection of non-null genes from a vast amount of null genes in microarray analysis. Bayesian classification results are obtained based on posterior expectation of the latent indicators. Typically, latent indicator variables are treated as missing data for an EM algorithm to be implemented in an empirical Bayes approach. Alternatively, fully Bayesian inference based on hierarchical structure of the model can be conducted via MCMC which approximates marginal posterior distributions. Finite-mixture-model-based classification allows strength to be borrowed across features in high-dimensional problems, in particular "large p small n" problems where p, the number of features to be classified, is several orders of magnitude greater than n, the sample size. However, in such high-dimensional problems it is difficult to assess convergence of fully Bayesian methods implemented using MCMC algorithms and the computational burden may be prohibitive.

In the context of finite mixture models, model complexity prevents exact marginal posteriors from being derived explicitly. For example, consider the following simple two-component mixture model in a fully Bayesian framework, with observed data $\{d_g\}$ and known hyperparameters $\tau_0, \sigma^2_{\tau_0}, \psi_0, \sigma^2_{\psi_0}, a_0, b_0, \alpha_1, \alpha_0$:

$$\begin{aligned} d_g|b_g, \tau, \psi, \sigma^2 &= (1 - b_g)N(\tau, \sigma^2) + b_g N(\tau + \psi, \sigma^2), \qquad (7.1) \\ b_g|p &\sim \text{Bernoulli}(p), \qquad \tau \sim N(\tau_0, \sigma^2_{\tau_0}), \qquad \psi \sim N(\psi_0, \sigma^2_{\psi_0}), \\ \sigma^2 &\sim \text{IG}(a_0, b_0), \qquad p \sim \text{Beta}(\alpha_1, \alpha_0), \end{aligned}$$

where IG(a, b) denotes an Inverse-Gamma distribution with shape a and scale b. Classification is conducted based on the magnitude of the posterior mean of the latent indicator b_g; a posterior mean close to 1 indicating membership in the non-null group for feature g. The Integrated Nested Laplace Approximations (INLA) framework is another popular approximate Bayesian method, introduced in Rue et al. (2009), whose implementation is available in the `R-INLA` package (Martino and Rue 2009). However, INLA does not fit any of the mixture models mentioned above because they are not members of the class of latent Gaussian models. In the context of model (1), numerical experiments show that MCMC produces believable approximate marginal posterior densities which can be regarded as proxies of the true marginal posteriors despite the label-switching phenomenon (Marin and Robert 2007) well-known for finite mixture models. In Sect. 7.2, we show that VB approximated densities are comparable to MCMC approximated ones in terms of both posterior mean and variance in this context, but VB achieves substantial gains in computational speed over MCMC.

More general finite mixture models are common for (empirical and fully) Bayesian inference in high-dimensional application areas, such as bioinformatics. Computations for the two-groups model in Smyth (2004), for example, are burdensome due to presence of the gene-specific error variance and different variances for the two-component distributions. Fully Bayesian inference via MCMC procedures is limited in practicality due to heavy computational burden resulting from the high-dimensionality of the data.

Our objective in this chapter is to investigate the feasibility and performance of VB for finite mixture models in sparse classification in a fully Bayes framework. We will see in later sections that there are significant computational issues with MCMC implementations in this context, making this approach impractical. In contrast, VB is fast, accurate, and easy to implement due to its deterministic nature. Moreover, the VB algorithm results in a very accurate classifier, despite the fact that it significantly underestimates the posterior variances of model parameters in some cases. To our knowledge, GBVA has not been evaluated before in this setting and our investigation indicates that it does not result in improved accuracy while adding substantially to the computational burden.

The plan for this chapter is as follows. Section 7.2 reviews theory underlying the VB method and approximation of marginal posterior densities via VB for the aforementioned simple two-component mixture model. Motivation behind using a hierarchical mixture model framework and implementation of VB for general finite mixture models are outlined in Sect. 7.3. We illustrate application of VB via examples involving simulated and real data in Section 7.4. Finally, we conclude with some discussion in Sect. 7.5. Many of the technical arguments are given in Appendix.

7.2 Variational Bayes

7.2.1 Overview of the VB Method

VB is a deterministic estimation methodology based on a factorization assumption on the approximate joint posterior distribution. This is a free-form approximation in that implementation of VB does not start with any assumed parametric form of the posterior distributions. The free-form factors are approximate marginal posterior distributions that one tries to obtain according to minimization of the Kullback–Leibler divergence between the approximate joint posterior and the true joint posterior. As a tractable solution to the minimization problem, the optimal factors depend on each other, which naturally leads to an iterative scheme that cycles through an update on each factor. The convergence of the algorithm is monitored via a single scalar criterion. Ormerod and Wand (2010) and Chapter 10 of Bishop (2006) present detailed introduction to VB in statistical terms. The VB approximation in this chapter refers to variational approximation to the joint and marginal posterior distributions under, in terminology used in Ormerod and Wand (2010), the product density (independence) restriction. The concept of a mean-field approximation has its roots in statistical physics and is a form of VB approximation under a more stringent product density restriction that the approximate joint posterior fully factorizes. The basic VB theory is outlined as follows.

Consider a Bayesian model with observed data $\mathbf{y}$, latent variables $\mathbf{x}$ and parameters $\boldsymbol{\Theta} = \{\boldsymbol{\theta}_1, \boldsymbol{\theta}_2, \ldots, \boldsymbol{\theta}_m\}$. Denote $H = \{\mathbf{x}, \boldsymbol{\Theta}\}$ as the collection of all unknown quantities. The posterior density $p(H|\mathbf{y}) = p(\mathbf{y}, H)/p(\mathbf{y})$ is not necessarily tractable due to the integral involved in computing the marginal data density $p(\mathbf{y})$. This intractability prevents computing marginal posterior densities such as $p(\boldsymbol{\theta}_i|\mathbf{y}) = \int p(H|\mathbf{y})d\{-\boldsymbol{\theta}_i\}$ where $\{-\boldsymbol{\theta}_i\}$ refers to $H \backslash \{\boldsymbol{\theta}_i\} = \{\mathbf{x}, \boldsymbol{\theta}_1, \boldsymbol{\theta}_2, \ldots, \boldsymbol{\theta}_{i-1}, \boldsymbol{\theta}_{i+1}, \ldots, \boldsymbol{\theta}_m\}$. Two key elements, namely minimization of the Kullback–Leibler divergence and employment of the product density restriction, underpin the VB algorithm. The Kullback–Leibler divergence gives rise to a scalar convergence criterion that governs the iterative algorithm. Through the product density restriction one assumes independence which allows derivation of tractable density functions.

For an arbitrary density for H, $q(H)$,

$$\log p(\mathbf{y}) = \int q(H) \log\left(\frac{p(\mathbf{y}, H)}{q(H)}\right) dH + \int q(H) \log\left(\frac{q(H)}{p(H|\mathbf{y})}\right) dH. \tag{7.2}$$

The second integral on the right-hand side of (7.2) is the Kullback–Leibler divergence between $q(H)$ and $p(H|\mathbf{y})$, i.e. $D_{\mathrm{KL}}(q(H)\|p(H|\mathbf{y}))$ which is nonnegative. Hence,

$$\log p(\mathbf{y}) \geq \int q(H) \log\left(\frac{p(\mathbf{y}, H)}{q(H)}\right) dH = C_q(\mathbf{y}),$$

where $C_q(\mathbf{y})$ denotes the lower bound on the log of the marginal data density. This lower bound depends on the density q. Given a data set, $\log p(\mathbf{y})$ is a fixed quantity. This means that if a density q is sought to minimize $D_{\mathrm{KL}}(q(H)\|p(H|\mathbf{y}))$, q also maximizes $C_q(\mathbf{y})$ making the lower bound a viable approximation to $\log p(\mathbf{y})$.

The product density restriction states that $q(H)$ factorizes into some partition of H, i.e. $q(H) = \prod_{i=1}^{k} q_i(\mathbf{h}_i)$, but the parametric form of the $q_i(\mathbf{h}_i)$ factors is not specified. It can be shown that, under the product density restriction, the optimal $q_i(\mathbf{h}_i)$ takes the following form (Ormerod and Wand 2010):

$$q_i(\mathbf{h}_i) \propto \exp\{E_{-\mathbf{h}_i}(\log p(\mathbf{y}, H))\} \tag{7.3}$$

for $i = 1, \ldots, k$, where expectation is taken over all unknown quantities except $\mathbf{h}_i$ with respect to the density $\prod_{j \neq i}^{k} q_j(\mathbf{h}_j)$.

Since each variational posterior $q_i(\mathbf{h}_i)$ depends on other variational posterior quantities, the algorithm involves iteratively updating $q_i(\mathbf{h}_i), 1 \leq i \leq k$ until the increase in $C_q(\mathbf{y})$, computed in every iteration after each of the updates $q_i(\mathbf{h}_i), 1 \leq$

$i \leq k$ has been made, is negligible. Upon convergence, approximate marginal posterior densities $q^*_i(\mathbf{h}_i), 1 \leq i \leq k$ are obtained, as well as an estimate $C_{q^*}(\mathbf{y})$ of the log marginal data density $\log p(\mathbf{y})$.

7.2.2 Practicality

As noted in Faes et al. (2011), under mild assumptions convergence of the VB algorithm is guaranteed (Luenberger and Ye 2008, p. 253). Bishop (2006) refers to Boyd and Vandenberghe (2004) and states that convergence is guaranteed because the lower bound $C_q(\mathbf{y})$ is convex with respect to each of the factors $q_i(\mathbf{h}_i)$. Convergence is easily monitored through the scalar quantity $C_q(\mathbf{y})$. Moreover, upon convergence, since $C_{q^*}(\mathbf{y})$ approximates $\log p(\mathbf{y})$, VB can be used to compare solutions produced under the same model but with different starting values or different orders of updating the q-densities. Because the optimal set of starting values corresponds to the largest $C_{q^*}(\mathbf{y})$ upon convergence, starting values can be determined empirically with multiple runs of the same VB algorithm. Experiments on starting values are generally undemanding due to high computational efficiency of the VB method, and are especially useful for problems where the true solutions are multimodal.

Computational convenience of VB is achieved if conjugate priors are assigned for model parameters and the complete-data distribution belongs to the exponential family. Beal (2003) describes models that satisfy these conditions as conjugate-exponential models. In these models the approximate marginal posterior densities $q_i(\mathbf{h}_i)$ take the conjugate form and the VB algorithm only requires finding the characterizing parameters. The induced conjugacy plays an important role in producing analytical solutions. Feasibility of VB for models with insufficient conjugacy is an open research area.

In practice, it suffices to impose some relaxed factorization of $q(H)$ as long as the factorization allows derivation of a tractable solution. Although, in many cases, the chosen product density restriction leads to an induced product form of factors of $q(H)$, the imposed factorization and induced factorization of $q(H)$ are generally not equivalent and may lead to different posterior independence structures.

As an introductory example, consider a simple fully Bayesian model with a $N(\mu, \sigma^2)$ data distribution and a bivariate Normal-Inverse-Gamma prior $p_{(\mu,\sigma^2)}(\mu, \sigma^2) = p_{\mu|\sigma^2}(\mu|\sigma^2)p_{\sigma^2}(\sigma^2)$ with $p_{\mu|\sigma^2}(\mu|\sigma^2) = N(\mu_0, \lambda_0\sigma^2)$ and $p_{\sigma^2}(\sigma^2) = \text{IG}(A_0, B_0)$. Imposing the product density restriction $q_{(\mu,\sigma^2)}(\mu, \sigma^2) = q_\mu(\mu)q_{\sigma^2}(\sigma^2)$ leads to independent Normal marginal posterior $q_\mu(\mu)$ and Inverse-Gamma marginal posterior $q_{\sigma^2}(\sigma^2)$. However, not imposing the product density restriction in this case leads to the bivariate Normal-Inverse-Gamma joint posterior $q_{\mu|\sigma^2}(\mu|\sigma^2)q_{\sigma^2}(\sigma^2)$ that reflects a different posterior independence structure. Hence, although choice of prior and product density restriction is problem-based, caution should be taken in examining possible correlation between model parameters, because imposing too much factorization than needed risks poor VB approximations if dependencies between the hidden quantities are necessary. This type of degradation in VB accuracy is noted in Ormerod and Wand (2010). There is a trade-off between tractability and accuracy with any type of problem simplification via imposed restrictions.

7.2.3 *Over-Confidence*

Underestimation of posterior variance of the VB approximate density has been observed in different settings such as those in Wang and Titterington (2005), Consonni and Marin (2007), and Rue et al. (2009). We review a possible reason why the variance of the approximate posterior given by VB tends to be smaller than that of the true posterior.

As explained previously by (2), a density $q(H)$ is sought to minimize the Kullback–Leibler divergence between itself and the true joint posterior $p(H|\mathbf{y})$. Upon examination of the form of the Kullback–Leibler divergence used in the VB method,

$$D_{\mathrm{KL}}(q(H)\|p(H|\mathbf{y})) = \int q(H)\left(-\log\left(\frac{p(H|\mathbf{y})}{q(H)}\right)\right) dH,$$

Bishop (2006) points out that, in regions of H space where density $p(H|\mathbf{y})$ is close to zero, $q(H)$, the minimizer, has to be close to zero also; otherwise the negative log term in the integrand would result in a large positive contribution to the Kullback–Leibler divergence. Thus, the optimal $q^*(H)$ tends to avoid regions where $p(H|\mathbf{y})$ is small. The factorized form of $q^*(H)$ implies that each of the factor q^*-densities in turn tends to avoid intervals where the true marginal posterior density is small. Thus, the marginal posterior densities approximated by VB are likely to be associated with underestimated posterior variance and a more compact shape than the true marginal posterior densities.

7.2.4 *Simple Two-Component Mixture Model*

For model (1) where $g = 1, 2, \ldots, G$, with known hyperparameters $\tau_0, \sigma^2_{\tau_0}, \psi_0, \sigma^2_{\psi_0}$, $a_0, b_0, \alpha_1, \alpha_0$, observed data $E = \{d_g\}$ and hidden quantities $H = \{\tau, \psi, \sigma^2, p, \{b_g\}\}$, imposing the product density restriction $q(H) = q_{\{b_g\}}(\{b_g\}) \times q_{(\tau,p)}(\tau, p) \times q_{\psi}(\psi) \times q_{\sigma^2}(\sigma^2)$ results in q-densities $q_{\{b_g\}}(\{b_g\}) = \prod_g q_{b_g}(b_g) = \prod_g \text{Bernoulli}\left(\frac{\exp(\hat{\rho}_{1g})}{\exp(\hat{\rho}_{1g}) + \exp(\hat{\rho}_{0g})}\right)$, $q_\tau(\tau) = N(\widehat{M_\tau}, \widehat{V_\tau})$, $q_p(p) = \text{Beta}(\hat{\alpha}_1, \hat{\alpha}_0)$, $q_\psi(\psi) = N(\widehat{M_\psi}, \widehat{V_\psi})$, and $q_{\sigma^2}(\sigma^2) = \text{IG}(\hat{a}, \hat{b})$. Optimal q-densities are found by employing the following iterative scheme for computing variational parameters.

1. Set $\hat{a} = \frac{G}{2} + a_0$. Initialize $\widehat{M_{\sigma^{-2}}} = 1$, $\hat{\rho}_{0g} = \hat{\rho}_{1g} = 0$ for each g, and

$$\widehat{M_{b_g}} = \begin{cases} 1 & \text{ifrank}(d_g) \geq 0.9G \\ 0 & \text{otherwise} \end{cases} \text{foreach} g,$$

$$\widehat{M_\psi} = \left| \sum_{g=1}^{G} d_g - \sum_{\{g:\text{rank}(d_g) \geq 0.9G\}} d_g \right|.$$

2. Update

$$\widehat{M_\tau} \leftarrow \frac{\sum_g \widehat{M_{\sigma^{-2}}} \left((1 - \widehat{M_{b_g}}) d_g + \widehat{M_{b_g}} (d_g - \widehat{M_\psi}) \right)}{G\widehat{M_{\sigma^{-2}}} + \frac{1}{\sigma_{\tau_0}^2}}$$

$$\widehat{M_\psi} \leftarrow \frac{\sum_g \widehat{M_{\sigma^{-2}}} \widehat{M_{b_g}} (d_g - \widehat{M_\tau})}{\sum_g \widehat{M_{b_g}} \widehat{M_{\sigma^{-2}}} + \frac{1}{\sigma_{\psi_0}^2}}$$

$$\hat{b} \leftarrow \frac{1}{2} \sum_g \left(\frac{1}{G\widehat{M_{\sigma^{-2}}} + \frac{1}{\sigma_{\tau_0}^2}} + \frac{\widehat{M_{b_g}}}{\sum_g \widehat{M_{b_g}} \widehat{M_{\sigma^{-2}}} + \frac{1}{\sigma_{\psi_0}^2}} \right)$$
$$+ \frac{1}{2} \sum_g \left((1 - \widehat{M_{b_g}})(d_g - \widehat{M_\tau})^2 + \widehat{M_{b_g}} (d_g - \widehat{M_\tau} - \widehat{M_\psi})^2 \right) + b_0$$

$$\widehat{M_{\sigma^{-2}}} \leftarrow \frac{\hat{a}}{\hat{b}}$$

$$\hat{\rho}_{1g} \leftarrow \frac{-\widehat{M_{\sigma^{-2}}}}{2} \left(\frac{1}{G\widehat{M_{\sigma^{-2}}} + \frac{1}{\sigma_{\tau_0}^2}} + \frac{1}{\sum_g \widehat{M_{b_g}} \widehat{M_{\sigma^{-2}}} + \frac{1}{\sigma_{\psi_0}^2}} + (d_g - \widehat{M_\tau} - \widehat{M_\psi})^2 \right)$$
$$+ \widehat{\log p}$$

$$\hat{\rho}_{0g} \leftarrow \frac{-\widehat{M_{\sigma^{-2}}}}{2} \left(\frac{1}{G\widehat{M_{\sigma^{-2}}} + \frac{1}{\sigma_{\tau_0}^2}} + (d_g - \widehat{M_\tau})^2 \right) + \widehat{\log(1-p)}$$

$$\widehat{M_{b_g}} \leftarrow \frac{\exp(\hat{\rho}_{1g})}{\exp(\hat{\rho}_{1g}) + \exp(\hat{\rho}_{0g})}.$$

3. Repeat (2) until the increase in

$$
\begin{aligned}
C_q(\mathbf{y}) = \sum_g \Bigg\{ & -\frac{1}{2}\log(2\pi) - \frac{\widehat{M_{\sigma^{-2}}}}{2} \times (1-\widehat{M_{b_g}}) \left(\frac{1}{G\widehat{M_{\sigma^{-2}}} + \frac{1}{\sigma_{\tau 0}^2}} + (d_g - \widehat{M_\tau})^2 \right) \\
& - \frac{\widehat{M_{\sigma^{-2}}}}{2} \times \widehat{M_{b_g}} \left(\frac{1}{G\widehat{M_{\sigma^{-2}}} + \frac{1}{\sigma_{\tau 0}^2}} + \frac{1}{\sum_g \widehat{M_{b_g}}\widehat{M_{\sigma^{-2}}} + \frac{1}{\sigma_{\psi 0}^2}} \right. \\
& \left. + (d_g - \widehat{M_\tau} - \widehat{M_\psi})^2 \right) \Bigg\} \\
& - \sum_g \left\{ \widehat{M_{b_g}} \log \widehat{M_{b_g}} + (1-\widehat{M_{b_g}}) \log(1-\widehat{M_{b_g}}) \right\} \\
& + \log\left(\mathrm{Beta}\left(\sum_g \widehat{M_{b_g}} + \alpha_1, \sum_g (1-\widehat{M_{b_g}}) + \alpha_0 \right) \right) - \log\left(\mathrm{Beta}(\alpha_1, \alpha_0)\right) \\
& + \frac{1}{2} \left(\log\left(\frac{1}{G\widehat{M_{\sigma^{-2}}} + \frac{1}{\sigma_{\tau 0}^2}} \right) - \log \sigma_{\tau 0}^2 + 1 \right. \\
& \left. - \frac{\frac{1}{G\widehat{M_{\sigma^{-2}}} + \frac{1}{\sigma_{\tau 0}^2}} + \left(\widehat{M_\tau} - \tau_0\right)^2}{\sigma_{\tau 0}^2} \right) \\
& + \frac{1}{2} \left(\log\left(\frac{1}{\sum_g \widehat{M_{b_g}}\widehat{M_{\sigma^{-2}}} + \frac{1}{\sigma_{\psi 0}^2}} \right) - \log \sigma_{\psi 0}^2 + 1 \right. \\
& \left. - \frac{\frac{1}{\sum_g \widehat{M_{b_g}}\widehat{M_{\sigma^{-2}}} + \frac{1}{\sigma_{\psi 0}^2}} + \left(\widehat{M_\psi} - \psi_0\right)^2}{\sigma_{\psi 0}^2} \right) \\
& + a_0 \log b_0 - \log \Gamma(a_0) - \frac{b_0 \hat{a}}{\hat{b}} - \hat{a}\log(\hat{b}) + \log\Gamma(\hat{a}) + \hat{a}
\end{aligned}
$$

from previous iteration becomes negligible.

4. Upon convergence, the remaining variational parameters are computed:

$$\widehat{V_\tau} \leftarrow \frac{1}{G\widehat{M_{\sigma^{-2}}} + \frac{1}{\sigma_{\tau 0}^2}} \quad \text{and} \quad \widehat{V_\psi} \leftarrow \frac{1}{\sum_g \widehat{M_{b_g}}\widehat{M_{\sigma^{-2}}} + \frac{1}{\sigma_{\psi 0}^2}}.$$

Here, we set starting values for $\widehat{M_{b_g}}$ such that $\widehat{M_{b_g}} = 1$ for genes that correspond to the top 10% of d_g values and $\widehat{M_{b_g}} = 0$ otherwise. Classification with a different set of starting values such that $\widehat{M_{b_g}} = 1$ for genes that correspond to the top 5% and bottom 5% of d_g values and $\widehat{M_{b_g}} = 0$ otherwise leads to almost the same results. In practice, we recommend using the "top-5%-bottom-5%" scheme of setting starting values for $\widehat{M_{b_g}}$ in the VB algorithm for classification with little prior information on location of the mixture component distributions rather than randomly generating values for $\widehat{M_{b_g}}$—empirical evidence suggests that VB with randomly generated labels does not produce reasonable solutions.

7.2.5 Marginal Posterior Approximation

To investigate we simulated 20,000 values from the simple two-component model (1) with true values $\tau = 0$, $\psi = 20$, $p = 0.2$, $b_g \sim \text{Bernoulli}(p)$ *i.i.d.* for each g, and $\sigma^2 = 36$. In the Bayesian analysis τ and ψ were assigned $N(0, 100)$ priors, σ^2 an $\text{IG}(0.1, 0.1)$ prior, and p a $\text{Beta}(0.1, 0.9)$ prior. Starting values and priors used in VB, MCMC, and the base VB of GBVA are kept the same for comparison. MCMC was implemented in WinBUGS via R with 1 chain of length 20,000, a burn-in 15,000, and a thinning factor of 10. Figure 7.1 shows that VB-approximated marginal posterior densities closely match the MCMC-approximates. In particular, the posterior mean of $\{b_g\}$ estimated by VB, when plotted against posterior mean of $\{b_g\}$ estimated by MCMC, almost coincides with the intercept zero slope one reference line, indicating little difference between VB and MCMC in terms of classification. In fact, with 3972 simulated non-null genes and a 0.8 cutoff for classification, MCMC identified 3303 genes as non-null with true positive rate 0.803 and false positive rate 0.00705, and VB also detected 3291 genes as non-null with true positive rate 0.800 and the same false positive rate as MCMC. Moreover, on an Intel Core i5-2430M 2.40 GHz, 6 GB RAM computer, it took VB 1.26 s to reach convergence with a 10^{-6} error tolerance, whereas MCMC procedure took about 13 min. Both methods are suitable for classification and posterior inference, with VB demonstrating an advantage in speed, in this example.

The grid-based variational approximations (GBVA) method detailed in Ormerod (2011) was developed to correct for over-confidence in VB densities. The method involves running a base VB algorithm and subsequent VB algorithms for density approximation with the aid of numerical interpolation and integration. For each marginal posterior density of interest, re-applying VB over a grid of values in GBVA

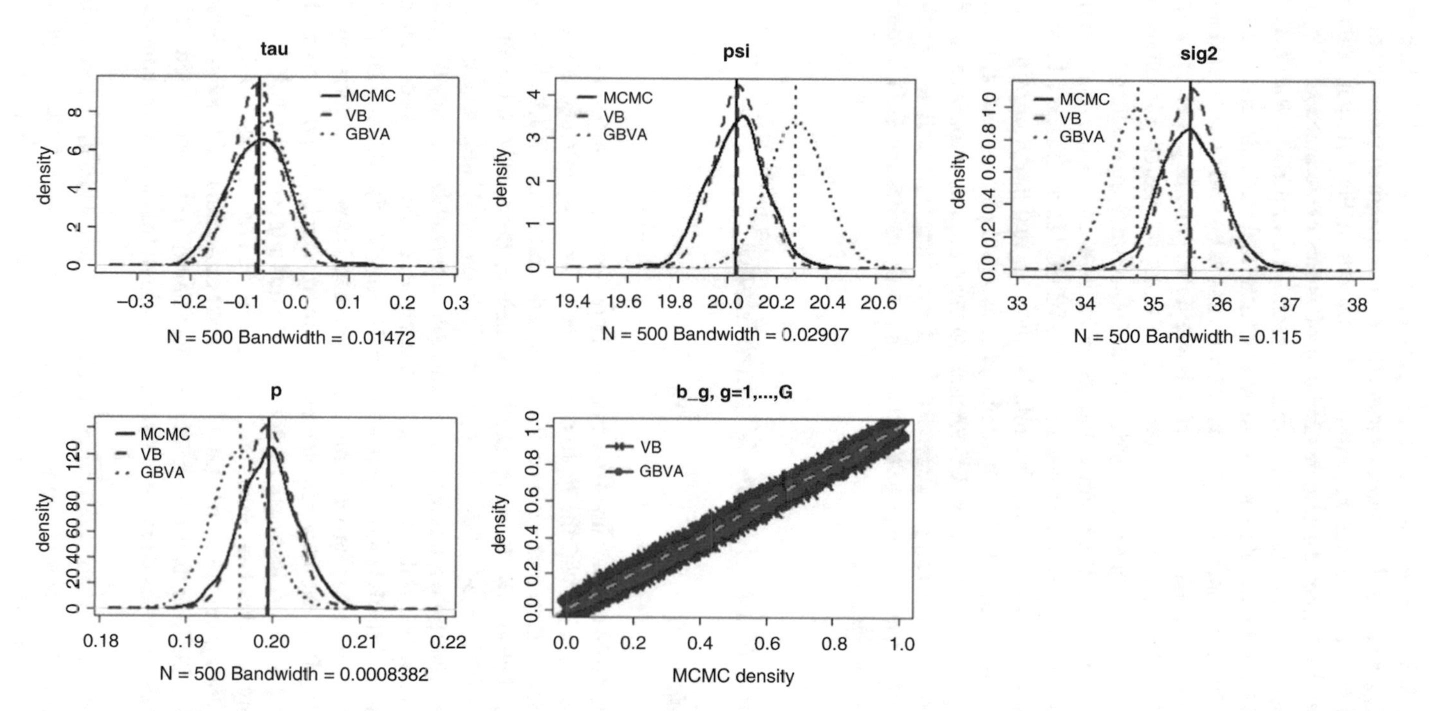

Fig. 7.1 Approximate marginal posteriors given by MCMC, VB and GBVA for simulated data under the simple two-component mixture model (1). In the last plot, the posterior mean of b_g, $g = 1, \ldots, G$ is compared between VB and MCMC, and between GBVA and MCMC

is similar in spirit to re-applying Laplace approximation in the INLA (Rue et al. 2009) method, and is the key step towards refining the marginal posterior density approximated by the base VB algorithm. We observe in Fig. 7.1 that, although GBVA appears to correct for underestimation of posterior variance of VB, loss in accuracy occurs in estimation of posterior mean in comparison with MCMC. Experiments with grid selection and precision of numerical methods did not correct the shift in posterior mean, and a systematic error caused by grid-based calculations of the unnormalized density is suspected. Although the shifts in posterior mean of GBVA densities observed in Fig. 7.1 are small in scale, for posterior inference the mismatch in mean may incur more serious loss in accuracy than underestimation of posterior variance of VB densities. Moreover, in the example, the total run time of GBVA (almost 6 h) is even larger than that of MCMC (13 min). The undesirable speed of GBVA is due to the fact that refining the marginal posterior of the latent variable $\{b_g\}$ must go through two iterations, i.e. two grid points 0 and 1, for each one of the G genes. Therefore, in GBVA implementation, this step alone requires $2 \times G$ iterations. This example suggests that for a "large p small n" problem, the approach of calculation over grid values can easily render the GBVA method computationally infeasible.

7.3 VB for a General Finite Mixture Model

7.3.1 Motivation

For large-scale problems that involve complex models on high-dimensional data, it is natural to consider MCMC for approximation of notoriously intractable marginal posterior densities. However, in many cases implementation of MCMC is impractical. MCMC, because of its sampling-based nature, becomes cumbersome for hierarchical "large p small n" models where high-dimensional unknown quantities on various hierarchies are inter-dependent. To achieve desirable accuracy of MCMC therefore requires (multiple) long chains. Yet convergence of high-dimensional MCMC chains is difficult to monitor (Cowles and Carlin 1996), and optimal MCMC chain settings often have to be determined based on heuristics. Moreover, in mixture models, an MCMC chain may produce evaluation of a marginal posterior density that is the same as other chains but associated with different component labels. The label-switching issue has been noted, for example in De Freitas et al. (2001), Marin and Robert (2007), and Grimmer (2011). Without correction of label-switching, misleading classification may arise in cases where multiple chains are used and the MCMC solution, taken as an average over chains with different or even contrasting labels, is wrong.

A common strategy for improving efficiency in approximation problems is to make simplifying assumptions so that plausible solutions for the complicated problem are obtained with satisfactory speed and accuracy. This is the motivation behind implementing VB for approximation of marginal posterior densities for finite mixture models in sparse classification problems. In this context, VB is promising in achieving comparable accuracy to MCMC given limited computational resources, with easy-to-monitor convergence and no label-switching due to its deterministic nature, in scenarios where MCMC implementation is unattractive.

7.3.2 The B-LIMMA Model

During the last decade, technological advances in molecular biology have led researchers toward high-throughput genomic studies, the simultaneous measurement of the expression levels of tens of thousands of genes has become a mainstay of biological and biomedical research. The use of microarrays to discover genes differentially expressed between two or more groups has been applied widely and has made fundamental breakthroughs in biomedical research possible.

The first approach to identify differentially expressed genes was fold-change estimation, however this approach did not take sampling variability into account. The t-test is likely the most popular naive method for assessing differential expression. Computing a t-statistic can be problematic because the variance estimates can be skewed by genes having a very low variance and its application for small sample sizes gives low statistical power. Consequently, the efficacy of a t-test along with the importance of variance modeling is problematic and has led to the development of many alternatives.

We illustrate a VB implementation for a general finite mixture models using a fully Bayesian version of the LIMMA model introduced by Smyth (2004). The LIMMA model is designed for testing significance of differential gene expression in microarray experiments via individual moderated t-tests and posterior log odds ratios. Model parameters can be estimated via an empirical Bayes approach, from which closed form test-statistics can be obtained. For a two-sample experimental design, let y_{ijg} denote the normalized log expression of gene g in sample j from treatment group i, where $g = 1, \ldots, G$, $i = 1, 2$, and $j = 1, \ldots, n_{ig}$. Then $d_g = \bar{y}_{2.g} - \bar{y}_{1.g}$ represents the observed differential expression for gene g, and $m_g = \sum_i \sum_j^{n_{ig}} (y_{ijg} - \bar{y}_{i.g})^2 / f_g$, where $f_g = n_{1g} + n_{2g} - 2$, is the mean squared error. A fully Bayesian version of the LIMMA model in this context is the following hierarchical mixture model:

$$d_g | b_g, \psi_g, \sigma^2_{\epsilon,g}, \tau = \tau + b_g \psi_g + \epsilon_g, \tag{7.4}$$

$$m_g | \sigma^2_{\epsilon,g} \sim \frac{\sigma^2_{\epsilon,g} \chi^2_{f_g}}{f_g}, \quad \text{where} f_g := n_{1g} + n_{2g} - 2,$$

$$\psi_g | \nu, \sigma^2_{\epsilon,g} \sim N(0, \nu \sigma^2_{\epsilon,g}),$$

$$\epsilon_g|\sigma^2_{\epsilon,g} \sim N(0, \sigma^2_g), \quad \text{where} \sigma^2_g := \sigma^2_{\epsilon,g} c_g, \quad c_g := \frac{1}{n_{1g}} + \frac{1}{n_{2g}},$$

$$b_g|p \sim \text{Bernoulli}(p) \quad i.i.d., \qquad \sigma^2_{\epsilon,g} \sim \text{IG}(A_\epsilon, B_\epsilon) \ i.i.d.,$$

$$\tau \sim N\left(\mu_{\tau_0}, \sigma^2_{\tau_0}\right), \qquad \nu \sim \text{IG}(A_\nu, B_\nu), \qquad p \sim \text{Bernoulli}(\alpha_1, \alpha_0).$$

In this model τ represents the overall mean treatment difference, ψ_g is the gene-specific effect of the treatment, and b_g is a latent indicator which takes the value 1 if gene g is non-null, i.e. is differentially expressed in the two treatment groups.

We refer to the above model as the B-LIMMA model. Derivation of the VB algorithm for the B-LIMMA model, which we refer to as VB-LIMMA, is as follows.

The set of observed data and the set of unobserved data are identified as $\mathbf{y}$ and H, respectively. For the B-LIMMA model, $\mathbf{y} = \{\{d_g\}, \{m_g\}\}$ and $H = \{\{b_g\}, \{\psi_g\}, \{\sigma^2_g\}, \tau, \nu, p\}$. The complete data log likelihood is

$$\log p(\mathbf{y}, H) = \sum_g \log p(d_g|b_g, \psi_g, \sigma^2_g, \tau) + \sum_g \log p(m_g|\sigma^2_g) + \sum_g \log p(\psi_g|\nu, \sigma^2_g)$$
$$+ \sum_g \log p(b_g|p) + \log p(\tau) + \log p(\nu) + \sum_g \log p(\sigma^2_g) + \log p(p).$$

In particular, the observed data log likelihood involves

$$\log p(d_g|b_g, \psi_g, \sigma^2_g, \tau) = \frac{1}{2} \log\left(2\pi\sigma_g^{-2}\right)$$
$$- \frac{1}{2\sigma^2_g}\left[(1-b_g)(d_g-\tau)^2 + b_g(d_g-\tau-\psi_g)^2\right],$$

$$\log p(m_g|\sigma^2_g) = -\log\left(\frac{\sigma^2_g}{c_g f_g}\right) - \log\left(2^{\frac{f_g}{2}}\Gamma\left(\frac{f_g}{2}\right)\right)$$
$$+ \left(\frac{f_g}{2} - 1\right)\log\left(\frac{m_g c_g f_g}{\sigma^2_g}\right) - \frac{1}{2}\left(\frac{m_g c_g f_g}{\sigma^2_g}\right).$$

The product density restriction is imposed such that $q(H) = q_{\{b_g\}}(\{b_g\}) \times q_{\{\psi_g\}}(\{\psi_g\}) \times q_{\{\sigma^2_g\}}(\{\sigma^2_g\}) \times q_{(\tau,\nu,p)}(\tau, \nu, p)$. Further factorizations are induced by applying (3):

$$q_{\{b_g\}}(\{b_g\}) = \prod_g q_{b_g}(b_g), \qquad q_{\{\psi_g\}}(\{\psi_g\}) = \prod_g q_{\psi_g}(\psi_g),$$

$$q_{\{\sigma^2_g\}}(\{\sigma^2_g\}) = \prod_g q_{\sigma^2_g}(\sigma^2_g), \quad \text{and} \quad q_{(\tau,\nu,p)}(\tau, \nu, p) = q_\tau(\tau) q_\nu(\nu) q_p(p).$$

Then, approximate marginal posterior distributions, with $\widehat{M_\theta}$ denoting the variational posterior mean $\int \theta q_\theta(\theta)d\theta$, and $\widehat{V_\theta}$ denoting the variational posterior variance $\int (\theta - \widehat{M_\theta})^2 q_\theta(\theta)d\theta$, are $q_\tau(\tau) = N(\widehat{M_\tau}, \widehat{V_\tau})$, $q_\nu(\nu) = \mathrm{IG}(\widehat{A_\nu}, \widehat{B_\nu})$, $q_p(p) = \mathrm{Beta}(\widehat{\alpha_1}, \widehat{\alpha_0})$, $q_{\psi_g}(\psi_g) = N(\widehat{M_{\psi_g}}, \widehat{V_{\psi_g}})$, $q_{\sigma_g^2}(\sigma_g^2) = \mathrm{IG}(\widehat{A_{\sigma_g^2}}, \widehat{B_{\sigma_g^2}})$, and $q_{b_g}(b_g) = \mathrm{Bernoulli}(\widehat{M_{b_g}})$. Updating the q-densities in an iterative scheme boils down to updating the variational parameters in the scheme. Convergence is monitored via the scalar quantity $C_q(\mathbf{y})$, the lower bound on the log of the marginal data density

$$\begin{aligned} C_q(\mathbf{y}) = E_{q(H)} \Bigg\{ & \sum_g \log p(d_g|b_g, \psi_g, \sigma_g^2, \tau) + \sum_g \log p(m_g|\sigma_g^2) \\ & + \sum_g \log p(\sigma_g^2) - \sum_g \log q(\sigma_g^2) + \log p(\tau) - \log q(\tau) \\ & + \sum_g \log p(\psi_g|\nu, \sigma_g^2) + \log p(\nu) - \sum_g \log q(\psi_g) - \log q(\nu) \\ & + \sum_g \log p(b_g|p) + \log p(p) - \sum_g \log q(b_g) - \log q(p) \Bigg\}. \end{aligned}$$

It is only necessary to update the variational posterior means $\widehat{M}.$ in the iterative scheme. Upon convergence, the other variational parameters are computed based on the converged value of those involved in the iterations. Therefore, the VB-LIMMA algorithm consists of the following steps:

Step 1: Initialize $\widehat{A_\nu}, \widehat{B_\nu}, \{\widehat{A_{\sigma_g^2}}\}, \{\widehat{B_{\sigma_g^2}}\}, \{\widehat{M_{\psi_g}}\}, \{\widehat{M_{b_g}}\}$.

Step 2: Cycle through $\widehat{M_\tau}, \widehat{B_\nu}, \{\widehat{B_{\sigma_g^2}}\}, \{\widehat{M_{\psi_g}}\}, \{\widehat{M_{b_g}}\}$ iteratively, until the increase in $C_q(\mathbf{y})$ computed at the end of each iteration is negligible.

Step 3: Compute $\widehat{V_\tau}, \{\widehat{V_{\psi_g}}\}, \widehat{\alpha_1}, \widehat{\alpha_0}$ using converged variational parameter values.

7.4 Numerical Illustrations

7.4.1 Simulation

The goal of this simulation study is to assess the performance of VB as a classifier when various model assumptions are correct, and to determine the accuracy of VB approximation of marginal posterior distributions.

7.4.1.1 The B-LIMMA Model

Under the B-LIMMA model (4), the difference in sample means $\{d_g\}$ and mean squared errors $\{m_g\}$ are simulated according to model (1) of Bar et al. (2010) with $\mu = 0$ and $\gamma_g \equiv 0$. VB-LIMMA approximates a marginal posterior density whose mean is then used as posterior estimate of the corresponding parameter for classification. Performance comparison is made between VB implemented in R and the widely used `limma` (Smyth 2005) package.

A data set containing $G = 5000$ genes was simulated under the B-LIMMA model with $n_{1g} \equiv n_{2g} \equiv 8, \tau = 0, \sigma_g^2 \sim \text{IG}(41, 400)$ *i.i.d.*, and ψ_g and b_g generated as specified in model (4). The shape and scale parameter values in the Inverse-Gamma distribution were chosen such that $\sigma_g^2, g = 1, \ldots, G$, were generated with mean 10 and moderate variation among the G genes. We set p, the non-null proportion, and ν, the variance factor such that $p \in \{0.05, 0.25\}, \nu \in \{\frac{1}{2}, 2\}$. Simulations with various sets of parameter values were investigated, but these values lead to representative results among our experiments. For classification, a $N(0, 100)$ prior for τ, IG(0.1, 0.1) prior for ν and for $\sigma_{\epsilon,g}^2$, and a Beta(1, 1) prior for the non-null proportion p were used. In the VB-LIMMA algorithm, the error tolerance was set to be 10^{-6}. The starting values were $\widehat{A_\nu} = \widehat{B_\nu} = 0.1, \sigma_{\epsilon,g}^2 \equiv 1, \widehat{M_{\psi_g}} \equiv 0$, and $\widehat{M_{b_g}} = 1$ for genes that are associated with the 5% largest values of d_g or the 5% smallest values of d_g, and $\widehat{M_{b_g}} = 0$ otherwise. For each gene g, $\widehat{M_{b_g}}$ was the VB-approximated posterior mean of the latent indicator b_g, i.e. the gene-specific posterior non-null probability. Hence, for detection of non-null genes, gene ranking based on $\widehat{M_{b_g}}$ can be compared with gene ranking based on the B-statistic, or equivalently gene-specific posterior non-null probability computable from the B-statistic, produced by `limma`.

We note that, varying the value of ν while keeping $\sigma_{\epsilon,g}^2$ unchanged in simulation varies the level of difficulty of the classification problem based on the simulated data. This is because, the B-LIMMA model specifies that the null component $N(\tau, \sigma_{\epsilon,g}^2 c_g)$ differs from the non-null component $N(\tau, \sigma_{\epsilon,g}^2 c_g + \nu\sigma_{\epsilon,g}^2)$ only in variance, and this difference is governed by the multiplicative factor ν. Therefore, data simulated with large ν tends to include non-null genes that are far away from mean τ and thus clearly distinguishable from null genes. The less overlap of the non-null and null component distributions, the easier the task of sparse classification for either method. This is reflected in improved ROC curves for both methods as ν is changed from 0.5 to 2 in Figs. 7.2 and 7.3. In fact, in Figs. 7.2 and 7.3 we also see that for simulated data with $p \in \{0.05, 0.25\}, \nu \in \{\frac{1}{2}, 2\}$, classification by VB-LIMMA is almost identical to classification by `limma` in terms of ROC curve performance. A cutoff value is defined such that if $\widehat{M_{b_g}} \geq$ cutoff then gene g is classified as non-null. Accuracy is defined as $\frac{TP+TN}{P+N}$ and false discovery rate (FDR) as $\frac{FP}{TP+FP}$, as in Sing et al. (2007). Focusing on a practical range of cutoff values, i.e. (0.5, 1), in Figs. 7.2 and 7.3, we see that accuracy and FDR of VB are both close to those of `limma`, and that VB has higher accuracy than `limma`. In summary,

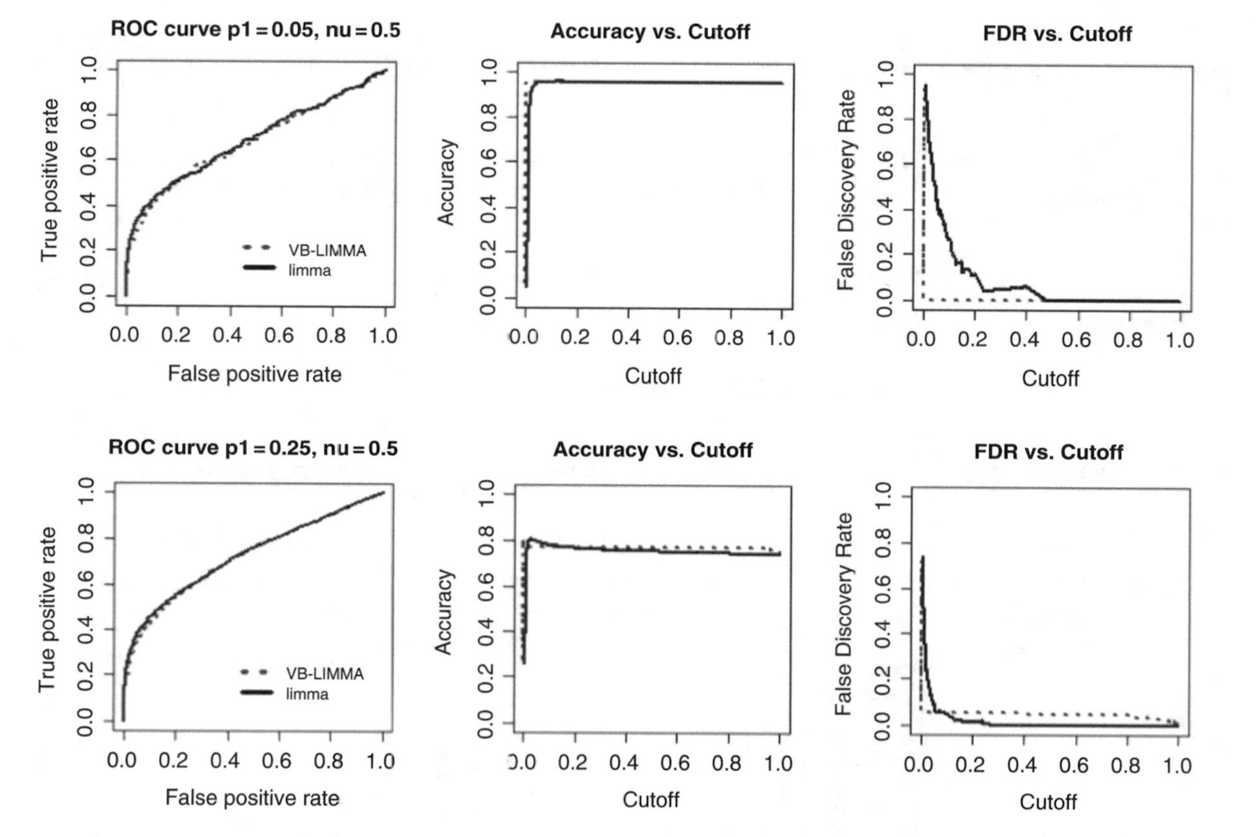

Fig. 7.2 `limma` and VB-LIMMA classification comparison on simulated two-sample data with $G = 5000, n_{1g} \equiv n_{2g} \equiv 8, \tau = 0, p \in \{0.05, 0.25\}, \nu = \frac{1}{2}, \sigma_g^2 \sim \text{IG}(41, 400)$ (*i.i.d.*)

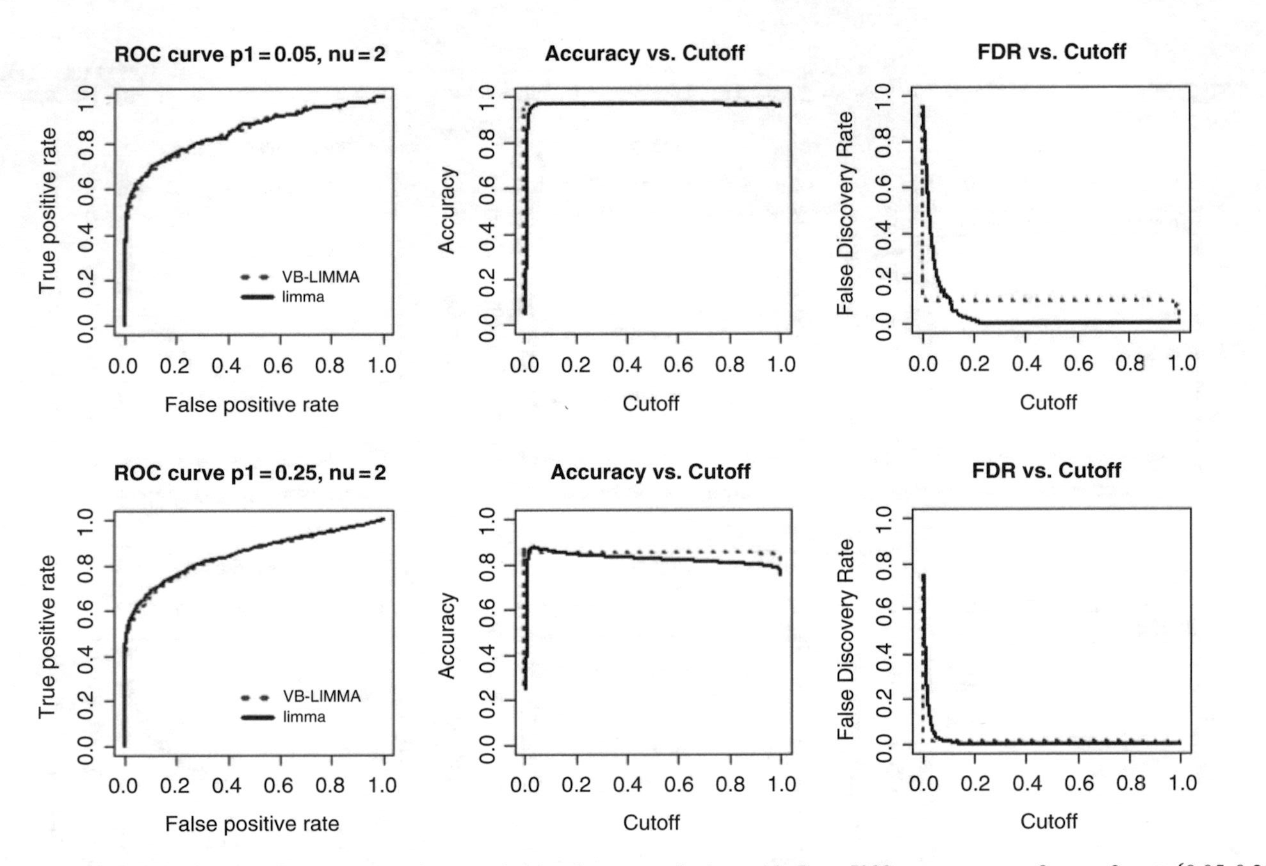

Fig. 7.3 `limma` and VB-LIMMA classification comparison on simulated two-sample data with $G = 5000, n_{1g} \equiv n_{2g} \equiv 8, \tau = 0, p \in \{0.05, 0.25\}, \nu = 2, \sigma_g^2 \sim \text{IG}(41, 400)$ (*i.i.d.*)

our experiment suggests that VB-LIMMA, the fully Bayesian classifier, acts as a comparable classifier to `limma`.

7.4.1.2 A Mixture Model Extended from the LIMMA Model

The two-component mixture model in Bar et al. (2010) differs from the LIMMA model in the assumption on the gene-specific treatment effect ψ_g, specifically

$$\psi_g|\psi, \sigma_\psi^2 \sim N(\psi, \sigma_\psi^2) \quad (i.i.d.) \tag{7.5}$$

where for the random effect the mean ψ is allowed to be non-zero and the variance σ_ψ^2 is assumed to be independent of the gene-specific error variance. This model is referred to as the LEMMA model in Bar et al. (2010). The LEMMA and LIMMA models both facilitate simultaneous testing of treatment effects on a large number of genes by "borrowing strength" across the genes. In Bar et al. (2010), an empirical Bayes approach is adopted in which parameter estimation of the prior on the error variance $\sigma_{\epsilon,g}^2$ is accomplished through maximum marginal likelihood and point estimates of the global parameters are obtained via the EM algorithm. In the EM algorithm, evaluation of the complete data likelihood involves integrating over the prior on the error variance $\sigma_{\epsilon,g}^2$ which is achieved via a Laplace approximation. In what follows we consider a natural extension to the LEMMA model: a fully Bayesian three-component mixture model, B-LEMMA. Formulation of the B-LEMMA model and details of the corresponding VB-LEMMA algorithm are similar to model (4) and what is outlined in Sect. 7.3.2, and are included in Appendix.

The computational method in Bar et al. (2010) for estimation via Laplace approximation and EM algorithm is implemented in the `lemma` R package (Bar and Schifano 2010). We consider an MCMC procedure for the B-LEMMA model so that the performance of VB-LEMMA for classification is assessed through comparison with MCMC-LEMMA with reference to `lemma`.

A data set containing $G = 5000$ genes was simulated under the B-LEMMA model using $n_{1g} \equiv 6, n_{2g} \equiv 8, \tau = 0, p_1 = p_2 = 0.1, \psi = 20, \sigma_\psi^2 = 2, \sigma_g^2 \sim \text{IG}(41, 400)$ *i.i.d.*, and ψ_g was generated based on (5). In the VB-LEMMA algorithm, the error tolerance was set to be 10^{-6}. Starting values were set under the same scheme as the VB-LIMMA algorithm, except that the parameter ν was replaced by ψ. The posterior mean of ψ, $\widehat{M_\psi}$, was initialized as the average of the absolute difference between the mean of d_g and the mean of the 5% largest d_g, and the absolute difference between the mean of d_g and the mean of the 5% smallest d_g. Non-informative independent priors were assigned to the global parameters, so that τ and ψ each had a $N(0, 100)$ prior, σ_ψ^2 and σ_g^2 each had an IG(0.1, 0.1) prior, and $(p_1, p_2, 1 - p_1 - p_2)$ had a Dirichlet(1, 1, 1) prior. For comparison of classification performance, we used the same prior distributions and starting values in MCMC-LEMMA as in VB-LEMMA. MCMC-LEMMA was run with 1 chain of length 1,200,000, a burn-in period 1,195,000, and a thinning factor 10. We made this choice of chain length to ensure chain convergence after experimenting with various chain settings.

Because point estimates suffice for classification, we present VB-LEMMA and MCMC-LEMMA classification results using posterior means computed from the approximate posterior marginal densities. Genes are classified with a prescribed cutoff value. That is, based on the mean from the approximate marginal posterior of (b_{1g}, b_{2g}) outputted by the VB-LEMMA algorithm, $\widehat{M_{b_{1g}}} \geq$ cutoff $\Longleftrightarrow$ gene g belongs to non-null 1 component, and $\widehat{M_{b_{2g}}} \geq$ cutoff $\Longleftrightarrow$ gene g belongs to non-null 2 component. The posterior mean of the (b_{1g}, b_{2g}) sample for any gene g is calculable from the MCMC samples and is used to classify gene g with the same cutoff. Point estimates outputted by `lemma` run with default starting values are used as a reference for accuracy of point estimates.

On an Intel Core i5-2430M 2.40 GHz, 6 GB RAM computer, it took `lemma`, VB-LEMMA, and MCMC-LEMMA 20.94 s, 29.24 s, and 25,353.88 s (7 h) to run, respectively. Table 7.1 shows the `lemma` point estimates and the approximate marginal posterior means from VB-LEMMA and from MCMC-LEMMA. The parameter estimates from the three methods are all close to the true values except for σ_ψ^2. The disagreement between the MCMC-LEMMA and VB-LEMMA estimates of σ_ψ^2 is also observable in Fig. 7.4 where the marginal posterior density plots are displayed. In Fig. 7.4, the posterior means of VB-LEMMA and MCMC-LEMMA agree well with each other and with `lemma`, since the vertical lines almost coincide for all parameters except for σ_ψ^2. Deviation of the MCMC-LEMMA estimate of σ_ψ^2 6.993 from the true value 2 renders chain convergence still questionable, despite the long chain we use. Yet we see in Fig. 7.5, which contains the mixture and component densities plots, that VB-LEMMA and MCMC-LEMMA appear to identify the correct non-null 1 and non-null 2 genes since the inward ticks and outward ticks on the x-axis locate at where the two non-null components truly are in the second and fourth plots. True positive rate, TPR, is defined as $\frac{TP1+TP2}{P1+P2}$ and accuracy defined as $\frac{TP1+TP2+TN}{P1+P2+N}$, where $TP1$, $TP2$, TN, $P1$, $P2$, and N are, respectively, number of correctly labeled non-null 1 genes, number of correctly labeled non-null 2 genes, number of correctly labeled null genes, number of true non-null 1 genes, number of true non-null 2 genes, and number of true null genes, such that in this example $P1+P2+N = G$. Figure 7.5 and Table 7.1 confirm that VB-LEMMA and MCMC-LEMMA classify the genes correctly with high true positive rate and accuracy on the simulated data.

The reason why we implement MCMC-LEMMA with only one chain is that we wish to avoid the label-switching issue noted in Sect. 7.3.1. For MCMC-LEMMA with one chain, should label-switching occur, ψ would have the opposite sign and the non-null proportions p_1 and p_2 would exchange positions, leading to classification results in which the labels of non-null components 1 and 2 are switched. Thus, with a single chain, the label switching issue is easily resolved. If multiple chains of ψ are run, some will have switched labels. The average of the MCMC simulated samples would then be misleading. Although using one chain requires a much longer chain than when multiple chains are used for reaching

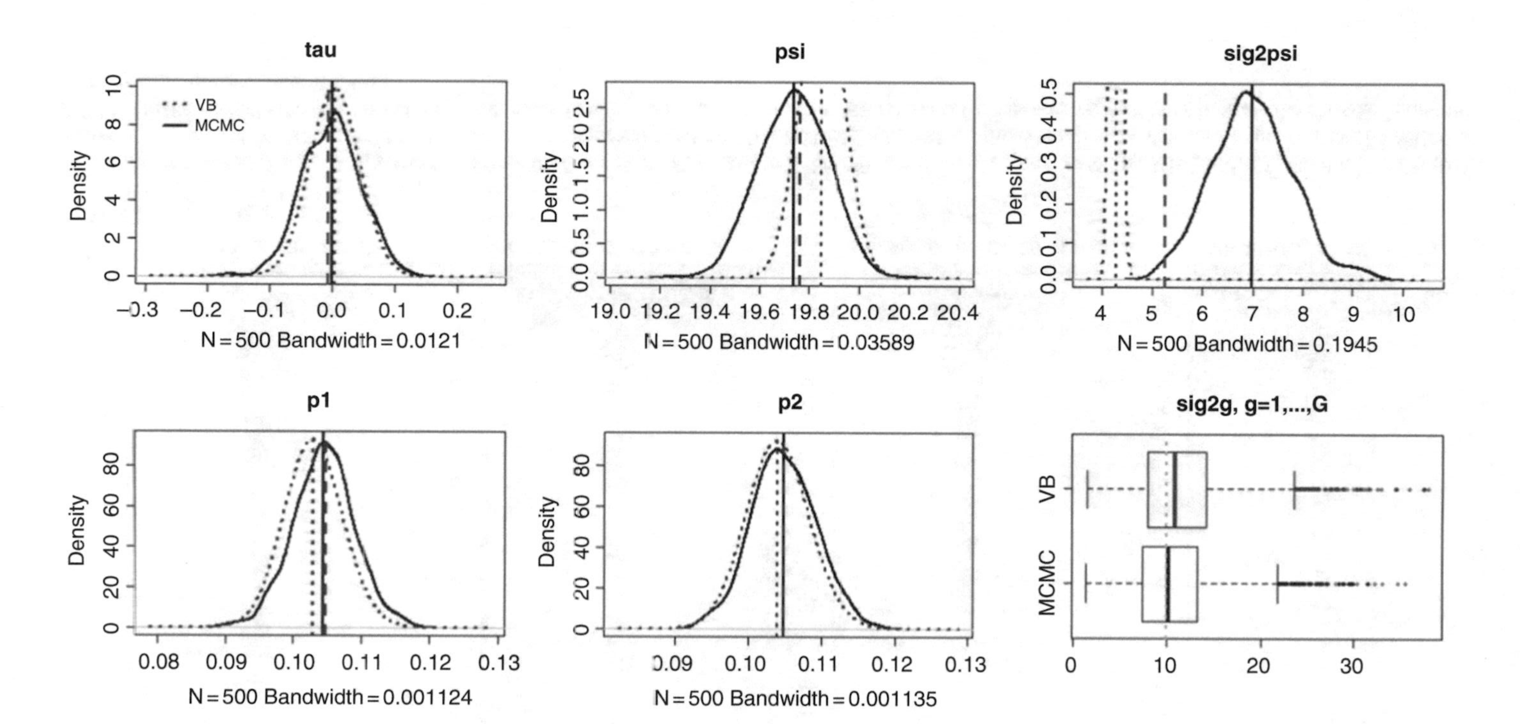

Fig. 7.4 Marginal posterior plots approximated by VB-LEMMA and by MCMC-LEMMA, for the parameters in the B-LEMMA model. For each parameter, the dotted and solid vertical lines represent the estimated posterior mean by VB and by MCMC respectively, and the dashed vertical line marks the estimated parameter value given by `lemma`. The bottom right plot shows the posterior mean of the gene-specific error variance σ_g^2 for each gene g, estimated by VB-LEMMA and by MCMC-LEMMA, and how this quantity varies across the G genes. The dotted line indicates the prior mean 10 of σ_g^2 since the prior distribution of σ_g^2 are *i.i.d.* IG(41, 400)

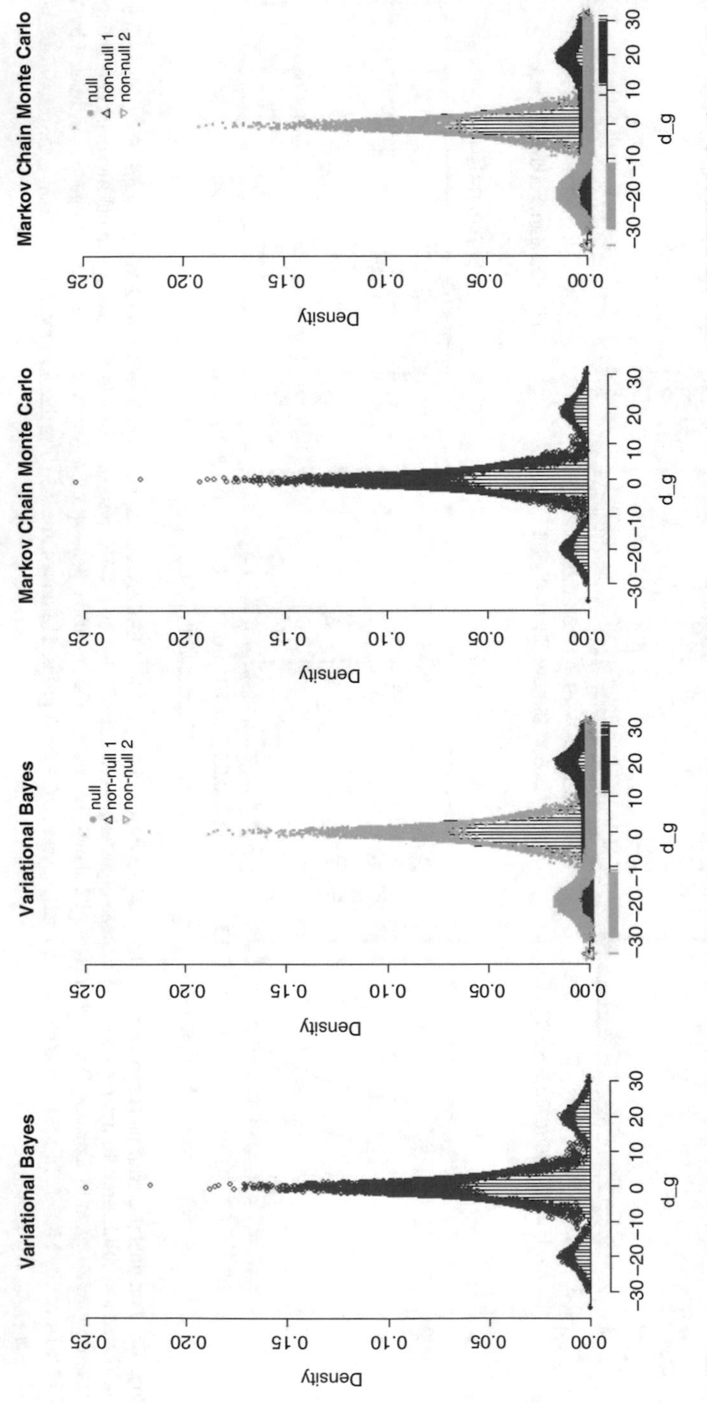

Fig. 7.5 Approximated densities on simulated data with $G = 5000$. From left to right: mixture density approximated by VB-LEMMA, the three component densities approximated by VB-LEMMA, mixture density approximated by MCMC-LEMMA, and the three component densities approximated by MCMC-LEMMA. On the x-axis in the second and the fourth plots, inward ticks indicate the classified non-null 1 genes and outward ticks indicate the classified non-null 2 genes by the corresponding method

Table 7.1 Posterior estimates, deduced number of non-null genes, and TPR and accuracy for `lemma`, VB-LEMMA, and MCMC-LEMMA on simulated B-LEMMA data with $G = 5000, n_{1g} \equiv 6, n_{2g} \equiv 8, \sigma_g^2 \sim \text{IG}(41, 400)$ (*i.i.d.*) and cutoff= 0.8

	True	`lemma`	VB-LEMMA	MCMC-LEMMA
τ	0	−0.00650	0.00432	−0.000682
ψ	20	19.764	19.850	19.737
σ_ψ^2	2	5.247	4.278	6.993
p_1	0.1	0.105	0.103	0.105
p_2	0.1	0.105	0.104	0.105
Number of genes in non-null 1	525	520	508	510
Number of genes in non-null 2	524	516	518	518
TPR	NA	0.987	0.977	0.978
Accuracy	NA	0.997	0.995	0.995

convergence, it allows us to distinguish label-switching from non-convergence and easily correct for label-switching in MCMC-LEMMA output.

One approach to address label-switching issue in mixture models is post-processing of MCMC solutions. In the B-LEMMA classification problem, once label-switching is detected, we can post-process MCMC-LEMMA classification results by giving the opposite sign to the MCMC sample of ψ, switching the p_1 sample for the p_2 sample and vice versa, and exchanging the non-null 1 and non-null 2 statuses. To make Fig. 7.6, where performance of VB-LEMMA and MCMC-LEMMA was recorded for 30 simulated data sets, the data simulation settings and the settings for the two methods were kept unchanged from the previous experiment, and MCMC-LEMMA results were post-processed according to the aforementioned scheme wherever label-switching occurred. Posterior means of σ_ψ^2 estimated by MCMC-LEMMA persisted as worse over-estimates than those estimated by VB-LEMMA. Nonetheless, the component densities plots for the 30 simulated data sets (not shown) with gene labels all appear similar to those in Fig. 7.5, indicating that both VB-LEMMA and MCMC-LEMMA are reliable classifiers in this context. However, VB-LEMMA is far superior in terms of computational efficiency.

Another way of avoiding label-switching issue in mixture models is assigning a prior distribution on the parameter that appropriately restricts the sampling process in MCMC, as Christensen et al. (2011) point out. Although the mixture and component densities plots (not shown) show that MCMC-LEMMA, when implemented with a half-normal prior on ψ, classifies the genes satisfactorily, restricting the mean of ψ to a positive value changes the marginal data distribution of the B-LEMMA model. Therefore, direct comparisons with the modified MCMC-LEMMA are no longer valid because VB-LEMMA and MCMC-LEMMA with a half-normal prior on ψ are based on two different models.

These B-LEMMA simulation experiments indicate that VB-LEMMA efficiently and accurately approximates fully Bayesian estimation and classification results. In contrast, MCMC-LEMMA, although theoretically capable of achieving extremely

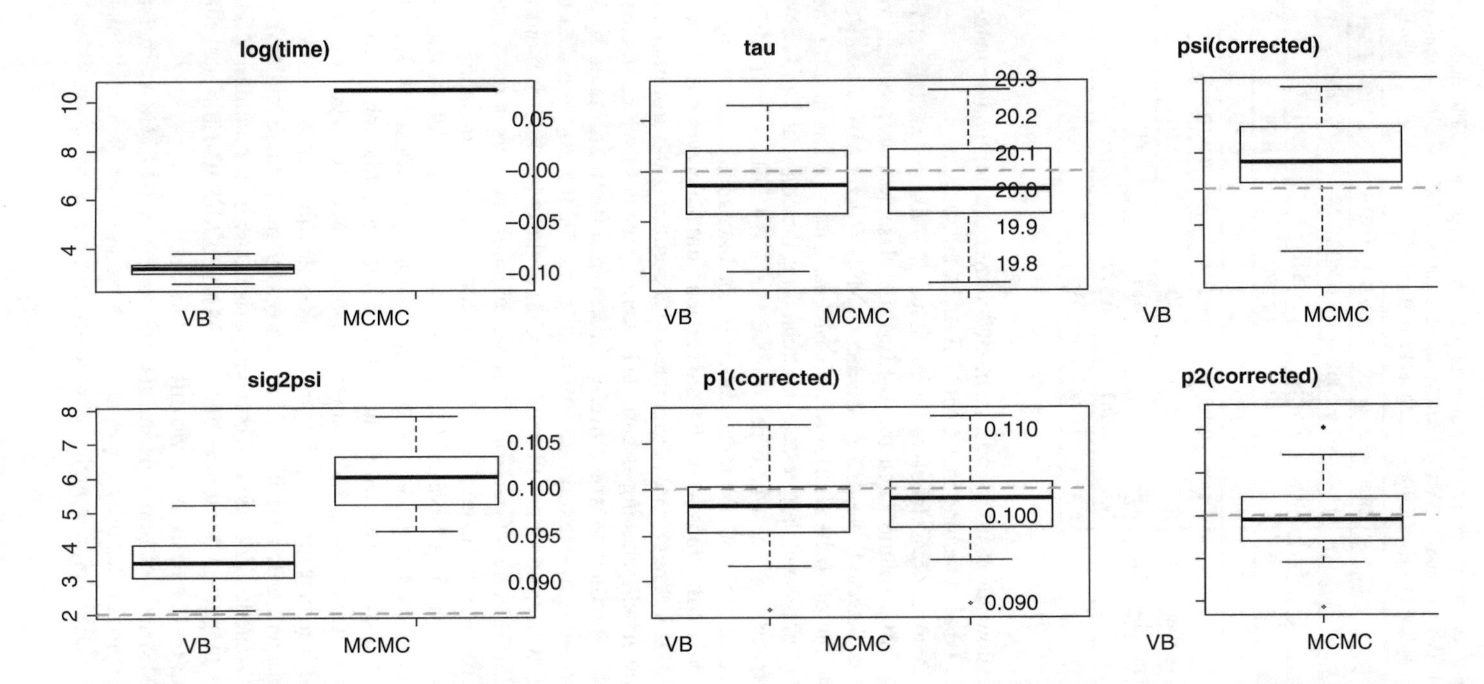

Fig. 7.6 Comparison between VB-LEMMA and MCMC-LEMMA with multiple simulation on an Intel Xeon L5410 2.33 GHz, 8 GB RAM computer. 30 data sets were simulated with same true parameter values in Table 7.1. MCMC-LEMMA results were post-processed so that results were corrected for label-switching among the simulated data sets. The dashed lines mark the true parameter values

high accuracy, requires substantially more work in overcoming issues such as non-convergence and label-switching.

7.4.1.3 A Mixture Model for Count Data

Mass-spectrometry-based shotgun proteomics has enabled large-scale identification and differential profiling of complex proteomes yielding significant insights into relevant biological systems. This approach typically involves liquid chromatography tandem mass spectrometry analysis and employs hybrid mass spectrometers with high data acquisition efficiency for intensity-based sampling of peptide ions. More recently, label-free techniques, such as peak intensity measurements and spectral counting, have emerged. Spectral counting involves measuring the abundance of a given protein based on the number of tandem mass spectral observations for all its constituent peptides. Spectral counts have been shown to correlate well with the abundance of the corresponding protein extending over a linear dynamic range of at least two orders of magnitude for complex protein mixtures. Most proteomics studies are interested in finding proteins, the abundance of which changes in different cellular states, under different conditions, or with respect to different treatments. Simple statistical methods have been employed to perform one protein at a time analysis using, for example, Wald or likelihood-ratio statistics, analogous to the t-test in the microarray context. In this subsection we propose an alternative VB approach for comparing spectral counts under two conditions, the approach allows for simultaneous testing of several thousand proteins. This two-group classification approach is analogous to the methods developed in Sect. 7.3.2.

Booth et al. (2011) propose a mixture of log-linear models with random protein specific factors for classifying proteins into null and non-null (differential abundance) categories. Their model, which is fully Bayesian, can be written as follows:

$$\begin{aligned}
y_{ij}|\mu_{ij} &\sim \text{Poisson}(\mu_{ij}), \qquad (7.6)\\
\log\mu_{ij}|I_i,\beta_0,\beta_1,b_{0i},b_{1i} &= \beta_0 + b_{0i} + \beta_1 T_j + b_{1i}I_iT_j + \beta_2 I_iT_j + \log L_i + \log N_j,\\
I_i|\pi_1 \sim \text{Bernoulli}(\pi_1)\ \ i.i.d., &\qquad b_{ki}|\sigma_k^2 \sim N(0,\sigma_k^2)\ \ i.i.d.,\ \ k=0,1,\\
\beta_m \sim N(0,\sigma_{\beta_m}^2),\ \ m=0,1,2, &\qquad \sigma_k^{-2} \sim \text{Gamma}(A_{\sigma_k^2},B_{\sigma_k^2}),\ \ k=0,1,\\
\pi_1 \sim \text{Beta}(\alpha,\beta),
\end{aligned}$$

where y_{ij} is the spectral count of protein i, $i = 1,\ldots,p$, and replicate j, $j = 1,\ldots,n$. L_i is the length of protein i, N_j is the average count for replicate j over all proteins, and

$$T_j = \begin{cases} 1 & \text{if replicate } j \text{ is in the treatment group} \\ 0 & \text{if replicate } j \text{ is in the control group.} \end{cases}$$

In fact, conjugacy in this Poisson GLMM is not sufficient for a tractable solution to be computed by VB. Therefore, a similar Poisson–Gamma HGLM where the parameters $\beta_m, m = 0, 1, 2$ and the latent variables $b_{ki}, k = 0, 1$ are transformed is used for the VB implementation:

$$
\begin{aligned}
&y_{ij}|\mu_{ij} \sim \text{Poisson}(\mu_{ij}), \qquad\qquad (7.7)\\
&\log \mu_{ij}|I_i, \beta_0, \beta_1, b_{0i}, b_{1i} = \beta_0 + b_{0i} + \beta_1 T_j + b_{1i} I_i T_j + \beta_2 I_i T_j + \log L_i + \log N_j,\\
&I_i|\pi_1 \sim \text{Bernoulli}(\pi_1)\ \textit{i.i.d.}, \qquad b_{ki}|\phi_{ki} = \log(\phi_{ki}^{-1}),\ k = 0, 1,\\
&\phi_{ki}|\delta_k \sim \text{IG}(\delta_k, \delta_k)\ \textit{i.i.d.},\ k = 0, 1, \qquad \delta_k \sim \text{Gamma}(A_{\delta_k}, B_{\delta_k}),\ k = 0, 1,\\
&\beta_m|\lambda_m = \log(\lambda_m^{-1}),\ m = 0, 1, 2, \qquad \lambda_m \sim \text{IG}(A_{\lambda_m}, B_{\lambda_m}),\ m = 0, 1, 2,\\
&\pi_1 \sim \text{Beta}(\alpha, \beta).
\end{aligned}
$$

As before classification is inferred from the posterior expectations of the latent binary indicators $I_i, i = 1, \ldots, p$. This fully Bayesian, two-component mixture model allows for derivation of a VB algorithm, VB-proteomics, the details of which are shown in Appendix.

For any real data, the exact model that the data distribution follows is unknown. To imitate challenges in sparse classification on real data sets, we simulated data from each of the two models: Poisson GLMM (6) and Poisson–Gamma HGLM (7), and applied VB-proteomics to both data sets. MCMC-proteomics, also derived based on Poisson–Gamma HGLM (7), was implemented via OpenBUGS in R with a chain of length 600,000, a burn-in period 595,000, and a thinning factor 5. Starting values implemented in VB-proteomics were consistent with those in MCMC-proteomics.

Two data sets, one following the Poisson GLMM (6) with $\beta_0 = -7.7009$, $\beta_1 = -0.1765$, $\beta_2 = 0$, $\sigma_0^2 = 1$, $\sigma_1^2 = 4$ and the other following the Poisson–Gamma HGLM (7) with $\lambda_0 = 2210.336$, $\lambda_1 = 1.193$, $\lambda_2 = 1$, $\delta_0 = 1$, $\delta_1 = 0.2$ were generated. In each data set there were four replicates under each treatment and $p = 1307$ proteins. The non-null indicator was generated with $\pi_1 = 0.2$. These values were chosen so that the simulated data sets were similar to the Synthetic twofold data analyzed in Booth et al. (2011).

Table 7.2 gives the classification and computational performance of VB-proteomics and MCMC-proteomics for the two simulated data sets on an Intel Core i5-2430M 2.40 GHz, 6 GB RAM computer. The longer running time and the larger deviation of posterior mean of π_1 from its true value associated with the simulated data under Poisson GLMM in Table 7.2 shows that it is indeed more difficult for VB-proteomics to perform classification on data from the "wrong" model. Nonetheless, VB-proteomics identifies the extreme proteins for both data sets with similar accuracy regardless of which model the data was generated from, and is much more desirable than MCMC-proteomics in terms of computational speed.

Table 7.2 Proteomics count data: simulation under the two models and VB-proteomics and MCMC-proteomics classifications

	Poisson GLMM			Poisson–Gamma HGLM		
	True	VB	MCMC	True	VB	MCMC
Number of non-nulls	259	109	96	262	103	90
π_1	0.2	0.0856	0.160	0.2	0.122	0.218
Accuracy		0.878	0.874		0.871	0.865
Time		54 s	7.5 h		10 s	7.4 h

Total number of proteins is 1307. Cutoff is 0.8. Accuracy is defined in Sect. 7.4.1.1

Figures 7.7 and 7.8 show plots of the log ratios of total counts (+1) in treatment and control groups against protein number, and ROC curves comparing performance of the model-based VB-proteomics, MCMC-proteomics approaches with one protein at a time Score tests. These plots clearly indicate that VB-proteomics acts as a better classifier than individual Score tests and as a comparable classifier to MCMC-proteomics in terms of ROC curve, accuracy, and false discovery rate performance.

7.4.2 *Real Data Examples*

In this section, we further illustrate classification performance of VB for fully Bayesian finite mixture models on real data. Two microarray examples are investigated, the APOA1 (Callow et al. 2000) and Colon Cancer (Alon et al. 1999) data sets.

7.4.2.1 APOA1 Data

The APOA1 data (Callow et al. 2000) contains 5548 genes associated with measurements from 8 control mice and 8 "knockout" mice. The data was originally obtained from a two-group design. The histogram of d_g in Fig. 7.9 indicates that the B-LIMMA model (4) is a good fit and that both `limma` and VB-LIMMA are applicable for classification, under the assumption that majority of the genes are null. The small proportion of the non-null genes is known since there are eight true non-null genes. Figure 7.10 shows the top 10 non-null genes by `limma`. VB-LIMMA was implemented, with non-informative priors and starting values such that $\widehat{M_{b_g}} = 1$ for genes that are associated with the 5% largest values of d_g or the 5% smallest values of d_g and $\widehat{M_{b_g}} = 0$ otherwise. VB-LIMMA identified the same top eight genes as `limma` in 53 s, indicated by close-to-one posterior non-null probability estimate of those genes in Fig. 7.9. The `convest` function in `limma` gave an over-estimate of the non-null proportion p, 0.130114, whereas VB-LIMMA produced a closer estimate 0.00162162 to the true value 0.00144.

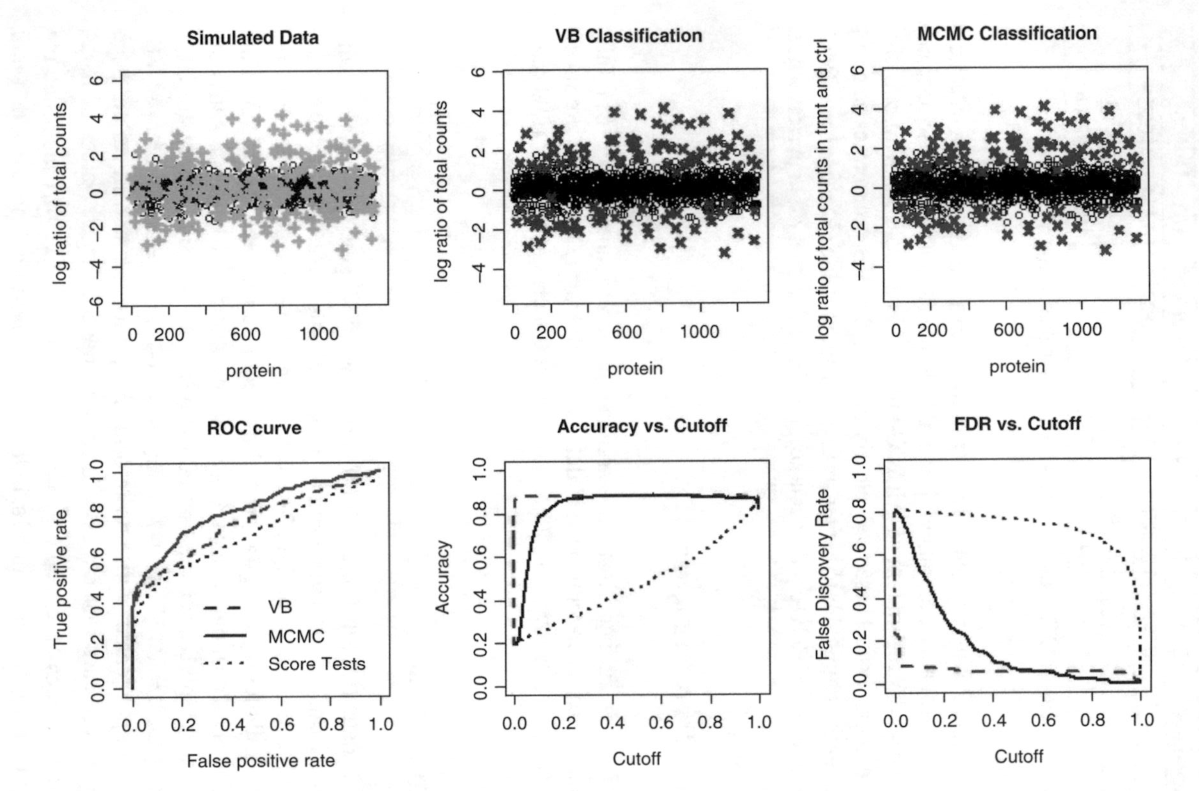

Fig. 7.7 Simulated data under the Poisson GLMM (6), i.e. the "wrong" model. Top row: plots of log ratios of total counts in treatment and control groups against protein number with true non-null proteins, VB-proteomics labeled proteins, and MCMC-proteomics labeled proteins indicated. Bottom row: ROC curves, accuracy vs. cutoff, and FDR vs. cutoff for VB-proteomics, MCMC-proteomics, and for one protein at a time score tests. Accuracy and FDR are as defined in Sect. 7.4.1.1

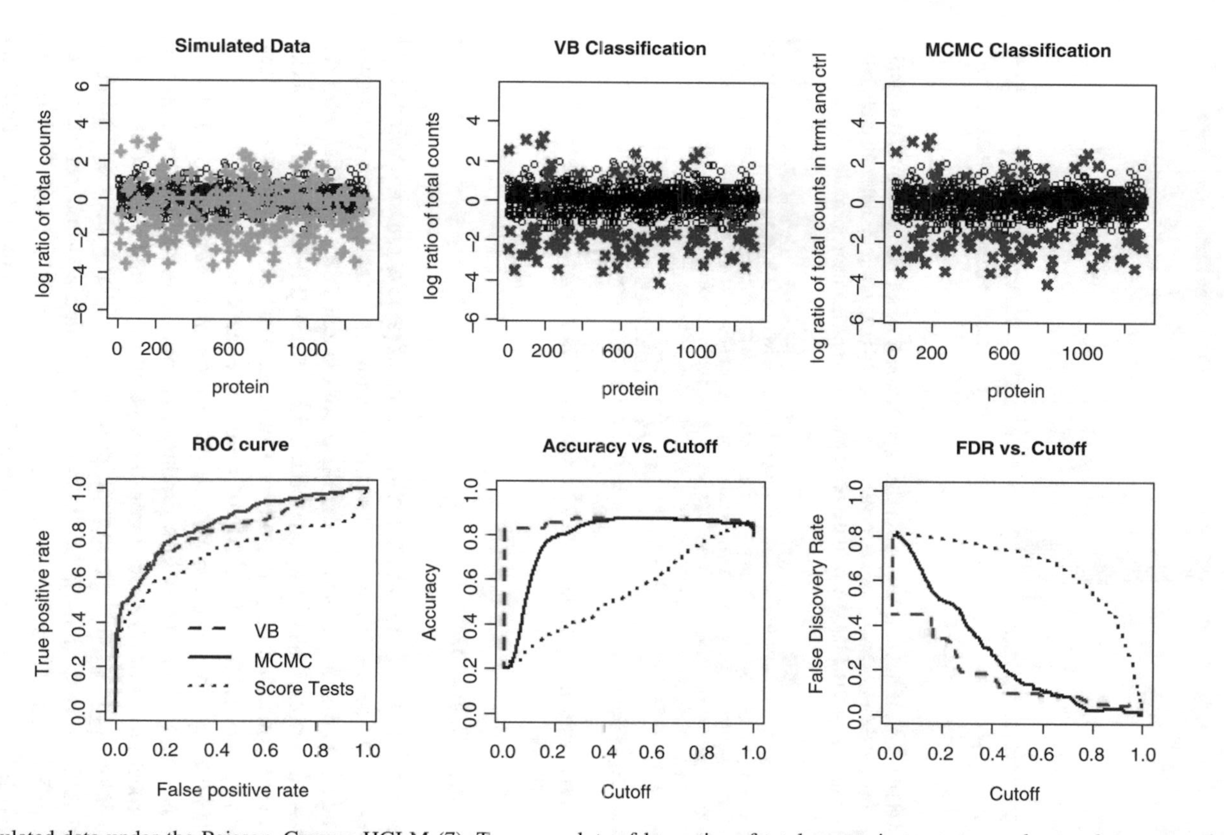

Fig. 7.8 Simulated data under the Poisson–Gamma HGLM (7). Top row: plots of log ratios of total counts in treatment and control groups against protein number with true non-null proteins, VB-proteomics labeled proteins, and MCMC-proteomics labeled proteins indicated. Bottom row: ROC curves, accuracy vs. cutoff, and FDR vs. cutoff for VB-proteomics, MCMC-proteomics, and for one protein at a time score tests. Accuracy and FDR are as defined in Sect. 7.4.1.1

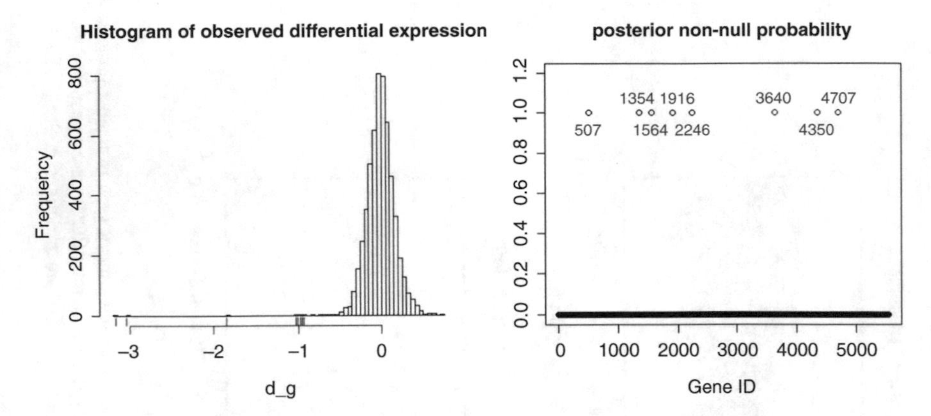

Fig. 7.9 APOA1 data: histogram of d_g with the eight non-null genes classified by VB-LIMMA indicated by inward ticks on the x-axis, and posterior mean of b_g (i.e., posterior non-null probability of gene g) estimated by VB-LIMMA with the eight non-null gene numbers identified

	NAME	logFC	t	P.Value	adj.P.Val	B
1916	ApoAI,lipid-Img	-3.1661645	-24.048796	5.442072e-15	3.019261e-11	15.960594
507	EST,HighlysimilartoA	-3.0485504	-12.938819	1.797129e-10	4.985235e-07	11.347701
4707	CATECHOLO-METHYLTRAN	-1.8481659	-12.458569	3.300689e-10	6.104074e-07	10.982250
3640	EST,WeaklysimilartoC	-1.0269537	-11.894827	6.905520e-10	9.577957e-07	10.524492
1564	ApoCIII,lipid-Img	-0.9325824	-9.927515	1.145390e-08	1.270925e-05	8.647893
2246	ESTs,Highlysimilarto	-1.0098117	-9.065277	4.461600e-08	3.579128e-05	7.667271
1354	est	-0.9774236	-9.057852	4.515842e-08	3.579128e-05	7.658359
4350	similartoyeaststerol	-0.9549693	-7.465666	7.068872e-07	4.902263e-04	5.546823

Fig. 7.10 `limma` output: Top 10 genes detected as non-null for APOA1 data. Gene numbers are shown as row names

7.4.2.2 Colon Cancer Data

The Colon Cancer data (Alon et al. 1999) consists of gene expression values from 2000 genes in 22 controls and 40 treatment samples. Since the two-component mixture model is a special case of the three-component mixture model with one of the non-null proportions being zero, both B-LIMMA and B-LEMMA models are suitable for classification of genes on the Colon Cancer data. Therefore, `limma`, VB-LIMMA, `lemma`, and VB-LEMMA were implemented.

A gene rankings table showing genes that are associated with the top 200 largest posterior non-null probabilities was produced according to results from each of the four procedures. For all procedures, the posterior non-null probability for any gene equals one minus the posterior null probability of that gene. Comparing the top genes tables reveals that out of the top 200 identified non-null genes, `limma` agrees with VB-LIMMA on 151 genes, `lemma` agrees with VB-LEMMA on 190 genes, and all four methods share 140 genes. This high level of agreement between the procedures shows that fully Bayesian classification via VB is comparable in accuracy to classification via empirical Bayes approaches on the Colon Cancer data.

MCMC-LEMMA was also implemented for the Colon Cancer data to illustrate performance of MCMC for classification on real data under the B-LEMMA model. A single chain of length 4,000,000, a burn-in period 3,995,000, and a thinning factor 10 was used to ensure convergence and avoid label-switching. MCMC diagnostics plots (not shown) suggests that convergence is tolerable, although it took MCMC-LEMMA 11.7 h to run on an Intel Xeon L5410 2.33 GHz, 8 GB RAM computer. There are 187 genes present in both top 200 genes tables based on MCMC-LEMMA and VB-LEMMA classifications, which shows accuracy of VB in classification is comparable with that of MCMC. From the component densities plots in Fig. 7.11 we see that the three-component mixture evaluated by VB-LEMMA and by MCMC-LEMMA are two different yet plausible solutions based on the B-LEMMA model. It is unknown which solution is closer to the truth. Nonetheless, with a 0.9 cutoff, MCMC-LEMMA did not detect any non-null genes, whereas VB-LEMMA identified 56 group 1 non-null genes and 30 group 2 non-null genes in 7 s.

7.5 Discussion

VB has been promoted in the statistical and computer science literature as an alternative to MCMC for Bayesian computation. However, the performance and feasibility of VB has not been widely investigated for hierarchical mixture models used in sparse classification problems. This chapter demonstrates the implementation of VB in that context and shows that it is capable of efficiently producing reliable solutions with classification performance comparable to much slower MCMC methods. A known issue with VB is its tendency to underestimate variability in marginal posteriors. GBVA is designed to correct this, but its performance in the context of hierarchical mixture models is not promising either in terms of accuracy or computational speed.

As the demand for high-dimensional data analysis tools grows the search for fast and accurate alternatives to MCMC continues to be an important open research area. Other deterministic alternatives to MCMC in posterior density approximation include EP (Minka 2001a,b) and INLA (Rue et al. 2009). In EP, while similar factorization of the joint posterior density to that in VB is assumed, the approximate posterior is found by moment matching implied by minimization of the reverse form of Kullback–Leibler divergence to that used in VB theory. However, as noted in Bishop (2006), drawbacks of EP include lack of guarantee of convergence and failure to capture any mode if the true posterior density is multimodal. For the class of latent Gaussian models, approximate Bayesian inference can be achieved via INLA (Rue et al. 2009). The INLA method relies on Gaussian approximations and numerical methods and bypasses any assumption of factorized density forms. A limitation of INLA is the fact that models that contain non-Gaussian latent variables in the linear predictor do not belong to the class of latent Gaussian models for which INLA is applicable.

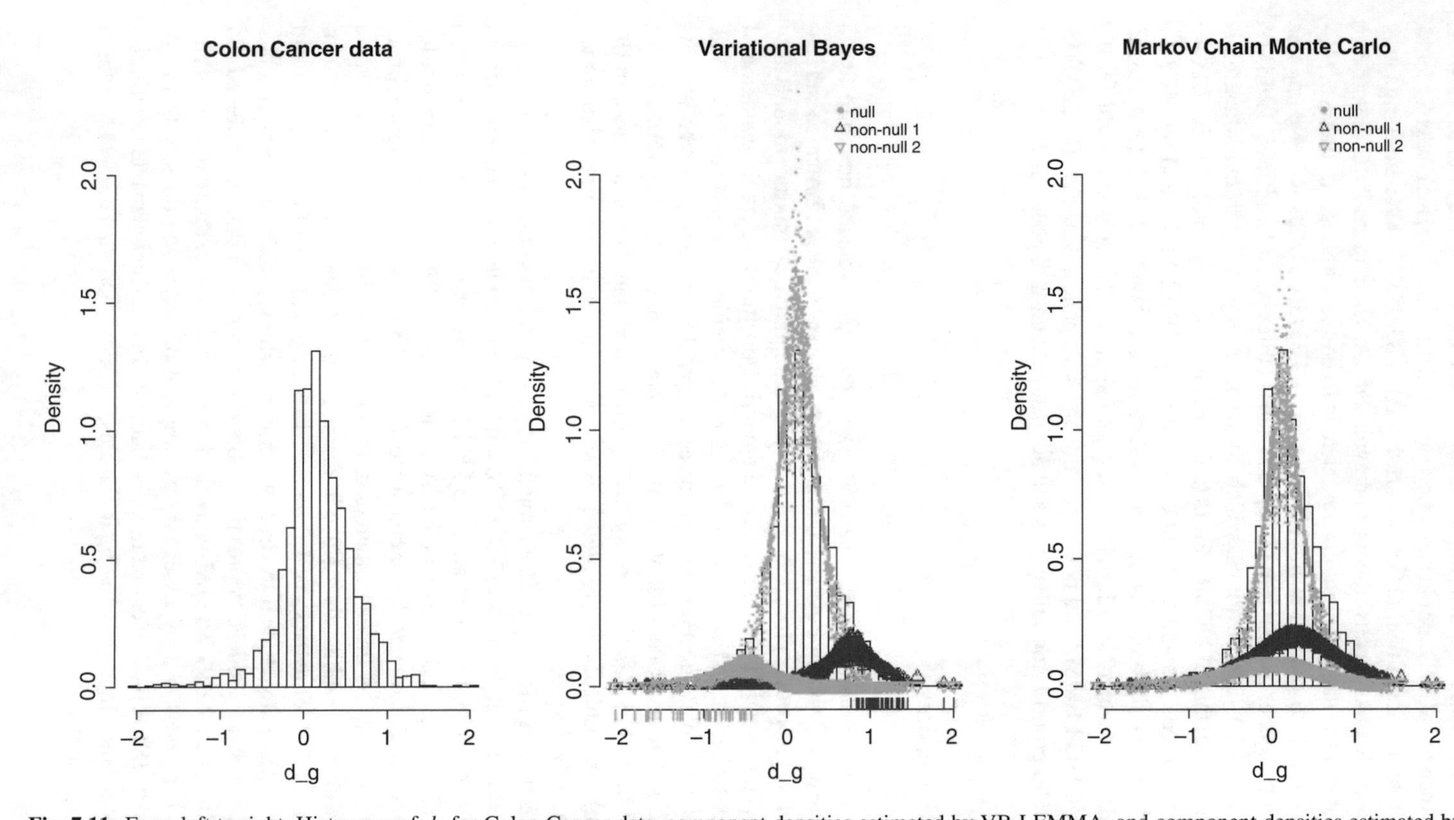

Fig. 7.11 From left to right: Histogram of d_g for Colon Cancer data, component densities estimated by VB-LEMMA, and component densities estimated by MCMC-LEMMA. On the *x*-axis, inward ticks indicate the classified non-null 1 genes and outward ticks indicate the classified non-null 2 genes with a 0.9 cutoff

VB has been criticized (Rue et al. 2009) for its applicability being restricted to the class of conjugate-exponential (Beal 2003) models. However, provided that VB is feasible for a reparameterized model whose inference is close to the original model (Beal 2003), VB is often a good approach for fully Bayesian inference because of its relative ease of implementation and computational speed.

Acknowledgements We would like to thank John T. Ormerod who provided supplementary materials for GBVA implementation in Ormerod (2011), and Haim Y. Bar for helpful discussions.

Professors Booth and Wells acknowledge the support of NSF-DMS 1208488 and NIH U19 AI111143.

Appendix: The VB-LEMMA Algorithm

The B-LEMMA Model

We consider a natural extension to the LEMMA model in Bar et al. (2010): a fully Bayesian three-component mixture model, B-LEMMA:

$$
\begin{aligned}
d_g|(b_{1g}, b_{2g}), \psi_g, \sigma^2_{\epsilon,g}, \tau &= \tau + (b_{1g} - b_{2g})\psi_g + \epsilon_g \\
m_g|\sigma^2_{\epsilon,g} &\sim \frac{\sigma^2_{\epsilon,g}\chi^2_{f_g}}{f_g}, \quad \text{where} f_g := n_{1g} + n_{2g} - 2 \\
\psi_g|\psi, \sigma^2_\psi &\sim N(\psi, \sigma^2_\psi) \quad i.i.d. \\
\epsilon_g|\sigma^2_{\epsilon,g} &\sim N(0, \sigma^2_g), \quad \text{where} \sigma^2_g := \sigma^2_{\epsilon,g} c_g, \\
c_g &:= \frac{1}{n_{1g}} + \frac{1}{n_{2g}} \\
(b_{1g}, b_{2g}, 1 - b_{1g} - b_{2g})|(p_1, p_2) &\sim \text{Multinomial}\,(1; p_1, p_2, 1 - p_1 - p_2) \quad i.i.d. \\
\tau &\sim N\left(\mu_{\tau_0}, \sigma^2_{\tau_0}\right) \\
\psi &\sim N\left(\mu_{\psi_0}, \sigma^2_{\psi_0}\right) \\
\sigma^2_\psi &\sim \text{IG}\left(A_\psi, B_\psi\right) \\
\sigma^2_{\epsilon,g} &\sim \text{IG}\left(A_\epsilon, B_\epsilon\right) \quad i.i.d. \\
(p_1, p_2, 1 - p_1 - p_2) &\sim \text{Dirichlet}\left(\alpha_1, \alpha_2, \alpha_0\right),
\end{aligned}
$$

where (b_{1g}, b_{2g}) takes values $(1, 0)$, $(0, 1)$, or $(0, 0)$, indicating that gene g is in non-null group 1, non-null group 2, or null group, respectively. p_1 and p_2 are proportions of non-null group 1 and non-null group 2 genes. Hence, the non-null proportion is $p_1 + p_2$. Each of τ and ψ_g represents the same quantity as in the B-LIMMA model.

Algorithm

The VB-LEMMA algorithm was derived based on an equivalent model to B-LEMMA. In the equivalent model, the gene-specific treatment effect is treated as the combination of a fixed global effect and a random zero-mean effect. That is, conditional distribution of d_g and that of ψ_g are replaced with

$$d_g|(b_{1g}, b_{2g}), u_g, \sigma^2_{\epsilon,g}, \tau = \tau + (b_{1g} - b_{2g})\psi + (b_{1g} + b_{2g})u_g + \epsilon_g$$
$$u_g|\sigma^2_\psi \sim N(0, \sigma^2_\psi) \quad i.i.d.$$

The set of observed data and the set of unobserved data are identified as

$$\mathbf{y} = \{\{d_g\}, \{m_g\}\}$$
$$H = \{\{\boldsymbol{b}_g\}, \{u_g\}, \{\sigma^2_g\}, \tau, \psi, \sigma^2_\psi, \boldsymbol{p}\}$$

where $\boldsymbol{b}_g = (b_{1g}, b_{2g})$ and $\boldsymbol{p} = (p_1, p_2)$.

Because of the similarities of the B-LEMMA model to the B-LIMMA model, derivation of VB-LEMMA was achieved by extending the derivation of VB-LIMMA that involves a gene-specific zero-mean random effect parameter. The VB algorithm based on the exact B-LEMMA model was also derived for comparison. However, little discrepancy in performance between the VB algorithm based on the exact model and VB-LEMMA was observed. Therefore, VB-LEMMA based on the equivalent model was adopted.

The product density restriction

$$q(H) = q_{\{\boldsymbol{b}_g\}}(\{\boldsymbol{b}_g\}) \times q_{\{\psi_g\}}(\{\psi_g\}) \times q_{\{\sigma^2_g\}}(\{\sigma^2_g\}) \times q_{(\tau,\boldsymbol{p})}(\tau, \boldsymbol{p}) \times q_{(\psi,\sigma^2_\psi)}(\psi, \sigma^2_\psi)$$

leads to q-densities

$$q_\tau(\tau) = N\left(\widehat{M_\tau}, \widehat{V_\tau}\right)$$
$$q_\psi(\psi) = N\left(\widehat{M_\psi}, \widehat{V_\psi}\right)$$
$$q_{\sigma^2_g}(\sigma^2_g) = \text{IG}\left(A_{\sigma^2_g}, \widehat{B_{\sigma^2_g}}\right)$$
$$q_{(b_{1g}, b_{2g}, 1-b_{1g}-b_{2g})}((b_{1g}, b_{2g}, 1 - b_{1g} - b_{2g})) = \text{Multinomial}\left(\widehat{M_{b_{1g}}}, \widehat{M_{b_{2g}}}, 1 - \widehat{M_{b_{1g}}} - \widehat{M_{b_{2g}}}\right)$$
$$q_{(p_1, p_2, 1-p_1-p_2)}((p_1, p_2, 1 - p_1 - p_2)) = \text{Dirichlet}\left(\widehat{\alpha_{p_1}}, \widehat{\alpha_{p_2}}, \widehat{\alpha_{p_0}}\right)$$
$$q_{\sigma^2_\psi}(\sigma^2_\psi) = \text{IG}\left(A_{\sigma^2_\psi}, \widehat{B_{\sigma^2_\psi}}\right).$$

It is only necessary to update the variational posterior means $\hat{M}.$ in VB-LEMMA. Upon convergence, the other variational parameters are computed based on the converged value of those involved in the iterations. The iterative scheme is as follows:

1. Initialize

$$\widehat{M_{\sigma_\psi^{-2}}} = 1$$

$$\widehat{M_{\sigma_g^{-2}}} = \frac{1}{c_g} \;\forall\, g$$

$$\widehat{M_{b_g}} = \begin{cases} (1,0,0) & \text{ifrank}(\mathrm{d_g}) \geq (1-0.05)\mathrm{G} \\ (0,1,0) & \text{ifrank}(\mathrm{d_g}) \leq 0.05\mathrm{G} \qquad \text{foreach} g \\ (0,0,1) & \text{otherwise} \end{cases}$$

$$\widehat{M_\psi} = \frac{1}{2}\left(\left|\sum_{\{g:\text{rank}(d_g)\geq(1-0.05)G\}} d_g - \sum_{g=1}^{G} d_g\right| + \left|\sum_{g=1}^{G} d_g - \sum_{\{g:\text{rank}(d_g)\leq 0.05G\}} d_g\right|\right)$$

$$\widehat{M_{u_g}} = 0 \;\forall\, g$$

$$\text{Set} A_{\sigma_\psi^2} = \frac{G}{2} + A_\psi \quad \text{and} \quad A_{\sigma_g^2} = \frac{1+f_g}{2} + A_\varepsilon \text{ for each } g.$$

2. Update

$$\widehat{M_\tau} \leftarrow \left\{\sum_g \widehat{M_{\sigma_g^{-2}}}\left[\left(1 - \widehat{M_{b_{1g}}} - \widehat{M_{b_{2g}}}\right) d_g + \widehat{M_{b_{1g}}}\left(d_g - \widehat{M_\psi} - \widehat{M_{u_g}}\right)\right.\right.$$
$$\left.\left. + \widehat{M_{b_{2g}}}\left(d_g + \widehat{M_\psi} - \widehat{M_{u_g}}\right)\right] + \frac{\mu_{\tau 0}}{\sigma_{\tau 0}^2}\right\} \times \frac{1}{\sum_g \widehat{M_{\sigma_g^{-2}}} + \frac{1}{\sigma_{\tau 0}^2}}$$

$$\widehat{M_\psi} \leftarrow \left\{\sum_g \widehat{M_{\sigma_g^{-2}}}\left(\widehat{M_{b_{1g}}} - \widehat{M_{b_{2g}}}\right)\left(d_g - \widehat{M_\tau} - \widehat{M_{u_g}}\right) + \frac{\mu_{\psi 0}}{\sigma_{\psi 0}^2}\right\}$$
$$\times \frac{1}{\sum_g \widehat{M_{\sigma_g^{-2}}}\left(\widehat{M_{b_{1g}}} + \widehat{M_{b_{2g}}}\right) + \frac{1}{\sigma_{\psi 0}^2}}$$

$$\widehat{M_{u_g}} \leftarrow \widehat{M_{\sigma_g^{-2}}}\left[\widehat{M_{b_{1g}}}\left(d_g - \widehat{M_\tau} - \widehat{M_\psi}\right) + \widehat{M_{b_{2g}}}\left(d_g - \widehat{M_\tau} + \widehat{M_\psi}\right)\right]$$
$$\times \frac{1}{\widehat{M_{\sigma_g^{-2}}}\left(\widehat{M_{b_{1g}}} + \widehat{M_{b_{2g}}}\right) + \widehat{M_{\sigma_\psi^{-2}}}}$$

$$\widehat{M_{\sigma_\psi^{-2}}} \leftarrow \frac{A_{\sigma_\psi^2}}{\frac{1}{2}\sum_g \left(\frac{1}{\widehat{M_{\sigma_g^{-2}}}\left(\widehat{M_{b_{1g}}}+\widehat{M_{b_{2g}}}\right)+\widehat{M_{\sigma_\psi^{-2}}}} + \widehat{M_{u_g}}^2 \right) + B_\psi}$$

$$\widehat{M_{\sigma_g^{-2}}} \leftarrow \frac{A_{\sigma_g^2}}{\widehat{B_{\sigma_g^2}}} \text{ where}$$

$$\widehat{B_{\sigma_g^2}} \leftarrow \frac{1}{2}\left\{ \frac{1}{\sum_g \widehat{M_{\sigma_g^{-2}}} + \frac{1}{\sigma_{\tau 0}^2}} + \left(\widehat{M_{b_{1g}}} + \widehat{M_{b_{2g}}}\right) \right.$$

$$\times \left(\frac{1}{\sum_g \widehat{M_{\sigma_g^{-2}}}\left(\widehat{M_{b_{1g}}} + \widehat{M_{b_{2g}}}\right) + \frac{1}{\sigma_{\psi 0}^2}} \right)$$

$$+ \left(1 - \widehat{M_{b_{1g}}} - \widehat{M_{b_{2g}}}\right)\left(d_g - \widehat{M_\tau}\right)^2$$

$$+ \widehat{M_{b_{1g}}}\left[\frac{1}{\widehat{M_{\sigma_g^{-2}}}\left(\widehat{M_{b_{1g}}} + \widehat{M_{b_{2g}}}\right) + \widehat{M_{\sigma_\psi^{-2}}}} + \left(d_g - \widehat{M_\tau} - \widehat{M_\psi} - \widehat{M_{u_g}}\right)^2 \right]$$

$$+ \widehat{M_{b_{2g}}}\left[\frac{1}{\widehat{M_{\sigma_g^{-2}}}\left(\widehat{M_{b_{1g}}} + \widehat{M_{b_{2g}}}\right) + \widehat{M_{\sigma_\psi^{-2}}}} \right.$$

$$\left.\left. + \left(d_g - \widehat{M_\tau} + \widehat{M_\psi} - \widehat{M_{u_g}}\right)^2 \right] \right\} + \frac{m_g f_g c_g}{2} + B_\varepsilon c_g$$

$$\widehat{M_{b_{1g}}} \leftarrow \frac{\exp\left(\widehat{\theta_{1g}}\right)}{\exp\left(\widehat{\theta_{1g}}\right) + \exp\left(\widehat{\theta_{2g}}\right) + 1} \text{ and}$$

$$\widehat{M_{b_{2g}}} \leftarrow \frac{\exp\left(\widehat{\theta_{2g}}\right)}{\exp\left(\widehat{\theta_{1g}}\right) + \exp\left(\widehat{\theta_{2g}}\right) + 1} \text{ where}$$

$$\widehat{\theta_{1g}} \leftarrow -\frac{\widehat{M_{\sigma_g^{-2}}}}{2}\left\{ \frac{1}{\sum_g \widehat{M_{\sigma_g^{-2}}}\left(\widehat{M_{b_{1g}}} + \widehat{M_{b_{2g}}}\right) + \frac{1}{\sigma_{\psi 0}^2}} \right.$$

$$+ \frac{1}{\widehat{M_{\sigma_g^{-2}}}\left(\widehat{M_{b_{1g}}} + \widehat{M_{b_{2g}}}\right) + \widehat{M_{\sigma_\psi^{-2}}}}$$

$$
\begin{aligned}
&+ \left(d_g - \widehat{M_\tau} - \widehat{M_\psi} - \widehat{M_{u_g}}\right)^2 - \left(d_g - \widehat{M_\tau}\right)^2 \Bigg\} \\
&+ \left[\text{digamma}\left(\sum_g \widehat{M_{b_{1g}}} + \alpha_1\right)\right. \\
&\left. -\text{digamma}\left(\sum_g \left(1 - \widehat{M_{b_{1g}}} - \widehat{M_{b_{2g}}}\right) + \alpha_0\right)\right]
\end{aligned}
$$

$$
\begin{aligned}
\widehat{\theta_{2g}} \leftarrow -\frac{\widehat{M_{\sigma_g^{-2}}}}{2} &\left\{ \frac{1}{\sum_g \widehat{M_{\sigma_g^{-2}}} \left(\widehat{M_{b_{1g}}} + \widehat{M_{b_{2g}}}\right) + \frac{1}{\sigma_{\psi 0}^2}} \right. \\
&+ \frac{1}{\widehat{M_{\sigma_g^{-2}}} \left(\widehat{M_{b_{1g}}} + \widehat{M_{b_{2g}}}\right) + \widehat{M_{\sigma_\psi^{-2}}}} \\
&\left. + \left(d_g - \widehat{M_\tau} + \widehat{M_\psi} - \widehat{M_{u_g}}\right)^2 - \left(d_g - \widehat{M_\tau}\right)^2 \right\} \\
&+ \left[\text{digamma}\left(\sum_g \widehat{M_{b_{2g}}} + \alpha_2\right)\right. \\
&\left. -\text{digamma}\left(\sum_g \left(1 - \widehat{M_{b_{1g}}} - \widehat{M_{b_{2g}}}\right) + \alpha_0\right)\right].
\end{aligned}
$$

3. Repeat (2) until the increase in

$$
\begin{aligned}
&\log \underline{p}\left(\mathbf{y}; \mathbf{q}\right) \\
&= \frac{-G}{2} \times \log\left(2\pi\right) - \sum_g \left[\widehat{M_{b_{1g}}} \log \widehat{M_{b_{1g}}} + \widehat{M_{b_{2g}}} \log \widehat{M_{b_{2g}}}\right. \\
&\quad \left. + \left(1 - \widehat{M_{b_{1g}}} - \widehat{M_{b_{2g}}}\right) \log\left(1 - \widehat{M_{b_{1g}}} - \widehat{M_{b_{1g}}}\right)\right] \\
&\quad + \log\left(\text{Beta}\left(\sum_g \widehat{M_{b_{1g}}} + \alpha_1, \sum_g \widehat{M_{b_{2g}}} + \alpha_2, \sum_g \left(1 - \widehat{M_{b_{1g}}} - \widehat{M_{b_{2g}}}\right) + \alpha_0\right)\right) \\
&\quad - \log\left(\text{Beta}\left(\alpha_1, \alpha_2, \alpha_0\right)\right) \\
&\quad + \left[\log\left(\frac{1}{\sum_g \widehat{M_{\sigma_g^{-2}}} + \frac{1}{\sigma_{\tau 0}^2}}\right)\right.
\end{aligned}
$$

$$-\log\sigma_{\tau0}^2 + 1 - \frac{\frac{1}{\sum_g \widehat{M_{\sigma_g^{-2}}} + \frac{1}{\sigma_{\tau0}^2}} + \left(\widehat{M_\tau} - \mu_{\tau0}\right)^2}{\sigma_{\tau0}^2}\Bigg] \times \frac{1}{2}$$

$$+ \left[\log\left(\frac{1}{\sum_g \widehat{M_{\sigma_g^{-2}}}\left(\widehat{M_{b_{1g}}} + \widehat{M_{b_{2g}}}\right) + \frac{1}{\sigma_{\psi0}^2}}\right) - \log\sigma_{\psi0}^2\right] \times \frac{1}{2}$$

$$+ \left[1 - \frac{\left(\frac{1}{\sum_g \widehat{M_{\sigma_g^{-2}}}\left(\widehat{M_{b_{1g}}} + \widehat{M_{b_{2g}}}\right) + \frac{1}{\sigma_{\psi0}^2}}\right) + \left(\widehat{M_\psi} - \mu_{\psi0}\right)^2}{\sigma_{\psi0}^2}\right] \times \frac{1}{2}$$

$$+ G \times A_\varepsilon \log B_\varepsilon + \sum_g \left[\left(\frac{f_g}{2} + A_\varepsilon\right)\log c_g \log\Gamma\left(A_{\sigma_g^2}\right) - \log\Gamma(A_\varepsilon)\right.$$

$$\left. + -\frac{f_g}{2}\log 2 - \log\Gamma\left(\frac{f_g}{2}\right) + \left(\frac{f_g}{2} - 1\right)\log m_g + \frac{f_g}{2}\log f_g\right]$$

$$+ \sum_g \left\{\frac{A_{\sigma_g^2}}{\widehat{B_{\sigma_g^2}}} \times \left[-\frac{1}{2} m_g f_g c_g - B_\varepsilon c_g\right.\right.$$

$$- \frac{1}{2}\left(\frac{1}{\sum_g \widehat{M_{\sigma_g^{-2}}} + \frac{1}{\sigma_{\tau0}^2}} + \left(1 - \widehat{M_{b_{1g}}} - \widehat{M_{b_{2g}}}\right)\left(d_g - \widehat{M_\tau}\right)^2\right)$$

$$- \frac{\widehat{M_{b_{1g}}} + \widehat{M_{b_{1g}}}}{2}\left(\frac{1}{\sum_g \widehat{M_{\sigma_g^{-2}}}\left(\widehat{M_{b_{1g}}} + \widehat{M_{b_{2g}}}\right) + \frac{1}{\sigma_{\psi0}^2}}\right)$$

$$- \frac{\widehat{M_{b_{1g}}}}{2}\left(\frac{1}{\widehat{M_{\sigma_g^{-2}}}\left(\widehat{M_{b_{1g}}} + \widehat{M_{b_{2g}}}\right) + \widehat{M_{\sigma_\psi^{-2}}}} + \left(d_g - \widehat{M_\tau} - \widehat{M_\psi} - \widehat{M_{u_g}}\right)^2\right)$$

$$\left.\left. - \frac{\widehat{M_{b_{2g}}}}{2}\left(\frac{1}{\widehat{M_{\sigma_g^{-2}}}\left(\widehat{M_{b_{1g}}} + \widehat{M_{b_{2g}}}\right) + \widehat{M_{\sigma_\psi^{-2}}}} + \left(d_g - \widehat{M_\tau} + \widehat{M_\psi} - \widehat{M_{u_g}}\right)^2\right)\right]\right\}$$

$$+ \sum_g A_{\sigma_g^2}\left(-\log\widehat{B_{\sigma_g^2}} + 1\right)$$

$$+\frac{1}{2}\sum_g\left[\log\left(\frac{1}{\widehat{M_{\sigma_g^{-2}}}\left(\widehat{M_{b_{1g}}}+\widehat{M_{b_{2g}}}\right)+\widehat{M_{\sigma_\psi^{-2}}}}\right)+1\right]$$

$$+A_\psi\log B_\psi-\log\Gamma(A_\psi)+A_{\sigma_\psi^2}\left(-\log\widehat{B_{\sigma_\psi^2}}\right)+\log\Gamma\left(A_{\sigma_\psi^2}\right)$$

from previous iteration becomes negligible.

4. Upon convergence, the remaining variational parameters are computed:

$$\widehat{V_\tau}\leftarrow\frac{1}{\sum_g\widehat{M_{\sigma_g^{-2}}}+\frac{1}{\sigma_{\tau_0}^2}}$$

$$\widehat{V_\psi}\leftarrow\frac{1}{\sum_g\widehat{M_{\sigma_g^{-2}}}\left(\widehat{M_{b_{1g}}}+\widehat{M_{b_{2g}}}\right)+\frac{1}{\sigma_{\psi_0}^2}}$$

$$\widehat{V_{u_g}}\leftarrow\frac{1}{\widehat{M_{\sigma_g^{-2}}}\left(\widehat{M_{b_{1g}}}+\widehat{M_{b_{2g}}}\right)+\widehat{M_{\sigma_\psi^{-2}}}}$$

$$\widehat{B_{\sigma_\psi^2}}\leftarrow\frac{1}{2}\sum_g\left[\frac{1}{\widehat{M_{\sigma_g^{-2}}}\left(\widehat{M_{b_{1g}}}+\widehat{M_{b_{2g}}}\right)+\widehat{M_{\sigma_\psi^{-2}}}}+\widehat{M_{u_g}}^2\right]+B_\psi$$

$$\widehat{\alpha_{p_1}}\leftarrow\sum_g\widehat{M_{b_{1g}}}+\alpha_1$$

$$\widehat{\alpha_{p_2}}\leftarrow\sum_g\widehat{M_{b_{2g}}}+\alpha_2$$

$$\widehat{\alpha_{p_0}}\leftarrow\sum_g\left(1-\widehat{M_{b_{1g}}}-\widehat{M_{b_{2g}}}\right)+\alpha_0.$$

The VB-Proteomics Algorithm

The Proteomics Model

As pointed out in Sect. 7.4.1.3, the fully Bayesian model in Booth et al. (2011) is as follows:

$$\begin{aligned}
y_{ij}|\mu_{ij} &\sim \text{Poisson}(\mu_{ij})\\
\log\mu_{ij}|I_i,\beta_0,\beta_1,b_{0i},b_{1i} &= \beta_0+b_{0i}+\beta_1T_j+b_{1i}I_iT_j+\beta_2I_iT_j+\log L_i+\log N_j\\
I_i|\pi_1 &\sim \text{Bernoulli}(\pi_1)\quad i.i.d.
\end{aligned}$$

$$
\begin{aligned}
b_{ki}|\sigma_k^2 &\sim N(0, \sigma_k^2) \quad i.i.d., \quad k = 0, 1 \\
\beta_m &\sim N(0, \sigma_{\beta_m}^2), \quad m = 0, 1, 2 \\
\sigma_k^{-2} &\sim \text{Gamma}(A_{\sigma_k^2}, B_{\sigma_k^2}), \quad k = 0, 1 \\
\pi_1 &\sim \text{Beta}(\alpha, \beta),
\end{aligned}
$$

where y_{ij} is the spectral count of protein i, $i = 1, \ldots, p$, and replicate j, $j = 1, \ldots, n$. L_i is the length of protein i, N_j is the average count for replicate j over all proteins, and

$$
T_j = \begin{cases} 1 & \text{if replicate } j \text{ is in the treatment group} \\ 0 & \text{if replicate } j \text{ is in the control group.} \end{cases}
$$

In fact, conjugacy in this Poisson GLMM is not sufficient for a tractable solution to be computed by VB. Therefore, a similar Poisson–Gamma HGLM where the parameters $\beta_m, m = 0, 1, 2$ and the latent variables $b_{ki}, k = 0, 1$ are transformed is used for the VB implementation:

$$
\begin{aligned}
y_{ij}|\mu_{ij} &\sim \text{Poisson}(\mu_{ij}) \\
\log \mu_{ij}|I_i, \beta_0, \beta_1, b_{0i}, b_{1i} &= \beta_0 + b_{0i} + \beta_1 T_j + b_{1i} I_i T_j + \beta_2 I_i T_j + \log L_i + \log N_j \\
I_i|\pi_1 &\sim \text{Bernoulli}(\pi_1) \quad i.i.d. \\
b_{ki}|\phi_{ki} &= \log(\phi_{ki}^{-1}), \quad k = 0, 1
\end{aligned}
$$

$$
\begin{aligned}
\phi_{ki}|\delta_k &\sim \text{IG}(\delta_k, \delta_k) \quad i.i.d., \quad k = 0, 1 \\
\delta_k &\sim \text{Gamma}(A_{\delta_k}, B_{\delta_k}), \quad k = 0, 1 \\
\beta_m|\lambda_m &= \log(\lambda_m^{-1}), \quad m = 0, 1, 2 \\
\lambda_m &\sim \text{IG}(A_{\lambda_m}, B_{\lambda_m}), \quad m = 0, 1, 2 \\
\pi_1 &\sim \text{Beta}(\alpha, \beta).
\end{aligned}
$$

As before classification is inferred from the posterior expectations of the latent binary indicators $I_i, i = 1, \ldots, p$.

Algorithm

The set of observed data and the set of unobserved data are identified as

$$
\begin{aligned}
\mathbf{y} &= \mathbf{y} \\
H &= \{\lambda_0, \lambda_1, \lambda_2, \boldsymbol{\phi}_\mathbf{0}, \boldsymbol{\phi}_\mathbf{1}, \delta_0, \delta_1, \mathbf{I}, \pi_1\}.
\end{aligned}
$$

The complete likelihood is

$$p(\mathbf{y}, H) = \prod_{i,j} p(y_{ij}|\lambda_0, \lambda_1, \lambda_2, \phi_{0i}, \phi_{1i}, I_i) \times \prod_i p(I_i|\pi_1) \times \prod_i p(\phi_{0i}|\delta_0)$$
$$\times \prod_i p(\phi_{1i}|\delta_1) \times p(\lambda_0)p(\lambda_1)p(\lambda_2)p(\delta_0)p(\delta_1)p(\pi_1),$$

in which the mixture density is

$$\begin{aligned}
&p(y_{ij}|\lambda_0, \lambda_1, \lambda_2, \phi_{0i}, \phi_{1i}, I_i) \\
&\quad= \text{Poisson}\left(y_{ij};\ \exp(\log \lambda_0^{-1} + \log \phi_{0i}^{-1} + \log(L_i N_j))\right)^{1-T_j} \\
&\quad\times \text{Poisson}\left(y_{ij};\ \exp(\log \lambda_0^{-1} + \log \phi_{0i}^{-1} + \log \lambda_1^{-1} + \log(L_i N_j))\right)^{(1-I_i)T_j} \\
&\quad\times \text{Poisson}\left(y_{ij};\ \exp(\log \lambda_0^{-1} + \log \phi_{0i}^{-1} + \log \lambda_1^{-1} + \log \phi_{1i}^{-1}\right. \\
&\quad\left. + \log \lambda_2^{-1} + \log(L_i N_j))\right)^{I_i T_j}.
\end{aligned}$$

The log densities that comprise the log complete likelihood are

$$\begin{aligned}
&\log p(y_{ij}|\lambda_0, \lambda_1, \lambda_2, \phi_{0i}, \phi_{1i}, I_i) \\
&\quad= (1 - T_j) \times \left[y_{ij}(\log \lambda_0^{-1} + \log \phi_{0i}^{-1}) - \lambda_0^{-1}\phi_{0i}^{-1}L_i N_j\right] \\
&\quad+ (1 - I_i)T_j \times \left[y_{ij}(\log \lambda_0^{-1} + \log \phi_{0i}^{-1} + \log \lambda_1^{-1}) - \lambda_0^{-1}\phi_{0i}^{-1}\lambda_1^{-1}L_i N_j\right] \\
&\quad+ I_i T_j \times \left[y_{ij}(\log \lambda_0^{-1} + \log \phi_{0i}^{-1} + \log \lambda_1^{-1} + \log \phi_{1i}^{-1} + \log \lambda_2^{-1})\right. \\
&\quad\left. - \lambda_0^{-1}\phi_{0i}^{-1}\lambda_1^{-1}\phi_{1i}^{-1}\lambda_2^{-1}L_i N_j\right] + y_{ij}\log(L_i N_j) - y_{ij}!
\end{aligned}$$

$$\begin{aligned}
&\log p(I_i|\pi_1) = I_i \log \pi_1 + (1 - I_i)\log(1 - \pi_1) \\
&\log p(\pi_1) = -\log(\text{Beta}(\alpha, \beta)) + (\alpha - 1)\log \pi_1 + (\beta - 1)\log(1 - \pi_1) \\
&\log p(\phi_{ki}|\delta_k) = \delta_k \log \delta_k - \log(\Gamma(\delta_k)) + (\delta_k + 1)\log \phi_k^{-1} - \delta_k \phi_k^{-1}, \quad \text{for } k = 0, 1 \\
&\log p(\delta_k) = -A_{\delta_k} \log B_{\delta_k} - \log(\Gamma(A_{\delta_k})) + (A_{\delta_k} - 1)\log \delta_k - \frac{\delta_k}{B_{\delta_k}}, \quad \text{for } k = 0, 1 \\
&\log p(\lambda_m) = A_{\lambda_m} \log B_{\lambda_m} - \log(\Gamma(A_{\lambda_m})) + (A_{\lambda_m} + 1)\log \lambda_m^{-1} - B_{\lambda_m}\lambda_m^{-1}, \\
&\qquad \text{for } m = 0, 1, 2.
\end{aligned}$$

The product density restriction

$$q(H) = q_{\lambda_0}(\lambda_0)q_{\lambda_1}(\lambda_1)q_{\lambda_2}(\lambda_2)q_{\boldsymbol{\phi}_0}(\boldsymbol{\phi}_0)q_{\boldsymbol{\phi}_1}(\boldsymbol{\phi}_1)q_{\delta_0}(\delta_0)q_{\delta_1}(\delta_1)q_{\mathbf{I}}(\mathbf{I})q_{\pi_1}(\pi_1)$$

leads to the following q-densities:

- Derivation of $q_{\lambda_0}(\lambda_0)$:

$$
\begin{aligned}
q_{\lambda_0}(\lambda_0) &\propto \exp \mathrm{E}_{-\lambda_0}\{\log p(\mathbf{y}, H)\} \\
&\propto \exp \mathrm{E}_{-\lambda_0}\left\{\sum_{i,j} \log p(y_{ij}|\lambda_0, \lambda_1, \lambda_2, \phi_{0i}, \phi_{1i}, I_i) + \log p(\lambda_0)\right\} \\
&\propto \exp\left\{\sum_{i,j}\Big[(1-T_j)(y_{ij}\log\lambda_0^{-1} - \lambda_0^{-1}\widehat{M_{\phi_{0i}^{-1}}}L_iN_j)\right. \\
&\quad +(1-\widehat{M_{I_i}})T_j(y_{ij}\log\lambda_0^{-1} - \lambda_0^{-1}\widehat{M_{\phi_{0i}^{-1}}}\widehat{M_{\lambda_1^{-1}}}L_iN_j) \\
&\quad +\widehat{M_{I_i}}T_j(y_{ij}\log\lambda_0^{-1} - \lambda_0^{-1}\widehat{M_{\phi_{0i}^{-1}}}\widehat{M_{\lambda_1^{-1}}}\widehat{M_{\phi_{1i}^{-1}}}\widehat{M_{\lambda_2^{-1}}}L_iN_j)\Big] \\
&\quad \left. +(A_{\lambda_0}+1)\log\lambda_0^{-1} - \frac{B_{\lambda_0}}{\lambda_0}\right\}
\end{aligned}
$$

The kernel of an Inverse-Gamma density is identified on the right hand side. Therefore, it can be deduced that

$$q_{\lambda_0}(\lambda_0) = \mathrm{IG}(\widehat{A_{\lambda_0}}, \widehat{B_{\lambda_0}})$$

with

$$
\begin{aligned}
\widehat{A_{\lambda_0}} &= \sum_{i,j} y_{ij} + A_{\lambda_0} \\
\widehat{B_{\lambda_0}} &= \sum_{i,j} L_iN_j\Big[(1-T_j)\widehat{M_{\phi_{0i}^{-1}}} + (1-\widehat{M_{I_i}})T_j\widehat{M_{\phi_{0i}^{-1}}}\widehat{M_{\lambda_1^{-1}}} \\
&\quad +\widehat{M_{I_i}}T_j\widehat{M_{\phi_{0i}^{-1}}}\widehat{M_{\lambda_1^{-1}}}\widehat{M_{\phi_{1i}^{-1}}}\widehat{M_{\lambda_2^{-1}}}\Big] + B_{\lambda_0}.
\end{aligned}
$$

Moreover, the posterior mean and posterior expected log of λ_0^{-1} are

$$
\begin{aligned}
\widehat{M_{\lambda_0^{-1}}} &= \frac{\widehat{A_{\lambda_0}}}{\widehat{B_{\lambda_0}}} \\
\widehat{\log\lambda_0^{-1}} &= \mathrm{digamma}(\widehat{A_{\lambda_0}}) - \log\widehat{B_{\lambda_0}}.
\end{aligned}
$$

- Derivation of $q_{\lambda_1}(\lambda_1)$ and $q_{\lambda_2}(\lambda_2)$ is similar to that of $q_{\lambda_0}(\lambda_0)$:

$$q(\lambda_1) \propto \exp \mathrm{E}_{-\lambda_1} \left\{ \sum_{i,j} \log p(y_{ij}|\lambda_0, \lambda_1, \lambda_2, \phi_{0i}, \phi_{1i}, I_i) + \log p(\lambda_1) \right\}$$

$$\begin{aligned}
\Rightarrow q(\lambda_1) &= \mathrm{IG}(\widehat{A_{\lambda_1}}, \widehat{B_{\lambda_1}}) \\
\widehat{A_{\lambda_1}} &= \sum_{i,j} T_j y_{ij} + A_{\lambda_1} \\
\widehat{B_{\lambda_1}} &= \sum_{i,j} L_i N_j [(1 - \widehat{M_{I_i}}) T_j \widehat{M_{\lambda_0^{-1}}} \widehat{M_{\phi_{0i}^{-1}}} + \widehat{M_{I_i}} T_j \widehat{M_{\lambda_0^{-1}}} \widehat{M_{\phi_{0i}^{-1}}} \widehat{M_{\phi_{1i}^{-1}}} \widehat{M_{\lambda_2^{-1}}}] \\
&\quad + B_{\lambda_1},
\end{aligned}$$

and

$$\begin{aligned}
\widehat{M_{\lambda_1^{-1}}} &= \frac{\widehat{A_{\lambda_1}}}{\widehat{B_{\lambda_1}}} \\
\widehat{\log \lambda_1^{-1}} &= \mathrm{digamma}(\widehat{A_{\lambda_1}}) - \log \widehat{B_{\lambda_1}}.
\end{aligned}$$

$$q_{\lambda_2}(\lambda_2) \propto \exp \mathrm{E}_{-\lambda_2} \left\{ \sum_{i,j} \log p(y_{ij}|\lambda_0, \lambda_1, \lambda_2, \phi_{0i}, \phi_{1i}, I_i) + \log p(\lambda_2) \right\}$$

$$\begin{aligned}
\Rightarrow q_{\lambda_2}(\lambda_2) &= \mathrm{IG}(\widehat{A_{\lambda_2}}, \widehat{B_{\lambda_2}}) \\
\widehat{A_{\lambda_2}} &= \sum_{i,j} \widehat{M_{I_i}} T_j y_{ij} + A_{\lambda_2} \\
\widehat{B_{\lambda_2}} &= \sum_{i,j} L_i N_j [\widehat{M_{I_i}} T_j \widehat{M_{\lambda_0^{-1}}} \widehat{M_{\phi_{0i}^{-1}}} \widehat{M_{\lambda_1^{-1}}} \widehat{M_{\phi_{1i}^{-1}}}] + B_{\lambda_2},
\end{aligned}$$

and

$$\begin{aligned}
\widehat{M_{\lambda_2^{-1}}} &= \frac{\widehat{A_{\lambda_2}}}{\widehat{B_{\lambda_2}}} \\
\widehat{\log \lambda_2^{-1}} &= \mathrm{digamma}(\widehat{A_{\lambda_2}}) - \log \widehat{B_{\lambda_2}}.
\end{aligned}$$

- Derivation of $q_{\boldsymbol{\phi_0}}(\boldsymbol{\phi_0})$ and $q_{\boldsymbol{\phi_1}}(\boldsymbol{\phi_1})$ is also similar to that of $q_{\lambda_0}(\lambda_0)$, with induced factorizations:

$$q_{\boldsymbol{\phi_0}}(\boldsymbol{\phi_0}) \propto \exp \mathrm{E}_{-\boldsymbol{\phi_0}} \left\{ \sum_{i,j} \log p(y_{ij}|\lambda_0, \lambda_1, \lambda_2, \phi_{0i}, \phi_{1i}, I_i) \right.$$

$$\left. + \sum_i \log p(\phi_{0i}|\delta_0) \right\}$$

$$\Rightarrow \quad q_{\boldsymbol{\phi_0}}(\boldsymbol{\phi_0}) = \prod_i q_{\phi_{0i}}(\phi_{0i}) \text{and}$$

$$q_{\phi_{0i}}(\phi_{0i}) \propto \exp \mathrm{E}_{-\boldsymbol{\phi_0}} \left\{ \sum_j \log p(y_{ij}|\lambda_0, \lambda_1, \lambda_2, \phi_{0i}, \phi_{1i}, I_i) + \log p(\phi_{0i}|\delta_0) \right\}.$$

Therefore, for each i,

$$\begin{aligned}
q_{\phi_{0i}}(\phi_{0i}) &= \mathrm{IG}(\widehat{A_{\phi_{0i}}}, \widehat{B_{\phi_{0i}}}) \\
\widehat{A_{\phi_{0i}}} &= \sum_j y_{ij} + \widehat{M_{\delta_0}} \\
\widehat{B_{\phi_{0i}}} &= \sum_j L_i N_j \widehat{M_{\lambda_0^{-1}}}[(1 - T_j) + (1 - \widehat{M_{I_i}}) T_j \widehat{M_{\lambda_1^{-1}}} \\
&\quad + \widehat{M_{I_i}} T_j \widehat{M_{\lambda_1^{-1}}} \widehat{M_{\phi_{1i}^{-1}}} \widehat{M_{\lambda_2^{-1}}}] + \widehat{M_{\delta_0}} \\
\widehat{M_{\phi_{0i}^{-1}}} &= \frac{\widehat{A_{\phi_{0i}}}}{\widehat{B_{\phi_{0i}}}} \\
\widehat{\log \phi_{0i}^{-1}} &= \mathrm{digamma}(\widehat{A_{\phi_{0i}}}) - \log \widehat{B_{\phi_{0i}}}.
\end{aligned}$$

$$q_{\boldsymbol{\phi_1}}(\boldsymbol{\phi_1}) \propto \exp \mathrm{E}_{-\boldsymbol{\phi_1}} \left\{ \sum_{i,j} \log p(y_{ij}|\lambda_0, \lambda_1, \lambda_2, \phi_{0i}, \phi_{1i}, I_i) \right.$$

$$\left. + \sum_i \log p(\phi_{1i}|\delta_1) \right\}$$

$$\Rightarrow \quad q_{\boldsymbol{\phi_1}}(\boldsymbol{\phi_1}) = \prod_i q_{\phi_{1i}}(\phi_{1i}) \text{and}$$

$$q_{\phi_{1i}}(\phi_{1i}) \propto \exp \mathrm{E}_{-\boldsymbol{\phi_1}} \left\{ \sum_j \log p(y_{ij}|\lambda_0, \lambda_1, \lambda_2, \phi_{0i}, \phi_{1i}, I_i) + \log p(\phi_{1i}|\delta_1) \right\}.$$

Therefore, for each i,

$$\begin{aligned}
q_{\phi_{1i}}(\phi_{1i}) &= \text{IG}(\widehat{A_{\phi_{1i}}}, \widehat{B_{\phi_{1i}}}) \\
\widehat{A_{\phi_{1i}}} &= \sum_j \widehat{M_{I_i}} T_j y_{ij} + \widehat{M_{\delta_1}} \\
\widehat{B_{\phi_{1i}}} &= \sum_j L_i N_j \widehat{M_{I_i}} T_j \widehat{M_{\lambda_0^{-1}}} \widehat{M_{\phi_{0i}^{-1}}} \widehat{M_{\lambda_1^{-1}}} \widehat{M_{\lambda_2^{-1}}} + \widehat{M_{\delta_1}} \\
\widehat{M_{\phi_{1i}^{-1}}} &= \frac{\widehat{A_{\phi_{1i}}}}{\widehat{B_{\phi_{1i}}}} \\
\widehat{\log \phi_{1i}^{-1}} &= \text{digamma}(\widehat{A_{\phi_{1i}}}) - \log \widehat{B_{\phi_{1i}}}.
\end{aligned}$$

- Derivation of $q_{\delta_0}(\delta_0)$:

$$\begin{aligned}
q_{\delta_0}(\delta_0) &\propto \exp \mathrm{E}_{-\delta_0} \left\{ \sum_i \log p(\phi_{0i}|\delta_0) + \log p(\delta_0) \right\} \\
&\propto \exp \left\{ \sum_i \left[\delta_0 \log \delta_0 - \log \Gamma(\delta_0) + (\delta_0 + 1) \widehat{\log \phi_{0i}^{-1}} - \delta_0 \widehat{M_{\phi_{0i}^{-1}}} \right] \right. \\
&\quad \left. + - A_{\delta_0} \log B_{\delta_0} - \log \Gamma(A_{\delta_0}) + (A_{\delta_0} - 1) \log \delta_0 - \frac{\delta_0}{B_{\delta_0}} \right\}.
\end{aligned}$$

The right-hand side does not contain the kernel of any standard distribution. Therefore, an approximation to $\log \Gamma(\delta_0)$ is used.

For complex number z with large $Re(z)$, because $\Gamma(z+1) = z! = z\Gamma(z)$,

$$\begin{aligned}
\log \Gamma(z) &= \log \Gamma(z+1) - \log z \\
&= \log z! - \log z \\
&\approx \left(\frac{1}{2} \log(2\pi z) + z \log z - z \right) - \log z \\
&\quad \text{by Stirling's approximation } n! \approx \sqrt{2\pi n} \left(\frac{n}{e} \right)^n \\
&\approx (z - \frac{1}{2}) \log z - z + \frac{1}{2} \log(2\pi).
\end{aligned}$$

Hence,

$$\delta_0 \log \delta_0 - \log \Gamma(\delta_0) \approx \frac{1}{2} \log \delta_0 + \delta_0 - \frac{1}{2} \log(2\pi) \text{ for large } \delta_0 > 0.$$

Substituting the above on the right-hand side of the formula for $q_{\delta_0}(\delta_0)$ leads to the kernel of a Gamma density. Therefore,

$$\begin{aligned} q_{\delta_0}(\delta_0) &\approx \text{Gamma}(\widehat{A_{\delta_0}}, \widehat{B_{\delta_0}}) \\ \widehat{A_{\delta_0}} &= \frac{p}{2} + A_{\delta_0} \\ \widehat{B_{\delta_0}} &= \frac{1}{-p - \sum_i \widehat{\log \phi_{0i}^{-1}} + \sum_i \widehat{M_{\phi_{0i}^{-1}}} + \frac{1}{B_{\delta_0}}}, \end{aligned}$$

and

$$\widehat{M_{\delta_0}} = \widehat{A_{\delta_0}} \widehat{B_{\delta_0}}.$$

- Derivation of $q_{\delta_1}(\delta_1)$ is similar to that of $q_{\delta_0}(\delta_0)$, with the same approximation employed:

$$\begin{aligned} q_{\delta_1}(\delta_1) &\approx \text{Gamma}(\widehat{A_{\delta_1}}, \widehat{B_{\delta_1}}) \\ \widehat{A_{\delta_1}} &= \frac{p}{2} + A_{\delta_1} \\ \widehat{B_{\delta_1}} &= \frac{1}{-p - \sum_i \widehat{\log \phi_{1i}^{-1}} + \sum_i \widehat{M_{\phi_{1i}^{-1}}} + \frac{1}{B_{\delta_1}}}, \end{aligned}$$

and

$$\widehat{M_{\delta_1}} = \widehat{A_{\delta_1}} \widehat{B_{\delta_1}}.$$

- Derivation of $q_{\mathbf{I}}(\mathbf{I})$:

$$q_{\mathbf{I}}(\mathbf{I}) \propto \exp \mathrm{E}_{-\mathbf{I}} \left\{ \sum_{i,j} \log p(y_{ij}|\lambda_0, \lambda_1, \lambda_2, \phi_{0i}, \phi_{1i}, I_i) + \sum_i \log p(I_i|\pi_1) \right\}$$

$$\Rightarrow \quad q_{\mathbf{I}}(\mathbf{I}) = \prod_i q_{I_i}(I_i) \text{and}$$

$$q_{I_i}(I_i) \propto \exp \mathrm{E}_{-\mathbf{I}} \left\{ \sum_j \log p(y_{ij}|\lambda_0, \lambda_1, \lambda_2, \phi_{0i}, \phi_{1i}, I_i) + \log p(I_i|\pi_1) \right\}.$$

Therefore, for each i,

$$q_{I_i}(I_i) = \text{Bernoulli}\left(\frac{\exp(\widehat{\theta_i})}{\exp(\widehat{\theta_i}) + 1}\right)$$

$$\widehat{\theta_i} = \sum_j y_{ij} T_j \left(\widehat{\log \phi_{1i}^{-1}} + \widehat{\log \lambda_2^{-1}}\right)$$

$$+ L_i \widehat{M_{\phi_{0i}^{-1}}} \left(\sum_j N_j T_j\right) \widehat{M_{\lambda_0^{-1}}} \widehat{M_{\lambda_1^{-1}}} \left(1 - \widehat{M_{\phi_{1i}^{-1}}} \widehat{M_{\lambda_2^{-1}}}\right)$$

$$+ \widehat{\log \pi_1} - \widehat{\log(1 - \pi_1)},$$

and

$$\widehat{M_{I_i}} = \frac{\exp(\widehat{\theta_i})}{\exp(\widehat{\theta_i}) + 1}.$$

- Derivation of $q_{\pi_1}(\pi_1)$:

$$q_{\pi_1}(\pi_1) \propto \exp \text{E}_{-\pi_1} \left\{ \sum_i \log p(I_i|\pi_1) + \log p(\pi_1) \right\}$$

$$\Rightarrow q_{\pi_1}(\pi_1) = \text{Beta}(\widehat{\alpha_{\pi_1}}, \widehat{\beta_{\pi_1}})$$

$$\widehat{\alpha_{\pi_1}} = \sum_i \widehat{M_{I_i}} + \alpha$$

$$\widehat{\beta_{\pi_1}} = \sum_i (1 - \widehat{M_{I_i}}) + \beta,$$

and

$$\widehat{\log \pi_1} = \text{digamma}(\widehat{\alpha_{\pi_1}}) - \text{digamma}(\widehat{\alpha_{\pi_1}} + \widehat{\beta_{\pi_1}})$$

$$\widehat{\log(1 - \pi_1)} = \text{digamma}(\widehat{\beta_{\pi_1}}) - \text{digamma}(\widehat{\alpha_{\pi_1}} + \widehat{\beta_{\pi_1}})$$

Updating the q-densities in an iterative scheme boils down to updating the variational parameters in the scheme. Convergence is monitored via the scalar quantity $C_q(\mathbf{y})$, the lower bound on the log of the marginal data density:

$$\log \underline{p}(\mathbf{y}; \mathbf{q})$$

$$= \text{E}_H \log p(\mathbf{y}, H) - \text{E}_H \log q(H)$$

$$
\begin{aligned}
&= \mathrm{E}_H \left\{ \sum_{i,j} \log p(y_{ij}|\lambda_0, \lambda_1, \lambda_2, \phi_{0i}, \phi_{1i}, I_i) - \sum_i \log q_{I_i}(I_i) \right\} \\
&+ \mathrm{E}_H \left\{ \sum_i \log p(I_i|\pi_1) + \log p(\pi_1) - \log q_{\pi_1}(\pi_1) \right\} \\
&+ \mathrm{E}_H \{ \log p(\lambda_0) + \log p(\lambda_1) + \log p(\lambda_2) - \log q_{\lambda_0}(\lambda_0) \\
&- \log q_{\lambda_1}(\lambda_1) - \log q_{\lambda_2}(\lambda_2) \} \\
&+ \mathrm{E}_H \left\{ \sum_i \log p(\phi_{0i}|\delta_0) + \sum_i \log p(\phi_{1i}|\delta_1) + \log p(\delta_0) + \log p(\delta_1) \right\} \\
&+ \mathrm{E}_H \left\{ - \sum_i \log q_{\phi_{0i}}(\phi_{0i}) - \sum_i \log q_{\phi_{1i}}(\phi_{1i}) - \log q_{\delta_0}(\delta_0) - \log q_{\delta_1}(\delta_1) \right\} \\
&= \sum_{i,j} \left\{ (1 - T_j) \left[y_{ij} \left(\widehat{\log \lambda_0^{-1}} + \widehat{\log \phi_{0i}^{-1}} + \log(L_i N_j) \right) - \widehat{M_{\lambda_0^{-1}}} \widehat{M_{\phi_{0i}^{-1}}} L_i N_j \right] \right\} \\
&+ \sum_{i,j} \left\{ (1 - \widehat{M_{I_i}}) T_j \left[y_{ij} \left(\widehat{\log \lambda_0^{-1}} + \widehat{\log \phi_{0i}^{-1}} + \widehat{\log \lambda_1^{-1}} + \log(L_i N_j) \right) \right. \right. \\
&\left. \left. - \widehat{M_{\lambda_0^{-1}}} \widehat{M_{\phi_{0i}^{-1}}} \widehat{M_{\lambda_1^{-1}}} L_i N_j \right] \right\} \\
&+ \sum_{i,j} \left\{ \widehat{M_{I_i}} T_j \left[y_{ij} \left(\widehat{\log \lambda_0^{-1}} + \widehat{\log \phi_{0i}^{-1}} + \widehat{\log \lambda_1^{-1}} \right. \right. \right. \\
&\left. + \widehat{\log \phi_{1i}^{-1}} + \widehat{\log \lambda_2^{-1}} + \log(L_i N_j) \right) \\
&\left. \left. - \widehat{M_{\lambda_0^{-1}}} \widehat{M_{\phi_{0i}^{-1}}} \widehat{M_{\lambda_1^{-1}}} \widehat{M_{\phi_{1i}^{-1}}} \widehat{M_{\lambda_2^{-1}}} L_i N_j \right] \right\} + \sum_{i,j} \{ -y_{ij}! \} \\
&- \sum_i \left(\widehat{M_{I_i}} \log \widehat{M_{I_i}} + (1 - \widehat{M_{I_i}}) \log(1 - \widehat{M_{I_i}}) \right) \\
&+ \left[- \log(\mathrm{Beta}(\alpha, \beta)) + \log(\mathrm{Beta}(\widehat{\alpha_{\pi_1}}, \widehat{\beta_{\pi_1}})) \right] \\
&+ A_{\lambda_0} \log B_{\lambda_0} - \log \Gamma(A_{\lambda_0}) - \widehat{A_{\lambda_0}} \log \widehat{B_{\lambda_0}} + \log \Gamma(\widehat{A_{\lambda_0}}) + \widehat{A_{\lambda_0}} \\
&- \left(\sum_{i,j} y_{ij} \right) \widehat{\log \lambda_0^{-1}} - B_{\lambda_0} \widehat{M_{\lambda_0^{-1}}} \\
&+ A_{\lambda_1} \log B_{\lambda_1} - \log \Gamma(A_{\lambda_1}) - \widehat{A_{\lambda_1}} \log \widehat{B_{\lambda_1}} + \log \Gamma(\widehat{A_{\lambda_1}}) + \widehat{A_{\lambda_1}} \\
&- \left(\sum_{i,j} T_j y_{ij} \right) \widehat{\log \lambda_1^{-1}} - B_{\lambda_1} \widehat{M_{\lambda_1^{-1}}}
\end{aligned}
$$

$$
\begin{aligned}
&+ A_{\lambda_2} \log B_{\lambda_2} - \log \Gamma(A_{\lambda_2}) - \widehat{A_{\lambda_2}} \log \widehat{B_{\lambda_2}} + \log \Gamma(\widehat{A_{\lambda_2}}) + \widehat{A_{\lambda_2}} \\
&- \left(\sum_{i,j} \widehat{M_{I_i}} T_j y_{ij} \right) \widehat{\log \lambda_2^{-1}} - B_{\lambda_2} \widehat{M_{\lambda_2^{-1}}} \\
&+ \sum_{k=0,1} \left\{ \sum_i (\widehat{M_{\delta_k}} - \widehat{A_{\phi_{ki}}}) \log \phi_{ki}^{-1} - \widehat{M_{\delta_k}} \left(\sum_i \widehat{M_{\phi_{ki}^{-1}}} - p \right) - \frac{p}{2} \log(2\pi) \right. \\
&\left. - \sum_i \left(\widehat{A_{\phi_{ki}}} \log \widehat{B_{\phi_{ki}}} - \log \Gamma(\widehat{A_{\phi_{ki}}}) - \widehat{A_{\phi_{ki}}} \right) \right\} \\
&+ \sum_{k=0,1} \left\{ -A_{\delta_k} \log B_{\delta_k} - \log(\Gamma(A_{\delta_k})) - \frac{\widehat{M_{\delta_k}}}{B_{\delta_k}} + \widehat{A_{\delta_k}} \log \widehat{B_{\delta_k}} + \log(\Gamma(\widehat{A_{\delta_k}})) + \widehat{A_{\delta_k}} \right\}.
\end{aligned}
$$

The VB-proteomics algorithm consists of the following steps:

Step 1: Initialize $\widehat{B_{\lambda_0}}, \widehat{B_{\lambda_1}}, \widehat{B_{\delta_0}}, \widehat{B_{\delta_1}}$, and $\widehat{A_{\phi_{0i}}}, \widehat{A_{\phi_{1i}}}, \widehat{B_{\phi_{0i}}}, \widehat{B_{\phi_{1i}}}$ for each i.
Step 2: Cycle through $\widehat{A_{\lambda_2}}, \widehat{B_{\lambda_2}}, \widehat{B_{\lambda_0}}, \widehat{B_{\lambda_1}}, \widehat{B_{\delta_0}}, \widehat{B_{\delta_1}}, \widehat{A_{\phi_{0i}}}, \widehat{B_{\phi_{0i}}}, \widehat{A_{\phi_{1i}}}, \widehat{B_{\phi_{1i}}}, \widehat{M_{I_i}}$ iteratively, until the increase in $C_q(\mathbf{y})$ computed at the end of each iteration is negligible.
Step 3: Compute $\widehat{\alpha_{\pi_1}}$ and $\widehat{\beta_{\pi_1}}$ using converged variational parameter values.

The values of the model parameters were chosen to form non-informative priors: The shape and scale parameters for the Inverse-Gamma priors were set to be 0.1, the shape and scale parameters for the Gamma priors were set to be 0.1 and 100 respectively, and the parameters for the Beta prior were both 1. Because a log-Normal distribution is approximately a Gamma distribution, the variance of $e^{b_{ki}}, k = 0, 1$ which follows a log-Normal distribution in the Poisson GLMM roughly equals to the variance of $\phi_{ki}^{-1}, k = 0, 1$ which follows a Gamma distribution in the Poisson–Gamma HGLM. That is, $(e^{\sigma_k^2} - 1)e^{\sigma_k^2} \approx 1/\delta_k,\ k = 0, 1$. Based on this we determined parameter values in the Gamma$(A_{\delta_k}, B_{\delta_k})$ prior for $\delta_k,\ k = 0, 1$.

Starting values of posterior mean of the latent indicator were $\widehat{M_{I_i}} = 1$ for proteins that are associated with the 20% smallest p-values from one protein at a time Score tests and $\widehat{M_{I_i}} = 0$ otherwise.

An approximation to the digamma function, digamma$(z) \approx \log z - \frac{1}{2z}$, was used wherever z was too small in VB implementation.

References

Alon U, Barkai N, Notterman D, Gish K, Ybarra S, Mack D, Levine A (1999) Broad patterns of gene expression revealed by clustering analysis of tumor and normal colon tissues probed by oligonucleotide arrays. Proc Natl Acad Sci 96(12):6745–6750

Attias H (2000) A variational Bayesian framework for graphical models. Adv Neural Inf Process Syst 12(1–2):209–215

Bar H, Schifano E (2010) Lemma: Laplace approximated EM microarray analysis. R package version 1.3-1. http://CRAN.R-project.org/package=lemma

Bar H, Booth J, Schifano E, Wells M (2010) Laplace approximated EM microarray analysis: an empirical Bayes approach for comparative microarray experiments. Stat Sci 25(3):388–407

Beal M (2003) Variational algorithms for approximate Bayesian inference. PhD thesis, University of London

Bishop C (1999) Variational principal components. In: Proceedings of ninth international conference on artificial neural networks, ICANN'99, vol 1. IET, pp 509–514

Bishop C (2006) Pattern recognition and machine learning. Springer Science+ Business Media, New York

Bishop C, Spiegelhalter D, Winn J (2002) VIBES: a variational inference engine for Bayesian networks. Adv Neural Inf Proces Syst 15:777–784

Blei D, Jordan M (2006) Variational inference for Dirichlet process mixtures. Bayesian Anal 1(1):121–143

Booth J, Eilertson K, Olinares P, Yu H (2011) A Bayesian mixture model for comparative spectral count data in shotgun proteomics. Mol Cell Proteomics 10(8):M110-007203

Boyd S, Vandenberghe L (2004) Convex optimization. Cambridge University Press, Cambridge

Callow M, Dudoit S, Gong E, Speed T, Rubin E (2000) Microarray expression profiling identifies genes with altered expression in HDL-deficient mice. Genome Res 10(12):2022–2029

Christensen R, Johnson WO, Branscum AJ, Hanson TE (2011) Bayesian ideas and data analysis: an introduction for scientists and statisticians. CRC, Boca Raton

Consonni G, Marin J (2007) Mean-field variational approximate Bayesian inference for latent variable models. Comput Stat Data Anal 52(2):790–798

Corduneanu A, Bishop C (2001) Variational Bayesian model selection for mixture distributions. In: Jaakkola TS, Richardson TS (eds) Artificial intelligence and statistics 2001. Morgan Kaufmann, Waltham, pp 27–34

Cowles MK Carlin BP (1996) Markov chain Monte Carlo convergence diagnostics: a comparative review. J Am Stat Assoc 91(434):883–904

De Freitas N, Højen-Sørensen P, Jordan M, Russell S (2001) Variational MCMC. In: Breese J, Koller D (eds) Proceedings of the seventeenth conference on uncertainty in artificial intelligence. Morgan Kaufmann, San Francisco, pp 120–127

Efron B (2008) Microarrays, empirical Bayes and the two-groups model. Stat Sci 23(1):1–22

Faes C, Ormerod J, Wand M (2011) Variational Bayesian inference for parametric and nonparametric regression with missing data. J Am Stat Assoc 106(495):959–971

Friston K, Ashburner J, Kiebel S, Nichols T, Penny W (2011) Statistical parametric mapping: the analysis of functional brain images. Academic, London

Gelman A, Carlin JB, Stern HS, Rubin DB (2003) Bayesian data analysis. Chapman & Hall/CRC, London/Boca Raton

Ghahramani Z, Beal M (2000) Variational inference for Bayesian mixtures of factor analysers. Adv Neural Inf Proces Syst 12:449–455

Goldsmith J, Wand M, Crainiceanu C (2011) Functional regression via variational Bayes. Electr J Stat 5:572

Grimmer J (2011) An introduction to Bayesian inference via variational approximations. Polit Anal 19(1):32–47

Honkela A, Valpola H (2005) Unsupervised variational Bayesian learning of nonlinear models. In: Saul LK, Weiss Y, Bottou L (eds) Advances in neural information processing systems, vol 17. MIT, Cambridge, pp 593–600

Jaakkola TS (2000) Tutorial on variational approximation methods. In: Opper M, Saad D (eds) Advanced mean field methods: theory and practice. MIT, Cambridge, pp 129–159

Li Z, Sillanpää M (2012) Estimation of quantitative trait locus effects with epistasis by variational Bayes algorithms. Genetics 190(1):231–249

Li J, Das K, Fu G, Li R, Wu R (2011) The Bayesian lasso for genome-wide association studies. Bioinformatics 27(4):516–523

Logsdon B, Hoffman G, Mezey J (2010) A variational Bayes algorithm for fast and accurate multiple locus genome-wide association analysis. BMC Bioinf 11(1):58

Luenberger D, Ye Y (2008) Linear and nonlinear programming. International series in operations research & management science, vol 116. Springer, New York

Marin J-M, Robert CP (2007) Bayesian core: a practical approach to computational Bayesian statistics. Springer, New York

Martino S, Rue H (2009) R package: INLA. Department of Mathematical Sciences, NTNU, Norway. Available at http://www.r-inla.org

McGrory C, Titterington D (2007) Variational approximations in Bayesian model selection for finite mixture distributions. Comput Stat Data Anal 51(11):5352–5367

McLachlan G, Peel D (2004) Finite mixture models. Wiley, New York

Minka T (2001a) Expectation propagation for approximate Bayesian inference. In: Breese J, Koller D (eds) Proceedings of the seventeenth conference on uncertainty in artificial intelligence. Morgan Kaufmann, San Francisco, pp 362–369

Minka T (2001b) A family of algorithms for approximate Bayesian inference. PhD thesis, Massachusetts Institute of Technology

Ormerod J (2011) Grid based variational approximations. Comput Stat Data Anal 55(1):45–56

Ormerod J, Wand M (2010) Explaining variational approximations. Am Stat 64(2):140–153

Rue H, Martino S, Chopin N (2009) Approximate Bayesian inference for latent Gaussian models by using integrated nested Laplace approximations. J R Stat Soc Ser B 71(2):319–392

Salter-Townshend M, Murphy T (2009) Variational Bayesian inference for the latent position and cluster model. In: NIPS 2009 (Workshop on analyzing networks & learning with graphs)

Sing T, Sander O, Beerenwinkel N, Lengauer T (2007) ROCR: visualizing the performance of scoring classifiers. R package version 1.0-2. http://rocr.bioinf.mpi-sb.mpg.de/ROCR.pdf/

Smídl V, Quinn A (2005) The variational Bayes method in signal processing. Springer, Berlin

Smyth G (2004) Linear models and empirical Bayes methods for assessing differential expression in microarray experiments. Stat Appl Genet Mol Biol 3(1):1–25. Article 3

Smyth G (2005) Limma: linear models for microarray data. In: Bioinformatics and computational biology solutions using R and bioconductor. Springer, New York, pp 397–420

Teschendorff A, Wang Y, Barbosa-Morais N, Brenton J, Caldas C (2005) A variational Bayesian mixture modelling framework for cluster analysis of gene-expression data. Bioinformatics 21(13):3025–3033

Tzikas D, Likas A, Galatsanos N (2008) The variational approximation for Bayesian inference. IEEE Signal Process Mag 25(6):131–146

Wand MP, Ormerod JT, Padoan SA, Frührwirth R (2011) Mean field variational Bayes for elaborate distributions. Bayesian Anal 6(4):1–48

Wang B, Titterington DM (2005) Inadequacy of interval estimates corresponding to variational Bayesian approximations. In: Cowell RG, Ghahramani Z (eds) Proceedings of the tenth international workshop on artificial intelligence and statistics. Society for Artificial Intelligence and Statistics, pp 373–380

Zhang M, Montooth K, Wells M, Clark A, Zhang D (2005) Mapping multiple quantitative trait loci by Bayesian classification. Genetics 169(4):2305–2318

Chapter 8
Hypothesis Testing for High-Dimensional Data

Wei Biao Wu, Zhipeng Lou, and Yuefeng Han

Abstract We present a systematic theory for tests for means of high-dimensional data. Our testing procedure is based on an invariance principle which provides distributional approximations of functionals of non-Gaussian vectors by those of Gaussian ones. Differently from the widely used Bonferroni approach, our procedure is dependence-adjusted and has an asymptotically correct size and power. To obtain cutoff values of our test, we propose a half-sampling method which avoids estimating the underlying covariance matrix of the random vectors. The latter method is shown via extensive simulations to have an excellent performance.

Keywords Gaussian approximation · Goodness-of-Fit Test · Half-sampling · High-dimensional data · Hypothesis testing · Large p small n · Rademacher weighted differencing

8.1 Introduction

With the advance of modern data collection techniques, high-dimensional data appear in various fields including physics, biology, healthcare, finance, marketing, social network, and engineering among others. A common feature in such datasets is that the data dimension or the number of involved parameters can be quite large. As a fundamentally important problem in the study of such data, one would like to perform statistical inference of those parameters such as multiple testing or construction of confidence regions. With that one is able to provide an answer to the question whether there is signal in the dataset, or whether the dataset consists only of random noises. Due to the high-dimensionality, the inferential procedures developed for low-dimensional problems may no longer be valid in the high-dimensional setting. Different approaches should be designed to account for high-dimensionality.

W. B. Wu (✉) · Z. Lou · Y. Han
Department of Statistics, University of Chicago, Chicago, IL, USA
e-mail: wbwu@galton.uchicago.edu; zplou@galton.uchicago.edu; yfhan@uchicago.edu

© Springer International Publishing AG, part of Springer Nature 2018

W. K. Härdle et al. (eds.), *Handbook of Big Data Analytics*, Springer Handbooks of Computational Statistics, https://doi.org/10.1007/978-3-319-18284-1_8

There exists a huge literature on multiple testing; see, for example, Dudiot and van der Laan (2008), Efron (2010) and Dickhaus (2014).

We now introduce the setting of our testing problem. Assume that $X_1, X_2, \ldots,$ are independent and identically distributed (i.i.d.) p-dimensional random vectors, with mean vector $\mu = (\mu_1, \ldots, \mu_p)^T = E(X_i)$ and covariance matrix $\Sigma = \mathrm{cov}(X_i) = (\sigma_{jk})_{j,k \le p}$. We are testing the hypothesis of existence of a signal

$$H_0 : \mu = 0 \text{ vs } H_A : \mu \neq 0 \tag{8.1}$$

based on the sample $X_1, \ldots, X_n$. This formulation is actually very general and its solution can be applied to many other problems; see Sect. 8.2. We can estimate μ by the sample mean vector $\hat{\mu} = \bar{X}_n = n^{-1} \sum_{i=1}^{n} X_i$. The classical Hotelling's T-squared test has the form

$$T = \bar{X}_n \hat{\Sigma}_n^{-1} \bar{X}_n, \tag{8.2}$$

where

$$\hat{\Sigma}_n = (n-1)^{-1} \sum_{i=1}^{n} (X_i - \bar{X}_n)(X_i - \bar{X}_n)^T \tag{8.3}$$

is the sample covariance matrix estimate of Σ. If p is small and fixed, by the Central Limit Theorem (CLT),

$$\sqrt{n}(\bar{X}_n - \mu) \Rightarrow N(0, \Sigma). \tag{8.4}$$

By the Law of Large Numbers, if Σ is non-singular,

$$\hat{\Sigma}_n^{-1} \to \Sigma^{-1} \text{ almost surely.} \tag{8.5}$$

Clearly (8.4) and (8.5) imply that under H_0, the Hotelling's T-squared statistic $nT \Rightarrow \chi_p^2$ (χ^2 distribution with degrees of freedom p). Thus we can reject H_0 at level $0 < \alpha < 1$ if $nT > \chi_{p,1-\alpha}^2$, the $(1-\alpha)$th quantile of χ_p^2.

In the high-dimensional situation in which p can be much larger than n, the CLT (8.4) is no longer valid; see Portnoy (1986). Furthermore, $\hat{\Sigma}_n$ is singular and thus T is not well-defined. Also the matrix convergence (8.5) may not hold, see Marčenko and Pastur (1967). In this chapter we shall apply a testing functional approach that does not use $\hat{\Sigma}_n^{-1}$ or the precision matrix Σ^{-1}. A function $g : \mathbb{R}^p \to [0, \infty)$ is said to be a testing functional if the following requirements are satisfied: (1) (monotonicity) for any $x = (x_1, \ldots, x_p)^T \in \mathbb{R}^p$ and $0 < c < 1$, $g(cx) \le g(x)$; (2) (identifiability) $g(x) = 0$ if and only if $x = 0$. We shall consider the test statistic

$$T_n = g(\sqrt{n}\bar{X}_n). \tag{8.6}$$

Examples of g include the L^2-based test with $g(x) = \sum_{j=1}^{p} x_j^2$, the L^∞-based test with $g(x) = \max_{j \le p} |x_j|$, the weighted empirical process $g(x) =$

$\sup_{u\geq 0}(\sum_{j=1}^{p} \mathbf{1}_{|x_j|\geq u} h(u))$, where $h(\cdot)$ is a nonnegative-valued non-decreasing function, among others. We reject H_0 in (8.1) if T_n is too big.

As a theoretical foundation, we base our testing procedure on the following invariance principle result

$$\sup_{t\in\mathbb{R}} |P[g(\sqrt{n}(\bar{X}_n - \mu)) \leq t] - P[g(\sqrt{n}\bar{Z}_n) \leq t]| \to 0, \tag{8.7}$$

where $Z, Z_1, Z_2, \ldots$ are i.i.d. $N(0, \Sigma)$ random vectors and $\bar{Z}_n = n^{-1}\sum_{i=1}^{n} Z_i =_{\mathcal{D}} n^{-1/2}Z$. Interestingly, though the CLT (8.4) does not generally hold in the high-dimensional setting, the testing functional form (8.7) may still be valid. Chernozhukov et al. (2014) proved (8.7) with the L^∞ norm $g(x) = \max_{j\leq p}|x_j|$, while Xu et al. (2014) consider the L^2 based test with $g(x) = \sum_{j=1}^{p} x_j^2$. In Sect. 8.5 we shall provide a sufficient condition so that (8.7) holds for certain testing functionals.

In applying (8.7) for testing (8.1), one needs to know the distribution of $g(\sqrt{n}\bar{Z}_n) =_{\mathcal{D}} g(Z)$ so that a suitable cutoff value can be obtained. The latter problem is highly nontrivial since the covariance matrix Σ, which is viewed as a nuisance parameter here, is typically not known and the associated estimation issue can be quite challenging. In Sect. 8.5 we shall propose a half-sampling technique which can avoid estimating the nuisance covariance matrix Σ.

8.2 Applications

Our paradigm (8.1) is actually quite general and it can be applied to testing of high-dimensional covariance matrices, testing of independence of high-dimensional data, analysis of variances with non-normal and heteroscedastic errors.

8.2.1 Testing of Covariance Matrices

There is a huge literature on testing covariance matrices such as uncorrelatedness, sphericity, or other patterns. For Gaussian data, tests for $\Sigma = \sigma^2 \mathrm{I}_p$, where I_p is the identity matrix, can be found in Ahmad (2010), Birke and Dette (2005), Chen et al. (2010), Fisher et al. (2010) and Ledoit and Wolf (2002). Tests for equality of covariance matrices are studied in Bai et al. (2009) and Jiang et al. (2012), and for sphericity is in Onatski et al. (2013). Minimax properties are considered in Cai and Ma (2013). For other contributions, see Qu and Chen (2012), Schott (2005, 2007), Srivastava (2005), Xiao and Wu (2013) and Zhang et al. (2013).

Assume that we have data matrix $\mathbf{Y}_n = (Y_{i,j})_{1\leq i\leq n, 1\leq j\leq p}$, where $(Y_{i,j})_{j=1}^{p}$, $i = 1, \ldots, n$, are i.i.d. p-dimensional random vectors. Let

$$\sigma_{jk} = \mathrm{cov}(Y_{1,j}, Y_{1,k}), \quad 1 \leq j, k \leq p, \tag{8.8}$$

be the covariance function. Consider testing hypothesis for uncorrelatedness:

$$H_0 : \sigma_{jk} = 0 \text{ for all } j \neq k. \tag{8.9}$$

For simplicity assume that $E(Y_{i,j}) = 0$. For a pair $a = (j, k)$ write $X_{i,a} = Y_{i,j}Y_{i,k}$, and $\bar{X}_a = n^{-1}\sum_{i=1}^n X_{i,a}$ and the $(p^2 - p)$-dimensional vector $\bar{X} = (\bar{X}_a)_{a\in\mathcal{A}}$, where $\mathcal{A} = \{(j, k) : j \neq k, j \leq p, k \leq p\}$. The hypothesis H_0 in (8.9) can be tested by using the test statistics $T = g(\sqrt{n}\bar{X})$. Xiao and Wu (2013) considered the L^∞ based test with $g(x) = \max_i |x_i|$, generalizing the result in Jiang (2004) which concerns the special case for i.i.d. vectors with independent entries. Han and Wu (2017) performed an L^2 based test for patterns of covariances with the test statistic

$$T = \sum_{a\in\mathcal{A}} \bar{X}_a^2 = \sum_{j\neq k} \hat{\sigma}_{jk}^2. \tag{8.10}$$

With slight modifications, one can also test the sphericity hypothesis

$$H_0 : \Sigma = \sigma^2 \mathrm{I}_p \text{ for some } \sigma^2 > 0, \tag{8.11}$$

where I_p is the $p \times p$ identity matrix. Let $\mathcal{A}_0 = \{(j, k) : j, k \leq p\}$ with diagonal entries added to $\mathcal{A}$. For $a = (j, j) \in \mathcal{A}_0$, let $X_{i,a} = Y_{i,j}^2 - \sigma^2$. If σ^2 is known, then H_0 in (8.11) can be rejected at level $\alpha \in (0, 1)$ if $T = g(\sqrt{n}\bar{X}) > t_{1-\alpha}$, where $t_{1-\alpha}$ is the $(1 - \alpha)$th quantile of $g(Z)$ and Z is a centered Gaussian vector with covariance structure $\mathrm{cov}(Z_a, Z_b) = E(X_{i,a}X_{i,b})$, $a, b \in \mathcal{A}_0$. In the case that σ^2 is not known, we shall use an estimate. For example, we can let $\hat{\sigma}^2 = n^{-1}\sum_{j=1}^n \hat{\sigma}_{jj}^2$, and consider $X_{i,a}^\circ = Y_{i,j}^2 - \hat{\sigma}^2$. Let $X_{i,a}^\circ = X_{i,a}$ if $a = (j, k)$ with $j \neq k$. The hypothesis H_0 in (8.11) can be tested by the statistic $T^\circ = g(\sqrt{n}\bar{X}^\circ)$.

8.2.2 *Testing of Independence*

Let $Y_i = (Y_{i,j})_{j=1}^p$, $i = 1, \ldots, n$, be i.i.d. p-dimensional random vectors with joint cumulative distribution function

$$F_{j_1,\ldots,j_d}(y_{j_1}, \ldots, y_{j_d}) = P(Y_{i,j_1} \leq y_{j_1}, \ldots, Y_{i,j_d} \leq y_{j_d}). \tag{8.12}$$

Consider the problem of testing whether entries of Y_i are independent. Assume that the marginal distributions are standard uniform[0, 1]. For $\mathbf{j} = (j_1, \ldots, j_d)$, write $F_{\mathbf{j}}(y_{\mathbf{j}}) = F_{j_1,\ldots,j_d}(y_{j_1}, \ldots, y_{j_d})$. For fixed d, the hypothesis of d-wise independence is

$$H_0 : \; F_{\mathbf{j}}(y_{\mathbf{j}}) = y_{j_1} \ldots y_{j_d} \text{ holds for all } y_1, \ldots, y_d \in (0, 1) \text{ and } \mathbf{j} \in \mathcal{A}_d, \tag{8.13}$$

where $\mathcal{A}_d = \{\mathbf{j} = (j_1, \ldots, j_d) : j_1 < \cdots < j_d \le p\}$. Pairwise and triple-wise independence correspond to $d = 2$ and $d = 3$, respectively. We estimate $F_{\mathbf{j}}(y_{\mathbf{j}})$ by the empirical cdf

$$\hat{F}_{\mathbf{j}}(y_{\mathbf{j}}) = \frac{1}{n} \sum_{i=1}^{n} \mathbf{1}_{Y_{i,\mathbf{j}} \le y_{\mathbf{j}}}, \tag{8.14}$$

where the notation $Y_{i,\mathbf{j}} \le y_{\mathbf{j}}$ means $Y_{i,j_h} \le y_{j_h}$ for all $h = 1, \ldots, d$. Let $y_{\mathbf{m}_1}, \ldots, y_{\mathbf{m}_N}$, $N \to \infty$, be a dense set of $[0, 1]^d$. For example, we can choose them to be the lattice set $\{1/K, \ldots, (K-1)/K\}^d$ with $N = (K-1)^d$. Let X_i, $1 \le i \le n$, be the $Np!/(d!(p-d)!)$-dimensional vector with the $(\ell\mathbf{j})$th component being $\mathbf{1}_{Y_{i,\mathbf{j}} \le y_{\mathbf{m}_\ell}} - \prod_{h \in \mathbf{m}_\ell} y_h$, $1 \le \ell \le N$, $\mathbf{j} \in \mathcal{A}_d$. Then the L^2-based test for (8.13) on the dense set $(y_{\mathbf{m}_\ell})_{\ell=1}^N$ has the form $n|\bar{X}|_2^2$.

8.2.3 *Analysis of Variance*

Consider the following two-way ANOVA model

$$Y_{ijk} = \mu + \alpha_i + \beta_j + \delta_{ij} + \varepsilon_{ijk}, i = 1, \ldots, I, j = 1, \ldots, J, k = 1, \ldots, K, \tag{8.15}$$

where μ is the grand mean, α_i and β_j are the main effects from the first and the second factors, respectively, and δ_{ij} are the interaction effect. Assume that $(Y_{ijk})_{i \le I, j \le J}$, $k = 1, \ldots, K$, are i.i.d. Consider the hypothesis of interaction:

$$H_0 : \delta_{ij} = 0 \text{ for all } i = 1, \ldots, I,\ j = 1, \ldots, J. \tag{8.16}$$

In the classical ANOVA procedure, one assumes that ε_{ijk}, $i \le I, j \le J$, are i.i.d. $N(0, \sigma^2)$ and makes use of the fact that the sum of squares

$$SS_I = \sum_{i=1}^{I} \sum_{j=1}^{J} (\bar{Y}_{ij\cdot} - \bar{Y}_{i\cdot\cdot} - \bar{Y}_{\cdot j\cdot} + \bar{Y}_{\cdots})^2 \tag{8.17}$$

is distributed as $\sigma^2 \chi^2_{(I-1)(J-1)}$. Here $\bar{Y}_{ij\cdot} = K^{-1} \sum_{k=1}^{K} Y_{ijk}$ and other sample averages $\bar{Y}_{i\cdot\cdot}$, $\bar{Y}_{\cdot j\cdot}$ and $\bar{Y}_{\cdots}$ are similarly defined. The null hypothesis H_0 is rejected at level $\alpha \in (0, 1)$ if

$$\frac{SS_I}{(I-1)(J-1)} > SS_E F_{(I-1)(J-1), IJ(K-1), 1-\alpha} \tag{8.18}$$

where $F_{(I-1)(J-1),IJ(K-1),1-\alpha}$ is the $(1-\alpha)$th quantile of the F-distribution $F_{(I-1)(J-1),IJ(K-1)}$ and

$$SS_E = \frac{\sum_{i=1}^{I}\sum_{j=1}^{J}(Y_{ijk} - \bar{Y}_{ij\cdot})^2}{IJ(K-1)} \tag{8.19}$$

is an estimate of σ^2.

The classical ANOVA procedure can be invalid when the assumption that ε_{ijk}, $i \le I, j \le J$ are i.i.d. $N(0,\sigma^2)$ is violated. In the latter case SS_I may no longer have a χ^2 distribution. However we can still approximate the distribution of SS_I in terms of (8.7). For $a = (i,j)$ let $X_{ak} = \bar{Y}_{ijk} - \bar{Y}_{i\cdot k} - \bar{Y}_{\cdot jk} + \bar{Y}_{\cdot\cdot k}$. Then $SS_I = \sum_{a\in\mathcal{A}}\bar{X}_{a\cdot}^2$, where $\bar{X}_{a\cdot} = K^{-1}\sum_{k=1}^{K}X_{ak}$.

8.3 Tests Based on L^∞ Norms

Fan et al. (2007) considered the L^∞ norm based test of (8.1) with the form

$$M_n = \max_{j\le p}\frac{\sqrt{n}|\hat{\mu}_j - \mu_j|}{\hat{\sigma}_j}, \text{ where } \hat{\sigma}_j^2 = \frac{1}{n}\sum_{i=1}^{n}(X_{ij} - \hat{\mu}_j)^2. \tag{8.20}$$

Assume that the dimension p satisfies

$$\log p = o(n^{1/3}) \tag{8.21}$$

and the uniform bounded third moment condition

$$\max_{j\le p} E|X_{ij} - \mu_j|^3 = O(1). \tag{8.22}$$

Let Φ be the standard normal cumulative distribution function and $z_\alpha = \Phi^{-1}(\alpha)$. Then

$$P(M_n \ge z_{1-\alpha/(2p)}) \le \alpha + o(1). \tag{8.23}$$

Namely, if we perform the test by rejecting H_0 of (8.1) whenever $M_n \ge z_{1-\alpha/(2p)}$, the familywise type I error of the latter test is asymptotically bounded by α. As a finite sample correction, the cutoff value $z_{1-\alpha/(2p)}$ in (8.23) can be replaced by the t-distribution quantile $t_{n-1,1-\alpha/(2p)}$ with degree of freedom $n-1$, noting that $(n-1)^{1/2}\hat{\mu}_j/\hat{\sigma}_j \sim t_{n-1}$ if X_{ij} are Gaussian. Due to the Bonferroni correction, the test by Fan et al. (2007) can be quite conservative if the dependence among entries of X_i is strong. For example, if $X_{i1} = X_{i2} = \cdots = X_{ip}$, then instead of using the cutoff value $z_{1-\alpha/(2p)}$, one should use $z_{1-\alpha/2}$, since the cutoff value $z_{1-\alpha/(2p)}$ leads to the extremely conservative type I error $\alpha/(2p)$. If entries of X_i are independent and X_i is Gaussian, then the type I error is $1-(1-\alpha/p)^p \to 1-e^{-\alpha}$ and it is slightly conservative. For example, when $\alpha = 0.05$, $1-e^{-\alpha} = 0.04877058$.

Liu and Shao (2013) obtained Gumbel convergence of M_n under the following conditions: (1) for some $r > 3$, the uniform bounded rth moment conditions $\max_{j\le p} E|X_{ij}-\mu_j|^r = O(1)$ holds, which is slightly stronger than (8.22) and (2) weak dependence among entries of X_i. For $\Sigma = (\sigma_{jk})_{j,k\le p}$, assume the correlation matrix $R = (r_{jk})_{j,k\le p}$ with $r_{jk} = \sigma_{jk}/(\sigma_{jj}^{1/2}\sigma_{kk}^{1/2})$ has the property: for some $\gamma > 0$,

$$\max \#\{j \le p : |r_{jk}| \ge (\log p)^{-1-\gamma}\} = O(p^{\rho}) \tag{8.24}$$

holds for all $\rho > 0$. Then under (8.21), Theorem 3.1 in Liu and Shao (2013) asserts the Gumbel convergence

$$M_n - 2\log p + \log\log p \Rightarrow \mathcal{G}, \tag{8.25}$$

where $\mathcal{G}$ follows the Gumbel distribution $P(\mathcal{G} \le y) = \exp(-e^{-y/2}/\pi^{1/2})$. By (8.25), one can reject H_0 in (8.1) at level $\alpha \in (0,1)$ based on the L^∞ norm test

$$\max_{j\le p} \frac{\sqrt{n}|\hat{\mu}_j|}{\hat{\sigma}_j} > 2\log p - \log\log p + g_{1-\alpha}, \tag{8.26}$$

where $g_{1-\alpha}$ is chosen such that $P(\mathcal{G} \le g_{1-\alpha}) = 1-\alpha$. Clearly the latter test has an asymptotically correct size.

Applying Theorem 2.2 in Chernozhukov et al. (2014), we can have the following Gaussian approximation result. Assume that there exist constants $c_1, c_2 > 0$ such that $c_1 \le E(X_{ij}-\mu_j)^2 \le c_2$ holds for all $j \le p$ and assume that $u = u_{n,p}$ satisfies

$$P\left[\max_{j\le p} |X_{1j}-\mu_j| \ge u\right] = o(n^{-1}) \tag{8.27}$$

Let $m_k = \max_{j\le p}(E|X_{1j}-\mu_j|^k)^{1/k}$ and further assume that

$$n^{-1/8}(m_3^{3/4} + m_4^{1/2})(\log(pn))^{7/8} + n^{-1/2}(\log(pn))^{3/2}u \to 0. \tag{8.28}$$

Let $Z \sim N(0,R)$. Then we have the Gaussian approximation result: as $n \to \infty$

$$\sup_t |P(M_n \ge t) - P(|Z|_\infty \ge t)| \to 0. \tag{8.29}$$

Let $t_{1-\alpha}$ be the $(1-\alpha)$th quantile of $|Z|_\infty$. The Gaussian approximation (8.29) leads to L^∞ norm based test: H_0 is rejected at level α if $\max_{j<p} \sqrt{n}|\hat{\mu}_j|/\hat{\sigma}_j \ge t_{1-\alpha}$. In comparison with the result in Fan et al. (2007), the latter test has an asymptotically correct size and it is dependence adjusted. To obtain an estimate for the cutoff value $t_{1-\alpha}$, Chernozhukov et al. (2014) proposed a Gaussian Multiplier Bootstrap (GMB)

method. Given $X_1, \dots, X_n$, let $\hat{t}_{1-\alpha}$ be such that

$$P\left(\max_{j\le p} n^{-1/2}\left|\sum_{i=1}^{n} X_{ij}e_i\right| \ge \hat{t}_{1-\alpha}|X_1, \dots, X_n\right) = \alpha, \tag{8.30}$$

where e_i are i.i.d. $N(0, 1)$ random variables independent of $(X_{ij})_{i\ge1,j\ge1}$. Note that $\hat{t}_{1-\alpha}$ can be numerically calculated by extensive Monté Carlo simulations. In Sect. 8.5 we shall propose a Hadamard matrix and a Rademacher weighted approaches. The simulation study in Sect. 8.6 shows that, for finite-sample performance, the latter approach gives a more accurate size than the method based on Gaussian Multiplier Bootstrap (8.30).

Chen et al. (2016) generalized Fan, Hall and Yao's L^∞ norm to high-dimensional dependent vectors. Assume that $(X_i)_{i\in\mathbb{Z}}$ is a p-dimensional stationary process of the form

$$X_t = G(\mathcal{F}_t) = (G_1(\mathcal{F}_t), \dots, G_p(\mathcal{F}_t))^T, \tag{8.31}$$

where ε_t, $t \in \mathbb{Z}$, are i.i.d. random variables, $\mathcal{F}_t = (\dots, \varepsilon_{t-1}, \varepsilon_t)$ and $G(\cdot)$ is a measurable function such that X_t is well-defined. Assume that the long-run covariance matrix

$$\Sigma_\infty = \sum_{i=-\infty}^{\infty} \operatorname{cov}(X_0, X_i) = (\omega_{jl})_{j,l\le p} \tag{8.32}$$

exists. Let $\varepsilon_i^*, \varepsilon_j, i, j \in \mathbb{Z}$, be i.i.d. random variables. Assume that X_t has finite rth moment, $r > 2$. Define the functional dependence measures (see, Wu 2005, 2011) as

$$\theta_r(m) = \max_{j\le p} \|X_{ij} - G_j(\dots, \varepsilon_{i-m-2}, \varepsilon_{i-m-1}, \varepsilon_{i-m}^*, \varepsilon_{i-m+1}, \dots, \varepsilon_i)\|_r. \tag{8.33}$$

If X_i are i.i.d., then $\Sigma_\infty = \Sigma$ and $\theta_r(m) = 0$ if $m \ge 1$. We say that (X_t) is *geometric moment contraction* (GMC; see Wu and Shao 2004) if there exist $\rho \in (0, 1)$ and $a_1 > 0$ such that

$$\theta_r(m) \le a_1\rho^m = a_1 e^{-a_2 m} \text{ with } a_2 = -\log\rho. \tag{8.34}$$

Let $\mu = EX_t$. To test the hypothesis H_0 in (8.1), Chen et al. (2016) introduced the following dependence-adjusted versions of Fan, Hall, and Yao's M_n. Let $n = mk$, where $m \asymp n^{1/4}$ and blocks $B_l = \{i : m(l-1) + 1 \le i \le ml\}$. Let $Y_{lj} = \sum_{i\in B_l} X_{ij}$, $1 \le j \le p$, $1 \le l \le k$, be the block sums. Define the block-normalized sum

$$M_n^\circ = \max_{j\le p} \frac{\sqrt{n}|\hat{\mu}_j - \mu_j|}{\hat{\sigma}_j^\circ}, \text{ where } (\hat{\sigma}_j^\circ)^2 = \frac{1}{mk}\sum_{l=1}^{k}(Y_{lj} - m\hat{\mu}_j)^2, \tag{8.35}$$

and the interlacing normalized sum: let $k^* = k/2$, $\mu_j^\dagger = (mk^*)^{-1}\sum_{l=1}^{k^*} Y_{2lj}$ and

$$M_n^\dagger = \max_{j\le p}\frac{\sqrt{n/2}|\mu_j^\dagger - \mu_j|}{\hat{\sigma}_j^\dagger}, \text{ where } (\hat{\sigma}_j^\dagger)^2 = \frac{1}{mk^*}\sum_{l=1}^{k^*}(Y_{2lj} - m\mu_j^\dagger)^2. \quad (8.36)$$

By Chen et al. (2016), we have the following result: Assume exists a constant $\zeta > 0$ such that the long-run variance $\omega_{jj} \ge \zeta$ for $j \le p$, (8.34) holds with $r = 3$, and

$$\log p = o(n^{1/4}). \quad (8.37)$$

Then (8.23) holds for both the block-normalized sum M_n° and the interlacing normalized sum $M_n^\dagger$. Note that, while (8.37) still allows ultra high dimensions, due to dependence, the allowed dimension p in condition (8.37) is smaller than the one in (8.21). Additionally, if the GMC (8.34) holds with some $r > 3$, (8.24) holds with the long-run correlation matrix $R = D^{-1/2}\Sigma_\infty D^{-1/2}$, where $D = \text{diag}(\Sigma_\infty)$, and for some $0 < \tau < 1/4$,

$$\log p = o(n^\tau), \quad (8.38)$$

then we have the Gumbel convergence for the interlacing normalized sum:

$$M_n^\dagger - 2\log p + \log\log p \Rightarrow \mathcal{G}, \quad (8.39)$$

where $\mathcal{G}$ is given in (8.25). Similarly as (8.26), one can perform the following test which has an asymptotically correct size: we reject H_0 in (8.1) at level $\alpha \in (0, 1)$ if

$$\max_{j\le p}\frac{\sqrt{n/2}|\mu_j^\dagger|}{\hat{\sigma}_j^\dagger} > 2\log p - \log\log p + g_{1-\alpha}. \quad (8.40)$$

8.4 Tests Based on L^2 Norms

In this section we shall consider the test which is based on the L^2 functional with $g(x) = \sum_{j=1}^p x_j^2$. Let $\lambda_1 \ge \cdots \ge \lambda_p \ge 0$ be the eigenvalues of Σ. For $Z \sim N(0, \Sigma)$, we have the distributional equality $g(Z) = Z^T Z =_{\mathcal{D}} \sum_{j=1}^p \lambda_j \eta_j^2$, where η_j are i.i.d. standard $N(0, 1)$ random variables. Let $f_k = (\sum_{j=1}^p \lambda_j^k)^{1/k}$, $k > 0$, and $f = f_2$. Then $Eg(Z) = f_1 = \text{tr}(\Sigma)$ and $\text{var}(g(Z)) = 2f^2$. Xu et al. (2014) provide a sufficient condition for the invariance principle (8.7) with the quadratic functional g. For some $0 < \delta \le 1$ let $q = 2 + \delta$.

Condition 1 *Let $\delta > 0$. Assume $EX_1 = 0$, $E|X_1|^{2q} < \infty$ and let*

$$K_\delta(X)^q := E\left|\frac{|X_1|_2^2 - f_1}{f}\right|^q < \infty \tag{8.41}$$

$$D_\delta(X)^q := E\left|\frac{X_1^T X_2}{f}\right|^q < \infty. \tag{8.42}$$

Observe that Condition 1, (8.41) and (8.42) are Lyapunov-type conditions. Assume that

$$\frac{K_0(X)^2}{n} + \frac{K_\delta(X)^q}{n^{q-1}} + \frac{E(X_1^T \Sigma X_1)^{q/2}}{n^{\delta/2} f^q} + \frac{D_\delta(X)^q}{n^\delta} \to 0 \text{ as } n \to \infty. \tag{8.43}$$

Then (8.7) holds (cf Xu et al. 2014). Consequently we have

$$\sup_{t \in \mathbb{R}} |P((n|\bar{X}_n|_2^2 - f_1)/f \le t) - P(V \le t)| \to 0, \text{ where } V = \sum_{j=1}^p f^{-1}\lambda_j(\eta_j^2 - 1). \tag{8.44}$$

In the literature, researchers primarily focus on developing the central limit theorem

$$R_n := \frac{n|\bar{X}_n|_2^2 - f_1}{f} = \frac{n\bar{X}_n^T \bar{X}_n - f_1}{f} \Rightarrow N(0, 2) \tag{8.45}$$

or its modified version; see, for example, Bai and Saranadasa (1996), Chen and Qin (2010) and Srivastava (2009). Xu et al. (2014) clarified an important issue on the CLT of R_n. By the Lindeberg–Feller central limit theorem, $V \Rightarrow N(0, 2)$ as $p \to \infty$ holds if and only if $\lambda_1/f \to 0$. The distributional approximation (8.44) indicates that, if λ_1/f does not go to 0, then the central limit theorem cannot hold for R_n.

Let $t_{1-\alpha}$ be the $(1-\alpha)$th quantile of $g(Z) = |Z|^2 = Z^T Z$. By (8.7) we can reject (8.1) at level $\alpha \in (0, 1)$ if

$$n|\bar{X}_n|^2 > t_{1-\alpha} \tag{8.46}$$

To calculate $t_{1-\alpha}$, one needs to know the eigenvalues $\lambda_1, \ldots, \lambda_p$. However, estimation of those eigenvalues is a very challenging problem, in particular if one does not impose certain structural assumptions on Σ. In Sect. 8.5.2 we shall propose a half-sampling based approach which does not need estimation of the covariance matrix Σ.

The L^∞ based tests discussed in Sect. 8.3 have a good power when the alternative consists of few large signals. If the signals are small and have a similar magnitude, then the L^2 test is more powerful. To this end, assume that there exists a constant $c > 0$ and a small $\delta > 0$ such that $c\delta \le \mu_j \le \delta/c$ holds for all $j = 1, \ldots, p$. We can interpret δ as the departure parameter (from the null H_0 with $\mu = 0$). For the L^∞-based test to have power approaching to 1, one necessarily requires that $\sqrt{n}\delta \to \infty$.

Elementary calculation shows that, under the much weaker condition $np^{1/2}\delta^2 \to \infty$, then the power of the L^2 based test, or the probability that event (8.46) occurs going to one. In the latter condition, larger dimension p is actually a blessing as it requires a smaller departure δ.

8.5 Asymptotic Theory

In Sects. 8.3 and 8.4, we discussed the classical L^∞ and L^2 functionals, respectively. For a general testing functional, we have the following invariance principle (cf Theorem 1), which asserts that functionals of sample means of non-Gaussian random vectors $X_1, X_2, \ldots$ can be approximated by those of Gaussian vectors $Z_1, Z_2, \ldots$ with same covariance structure. Assume $g \in \mathbb{C}^3(\mathbb{R}^p)$. For $\mathbf{x} = (x_1, \ldots, x_p)^T$ write $g_j = g_j(\mathbf{x}) = \partial g(\mathbf{x})/\partial x_j$. Similarly we define the partial derivatives g_{jk} and g_{jkl}. For all $j, k, l = 1, \ldots, p$, assume that

$$\kappa_{jkl} := \sup_{\mathbf{x}\in\mathbb{R}^p} (|g_j g_k g_l| + |g_{jk} g_l| + |g_{jl} g_k| + |g_{kl} g_j| + |g_{jkl}|) < \infty. \tag{8.47}$$

For $Z_1 \sim N(0, \Sigma)$ write $Z_1 = (Z_{11}, \ldots, Z_{1p})^T$. Define

$$\mathcal{K}_p = \sum_{j,k,l=1}^{p} \kappa_{jkl}(E|X_{1j}X_{1k}X_{1l}| + E|Z_{1j}Z_{1k}Z_{1l}|). \tag{8.48}$$

For $g(Z_1) =_{\mathcal{D}} g(\sqrt{n}\bar{Z}_n)$, we assume that its c.d.f. $F(t) = P[g(Z) \le t]$ is Hölder continuous: there exists $\ell_p > 0$, index $\alpha > 0$, such that for all $\psi > 0$, the concentration function

$$\sup_{t\in\mathbb{R}} P(t \le g(Z_1) \le t + \psi) \le \ell_p \psi^\alpha. \tag{8.49}$$

Theorem 1 (Lou and Wu (2018)) *Assume (8.47), (8.49) and* $\mathcal{K}_p \ell_p^{3/\alpha} = o(\sqrt{n})$. *Then*

$$\sup_{t\in\mathbb{R}} |P[g(\sqrt{n}(\bar{X}_n - \mu)) \le t] - P[g(\sqrt{n}\bar{Z}_n) \le t]| = O(\ell_p^3 \mathcal{K}_p^\alpha n^{-\alpha/2}) \to 0. \tag{8.50}$$

To apply Theorem 1 for hypothesis testing, we need to know the c.d.f. $F(t) = P[g(Z) \le t]$. Note that $F(\cdot)$ depends on g and the covariance matrix Σ. Thus we can also write $F(\cdot) = F_{g,\Sigma}(\cdot)$. If Σ is known, the distribution of $g(Z)$ is completely known and its cdf $F(t) = P[g(Z) \le t]$ can be calculated either analytically or by extensive Monte Carlo simulations. Let $t_{1-\alpha}$, $0 < \alpha < 1$, be the $(1-\alpha)$th quantile of $g(Z)$. Namely

$$P[g(Z) > t_{1-\alpha}] = \alpha. \tag{8.51}$$

Then the null hypothesis H_0 in (8.1) is rejected at level α if the test statistic $T_n = g(\sqrt{n}\bar{X}_n) > t_{1-\alpha}$. This test has asymptotically correct size α. Additionally, the $(1-\alpha)$ confidence region for μ can be constructed as

$$\{\mu \in \mathbb{R}^p : g(\sqrt{n}(\bar{X}_n - \mu)) \leq t_{1-\alpha}\} = \{\bar{X}_n + \nu \in \mathbb{R}^p : g(\sqrt{n}\nu) \leq t_{1-\alpha}\}. \quad (8.52)$$

If Σ is not known, as a straightforward way to approximate $F(t) = F_{g,\Sigma}(t)$, one may use an estimate $\tilde{\Sigma}$ so that $F_{g,\Sigma}(t)$ can be approximated by $F_{g,\tilde{\Sigma}}(t)$. Here we do not adopt this approach for the following two reasons. First, it can be quite difficult to consistently estimate Σ without assuming sparseness or other structural conditions. The latter assumptions are widely used in the literature; see, for example, Bickel and Levina (2008a), Bickel and Levina (2008b), Cai et al. (2011) and Fan et al. (2013). Second, it is difficult to quantify the difference $F_{g,\tilde{\Sigma}}(\cdot) - F(\cdot)$ based on operator norm or other type of matrix convergence of the estimate $\tilde{\Sigma}$. Xu et al. (2014) argued that, for the L^2 test with $g(x) = \sum_{j=1}^p x_j^2$, one needs to use the normalized consistency of $\tilde{\Sigma}$, instead of the widely used operator norm consistency. We propose using half-sampling and balanced Rademacher schemes.

8.5.1 Preamble: i.i.d. Gaussian Data

In practice, however, the covariance matrix Σ is typical unknown. Assume at the outset that $X_1, \ldots, X_n$ are i.i.d. $N(\mu, \Sigma)$ vectors. Assume that $n = 4m$, where m is a positive integer. Then we can estimate the cumulative distribution function $F(t) = P[g(Z) \leq t]$ by using Hadamard matrices (see, Georgiou et al. 2003; Hedayat and Wallis 1978; Yarlagadda and Hershey 1997). We say that H is an $n \times n$ Hadamard matrix if its first row consisting all 1s, and all its entries taking values 1 or -1 such that

$$HH^T = nI_n, \quad (8.53)$$

where I_n is the $n \times n$ identity matrix. Let

$$Y_j = \frac{1}{\sqrt{n}} \sum_{i=1}^n H_{ji} X_i, \; j = 1, \ldots, n. \quad (8.54)$$

By (8.53), we have $\sum_{i=1}^n H_{ji} = 0$ for $2 \leq j \leq n$ and $\sum_{i=1}^n H_{ji} H_{j'i} = 0$ if $j \neq j'$. Since $X_1, \ldots, X_n$ are i.i.d. $N(\mu, \Sigma)$, it is clear that $Y_2, \ldots, Y_n$ are also i.i.d. $N(0, \Sigma)$ vectors. Hence the random variables $g(Y_2), \ldots, g(Y_n)$ are independent and identically distributed as $g(Z)$. Therefore we can construct the empirical cumulative distribution function

$$\hat{F}_n(t) = \frac{1}{n-1} \sum_{j=2}^n \mathbf{1}_{g(Y_j) \leq t}, \quad (8.55)$$

which converges uniformly to $F(t)$ as $n \to \infty$, and $t_{1-\alpha}$ can be estimated by $\hat{t}_{1-\alpha} = \hat{F}_n^{-1}(1-\alpha)$, the $(1-\alpha)$th empirical quantile of $\hat{F}_n(\cdot)$. As an important feature of the latter method, one does not need to estimate the covariance matrix Σ, the nuisance parameter. In combinatorial experiment design, however, it is highly nontrivial to construct Hadamard matrices. If n is a power of 2, then one can simply apply Sylvester's construction. The Hadamard conjecture states that a Hadamard matrix of order n exists when $4|n$. The latter problem is still open. For example, it is unclear whether a Hadamard matrix exists when $n = 668$ (see Brent et al. 2015).

8.5.2 Rademacher Weighted Differencing

To circumvent the existence problem of Hadamard matrices in Sect. 8.5.1, we shall construct asymptotically independent realizations by using Rademacher random variables. Let $\varepsilon_{jk}, j, k \in \mathbb{Z}$, independent of $(X_i)_{i \geq 1}$, be i.i.d. Bernoulli random variables with $P(\varepsilon_{jk} = 1) = P(\varepsilon_{jk} = -1) = 1/2$. Define the Rademacher weighted differences

$$Y_j = D(A_j), \text{ where } D(A) = \frac{|A|^{1/2}(n-|A|)^{1/2}}{n^{1/2}} \left(\frac{\sum_{i \in A} X_i}{|A|} - \frac{\sum_{i \in \{1,\ldots,n\}-A} X_i}{n-|A|} \right), \tag{8.56}$$

where the random set

$$A_j = \{1 \leq i \leq n : \varepsilon_{ji} = 1\}. \tag{8.57}$$

When defining Y_j, we require that A_j satisfies $|A_j| \neq 0$ and $|A_j| \neq n$. By the Hoeffding inequality, $|A_j|$ concentrates around $n/2$ in the sense that, for $u \geq 0$, $P(||A_j| - n/2| \geq u) \leq 2\exp(-2u^2/n)$. Alternatively, we consider the balanced Rademacher weighted differencing: let $A_1^\circ, A_2^\circ, \ldots$ be simple random sample drawn equally likely from $\mathcal{A}_m = \{A \subset \{1,\ldots,n\} : |A| = m\}$, where $m = \lfloor n/2 \rfloor$. Similarly as Y_j in (8.56), we define

$$Y_j^\circ = D(A_j^\circ). \tag{8.58}$$

Clearly, given A_j (resp. A_j°), Y_j (resp. Y_j°) has mean 0 and covariance matrix Σ. Based on Y_j in (8.56) (resp. Y_j° in (8.58)), define the empirical distribution functions

$$\hat{F}_N(t) = \frac{1}{N} \sum_{j=1}^{N} \mathbf{1}_{g(Y_j) \leq t}, \tag{8.59}$$

where $N \to \infty$ and

$$\hat{F}_N^\circ(t) = \frac{1}{N}\sum_{j=1}^{N} \mathbf{1}_{g(Y_j^\circ)\le t}. \tag{8.60}$$

For sets $A, B \subset \{1,\ldots,n\}$, let $A^c = \{1,\ldots,n\} - A$, $B^c = \{1,\ldots,n\} - B$ and

$$d(A,B) = \max\left\{||A \cap B| - \frac{n}{4}|,\ ||A^c \cap B| - \frac{n}{4}|,\ ||A \cap B^c| - \frac{n}{4}|,\ ||A^c \cap B^c| - \frac{n}{4}|\right\}.$$

If A, B are chosen according to a Hadamard matrix, then $d(A,B) = 0$. Assume that

$$d(A,B) \le 0.1n. \tag{8.61}$$

Then there exists an absolute constant $c > 0$ such that

$$\operatorname{cov}(D(A), D(B)) = \delta\Sigma,\ \text{where } |\delta| \le c\frac{d(A,B)}{n}. \tag{8.62}$$

Again by the Hoeffding inequality, if we choose A_1, A_2 according to (8.57), there exists absolute constants $c_1, c_2 > 0$ such that $P(d(A_1,A_2) \ge u) \le c_1 \exp(-c_2 u^2/n)$, indicating that (8.61) holds with probability close to 1, $d(A_1,A_2) = O_P(n^{1/2})$ and hence the weak orthogonality with $\delta(A_1,A_2) = O_P(n^{-1/2})$.

Theorem 2 (Lou and Wu (2018)) *Under conditions of Theorem 1, we have* $\sup_t |\hat{F}_N^\circ(t) - F(t)| \to 0$ *in probability as* $N \to \infty$.

8.5.3 Calculating the Power

The asymptotic power expression is

$$B(\mu) = P[g(Z + \sqrt{n}\mu) \ge t_{1-\alpha}]. \tag{8.63}$$

Given the sample $X_1, \ldots, X_n$ whose mean vector μ may not necessarily be 0, based on the estimated $\hat{t}_{1-\alpha}$ from the empirical cumulative distribution functions (8.59) and (8.60), we can actually estimate the power function by the following:

$$\begin{aligned}\hat{B}(\nu) &= \hat{P}(g(D(A_j^\circ) + \sqrt{n}\nu) \ge \hat{t}_{1-\alpha}|X_1,\ldots,X_n)\\ &= \frac{1}{N}\sum_{j=1}^{N} \mathbf{1}_{g(D(A_j^\circ)+\sqrt{n}\nu)\ge\hat{t}_{1-\alpha}}.\end{aligned} \tag{8.64}$$

8.5.4 An Algorithm with General Testing Functionals

For ease of application, we shall in this section provide details of testing the hypothesis H_0 in (8.1) using the Rademacher weighting scheme described in Sect. 8.5.2.

To construct a confidence region for μ, one can use (8.52) with $t_{1-\alpha}$ therein replaced by the empirical quantile $\hat{t}^{\circ}_{1-\alpha}$.

8.6 Numerical Experiments

In this section, we shall perform a simulation study and evaluate the finite-sample performance of our Algorithm 1 with $\hat{F}^{\circ}_{N}(t)$ defined in (8.60). Tests for mean vectors and covariance matrices are considered in Sects. 8.6.1 and 8.6.2, respectively. Section 8.6.3 contains a real data application on testing correlations between different pathways of a pancreatic ductal adenocarcinoma dataset.

8.6.1 Test of Mean Vectors

We consider three different testing functionals: for $x = (x_1, \ldots, x_p)^\top \in \mathbb{R}^p$, let

$$g_1(x) = \max_{j \le p} |x_j|, \quad g_2(x) = \sum_{j=1}^{p} |x_j|^?, \quad g_3(x) = \sup_{c \ge 0} \left\{ c^? \sum_{j=1}^{p} |x_j|^? \mathbf{1}_{|x_j| \ge c} \right\}.$$

For the L^∞ form $g_1(x)$, four different testing procedures are compared: the procedure using our Algorithm 1 with $\hat{F}^{\circ}_{N}(\cdot)$ replaced by $\hat{F}_N(\cdot)$; cf (8.59); or by

$$\hat{F}^{\dagger}_{N}(t) = \frac{1}{N} \sum_{j=1}^{N} \mathbf{1}_{g(Y^{\dagger}_j) \le t}, \quad \text{where } Y^{\dagger}_j = \frac{1}{\sqrt{n}} \sum_{i=1}^{n} \varepsilon_{ji}(X_i - \bar{X}) \tag{8.65}$$

Algorithm 1: Rademacher weighted testing procedure

1. Input $X_1, \ldots, X_n$;
2. Compute the average $\bar{X}_n$ and the test statistic $T = g(\sqrt{n}\bar{X}_n)$;
3. Choose a large N in (8.60) and obtain the empirical quantile $\hat{t}^{\circ}_{1-\alpha}$;
4. Reject H_0 at level α if $T > \hat{t}^{\circ}_{1-\alpha}$;
5. Report the p-value as $\hat{F}^{\circ}_{N}(T)$.

---

and ε_{ji} are i.i.d. Bernoulli(1/2) independent of (X_{ij}); the test of Fan et al. (2007) (FHY, see (8.20) and (8.23)) and the Gaussian Multiplier Bootstrap method in Chernozhukov et al. (2014) (CCK, see (8.30)).

For $g_2(x)$, we compare the performance of our Algorithm 1 with $\hat{F}_N^\circ(\cdot)$, $\hat{F}_N(\cdot)$ and $\hat{F}_N^\dagger(\cdot)$, and also the CLT-based procedure of Chen and Qin (2010) (CQ), which is a variant of (8.45) with the numerator $n\bar{X}_n^T\bar{X}_n - f_1$ therein replaced by $n^{-1}\sum_{i\neq j} X_i^\top X_j$.

The portmanteau testing functional $g_3(x)$ is a marked weighted empirical process.

For our Algorithm 1 and the Gaussian Multiplier Bootstrap method, we calculate the empirical cutoff values with $N = 4000$. For each functional, we consider two models and use $n = 40, 80$ and $p = 500, 1000$. The empirical sizes for each case are calculated based on 1000 simulations.

Example 1 (Factor Model) Let Z_{ij} be i.i.d. $N(0, 1)$ and consider

$$X_i = (Z_{i1}, \ldots, Z_{ip})^\top + p^\delta (Z_{i0}, \ldots, Z_{i0})^\top, \quad i = 1, \ldots, n, \tag{8.66}$$

Then X_i are i.i.d. $N(0, \Sigma)$ with $\Sigma = \mathrm{I}_p + p^{2\delta}\mathbf{1}\mathbf{1}^\top$, where $\mathbf{1} = (1, \ldots, 1)^\top$. Larger δ implies stronger correlation among the entries $X_{i1}, \ldots, X_{ip}$.

Table 8.1 reports empirical sizes for the factor model with $g_1(\cdot)$ at the 5% significance level. For each choice of p, n, and δ, our Algorithm 1 with $\hat{F}_N^\circ(\cdot)$ and $\hat{F}_N(\cdot)$ perform reasonably well, while the empirical sizes using $\hat{F}_N^\dagger(\cdot)$ are generally slightly larger than 5%. The empirical sizes using Chernozhukov et al.'s (8.30) or Fan et al.'s (8.23) are substantially different from the nominal level 5%. For large δ, as expected, the procedure of Fan, Hall, and Yao can be very conservative.

The empirical sizes for the factor model using $g_2(\cdot)$ are summarized in Table 8.2. Our Algorithm 1 with $\hat{F}_N^\circ(\cdot)$ and $\hat{F}_N(\cdot)$ perform quite well. The empirical sizes for Chen and Qin's procedure deviate significantly from 5%. This can be explained by the fact that CLT of type (8.45) is no longer valid for model (8.66); see the discussion following (8.45) and Theorem 2.2 in Xu et al. (2014).

When using functional $g_3(x)$, our Algorithm 1 with $\hat{F}_N^\circ(\cdot)$ and $\hat{F}_N(\cdot)$ perform slightly better than $\hat{F}_N^\dagger(\cdot)$ and approximate the nominal 5% level well (Table 8.3).

Table 8.1 Empirical sizes for the factor model (8.66) with $g_1(\cdot)$

		$n = 40$					$n = 80$				
p	δ	$\hat{F}_N^\circ$	CCK	$\hat{F}_N$	FHY	$\hat{F}_N^\dagger$	$\hat{F}_N^\circ$	CCK	$\hat{F}_N$	FHY	$\hat{F}_N^\dagger$
500	−0.05	0.053	0.028	0.052	0.028	0.059	0.053	0.037	0.052	0.031	0.055
1000		0.052	0.023	0.052	0.035	0.057	0.051	0.036	0.051	0.034	0.053
500	0.05	0.051	0.034	0.054	0.014	0.064	0.047	0.030	0.044	0.018	0.047
1000		0.057	0.035	0.058	0.011	0.063	0.053	0.044	0.055	0.015	0.056
500	0.1	0.046	0.026	0.048	0.009	0.055	0.053	0.042	0.054	0.007	0.056
1000		0.059	0.041	0.059	0.007	0.063	0.052	0.045	0.054	0.008	0.056

Table 8.2 Empirical sizes for the factor model (8.66) using functional $g_2(x)$

		$n = 40$				$n = 80$			
p	δ	CQ	$\widehat{F}_N^{\circ}$	$\widehat{F}_N$	$\widehat{F}_N^{\dagger}$	CQ	$\widehat{F}_N^{\circ}$	$\widehat{F}_N$	$\widehat{F}_N^{\dagger}$
500	−0.05	0.078	0.055	0.061	0.066	0.063	0.048	0.047	0.048
1000		0.081	0.063	0.066	0.072	0.066	0.050	0.049	0.053
500	0.05	0.074	0.054	0.054	0.059	0.067	0.052	0.053	0.054
1000		0.075	0.054	0.052	0.056	0.076	0.058	0.057	0.059
500	0.1	0.067	0.049	0.051	0.052	0.068	0.055	0.052	0.056
1000		0.083	0.064	0.064	0.067	0.068	0.048	0.051	0.051

Table 8.3 Empirical sizes for the factor model (8.66) using functional $g_3(x)$

		$n = 40$			$n = 80$		
p	δ	$\widehat{F}_N^{\circ}$	$\widehat{F}_N$	$\widehat{F}_N$	$\widehat{F}_N^{\circ}$	$\widehat{F}_N$	$\widehat{F}_N^{\dagger}$
500	−0.05	0.061	0.059	0.066	0.049	0.048	0.049
1000		0.062	0.064	0.073	0.058	0.059	0.063
500	0.05	0.054	0.058	0.060	0.053	0.053	0.055
1000		0.053	0.054	0.057	0.059	0.059	0.060
500	0.1	0.049	0.049	0.051	0.053	0.053	0.055
1000		0.053	0.054	0.057	0.059	0.059	0.060

Example 2 (Multivariate t-Distribution) Consider the multivariate t_ν vector

$$X_i = (X_{i1}, \ldots, X_{ip})^\top = Y_i\sqrt{\nu/W_i} \sim t_\nu(0, \Sigma), \quad i = 1, \ldots, n \tag{8.67}$$

where the degrees of freedom $\nu = 4$, $\Sigma = (\sigma_{jk})_{j,k=1}^{p}$, $\sigma_{jj} = 1$ for $j = 1, \ldots, p$ and

$$\sigma_{jk} = c|j-k|^{-d}, \quad 1 \leq j \neq k \leq p,$$

and $Y_i \sim N(0, \Sigma)$, $W_i \sim \chi^2_\nu$ are independent. The above covariance structure allows long-range dependence among $X_{i1}, \ldots, X_{ip}$; see Veillette and Taqqu (2013).

We summarize the simulated sizes for model (8.67) in Tables 8.4, 8.5, and 8.6. As in Example 1, similar conclusions apply here. Due to long-range dependence, the procedure of Fan, Hall, and Yao appears conservative. The Gaussian Multiplier Bootstrap (8.30) yields empirical sizes that are quite different from 5%. The CLT-based procedure of Chen and Qin is severely affected by the dependence. In practice we suggest using Algorithm 1 with $\widehat{F}_N^{\circ}(\cdot)$ which has a good size accuracy.

Table 8.4 Empirical sizes for multivariate t-distribution using functional $g_1(x)$

t_4		$n=40$					$n=80$				
c	d	$\hat{F}_N^{\dagger}$	CCK	$\hat{F}_N$	FHY	$\hat{F}_N^{\dagger}$	$\hat{F}_N^{\circ}$	CCK	$\hat{F}_N$	FHY	$\hat{F}_N^{\dagger}$
0.5	1/8	0.047	0.011	0.044	0.016	0.053	0.051	0.017	0.045	0.013	0.049
		0.059	0.015	0.056	0.014	0.061	0.055	0.017	0.055	0.022	0.059
0.5	1/4	0.057	0.010	0.055	0.023	0.061	0.050	0.016	0.050	0.022	0.053
		0.051	0.005	0.048	0.018	0.058	0.054	0.014	0.055	0.022	0.060
0.8	1/8	0.054	0.020	0.050	0.017	0.061	0.052	0.030	0.051	0.016	0.053
		0.049	0.019	0.044	0.012	0.049	0.049	0.022	0.048	0.017	0.051
0.8	1/4	0.048	0.013	0.050	0.022	0.053	0.046	0.019	0.042	0.036	0.044
		0.054	0.008	0.053	0.017	0.057	0.051	0.018	0.050	0.018	0.052

For each choice of c and d, the upper line corresponding to $p=500$ and the second for $p=1000$

Table 8.5 Empirical sizes for multivariate t-distribution using functional $g_2(x)$

t_4		$n=40$				$n=80$			
c	d	CQ	$\hat{F}_N^{\circ}$	$\hat{F}_N$	$\hat{F}_N^{\dagger}$	CQ	$\hat{F}_N^{\circ}$	$\hat{F}_N$	$\hat{F}_N^{\dagger}$
0.5	1/8	0.074	0.053	0.053	0.056	0.076	0.060	0.052	0.058
		0.073	0.055	0.050	0.054	0.077	0.062	0.061	0.064
0.5	1/4	0.067	0.052	0.044	0.051	0.073	0.055	0.054	0.057
		0.072	0.057	0.054	0.060	0.070	0.056	0.055	0.060
0.8	1/8	0.074	0.059	0.062	0.066	0.070	0.047	0.051	0.052
		0.064	0.052	0.053	0.057	0.075	0.052	0.054	0.055
0.8	1/4	0.081	0.063	0.058	0.063	0.080	0.055	0.056	0.061
		0.067	0.052	0.051	0.059	0.068	0.053	0.052	0.056

For each choice of c and d, the upper line corresponding to $p=500$ and the second for $p=1000$

Table 8.6 Empirical sizes for multivariate t-distribution using functional $g_3(x)$

t_4		$n=40$			$n=80$		
c	d	$\hat{F}_N^{\circ}$	$\hat{F}_N$	$\hat{F}_N^{\dagger}$	$\hat{F}_N^{\circ}$	$\hat{F}_N$	$\hat{F}_N^{\dagger}$
0.5	1/8	0.053	0.050	0.056	0.055	0.051	0.054
		0.050	0.049	0.056	0.059	0.055	0.060
0.5	1/4	0.052	0.048	0.053	0.056	0.056	0.060
		0.056	0.048	0.060	0.055	0.056	0.061
0.8	1/8	0.059	0.059	0.066	0.049	0.049	0.052
		0.048	0.048	0.054	0.051	0.051	0.058
0.8	1/4	0.067	0.063	0.069	0.053	0.056	0.061
		0.049	0.048	0.052	0.048	0.048	0.051

For each choice of c and d, the upper line corresponding to $p=500$ and the second for $p=1000$

8.6.2 *Test of Covariance Matrices*

8.6.2.1 Sizes Accuracy

We first consider testing for $H_{0a} : \Sigma = I$ for the following model:

$$X_{ij} = \varepsilon_{i,j}\varepsilon_{i,j+1},\ 1 \le i \le n,\ 1 \le j \le p, \tag{8.68}$$

where ε_{ij} are i.i.d. (1) standard normal; (2) centralized Gamma(4,1); and (3) the student t_5. We then study the second test $H_{0b} : \Sigma_{1,2} = 0$, by partitioning equally the entire random vector $X_i = (X_{i1}, \ldots, X_{ip})^T$ into two subvectors of $p_1 = p/2$ and $p_2 = p - p_1$. In the simulation, we generate samples of two subvectors independently according to model (8.68). We shall use Algorithm 1 with L^2 functional. Tables 8.7 and 8.8 report the simulated sizes based on 1000 replications with $N = 1000$ half-sampling implementations, and they are reasonably closed to the nominal level 5%.

8.6.2.2 Power Curve

To access the power for testing $H_0 : \Sigma = \mathrm{I}_p$ using the L^2 test, we consider the model

$$X_{ij} = \varepsilon_{i,j}\varepsilon_{i,j+1} + \rho\zeta_i,\ 1 \le i \le n,\ 1 \le j \le p, \tag{8.69}$$

where ε_{ij} and ζ_i are i.i.d. Student t_5 and ρ is chosen to be $0, 0.02, 0.04, \ldots, 0.7$. The power curve is shown in Fig. 8.1. As expected, the power increases with n.

Table 8.7 Simulated sizes of the L^2 test for H_{0a}

	$N(0,1)$		$\Gamma(4,1)$		t_5	
	p					
n	64	128	64	128	64	128
20	0.045	0.054	0.046	0.047	0.053	0.048
50	0.044	0.045	0.055	0.045	0.046	0.050
100	0.050	0.054	0.047	0.053	0.051	0.049

Table 8.8 Simulated sizes of the L^2 test for H_{0b}

	$N(0,1)$		$\Gamma(4,1)$		t_5	
	p					
n	64	128	64	128	64	128
20	0.044	0.050	0.043	0.055	0.045	0.043
50	0.045	0.043	0.049	0.044	0.053	0.045
100	0.053	0.053	0.053	0.045	0.050	0.050

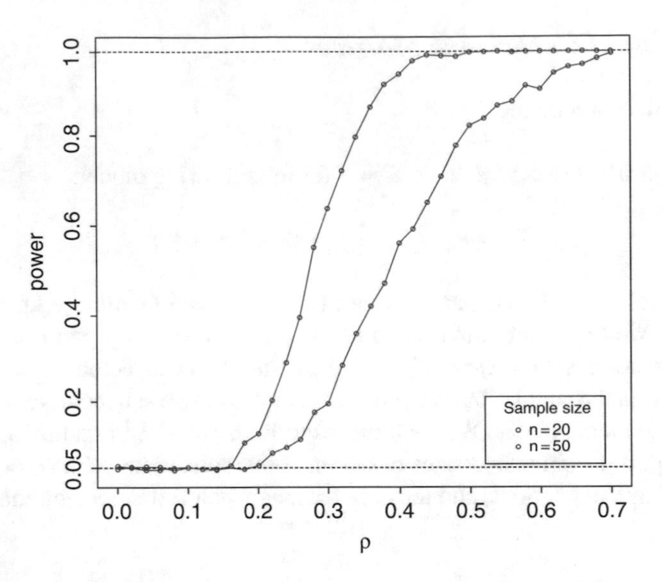

Fig. 8.1 Power curve for testing $H_0 : \Sigma = \mathrm{I}_p$ with model (8.69), and $n = 20, 50$, using the L^2 test

8.6.3 A Real Data Application

We now apply our testing procedures to a pancreatic ductal adenocarcinoma (PDAC) dataset, preprocessed from NCBI's Gene Expression Omnibus, accessible through GEO Series accession number GSE28735 (http://www.ncbi.nlm.nih.gov/geo/query/acc.cgi?acc=GSE28735). The dataset consists of two classes of gene expression levels that came from 45 pancreatic tumor patients and 45 pancreatic normal patients. There are a total of 28,869 genes. We shall test existence of correlations between two subvectors, which can be useful for identifying sets of genes which are significantly correlated.

We consider genetic pathways of the PDAC dataset. Pathways are found to be highly significantly associated with the disease even if they harbor a very small amount of individually significant genes. According to the KEGG database, the pathway "hsa05212" is relevant to pancreatic cancer. Among the 28,869 genes, 66 are mapped to this pathway. We are interested in testing whether the pathway to pancreatic cancer is correlated with some common pathways, "hsa04950" (21 genes, with name "Maturity onset diabetes of the young"), "hsa04940" (59 genes, with name "Type I diabetes mellitus"), "hsa04972" (87 genes, with name "Pancreatic secretion"). Let W_i, X_i, Y_i, and Z_i be the expression levels of individual i from the tumor group for pathways "hsa05212," "hsa04950," "hsa04940," and "hsa04972," respectively. The null hypotheses are $H_{01}^T : \mathrm{cov}(W_i, X_i) = 0_{66\times 21}$, $H_{02}^T : \mathrm{cov}(W_i, Y_i) = 0_{66\times 59}$ and $H_{03}^T : \mathrm{cov}(W_i, Z_i) = 0_{66\times 87}$. Similar null hypothesis $H_{01}^N, H_{01}^N, H_{01}^N$ can be formulated for the normal group. Our L^2 test of Algorithm 1 is compared with the Gaussian multiplier bootstrap (8.30). The results are summarized

Table 8.9 Estimated p-values of tests for covariances between pathway "pancreatic cancer" and other different pathways, based on $N = 10^6$ half-sampling implementations

Pathway	Name	Tumor patients		Normal patients	
		CCK	L^2 test	CCK	L^2 test
hsa04950	Maturity onset diabetes of the young	0.013116	0.000000	0.006618	0.000000
hsa04940	Type I diabetes mellitus	0.066270	0.000000	0.074014	0.002327
hsa04972	Pancreatic secretion	0.063291	0.000003	0.095358	0.001189

in Table 8.9. The CCK test is not able to reject the null hypothesis H_{03} at 5% level since it gives a p-value of 0.063291. However using the L^2 test, H_{03} is rejected, suggesting that there is a substantial correlation between pathways "hsa05212" and "hsa04972." Similar claims can be made for other cases. The L^2 test also suggests that, at 0.1% level, for the tumor group, the hypotheses H_{02}^T and H_{03}^T are rejected, while for the normal group, the hypotheses H_{02}^N and H_{03}^N are not rejected.

References

Ahmad MR (2010) Tests for covariance matrices, particularly for high dimensional data. Technical Reports, Department of Statistics, University of Munich. http://epub.ub.uni-muenchen.de/11840/1/tr091.pdf. Accessed 3 Apr 2018

Bai ZD, Saranadasa H (1996) Effect of high dimension: by an example of a two sample problem. Stat Sin 6:311–329

Bai ZD, Jiang DD, Yao JF, Zheng SR (2009) Corrections to LRT on large-dimensional covariance matrix by RMT. Ann Stat 37:3822–3840

Bickel PJ, Levina E (2008a) Regularized estimation of large covariance matrices. Ann Stat 36:199–227

Bickel PJ, Levina E (2008b) Covariance regularization by thresholding. Ann Stat 36:2577–2604

Birke M, Dette H (2005) A note on testing the covariance matrix for large dimension. Stat Probab Lett 74:281–289

Brent RP, Osborn JH, Smith WD (2015) Probabilistic lower bounds on maxima determinants of binary matrices. Available at arxiv.org/pdf/1501.06235. Accessed 3 Apr 2018

Cai Y, Ma ZM (2013) Optimal hypothesis testing for high dimensional covariance matrices. Bernoulli 19:2359–2388

Cai T, Liu WD, Luo X (2011) A constrained l_1 minimization approach to sparse precision matrix estimation. J Am Stat Assoc 106:594–607

Chen SX, Qin Y-L (2010) A two-sample test for high-dimensional data with applications to gene-set testing. Ann Stat 38:808–835

Chen SX, Zhang L-X, Zhong P-S (2010) Tests for high-dimensional covariance matrices. J Am Stat Assoc 105:810–819

Chen XH, Shao QM, Wu WB, Xu LH (2016) Self-normalized Cramér type moderate deviations under dependence. Ann Stat 44:1593–1617

Chernozhukov V, Chetverikov D, Kato K (2014) Gaussian approximations and multiplier bootstrap for maxima of sums of high-dimensional random vectors. Ann Stat 41:2786–2819

Dickhaus T (2014) Simultaneous statistical inference: with applications in the life sciences. Springer, Heidelberg

Dudiot S, van der Laan M (2008) Multiple testing procedures with applications to genomics. Springer, New York

Efron B (2010) Large-scale inference: empirical Bayes methods for estimation, testing, and prediction. Cambridge University Press, Cambridge
Fan J, Hall P, Yao Q (2007) To how many simultaneous hypothesis tests can normal, Student's t or bootstrap calibration be applied. J Am Stat Assoc 102:1282–1288
Fan J, Liao Y, Mincheva M (2013) Large covariance estimation by thresholding principal orthogonal complements. J R Stat Soc Ser B Stat Methodol 75:603–680
Fisher TJ, Sun XQ, Gallagher CM (2010) A new test for sphericity of the covariance matrix for high dimensional data. J Multivar Anal 101:2554–2570
Georgiou S, Koukouvinos C, Seberry J (2003) Hadamard matrices, orthogonal designs and construction algorithms. In: Designs 2002: further computational and constructive design theory, vols 133–205. Kluwer, Boston
Han YF, Wu WB (2017) Test for high dimensional covariance matrices. Submitted to Ann Stat
Hedayat A, Wallis WD (1978) Hadamard matrices and their applications. Ann Stat 6:1184–1238
Jiang TF (2004) The asymptotic distributions of the largest entries of sample correlation matrices. Ann Appl Probab 14:865–880
Jiang DD, Jiang TF, Yang F (2012) Likelihood ratio tests for covariance matrices of high-dimensional normal distributions. J Stat Plann Inference 142:2241–2256
Ledoit O, Wolf M (2002) Some hypothesis tests for the covariance matrix when the dimension is large compared to the sample size. Ann Stat 30:1081–1102
Liu WD, Shao QM (2013) A Cramér moderate deviation theorem for Hotelling's T^2-statistic with applications to global tests. Ann Stat 41:296–322
Lou ZP, Wu WB (2018) Construction of confidence regions in high dimension (Paper in preparation)
Marčenko VA, Pastur LA (1967) Distribution of eigenvalues for some sets of random matrices. Math U S S R Sbornik 1:457–483
Onatski A, Moreira MJ, Hallin M (2013) Asymptotic power of sphericity tests for high-dimensional data. Ann Stat 41:1204–1231
Portnoy S (1986) On the central limit theorem in $\mathbb{R}^p$ when $p \to \infty$. Probab Theory Related Fields 73:571–583
Qu YM, Chen SX (2012) Test for bandedness of high-dimensional covariance matrices and bandwidth estimation. Ann Stat 40:1285–1314
Schott JR (2005) Testing for complete independence in high dimensions. Biometrika 92:951–956
Schott JR (2007) A test for the equality of covariance matrices when the dimension is large relative to the sample size. Comput Stat Data Anal 51:6535–6542
Srivastava MS (2005) Some tests concerning the covariance matrix in high-dimensional data. J Jpn Stat Soc 35:251–272
Srivastava MS (2009) A test for the mean vector with fewer observations than the dimension under non-normality. J Multivar Anal 100:518–532
Veillette MS, Taqqu MS (2013) Properties and numerical evaluation of the Rosenblatt distribution. Bernoulli 19:982–1005
Wu WB (2005) Nonlinear system theory: another look at dependence. Proc Natl Acad Sci USA 102:14150–14154 (electronic)
Wu WB (2011) Asymptotic theory for stationary processes. Stat Interface 4:207–226
Wu WB, Shao XF (2004) Limit theorems for iterated random functions. J Appl Probab 41:425–436
Xiao H, Wu WB (2013) Asymptotic theory for maximum deviations of sample covariance matrix estimates. Stoch Process Appl 123:2899–2920
Xu M, Zhang DN, Wu WB (2014) L^2 asymptotics for high-dimensional data. Available at arxiv.org/pdf/1405.7244v3. Accessed 3 Apr 2018
Yarlagadda RK, Hershey JE (1997) Hadamard matrix analysis and synthesis. Kluwer, Boston
Zhang RM, Peng L, Wang RD (2013) Tests for covariance matrix with fixed or divergent dimension. Ann Stat 41:2075–2096

Chapter 9
High-Dimensional Classification

Hui Zou

Abstract There are three fundamental goals in constructing a good high-dimensional classifier: high accuracy, interpretable feature selection, and efficient computation. In the past 15 years, several popular high-dimensional classifiers have been developed and studied in the literature. These classifiers can be roughly divided into two categories: sparse penalized margin-based classifiers and sparse discriminant analysis. In this chapter we give a comprehensive review of these popular high-dimensional classifiers.

Keywords Classifier · Bayes rule · High dimensional data · Regularization · Sparsity · Variable selection

9.1 Introduction

Classification is a fundamental topic in modern statistical learning and statistical classification methods have found numerous successful applications in science and engineering fields. Let $\{(X_i, y_i\}_{i=1}^n\}$ be a random sample from some distribution $P(y, X)$ where y denotes the class label and X represents the predictor vector of dimension p. A classifier predicts the class label at X as $\widehat{y}(X)$. Under the standard 0–1 loss, the misclassification error is $\Pr(y \neq \widehat{y}(X))$ which is lower bounded by the misclassification error of Bayes rule whose predicted label is $\text{argmax}_c \Pr(y = c|X)$ where c is the possible values of y.

In the classical domain where n is much larger than p, there are several off-the-shelf classifiers that can achieve very competitive performance in general, including the famous support vector machines (SVM) (Vapnik 1996), AdaBoost (Freund and Schapire 1997), MART (Friedman 2001), and random forests (Breiman 2001). The more traditional classifiers such as linear discriminant analysis (LDA) and

---

H. Zou (✉)
School of Statistics, University of Minnesota, Minneapolis, MN, USA
e-mail: zouxx019@umn.edu

© Springer International Publishing AG, part of Springer Nature 2018

W. K. Härdle et al. (eds.), *Handbook of Big Data Analytics*, Springer Handbooks of Computational Statistics, https://doi.org/10.1007/978-3-319-18284-1_9

linear logistic regression often perform satisfactorily in practice, and when they do not perform well, one can use their kernel counterparts, kernel discriminant analysis (Mika et al. 1999), and kernel logistic regression (Zhu and Hastie 2005), to improve the classification accuracy. In short, the practitioner can choose from a rich collection of powerful methods for solving low or moderate dimensional classification problems.

The focus in this chapter is on high-dimensional classification where the ambient dimension p is comparable, or even much larger than the sample size n. High-dimensionality brings two great challenges. First, many successful classifiers in the traditional lower-dimension setting may not perform satisfactorily under high-dimensions, because those classifiers are built on all predictors and hence fail to explore the sparsity structure in the high-dimensional model. If there are many noisy features in the data, it is not wise to use all predictors in classification. Second, feature selection has strong scientific foundations. In many scientific studies, the hypothesis is that only a few important predictors determine the response value, and the goal of analysis is to identify these important variables. Such applications naturally require sparsity in the statistical model. For example, many have reported that kernel SVMs can provide accurate classification results in microarray applications, but it does not tell which genes are responsible to classification. Hence, many prefer to use more interpretable methods such as nearest shrunken centroids that select a small subset of genes and also deliver competitive classification performance.

The most popular classifiers with high-dimensional data include sparse penalized svm/large margin classifiers and sparse discriminant analysis. Sparse versions of AdaBoost/boosting and random forests are yet to be developed. In this chapter we present the following high-dimensional classifiers:

- the lasso and elastic-net penalized SVMs;
- the lasso and elastic-net penalized logistic regression;
- the elastic-net penalized Huberized SVMs;
- concave penalized margin-based classifiers;
- independence rules: nearest shrunken centroids (NSC) and FAIR;
- linear programming discriminant (LPD);
- direct sparse discriminant analysis (DSDA);
- sparse semiparametric discriminant analysis;
- sparse additive margin-based classifiers.

Some sparse binary classifier can have a rather straightforward multiclass generalization, while it is much more challenging to derive the multiclass generalizations of other sparse binary classifiers. In this chapter, we will primarily focus on methods for high-dimensional binary classification. Some multiclass sparse classifiers will be briefly mentioned.

9.2 LDA, Logistic Regression, and SVMs

9.2.1 LDA

Linear discriminant analysis (LDA) is perhaps the oldest classification method in the literature and yet is still being routinely used in modern applications (Michie et al. 1994; Hand 2006). The LDA model starts with the assumption that the conditional distribution of the predictor vector given the class label is a multivariate normal distribution:

$$X|y = k \sim N(\mu_k, \boldsymbol{\Sigma}), \quad k = 1, \ldots, K. \tag{9.1}$$

Define $\pi_k = \Pr(Y = k)$, the LDA Bayes rule can be written as

$$\operatorname{argmax}_k[\log \pi_k + \mu_k^T \boldsymbol{\Sigma}^{-1}(X - \mu_k)]. \tag{9.2}$$

The LDA rule replaces these unknown parameters in the Bayes rule with their natural estimates. Let n_k be the number of observations in class k. Define $\widehat{\pi}_k = \frac{n_k}{n}$, $\widehat{\mu}_k = \frac{\sum_{y_i=k} X_i}{n_k}$, and

$$\widehat{\boldsymbol{\Sigma}} = \frac{\sum_{k=1}^{K} \sum_{y_i=k}(X_i - \widehat{\mu}_k)(X_i - \widehat{\mu}_k)^T}{n - K}. \tag{9.3}$$

For the binary case, the LDA classifies an observation to Class 2 if and only if $X^T \widehat{\boldsymbol{\beta}} + \widehat{\boldsymbol{\beta}}_0 > 0$ where

$$\widehat{\boldsymbol{\beta}} = \widehat{\boldsymbol{\Sigma}}^{-1}(\widehat{\mu}_2 - \widehat{\mu}_1), \tag{9.4}$$

$$\widehat{\boldsymbol{\beta}}_0 = -\frac{1}{2}(\widehat{\mu}_1 + \widehat{\mu}_2)^T \widehat{\boldsymbol{\Sigma}}^{-1}(\widehat{\mu}_2 - \widehat{\mu}_1) + \log \frac{\widehat{\pi}_2}{\widehat{\pi}_1}. \tag{9.5}$$

There is also a nice geometric view of LDA, originally suggested by Fisher (1936). Fisher considered the problem of finding an optimal linear projection of the data in which the optimality measure is the ratio of the between-class variance to the within-class variance on the projected data. Fisher's solution is equivalent to the LDA. In Fisher's approach, the normality assumption is not needed but the equal covariance assumption is important.

9.2.2 Logistic Regression

Here we only discuss the logistic regression model for binary classification. The multi-class version of logistic regression is known as multinomial logistic regression. For notation convenience, write $p(X) = \Pr(y = 1|X)$, $1 - p(x) = \Pr(y = 0|X)$. The log-odds function is defined as $f(X) = \log\left(\frac{\Pr(y=1|X)}{\Pr(y=0|X)}\right) = \log\left(p(X)/(1-p(X))\right)$. The linear logistic regression model assumes that $f(X)$ is a linear function of X, i.e.,

$$f(X) = \beta_0 + X^T \boldsymbol{\beta} \tag{9.6}$$

which is equivalent to assuming that

$$p(X) = \frac{\exp(\beta_0 + X^T \boldsymbol{\beta})}{1 + \exp(\beta_0 + X^T \boldsymbol{\beta})}. \tag{9.7}$$

The parameters $(\boldsymbol{\beta}, \beta_0)$ are estimated by maximum conditional likelihood

$$(\widehat{\beta}_0, \widehat{\boldsymbol{\beta}}) = \operatorname{argmax}_{\beta_0, \boldsymbol{\beta}} \ell(\beta_0, \boldsymbol{\beta}). \tag{9.8}$$

where

$$\ell(\beta_0, \boldsymbol{\beta}) = \sum_{i=1}^{n} y_i(\beta_0 + X_i^T \boldsymbol{\beta}) - \sum_{i=1}^{n} \log\left(1 + \exp(\beta_0 + X_i^T \boldsymbol{\beta})\right). \tag{9.9}$$

We classify X to class $I(\widehat{\beta}_0 + X^T \widehat{\boldsymbol{\beta}} > 0)$.

The linear logistic regression model is often fitted via an iteratively re-weighted least squares (IRWLS) algorithm. It turns out that IRWLS can be a part of the algorithm used for solving the sparse penalized linear logistic regression (Friedman et al. 2010), which is discussed in Sect. 9.4.

If we use $y = +1, -1$ to code the class label, then it is easy to check that (9.8) is equivalent to

$$\operatorname{argmin}_{\beta_0, \boldsymbol{\beta}} \sum_{i=1}^{n} \phi(y_i(\beta_0 + X_i^T \boldsymbol{\beta})) \tag{9.10}$$

where $\phi(t) = \log(1 + e^{-t})$ is called the logistic regression loss. This loss-based formulation is useful when we try to unify logistic regression and the support vector machine.

9.2.3 The Support Vector Machine

The support vector machine (SVM) is one of the most successful modern classification techniques (Vapnik 1996). There are a lot of nice introductory books (e.g., Hastie et al. 2009) on the SVM and its kernel generalizations. Our review begins with a nice geometric view of the SVM: it finds the optimal separating hyperplane when the training data can be perfectly separated by a hyperplane. A hyperplane is defined by $\{X : \boldsymbol{\beta}_0 + X^T\boldsymbol{\beta} = 0\}$, where $\boldsymbol{\beta}$ is a unit vector ($\|\boldsymbol{\beta}\|_2 = 1$). We use $1, -1$ to code the class labels. If the training data are linearly separable, then there exists a hyperplane such that

$$y_i(\boldsymbol{\beta}_0 + X_i^T\boldsymbol{\beta}) > 0, \quad \forall i. \tag{9.11}$$

The margin is defined as the smallest distance from the training data to the hyperplane. The SVM uses the separating hyperplane with the largest margin. So the SVM problem is formulated as

$$\max_{\boldsymbol{\beta},\boldsymbol{\beta}_0,\|\boldsymbol{\beta}\|_2=1} \quad C, \tag{9.12}$$

$$\text{subjectto} \;\; y_i(\beta_0 + X_i^T\boldsymbol{\beta}) \geq C, \;\; i = 1, \ldots, n, \tag{9.13}$$

where C is the margin. See Fig. 9.1 for a graphical illustration.

Of course, the training data are rarely linearly separable. Then the general SVM problem is defined by

$$\max_{\boldsymbol{\beta},\beta_0,\|\boldsymbol{\beta}\|_2=1} \quad C, \tag{9.14}$$

$$\text{subjectto} \;\; y_i(\beta_0 + X_i^T\boldsymbol{\beta}) \geq C(1 - \xi_i), \;\; i = 1, \ldots, n, \tag{9.15}$$

$$\xi_i \geq 0, \sum \xi_i \leq B, \tag{9.16}$$

where $\xi_i, \xi_i \geq 0$ are slack variables, and B is a pre-specified positive number which can be regarded as a *tuning parameter*. Notice that the use of slack variables allow

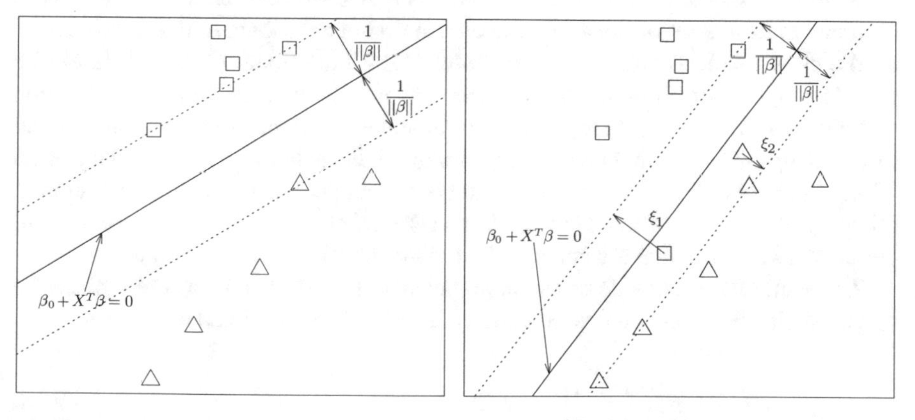

Fig. 9.1 The left panel shows the geometry of the SVM for separable data; the right panel shows the case with non-separable data. Slack variables are introduced in the right panel

some data points are on the wrong side of the hyperplane. See Fig. 9.1 (right panel) for a graphical illustration.

9.3 Lasso and Elastic-Net Penalized SVMs

9.3.1 The ℓ_1 SVM

It can be shown that the SVM defined in (9.14)–(9.16) is the solution to the following optimization problem:

$$\min_{\beta_0,\boldsymbol{\beta}} \frac{1}{n}\sum_{i=1}^{n}\left[1 - y_i(\beta_0 + X_i^T\boldsymbol{\beta})\right]_+ + \lambda\|\boldsymbol{\beta}\|_2^2, \tag{9.17}$$

where z_+ indicates the positive part of a real valued number z, and λ has a one-to-one correspondence to B in (9.16). The loss function $(1-t)_+$ is called the *hinge* loss.

From (9.17) we can easily obtain the so-called kernel SVM in a reproducing kernel Hilbert space, and it turns out the kernel SVM is also the solution to a quadratically penalized hinge loss problem. The readers are referred to Chapter 12 of Hastie et al. (2009) for the detailed discussion. The use of the ℓ_2 penalty in the SVM is critically important for the success of the SVM in the low-dimensional classification problems but is also responsible for its failure in the high-dimensional cases. The main reason is that the solution to (9.17) is dense: all $\widehat{\beta}_j$s are nonzero in general. Therefore, the resulting SVM classifier uses all predictors. A simulation study in Chapter 12 of Hastie et al. (2009) clearly shows that the classification accuracy of the SVM will be greatly degraded if there are more and more noise predictors included in the predictor set. With high-dimensional data it is very common to have a lot of noise covariates that do not contribute to the classification of the class label. Therefore, a successful high-dimensional classifier should be able to suppress or eliminate the influence of noise variables. Many penalization techniques have been developed for simultaneous variable selection and model fitting in high-dimensional linear regression models, including the lasso (Tibshirani 1996), the SCAD (Fan and Li 2001), the elastic-net (Zou and Hastie 2005), among others. These penalization techniques can be combined with the hinge loss to generate new SVMs that are suitable for high-dimensional classification.

Zhu et al. (2003) considered the lasso penalized SVM. The idea is to replace the ℓ_2 penalty in (9.17) with the lasso penalty. The ℓ_1 SVM is defined as follows:

$$\min_{\beta_0,\boldsymbol{\beta}} \frac{1}{n}\sum_{i=1}^{n}\left[1 - y_i(\beta_0 + X_i^T\boldsymbol{\beta})\right]_+ + \lambda\|\boldsymbol{\beta}\|_1. \tag{9.18}$$

where $\|\boldsymbol{\beta}\|_1 = \sum_j |\beta_j|$ is called the lasso penalty or the ℓ_1 penalty. A fundamental difference between the ℓ_2 penalty and the ℓ_1 penalty is that the latter is singular at zero. Thus, the solution to (9.18) could have some exact zero $\widehat{\beta}_j$s. The regularization parameter λ controls the sparsity as well as the classification accuracy of the SVM. It can be easily shown that when λ is sufficiently large, the solution to (9.18) must have $\widehat{\beta}_j = 0$ for all js. When λ is relaxed to smaller values, more $\widehat{\beta}_j$s become nonzero, which means more predictors enter the ℓ_1 SVM model. The ℓ_1 penalized SVM was proposed in Bradley and Mangasarian (1998). The major contribution in Zhu et al. (2003) is a piecewise linear path-following algorithm for computing the entire solution paths of the ℓ_1 SVM. Wang and Shen (2006) developed a multicategory ℓ_1 support vector machine.

9.3.2 The DrSVM

Although the ℓ_1 SVM is better than the standard ℓ_2 SVM for high-dimensional classification in general, the ℓ_1 SVM also has some serious drawbacks due to the lasso penalty. When there are a group of highly correlated variables, the lasso tends to randomly pick one and ignore the rest from the group. In addition, the lasso solution suffers from high variability from collinearity. The elastic net (Zou and Hastie 2005) offers a nice fix to the above issues by adaptively mixing the ℓ_2 penalty and the ℓ_1 penalty. The elastic-net penalty is defined as

$$\frac{\lambda_2}{2}\|\beta\|_2^2 + \lambda_1\|\beta\|_1, \tag{9.19}$$

$$\text{or } \lambda[(1-\alpha)\frac{\|\beta\|_2^2}{2} + \alpha\|\beta\|_1], \quad 0 < \alpha < 1 \tag{9.20}$$

The elastic-net penalty is able to enjoy the good properties of the ℓ_2 and the ℓ_1 penalties. Similar to the ℓ_1 penalty, the elastic-net penalty is singular at zero, which allows it to produce sparse solutions automatically. The ℓ_2 component of the elastic-net penalty further regularizes the solution path to make it more stable. Thus, the elastic-net usually outperforms the lasso in prediction. Furthermore, the ℓ_2 component encourages a grouping effect such that highly correlated variables tend to be selected together.

Wang et al. (2008) applied the elastic-net penalty to the SVM and defined the so-called DrSVM as follows:

$$\min_{\beta_0,\beta} \sum_{i=1}^{n} [1 - y_i(\beta_0 + x_i^T\beta)]_+ + \frac{\lambda_2}{2}\|\beta\|_2^2 + \lambda_1\|\beta\|_1. \tag{9.21}$$

Notice that we use (9.19) for the elastic-net penalty in order to gain a nice computational advantage, which will be explained later.

Theorem 1 (Wang et al. 2008) *Denote the solution to (9.21) as $\widehat{\boldsymbol{\beta}}$. Then for any pair (j,ℓ), we have*

$$\left|\widehat{\beta}_j - \widehat{\beta}_\ell\right| \leq \frac{1}{\lambda_2}\|x_j - x_\ell\|_1. \tag{9.22}$$

Furthermore, if the input variable x_j, x_ℓ are centered and normalized, then

$$\left|\widehat{\beta}_j - \widehat{\beta}_\ell\right| \leq \frac{\sqrt{n}}{\lambda_2}\sqrt{2(1-\rho_{j\ell})}, \tag{9.23}$$

where $\rho_{j\ell}$ is the sample correlation between x_j and x_ℓ.

Theorem 1 gives a quantitative explanation of the grouping effect of the DrSVM. The theorem is actually established for a whole class of elastic-net penalized margin-based loss classifiers (Wang et al. 2008).

The DrSVM enjoys a piecewise linear solution path property, which is the foundation of an efficient path-following algorithm for solving the DrSVM.

Theorem 2 (Wang et al. 2008) *Write the solution $\widehat{\boldsymbol{\beta}}$ for (9.21) as $\widehat{\boldsymbol{\beta}}_{\lambda_2}(\lambda_1)$. For each fixed $\lambda_2 \geq 0$, $\widehat{\boldsymbol{\beta}}_{\lambda_2}(\lambda_1)$ is a piecewise linear function of λ_1.*

Notice that when $\lambda_2 = 0$, the DrSVM reduces to the ℓ_1 SVM. Thus, the above theorem shows the ℓ_1 SVM has a piecewise linear solution path. In fact, Zhu et al. (2003) already proved this property and derive a path-following algorithm for computing the entire solution paths of the ℓ_1 SVM. Wang et al. (2008) further generalize the arguments used in Zhu et al. (2003) to develop an efficient path-following algorithm for the DrSVM. By the piecewise linear property, we only need to compute the solutions at change points and the solution path between two change points can be linearly extrapolated. The main idea of the algorithm follows the least angle regression algorithm by Efron et al. (2004) for solving the lasso penalized linear regression. The R code for fitting the DrSVM can be downloaded from

```
http://dept.stat.lsa.umich.edu/~jizhu/code/drsvm/
```

9.4 Lasso and Elastic-Net Penalized Logistic Regression

Under the linear logistic regression model, the log-likelihood function is

$$\ell(\beta_0, \boldsymbol{\beta}) = \sum_{i=1}^{n} y_i(\beta_0 + X_i^T\boldsymbol{\beta}) - \sum_{i=1}^{n} \log\left(1 + \exp(\beta_0 + X_i^T\boldsymbol{\beta})\right). \tag{9.24}$$

Tibshirani (1996) mentioned the lasso penalized logistic regression:

$$\operatorname{argmin}_{\beta_0,\boldsymbol{\beta}} \left\{ -\frac{1}{n}\ell(\beta_0,\boldsymbol{\beta}) + \lambda \sum_{j=1}^{p} |\beta_j| \right\}. \tag{9.25}$$

However, Tibshirani did not provide an efficient algorithm for fitting the lasso penalized logistic regression model. Recently, Friedman et al. (2010) developed an R package `glmnet` (available from CRAN) for solving the more general elastic-net penalized logistic regression model:

$$\operatorname{argmin}_{\beta_0,\boldsymbol{\beta}} \left\{ -\frac{1}{n}\ell(\beta_0,\boldsymbol{\beta}) + \lambda \sum_{j=1}^{p} \left(\frac{1-\alpha}{2}|\beta_j|^2 + \alpha|\beta_j| \right) \right\}. \tag{9.26}$$

When $\alpha = 1$, `glmnet` solves the lasso penalized logistic regression. `glmnet` also handles the elastic-net penalized multinomial regression for multiclass problems.

The algorithm used in `glmnet` is very simple, which is referred to FHT algorithm in what follows. First, we follow the Newton-Raphson algorithm for solving the ordinary logistic regression to get a quadratic approximation of the logistic regression loss. Then, a very fast coordinate descent algorithm can be used to find the minimizer of the approximate objective function. The final solution is obtained by iterating between the Newton-Raphson step and a cyclic coordinate descent loop.

Let $(\tilde{\beta}_0, \widetilde{\boldsymbol{\beta}})$ be the current solution to (9.26). Define

$$\tilde{p}(X_i) = \frac{\exp(\tilde{\beta}_0 + X_i^T \widetilde{\boldsymbol{\beta}})}{1 + \exp(\tilde{\beta}_0 + X_i^T \widetilde{\boldsymbol{\beta}})} \tag{9.27}$$

$$w_i = \tilde{p}(X_i)(1 - \tilde{p}(X_i)) \tag{9.28}$$

$$z_i = \tilde{\beta}_0 + X^T \widetilde{\boldsymbol{\beta}} + \frac{y_i - \tilde{p}(X_i)}{w_i}. \tag{9.29}$$

Friedman et al. (2010) considered a quadratic approximation of $\ell(\beta_0, \boldsymbol{\beta})$ and then turned (9.26) into a penalized quadratic problem. Specifically, FHT algorithm updates $(\tilde{\beta}_0, \widetilde{\boldsymbol{\beta}})$ by solving the following penalized weighted least squares problem:

$$\operatorname{argmin}_{\beta_0,\boldsymbol{\beta}} \frac{1}{2n} \sum_{i=1}^{n} w_i (z_i - \beta_0 - X_i^T \boldsymbol{\beta})^2 + \lambda \sum_{j=1}^{p} \left(\frac{1-\alpha}{2}|\beta_j|^2 + \alpha|\beta_j| \right). \tag{9.30}$$

Notice that if $\lambda = 0$, (9.30) reduces to the IRWLS for solving the linear logistic regression.

FHT algorithm uses a coordinate descent algorithm to solve (9.30). Suppose that we have the solution to (9.30) except β_j. Then we have

$$\beta_j = \text{argmin}_{\beta_j} \frac{1}{2n} \sum_{i=1}^{n} w_i (z_i - \tilde{y}_i - x_{ij}\beta_j)^2 + \lambda \left(\frac{1-\alpha}{2} |\beta_j|^2 + \alpha |\beta_j| \right). \tag{9.31}$$

where $\tilde{y}_i = \beta_0 + \sum_{\ell \neq j} x_\ell \tilde{\beta}_\ell$. The solution to (9.31) is given by the elastic-net thresholding rule (Zou and Hastie 2005):

$$\beta_j = \frac{S(\sum_{i=1}^{n} \frac{w_i}{n} x_{ij}(z_i - \tilde{y}_i), \lambda\alpha)}{\sum_{i=1}^{n} \frac{w_i}{n} x_{ij}^2 + \lambda(1-\alpha)} \tag{9.32}$$

where S is the so-called soft-thresholding operator: $S(t, s) = \text{Sign}(z)(|z| - s)_+$. We repeatedly apply (9.32) to update the coefficient of each coordinate $j = 1, 2, \ldots, p, 1, 2, \ldots, p, 1, 2, \ldots,$ and the cycle continues until convergence. The simplicity of (9.32) makes cyclic coordinate descent a very attractive algorithm for solving (9.31). In fact, Friedman et al. (2010) demonstrated that cyclic coordinate descent can be faster than the least angle regression algorithm for solving the lasso penalized least squares.

Several implementation tricks, such as warm start, active set update, the strong rule, are used in the `glmnet` package to optimize the computation speed of FHT algorithm. The readers are referred to Friedman et al. (2010) for the details.

Algorithm 1: FHT algorithm for the elastic-net penalized logistic regression

---

1. Initialize $(\beta_0^{(0)}, \boldsymbol{\beta}^{(0)})$.
2. for $k = 0, 1, \ldots,$ do

   (2.a) $\tilde{\beta}_0 = \beta_0^{(k)}, \tilde{\boldsymbol{\beta}}_0 = \boldsymbol{\beta}^{(k)}$

   (2.b) compute $\tilde{p}(X_i), w_i, z_i$ as defined in (1.30)–(1.32)

   (2.c) let $(\beta_0^{(k)}, \boldsymbol{\beta}^{(k)})$ be the solution to (9.30), which is solved by cyclic coordinate descent based on (9.32).

3 repeat steps (2.a), (2.b), (2.c) till convergence.

9.5 Huberized SVMs

Wang et al. (2008) proposed a smoothed SVM for high-dimensional classification where the hinge loss underlying the standard SVM is replaced with a Huberized hinge loss. The motivation is to use a smoother loss function in order to achieve gains in computational efficiency. See Fig. 9.2 for a plot of the Huberized hinge loss.

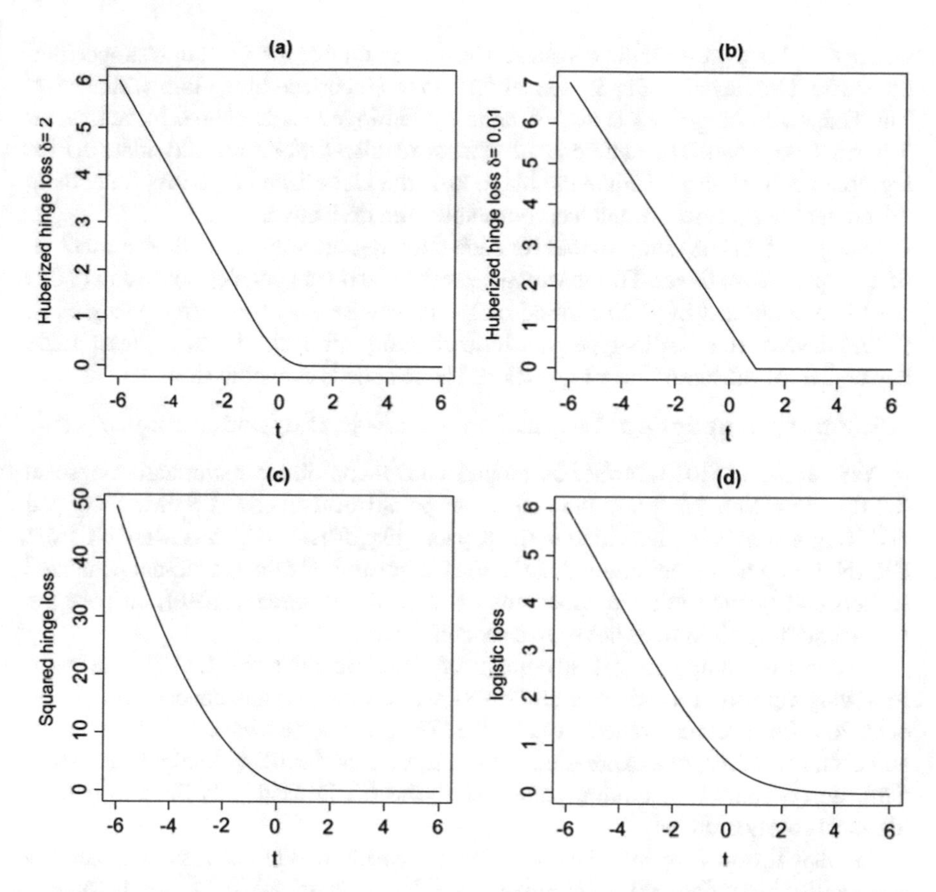

Fig. 9.2 (**a**) The Huberized hinge loss function (with $\delta = 2$). (**b**) The Huberized hinge loss function (with $\delta = 0.01$). (**c**) The squared hinge loss function. (**d**) The logistic loss function

The elastic-net penalized Huberized SVM is defined as

$$\min_{(\beta_0,\boldsymbol{\beta})} \frac{1}{n} \sum_{i=1}^{n} \phi_\delta(y_i(\beta_0 + \mathbf{x}_i^\top \boldsymbol{\beta})) + P_{\lambda_1,\lambda_2}(\boldsymbol{\beta}). \tag{9.33}$$

where $P_{\lambda_1,\lambda_2}(\boldsymbol{\beta})$ is the elastic net penalty:

$$P_{\lambda_1,\lambda_2}(\boldsymbol{\beta}) = \sum_{j=1}^{p} p_{\lambda_1,\lambda_2}(\beta_j) = \sum_{j=1}^{p} \left(\lambda_1 |\beta_j| + \frac{\lambda_2}{2} \beta_j^2 \right), \tag{9.34}$$

Note that $\phi(\cdot)$ in (9.33) is the Huberized hinge loss

$$\phi_\delta(t) = \begin{cases} 0, & t > 1 \\ (1-t)^2/2\delta, & 1-\delta < t \leq 1 \\ 1-t-\delta/2, & t \leq 1-\delta, \end{cases}$$

where $\delta > 0$ is a pre-specific constant. The default choice for δ is 2 unless specified otherwise. Displayed in Fig. 9.2(panel (a)) is the Huberized hinge loss with $\delta = 2$. The Huberized hinge loss is very similar to the hinge loss in shape. In fact, when δ is small, the two loss functions are almost identical. See Fig. 9.2(panel (b)) for a graphical illustration. Unlike the hinge loss, the Huberized hinge loss function is differentiable everywhere and has continuous first derivative.

Wang et al. (2008) showed that for each fixed λ_2, the solution path as a function of λ_1 is piecewise linear. This is why we use the elastic net penalty defined in (9.34) not the one defined in (9.20). Based on the piecewise linear property, Wang et al. (2008) developed a LARS-type path following algorithm similar to the least angle regression for the lasso linear regression. The R code is available from

`http://dept.stat.lsa.umich.edu/~jizhu/code/hhsvm/`

Yang and Zou (2013) further developed a novel coordinate-majorization-descent (CMD) algorithm for fitting the elastic-net penalized Huberized SVM. Yang and Zou's algorithm is implemented in the R package `gcdnet` which is available from CRAN. It is worth mentioning that the FHT algorithm for the elastic-net penalized Huberized logistic regression does not work for the Huberized SVM, because the Huberized hinge loss does not have a second derivative.

To see the computational advantage of `gcdnet` over the LARS type path-following algorithm in Wang et al. (2008), we use the prostate cancer data (Singh et al. 2002) as a demonstration. See Fig. 9.3. The prostate data have 102 observations and each has 6033 gene expression values. It took the LARS-type algorithm about 5 min to compute the solution paths, while the CMD used only 3.5 s to get the identical solution paths.

In what follows we introduce the CMD algorithm. Without loss of generality assume the input data are standardized: $\frac{1}{n}\sum_{i=1}^{n} x_{ij} = 0$, $\frac{1}{n}\sum_{i=1}^{n} x_{ij}^2 = 1$, for $j = 1, \ldots, p$. The standard coordinate descent algorithm proceeds as follows. Define the current margin $r_i = y_i(\tilde{\beta}_0 + \mathbf{x}_i^{\mathsf{T}}\tilde{\boldsymbol{\beta}})$ and

$$F(\beta_j|\tilde{\beta}_0, \tilde{\boldsymbol{\beta}}) = \frac{1}{n}\sum_{i=1}^{n} \phi_\delta(r_i + y_i x_{ij}(\beta_j - \tilde{\beta}_j)) + p_{\lambda_1,\lambda_2}(\beta_j). \tag{9.35}$$

For fixed λ_1 and λ_2, the standard coordinate descent algorithm (Tseng 2001) proceeds as follows:

1. Initialization: $(\tilde{\beta}_0, \tilde{\boldsymbol{\beta}})$
2. Cyclic coordinate descent: for $j = 0, 1, 2, \ldots, p$: update $\tilde{\beta}_j$ by minimizing the objective function

$$\tilde{\beta}_j = \underset{\beta_j}{\operatorname{argmin}} F(\beta_j|\tilde{\beta}_0, \tilde{\boldsymbol{\beta}}). \tag{9.36}$$

3. Repeat Step 2 till convergence.

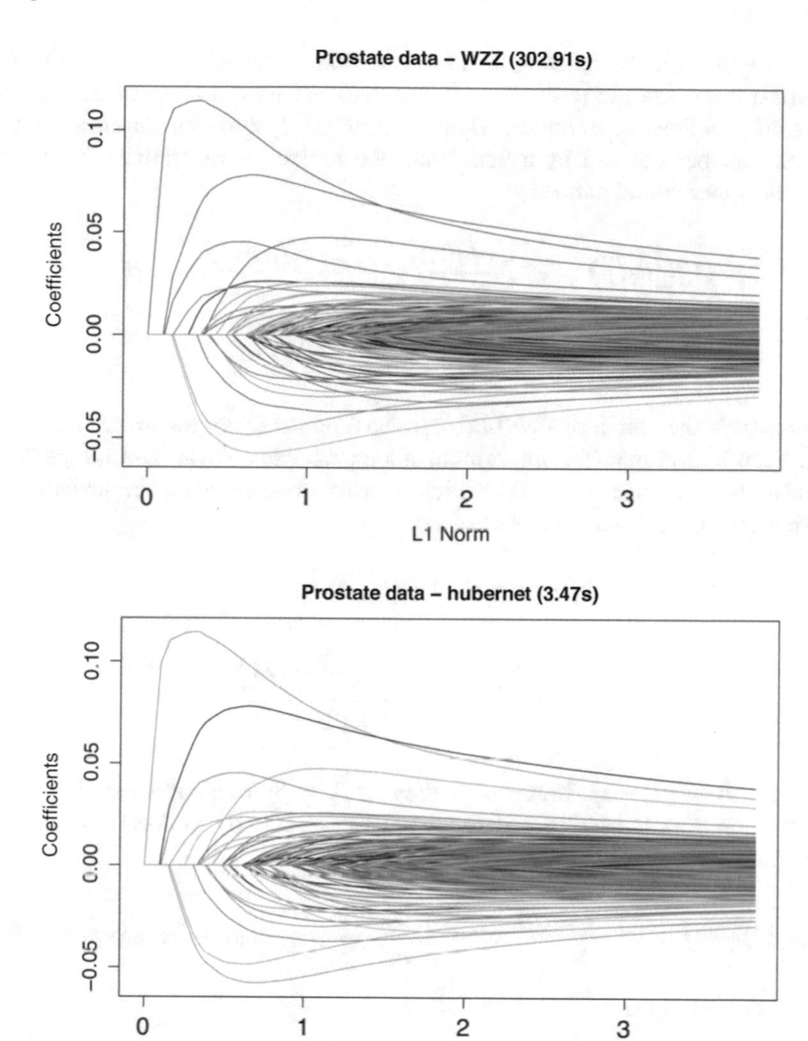

Fig. 9.3 Solution paths and timings of the elastic-net penalized Huberized SVM on the prostate cancer data with 102 observations and 6033 predictors. The top panel shows the solution paths computed by the LARS-type algorithm in Wang et al. (2008); the bottom panel shows the solution paths computed by CMD. CMD is 87 times faster

The major difficulty in using the above coordinate descent procedure is that the univariate minimization problem in (9.36) does not have a closed form solution, unlike the penalized least squares. Yang and Zou (2013) show that the computational obstacle can be resolved by a neat trick. We replace F function in (9.35) by a quadratic upper bound defined as

$$Q(\beta_j|\tilde{\beta}_0, \tilde{\boldsymbol{\beta}}) = \frac{\sum_{i=1}^n \phi_\delta(r_i)}{n} + \frac{\sum_{i=1}^n \phi'_\delta(r_i) y_i x_{ij}}{n}(\beta_j - \tilde{\beta}_j) + \frac{1}{\delta}(\beta_j - \tilde{\beta}_j)^2 + P_{\lambda_1,\lambda_2}(\beta_j), \tag{9.37}$$

where $\phi_\delta'(t)$ is the first derivative of $\phi_\delta(t)$. The function Q is a majorization function of F. Then by the majorization-minimization principle (Hunter and Lange 2004), we solve the minimizer of (9.37), which is easily obtained via a simple elastic-net thresholding rule (Zou and Hastie 2005):

$$\widehat{\beta}_j^{\mathrm{C}} = \underset{\beta_j}{\operatorname{argmin}} Q(\beta_j|\tilde{\beta}_0, \tilde{\boldsymbol{\beta}}) = \frac{S\left(\frac{2}{\delta}\tilde{\beta}_j - \frac{\sum_{i=1}^n \phi'_c(r_i) y_i x_{ij}}{n}, \lambda_1\right)}{\frac{2}{\delta} + \lambda_2}, \tag{9.38}$$

where $S(z, s) = (|z| - s)_+ \mathrm{Sign}(z)$. We then set $\tilde{\beta}_j = \widehat{\beta}_j^{\mathrm{C}}$ as the new estimate.

The same trick is used to update intercept β_0. Similarly to (9.37), we consider minimizing a quadratic approximation

$$Q(\beta_0|\tilde{\beta}_0, \tilde{\boldsymbol{\beta}}) = \frac{\sum_{i=1}^n \phi_c(r_i)}{n} + \frac{\sum_{i=1}^n \phi'_c(r_i) y_i}{n}(\beta_0 - \tilde{\beta}_0) + \frac{1}{\delta}(\beta_0 - \tilde{\beta}_0)^2, \tag{9.39}$$

which has a minimizer

$$\widehat{\beta}_0^{\mathrm{C}} = \tilde{\beta}_0 - \frac{\delta}{2}\frac{\sum_{i=1}^n \phi'_c(r_i) y_i}{n}. \tag{9.40}$$

We set $\tilde{\beta}_0 = \widehat{\beta}_0^{\mathrm{C}}$ as the new estimate.

Algorithm 2 has the complete details of the CMD algorithm for solving the elastic-net penalized Huberized SVM. The beauty of Algorithm 2 is that it is remarkably simple and almost identical to the coordinate descent algorithm for computing the elastic net penalized regression. Moreover, Yang and Zou (2013) showed that the CMD algorithm can solve a whole class of elastic-net penalized large margin classifiers. Let $\phi(t)$ be a convex, differentiable decreasing function such that $\phi'(0) < 0$. Then $\phi(t)$ is a classification calibrated margin-based loss function for binary classification (Bartlett et al. 2006). The Huberized hinge loss is a smooth margin-based loss function. Other smooth margin-based loss functions

---

Algorithm 2: The CMD algorithm for solving the elastic-net penalized Huberized SVM

- Initialize $(\widetilde{\beta}_0, \widetilde{\boldsymbol{\beta}})$.
- Iterate 2(a)–2(b) until convergence:
  - 2(a) Cyclic coordinate descent: for $j = 1, 2, \ldots, p$,
    * (2.a.1) Compute $r_i = y_i(\widetilde{\beta}_0 + \mathbf{x}_i^{\mathsf{T}} \widetilde{\boldsymbol{\beta}})$.
    * (2.a.2) Compute

$$\widehat{\beta}_j^{\mathrm{C}} = \frac{S\left(\frac{2}{\delta}\widetilde{\beta}_j - \frac{\sum_{i=1}^{n} \phi_c'(r_i) y_i x_{ij}}{n}, \lambda_1\right)}{\frac{2}{\delta} + \lambda_2}.$$

    * (2.a.3) Set $\widetilde{\beta}_j = \widehat{\beta}_j^{\mathrm{C}}$.
  - 2(b) Update the intercept term
    * (2.b.1) Re-compute $r_i = y_i(\widetilde{\beta}_0 + \mathbf{x}_i^{\mathsf{T}} \widetilde{\boldsymbol{\beta}})$.
    * (2.b.2) Compute

$$\widehat{\beta}_0^{\mathrm{C}} = \widetilde{\beta}_0 - \frac{\sum_{i=1}^{n} \phi_c'(r_i) y_i}{\frac{2}{\delta} n}.$$

    * (2.b.3) Set $\widetilde{\beta}_0 = \widehat{\beta}_0^{\mathrm{C}}$.

---

include the logistic regression loss and the squared hinge loss. See Fig. 9.2(panel (c) and panel (d)).

$$\text{logistic regression loss}: \ \phi(t) = \log(1 + e^{-t}) \tag{9.41}$$

$$\text{squared hinge loss}: \ \phi(t) = ([1 - t]_+)^2 \tag{9.42}$$

Given the loss function, the elastic-net penalized ϕ margin classifier is defined as

$$\min_{\beta_0, \boldsymbol{\beta}} \frac{1}{n} \sum_{i=1}^{n} \phi\left(y_i(\beta_0 + X_i^T \boldsymbol{\beta})\right) + \sum_{j=1}^{p} P_{\lambda_1, \lambda_2}(|\beta_j|). \tag{9.43}$$

It is shown that the CMD algorithm can be used to solve (9.43) as long as ϕ satisfies the following *quadratic majorization condition*

$$\phi(t + a) \leq \phi(t) + \phi'(t)a + \frac{M}{2}a^2 \quad \forall t, a. \tag{9.44}$$

It is easy to check that $m = \frac{1}{4}$ for the logistic regression loss, $M = 4$ for the squared hinge loss, and $M = \frac{2}{\delta}$ for the Huberized hinge loss. We can simply replace $\frac{2}{\delta}$ with M in Algorithm 2 to solve for (9.43). See Yang and Zou (2013) for details.

9.6 Concave Penalized Margin-Based Classifiers

In the literature concave penalty functions have also received a lot of attention as ways to produce sparse models. The most famous examples of concave penalty functions are the smoothly clipped absolute deviation (SCAD) penalty proposed by Fan and Li (2001) and the minimax concave penalty (MCP) proposed by Zhang (2010). Another popular concave penalty function is the ℓ_q penalty with $0 < q < 1$. Fan and Li (2001) argued that a good penalty function should produce a penalized estimate that possesses the properties of sparsity, continuity, and unbiasedness. The lasso penalty does not have the unbiasedness property because it also shrinks large coefficients that are very likely to be significant. The original motivation for using the concave penalty is to correct the shrinkage bias by the lasso. To this end, Fan and Li (2001) proposed the SCAD penalty defined by $P_\lambda^{\text{scad}}(0) = 0$, $P_\lambda(t)$ is symmetric and for $t > 0$

$$\frac{dP_\lambda^{\text{scad}}(t)}{dt} = \lambda I(t \le \lambda) + \frac{(a\lambda - t)_+}{a-1} I(t > \lambda), \quad a > 2$$

Note that z_+ stands for the positive part of z. The default value for a is 3.7 (Fan and Li 2001). Hence the SCAD behaves like the lasso for small coefficients and like the hard thresholding for large coefficients. The MCP penalty is defined by $P_\lambda^{\text{mcp}}(0) = 0$, $P_\lambda^{\text{mcp}}(t)$ is symmetric and for $t > 0$. The derivative of MCP is

$$\frac{dP_\lambda^{\text{mcp}}(t)}{dt} = \left(\lambda - \frac{t}{a}\right)_+.$$

Figure 9.4 shows the plots of the lasso, SCAD, and MCP penalty functions.

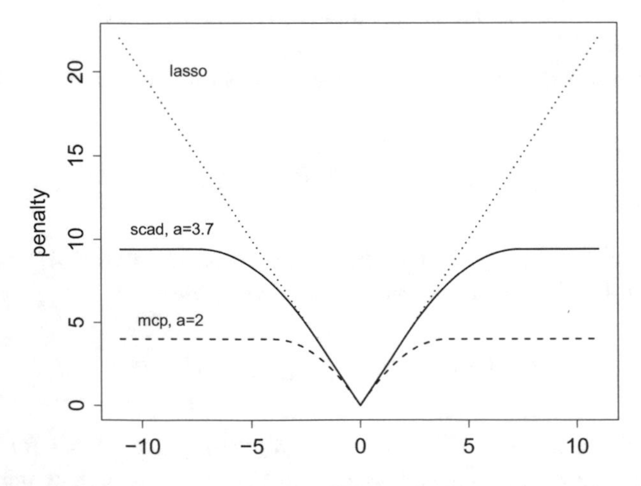

Fig. 9.4 Three penalty functions: lasso ($\lambda = 2$), SCAD ($\lambda = 2, a = 3.7$), MCP ($\lambda = 2, a = 2$)

In this section we discuss concave penalized large margin classifiers. We use $y = 1+, -1$ to code the class label and consider a margin-based loss function $\phi(\cdot)$. A concave penalized ϕ-classifier is defined as

$$(\widehat{\beta}_0, \widehat{\boldsymbol{\beta}}) = \operatorname{argmin}_{\boldsymbol{\beta}} \frac{1}{n} \sum_{i=1}^{n} \phi(y_i(\beta_0 + X_i^T \boldsymbol{\beta})) + \sum_{j=1}^{p} P_\lambda(|\beta_j|), \tag{9.45}$$

where P_λ is a concave penalty such as the SCAD or the MCP. The classifier is $\text{Sign}(\widehat{\beta}_0 + X^T \widehat{\boldsymbol{\beta}})$. Zhang et al. (2006) used the SCAD penalized SVM for microarray classification problems and reported good performance.

The computation of a concave penalized problem is more challenging than the lasso problem. Fan and Li (2001) proposed a local quadratic approximation (LQA) algorithm for solving both lasso and concave penalized likelihood models. The main idea is to approximate the penalty function by a quadratic function and then iteratively solve the quadratic penalized likelihood model. The LQA works reasonably well for moderate dimensions but it cannot handle high-dimensions, and the LQA algorithm has a technical difficulty at zero: once the current estimate is very close to zero, the LQA will truncate it to zero and remove it from the latter iterations. Zou and Li (2008) proposed a novel local linear approximation (LLA) algorithm to eliminate the drawbacks of LQA. The LLA algorithm is remarkably simple. For a concave penalty function, we have

$$P_\lambda(|\beta_j|) \leq P_\lambda(|\tilde{\beta}_j|) + P'_\lambda(|\tilde{\beta}_j|)(|\beta_j| - |\tilde{\beta}_j|) \tag{9.46}$$

This inequality suggests us to revoke the MM principle (Hunter and Lange 2004). Hence we update the current solution by solving

$$\operatorname{argmin}_{\boldsymbol{\beta}} \frac{1}{n} \sum_{i=1}^{n} \phi(y_i(\beta_0 + X_i^T \boldsymbol{\beta})) + \lambda \sum_{j=1}^{p} \{P_\lambda(|\tilde{\beta}_j|) + P'_\lambda(|\tilde{\beta}_j|)(|\beta_j| - |\tilde{\beta}_j|)\} \tag{9.47}$$

which is equivalent to

$$\operatorname{argmin}_{\boldsymbol{\beta}} \frac{1}{n} \sum_{i=1}^{n} \phi(y_i(\beta_0 + X_i^T \boldsymbol{\beta})) + \lambda \sum_{j=1}^{p} w_j |\beta_j|. \tag{9.48}$$

where $w_j = P'_\lambda(|\tilde{\beta}_j|)$. Algorithm 3 summarizes the details of LLA algorithm. It is important to see that the LLA algorithm turns the original concave penalization problem into a sequence of weighted ℓ_1 penalization problems via the MM principle. At each LLA step, the weighted ℓ_1 penalization problem has an exact sparse presentation without using any artificial truncation and the weighted ℓ_1 penalization problem can be solved very efficiently. These advantages make LLA the standard algorithm for solving concave penalized problems.

Algorithm 3: The local linear approximation (LLA) algorithm

1. Initialize $\widehat{\boldsymbol{\beta}}^{(0)} = \widehat{\boldsymbol{\beta}}^{initial}$ and compute the adaptive weight

$$\widehat{\mathbf{w}}^{(0)} = (\widehat{w}_1^{(0)}, \cdots, \widehat{w}_p^{(0)})' = \left(P'_\lambda(|\widehat{\beta}_1^{(0)}|), \cdots, P'_\lambda(|\widehat{\beta}_p^{(0)}|)\right)'.$$

2. For $m = 1, 2, \ldots$, repeat the LLA iteration till convergence

   (2.a) Obtain $\widehat{\boldsymbol{\beta}}^{(m)}$ by solving the following optimization problem

$$\widehat{\boldsymbol{\beta}}^{(m)} = \min_{\boldsymbol{\beta}} \frac{1}{n} \sum_{i=1}^{n} \phi(y_i(\beta_0 + X_i^T \boldsymbol{\beta})) + \sum_{j=1}^{p} \widehat{w}_j^{(m-1)} \cdot |\beta_j|,$$

   (2.b) Update the adaptive weight vector $\widehat{\mathbf{w}}^{(m)}$ with $\widehat{w}_j^{(m)} = P'_\lambda(|\widehat{\beta}_j^{(m)}|)$.

We can use Algorithm 3 to solve a class of concave penalized margin-based classifiers. We only need to specify the algorithm used for solving step (2.a). For example, if ϕ is the logistic regression loss, then step (2.a) can be solved by using `glmnet` or `gcdnet`. When ϕ is the Huberized hinge loss or the squared hinge loss, `gcdnet` can be used for step (2.a) When ϕ is the hinge loss, then one can use the algorithm by Zhu et al. (2003) to compute step (2.a).

There has been a lot of theoretical work on the concave penalization. A strong theoretical result states that the concave penalized estimator is able to produce the so-called oracle estimator. We use the logistic regression to illustrate the oracle property of concave penalization. Our discussion here follows Fan et al. (2014). Given X_i, y_i has a Bernoulli distribution with

$$P(y_i = 1|\mathbf{x}_i, \boldsymbol{\beta}^\star) = \frac{\exp(\mathbf{x}_i'\boldsymbol{\beta}^\star)}{1 + \exp(\mathbf{x}_i'\boldsymbol{\beta}^\star)}$$

where $\boldsymbol{\beta}^\star$ is a sparse parameter vector. Let s be the number of nonzero elements of $\boldsymbol{\beta}^\star$. Define $\|\boldsymbol{\beta}^\star_{\mathcal{A}}\|_{\min} = \min |\beta_j|, \beta_j \neq 0$. $\mathcal{A} = \{j : \beta_j \neq 0\}$. The oracle estimator is defined as

$$\widehat{\boldsymbol{\beta}}^{\text{oracle}}_{\text{Logit}} = \text{argmin}_{\boldsymbol{\beta}}\ \ell_n^{\text{Logit}}(\boldsymbol{\beta}) \quad \text{subjectto} \quad \boldsymbol{\beta}_{\mathcal{A}^c} = \mathbf{0}. \tag{9.49}$$

where $\ell_n^{\text{Logit}}(\boldsymbol{\beta})$ denotes the negative log-Bernoulli likelihood. The lasso logistic regression estimator is

$$\widehat{\boldsymbol{\beta}}^{\text{lasso}}_{\text{Logit}} = \arg\min_{\boldsymbol{\beta}} \ \ell_n^{\text{Logit}}(\boldsymbol{\beta}) + \lambda_{\text{lasso}} \|\boldsymbol{\beta}\|_{\ell_1} \tag{9.50}$$

To present the theorem we also need the following notation. Define $\psi(t) = \log(1 + \exp(t))$ and

$$\boldsymbol{\mu}(\boldsymbol{\beta}) = (\psi'(\mathbf{x}_1'\boldsymbol{\beta}), \ldots, \psi'(\mathbf{x}_n'\boldsymbol{\beta}))'$$
$$\boldsymbol{\Sigma}(\boldsymbol{\beta}) = \mathrm{diag}\{\psi''(\mathbf{x}_1'\boldsymbol{\beta}), \ldots, \psi''(\mathbf{x}_n'\boldsymbol{\beta})\}$$
$$Q_1 = \max_j \lambda_{\max}\left(\frac{1}{n}\mathbf{X}_{\mathcal{A}}' \,\mathrm{diag}\{|\mathbf{x}_{(j)}|\}\mathbf{X}_{\mathcal{A}}\right)$$
$$Q_2 = \left\| \left(\frac{1}{n}\mathbf{X}_{\mathcal{A}}'\boldsymbol{\Sigma}(\boldsymbol{\beta}^{\star})\mathbf{X}_{\mathcal{A}}\right)^{-1} \right\|_{\ell_\infty}$$
$$Q_3 = \|\mathbf{X}_{\mathcal{A}^c}'\boldsymbol{\Sigma}(\boldsymbol{\beta}^{\star})\mathbf{X}_{\mathcal{A}}(\mathbf{X}_{\mathcal{A}}'\boldsymbol{\Sigma}(\boldsymbol{\beta}^{\star})\mathbf{X}_{\mathcal{A}})^{-1}\|_{\ell_\infty}.$$

in which $\mathrm{diag}\{|\mathbf{x}_{(j)}|\}$ is a diagonal matrix with elements $\{|x_{ij}|\}_{i=1}^n$.

Theorem 3 (Fan et al. 2014) *Assume the restricted eigenvalue condition:*

$$\kappa_{Logit} = \min_{\mathbf{u}\neq\mathbf{0}:\|\mathbf{u}_{\mathcal{A}^c}\|_{\ell_1}\leq 3\|\mathbf{u}_{\mathcal{A}}\|_{\ell_1}} \frac{\mathbf{u}'\nabla^2\ell_n^{Logit}(\boldsymbol{\beta}^{\star})\mathbf{u}}{\mathbf{u}'\mathbf{u}} \in (0,\infty) > 0.$$

and $\|\boldsymbol{\beta}^{\star}_{\mathcal{A}}\|_{\min} > (a+1)\lambda$. *Let* $\lambda \geq \frac{5s^{1/2}}{a_0\kappa_{Logit}}\lambda_{lasso}$, *with* $a_0 = 1$ *for SCAD or MCP.*

Using the lasso logistic regression in (9.50) as its initial value, the LLA algorithm for the SCAD or MCP penalized logistic regression converges to the oracle estimator $\widehat{\boldsymbol{\beta}}^{oracle}_{Logit}$ *in (9.49) after two iterations with a probability at least* $1 - \delta_0^{Logit} - \delta_1^{Logit} - \delta_2^{Logit}$, *where*

$$\delta_0^{Logit} = 2p\exp\left(-\frac{1}{2M}n\lambda_{lasso}^2\right)$$
$$\delta_1^{Logit} = 2s\cdot\exp\left(-\frac{n}{M}\cdot\min\left\{\frac{2}{Q_1^2Q_2^4s^2}, \frac{a_1^2\lambda^2}{2(1+2Q_3)^2}\right\}\right) + 2(p-s)\cdot\exp\left(-\frac{a_1^2n\lambda^2}{2M}\right)$$
$$\delta_2^{Logit} = 2s\cdot\exp\left(-\frac{n}{MQ_2^2}\cdot\min\left\{\frac{2}{Q_1^2Q_2^2s^2}, \frac{1}{2}(\|\boldsymbol{\beta}^{\star}_{\mathcal{A}}\|_{\min} - a\lambda)^2\right\}\right)$$

with $M = \max_j \frac{1}{n}\sum_{i=1}^n x_{ij}^2$, *the constant* $a_1 = 1$ *for SCAD and* $a_1 = 1 - a^{-1}$ *for MCP.*

Fan et al. (2014) actually developed a general theory for a wide class of folded concave penalization problems. The above theorem is just a special case of their general theory. Their theory also suggests that zero can be a good initial value for the LLA algorithm if the following condition holds:

$$\kappa_{\mathrm{Logit}} > 5s^{1/2}$$

Then they showed that the LLA algorithm with zero as its initial value converges to the oracle estimator in (9.49) after three iterations with a high probability. Fan et al. (2014) also provided numeric demonstrations of the 2-step LLA and 3-step LLA estimator of the concave penalized logistic regression. Their numeric examples also showed that the coordinate descent algorithm generally provides an inferior solution to the LLA solution.

9.7 Sparse Discriminant Analysis

The LDA classifier is very simple and yet has many successful applications in the classical low-dimension-larger-sample-size setting (Michie et al. 1994; Hand 2006). Because of that, many researchers have tried to extend LDA to handle high-dimensional classification problems. An obvious difficulty in the LDA classifier is that the usual estimator of the common within-class covariance matrix, $\widehat{\boldsymbol{\Sigma}}$, does not have a proper inverse. Hence, many have tried to tackle this singular matrix issue in their ways to modify LDA under high dimensionality. However, this is the wrong approach to high-dimensional LDA for at least two reasons. First, the Bayes rule under the LDA model is linear, which means that in principle we only need to estimate $p+1$ parameters not a $p \times p$ covariance matrix. Estimating a $p \times p$ covariance matrix is a much harder problem than estimating the linear Bayes rule. Second, modifying the covariance matrix estimator does not directly fix the fundamental issue that is responsible for the failure of the ordinary LDA under high-dimensions. Consider an ideal situation in which $\boldsymbol{\Sigma}$ is the identity matrix and we are aware of this fact. Then we do not need to estimate $\boldsymbol{\Sigma}$ at all, which means that we do not have the difficulty of estimating a large matrix $\boldsymbol{\Sigma}$ or its inverse. Even under this ideal LDA model, the noise accumulation in estimating the mean vectors may degrade the LDA classifier to no better than random guessing (Fan and Fan 2008). This theoretical result reveals the fundamental role of sparsity in high-dimensional classification and suggests that a successful high-dimensional generalization of LDA must incorporate sparsity.

There are many sparse LDA proposals in the literature. Two popular methods are nearest shrunken centroids classifier (NSC) (Tibshirani et al. 2002) and features annealed independent rule (FAIR) (Fan and Fan 2008). Both NSC and FAIR are the independent rule methods that ignore the correlation between features. They are straightforward, computationally efficient and can have surprisingly good performance in applications. However, ignoring correlation leads to biased feature selection, which in turn implies that the independent rule methods may be theoretically sub-optimal. There are also sparse LDA methods that respect the possible correlation structure between features. An incomplete list includes Wu et al. (2008), Clemmensen et al. (2011), Mai et al. (2012), Witten and Tibshirani (2011), Cai and Liu (2011), and Fan et al. (2012).

9.7.1 *Independent Rules*

An independence rule tries to bypass the difficulty in estimating $\mathbf{\Sigma}$ and $\mathbf{\Sigma}^{-1}$ in the LDA model by using a very simple (but biased) estimator: it only takes the diagonals of the usual sample estimator of $\mathbf{\Sigma}$. In other words, an independent rule ignores the correlation structure between variables. If the independent rule also uses the usual sample estimators of the mean vectors, then it classifies an observation to Class 2 if and only if

$$\sum_{j=1}^{p} \frac{(\widehat{\mu}_{2j} - \widehat{\mu}_{1j})^T (X_j - (\widehat{\mu}_{1j} + \widehat{\mu}_{2j})/2)}{\widehat{\sigma}_{jj}} > 0. \tag{9.51}$$

The next step in an independence rule is to incorporate sparsity in the above classifier. Different independence rules use different feature selection methods.

Tibshirani et al. (2002) proposed the nearest shrunken centroids classifier (NSC) which is implemented in an R package `pamr`. NSC handles a general multi-class classification problem. Let

$$d_{kj} = \frac{\widehat{\mu}_{kj} - \widehat{\mu}_j}{m_k(s_j + s_0)}, \tag{9.52}$$

where

$$s_j^2 = \frac{1}{n-K} \sum_k \sum_{y_i = c_k} (X_{ij} - \widehat{\mu}_{kj})^2, \tag{9.53}$$

and $m_k = \sqrt{1/n_k - 1/n}$ so that $m_k s_j$ gives the standard error of $\widehat{\mu}_{kj} - \widehat{\mu}_j$. The number s_0 is a positive constant across all features, added to stabilize the algorithm numerically. In theory s_0 can be zero. Tibshirani et al. (2002) recommended using s_0 as the median value of s_is in practice. NSC shrinks d_{jk} to d'_{jk} by soft-thresholding

$$d'_{kj} = \operatorname{sign}(d_{jk})(|d_{jk} - \Delta|)_+, \tag{9.54}$$

where Δ is a positive constant that determines the amount of shrinkage. In practice Δ is chosen by cross-validation. Then the shrunken centroids are

$$\widehat{\mu}'_{kj} = \widehat{\mu}_j + m_k(s_i + s_0)d'_{jk}. \tag{9.55}$$

For an observation X, define the discriminant score for class c_k as

$$\delta_k(X) = \sum_{j=1}^{p} \frac{(X_j - \widehat{\mu}'_{kj})^2}{(s_i + s_0)^2} - 2\log \widehat{\pi}_k. \tag{9.56}$$

Then NSC classification rule is

$$\delta(X) = \text{argmin}_k \delta_k(X). \tag{9.57}$$

Notice that in (9.54) if Δ is sufficiently large such that $d_{jk} = 0$, then variable X_j does not contribute to the NSC classifier in (9.57). In other words, only the features with unequal shrunken centroids are selected for classification. This treatment effectively decreases the dimension of the final classifier.

Features annealed independent rule (FAIR) (Fan and Fan 2008) inherits the reasoning behind NSC that only the features with distinct means should be responsible for classification, except that it uses hard-thresholding rather than soft-thresholding for feature selection. The main focus in Fan and Fan (2008) is on the theoretical understanding of sparse independent rules in binary classification. For the jth variable, FAIR computes its two-sample t-statistic

$$t_j = \frac{\widehat{\mu}_{1j} - \widehat{\mu}_{2j}}{\sqrt{s_{1j}^2 + s_{2j}^2}}, \tag{9.58}$$

where s_{1j}^2, s_{2j}^2 are with-in group sample variance. The magnitudes of t_j can be viewed as the jth variable's importance. Therefore, for a given threshold $\Delta > 0$, FAIR classifies an observation to Class 2 if and only if

$$\sum_{j=1}^{p} \frac{(\widehat{\mu}_{2j} - \widehat{\mu}_{ij})\left(X_j - \dfrac{\widehat{\mu}_{1j} + \widehat{\mu}_{2j}}{2}\right)}{\widehat{\Sigma}_{jj}} 1(|t_j| > \Delta) > 0. \tag{9.59}$$

Under suitable regularity conditions, FAIR is proven to be capable of separating the features with distinct means from those without, even when the equal-variance and the normality assumptions are not satisfied. More explicitly, define $S = \{j : \mu_{1j} \neq \mu_{2j}\}$ and s is the cardinality of the set S. Fan and Fan (2008) proved the following results.

Theorem 4 (Fan and Fan 2008) *If the following conditions are satisfied:*

1. *there exist constants* ν_1, ν_2, M_1, M_2, *such that* $E|X_{kj}|^m \leq m!M_1^{m-2}\nu_1/2$ *and* $E|X_{kj}^2 - \text{Var}(X_{kj})|^m \leq m!M_2\nu_2/2$;
2. $\Sigma_{k,jj}$ *are bounded away from 0;*
3. *there exists* $0 < \gamma < 1/3$ *and a sequence* $b_n \to$ *such that* $\log(p - s) = o(n^\gamma)$ *and* $\log s = o(n^{1/2-\gamma} b_n)$;

then for $\Delta \sim cn^{\gamma/2}$, *where* c *is a positive constant, we have*

$$\Pr\left(\min_{j\in S} |t_j| \geq \Delta, \text{ and } \max_{j\notin S} |t_j| < \Delta\right) \to 1 \tag{9.60}$$

Although the independence rules are simple to implement, they may produce sub-optimal or even misleading results if the correlations should not be ignored. Mai et al. (2012) carefully analyzed this issue. The Bayes rule in the context of LDA is a linear classifier with the classification coefficient vector $\boldsymbol{\beta}^{\text{Bayes}} = \boldsymbol{\Sigma}^{-1}(\mu_2 - \mu_1)$. Define $D = \{j : \boldsymbol{\beta}_j^{\text{Bayes}} \neq 0\}$ which should be the target of variable selection. NSC and FAIR, on the other hand, target at $\mathcal{S} = \{j : \mu_{1j} \neq \mu_{2j}\}$. The following theorem given in Mai et al. (2012) shows that D and S need not be identical, and hence NSC and FAIR may select a wrong set of variables.

Theorem 5 (Mai et al. 2012) *Let*

$$\boldsymbol{\Sigma} = \begin{pmatrix} \boldsymbol{\Sigma}_{\mathcal{S},\mathcal{S}} & \boldsymbol{\Sigma}_{\mathcal{S},\mathcal{S}^c} \\ \boldsymbol{\Sigma}_{\mathcal{S}^c,\mathcal{S}} & \boldsymbol{\Sigma}_{\mathcal{S}^c,\mathcal{S}^c} \end{pmatrix}, \boldsymbol{\Sigma} = \begin{pmatrix} \boldsymbol{\Sigma}_{\mathcal{D},\mathcal{D}} & \boldsymbol{\Sigma}_{\mathcal{D},\mathcal{D}^c} \\ \boldsymbol{\Sigma}_{\mathcal{D}^c,\mathcal{D}} & \boldsymbol{\Sigma}_{\mathcal{D}^c,\mathcal{D}^c} \end{pmatrix}.$$

1. *If and only if* $\boldsymbol{\Sigma}_{\mathcal{S}^c,\mathcal{S}}\boldsymbol{\Sigma}_{\mathcal{S},\mathcal{S}}^{-1}(\mu_{2,\mathcal{S}} - \mu_{1,\mathcal{S}}) = 0$, *we have* $\mathcal{D} \subseteq \mathcal{S}$;
2. *If and only if* $\mu_{2,\mathcal{D}^c} = \mu_{1,\mathcal{D}^c}$ *or* $\boldsymbol{\Sigma}_{\mathcal{D}^c,\mathcal{D}}\boldsymbol{\Sigma}_{\mathcal{D},\mathcal{D}}^{-1}(\mu_{2,\mathcal{D}} - \mu_{1,\mathcal{D}}) = 0$, *we have* $\mathcal{S} \subseteq \mathcal{D}$.

With Theorem 5, examples can be easily constructed that D and S are not identical. Consider an LDA model with $p = 25$, $\mu_1 = 0_p$, $\sigma_{ii} = 1$, and $\sigma_{ij} = 0.5, i \neq j$. If $\mu_2 = (1_5^T, 0_{20}^T)^T$, then $\mathcal{S} = \{1, 2, 3, 4, 5\}$ and $\mathcal{D} = \{j : j = 1, \ldots, 25\}$, since $\boldsymbol{\beta}^{\text{Bayes}} = (1.62 \times 1_5^T, -0.38 \times 1_{20}^T)^T$. On the other hand, if we let $\mu_2 = (3 \times 1_5^T, 2.5 \times 1_5^T)^T$, then $\mathcal{S} = \{1, \ldots, 25\}$ but $\mathcal{D} = \{1, 2, 3, 4, 5\}$, because $\boldsymbol{\beta}^{\text{Bayes}} = (1_5^T, 0_{20}^T)^T$. These two examples show that ignoring correlation can yield inconsistent variable selection.

9.7.2 Linear Programming Discriminant Analysis

Cai and Liu (2011) suggested a linear programming discriminant (LPD) for deriving a sparse LDA method. LPD begins with the observation that the Bayes rule classification vector $\boldsymbol{\beta}^{\text{Bayes}} = \boldsymbol{\Sigma}^{-1}(\mu_2 - \mu_1)$ satisfies the linear equation $\boldsymbol{\Sigma}\boldsymbol{\beta} = \mu_2 - \mu_1$. Then LPD estimates β^{Bayes} by solving the following linear programming problem:

$$\widehat{\boldsymbol{\beta}} = \text{argmin}_{\boldsymbol{\beta}} \sum_{j=1}^{p} \|\boldsymbol{\beta}\|_1, \text{ s.t. } \|\widehat{\boldsymbol{\Sigma}}\boldsymbol{\beta} - (\widehat{\mu}_2 - \widehat{\mu}_1)\|_\infty \leq \lambda_n. \tag{9.61}$$

Cai and Liu (2011) showed that the misclassification error of LPD can approach the Bayes error under the LDA model as $n, p \to \infty$. For simplicity, Cai and Liu (2011) assumed the equal class probability $P(Y = 1) = P(Y = 2) = 1/2$ and LPD classifies an observation X to class 2 if

$$\widehat{\beta}^T \left(X - \frac{\widehat{\mu}_2 + \widehat{\mu}_1}{2} \right) > 0. \tag{9.62}$$

Theorem 6 (Cai and Liu 2011) *Define* $\Delta = (\mu_2 - \mu_1)^T \boldsymbol{\Sigma}^{-1}(\mu_2 - \mu_1)$. *Let* R_n *denote the misclassification error rate of LPD conditioning on the observed data. Suppose* $\log p = o(n)$, $\Sigma_{jj} < K, j = 1, \ldots, p$ *and* $\Delta > c_1$ *for some positive constants* K, c_1. *Let* $\lambda_n = O(\sqrt{\Delta \log p/n})$. *Then we have*

1. *if*

$$\frac{\|\boldsymbol{\beta}^{Bayes}\|_1}{\Delta^{1/2}} + \frac{\|\boldsymbol{\beta}^{Bayes}\|_1^2}{\Delta^2} = o\left(\sqrt{\frac{n}{\log p}}\right), \tag{9.63}$$

then $R_n - R(Bayes) \to 0$ *in probability as* $n, p \to \infty$;

2. *if*

$$\|\boldsymbol{\beta}^{Bayes}\|_1 \Delta^{1/2} + \|\beta^{Bayes}\|_1^2 = o\left(\sqrt{\frac{n}{\log p}}\right), \tag{9.64}$$

then

$$\frac{R_n}{R(Bayes)} - 1 = O\left((\|\boldsymbol{\beta}^{Bayes}\|_1 \Delta^{1/2} + \|\boldsymbol{\beta}^{Bayes}\|_1^2) \sqrt{\frac{n}{\log p}} \right) \tag{9.65}$$

with probability greater than $1 - O(p^{-1})$.

9.7.3 Direct Sparse Discriminant Analysis

Mai et al. (2012) proposed Direct Sparse Discriminant Analysis (DSDA) by utilizing a least squares representation of LDA in the binary case. Let $\widehat{\boldsymbol{\beta}}^{\text{Bayes}}$ be the LDA estimator of the Bayes rule vector $\boldsymbol{\beta}^{\text{Bayes}} = \boldsymbol{\Sigma}^{-1}(\mu_2 - \mu_1)$. Suppose that the class label is coded by two distinct numeric values and write y as the numeric vector of the class labels. Define

$$(\widehat{\beta}_0^{\text{ols}}, \widehat{\boldsymbol{\beta}}^{\text{ols}}) = \text{argmin}_{(\beta_0, \boldsymbol{\beta})} \sum_{i=1}^{n} (y_i - \beta_0 - X_i^T \beta)^2. \tag{9.66}$$

It can be shown that $\widehat{\boldsymbol{\beta}}^{\text{ols}} = c\widehat{\beta}^{\text{Bayes}}$ for some positive constant c, which means that the LDA classification direction can be exactly recovered by doing the least squares computation.

Mai et al. (2012) showed that we can consider a penalized least squares formulation to produce a properly modified sparse LDA method. DSDA is defined by

$$\widehat{\boldsymbol{\beta}}^{\text{DSDA}} = \text{argmin}_{\boldsymbol{\beta}} n^{-1} \sum_{i=1}^{n} (y_i - \beta_0 - X_i^T \boldsymbol{\beta})^2 + P_\lambda(\boldsymbol{\beta}), \tag{9.67}$$

where $P_\lambda(\cdot)$ is a penalty function, such as LASSO and SCAD penalty. The tuning parameter λ can be chosen by cross validation. DSDA classifies X to class 2 if

$$X^T \widehat{\beta}^{\text{DSDA}} + \widehat{\beta}_0 > 0, \tag{9.68}$$

with

$$\widehat{\beta}_0 = -(\widehat{\mu}_2 + \widehat{\mu}_1)^T \widehat{\boldsymbol{\beta}}^{\text{DSDA}}/2 + (\widehat{\boldsymbol{\beta}}^{\text{DSDA}})^T \widehat{\boldsymbol{\Sigma}} \widehat{\boldsymbol{\beta}}^{\text{DSDA}} \{(\widehat{\mu}_2 - \widehat{\mu}_1)^T \widehat{\boldsymbol{\beta}}^{\text{DSDA}}\}^{-1} \log\left(\frac{n_2}{n_1}\right), \tag{9.69}$$

and n_1 (n_2) is the number of observations in class 1 (class 2).

DSDA is computationally very efficient, because we can take advantage of the fast algorithms for computing the penalized least squares. For example, one could use the least angle regression implemented in `lars` or coordinate descent implemented in `glmnet` to compute DSDA with the LASSO penalty. An R package `dsda` for DSDA is available from

`http://stat.fsu.edu/~mai/dsda_1.11.tar.gz`

Strong theoretical results have been established for DSDA (Mai et al. 2012). Let $\beta^{\text{Bayes}} = \Sigma^{-1}(\mu_2 - \mu_1)$ represent the Bayes classifier coefficient vector. Write $\mathcal{D} = \{j : \beta_j^{\text{Bayes}} \neq 0\}$ and let s be the cardinality of $\mathcal{D}$. We use C to represent the marginal covariance matrix of the predictors and partition C as

$$C = \begin{pmatrix} C_{\mathcal{D}\mathcal{D}} & C_{\mathcal{D}\mathcal{D}^c} \\ C_{\mathcal{D}^c\mathcal{D}} & C_{\mathcal{D}^c\mathcal{D}^c} \end{pmatrix}.$$

Denote $\boldsymbol{\beta}^* = (C_{\mathcal{D}\mathcal{D}})^{-1}(\mu_{2\mathcal{D}} - \mu_{1\mathcal{D}})$ and define $\tilde{\boldsymbol{\beta}}^{\text{Bayes}}$ by letting $\tilde{\boldsymbol{\beta}}_{\mathcal{D}}^{\text{Bayes}} = \boldsymbol{\beta}^*$ and $\tilde{\boldsymbol{\beta}}_{\mathcal{D}^c}^{\text{Bayes}} = 0$.

Theorem 7 (Mai et al. 2012) *The quantities $\tilde{\boldsymbol{\beta}}^{Bayes}$ and $\boldsymbol{\beta}^{Bayes}$ are equivalent in the sense that $\tilde{\boldsymbol{\beta}}^{Bayes} = c\boldsymbol{\beta}^{Bayes}$ for some positive constant c and the Bayes classifier can be written as assigning X to class 2 if*

$$\{X - (\mu_1 + \mu_2)/2\}^T \tilde{\boldsymbol{\beta}}^{Bayes} + (\tilde{\boldsymbol{\beta}}^{Bayes})^T \boldsymbol{\Sigma} \tilde{\boldsymbol{\beta}}^{Bayes} \{(\mu_2 - \mu_1)^T \tilde{\boldsymbol{\beta}}^{Bayes}\}^{-1} \log \frac{\pi_2}{\pi_1} > 0.$$

According to the above theorem it suffices to show that DSDA can consistently recover the support of $\tilde{\beta}^{\text{Bayes}}$ and estimate β^*. Non-asymptotic analysis is given in Mai et al. (2012). Here we only present the corresponding asymptotic results to highlight the main points of the theory for DSDA.

Assume that Δ, κ, φ are constants. In addition, we need the following regularity conditions:

(1) $n, p \to \infty$ and $\log(ps)s^2/n \to 0$;

(2) $\min_{j\in A} |\beta_j^*| \gg \{\log(ps)s^2/n\}^{1/2}$.
(3) $\|(C_{\mathcal{DD}})^{-1}\|_\infty$ and $\|\mu_{2\mathcal{D}} - \mu_{1\mathcal{D}}\|_\infty$ are bounded.

Theorem 8 (Mai et al. 2012) *Let $\widehat{\boldsymbol{\beta}}^{lasso}$ be the DSDA estimator with LASSO penalty and $\widehat{\mathcal{D}}^{lasso} = \{j : \widehat{\beta}_j^{lasso} \neq 0\}$. If we choose the LASSO penalty parameter $\lambda = \lambda_n$ such that $\lambda_n \ll \min_{j\in\mathcal{D}} |\beta_j^*|$ and $\lambda_n \gg \{\log(ps)s^2/n\}^{1/2}$, and further assume*

$$\kappa = \|C_{\mathcal{D}^c\mathcal{D}}(C_{\mathcal{DD}})^{-1}\|_\infty < 1, \tag{9.70}$$

then with probability tending to 1, $\widehat{\mathcal{D}}^{lasso} = \mathcal{D}$ and $\|\widehat{\beta}_{\mathcal{D}}^{lasso} - \beta^*\|_\infty \leq 4\varphi\lambda_n$.

Mai et al. (2018) developed a multiclass sparse discriminant analysis and proved its strong theoretical and computational properties. They also provided an R package `msda` for their method, which is available from CRAN.

9.8 Sparse Semiparametric Discriminant Analysis

Motivated by the Box-Cox model in regression analysis, we can consider the following semiparametric linear discriminant analysis (SeLDA) model

$$\left(h_1(X_1), \ldots, h_p(X_p)\right) \mid Y \sim N(\mu_Y, \Sigma), \tag{9.71}$$

where $h = (h_1, \ldots, h_p)$ is a set of strictly monotone univariate transformations. It is important to note that the SeLDA model does not assume that these univariate transformations are known or have any parametric forms. This is the nonparametric component of the semiparametric model. It is clear that each transformation function h is only unique up to location and scale shifts. Thus, for identifiability we assume that $n_+ \geq n_-$, $\mu_+ = 0$, $\Sigma_{jj} = 1, 1 \leq j \leq p$. The Bayes rule of the SeLDA model is

$$\widehat{Y}^{\text{Bayes}} = \text{Sign}\left(\left(h(X) - \frac{1}{2}(\mu_+ + \mu_-)\right)^T \Sigma^{-1}(\mu_+ - \mu_-) + \log\frac{\pi_+}{\pi_-}\right). \tag{9.72}$$

The SeLDA model was studied in Lin and Jeon (2003) under the classical low-dimensional setting when p is fixed and n goes to infinity. Mai and Zou (2015) established the high-dimensional estimation method and theory of SeLDA when p is allowed to grow faster than a polynomial order of n.

Note that for any continuous univariate random variable X, we have

$$\Phi^{-1} \circ F(X) \sim N(0, 1), \tag{9.73}$$

where F is the cumulative probability function (CDF) of W and Φ is the CDF of the standard normal distribution. Based on this fact, we have

$$h_j = \Phi^{-1} \circ F_{+j}, \tag{9.74}$$

where $F_{+,j}$ is the conditional CDF of X_j given $Y = +1$. To fix idea, let us use a simple estimator of h_j that can be obtained by plugging a good estimator of F_{+j} into (9.74). Let $\tilde{F}_{+j}$ be the empirical conditional CDF. Given a pair of numbers (a, b), define

$$\widehat{F}_{+j}^{a,b}(x) = \begin{cases} b & \text{if } \tilde{F}_{+j}(x) > b; \\ \tilde{F}_{+j}(x) & \text{if } a \le \tilde{F}_{+j}(x) \le b; \\ a & \text{if } \tilde{F}_{+j}(x) < a. \end{cases} \tag{9.75}$$

Then

$$\widehat{h}_j = \Phi^{-1} \circ \widehat{F}_{+j}^{a,b}. \tag{9.76}$$

With $\widehat{h}_j$, the covariance matrix Σ is estimated by the pooled sample covariance matrix of $\widehat{h}(X^i)$. The mean vector μ_{-j} is estimated by

$$\widehat{\mu}_{-j} = q^{-1}\Bigg(\frac{1}{n_-}\sum_{i=1}^{n_-}\widehat{h}(X^i_{-j})1_{\tilde{F}(X^i_{-j})\in(a,b)} \tag{9.77}$$
$$+\phi(\Phi^{-1}\circ\tilde{F}_{-j}\circ\tilde{F}_{+j}^{-1}(b)) - \phi(\Phi^{-1}\circ\tilde{F}_{-j}\circ\tilde{F}_{+j}^{-1}(a))\Bigg)$$

where

$$q = \frac{1}{n_-}\sum_{i=1}^{n_-} 1_{\tilde{F}_{+j}(X^i)\in(a,b)}. \tag{9.78}$$

The complicated form of (9.77) is due to the Winsorization in (9.75).

After estimating the transformation functions, one can substitute h_j with its estimator $\widehat{h}_j$ and apply any good sparse LDA method to the pseudo data $(Y, \widehat{h}(X))$. For example, let us apply DSDA by solving for

$$\widehat{\boldsymbol{\beta}} = \text{argmin}_{\boldsymbol{\beta}} n^{-1}\sum_{i=1}^{n}(y_i - \beta_0 - \widehat{h}(X_i)^T\boldsymbol{\beta})^2 + \sum_{j=1}^{p} P_\lambda(|\beta_j|) \tag{9.79}$$

$$\widehat{\beta}_0 = -\frac{(\widehat{\mu}_+ + \widehat{\mu}_-)^T\widehat{\boldsymbol{\beta}}}{2} + \frac{\widehat{\boldsymbol{\beta}}^T\widehat{\Sigma}\widehat{\boldsymbol{\beta}}}{(\widehat{\mu}_+ - \widehat{\mu}_-)^T\widehat{\boldsymbol{\beta}}}\log\frac{\widehat{\pi}_+}{\widehat{\pi}_-}.$$

Then the SeSDA classification rule is $\text{Sign}\left(\widehat{\beta}_0 + \widehat{h}(X)^T\widehat{\boldsymbol{\beta}}\right)$.

For the classical low dimensional case the penalty term in (9.79) is not needed. Lin and Jeon (2003) proved the asymptotic consistency of the low-dimensional SeLDA estimator when fixing a and b as constants as long as $0 < a < b < 1$. In particular, they suggested choosing a and b such that a, b, $\tilde{F}_-(\tilde{F}_+(a))$ and $\tilde{F}_-(\tilde{F}_+(b))$ are all between 2.5% and 97.5%. However, it is unclear whether their choice of Winsorization parameters (a, b) is still valid when p is much larger than n. In order to handle the high-dimensional SeLDA model and the SeSDA classifier, Mai and Zou (2015) proposed using

$$(a, b) = (a_n, b_n) = \left(\frac{1}{n_+^2}, 1 - \frac{1}{n_+^2}\right) \tag{9.80}$$

and established valid asymptotic theory as long as $\log(p) \ll n^{1/3-\rho}$.

We need to define some notation for presenting the asymptotic theory. Define $\beta^\star = C^{-1}(\mu_+ - \mu_-)$. Let C be the marginal covariance matrix of $(h_1(X_1), \ldots, h_p(X_p))$. Recall that $\beta^\star$ is equal to $c\Sigma^{-1}(\mu_+ - \mu_-) = c\beta^{\text{Bayes}}$ for some positive constant (Mai et al. 2012). Therefore we can write the Bayes error rate as

$$R = \Pr(Y \neq \text{Sign}(h(X)^T \boldsymbol{\beta}^* + \beta_0))$$

The HD-SeLDA classifier error rate is

$$R_n = \Pr(Y \neq \text{Sign}(\widehat{h}(X)^T \widehat{\boldsymbol{\beta}} + \widehat{\beta}_0)).$$

A variable has contribution to the SeLDA model if and only if $\beta_j^{\text{Bayes}} \neq 0$. We can write $A = \{j : \beta_j^{\text{Bayes}} \neq 0\} = \{j : \beta_j^\star \neq 0\}$. Let s be the cardinality of A. We partition C into four submatrices: $C_{AA}, C_{A^cA}, C_{AA^c}, C_{A^cA^c}$.

Theorem 9 (Mai and Zou 2015) *Suppose that we use the lasso penalty in (9.79). Assume the irrepresentable condition $\kappa = \|C_{A^cA}(C_{AA})^{-1}\|_\infty < 1$ and two regularity conditions*

(C1) $n, p \to \infty$ *and* $\dfrac{s^2 \log(ps)}{n^{\frac{1}{3}-\rho}} \to 0$, *for some ρ in $(0, 1/3)$;*

(C2) $\min_{j\in A} |\beta_j| \gg \max\left\{ sn^{-1/4}, \sqrt{\log(ps)\dfrac{s^2}{n^{\frac{1}{3}-\rho}}} \right\}$ *for some ρ in $(0, 1/3)$.*

Let $\widehat{A} = \{j : \widehat{\beta}_j \neq 0\}$. Under conditions (C1) and (C2), if we choose $\lambda = \lambda_n$ such that $\lambda_n \ll \min_{j\in A} |\beta_j|$ and $\lambda_n \gg \sqrt{\log(ps)\dfrac{s^2}{n^{\frac{1}{3}-\rho}}}$, then $\Pr(\widehat{A} = A) \to 1$ and $\Pr\left(\|\widehat{\beta}_A - \beta_A\|_\infty \leq 4\varphi\lambda_n\right) \to 1$. Moreover, $R_n - R \to 0$ in probability.

In practice, we can further improve the estimator of $\widehat{h}_j$ by combining the data from both classes. The estimator $\widehat{h}_j$ in (9.76) only uses data in class $+1$ to estimate $\widehat{h}_j$. In theory, the two classes are symmetric and one can also use data in class -1 to estimate $\widehat{h}_j$. Then we can consider a combined estimator from both classes. The pooled estimator of SeLDA is presented in Mai and Zou (2015) where numeric examples are given to show the pooled SeLDA has better classification performance.

R code for doing SeLDA is in the R package `dsda` which is available from

```
http://stat.fsu.edu/~mai/dsda_1.11.tar.gz
```

9.9 Sparse Penalized Additive Models for Classification

Generalized additive models (Hastie and Tibshirani 1990) are widely used in statistics because they offer a nice way to let covariates have nonlinear contributions to the model while enjoying the nice interpretation of a linear model. For example, consider the generalized additive model for logistic regression. The log-odd function is written as $P(Y = 1|X) = \frac{e^{f(X)}}{1+e^{f(X)}}$ where $f(X) = \beta_0 + \sum_{k=1}^{p} f_j(X_j)$. Assuming each univariate function $f_j(X_j)$ is smooth, then the above additive model can be estimated by back-fitting and the commonly used smoothing techniques such as regression spline, local linear smoother or smoothing spline. See Hastie and Tibshirani (1990) and Hastie et al. (2009) for the details.

Generalized additive models can be extended to the high-dimensional setting straightforwardly, if we model each univariate function $f_j(X_j)$ by a linear combination of basis functions such as B-splines. In the additive logistic regression model, let

$$f_k(X_k) = \sum_{m}^{p_k} \beta_{km} B_m(X_k)$$

where $B_k(X_j)$ represents a basis function of X_j. We use $+1, -1$ to code each label and adopt the margin-based logistic regression loss $\phi(t) = \log(1 + e^{-t})$. The penalized additive logistic regression estimator is defined as

$$\operatorname{argmin}_{\boldsymbol{\beta}} \frac{1}{n} \sum_{i=1}^{n} \log(1 + e^{-y_i f(X_i)}) + \lambda \sum_{k=1}^{p} \sqrt{p_k} \|\boldsymbol{\beta}^{(k)}\|_2, \qquad \lambda > 0, \qquad (9.81)$$

where $\|\boldsymbol{\beta}^{(k)}\|_2 = \sqrt{\sum_m^{p_k} \beta_{km}^2}$ is the group-lasso penalty (Yuan and Lin 2006). Notice that we do not use the lasso penalty $\lambda \sum_{k,m} |\beta_{km}|$ in this case. The reason is simple. In the generalized additive model, variable X_j is removed from the model if and only if the univariate function f_j is zero. Hence, we need to simultaneously set $\beta_{km} = 0$ for all $m = 1, \ldots, p_k$. The lasso penalty cannot guarantee group elimination, while the group-lasso can. Yuan and Lin (2006) first proposed the group-lasso for doing groupwise selection in the linear regression model. Meier et al. (2008) extended the group-lasso idea to penalized logistic regression.

The computation of a group-lasso problem is generally more challenging than a lasso problem. Under a strong group-wise orthonormal condition, Yuan and Lin (2006) suggested a blockwise descent algorithm for solving the group-lasso penalized linear regression problem. Meier et al. (2008) extended the blockwise descent algorithm to solve the group-lasso penalized logistic regression. Meier's algorithm is implemented in an R package `grplasso` available from CRAN. Liu et al. (2009) implemented Nesterov's method (Nesterov 2004, 2007) for solving the group-lasso penalized linear regression and logistic regression.

Yang and Zou (2014) proposed a very efficient unified algorithm named groupwise-majorization-descent (GMD) for solving a whole class of group-lasso learning problems, including the group-lasso penalized logistic regression, Huberized SVM, and squared SVM. GMD works for general design matrices, without requiring the group-wise orthogonal assumption. Yang and Zou (2014) have implemented GMD in an R package `gglasso` available from CRAN. To appreciate the speed advantage of `gglasso` over `grplasso` and `SLEP`, let us consider fitting a group-lasso penalized logistic regression model on breast cancer data (Graham et al. 2010) where $n = 42$ and $p = 22{,}283$. Each variable contributes an additive component that is expressed by five B-spline basis functions. The group-lasso penalty is imposed on the coefficients of five B-spline basis functions for each variable. So the corresponding group-lasso logistic regression model has 22,283 groups and each group has five coefficients to be estimated. Displayed in Fig. 9.5 are three solution path plots produced by `grplasso`, `SLEP`, and `gglasso`. It took `SLEP` about 450 and `grplasso` about 360 s to compute the logistic regression paths, while `gglasso` used only about 10 s.

Figures 9.6 and 9.7 show the solution paths computed by using `gglasso` of the sparse additive Huberized SVM and the sparse additive squared SVM on the breast cancer data (Graham et al. 2010), respectively.

In what follows we introduce the GMD algorithm for solving the group-lasso penalized large margin classifier. For simplicity, we use $X_1, \ldots, X_p$ to denote the generic predictors used to fit the group-lasso model and use $Z_1, \ldots, Z_q$ to denote the original variables in the data. For example, we generate $\mathbf{x}$ variables by using basis functions of $z_1, \ldots, z_q$. For instance, $X_1 = Z_1, X_2 = Z_1^2, X_3 = Z_1^3$, $X_4 = Z_2, X_5 = Z_2^2$, etc. It is important that the user has defined the $\mathbf{x}$ variables before fitting the group-lasso model. Let $\mathbf{X}$ be the design matrix with n rows and p columns. If an intercept is included, we let the first column of $\mathbf{X}$ be a vector of 1. It is also assumed that the group membership is already defined such that $(1, 2, \ldots, p) = \bigcup_{k=1}^{K} I_k$ and the cardinality of index set I_k is p_k, $I_k \bigcap I_{k'} = \emptyset$ for $k \neq k', 1 \leq k, k' \leq K$. Group k contains $x_j, j \in I_k$, for $1 \leq k \leq K$. If an intercept is included, then $I_1 = \{1\}$. We use $\boldsymbol{\beta}^{(k)}$ to denote the segment of $\boldsymbol{\beta}$ corresponding to group k. This notation is used for any p-dimensional vector.

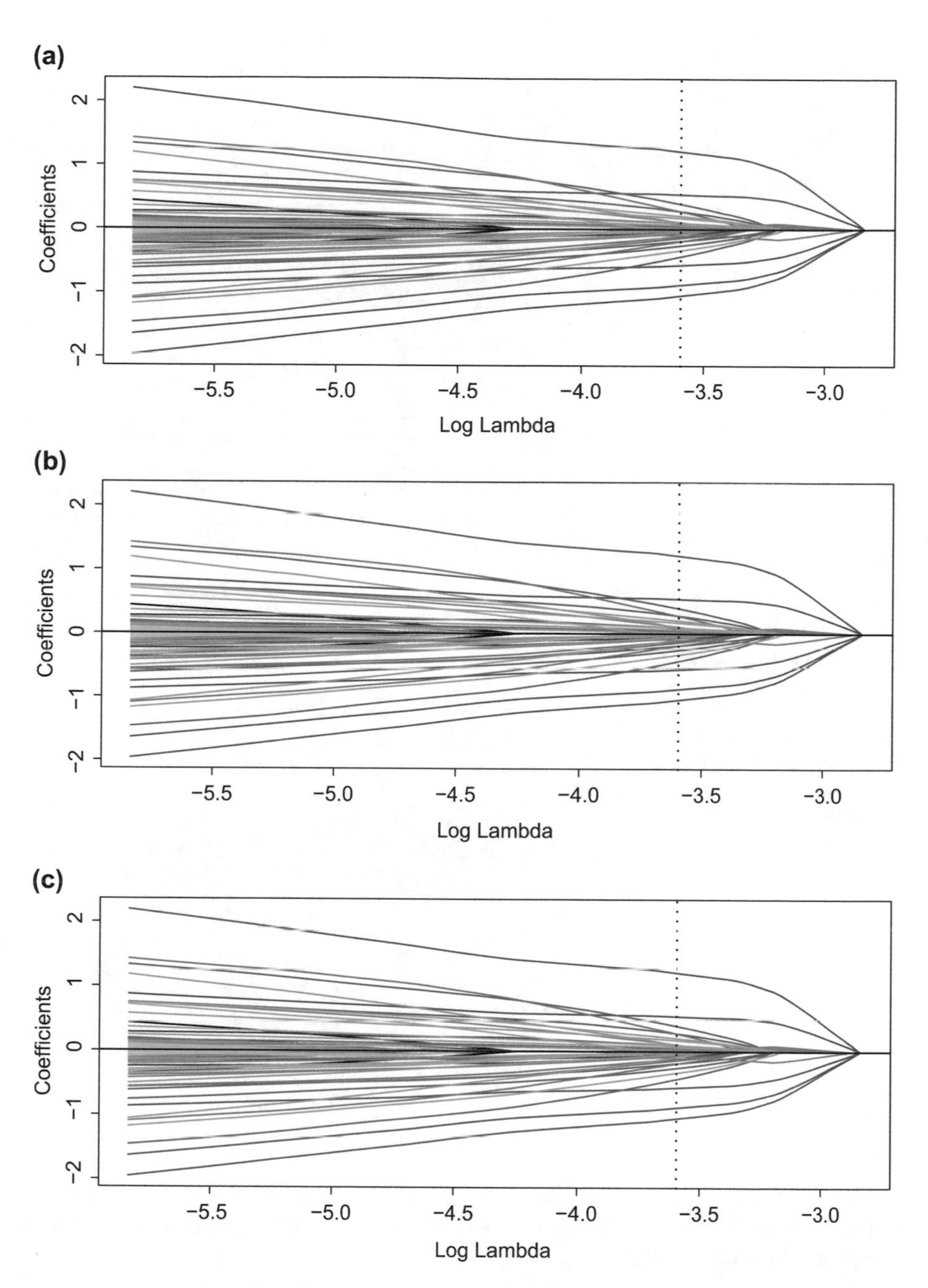

Fig. 9.5 Fit a sparse additive logistic regression model using the group-lasso on the breast cancer data with $n = 42$ patients and $22,283$ genes (groups). Each gene's contribution is modeled by 5 B-Spline basis functions. The solution paths are computed at 100 λ values on an Intel Xeon X5560 (Quad-core 2.8 GHz) processor. The vertical dotted lines indicate the selected λ ($\log \lambda = -3.73$), which selects eight genes. (**a**) SLEP—Liu et al. (2009). Breast Cancer Data (approximately 450 s). (**b**) grplasso—Meier et al. (2008). Breast Cancer Data (approximately 360 s). (**c**) `gglasso`—BMD Algorithm. Breast Cancer Data (approximately 10 s)

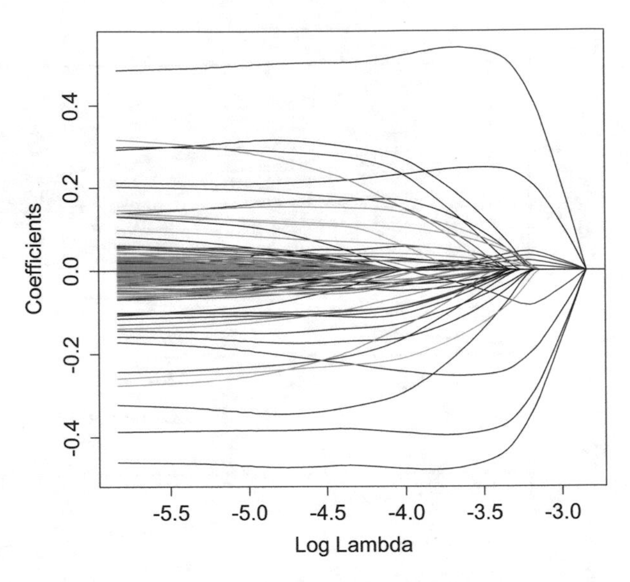

Fig. 9.6 Fit a sparse additive Huberized SVM using the group-lasso on the breast cancer data (Graham et al. 2010) with $n = 42$ patients and 22,283 genes (groups). Each gene's contribution is modeled by 5 B-Spline basis functions. The solution paths are computed at 100 λ values

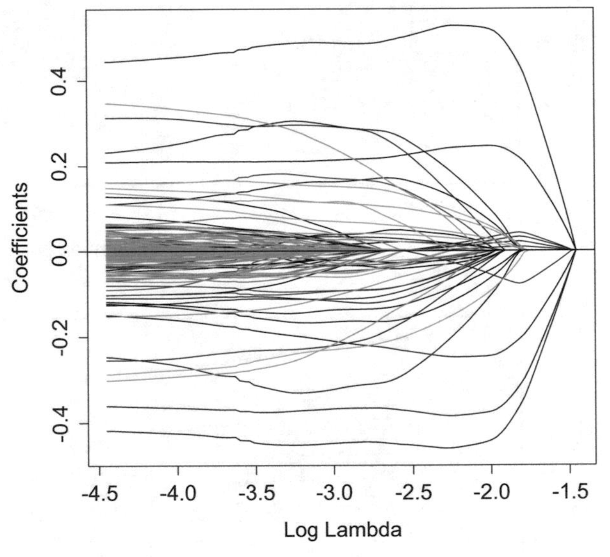

Fig. 9.7 Fit a sparse additive squared SVM using the group-lasso on the breast cancer data (Graham et al. 2010) with $n = 42$ patients and 22,283 genes (groups). Each gene's contribution is modeled by 5 B-Spline basis functions. The solution paths are computed at 100 λ values

Let $Y = 1, -1$ denote the class label. Consider a smooth convex loss function $\phi(\cdot)$. The classifier is $\mathrm{Sign}(X^T\widehat{\boldsymbol{\beta}})$, where $\widehat{\boldsymbol{\beta}}$ is computed via the group-lasso penalized empirical loss formulation:

$$\mathrm{argmin}_{\boldsymbol{\beta}} \frac{1}{n}\sum_{i=1}^{n} \phi(y_i, X_i^T\boldsymbol{\beta}) + \lambda w_k \sum_{k=1}^{K} \|\boldsymbol{\beta}^{(k)}\|_2, \tag{9.82}$$

We use weights w_ks in order to make a more flexible group-lasso model. The default choice for w_k is $\sqrt{p_k}$. If we do not want to penalize a group of predictors, simply let the corresponding weight be zero. For example, the intercept is typically not penalized so that $w_1 = 0$. Following the adaptive lasso idea (Zou 2006), one could define the adaptively weighted group-lasso which often has better estimation and variable selection performance than the un-weighted group-lasso.

Let the empirical ϕ loss be

$$L(\boldsymbol{\beta} \mid \mathbf{D}) = \frac{1}{n}\sum_{i=1}^{n}\phi(y_i, \boldsymbol{\beta}^T\mathbf{x}_i).$$

The loss function ϕ is said to satisfy the quadratic majorization (QM) condition, if and only if the following two assumptions hold:

(i) $L(\boldsymbol{\beta} \mid \mathbf{D})$ is differentiable as a function of $\boldsymbol{\beta}$, i.e., $\nabla L(\boldsymbol{\beta}|\mathbf{D})$ exists everywhere.
(ii) There exists a $p \times p$ matrix $\mathbf{H}$, which may only depend on the data $\mathbf{D}$, such that for all $\boldsymbol{\beta}, \boldsymbol{\beta}^*$,

$$L(\boldsymbol{\beta} \mid \mathbf{D}) \le L(\boldsymbol{\beta}^* \mid \mathbf{D}) + (\boldsymbol{\beta}-\boldsymbol{\beta}^*)^T\nabla L(\boldsymbol{\beta}^*|\mathbf{D}) + \frac{1}{2}(\boldsymbol{\beta}-\boldsymbol{\beta}^*)^T\mathbf{H}(\boldsymbol{\beta}-\boldsymbol{\beta}^*). \quad (9.83)$$

It has been verified that the QM condition holds for the logistic regression loss, squared hinge loss, and Huberized hinge loss (Yang and Zou 2014). See Table 9.1.

Once the QM condition is verified, we can derive the GMD algorithm for computing the solution of (9.82). GMD is a cyclic groupwise updating procedure that continues until convergence. Let $\widetilde{\boldsymbol{\beta}}$ denote the current solution of $\boldsymbol{\beta}$. Without loss of generality, we derive the GMD update of $\widetilde{\boldsymbol{\beta}}^{(k)}$, the coefficients of group k. Define $\mathbf{H}^{(k)} = [H_{ij}], i, j \in \text{group}k$. Write $\boldsymbol{\beta}$ such that $\boldsymbol{\beta}^{(k')} = \widetilde{\boldsymbol{\beta}}^{(k')}$ for $k' \neq k$. Given $\boldsymbol{\beta}^{(k')} = \widetilde{\boldsymbol{\beta}}^{(k')}$ for $k' \neq k$, the optimal $\boldsymbol{\beta}^{(k)}$ is defined as

$$\text{argmin}_{\boldsymbol{\beta}^{(k)}} L(\boldsymbol{\beta} \mid \mathbf{D}) + \lambda w_k \|\boldsymbol{\beta}^{(k)}\|_2. \quad (9.84)$$

Unfortunately, there is no closed form solution to (9.84) for a general loss function with a general design matrix.

We overcome the computational obstacle by taking advantage of the QM condition and the MM principle (Hunter and Lange 2004). From (9.83) we have

$$L(\boldsymbol{\beta} \mid \mathbf{D}) \le L(\widetilde{\boldsymbol{\beta}} \mid \mathbf{D}) + (\boldsymbol{\beta} - \widetilde{\boldsymbol{\beta}})^T\nabla L(\widetilde{\boldsymbol{\beta}}|\mathbf{D}) + \frac{1}{2}(\boldsymbol{\beta} - \widetilde{\boldsymbol{\beta}})^T\mathbf{H}(\boldsymbol{\beta} - \widetilde{\boldsymbol{\beta}}).$$

Table 9.1 The QM condition is verified for the logistic regression loss, squared hinge loss, and Huberized hinge loss

Loss	$-\nabla L(\boldsymbol{\beta} \mid \mathbf{D})$	$\mathbf{H}$
Logistic regression	$\frac{1}{n}\sum_{i=1}^{n} y_i\mathbf{x}_i \frac{1}{1+\exp(y_i\mathbf{x}_i^T\boldsymbol{\beta})}$	$\frac{1}{4}\mathbf{X}^T\mathbf{X}/n$
Squared hinge loss	$\frac{1}{n}\sum_{i=1}^{n} 2y_i\mathbf{x}_i(1 - y_i\mathbf{x}_i^T\boldsymbol{\beta})_+$	$4\mathbf{X}^T\mathbf{X}/n$
Huberized hinge loss	$\frac{1}{n}\sum_{i=1}^{n} y_i\mathbf{x}_i\text{hsvm}'(y_i\mathbf{x}_i^T\boldsymbol{\beta})$	$\frac{2}{8}\mathbf{X}^T bX/n$

Write $U(\widetilde{\boldsymbol{\beta}}) = -\nabla L(\widetilde{\boldsymbol{\beta}}|\mathbf{D})$. Using

$$\boldsymbol{\beta} - \widetilde{\boldsymbol{\beta}} = (\underbrace{0, \ldots, 0}_{k-1}, \boldsymbol{\beta}^{(k)} - \widetilde{\boldsymbol{\beta}}^{(k)}, \underbrace{0, \ldots, 0}_{K-k}),$$

we can write

$$L(\boldsymbol{\beta} \mid \mathbf{D}) \leq L(\widetilde{\boldsymbol{\beta}} \mid \mathbf{D}) - (\boldsymbol{\beta}^{(k)} - \widetilde{\boldsymbol{\beta}}^{(k)})^T U^{(k)} + \frac{1}{2}(\boldsymbol{\beta}^{(k)} - \widetilde{\boldsymbol{\beta}}^{(k)})^T \mathbf{H}^{(k)} (\boldsymbol{\beta}^{(k)} - \widetilde{\boldsymbol{\beta}}^{(k)}). \quad (9.85)$$

Let η_k be the largest eigenvalue of $\mathbf{H}^{(k)}$. We set $\gamma_k = (1 + \varepsilon^*)\eta_k$, where $\varepsilon^* = 10^{-6}$. Then we can further relax the upper bound in (9.85) as

$$L(\boldsymbol{\beta} \mid \mathbf{D}) \leq L(\widetilde{\boldsymbol{\beta}} \mid \mathbf{D}) - (\boldsymbol{\beta}^{(k)} - \widetilde{\boldsymbol{\beta}}^{(k)})^T U^{(k)} + \frac{1}{2}\gamma_k(\boldsymbol{\beta}^{(k)} - \widetilde{\boldsymbol{\beta}}^{(k)})^T (\boldsymbol{\beta}^{(k)} - \widetilde{\boldsymbol{\beta}}^{(k)}). \quad (9.86)$$

It is important to note that the inequality strictly holds unless for $\boldsymbol{\beta}^{(k)} = \widetilde{\boldsymbol{\beta}}^{(k)}$. Instead of minimizing (9.84) we solve

$$\operatorname{argmin}_{\boldsymbol{\beta}^{(k)}} L(\widetilde{\boldsymbol{\beta}} \mid \mathbf{D}) - (\boldsymbol{\beta}^{(k)} - \widetilde{\boldsymbol{\beta}}^{(k)})^T U^{(k)} + \frac{1}{2}\gamma_k(\boldsymbol{\beta}^{(k)} - \widetilde{\boldsymbol{\beta}}^{(k)})^T (\boldsymbol{\beta}^{(k)} - \widetilde{\boldsymbol{\beta}}^{(k)}) + \lambda w_k \|\boldsymbol{\beta}^{(k)}\|_2. \quad (9.87)$$

Denote by $\widetilde{\boldsymbol{\beta}}^{(k)}(\text{new})$ the solution to (9.87). It is straightforward to see that $\widetilde{\boldsymbol{\beta}}^{(k)}(\text{new})$ has a simple closed-from expression

$$\widetilde{\boldsymbol{\beta}}^{(k)}(\text{new}) = \frac{1}{\gamma_k}\left(U^{(k)} + \gamma_k \widetilde{\boldsymbol{\beta}}^{(k)}\right)\left(1 - \frac{\lambda w_k}{\|U^{(k)} + \gamma_k \widetilde{\boldsymbol{\beta}}^{(k)}\|_2}\right)_+. \quad (9.88)$$

Algorithm 4 summarizes the details of GMD.

Algorithm 4: The GMD algorithm for a general group-lasso penalized problem

---

1. For $k = 1, \ldots, K$, compute γ_k, the largest eigenvalue of $\mathbf{H}^{(k)}$.
2. Initialize $\widetilde{\boldsymbol{\beta}}$.
3. Repeat the following cyclic groupwise updates until convergence:
   - for $k = 1, \ldots, K$, do step (3.1)–(3.3)

     3.1 Compute $U(\widetilde{\boldsymbol{\beta}}) = -\nabla L(\widetilde{\boldsymbol{\beta}}|\mathbf{D})$.

     3.2 Compute $\widetilde{\boldsymbol{\beta}}^{(k)}(\text{new}) = \frac{1}{\gamma_k}\left(U^{(k)} + \gamma_k \widetilde{\boldsymbol{\beta}}^{(k)}\right)\left(1 - \frac{\lambda w_k}{\|U^{(k)} + \gamma_k \widetilde{\boldsymbol{\beta}}^{(k)}\|_2}\right)_+$.

     3.3 Set $\widetilde{\boldsymbol{\beta}}^{(k)} = \widetilde{\boldsymbol{\beta}}^{(k)}(\text{new})$.

One may also consider a concave group-lasso penalized additive model.

$$\text{argmin}_{\boldsymbol{\beta}} \frac{1}{n} \sum_{i=1}^{n} \phi(y_i f(X_i)) + \lambda \sum_{k=1}^{K} P_\lambda(\|\boldsymbol{\beta}^{(k)}\|_2), \tag{9.89}$$

where P_λ can be the SCAD or the MCP penalty. The above problem can be solved by combining the LLA algorithm and `gglasso`. Let $\widetilde{\boldsymbol{\beta}}$ be the current estimate, then we update the estimate by solving

$$\min_{\boldsymbol{\beta}} \frac{1}{n} \sum_{i=1}^{n} \phi(y_i f(X_i)) + \sum_{k=1}^{K} \tilde{w}_k \|\boldsymbol{\beta}^{(k)}\|_2$$

where

$$\tilde{w}_k = P_\lambda'(\|\widetilde{\boldsymbol{\beta}}^{(k)}\|_2).$$

References

Bartlett P, Jordan M, McAuliffe J (2006) Convexity, classification and risk bounds. J Am Stat Assoc 101:138–156

Bradley P, Mangasarian O (1998) Feature selection via concave minimization and support vector machines. In: Machine learning proceedings of the fifteenth international conference (ICML'98), Citeseer, pp 82–90

Breiman L (2001) Random forests. Mach Learn 45:5–32

Cai T, Liu W (2011) A direct estimation approach to sparse linear discriminant analysis. J Am Stat Assoc 106:1566–1577

Clemmensen L, Hastie T, Witten D, Ersboll B (2011) Sparse discriminant analysis. Technometrics 53:406–413

Efron B, Hastie T, Johnstone I, Tibshirani R (2004) Least angle regression (with discussion). Ann Stat 32:407–499

Fan J, Fan Y (2008) High dimensional classification using features annealed independence rules. Ann Stat 36:2605–2637

Fan J, Li R (2001) Variable selection via nonconcave penalized likelihood and its oracle properties. J Am Stat Assoc 96:1348–1360

Fan J, Feng Y, Tong X (2012) A ROAD to classification in high dimensional space. J R Stat Soc Ser B 74:745–771

Fan J, Xue L, Zou H (2014) Strong oracle optimality of folded concave penalized estimation. Ann Stat 42:819–849

Fisher R (1936) The use of multiple measurements in taxonomic problems. Ann Eugen 7:179–188

Friedman J (2001) Greedy function approximation: a gradient boosting machine. Ann Stat 29:1189–1232

Friedman J, Hastie T, Tibshirani R (2010) Regularization paths for generalized linear models via coordinate descent. J Stat Softw 33:1–22

Freund Y, Schapire R, (1997) A Decision-Theoretic Generalization of On-Line Learning and an Application to Boosting. J Comput Syst Sci 55(1):119–139

Graham K, de LasMorenas A, Tripathi A, King C, Kavanah M, Mendez J, Stone M, Slama J, Miller M, Antoine G et al (2010) Gene expression in histologically normal epithelium from breast cancer patients and from cancer-free prophylactic mastectomy patients shares a similar profile. Br J Cancer 102:1284–1293

Hand DJ (2006) Classifier technology and the illusion of progress. Stat Sci 21:1–14

Hastie T, Tibshirani R (1990) Generalized additive models. Chapman and Hall/CRC, Boca Raton

Hastie T, Tibshirani R, Friedman J (2009) The elements of statistical learning: data mining, inference, and prediction, 2nd edn. Springer, New York

Hunter D, Lange K (2004) A tutorial on MM algorithms. Am Stat 58:30–37

Lin Y, Jeon Y (2003) Discriminant analysis through a semiparametric model. Biometrika 90:379–392

Liu J, Ji S, Ye J (2009) SLEP: sparse learning with efficient projections. Arizona State University. http://www.public.asu.edu/jye02/Software/SLEP

Mai Q, Zou H (2015) Sparse semiparametric discriminant analysis. J Multivar Anal 135:175–188

Mai, Q, Zou H, Yuan M (2012) A direct approach to sparse discriminant analysis in ultra-high dimensions. Biometrika 99:29–42

Mai Q, Yang Y, Zou H (2018, to appear) Multiclass sparse discriminant analysis. Stat Sin

Meier L, van de Geer S, Buhlmann P (2008) The group lasso for logistic regression. J R Stat Soc Ser B 70:53–71

Michie D, Spiegelhalter D, Taylor C (1994) Machine learning, neural and statistical classification, 1st edn. Ellis Horwood, Upper Saddle River

Mika S, Räsch G, Weston J, Schölkopf B, Müller KR (1999) Fisher discriminant analysis with kernels. Neural Netw Signal Process IX:41–48

Nesterov Y (2004) Introductory lectures on convex optimization: a basic course. Springer, New York

Nesterov Y (2007) Gradient methods for minimizing composite objective function. Technical Report, Center for Operations Research and Econometrics (CORE), Catholic University of Louvain (UCL)

Singh D, Febbo P, Ross K, Jackson D, Manola J, Ladd C, Tamayo P, Renshaw AA, D'Amico AV, Richie J, Lander E, Loda M, Kantoff P, Golub T, Sellers W (2002) Gene expression correlates of clinical prostate cancer behavior. Cancer Cell 1:203–209

Tibshirani R (1996) Regression shrinkage and selection via the lasso. J R Stat Soc Ser B 58: 267–288

Tibshirani R, Hastie T, Narasimhan B, Chu G (2002) Diagnosis of multiple cancer types by shrunken centroids of gene expression. Proc Natl Acad Sci 99:6567–6572

Tseng P (2001) Convergence of a block coordinate descent method for nondifferentiable minimization. J Optim Theory Appl 109:475–494

Vapnik V (1996) The nature of statistical learning. Springer, New York

Wang L, Shen X (2006) Multicategory support vector machines, feature selection and solution path. Stat Sin 16:617–634

Wang L, Zhu J, Zou H (2008) Hybrid huberized support vector machines for microarray classification and gene selection. Bioinformatics 24:412–419

Witten D, Tibshirani R (2011) Penalized classification using Fisher's linear discriminant. J R Stat Soc Ser B 73:753–772

Wu M, Zhang L, Wang Z, Christiani D, Lin X (2008) Sparse linear discriminant analysis for simultaneous testing for the significance of a gene set pathway and gene selection. Bioinformatics 25:1145–1151

Yang Y, Zou H (2013) An efficient algorithm for computing the HHSVM and its generalizations. J Comput Graph Stat 22:396–415

Yang Y, Zou H (2015) A fast unified algorithm for solving group-lasso penalized learning problems. Stat Comput 25(6):1129–1141

Yuan M, Lin Y (2006) Model selection and estimation in regression with grouped variables. J R Stat Soc Ser B 68:49–67

Zhang C-H (2010) Nearly unbiased variable selection under minimax concave penalty. Ann Stat 38:894–942

Zhang HH, Ahn J, Lin X, Park C (2006) Gene selection using support vector machines with nonconvex penalty. Bioinformatics 2:88–95

Zhu J, Hastie T (2005) Kernel logistic regression and the import vector machine. J Comput Graph Stat 14:185–205

Zhu J, Rosset S, Hastie T, Tibshirani R (2003) 1-Norm support vector machine. In: Neural information processing systems. MIT Press, Cambridge, p 16

Zou H (2006) The adaptive lasso and its oracle properties. J Am Stat Assoc 101:1418–1429

Zou H, Hastie T (2005) Regularization and variable selection via the elastic net. J R Stat Soc Ser B 67:301–320

Zou H, Li R (2008) One-step sparse estimates in nonconcave penalized likelihood models (with discussion). Ann Stat 36:1509–1533

Chapter 10
Analysis of High-Dimensional Regression Models Using Orthogonal Greedy Algorithms

Hsiang-Ling Hsu, Ching-Kang Ing, and Tze Leung Lai

Abstract We begin by reviewing recent results of Ing and Lai (Stat Sin 21:1473–1513, 2011) on the statistical properties of the orthogonal greedy algorithm (OGA) in high-dimensional sparse regression models with independent observations. In particular, when the regression coefficients are absolutely summable, the conditional mean squared prediction error and the empirical norm of OGA derived by Ing and Lai (Stat Sin 21:1473–1513, 2011) are introduced. We then explore the performance of OGA under more general sparsity conditions. Finally, we obtain the convergence rate of OGA in high-dimensional time series models, and illustrate the advantage of our results compared to those established for Lasso by Basu and Michailidis (Ann Stat 43:1535–1567, 2015) and Wu and Wu (Electron J Stat 10:352–379, 2016).

Keywords Conditional mean squared prediction errors · Empirical norms · High-dimensional models · Lasso · Orthogonal greedy algorithms · Sparsity · Time series

10.1 Introduction

Consider a high-dimensional linear regression model

$$y_t = \alpha + \sum_{j=1}^{p} \beta_j x_{tj} + \varepsilon_t = y(\mathbf{x}_t) + \varepsilon_t, t = 1, 2, \ldots, n \qquad (10.1.1)$$

H.-L. Hsu
National University of Kaohsiung, Kaohsiung, Taiwan
e-mail: hsuhl@nuk.edu.tw

C.-K. Ing (✉)
National Tsing Hua University, Hsinchu, Taiwan
e-mail: cking@stat.nthu.edu.tw

T. L. Lai
Stanford University, Stanford, CA, USA
e-mail: lait@stanford.edu

© Springer International Publishing AG, part of Springer Nature 2018

W. K. Härdle et al. (eds.), *Handbook of Big Data Analytics*, Springer Handbooks of Computational Statistics, https://doi.org/10.1007/978-3-319-18284-1_10

where $p \gg n$ and the predictor vector, $\mathbf{x}_t = (x_{t1}, \ldots, x_{tp})^\top$, is uncorrelated with the zero-mean random disturbance ε_t. Since $p \gg n$, it is difficult to apply the conventional statistical procedures, such as the maximum likelihood estimate and the least squares estimate, to estimate the unknown regression function

$$y(\mathbf{x}) = \alpha + \boldsymbol{\beta}^\top \mathbf{x}, \tag{10.1.2}$$

where $\boldsymbol{\beta} = (\beta_1, \ldots, \beta_p)^\top$ and $\mathbf{x} = (x_1, \ldots, x_p)^\top$ denotes an independent replicate of $\mathbf{x}_t$. A major advance to resolve this difficulty is the lasso (least absolute shrinkage and selection operator, Tibshirani (1996)), which is the l_1-penalized least squares method. In particular, lasso can consistently estimate (10.1.2) under a sparsity condition and some additional regularity conditions on the regression coefficients; see, e.g., Bickel et al. (2009). On the other hand, Fan and Li (2001) have argued that the l_1-penalty used by lasso may result in severe bias for large regression coefficients and have proposed a SCAD (smoothly clipped absolute deviation) penalty to address this problem. Because the associated minimization problem is non-convex and direct computation is infeasible for large p, multi-step procedures in which each step involves convex optimization, such as local linear and local quadratic approximations, have been introduced. One such procedure is Zou's (2006) adaptive lasso, which uses lasso as an initial estimate to determine the weights for a second-stage weighted lasso.

An alternative method for handling a large number of input variables is stepwise least squares regression. In Sect. 10.2, we introduce recent results of Ing and Lai (2011) on a fast stepwise regression method, called the orthogonal greedy algorithm (OGA), that selects input variables to enter a p-dimensional linear regression model sequentially so that the selected variable at each step minimizes the residual sum squares. When $(\mathbf{x}_t^\top, \varepsilon_t)$ are i.i.d. and the regression coefficients obey the following sparse condition,

$$\sup_n \sum_{j=1}^{p_n} |\beta_j| \sigma_j < \infty, \tag{10.1.3}$$

where $\sigma_j^2 = \mathrm{Var}(x_j)$, they have derived the convergence rate of the conditional mean squared prediction error (CMSPE),

$$E\{(y(\mathbf{x}) - \hat{y}_m(\mathbf{x}))^2 | \, y_1, \mathbf{x}_1, \ldots, y_n, \mathbf{x}_n\}, \tag{10.1.4}$$

of the OGA predictor $\hat{y}_m(\mathbf{x})$ of $y(\mathbf{x})$ after m iterations are performed. They have also developed a consistent model selection procedure along the OGA path that can adjust for potential spuriousness of the greedily chosen regressors among a large number of candidate variables. The resultant regression estimate is shown to have the oracle property of being equivalent to least squares regression on an asymptotically minimal set of relevant regressors under the strong sparsity condition, which assumes that

$$\sharp(J_n^*) \ll n, \tag{10.1.5}$$

where $J_n^* = \{1 \le j \le p_n : \beta_j \neq 0\}$ denotes the set of relevant predictor variables and $\sharp(J)$ denotes the cardinality of J. Section 10.2 also reviews Ing and Lai's (2011) result on the performance of OGA in situations where $\mathbf{x}_t$ is non-random, which is referred to as the fixed design case. Since in this case, the more commonly used performance measure is the empirical norm,

$$\|\hat{y}_m(\cdot) - y(\cdot)\|_n^2 = n^{-1} \sum_{t=1}^{n} (\hat{y}_m(\mathbf{x}_t) - y(\mathbf{x}_t))^2, \tag{10.1.6}$$

where $\hat{y}_m(\mathbf{x}_t)$ is the OGA predictor of $y(\mathbf{x}_t)$ after m iterations, we focus on their upper bound established for (10.1.6) under the sparse condition (10.1.3). This upper bound provides an analog of the oracle inequalities of Candés and Tao (2007), Bunea et al. (2007), Bickel et al. (2009), and Candés and Plan (2009) for lasso and Dantzig selector.

In Sect. 10.3, we introduce the results of Gao et al. (2013) and Ing and Lai (2015) on the rate of convergence of OGA under sparse conditions stronger than (10.1.3), but weaker than (10.1.5). We also report simulation studies of the performance of OGA+HDAIC relative to that of lasso and Bühlmann's (2006) PGA+AIC$_c$, where HDAIC, defined in Sect. 10.2.1, is the abbreviation for high-dimensional Akaike's information criterion. Section 10.4 is concerned with the performance of OGA under (10.1.1) with $(\mathbf{x}_t, \varepsilon_t)$ being stationary time series. We derive the convergence rate of OGA in a general time series setting, and illustrate the advantage of our results compared to those established for lasso by Basu and Michailidis (2015) and Wu and Wu (2016).

10.2 Convergence Rates of OGA

10.2.1 Random Regressors

We begin this section by furnishing a formal definition of OGA. Replacing y_t by $y_t - \bar{y}$ and x_{tj} by $x_{tj} - \bar{x}_j$, where $\bar{x}_j = n^{-1} \sum_{t=1}^n x_{tj}$ and $\bar{y} = n^{-1} \sum_{t=1}^n y_t$, it will be assumed throughout the rest of the chapter that $\alpha = 0, E(y_t) = 0$, and $E(x_{tj}) = 0$ for all t and j. Letting $\hat{J}_0 = \emptyset$, $\mathbf{Y} = (y_1, \ldots, y_n)'$ and $\mathbf{X}_j = (x_{1j}, \ldots, x_{nj})'$, $1 \le j \le p$, OGA chooses $\mathbf{X}_{\hat{j}_1}, \mathbf{X}_{\hat{j}_2}, \ldots$ in the following sequential manner:

Step 1. $\hat{j}_1 = \arg\max_{1 \le j \le p} |\mathbf{Y}'\mathbf{X}_j / \|\mathbf{X}_j\||$ and $\hat{J}_1 = \hat{J}_0 \bigcup \{\hat{j}_1\}$.

$\vdots$

Step m. $\hat{j}_m = \arg\max_{1 \le j \le p} |\mathbf{Y}'(\mathbf{I} - \mathbf{H}_{\hat{J}_{m-1}})\mathbf{X}_j / \|\mathbf{X}_j\||$ and $\hat{J}_m = \hat{J}_{m-1} \bigcup \{\hat{j}_m\}$, where for $J \subset \{1, \ldots, p\}$, $\mathbf{H}_J$ is the orthogonal projection matrix on span$\{\mathbf{X}_j, j \in J\}$.

When iterations stop at step m, we predict $y(\mathbf{x})$ and $y(\mathbf{x}_t)$ by $\hat{y}_m(\mathbf{x}) = \mathbf{x}'(\hat{J}_m)\hat{\boldsymbol{\beta}}(\hat{J}_m)$ and $\hat{y}_m(\mathbf{x}_t) = \mathbf{x}_t'(\hat{J}_m)\hat{\boldsymbol{\beta}}(\hat{J}_m)$, respectively, where $\mathbf{x}(J) = (x_t, i \in J)$, $\mathbf{x}_t(J) = (x_{ti}, i \in J)$ and

$$\hat{\boldsymbol{\beta}}(J) = \left(\sum_{t=1}^{n} \mathbf{x}_t(J)\mathbf{x}_t'(J)\right)^{-1} \sum_{t=1}^{n} \mathbf{x}_t(J) y_t$$

is the least squares estimate of the regression coefficient vector of model J. Let $\boldsymbol{\Gamma}(J) = E\left\{\mathbf{z}(J)\mathbf{z}^\top(J)\right\}$ and $\mathbf{g}_i(J) = E(z_i\mathbf{z}(J))$, where $z_j = x_j/\sigma_j$ and $\mathbf{z}(J) = (z_j, j \in J)$. Ing and Lai (2011) assume that for some $\theta_0, \theta_1 > 0$,

$$E\{\exp(\theta\varepsilon)\} < \infty, |\theta| \le \theta_0, \tag{10.2.1}$$

$$\limsup_{n\to\infty} \max_{1\le j\le p_n} E\left\{\exp(\theta_1 z_j^2)\right\} < \infty, \tag{10.2.2}$$

and for some $\delta > 0$, $M > 0$ and all large n,

$$\min_{1\le \sharp(J)\le K_n} \lambda_{\min}(\boldsymbol{\Gamma}(J)) > \delta, \quad \max_{1\le \sharp(J)\le K_n,\, i\notin J} \| \boldsymbol{\Gamma}^{-1}(J)\mathbf{g}_i(J) \|_1 < M, \tag{10.2.3}$$

where K_n is a prescribed upper bound for the number of OGA iterations and $\| \mathbf{v} \|_1 = \sum_{j=1}^{k} |v_j|$ for $\mathbf{v} = (v_1, \ldots, v_k)^\top$. By making use of these assumptions together with the sparsity condition (10.1.3), Theorem 3.1 of Ing and Lai (2011) provides the following uniform bound for the CMSPE of OGA.

Theorem 10.2.1 *Suppose* $(\mathbf{x}_t, \varepsilon_t)$ *are i.i.d.,* $p = p_n \to \infty$ *and* $\log p = o(n)$. *Assume* (10.1.3), (10.2.1)– (10.2.3) *and* $K_n \to \infty$ *such that* $K_n = O((n/\log p_n)^{1/2})$. *Then,*

$$\max_{1\le m\le K_n} \left(\frac{E[\{y(\mathbf{x}) - \hat{y}_m(\mathbf{x})\}^2 | y_1, \mathbf{x}_1, \ldots, y_n, \mathbf{x}_n]}{m^{-1} + n^{-1} m \log p_n}\right) = O_p(1). \tag{10.2.4}$$

Some comments are in order. Note first that by Theorem 3 of Temlyakov (2000), the squared bias in approximating $y(\mathbf{x})$ by $y_{J_m}(\mathbf{x})$ is $E(y(\mathbf{x}) - y_{J_m}(\mathbf{x}))^2 = O(m^{-1})$, where $y_J(\mathbf{x})$ is the best linear predictor of $y(\mathbf{x})$ based on $\{x_j, j \in J\}$ and J_k be the set of predictor variables selected by the population version of OGA at the end of stage k. Moreover, since OGA uses $\hat{y}_m(\cdot)$ instead of $y_{J_m}(\cdot)$, it has not only larger squared bias but also variance in the least squares estimates whose order of magnitude is $O(n^{-1} m \log p_n)$. Note also that m is the number of estimated regression coefficients, $O(n^{-1})$ is the variance per coefficient, and $O(\log p_n)$ is the variance inflation factor due to data-dependent selection of $\hat{j}_i$ from $\{1, \ldots, p_n\}$. Combining the squared bias with the variance suggests that $O(m^{-1} + n^{-1} m \log p_n)$ is the smallest order one can expect for $E_n(\{y(\mathbf{x}) - \hat{y}_m(\mathbf{x})\}^2)$. Equation (10.2.4) reveals that uniformly in $m = O((n/\log p_n)^{1/2})$, OGA can attain this best order of $m^{-1} + n^{-1} m \log p_n$ for $E_n(\{y(\mathbf{x}) - \hat{y}_m(\mathbf{x})\}^2)$, where $E_n(\cdot) \equiv E[\cdot | y_1, \mathbf{x}_1, \ldots, y_n, \mathbf{x}_n]$.

Whereas (10.2.4) suggests ceasing OGA iterations after $O((n/\log p)^{1/2})$ steps, Ing and Lai (2011) have proposed to choose along the OGA path the model that has the smallest value of a suitably chosen model selection criterion, which is called a "high-dimensional information criterion" (HDIC) and defined by

$$\mathrm{HDIC}(J) = n \log \hat{\sigma}^2_{n,J} + \sharp(J) w_n \log p, \tag{10.2.5}$$

where for a non-empty subset $J \subset \{1, \ldots, p\}$, $\hat{\sigma}^2_{n,J} = n^{-1} \sum_{t=1}^{n} (y_t - \hat{y}_{t;J})^2$ with $\hat{y}_{t;J}$ denoting the fitted value of y_t when $\mathbf{Y}$ is projected into the linear space spanned by $\mathbf{X}_j, j \in J \neq \emptyset$. It is worth mentioning that different information criteria correspond to different choices of w_n. In particular, $w_n = \log n$ corresponds to HDBIC; without the $\log p$ factor, (10.2.5) reduces to the usual BIC. The case $w_n = c$ corresponds to high-dimensional Akaike's information criterion (HDAIC). Ing and Lai (2011) have suggested choosing model $\hat{J}_{\hat{k}_n}$, where

$$\hat{k}_n = \arg \min_{1 \leq k \leq K_n} \mathrm{HDIC}(\hat{J}_k). \tag{10.2.6}$$

Under the strong sparsity assumption that there exists $0 \leq \gamma < 1/2$ such that $n^{\gamma} = o((n/\log p_n)^{1/2})$ and

$$\liminf_{n \to \infty} n^{\gamma} \min_{j \in J_n^*} \beta_j^2 \sigma_j^2 > 0, \tag{10.2.7}$$

Theorem 4.1 of Ing and Lai (2011) shows that if $K_n/n^{\gamma} \to \infty$ and $K_n = O((n/\log p_n)^{1/2})$, then

$$\lim_{n \to \infty} P(J_n^* \subset \hat{J}_{K_n}) = 1. \tag{10.2.8}$$

Thus, with probability approaching 1 as $n \to \infty$, the OGA path includes all relevant regressors when the number of iterations is large enough. Define the minimal number of relevant regressors along an OGA path by

$$\widetilde{k}_n = \min\{k : 1 \leq k \leq K_n, J_n^* \subseteq \hat{J}_k\} \ (\min \emptyset = K_n). \tag{10.2.9}$$

Under (10.2.7), $K_n/n^{\gamma} \to \infty$ and $K_n = O((n/\log p_n)^{1/2})$, Theorem 4.2 of Ing and Lai (2011) further proves that

$$\lim_{n \to \infty} P(\hat{k}_n = \tilde{k}_n) = 1, \tag{10.2.10}$$

provided w_n in (10.2.5) satisfying

$$w_n \to \infty, \ w_n \log p_n = o(n^{1-2\gamma}). \tag{10.2.11}$$

Therefore, consistency of variable selection along OGA paths by HDIC is established.

Although (10.2.11) shows that $\tilde{k}_n$ can be consistently estimated by $\hat{k}_n$, $\hat{J}_{\hat{k}_n}$ may still contain irrelevant variables that are included along the OGA path, as shown in Example 3 of Section 5 in Ing and Lai (2011). To exclude irrelevant variables, Ing and Lai (2011) have defined a subset $\hat{J}_n^*$ of $\hat{J}_{\hat{k}_n}$ by

$$\hat{J}_n^* = \{\hat{j}_l : \text{HDIC}(\hat{J}_{\hat{k}_n} - \{\hat{j}_l\}) > \text{HDIC}(\hat{J}_{\hat{k}_n}), 1 \le l \le \hat{k}_n\} \text{ if } \hat{k}_n > 1, \quad (10.2.12)$$

and $\hat{J}_n^* = \{\hat{j}_1\}$ if $\hat{k}_n = 1$. Theorem 4.3 of Ing and Lai (2011) establishes the oracle property of $\hat{J}_n^*$ and shows that this simple procedure, which is denoted by OGA+HDIC+Trim, achieves variable selection consistency in the sense that $\lim_{n\to\infty} P(\hat{J}_n^* = J_n^*) = 1$. For strongly sparse regression models, Zhao and Yu (2006) have shown that lasso is variable-selection consistent under a 'strong irrepresentability' condition of the design matrix and additional regularity conditions. They have also shown that lasso can fail to distinguish irrelevant predictors that are highly correlated with relevant predictors, and the strong irrepresentable condition is used to rule out such cases. On the other hand, OGA+HDIC+Trim can achieve variable selection consistency without this type of condition.

10.2.2 The Fixed Design Case

As mentioned previously, the empirical norm defined in (10.1.6) is the more commonly used performance measure in fixed designs. When ε_t in (10.1.1) are assumed to be either zero or nonrandom, upper bounds for (10.1.6) have been discussed in the approximation theory literature; see, e.g., Tropp (2004) and Donoho et al. (2006). When the ε_t in (10.1.1) are zero-mean random variables, an upper bound for (10.1.6) should involve the variance besides the bias of the regression estimate. Since the regression function in (10.1.1) has infinitely many representations when $p > n$, Ing and Lai (2011) have introduced the sets

$$\mathbf{B} = \{\mathbf{b} : \mathbf{X}\mathbf{b} = (y(\mathbf{x}_1), \ldots, y(\mathbf{x}_n))^\top\}, \quad (10.2.13)$$

where $\mathbf{X} = (\mathbf{X}_1, \ldots, \mathbf{X}_p)$ is $n \times p$, and $\mathbf{B}_{J,i} = \{\boldsymbol{\theta}_{J,i} : \mathbf{X}_J^\top \mathbf{X}_i = \mathbf{X}_J^\top \mathbf{X}_J \boldsymbol{\theta}_{J,i}\}$, where $J \subseteq \{1, \ldots, p\}$ and $1 \le i \le p$ with $i \notin J$, and defined

$$r_p = \arg \min_{0<r<1/2} \{1 + (\log \sqrt{1/(1-2r)}/ \log p)\}/r, \; \tilde{r}_p = 1/(1-2r_p). \quad (10.2.14)$$

With these notations, Ing and Lai (2011) have provided the following upper bound for (10.1.6).

Theorem 10.2.2 *Suppose ε_t are i.i.d. normal random variables with $E(\varepsilon_t) = 0$ and $E(\varepsilon_t^2) = \sigma^2$. Assume that x_{tj} are nonrandom constants, normalized so that $n^{-1}\sum_{t=1}^n x_{tj}^2 = 1$, and satisfying*

$$\max_{1\leq \sharp(J)\leq \lfloor n/\log p\rfloor, i\notin J} \inf_{\boldsymbol{\theta}_{J,i}\in \mathbf{B}_{J,i}} \|\boldsymbol{\theta}_{J,i}\|_1 < M \text{ for some } M > 0. \tag{10.2.15}$$

Let $0 < \xi < 1$, $C > \sqrt{2}(1+M)$, $s > \{1 + (2\log p)^{-1}\log \tilde{r}_p\}/r_p$, where r_p and $\tilde{r}_p$ are defined by (10.2.14), *and*

$$\omega_{m,n} = (\inf_{\mathbf{b}\in\mathbf{B}} \|\mathbf{b}\|_1) \max\left\{\frac{\inf_{\mathbf{b}\in\mathbf{B}}\|\mathbf{b}\|_1}{1+m\xi^2}, \frac{2C\sigma}{1-\xi}(\frac{\log p}{n})^{1/2}\right\}. \tag{10.2.16}$$

Then for all $p \geq 3$, $n \geq \log p$, and $1 \leq m \leq \lfloor n/\log p\rfloor$,

$$\|\hat{y}_m(\cdot) - y(\cdot)\|_n^2 \leq \omega_{m,n} + s\sigma^2 m(\log p)/n \tag{10.2.17}$$

with probability at least

$$1 - p\exp\left\{-\frac{C^2\log p}{2(1+M)^2}\right\} - \frac{\tilde{r}_p^{1/2}p^{-(sr_p-1)}}{1-\tilde{r}_p^{1/2}p^{-(sr_p-1)}}.$$

Some comments on Theorem 10.2.2 are as follows. The upper bound (10.2.17) for the prediction risk of OGA is a sum of a variance term, $s\sigma^2 m(\log p)/n$, and a squared bias term, $\omega_{m,n}$. The variance term is the usual least squares risk $m\sigma^2/n$ multiplied by a risk inflation factor $s\log p$; see Foster and George (1994) for a detailed discussion of the idea of risk inflation. The squared bias term is the maximum of $(\inf_{\mathbf{b}\in\mathbf{B}}\|\mathbf{b}\|_1)^2/(1+m\xi^2)$, which is the approximation error of the noiseless OGA, and $2C\sigma(1-\xi)^{-1}\inf_{\mathbf{b}\in\mathbf{B}}\|\mathbf{b}\|_1(n^{-1}\log p)^{1/2}$, which can be viewed as the error caused by the discrepancy between the noiseless OGA and the sample OGA. The $\|\boldsymbol{\theta}_{J,i}\|_1$ in (10.2.15) is closely related to the cumulative coherence function introduced by Tropp (2004). Since Theorem 10.2.2 does not put any restriction on M and $\inf_{\mathbf{b}\in\mathbf{B}}\|\mathbf{b}\|_1$, the theorem can be applied to any design matrix although a large value of M or $\inf_{\mathbf{b}\in\mathbf{B}}\|\mathbf{b}\|_1$ will result in a large bound on the right-hand side of (10.2.17). Note that the population analog of $\|\boldsymbol{\theta}_{J,i}\|_1$ for random regressors is $\|\boldsymbol{\Gamma}^{-1}(J)\mathbf{g}_i(J)\|_1$, which appears in the second part of (10.2.3). In addition, although (10.2.3) makes an assumption on $\lambda_{\min}(\boldsymbol{\Gamma}(J))$, it does not make assumptions on $\lambda_{\max}(\boldsymbol{\Gamma}(J))$. This is similar to the restricted eigenvalue (RE) assumption introduced by Bickel et al. (2009) but differs from the sparse Riesz condition introduced by Zhang and Huang (2008).

10.3 The Performance OGA Under General Sparse Conditions

10.3.1 *Rates of Convergence*

In this section, we first investigate the performance of OGA under the following sparse condition,

$$\sup_{n\geq 1}\sum_{j=1}^{p_n}|\beta_j\sigma_j|^{1/\gamma} < \infty \text{ for some } \gamma \geq 1, \tag{10.3.1}$$

which we call the *algebraic decay case*. Obviously, (10.3.1) includes (10.1.3) as a special case. The parameter γ in (10.3.1) can be viewed as an index to describe the degree of sparseness in the underlying high-dimensional regression model. The larger the γ is, the sparser the model is. In fact, an assumption similar to (10.3.1) has also been made by Bickel and Levian (2008) and Cai et al. (2010) in the context of estimating high-dimensional covariance matrices. We start by defining the population version of the weak orthogonal greedy algorithm (WOGA). Let $0 < \xi \leq 1$, $J_{\xi,0} = \emptyset$ and $u_0 = y(\mathbf{x})$.

Step 1: Let $j_{\xi,m}$ be any integer $j \in \{1,\ldots,p\}$ such that $|E(u_0 z_j)| \geq \xi \max_{1\leq i\leq p}|E(u_0 z_i)|$. Define $J_{\xi,1} = J_{\xi,0}\bigcup\{j_{\xi,1}\}$ and $u_1 = y(\mathbf{x}) - y_{J_{\xi,1}}(\mathbf{x})$, where $y_J(\mathbf{x})$ is defined after Theorem 10.2.1.

$\vdots$

Step m: Let $j_{\xi,m}$ be any integer $j \in \{1,\ldots,p\}$ satisfying $|E(u_{m-1}z_j)| \geq \xi \max_{1\leq i\leq p}|E(u_{m-1}z_i)|$. Define $J_{\xi,m} = J_{\xi,m-1}\bigcup\{j_{\xi,m}\}$ and $u_m = y(\mathbf{x}) - y_{J_{\xi,m}}(\mathbf{x})$.

Gao et al. (2013) have established a rate of convergence of $E(u_m^2)$.

Lemma 10.3.1 *Assume* (10.3.1) *and*

$$\lambda_{\min}(\boldsymbol{\Gamma}) := \lambda_{\min}(E(\mathbf{z}\mathbf{z}^\top)) \geq \lambda_1 > 0, \tag{10.3.2}$$

where $\mathbf{z} = (z_1,\ldots,z_p)^\top$. *Then, there exists* $C_1 > 0$ *such that*

$$E(y(\mathbf{x}) - y_{J_{\xi,m}}(\mathbf{x}))^2 \leq C_1 m^{-2\gamma+1}.$$

By making use of Lemma 10.3.1, Ing and Lai (2015) have provided the following extension of Theorem 10.2.1.

Theorem 10.3.2 *Suppose* $(\mathbf{x}_t, \varepsilon_t)$ *are i.i.d.,* $p = p_n \to \infty$ *and* $\log p = o(n)$. *Assume (10.2.1)–(10.2.3), (10.3.1), (10.3.2) and* $K_n \to \infty$ *such that* $K_n = O((n/\log p_n)^{1/2})$. *Then*

$$\max_{1\leq m\leq K_n}\left(\frac{E[\{y(\mathbf{x}) - \hat{y}_m(\mathbf{x})\}^2 | y_1, \mathbf{x}_1, \ldots, y_n, \mathbf{x}_n]}{m^{-2\gamma+1} + n^{-1}m\log p_n}\right) = O_p(1). \tag{10.3.3}$$

In view of (10.3.3), to strike a suitable balance between squared bias and variance, one should choose $m \asymp (n/\log p_n)^{1/2\gamma}$, which yields a rate of convergence, $(n^{-1}\log p_n)^{1-(2\gamma)^{-1}}$. Indeed, recent papers by Raskutti et al. (2011) and Wang et al. (2014) have shown that under certain regularity conditions, $(n^{-1}\log p_n)^{1-(2\gamma)^{-1}}$ is also the minimax optimal rate for fixed and random designs, respectively. Moreover, Negahban et al. (2012) have shown that this rate is achievable by the lasso estimate when $\mathbf{x}_t$ are non-random.

Ing and Lai (2015) have also considered a coefficient condition sparser than (10.3.1):

$$\sum_{j\in J}|\beta_j\sigma_j| \le M_1 \max_{j\in J}|\beta_j\sigma_j| \text{ for any } J \subseteq \{1,\ldots,p\}, \tag{10.3.4}$$

where $M_1 > 1$. It is easy to see that (10.3.4) is fulfilled by any exponential decay function defined on the set of positive integers. We therefore call it the *exponential decay case*. Based on an argument similar to that used to prove Lemma 10.3.1 and Theorem 10.3.2, Ing and Lai (2015) have obtained the next theorem.

Theorem 10.3.3 *Suppose* $(\mathbf{x}_t, \varepsilon_t)$ *are i.i.d.,* $p = p_n \to \infty$ *and* $\log p = o(n)$. *Assume* (10.2.1)–(10.2.3), (10.3.4), (10.3.2) *and* $K_n \to \infty$ *such that* $K_n = O((n/\log p_n)^{1/2})$. *Then, there exist* $C_2, C_3 > 0$ *such that*

$$E(y(\mathbf{x}) - y_{J_{\xi,m}}(\mathbf{x}))^2 \le C_2\exp(-C_3 m), \tag{10.3.5}$$

and

$$\max_{1\le m\le K_n}\left(\frac{E[\{y(\mathbf{x}) - \hat{y}_m(\mathbf{x})\}^2 | y_1, \mathbf{x}_1, \ldots, y_n, \mathbf{x}_n]}{\exp(-C_3 m) + n^{-1} m \log p_n}\right) = O_p(1). \tag{10.3.6}$$

In view of (10.3.6), it can be shown that the optimal convergence rate of OGA under (10.3.4) is $n^{-1}\log n \log p_n$, which is achieved by taking $m \approx S\log n$ for some S large enough. This rate is faster than the one obtained in Theorem 3.1 because a sparser condition on the regression coefficients is imposed.

10.3.2 Comparative Studies

In this section, we report simulation studies of the performance of OGA+HDAIC relative to that of lasso and Bühlmann's (2006) PGA+AIC$_c$. These studies consider the regression model

$$y_t = \sum_{j=1}^{p} \beta_j x_{tj} + \varepsilon_t,\ t = 1,\ldots,n, \tag{10.3.7}$$

where $p \gg n$, ε_t are i.i.d. $N(0, \sigma^2)$ and are independent of the x_{tj}, and x_{tj} are generated in the following way:

$$x_{tj} = d_{tj} + \eta w_t, \tag{10.3.8}$$

in which $\eta \geq 0$ and $(d_{t1}, \ldots, d_{tp}, w_t)^\top, 1 \leq t \leq n$, are i.i.d. normal with mean $(0, \ldots, 0)^\top$ and covariance matrix $\mathbf{I}$. Since under (10.3.8), $\lambda_{\min}(\mathbf{\Gamma}) = 1/(1+\eta^2) > 0$ and for any $J \subseteq \{1, \ldots, p\}$ and $1 \leq i \leq p$ with $i \notin J$, $\|\mathbf{\Gamma}^{-1}(J)\mathbf{g}_i(J)\|_1 \leq 1$, (10.2.3) are satisfied; moreover, $\mathrm{Corr}(x_{tj}, x_{tk}) = \eta^2/(1+\eta^2)$ increases as $|\eta|$ does.

Example 1 (The Algebraic Decay Case) Consider (10.3.7) with $\beta_j = s \times j^{-a}$, $p = 2000$ and $n = 400$. In addition, assume that $\eta = 0$ and 2, $\sigma^2 = 0.01, 0.1$ and 1, $s = 1, 2$ and 5, and $a = 1.5$ and 2. In light of Theorem 10.3.2 which requires the number K_n of iterations to satisfy $K_n = O((n/\log p_n)^{1/2})$, we choose $K_n = \lfloor 5(n/\log p_n)^{1/2} \rfloor$. We have also allowed D in $K_n = \lfloor D(n/\log p_n)^{1/2} \rfloor$ to vary between 3 and 10, and the results are similar to those for $D = 5$. Table 10.1 summarizes the mean squared prediction errors (MSPE) of the OGA+HDAIC, lasso and PGA+AIC$_c$ based 1000 simulation replications, where the MSPE is defined by

$$\mathrm{MSPE} = \frac{1}{1000} \sum_{l=1}^{1000} \left(\sum_{j=1}^{p} \beta_j x_{n+1,j}^{(l)} - \hat{y}_{n+1}^{(l)} \right)^2, \tag{10.3.9}$$

in which $x_{n+1,1}^{(l)}, \ldots, x_{n+1,p}^{(l)}$ are the regressors associated with $y_{n+1}^{(l)}$, the new outcome in the lth simulation run, and $\hat{y}_{n+1}^{(l)}$ denotes the predictor of $y_{n+1}^{(l)}$. Here and in the sequel, we choose $c = 2.01$ for HDAIC. We have allowed c in HDAIC to vary among 2.01, 2.51, 3.01, 3.51, and 4.01, but the results are quite similar for the different choices of c. To implement lasso, we use the `Glmnet` package in `R` (Friedman et al. 2010) that conducts fivefold cross-validation to select the optimal penalty r. In addition, the step size ν of PGA is chosen to be $\nu = 0.1$ and the number of PGA iterations is set to 1000.

First note that all MSPEs increase as σ^2 and s increase or as a decreases. The MSPEs of lasso are about two times those of OGA+HDAIC in the case of $a = 1.5$, and about three times those of OGA+HDAIC in the case of $a = 2$. This observation applies to all $\sigma^2 = 0.01, 0.1$ and 1, $s = 1, 2$ and 5 and $\eta = 0$ and 2. The MSPEs of PGA+AIC$_c$ are smaller than those of lasso but larger than those of OGA+HDAIC.

Example 2 (The Exponential Decay Case) Consider (10.3.7) with $\beta_j = s \times \exp(-ja)$, $p = 2000$ and $n = 400$. In addition, assume the same values for η, σ^2, s, and a as those used in Example 1. Like the algebraic decay case, the MSPEs of OGA+HDAIC in the current example are also smaller than those of lasso and PGA+AIC$_c$. In particular, for both $\eta = 0$ and 2, the MSPEs of lasso are about 5–10 times those of OGA+HDAIC when $\sigma^2 = 0.01$ and 0.1, and about 1.1–6 times

Table 10.1 MSPEs of OGA+HDAIC, lasso and PGA+AIC$_c$ in the algebraic-decay case

η	σ^2	a	s	OGA+HDAIC	Lasso	PGA+AIC$_c$
0	0.01	1.5	1	0.250	0.468	0.304
			2	0.640	1.141	0.850
			5	2.291	5.001	3.639
		2	1	0.021	0.065	0.037
			2	0.041	0.122	0.070
			5	0.102	0.616	0.184
	0.1	1.5	1	0.563	1.143	0.670
			2	1.369	2.690	1.654
			5	4.671	8.378	5.916
		2	1	0.068	0.234	0.141
			2	0.133	0.427	0.249
			5	0.309	0.956	0.550
	1	1.5	1	1.334	2.863	1.865
			2	3.007	6.579	3.794
			5	10.234	19.940	12.171
		2	1	0.235	0.820	0.737
			2	0.448	1.544	1.140
			5	1.056	3.393	1.935
2	0.01	1.5	1	0.254	0.478	0.313
			2	0.630	1.122	0.831
			5	2.282	5.059	3.656
		2	1	0.021	0.064	0.036
			2	0.040	0.121	0.068
			5	0.100	0.613	0.184
	0.1	1.5	1	0.565	1.150	0.666
			2	1.405	2.706	1.673
			5	4.689	8.465	6.063
		2	1	0.068	0.225	0.148
			2	0.128	0.417	0.240
			5	0.323	1.006	0.571
	1	1.5	1	1.292	2.816	1.856
			2	3.202	6.693	4.080
			5	10.180	20.260	12.313
		2	1	0.236	0.845	0.815
			2	0.440	1.527	1.077
			5	1.036	3.416	2.110

those of the latter when $\sigma^2 = 1$. On the other hand, unlike the algebraic decay case, the MSPEs of PGA+AIC$_c$ are larger than those of lasso in the exponential decay case. The difference between these two methods is particularly notable as $\sigma^2 = 1$ (Table 10.2).

Table 10.2 MSPEs of OGA+HDAIC, lasso and PGA+AIC$_c$ in the exponential-decay case

η	σ^2	a	s	OGA+HDAIC	Lasso	PGA+AIC$_c$
0	0.01	1.5	1	0.0005	0.0046	0.0068
			2	0.0008	0.0063	0.0061
			5	0.0008	0.0074	0.0071
		2	1	0.0003	0.0034	0.0058
			2	0.0007	0.0038	0.0056
			5	0.0009	0.0048	0.0061
	0.1	1.5	1	0.0046	0.0312	0.0599
			2	0.0080	0.0418	0.0588
			5	0.0090	0.0577	0.0690
		2	1	0.0033	0.0218	0.0539
			2	0.0041	0.0286	0.0558
			5	0.0061	0.0366	0.0596
	1	1.5	1	0.0526	0.1381	0.5587
			2	0.0585	0.2596	0.5419
			5	0.0624	0.4090	0.5692
		2	1	0.0367	0.0458	0.5633
			2	0.0434	0.1613	0.5304
			5	0.0915	0.2593	0.5812
2	0.01	1.5	1	0.0007	0.0049	0.0063
			2	0.0008	0.0061	0.0064
			5	0.0008	0.0073	0.0067
		2	1	0.0004	0.0034	0.0058
			2	0.0006	0.0039	0.0058
			5	0.0008	0.0046	0.0062
	0.1	1.5	1	0.0043	0.0351	0.0564
			2	0.0084	0.0411	0.0608
			5	0.0084	0.0565	0.0700
		2	1	0.0035	0.0216	0.0554
			2	0.0037	0.0261	0.0516
			5	0.0055	0.0359	0.0536
	1	1.5	1	0.0564	0.1362	0.5729
			2	0.0596	0.2854	0.5531
			5	0.0651	0.4156	0.5979
		2	1	0.0406	0.0456	0.5415
			2	0.0522	0.1551	0.5077
			5	0.0991	0.2623	0.5844

10.4 The Performance of OGA in High-Dimensional Time Series Models

In this section, we develop an upper bound for the empirical norm $\|\hat{y}_m(\cdot) - y(\cdot)\|_n^2$ of OGA under (10.1.1) with $(\mathbf{x}_t, \varepsilon_t)$ being dependent series. Specifically, we assume that $(\mathbf{x}_t, \varepsilon_t)$ are stationary time series, and $0 < \underline{\tau} < \min_{1\le i\le p}\sigma_i \le \max_{1\le i\le p}\sigma_i < \bar{\tau} < \infty$. Moreover,

(A1) $\max_{1\le j\le p} E|n^{-1/2}\sum_{t=1}^n z_{tj}\varepsilon_t|^q = O(1)$ for some $q \ge 2$.
(A2) $\max_{1\le i,j\le p} E|n^{-1/2}\sum_{t=1}^n (z_{ti}z_{tj} - \rho_{ij})|^{2q_1} = O(1)$, where $q_1 > q$ and $\rho_{ij} = E(z_i z_j)$.
(A3) $p = p_n \to \infty$ and $p_n^{2/q}/n = o(1)$.

The following examples show that (A1) and (A2) are fulfilled by a broad class of time series models.

Example 3 Let $\{\varepsilon_t\}$ be a sequence of martingale differences with respect to an increasing sequence of σ-fields $\{\mathcal{F}_t\}$ such that supt $\sup_{-\infty<t<\infty} E[|\varepsilon_t|^q|\mathcal{F}_{t-1}] < D < \infty$ a.s. for some $q \ge 2$. Assume also that $\{\mathbf{z}_t\}$ is $\mathcal{F}_{t-1}$-measurable and obeys $\max_{1\le t\le n, 1\le j\le p} E|z_{tj}|^q = O(1)$. Then (A1) holds true. To see this, by Lemma 2 of Wei (1987),

$$E|n^{-1/2}\sum_{t=1}^n z_{tj}\varepsilon_t|^q \le D_q E|n^{-1}\sum_{t=1}^n z_{tj}^2|^{q/2},$$

where D_q is a positive constant dependent only on D and q. Moreover, by the convexity of $|x|^{q/2}$, it holds that for all $1 \le j \le p$,

$$\max_{1\le j\le p} E|n^{-1}\sum_{t=1}^n z_{tj}^2|^{q/2} \le \max_{1\le j\le p} n^{-1}\sum_{t=1}^n E|z_{tj}|^q \le \max_{1\le t\le n, 1\le j\le p} E|z_{tj}|^q = O(1).$$

As a result, (A1) follows.

Example 4 Let $\{\varepsilon_t\}$ and $\{\mathbf{z}_t\}$ be independent time series. Assume $\{\mathbf{z}_t\}$ satisfies the same assumptions as in Example 3, and $\varepsilon_t = \sum_{i=0}^{\infty} a_j\delta_{t-j}$, where $\sum_{j=0}^{\infty} a_j^2 < \infty$, $\{\delta_t, \mathcal{F}_t\}$ is a martingale difference sequence obeying $E(\delta_t^2) = \sigma_\delta^2 < \infty$ for all t, and $\sup_{-\infty<t<\infty} E[|\delta_t|^q|\mathcal{F}_{t-1}] < H < \infty$ a.s. for some $q \ge 2$. In addition, assume that the spectral density, $f_\varepsilon(\cdot)$, of $\{\varepsilon_t\}$ follows

$$\sup_{-\pi\le\lambda\le\pi} f_\varepsilon(\lambda) < \infty.$$

Let $\gamma_\varepsilon(k) = E(\varepsilon_t\varepsilon_{t+k})$. By Wei (1987, Lemma 2) and the convexity of $|x|^{q/2}$, there exist positive constants H_q and H_q' dependent only on H and q such that

$$E|n^{-1/2}\sum_{t=1}^n z_{tj}\varepsilon_t|^q \le H_q E|n^{-1}\sum_{1\le k,l\le n} \gamma_\varepsilon(k-l) z_{kj}z_{lj}|^{q/2}$$

$$\le H_q'(\sup_{-\pi\le\lambda\le\pi} f_\varepsilon(\lambda))^{q/2} E|n^{-1}\sum_{t=1}^n z_t^2|^{q/2} \le H_q'(\sup_{-\pi\le\lambda\le\pi} f_\delta(\lambda))^{q/2} \max_{1\le t\le n, 1\le j\le p} E|z_{tj}|^q,$$

and hence (A1) follows.

Example 5 Assume $\{z_{tj}\}$ has a linear representation,

$$z_{tj} = \sum_{l=-\infty}^{\infty} a_l^{(j)} \alpha_{t-l}^{(j)}, \tag{10.4.1}$$

where $\{\alpha_t^{(j)}, \mathcal{F}_t\}$ is a martingale difference sequence, $E[(\alpha_t^{(j)})^2] = \omega_j$, and $\omega_j \sum_{l=-\infty}^{\infty} (a_l^{(j)})^2 = 1$ for all t and $1 \leq j \leq p$. Also assume that $\sup_{-\infty<t<\infty} E[|\alpha_t^{(j)}|^{4q_1}|\mathcal{F}_{t-1}] < L_j < \infty$ a.s. for some $q_1 > q \geq 2$, $\max_{1\leq j\leq p} L_j = O(1)$, and $\max_{1\leq j\leq p} \sum_{k=-\infty}^{\infty} \gamma_j^2(k) = O(1)$, where $\gamma_j(k) = E(z_{tj}z_{t+k,j})$. By the First Moment Bound Theorem of Findley and Wei (1993) and an argument similar to that used in Lemma 2 of Ing and Wei (2003), it can be shown that (A2) is fulfilled by (10.4.1). It is worth mentioning that (10.4.1) includes not only short-memory autoregressive moving average (ARMA) models, but also some long-memory processes; see Findley and Wei (1993) for more details.

We first analyze the "noiseless" OGA that replaces y_t in OGA by its mean $y(\mathbf{x}_t)$. Let $\boldsymbol{\mu} = (y(\mathbf{x}_1), \ldots, y(\mathbf{x}_n))^\top$, $\mathbf{U}^{(0)} = \boldsymbol{\mu}$, $\tilde{j}_1 = \arg\max_{1\leq j\leq p} |(\mathbf{U}^{(0)})^\top \mathbf{X}_j| / \|\mathbf{X}_j\|$ and $\mathbf{U}^{(1)} = (\mathbf{I} - \mathbf{H}_{\{\tilde{j}_1\}})\boldsymbol{\mu}$. Proceeding inductively yields

$$\tilde{j}_m = \arg\max_{1\leq j\leq p} |(\mathbf{U}^{(m-1)})^\top \mathbf{X}_j| / \|\mathbf{X}_j\|, \ \mathbf{U}^{(m)} = (\mathbf{I} - \mathbf{H}_{\{\tilde{j}_1,\ldots,\tilde{j}_m\}})\boldsymbol{\mu}.$$

When the procedure stops after m iterations, the noiseless OGA determines an index set $\tilde{J}_m = \{\tilde{j}_1, \ldots, \tilde{j}_m\}$ and approximates $\boldsymbol{\mu}$ by $\mathbf{H}_{\tilde{J}_m}\boldsymbol{\mu}$. A generalization of noiseless OGA takes $0 < \xi \leq 1$ and replaces $\tilde{j}_i$ by $\tilde{j}_{i,\xi}$, where $\tilde{j}_{i,\xi}$ is any $1 \leq l \leq p$ satisfying

$$|(\mathbf{U}^{(i-1)})^\top \mathbf{X}_l| / \|\mathbf{X}_l\| \geq \xi \max_{1\leq j\leq p} |(\mathbf{U}^{(i-1)})^\top \mathbf{X}_j| / \|\mathbf{X}_j\|. \tag{10.4.2}$$

The next lemma gives a rate of convergence of $n^{-1}\|\mathbf{U}^{(m)}\|^2$.

Lemma 10.4.1 *Assume* (10.1.3) *and* (A2). *Then, there exists a sequence of positive random variables* $\{G_n\}$ *satisfying* $G_n = O_p(1)$ *such that for any* $m \geq 1$,

$$n^{-1}\|(\mathbf{I} - \mathbf{H}_{\tilde{J}_{m,\xi}})\boldsymbol{\mu}\|^2 \leq \frac{G_n}{1 + m\xi^2}, \tag{10.4.3}$$

where $\tilde{J}_{m,\xi} = \{\tilde{j}_{1,\xi}, \ldots, \tilde{j}_{m,\xi}\}$.

Proof Define $\hat{\rho}_{jj} = n^{-1}\sum_{t=1}^n z_{tj}^2$ and $\nu_{J,i} = (\mathbf{X}_i)^\top(\mathbf{I} - \mathbf{H}_J)\boldsymbol{\mu}/(n^{1/2}\|\mathbf{X}_i\|)$ for $J \subseteq \{1, \ldots, p\}$ and $i \in \{1, \ldots, p\}$. Then for $m \geq 1$,

$$\begin{aligned}
&n^{-1}\|(\mathbf{I} - \mathbf{H}_{\tilde{J}_{m,\xi}})\boldsymbol{\mu}\|^2 \\
&\leq n^{-1}\|(\mathbf{I} - \mathbf{H}_{\tilde{J}_{m-1,\xi}})\boldsymbol{\mu} - \frac{\boldsymbol{\mu}^\top(\mathbf{I} - \mathbf{H}_{\tilde{J}_{m-1,\xi}})\mathbf{X}_{\tilde{j}_{m,\xi}}}{\|\mathbf{X}_{\tilde{j}_{m,\xi}}\|^2}\mathbf{X}_{\tilde{j}_{m,\xi}}\|^2 \\
&\leq n^{-1}\|(\mathbf{I} - \mathbf{H}_{\tilde{J}_{m-1,\xi}})\boldsymbol{\mu}\|^2 - \nu^2_{\tilde{J}_{m-1,\xi},\tilde{j}_{m,\xi}} \leq n^{-1}\|(\mathbf{I} - \mathbf{H}_{\tilde{J}_{m-1,\xi}})\boldsymbol{\mu}\|^2 - \xi^2 \max_{1\leq j\leq p} \nu^2_{\tilde{J}_{m-1,\xi},j},
\end{aligned} \tag{10.4.4}$$

in which $\mathbf{H}_{\tilde{J}_{0,\xi}} = \mathbf{0}$. Moreover, we have

$$n^{-1}\|(\mathbf{I}-\mathbf{H}_{\tilde{J}_{m-1,\xi}})\boldsymbol{\mu}\|^2 = \sum_{j=1}^{p} \beta_j\sigma_j\hat{\rho}_{jj}^{1/2}\nu_{\tilde{J}_{m-1,\xi},j} \leq (\max_{1\leq j\leq p}|\nu_{\tilde{J}_{m-1,\xi},j}|)\sum_{j=1}^{p}|\beta_j\sigma_j\hat{\rho}_{jj}^{1/2}|. \tag{10.4.5}$$

Let $G_n = (\sum_{j=1}^{p}|\beta_j\sigma_j\hat{\rho}_{jj}^{1/2}|)^2$. It follows from (10.4.4) and (10.4.5) that

$$n^{-1}\|(\mathbf{I}-\mathbf{H}_{\tilde{J}_{m,\xi}})\boldsymbol{\mu}\|^2 \leq n^{-1}\|(\mathbf{I}-\mathbf{H}_{\tilde{J}_{m-1,\xi}})\boldsymbol{\mu}\|^2\{1-\xi^2 n^{-1}\|(\mathbf{I}-\mathbf{H}_{\tilde{J}_{m-1,\xi}})\boldsymbol{\mu}\|^2/G_n\}. \tag{10.4.6}$$

Since $n^{-1}\|(\mathbf{I}-\mathbf{H}_{\tilde{J}_{0,\xi}})\boldsymbol{\mu}\|^2 = n^{-1}\|\boldsymbol{\mu}\|^2 \leq G_n$, (10.4.6) and Lemma 10.3.1 of Temlyakov (2000) yield (10.4.3). It remains to show that $G_n = O_p(1)$, which follows immediately from

$$E(G_n) \leq \left\{\sum_{j=1}^{p}|\beta_j\sigma_j|E^{1/2}(\hat{\rho}_{jj})\right\}^2 \leq \max_{1\leq j\leq p}E(\hat{\rho}_{jj})\left(\sum_{j=1}^{p}|\beta_j\sigma_j|\right)^2 = O(1)$$

noting that the first relationship is ensured by Minkowski's inequality, whereas the last one is guaranteed by (A2) and (10.1.3). □

Define the sample counterpart $\hat{\mu}_{J,i}$ of $\nu_{J,i}$, where $\hat{\mu}_{J,i} = (\mathbf{X}_i)^\top(\mathbf{I}-\mathbf{H}_J)\mathbf{Y}/(n^{1/2}\|\mathbf{X}_i\|)$. Lemma 10.4.2 provides a uniform bound for the difference between $\hat{\mu}_{J,i}$ and $\nu_{J,i}$ over $1 \leq i \leq p$ and $J \subset \{1,\ldots,p\}$ with $i \notin J$ and $\sharp(J) \leq K_n = O(n^{1/2}/p^{1/q})$.

Lemma 10.4.2 *Assume* (A1)–A(3) *and* (1.7). *Suppose* $K_n = O(n^{1/2}/p_n^{1/q})$. *Then*

$$\max_{(J,i):\sharp(J)\leq K_n, i\notin J}|\hat{\mu}_{J,i}-\nu_{J,i}| = O_p\left(\frac{p_n^{1/q}}{n^{1/2}}\right) = o_p(1). \tag{10.4.7}$$

Proof Note first that for any $J \subseteq \{1,\ldots,p_n\}$, $1 \leq i \leq p_n$ and $i \notin J$,

$$|\hat{\mu}_{J,i}-\nu_{J,i}| = \left(n^{-1}\sum_{t=1}^{n}z_{ti}^2\right)^{-1/2} n^{-1}(\mathbf{Z}_i'(\mathbf{I}-\mathbf{H}_J)\boldsymbol{\varepsilon}),$$

where $\mathbf{Z}_i = (z_{1i},\ldots,z_{ni})^\top$ and $\boldsymbol{\varepsilon} = (\varepsilon_1,\ldots,\varepsilon_n)^\top$. The desired result (10.4.7) follows immediately from

$$\max_{1\leq i\leq p_n}\left(n^{-1}\sum_{t=1}^{n}z_{ti}^2\right)^{-1} = O_p(1), \tag{10.4.8}$$

$$\max_{1\leq i\leq p} n^{-1}|\mathbf{Z}_i'\boldsymbol{\varepsilon}| = O_p\left(\frac{p_n^{1/q}}{n^{1/2}}\right), \tag{10.4.9}$$

and

$$\max_{(J,i):\sharp(J)\le K_n, i\notin J} n^{-1}|\mathbf{Z}_i'\mathbf{H}_J\boldsymbol{\varepsilon}| = O_p\left(\frac{p_n^{1/q}}{n^{1/2}}\right). \tag{10.4.10}$$

By (A2), it can be shown that $\max_{1\le i\le p_n}|n^{-1}\sum_{t=1}^n(z_{ti}^2-1)| = O(p_n^{1/2q_1}/n^{1/2}) = o_p(1)$, which leads to (10.4.8). Similarly, (10.4.9) is ensured by (A1). To show (10.4.10), we have

$$\begin{aligned}
&\max_{(J,i):\sharp(J)\le K_n, i\notin J} n^{-1}|\mathbf{Z}_i'\mathbf{H}_J\boldsymbol{\varepsilon}| \\
&\le \max_{1\le\sharp(J)\le K_n, i\notin J}\left|\left(n^{-1}\sum_{t=1}^n z_{ti:J}^{\perp}\mathbf{z}_t(J)\right)^{\top}\hat{\boldsymbol{\Gamma}}^{-1}(J)\left(n^{-1}\sum_{t=1}^n \varepsilon_t\mathbf{z}_t(J)\right)\right| \\
&\quad + \max_{1\le\sharp(J)\le K_n, i\notin J}\left|\mathbf{g}_i^{\top}(J)\boldsymbol{\Gamma}^{-1}(J)\left(n^{-1}\sum_{t=1}^n \varepsilon_t\mathbf{z}_t(J)\right)\right| := Q_{1,n} + Q_{2,n},
\end{aligned} \tag{10.4.11}$$

where $z_{ti:J}^{\perp} = z_{ti} - \mathbf{g}_i^{\top}(J)\boldsymbol{\Gamma}^{-1}(J)\mathbf{z}_t(J)$, $\mathbf{z}_t(J)$ denotes a subvector of $\mathbf{z}_t = (z_{t1},\ldots,z_{tp})^{\top}$, with J being the corresponding subset of indices, and $\hat{\boldsymbol{\Gamma}}(J) = n^{-1}\sum_{t=1}^n \mathbf{z}_t(J)\mathbf{z}_t'(J)$. By (10.2.3) and (10.4.9), it holds that

$$Q_{2,n} \le \max_{1\le i\le p_n}\left|n^{-1}\sum_{t=1}^n \varepsilon_t x_{ti}\right| \max_{1\le\sharp(J)\le K_n-1, i\notin J}\|\boldsymbol{\Gamma}^{-1}(J)\mathbf{g}_i(J)\|_1 = O_p\left(\frac{p_n^{1/q}}{n^{1/2}}\right). \tag{10.4.12}$$

Moreover, since $\max_{1\le\sharp(J)\le K_n, i\notin J}\|n^{-1}\sum_{t=1}^n \varepsilon_t\mathbf{z}_t(J)\| \le K_n^{1/2}\max_{1\le i\le p_n}|n^{-1}\sum_{t=1}^n \varepsilon_t z_{ti}|$ and

$$\begin{aligned}
&\max_{1\le\sharp(J)\le K_n, i\notin J}\left\|n^{-1}\sum_{t=1}^n z_{ti;J}^{\perp}\mathbf{z}_t(J)\right\| \\
&\le K_n^{1/2}\max_{1\le i,j\le p_n}\left|n^{-1}\sum_{t=1}^n z_{ti}z_{tj} - \rho_{ij}\right|(1 + \max_{1\le\sharp(J)\le K_n, i\notin J}\|\boldsymbol{\Gamma}^{-1}(J)\mathbf{g}_i(J)\|_1),
\end{aligned}$$

it follows from (10.2.3) that for all large n,

$$\begin{aligned}
Q_{1,n} \le & (\max_{1\le\sharp(J)\le K_n}\|\hat{\boldsymbol{\Gamma}}^{-1}(J)\|)K_n(1+M) \\
&\times\left(\max_{1\le i,j\le p_n}|n^{-1}\sum_{t=1}^n z_{ti}z_{tj} - \rho_{ij}|\right)\left(\max_{1\le i\le p_n}|n^{-1}\sum_{t=1}^n \varepsilon_t z_{ti}|\right).
\end{aligned} \tag{10.4.13}$$

By (A2),

$$\max_{1\le i,j\le p_n} |n^{-1}\sum_{t=1}^{n} z_{ti}z_{tj} - \rho_{ij}| = O_p\left(\frac{p_n^{1/q_1}}{n^{1/2}}\right) = o_p\left(\frac{p_n^{1/q}}{n^{1/2}}\right), \tag{10.4.14}$$

which, together with (10.2.3), yields

$$\begin{aligned}
&\max_{1\le\sharp(J)\le K_n} \|\hat{\mathbf{\Gamma}}^{-1}(J) - \mathbf{\Gamma}^{-1}(J)\| \\
&\le \delta^{-1}\left(\max_{1\le\sharp(J)\le K_n} \|\hat{\mathbf{\Gamma}}^{-1}(J) - \mathbf{\Gamma}^{-1}(J)\| + \delta^{-1}\right) \max_{1\le\sharp(J)\le K_n} \|\hat{\mathbf{\Gamma}}(J) - \mathbf{\Gamma}(J)\| \\
&\le \delta^{-1}\left(\max_{1\le\sharp(J)\le K_n} \|\hat{\mathbf{\Gamma}}^{-1}(J) - \mathbf{\Gamma}^{-1}(J)\| + \delta^{-1}\right) K_n \max_{1\le i,j\le p_n} |n^{-1}\sum_{t=1}^{n} z_{ti}z_{tj} - \rho_{ij}| \\
&= o_p\left(\max_{1\le\sharp(J)\le K_n} \|\hat{\mathbf{\Gamma}}^{-1}(J) - \mathbf{\Gamma}^{-1}(J)\|\right) + o_p(1),
\end{aligned}$$

and hence

$$\max_{1\le\sharp(J)\le K_n} \|\hat{\mathbf{\Gamma}}^{-1}(J)\| \le \max_{1\le\sharp(J)\le K_n} \|\hat{\mathbf{\Gamma}}^{-1}(J) - \mathbf{\Gamma}^{-1}(J)\| + \delta^{-1} = O_p(1). \tag{10.4.15}$$

In view of (10.4.9), (10.4.13)–(10.4.15), and the restriction on K_n, one obtains $Q_{1,n} = o_p(p_n^{1/q}/n^{1/2})$. Combining this with (10.4.11) and (10.4.12) gives (10.4.10). Thus the proof is complete. □

The main result of this section is given as follows.

Theorem 10.4.3 *Assume* (10.1.3), (10.2.3), *and* (A1)–(A3). *Suppose* $K_n = O(n^{1/2}/p_n^{1/q})$. *Then*

$$\max_{1\le m\le K_n} \frac{\|\hat{y}_m(\cdot) - y(\cdot)\|_n^2}{m^{-1} + \frac{mp_n^{2/q}}{n}} = O_p(1). \tag{10.4.16}$$

Proof By Lemma 10.4.2, for arbitrarily small $\iota > 0$, there exists a positive number S^* such that $P(A_{K_n}^c) < \iota$, where

$$A_m = \left\{\max_{(J,i):\sharp(J)\le m-1, i\notin J} |\hat{\mu}_{J,i} - \nu_{J,i}| \le S^* p_n^{1/q}/n^{1/2}\right\}.$$

Let $0 < \xi < 1$ and $\tilde{\xi} = 2/(1-\xi)$. Define

$$B_m = \left\{ \min_{0 \le i \le m-1} \max_{1 \le j \le p_n} |\nu_{\hat{J}_i, j}| > \tilde{\xi} S^* p_n^{1/q} / n^{1/2} \right\}.$$

Then on the set $A_m \bigcap B_m$, we have for all $1 \le q \le m$,

$$\begin{aligned}
|\nu_{\hat{J}_{q-1}\hat{j}_q}| &\ge -|\hat{\mu}_{\hat{J}_{q-1}\hat{j}_q} - \nu_{\hat{J}_{q-1}\hat{j}_q}| + |\hat{\mu}_{\hat{J}_{q-1}\hat{j}_q}| \\
&\ge - \max_{(J,i):\sharp(J) \le m-1, i \notin J} |\hat{\mu}_{J,i} - \nu_{J,i}| + |\hat{\mu}_{\hat{J}_{q-1}\hat{j}_q}| \\
&\ge -S^* p_n^{1/q}/n^{1/2} + \max_{1 \le j \le p_n} |\hat{\mu}_{\hat{J}_{q-1}j}| \\
&\ge -2S^* p_n^{1/q}/n^{1/2} + \max_{1 \le j \le p_n} |\nu_{\hat{J}_{q-1}j}| \ge \xi \max_{1 \le j \le p_n} |\nu_{\hat{J}_{q-1}j}|,
\end{aligned}$$

implying that on the set $A_m \bigcap B_m$, $\hat{J}_q, 1 \le q \le m$ obeys (10.4.2). Therefore, it follows from Lemma 10.4.1 that

$$n^{-1} \|(\mathbf{I} - \mathbf{H}_{\hat{J}_m})\boldsymbol{\mu}\|^2 I_{A_m \bigcap B_m} \le \frac{G_n}{1 + m\xi^2}, \tag{10.4.17}$$

where G_n is defined in the proof of Lemma 10.4.1. Moreover, for $0 \le i \le m-1$, $\|(\mathbf{I} - \mathbf{H}_{\hat{J}_m})\boldsymbol{\mu}\|^2 \le \|(\mathbf{I} - \mathbf{H}_{\hat{J}_i})\boldsymbol{\mu}\|^2$, and hence

$$\begin{aligned}
n^{-1} \|(\mathbf{I} - \mathbf{H}_{\hat{J}_m})\boldsymbol{\mu}\|^2 &\le \min_{0 \le i \le m-1} n^{-1} \sum_{j=1}^{p_n} \beta_j \mathbf{X}_j^\top (\mathbf{I} - \mathbf{H}_{\hat{J}_i})\boldsymbol{\mu} \\
&\le \left(\min_{0 \le i \le m-1} \max_{1 \le j \le p_n} |\nu_{\hat{J}_i, j}| \right) G_n^{1/2} \le \tilde{\xi} S^* p_n^{1/q} G_n^{1/2} / n^{1/2} \text{ on } B_m^c.
\end{aligned} \tag{10.4.18}$$

Since A_m decreases as m increases and $m \le K_n = O(n^{1/2}/p_n^{1/q})$, it follows from (10.4.17) and (10.4.18) that there exists a positive constant C^* such that

$$\max_{1 \le m \le K_n} m n^{-1} \|(\mathbf{I} - \mathbf{H}_{\hat{J}_m})\boldsymbol{\mu}\|^2 \le C^*(G_n + G_n^{1/2}) \text{ on } A_{K_n}. \tag{10.4.19}$$

By noticing that $G_n = O_p(1)$ and the probability ι of $A_{k_n}^c$ can be arbitrarily small, (10.4.19) further yields

$$\max_{1 \le m \le K_n} m n^{-1} \|(\mathbf{I} - \mathbf{H}_{\hat{J}_m})\boldsymbol{\mu}\|^2 = O_p(1). \tag{10.4.20}$$

Utilizing (10.4.9) and (10.4.15), one obtains

$$\max_{1\le m\le K_n} \frac{\boldsymbol{\varepsilon}^\top \mathbf{H}_{\hat{J}_m}\boldsymbol{\varepsilon}}{m p_n^{2/q}/n} \le \max_{1\le \sharp(J)\le K_n} \|\hat{\boldsymbol{\Gamma}}^{-1}(J)\| \max_{1\le j\le p_n} (n^{-1/2}\sum_{t=1}^{n} z_{tj}\varepsilon_t)^2 p_n^{-2/q} = O_p(1). \tag{10.4.21}$$

Now the desired conclusion (10.4.16) follows from (10.4.20), (10.4.21) and $\|\hat{y}_m(\cdot) - y(\cdot)\|_n^2 = n^{-1}(\|(\mathbf{I}-\mathbf{H}_{\hat{J}_m})\boldsymbol{\mu}\|^2 + \boldsymbol{\varepsilon}^\top \mathbf{H}_{\hat{J}_m}\boldsymbol{\varepsilon})$. □

Some comments on Theorem 10.4.3 are in order.

1. Instead of the sub-exponential and sub-Gaussian conditions described in (10.2.1) and (10.2.2), we assume (A1) and (A2), which substantially broaden the applicability of Theorem 10.4.3. Moreover, Examples 3–5 reveal that (A1) and (A2) hold not only for high-dimensional regression models with time series errors in which $\{\mathbf{x}_t\}$ and $\{\varepsilon_t\}$ are independent, but also for high-dimensional autoregressive exogenous (ARX) models in which ε_t and $\mathbf{x}_j, j > t$, are correlated.
2. It is clear from (10.4.21) that the variance inflation factor $p_n^{2/q}$ in the denominator of (10.4.16) is contributed by the order of magnitude of $\max_{1\le j\le p_n}(n^{-1/2}\sum_{t=1}^{n} z_{tj}\varepsilon_t)^2$, and is derived under condition (A1) that $\max_{1\le j\le p_n} E|n^{-1/2}\sum_{t=1}^{n} z_{tj}\varepsilon_t|^q = O(1)$. Let Z_i be independent random variables following $E|Z_i|^q = 1$ for all i, Ing and Lai (2016) have recently constructed an example showing that for any small $\varsigma > 0$, $P(p_n^{-1/q}\max_{1\le i\le p_n}|Z_i| < \varsigma) = o(1)$, and hence the divergence rate of $\max_{1\le i\le p_n} Z_i^2$ cannot be slower than $p_n^{2/q}$. Their example suggests that the variance inflation factor $p_n^{2/q}$ seems difficult to improve. By (10.4.16), OGA's empirical norm has the optimal rate $p_n^{1/q}/n^{1/2}$, which is achieved by choosing $m \asymp n^{1/2}/p_n^{1/q}$.

We close this section by mentioning some recent developments of lasso in high-dimensional time series models. Basu and Michailidis (2015) have derived an upper bound for the empirical norm of the lasso estimate under (10.1.1) with $\{\mathbf{x}_t\}$ and $\{\varepsilon_t\}$ being independent and stationary Gaussian processes. They show that when $r \ge \underline{c}(\log p_n/n)^{1/2}$ for some $\underline{c} > 0$,

$$n^{-1}\sum_{t=1}^{n}(\hat{y}_{\text{lasso}(r)}(\mathbf{x}_t) - y(\mathbf{x}_t))^2 = O_p(k\log p_n/n), \tag{10.4.22}$$

provided (10.3.2) holds true and the regression coefficients are k-sparse, namely $\sharp(J_n^*) = k$, with k satisfying

$$k\{\max_{\sharp(J)=k} \lambda_{\max}(\boldsymbol{\Gamma}(J))\}^2 \lesssim n/\log p_n. \tag{10.4.23}$$

Let $\{a_n\}$ and $\{b_n\}$ be sequences of positive numbers. We say $a_n \lesssim b_n$ if there exists positive constant M such that $a_n \le M b_n$. The bound on the right-hand side of (10.4.22) seems satisfactory because it is the same as the one obtained in the fixed design model with i.i.d. Gaussian error; see (2.20). However, while (10.3.2) is not an uncommon assumption for time series models, it may seem restrictive compared to (10.2.3) or the restricted eigenvalue assumption. Assumption (10.4.23) may also lead to a stringent limitation on k. For example, (10.4.23) becomes $k \lesssim (n/\log p_n)^{1/3}$ in the case of $0 < E(z_{ti}z_{tj}) = \rho < 1$ for all $1 \leq i \neq j \leq p_n$, which is a benchmark case in high-dimensional data analysis; see also Sect. 10.3.2. In addition, non-Gaussian time series or ARX models are precluded by the assumption that $\{\mathbf{x}_t\}$ and $\{\varepsilon_t\}$ are independent Gaussian processes. Alternatively, Wu and Wu (2016) have considered lasso estimation with a fixed design matrix. They assume that an RE condition is satisfied and $\{\varepsilon_t\}$ is a stationary time series having a finite q-th moment. When the model is k-sparse, they derive a sharp upper bound for the empirical norm of lasso. However, it seems difficult to extend their results to high-dimensional ARX models.

References

Basu S, Michailidis G (2015) Regularized estimation in sparse high-dimensional time series models. Ann Stat 43:1535–1567

Bickel PJ, Levina E (2008) Regularized estimation of large covariance matrices. Ann Stat 36: 199–227

Bickel PJ, Ritov Y, Tsybakov AB (2009) Simultaneous analysis of Lasso and Dantzig selector. Ann Stat 37:1705–1732

Bühlmann P (2006) Boosting for high-dimensional linear models. Ann Stat 34:559–583

Bunea F, Tsybakov AB, Wegkamp MH (2007) Sparsity oracle inequalities for the Lasso. Electr J Stat 1:169–194

Candés EJ, Plan Y (2009) Near-ideal model selection by ℓ_1 minimization. Ann Stat 37:2145–2177

Candés EJ, Tao T (2007) The Dantzig selector: statistical estimation when p is much larger than n. Ann Stat 35:2313–2351

Cai T, Zhang C-H, Zhou HH (2010) Optimal rates of convergence for covariance matrix estimation. Ann Stat 38:2118–2144

Donoho DL, Elad M, Temlyakov VN (2006) Stable recovery of sparse overcomplete representations in the presence of noise. IEEE Trans Inform Theory 52:6–18

Fan J, Li R (2001) Variable selection via nonconcave penalized likelihood and its oracle properties. J Am Stat Assoc 96:1348–1360

Findley DF, Wei C-Z (1993) Moment bounds for deriving time series CLT's and model selection procedures. Stat Sin 3:453–470

Foster DP, George EI (1994) The risk inflation criterion for multiple regression. Ann Stat 22:1947–1975

Friedman J, Hastie T, Tibshirani R (2010) glmnet: Lasso and elastic-net regularized generalized linear models. R package version 1.1-5. http://cran.r-project.org/web/packages/glmnet/index.html. Accessed 10 Dec 2012

Gao F, Ing C-K, Yang Y (2013) Metric entropy and sparse linear approximation of l_q-Hulls for $0 < q \leq 1$. J Approx Theory 166:42–55

Ing C-K, Lai TL (2011) A stepwise regression method and consistent model selection for high-dimensional sparse linear models. Stat Sin 21:1473–1513

Ing C-K, Lai TL (2015) An efficient pathwise variable selection criterion in weakly sparse regression models. Technical Report, Academia Sinica

Ing C-K, Lai TL (2016) Model selection for high-dimensional time series. Technical Report, Academia Sinica

Ing C-K, Wei C-Z (2003) On same-realization prediction in an infinite-order autoregressive process. J Multivar Anal 85:130–155

Negahban SN, Ravikumar P, Wainwright MJ, Yu B (2012) A unified framework for high-dimensional analysis of m-estimators with decomposable regularizers. Stat Sci 27:538–557

Raskutti G, Wainwright MJ, Yu B (2011) Minimax rates of estimation for high-dimensional linear regression over l_q-balls. IEEE Trans Inform Theory 57:6976–6994

Temlyakov VN (2000) Weak greedy algorithms. Adv Comput Math 12:213–227

Tibshirani R (1996) Regression shrinkage and selection via the Lasso. J R Stat Soc Ser B 58:267–288

Tropp JA (2004) Greed is good: algorithmic results for sparse approximation. IEEE Trans Inform Theory 50:2231–2242

Wang Z, Paterlini S, Gao F, Yang Y (2014) Adaptive minimax regression estimation over sparse hulls. J Mach Learn Res 15:1675–1711

Wei C-Z (1987) Adaptive prediction by least squares predictors in stochastic regression models with applications to time series. Ann Stat 15:1667–1682

Wu WB, Wu YN (2016) Performance bounds for parameter estimates of high-dimensional linear models with correlated errors. Electron J Stat 10:352–379

Zhang C-H, Huang J (2008) The sparsity and bias of the Lasso selection in highdimensional linear regression. Ann Stat 36:1567–1594

Zhao P, Yu B (2006) On model selection consistency of Lasso. J Mach Learn Res 7:2541–2563

Zou H (2006) The adaptive Lasso and its oracle properties. J Am Stat Assoc 101:1418–1429

Chapter 11
Semi-supervised Smoothing for Large Data Problems

Mark Vere Culp, Kenneth Joseph Ryan, and George Michailidis

Abstract This book chapter is a description of some recent developments in non-parametric semi-supervised regression and is intended for someone with a background in statistics, computer science, or data sciences who is familiar with local kernel smoothing (Hastie et al., The elements of statistical learning (data mining, inference and prediction), chapter 6. Springer, Berlin, 2009). In many applications, response data often require substantially more effort to obtain than feature data. Semi-supervised learning approaches are designed to explicitly train a classifier or regressor using all the available responses and the full feature data. This presentation is focused on local kernel regression methods in semi-supervised learning and provides a good starting point for understanding semi-supervised methods in general.

Keywords Computational statistics · Machine learning · Non-parametric regression

11.1 Introduction

This book chapter is a description of recent developments in non-parametric semi-supervised regression and is intended for someone with a background in statistics, computer science, or data sciences who is familiar with local kernel smoothing (Hastie et al. 2009, Chapter 6). In many applications, response data often require

---

M. V. Culp (✉) · K. J. Ryan
West Virginia University, Department of Statistics, Morgantown, WV, USA
e-mail: mvculp@mail.wvu.edu; kjryan@mail.wvu.edu

G. Michailidis
University of Florida, Department of Statistics, Gainesville, FL, USA
e-mail: gmichail@umich.edu

© Springer International Publishing AG, part of Springer Nature 2018

W. K. Härdle et al. (eds.), *Handbook of Big Data Analytics*, Springer Handbooks of Computational Statistics, https://doi.org/10.1007/978-3-319-18284-1_11

substantially more effort to obtain than feature data. For example, in some non-clinical pharmaceutical applications, the feature data are often measurements on compounds while the response is a hard to determine attribute such as whether or not the drug will have a particular side-effect (Lundblad 2004). The response information is expensive to obtain, and thus a semi-supervised approach that uses all the available data may be preferred. Other applications are in text analysis (McCallum et al. 2000), spam in email detection problems (Koprinska et al. 2007), chemogenomics (Bredel and Jacoby 2004), and proteomics (Yamanishi et al. 2004).

A common theme in these big data applications is that large feature data sets are available with only some observed responses. Handling such data mart problems has been of recent interest in machine learning, and semi-supervised learning addresses this specific class of problems. This book chapter provides a starting point for grasping some of the popular semi-supervised approaches designed to address these challenging and practical problems.

Semi-supervised learning approaches are designed to explicitly train a classifier or regressor using all the available responses and the full feature data (Chapelle et al. 2006; Zhu 2008; Abney 2008). This chapter is organized as follows. First, supervised kernel regression is extended into semi-supervised learning. Then some properties of this estimator are provided. Several, other well-known semi-supervised approaches can be viewed as generalizations of semi-supervised kernel regression. This chapter closes with a brief summary of these types of interesting extensions.

11.2 Semi-supervised Local Kernel Regression

In semi-supervised learning, one starts with both partially observed responses and a complete feature data set. Formally, assume that $(y_i, r_i, \boldsymbol{x}_i)$ are instances from the joint distribution of random variables $\boldsymbol{Y}$, $\boldsymbol{R}$, and random vector $\boldsymbol{X}$. The r_i denote the indicator of response availability. The responses are assumed to be Missing Completely at Random (MCAR). The MCAR assumption means that $\boldsymbol{R} \perp\!\!\!\perp \boldsymbol{Y} \mid \boldsymbol{X}$ and that $\boldsymbol{R}|\boldsymbol{X}$ is Bernoulli distributed with success probability $p(\boldsymbol{X}) = p$. This assumption underpins nearly every semi-supervised approach (Lafferty and Wasserman 2007).

To begin, assume that n observations $(y_i, r_i, \boldsymbol{x}_i)$ are obtained. In supervised learning, all training is done with responses $r_i y_i$ and data $r_i \boldsymbol{x}_i$. This is commonly referred to as the labeled data which is indexed by set $L = \{i|r_i = 1\}$. The focus of this work is to obtain a prediction for the point $\boldsymbol{x}_0$. In machine learning, a distinction is made between prediction of an arbitrary $\boldsymbol{x}_0$ versus the so-called transductive prediction problem of predicting $(1-r_i)\boldsymbol{x}_i$. Observations corresponding to the transductive prediction problem are referred to as unlabeled with index set $U = \{i|r_i = 0\}$. The latent response vector for the unlabeled data is denoted by $\boldsymbol{Y}_U$ and is unavailable for training. For finite samples with sets L and U fixed and for simplicity, we assume that the first m observations are labeled and the remaining $n-m$ are unlabeled; the data are then partitioned as:

$$Y(Y_U) = \begin{pmatrix} Y_L \\ Y_U \end{pmatrix} \quad X = \begin{pmatrix} X_L \\ X_U \end{pmatrix}, \tag{11.1}$$

where Y_U is arbitrary in $\mathbb{R}^{|U|}$.

11.2.1 Supervised Kernel Regression

The local kernel smoother operates on a kernel matrix involving distances between observations x_i, x_j, e.g. $D_{ij} = ||x_i^T - x_j^T||_2$ (Euclidean) or $D_{ij} = ||x_i^T - x_j^T||_1^1$ (Manhattan). Using the distance function, the local kernel function is denoted as: $K_h(x_i, x_0) = K(D_{i0}/h)$ with $x_i^T, x_0^T \in \mathbb{R}^p$, e.g., the Gaussian kernel is most common with $K_h(x_i, x_0) = \exp\left(-||x_i^T - x_0^T||_2^2/(2h)\right)$. The generic notation using an arbitrary response y of length k and data X of dimensions $k \times p$ for prediction of observation x_0 is given by

$$\widehat{m}_y(x_0) = \frac{\sum_{i=1}^k K_h(x_i, x_0) y_i}{\sum_{i=1}^k K_h(x_i, x_0)}. \tag{11.2}$$

In supervised local kernel regression this estimator is the local average

$$\widehat{m}_{Y_L}(x_0) = \frac{\sum_{i \in L} K_h(x_i, x_0) y_i}{\sum_{i \in L} K_h(x_i, x_0)} = \frac{\sum_{i=1}^n K_h(x_i, x_0) r_i y_i}{\sum_{i=1}^n K_h(x_i, x_0) r_i}. \tag{11.3}$$

The parameter h is to be estimated by cross-validation using the labeled data. The supervised local kernel estimator has a long history in statistics and is fairly deeply understood (Hastie et al. 2009, Chapter 6). The purpose of this book chapter is to articulate how this local smoothing concept is used in the semi-supervised setting.

Predictions (11.3) applied to x_i with $i \in L \cup U$ are of particular note. The kernel Gram matrix $W_{ij} = K\left(D_{ij}/h\right)$ partitions

$$W = \begin{pmatrix} W_{LL} & W_{LU} \\ W_{UL} & W_{UU} \end{pmatrix}, \tag{11.4}$$

where W_{LL} corresponds to the similarities between labeled observations, $W_{UL} = W_{LU}^T$ corresponds to similarities between labeled and unlabeled observations, and W_{UU} corresponds to similarities between unlabeled observations. In vector form the predictions from local kernel smoother (11.3) applied to observations in $L \cup U$ are given by

$$\widetilde{\mathbf{m}} = \begin{pmatrix} \widetilde{\mathbf{m}}_L \\ \widetilde{\mathbf{m}}_U \end{pmatrix} = \begin{pmatrix} D_{LL}^{-1} W_{LL} Y_L \\ D_{UL}^{-1} W_{UL} Y_L \end{pmatrix} = \begin{pmatrix} T_{LL} Y_L \\ T_{UL} Y_L \end{pmatrix}, \tag{11.5}$$

where $\boldsymbol{D}_{AB}$ is the diagonal row sum matrix of the $\boldsymbol{W}_{AB}$ partition of $\boldsymbol{W}$ for index sets A, B. The matrices $\boldsymbol{T}_{LL}, \boldsymbol{T}_{UL}$ are right stochastic kernel matrices (i.e., $\boldsymbol{T}_{LL}\mathbf{1} = \mathbf{1}$ and $(\boldsymbol{T}_{LL})_{ij} \geq 0$).

The inverse of matrix $\boldsymbol{D}_{UL}$ is necessary to obtain unique predictions (11.5). This inverse is non-unique whenever unlabeled cases have zero weight to all labeled cases. In a sense, the assumption of the supervised kernel estimator is that each unlabeled case has a non-zero adjacency to a labeled case. This is a very strong assumption and as we will see the semi-supervised estimators use the unlabeled connectivity in $\boldsymbol{W}_{UU}$ to generalize this restriction. The supervised prediction border for the two moons data set is provided in Fig. 11.1.

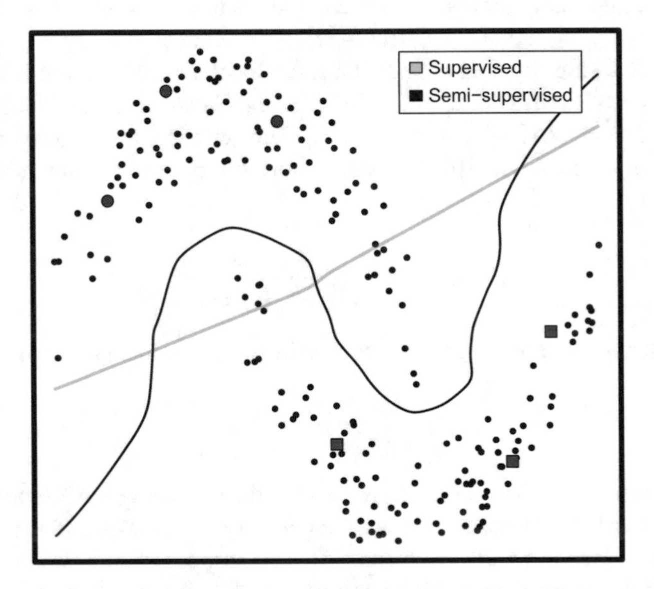

Fig. 11.1 The two moons cluster example. The supervised local kernel smoother (gray) classifies in the logical way when only using the labeled observations (red for class one, and blue for class two). The unlabeled data presents two structured moons. The semi-supervised local kernel estimator passes between these moons. Although this example justifies the use of semi-supervised estimation, it may be misleading in the sense that real data do not often conform to perfect classification. The preponderance of empirical evidence along with much practical experience suggests that semi-supervised local kernel regression will typically outperform supervised local kernel when the labeled data are small relative to the size of the unlabeled data. It is not fully understood why these estimators work so well in practice given that this type of perfect clustering does not usually occur

Example 1 Suppose there are six observations with four observed responses. In this case, $L = \{1, 2, 3, 4\}$ and $U = \{5, 6\}$. Also, suppose the observed responses and feature data are

$$\boldsymbol{Y}_L = \begin{pmatrix} 1 \\ 1 \\ -1 \\ -1 \end{pmatrix} \text{ and } \boldsymbol{X} = \begin{pmatrix} 0.2 \\ 0.9 \\ 1.8 \\ 4.3 \\ 6.1 \\ 7.8 \end{pmatrix}.$$

A Gaussian kernel is used with $h = 0.9$. To aid in presentation the output of the kernel function was rounded to one decimal point. The weight matrix is given by

$$\boldsymbol{W} = \left(\begin{array}{cccc|cc} 1.0 & 0.8 & 0.2 & 0.0 & 0.0 & 0.0 \\ 0.8 & 1.0 & 0.6 & 0.0 & 0.0 & 0.0 \\ 0.2 & 0.6 & 1.0 & 0.0 & 0.0 & 0.0 \\ 0.0 & 0.0 & 0.0 & 1.0 & 0.2 & 0.0 \\ \hline 0.0 & 0.0 & 0.0 & 0.2 & 1.0 & 0.2 \\ 0.0 & 0.0 & 0.0 & 0.0 & 0.2 & 1.0 \end{array}\right).$$

From this,

$$\boldsymbol{W}_{LL} = \begin{pmatrix} 1.0 & 0.8 & 0.2 & 0.0 \\ 0.8 & 1.0 & 0.6 & 0.0 \\ 0.2 & 0.6 & 1.0 & 0.0 \\ 0.0 & 0.0 & 0.0 & 1.0 \end{pmatrix}, \boldsymbol{W}_{LU} = \boldsymbol{W}_{UL}^T = \begin{pmatrix} 0.0 & 0.0 \\ 0.0 & 0.0 \\ 0.0 & 0.0 \\ 0.2 & 0.0 \end{pmatrix}, \text{ and } \boldsymbol{W}_{UU} = \begin{pmatrix} 1.0 & 0.2 \\ 0.2 & 1.0 \end{pmatrix}.$$

The labeled 4×4 diagonal row-sum matrices are given by

$$\boldsymbol{D}_{LL} = \begin{pmatrix} 2.0 & 0.0 & 0.0 & 0.0 \\ 0.0 & 2.4 & 0.0 & 0.0 \\ 0.0 & 0.0 & 1.8 & 0.0 \\ 0.0 & 0.0 & 0.0 & 1.0 \end{pmatrix}, \boldsymbol{D}_{LU} = \begin{pmatrix} 0.0 & 0.0 & 0.0 & 0.0 \\ 0.0 & 0.0 & 0.0 & 0.0 \\ 0.0 & 0.0 & 0.0 & 0.0 \\ 0.0 & 0.0 & 0.0 & 0.2 \end{pmatrix},$$

and the corresponding unlabeled 2×2 matrices are

$$\boldsymbol{D}_{UL} = \begin{pmatrix} 0.2 & 0.0 \\ 0.0 & 0.0 \end{pmatrix}, \text{ and } \boldsymbol{D}_{UU} = \begin{pmatrix} 1.2 & 0.0 \\ 0.0 & 1.2 \end{pmatrix}.$$

(continued)

Example 1 (continued)
The supervised estimator will not be unique in this case since $(\boldsymbol{D}_{UL})_{1,1} = 0.0$. A generalized inverse $\boldsymbol{D}_{UL}^{-}$ with 0.0 for this case is used. This yields supervised stochastic matrices

$$\boldsymbol{T}_{LL} = \begin{pmatrix} 0.50 & 0.40 & 0.10 & 0.00 \\ 0.33 & 0.42 & 0.25 & 0.00 \\ 0.11 & 0.33 & 0.56 & 0.00 \\ 0.00 & 0.00 & 0.00 & 1.00 \end{pmatrix}, \text{ and } \boldsymbol{T}_{UL} = \begin{pmatrix} 0.0 & 0.0 & 0.0 & 1.0 \\ 0.0 & 0.0 & 0.0 & 0.0 \end{pmatrix}.$$

The transductive predictions (11.5) for the supervised estimator is then

$$\widetilde{\mathbf{m}} = \begin{pmatrix} \boldsymbol{T}_{LL}\boldsymbol{Y}_L \\ \boldsymbol{T}_{UL}\boldsymbol{Y}_L \end{pmatrix} = \begin{pmatrix} 0.80 \\ 0.50 \\ -0.11 \\ -1.00 \\ -1.00 \\ 0.00 \end{pmatrix}.$$

The supervised prediction for observation 6 is $\widetilde{\mathbf{m}}_6 = 0.0$. This observation is close in proximity to unlabeled observation 5 which is strongly classified as $\widetilde{\mathbf{m}}_5 = -1.0$. It seems reasonable that observation 6 should also be classified as -1.0. This is the issue that the semi-supervised estimator will address by using the unlabeled data in training. This example will be revisited with the semi-supervised estimator introduced next.

11.2.2 Semi-supervised Kernel Regression with a Latent Response

In semi-supervised kernel regression, the approach treats the problem as "supervised" with a latent response variable $\boldsymbol{Y}_U \in \mathbb{R}^{|U|}$. In other words, estimator $\widehat{m}_{\boldsymbol{Y}(\boldsymbol{Y}_U)}(\boldsymbol{x}_0)$ is a function of $\boldsymbol{Y}_U$. The estimation problem then boils down to two parts: (1) employ supervised kernel regression with $\boldsymbol{Y}_U$ fixed to get the estimator $\widehat{m}_{\boldsymbol{Y}(\boldsymbol{Y}_U)}(\boldsymbol{x}_0)$ and (2) determine the estimate $\widehat{\boldsymbol{Y}}_U$. The final semi-supervised kernel estimator is then $\widehat{m}_{\boldsymbol{Y}(\widehat{\boldsymbol{Y}}_U)}(\boldsymbol{x}_0)$.

The semi-supervised local kernel estimator for any unlabeled latent response $\boldsymbol{Y}_U \in \mathbb{R}^{|U|}$ is

$$\widehat{m}_{\boldsymbol{Y}(\boldsymbol{Y}_U)}(\boldsymbol{x}_0) = \frac{\sum_{i \in L \cup U} K_h(\boldsymbol{x}_i, \boldsymbol{x}_0) \boldsymbol{Y}_i(\boldsymbol{Y}_U)}{\sum_{i \in L \cup U} K_h(\boldsymbol{x}_i, \boldsymbol{x}_0)}. \tag{11.6}$$

The estimator $\widehat{m}_{Y(Y_U)}(\boldsymbol{x}_0)$ is designed to be exactly the supervised kernel regression estimator if $\boldsymbol{Y}_U$ were known during training.

Now, define $\widehat{\boldsymbol{m}}\left(\boldsymbol{Y}_U\right) = \left(\widehat{m}_{Y(Y_U)}(\boldsymbol{x}_i)\right)_{i \in L \cup U}$, which is the estimator applied to the observed labeled and unlabeled feature data. From this, define $\boldsymbol{S}_{ij} = \frac{K_h(\boldsymbol{x}_i, \boldsymbol{x}_j)}{\sum_{\ell \in L \cup U} K_h(\boldsymbol{x}_i, \boldsymbol{x}_\ell)}$. The matrix

$$\boldsymbol{S} = \begin{pmatrix} \boldsymbol{S}_{LL} & \boldsymbol{S}_{LU} \\ \boldsymbol{S}_{UL} & \boldsymbol{S}_{UU} \end{pmatrix}$$

is a right stochastic kernel smoother matrix. From this, for any $\boldsymbol{Y}_U$, the estimator applied to the observed feature data is written as:

$$\widehat{\boldsymbol{m}}\left(\boldsymbol{Y}_U\right) = \boldsymbol{S}\boldsymbol{Y}\left(\boldsymbol{Y}_U\right) = \begin{pmatrix} \boldsymbol{S}_{LL}\boldsymbol{Y}_L + \boldsymbol{S}_{LU}\boldsymbol{Y}_U \\ \boldsymbol{S}_{UL}\boldsymbol{Y}_L + \boldsymbol{S}_{UU}\boldsymbol{Y}_U \end{pmatrix} = \begin{pmatrix} \widehat{\boldsymbol{m}}_L\left(\boldsymbol{Y}_U\right) \\ \widehat{\boldsymbol{m}}_U\left(\boldsymbol{Y}_U\right) \end{pmatrix}. \tag{11.7}$$

This estimator is well-defined for any $\boldsymbol{Y}_U \in \mathbb{R}^{|U|}$.

The issue of choosing an optimal $\boldsymbol{Y}_U$ is of interest. For this, consider the unlabeled fitted residual vector as a function of the unlabeled response $\boldsymbol{Y}_U$:

$$\widehat{\boldsymbol{\epsilon}}_U\left(\boldsymbol{Y}_U\right) = \boldsymbol{Y}_U - \widehat{\boldsymbol{m}}_U\left(\boldsymbol{Y}_U\right).$$

Since we do not know $\boldsymbol{Y}_U$, it seems reasonable to choose $\widehat{\boldsymbol{Y}}_U$ as the minimizer of unlabeled residual squared error, i.e., choose $\widehat{\boldsymbol{Y}}_U$ such that

$$\mathrm{RSS}_U\left(\boldsymbol{Y}_U\right) = \widehat{\boldsymbol{\epsilon}}_U\left(\widehat{\boldsymbol{Y}}_U\right)^T \widehat{\boldsymbol{\epsilon}}_U\left(\widehat{\boldsymbol{Y}}_U\right) = \sum_{j \in U} \widehat{\epsilon}_j^{\,2}\left(\boldsymbol{Y}_U\right) = 0.$$

The solution is to force each unlabeled fitted residual to zero or equivalently solve

$$\begin{aligned} \widehat{\boldsymbol{Y}}_U &= \widehat{\boldsymbol{m}}_U\left(\boldsymbol{Y}_U\right) = \boldsymbol{S}_{UL}\boldsymbol{Y}_L + \boldsymbol{S}_{UU}\widehat{\boldsymbol{Y}}_U \\ &= \left(\boldsymbol{I} - \boldsymbol{S}_{UU}\right)^{-1}\boldsymbol{S}_{UL}\boldsymbol{Y}_L. \end{aligned} \tag{11.8}$$

This estimator $\widehat{\boldsymbol{Y}}_U$ is unique as long as all the eigenvalues of $\boldsymbol{S}_{UU}$, which are real, are strictly less than one (Culp and Ryan 2013). Using this predicted latent response, the estimator is

$$\widehat{m}_{Y(\widehat{Y}_U)}\left(\boldsymbol{x}_0\right) = \frac{\sum_{i \in L \cup U} K_h(\boldsymbol{x}_i, \boldsymbol{x}_0) Y_i\left(\widehat{\boldsymbol{Y}}_U\right)}{\sum_{i \in L \cup U} K_h(\boldsymbol{x}_i, \boldsymbol{x}_0)}, \tag{11.9}$$

and restricted to the $L \cup U$ cases we get the following fitted response vector:

$$\widehat{\boldsymbol{m}}\left(\widehat{\boldsymbol{Y}}_U\right) = \begin{pmatrix} \boldsymbol{S}_{LL}\boldsymbol{Y}_L + \boldsymbol{S}_{LU}\left(\boldsymbol{I} - \boldsymbol{S}_{UU}\right)^{-1}\boldsymbol{S}_{UL}\boldsymbol{Y}_L \\ \left(\boldsymbol{I} - \boldsymbol{S}_{UU}\right)^{-1}\boldsymbol{S}_{UL}\boldsymbol{Y}_L \end{pmatrix}. \tag{11.10}$$

The matrix $(\boldsymbol{I}-\boldsymbol{S}_{UU})^{-1}\boldsymbol{S}_{UL}$ is the product of the $|U|\times|U|$ right stochastic matrix $\boldsymbol{P}=(\boldsymbol{I}-\boldsymbol{S}_{UU})^{-1}(\boldsymbol{D}_{UU}+\boldsymbol{D}_{UL})^{-1}\boldsymbol{D}_{UL}$ and the $|U|\times|L|$ supervised prediction matrix $\boldsymbol{T}_{UL}=\boldsymbol{D}_{UL}^{-1}\boldsymbol{W}_{UL}$ (generalizes naturally for cases when $\boldsymbol{D}_{UL}^{-1}$ is not unique), that is,

$$\begin{aligned}\widehat{\boldsymbol{m}}_U\left(\widehat{\boldsymbol{Y}}_U\right) &= (\boldsymbol{I}-\boldsymbol{S}_{UU})^{-1}\boldsymbol{S}_{UL}\boldsymbol{Y}_L\\ &= (\boldsymbol{I}-\boldsymbol{S}_{UU})^{-1}(\boldsymbol{D}_{UU}+\boldsymbol{D}_{UL})^{-1}\boldsymbol{D}_{UL}\boldsymbol{T}_{UL}\boldsymbol{Y}_L\\ &= \boldsymbol{P}\widetilde{\mathbf{m}}_U.\end{aligned}$$

Each unlabeled semi-supervised estimate is a probability weighted linear combination of the local supervised kernel smoother applied to observations in U. The semi-supervised prediction border for the two moons data set is provided in Fig. 11.1.

Example 1 (continued) The smoother is

$$\boldsymbol{S}=\left(\begin{array}{cccc|cc} 0.50 & 0.40 & 0.10 & 0.00 & 0.00 & 0.00\\ 0.33 & 0.42 & 0.25 & 0.00 & 0.00 & 0.00\\ 0.11 & 0.33 & 0.56 & 0.00 & 0.00 & 0.00\\ 0.00 & 0.00 & 0.00 & 0.83 & 0.17 & 0.00\\ \hline 0.00 & 0.00 & 0.00 & 0.14 & 0.71 & 0.14\\ 0.00 & 0.00 & 0.00 & 0.00 & 0.17 & 0.83\end{array}\right).$$

From this, $\widehat{\boldsymbol{Y}}_U=(\boldsymbol{I}-\boldsymbol{S}_{UU})^{-1}\boldsymbol{S}_{UL}\boldsymbol{Y}_L=\mathbf{-1}$ and the semi-supervised kernel smoother is

$$\widehat{\boldsymbol{m}}\left(\widehat{\boldsymbol{Y}}_U\right)=\begin{pmatrix}\boldsymbol{S}_{LL}\boldsymbol{Y}_L+\boldsymbol{S}_{LU}\widehat{\boldsymbol{Y}}_U\\ \widehat{\boldsymbol{Y}}_U\end{pmatrix}=\begin{pmatrix}0.80\\ 0.50\\ -0.11\\ -1.00\\ -1.00\\ -1.00\end{pmatrix}.$$

Notice that $\widehat{\boldsymbol{m}}\left(\widehat{\boldsymbol{Y}}_U\right)_5=\widehat{\boldsymbol{m}}\left(\widehat{\boldsymbol{Y}}_U\right)_6=-1.0$. The semi-supervised estimator exploits the proximity of the unlabeled feature data to classify more distant unlabeled observations.

11.2.3 Adaptive Semi-supervised Kernel Regression

A semi-supervised estimator exploits the structures within the unlabeled feature data to improve performance. Such structures are rendered unusable if large amounts of error are added to the feature data, and the semi-supervised estimator will perform poorly. The supervised estimator will also perform poorly, but may arguably have an advantage in these circumstances. It is reasonable to adapt between the purely semi-supervised estimator (11.9) and the purely supervised estimator (11.3) to optimize the bias/variance trade-off on prediction performance.

One way to compromise between these two estimators is to consider the function $\widehat{\boldsymbol{m}}(\mathbf{0})$ given in Display (11.6) with latent response estimate $\widehat{\boldsymbol{Y}}_U = \mathbf{0}$. Employing the zero vector for the latent response has a nice connection to the supervised estimator. To see this, notice that for prediction $\boldsymbol{x}_0$ the semi-supervised estimator is decomposed as:

$$\widehat{m}_{Y(Y_U)}(\boldsymbol{x}_0) = p(\boldsymbol{x}_0)\widehat{m}_{Y_L}(\boldsymbol{x}_0) + (1 - p(\boldsymbol{x}_0))\,\widehat{m}_{Y_U}(\boldsymbol{x}_0), \tag{11.11}$$

with $p(\boldsymbol{x}_0) = \frac{\sum_{i \in L} K_h(\boldsymbol{x}_i, \boldsymbol{x}_0)}{\sum_{i \in L \cup U} K_h(\boldsymbol{x}_i, \boldsymbol{x}_0)}$. Plugging $\widehat{\boldsymbol{Y}}_U = \mathbf{0}$ results in a shrunken version of supervised estimation with $|\widehat{\boldsymbol{m}}_{Y(\mathbf{0})}(\boldsymbol{x}_0)| = |p(\boldsymbol{x}_0)\widehat{m}_{Y_L}(\boldsymbol{x}_0)| \leq |\widehat{m}_{Y_L}(\boldsymbol{x}_0)|$. In practice, these estimators are usually somewhat close (e.g., they have the same sign in classification with a ± 1 response). To force the semi-supervised estimator to be adaptive we consider a set of shrinking responses indexed by a parameter γ, where $\gamma = 0$ results in pure semi-supervised and $\gamma = \infty$ results in approximate supervised. The new estimator $\widehat{m}_{Y(\widehat{Y}_{U_\gamma})}(\boldsymbol{x}_0)$ with updated unlabeled response

$$\widehat{\boldsymbol{Y}}_{U_\gamma} = \tfrac{1}{1+\gamma}\left(\boldsymbol{I} - \tfrac{1}{1+\gamma}\boldsymbol{S}_{UU}\right)^{-1} \boldsymbol{S}_{UL}\boldsymbol{Y}_L \tag{11.12}$$

adapts between the supervised and semi-supervised extremes. Consider the two moons data example in Fig. 11.2. As γ is increased the classification border approaches the supervised classification border. In practice this type of adaptive estimator has some advantages over the extreme versions. It allows for using the unlabeled data structure, but also allows for it to adjust to noise. To make this point more clear, consider the simulation study presented in Fig. 11.3. Here each observation from the two moons data was generated with noise $\boldsymbol{x}_i^\star \sim N(\boldsymbol{x}_i, \sigma^2)$. The truth was taken as the constant labeling over the moon, and the unlabeled error was plotted as function of σ. The optimal technique with minimum unlabeled error is purely semi-supervised for small σ, purely supervised for large σ, and an adaptive solution for intermediate σ.

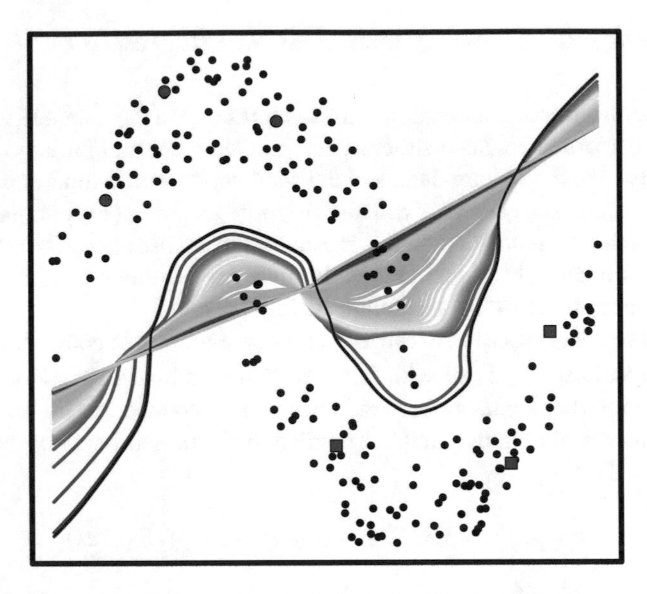

Fig. 11.2 The “two moons” data with regularized classification boundary curves. Rainbow spectrum: ordered by $\gamma \in (0, \infty)$

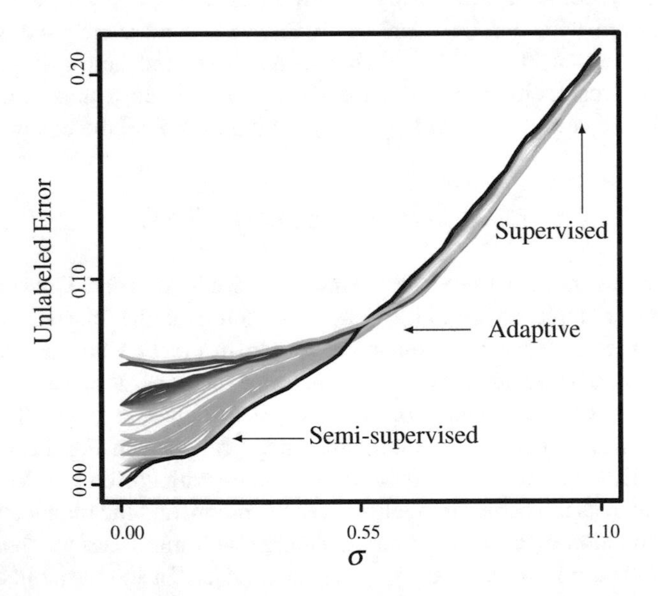

Fig. 11.3 Noise degradation study for the “two moons” data. Black: $\gamma = 0$ (harmonic extreme). Gray: $\gamma = \infty$ (supervised extreme). Rainbow spectrum: ordered by $\gamma \in (0, \infty)$

11.2.4 Computational Issues for Large Data

Several of the practical applications discussed above typically occur with small labeled but large unlabeled sets. In such a case the semi-supervised estimator $\widehat{m}_{\boldsymbol{Y}(\widehat{Y}_{U_\gamma})}(\boldsymbol{x}_0)$ requires tuning parameters (h, γ). Use $(\hat{h}, \hat{\gamma})$ to get unlabeled response $\widehat{\boldsymbol{Y}}_{U_\gamma}$ in (11.12). Then one can predict any $\boldsymbol{x}_0$ directly using the adaptive semi-supervised estimator $\widehat{m}_{\boldsymbol{Y}(\widehat{Y}_{U_\gamma})}(\boldsymbol{x}_0)$. If the unlabeled data is particularly large, one could obtain a representative set $\widetilde{U} \subseteq U$ and use data $(\boldsymbol{Y}_L, \boldsymbol{X}_L, \boldsymbol{X}_{\widetilde{U}})$ to estimate (h, γ) by cross-validation to obtain $\widehat{\boldsymbol{Y}}_{\widetilde{U}_{\hat{\gamma}}}$. Then prediction $\widehat{m}_{\boldsymbol{Y}\left(\widehat{Y}_{\widetilde{U}_{\hat{\gamma}}}\right)}(\boldsymbol{x}_0)$ can be applied with $\boldsymbol{x}_0$ from the entire unlabeled data or a new observation, and this prediction problem is embarrassingly parallel.

Much work in semi-supervised learning has been concerned with the construction of these local kernel estimators for cases where both $|L \cup U|$ and p are large. The main idea is to induce sparsity into the local kernel metric. Often $\boldsymbol{W}$ is treated as an $n \times n$ adjacency matrix for a graph. As such, each observation is a node on the graph, and weighted edges are defined by non-zero off-diagonal elements in $\boldsymbol{W}$. Much work is on computing k-NN or ϵ thresholds for the elements in the graph. Graph construction often has a substantial effect on the performance of kernel based semi-supervised learning methods (Jebara et al. 2009). Computing the graph quickly for large data sets often involves anchor point methods (Liu et al. 2010). Conceptually, one needs a representative sample of points to construct the central nodes of the graph and then proceed to construct the graph much quicker. This usually leads to fast algorithms for constructing $\boldsymbol{W}$. This problem of graph construction from feature data is sometimes referred to as the learning the graph problem. Earlier work in the area incorporated estimating the parameter h as part of the learning the graph problem; however, currently it is accepted that estimating this parameter by cross-validation is the more sound approach.

11.3 Optimization Frameworks for Semi-supervised Learning

The semi-supervised local kernel smoother presented above provides a good starting point for understanding how unlabeled data are used in semi-supervised learning. This approach is now used to motivate the general classes of techniques found in the semi-supervised literature. Most semi-supervised approaches are based on optimization frameworks which use the unlabeled data in either the loss function or penalty or both. The $\boldsymbol{W}$ matrix is often regarded as an $n \times n$ adjacency matrix for a graph with n nodes as discussed previously. Now, the graph is not fixed, and new nodes unavailable during training can be predicted by nearly all techniques. Next key literature based approaches are presented with their main ideas derived from the supervised local kernel technique.

It is well known that the supervised Nadaraya Watson kernel estimator solves optimization problem:

$$\min_{f(\boldsymbol{x}_0)\in\mathbb{R}} \sum_{i\in L} K_h(\boldsymbol{x}_i, \boldsymbol{x}_0)\,(Y_{L_i} - f(\boldsymbol{x}_0))^2 .$$

Instead of focusing on prediction, consider training only the labeled function and treating $\boldsymbol{W}_{LL}$ as a graph. The discrete combinatorial Laplacian operator for this graph is given by $\widetilde{\boldsymbol{\Delta}}_{LL} = \boldsymbol{D}_{LL} - \boldsymbol{W}_{LL}$. It turns out that the estimator for the labeled function vector $\widetilde{\boldsymbol{m}}_L$ is the solution to the optimization problem

$$\min_{\boldsymbol{f}_L} (\boldsymbol{Y}_L - \boldsymbol{f}_L)^T \boldsymbol{W}_{LL} (\boldsymbol{Y}_L - \boldsymbol{f}_L) + \boldsymbol{f}_L^T \widetilde{\boldsymbol{\Delta}}_{LL} \boldsymbol{f}_L. \tag{11.13}$$

One can solve this to see that the optimal solution is indeed $\widetilde{\boldsymbol{m}}_L$. This optimization problem provides no explanation for how to predict the unlabeled observations, but it does provide the starting point for how one can view several semi-supervised approaches in the literature.

The first class of techniques are based on the labeled loss criterion. The main idea is to penalize the unlabeled prediction vector against some penalty matrix defined on the full graph. Usually the penalty matrix is a variation of the discrete combinatorial Laplacian matrix $\boldsymbol{\Delta} = \boldsymbol{D} - \boldsymbol{W}$ where $\boldsymbol{D}$ is the row sum matrix of $\boldsymbol{W}$. It is natural to generalize the penalty in optimization problem (11.13) to $\boldsymbol{f}^T \boldsymbol{\Delta} \boldsymbol{f}$. Belkin et al. (2004) proposed optimization problem

$$\min_{\boldsymbol{f}} (\boldsymbol{Y}_L - \boldsymbol{f}_L)^T (\boldsymbol{Y}_L - \boldsymbol{f}_L) + \boldsymbol{f}^T \boldsymbol{\Delta} \boldsymbol{f},$$

which is an example of this type of generalization.[1] In semi-supervised learning, the *labeled loss* optimization framework is

$$\min_{\boldsymbol{f}} L(\boldsymbol{Y}_L, \boldsymbol{f}_L) + \eta_1 \boldsymbol{f}^T \boldsymbol{\Delta} \boldsymbol{f} + \eta_2 \boldsymbol{f}^T \boldsymbol{f}, \tag{11.14}$$

where $L(\cdot,\cdot)$ is a loss function and both $\eta_1 \geq 0$ and $\eta_2 \geq 0$. Several semi-supervised graph-based learning techniques are based off of this general criterion including manifold regularization where the problem is adapted to Hilbert space optimization (Belkin et al. 2006), energy diffusion with the normalized Laplacian (Zhou et al. 2004), and a deformed Laplacian approach (Gong et al. 2015). The energy optimization viewpoint for semi-supervised learning is discussed extensively in Abney (2008). Also, several of these authors make connections to graph-cutting routines (Wang et al. 2013) and harmonic functions (Zhu 2008). The harmonic

[1] Note, one could weight the loss function as $(\boldsymbol{Y}_L - \boldsymbol{f}_L)^T \boldsymbol{W}_{LL} (\boldsymbol{Y}_L - \boldsymbol{f}_L)$, however, to our knowledge this specific loss function in the context of semi-supervised learning with a labeled loss approach has not been studied.

property produces a general averaging estimator over the graph, where any function f is harmonic whenever it satisfies

$$\widehat{\boldsymbol{f}} = \begin{pmatrix} \widehat{\boldsymbol{f}}_L \\ \widehat{\boldsymbol{f}}_U \end{pmatrix} = \begin{pmatrix} \widehat{\boldsymbol{f}}_L \\ (\boldsymbol{I} - \boldsymbol{S}_{UU})^{-1} \boldsymbol{S}_{UL} \widehat{\boldsymbol{f}}_L \end{pmatrix} \tag{11.15}$$

for semi-supervised learning. Notice the similarity with the unlabeled estimator (11.10). Culp and Ryan (2013) showed that all solutions to labeled loss with $\eta_2 = 0$ are harmonic (11.15) for any loss function, that is, $\widehat{\boldsymbol{f}}_L$ is the solution to $\min_{\boldsymbol{f}_L} L(\boldsymbol{Y}_L, \boldsymbol{f}_L) + \eta \boldsymbol{f}_L^T \boldsymbol{\Delta}_{LL}^{\star} \boldsymbol{f}_L$ where $\boldsymbol{\Delta}_{LL}^{\star} = \boldsymbol{\Delta}_{LL} - \boldsymbol{\Delta}_{LU} (\boldsymbol{\Delta}_{UU})^{-1} \boldsymbol{\Delta}_{UL}$. This establishes that the loss function directly determines the estimate for the few labeled observations and results in the harmonic estimator for the unlabeled data.

Joint training provides another class of semi-supervised approaches. In this case, similar to the semi-supervised local kernel estimator (11.9), a latent response is required. The joint training optimization functional

$$\min_{\boldsymbol{f}, \boldsymbol{Y}_U} L(\boldsymbol{Y}(\boldsymbol{Y}_U), \boldsymbol{f}) + \eta_1 J_1(\boldsymbol{f}) + \eta_2 J_2(\boldsymbol{Y}_U), \tag{11.16}$$

yields both an unlabeled response estimate and function, simultaneously. The connection to semi-supervised kernel regression (11.3) is demonstrated by the natural generalization of (11.13), i.e.,

$$\min_{\boldsymbol{f}, \boldsymbol{Y}_U} (\boldsymbol{Y}(\boldsymbol{Y}_U) - \boldsymbol{f})^T \boldsymbol{W} (\boldsymbol{Y}(\boldsymbol{Y}_U) - \boldsymbol{f}) + \boldsymbol{f}^T \boldsymbol{\Delta} \boldsymbol{f} + \gamma \boldsymbol{Y}_U^T \boldsymbol{Y}_U, \tag{11.17}$$

which results in

$$\widehat{m}_{\boldsymbol{Y}(\widetilde{\boldsymbol{Y}}_{U_\gamma})}(\boldsymbol{x}_0) = \frac{\sum_{i \in L \cup U} K_h(\boldsymbol{x}_i, \boldsymbol{x}_0) \boldsymbol{Y}_i\left(\widetilde{\boldsymbol{Y}}_{U_\gamma}\right)}{\sum_{i \in L \cup U} K_h(\boldsymbol{x}_i, \boldsymbol{x}_0)} \text{ where}$$
$$\widetilde{\boldsymbol{Y}}_{U_\gamma} = ((\boldsymbol{\Delta S})_{UU} + \gamma \boldsymbol{I})^{-1} (\boldsymbol{\Delta S})_{UL} \boldsymbol{Y}_L.$$

This estimator has several properties similar to the local semi-supervised kernel smoother (11.9) and was studied in Culp and Ryan (2013). Other versions of the joint optimization problem are also of note, including the S^3VM which uses hinge loss (Chapelle et al. 2008) and ψ-learning approaches (Wang and Shen 2007). The joint optimization problem is a worthwhile framework and has been of recent interest in the literature.

Self-training provides another class of techniques to fit semi-supervised estimators (Zhu 2008). The main idea is to treat the latent response $\boldsymbol{Y}_U$ as if it were known and then iteratively update. For example, a kernel smoother is extended to semi-supervised learning in the following way: set $\widehat{\boldsymbol{f}}_U = \boldsymbol{0}$ and iterate $\widehat{\boldsymbol{f}} = \boldsymbol{SY}\left(\widehat{\boldsymbol{f}}_U\right)$. The convergent solution to this problem results into $\widehat{\boldsymbol{f}} = \widehat{\boldsymbol{m}}_U\left(\widehat{\boldsymbol{Y}}_U\right)$, i.e., the solution is exactly the semi-supervised kernel smoother (11.9). This algorithm is a simplification of common self-training algorithms including: Yarowski's algorithm

(Abney 2004) and fitting the fits (Culp and Michailidis 2008). The pure generality of the approach is of particular note, because one can extend any supervised technique using self-training. Unfortunately, not much is currently known about the general behavior of self-training algorithms when using generic supervised functions.

There are many other techniques in the literature motivated from different perspectives including energy optimization and physics (Zhu 2008), learner agreement approaches (Blum and Mitchell 1998; Shilang 2013), and generalized EM approaches (Zhu 2008; Chapelle et al. 2006). The goal of this chapter was to demonstrate key semi-supervised research areas in terms of extending local kernel smoothing into semi-supervised learning.

Acknowledgements NSF CAREER/DMS-1255045 grant supported the work of Mark Vere Culp. The opinions and views expressed in this chapter are those of the authors and do not reflect the opinions or views at the NSF.

References

Abney S (2004) Understanding the Yarowsky algorithm. Comput Linguist 30(3):365–395

Abney S (2008) Semisupervised learning for computational linguistics. Chapman and Hall, CRC, Boca Raton

Belkin M, Matveeva I, Niyogi P (2004) Regularization and semi-supervised learning on large graphs. In: COLT, pp 624–638

Belkin M, Niyogi P, Sindhwani V (2006) Manifold regularization: a geometric framework for learning from labeled and unlabeled examples. J Mach Learn Res 7:2399–2434

Blum A, Mitchell T (1998) Combining labeled and unlabeled data with co-training. In: Computational learning theory, pp 92–100

Bredel M, Jacoby E (2004) Chemogenomics: an emerging strategy for rapid target and drug discovery. Nat Rev Genet 5(4):262–275

Chapelle O, Schölkopf B, Zien A (2006) Semi-supervised learning. MIT Press, Cambridge. http://www.kyb.tuebingen.mpg.de/ssl-book

Chapelle O, Sindhwani V, Keerthi S (2008) Optimization techniques for semi-supervised support vector machines. J Mach Learn Res 9:203–233

Culp M, Michailidis G (2008) An iterative algorithm for extending learners to a semi-supervised setting. J Comput Graph Stat 17(3):545–571

Culp M, Ryan K (2013) Joint harmonic functions and their supervised connections. J Mach Learn Res 14:3721–3752

Gong C, Liu T, Tao D, Fu K, Tu E, Yang J (2015) Deformed graph Laplacian for semisupervised learning. IEEE Trans Neural Nets Learn Syst 26:2261–2274

Hastie T, Tibshirani R, Friedman J (2009) The elements of statistical learning (data mining, inference and prediction). Springer, Berlin

Jebara T, Wang J, Chang S (2009) Graph construction and *b*-matching for semi-supervised learning. In: International conference of machine learning

Koprinska I, Poon J, Clark J, Chan J (2007) Learning to classify e-mail. Inf Sci 177(10):2167–2187

Lafferty J, Wasserman L (2007) Statistical analysis of semi-supervised regression. In: Advances in NIPS. MIT Press, Cambridge, pp 801–808

Liu W, He J, Chang S (2010) Large graph construction for scalable semi-supervised learning. In: International conference of machine learning

Lundblad R (2004) Chemical reagents for protein modification. CRC Press, Boca Raton

McCallum A, Nigam K, Rennie J, Seymore K (2000) Automating the construction of internet portals with machine learning. Inf Retr J 3:127–163

Shilang S (2013) A survey of multi-view machine learning. Neural Comput Appl 7–8(28):2013–2038

Wang J, Shen X (2007) Large margin semi-supervised learning. J Mach Learn Res 8:1867–1897

Wang J, Jebara T, Chang S (2013) Semi-supervised learning using greedy max-cut. J Mach Learn Res 14:771–800

Yamanishi Y, Vert J, Kanehisa M (2004) Protein network inference from multiple genomic data: a supervised approach. Bioinformatics 20:363–370

Zhou D, Bousquet O, Lal TN, Weston J, Schölkopf B (2004) Learning with local and global consistency. In: Advances in neural information processing systems 16

Zhu X (2008) Semi-supervised learning literature survey. Technical report, Computer Sciences, University of Wisconsin-Madison

Chapter 12
Inverse Modeling: A Strategy to Cope with Non-linearity

Qian Lin, Yang Li, and Jun S. Liu

Abstract In the big data era, discovering and modeling potentially non-linear relationships between predictors and responses might be one of the toughest challenges in modern data analysis. Most forward regression modeling procedures are seriously compromised due to the curse of dimension. In this chapter, we show that the inverse modeling idea, originated from the *Sliced Inverse Regression* (SIR), can help us detect nonlinear relations effectively, and survey a few recent advances, both algorithmically and theoretically, in which the inverse modeling idea leads to unforeseeable benefits in nonlinear variable selection and nonparametric screening.

Keywords Correlation pursuit · Multiple index models · Nonparametric screening · Sliced inverse regression · Sufficient dimension reduction · Sub-Gaussian

12.1 Introduction

Studying relationships among random variables is one of the main focuses in statistics. In the past century and perhaps until nowadays, linear regression/OLS has been playing the dominant role in real-world data analysis. Characteristics of modern data, however, challenge this prevailing framework in at least two aspects: (1) the linear system could be under-determined and over-fitting, i.e., '*large p, small n*'; and (2) the linear assumption might be inappropriate and under-fitting, i.e.,

---

Q. Lin
Center for Statistical Science, Department of Industrial Engineering, Tsinghua University, Beijing, China

Y. Li
Vatic Labs, New York City, NY, USA

J. S. Liu (✉)
Department of Statistics, Harvard University, Cambridge, MA, USA

Center for Statistical Science, Tsinghua University, Beijing, China
e-mail: jliu@stat.harvard.edu

© Springer International Publishing AG, part of Springer Nature 2018

W. K. Härdle et al. (eds.), *Handbook of Big Data Analytics*, Springer Handbooks of Computational Statistics, https://doi.org/10.1007/978-3-319-18284-1_12

nonlinearity. Significant progresses have been made for the high dimensional linear regression in the last two decades. By assuming that only a very few among many candidate predictors are related to the response in a regression (i.e., the sparsity assumption), researchers have proposed the LASSO (Tibshirani 1996) and Danztig Selector (Candes and Tao 2007) to effectively select important predictors. Successes of LASSO and its follow-up work provide us with certain confidence that linear regression in high dimension is no longer a formidable task. However, detecting nonlinear relationships in high-dimension is still a challenge.

Forward regressions, in which people interpret the response as certain function (parametric or non-parametric) of the predictors with some random perturbations, have been widely adopted by almost all data scientists. To be precise, let Y be the response and $\boldsymbol{X} \in \mathbb{R}^p$ be the predictor vector. Forward regressions construct models for the conditional density $p(Y|\boldsymbol{X})$. In contrast, inverse modeling studies the distribution of predictors conditional on a particular value of the response variable, i.e., focusing on the density function $p(\boldsymbol{X}|\, Y)$. Although it is a little counter-intuitive, this formulation is natural from a Bayesian point of view. A celebrated nontrivial application of the inverse modeling idea dates back to Li (1991), where Li proposed the sliced inverse regression (SIR) for dimension reductions. SIR inspired the so-called *Sufficient Dimension Reduction* (SDR) formulation, which postulates the existence of the central subspace—the minimal subspace $\mathcal{S}$ such that $Y \perp\!\!\!\perp \boldsymbol{X}|\mathbb{P}_{\mathcal{S}}\boldsymbol{X}$ and enables people to find such a space, at least approximately, in real data. In the last two decades, the SDR has gained successes in various scientific fields, but the inverse modeling idea of SIR is somehow overlooked. Modeling from an inverse perspective has not been explicitly stated until recently, where Cook (2007) proposed the principal fitted components (PFC). Inverse modeling not only clarifies various SDR algorithms, but also brings us the desired theoretical treatment of SIR/SDR in high dimension. In fact, since its birth, SIR has been advocated as a nonlinear alternative to the multiple linear regression (Chen and Li 1998). However, the lack of the thorough theoretical treatment prevents SIR to be a nonlinear counterpart of the multiple linear regression in high dimensional settings. Some recent theoretical advances for SIR/SDR will be briefly reviewed in Sects. 12.2 and 12.4.

In high-dimensional settings, it is scientifically desirable to look for a small number of predictors that can explain the response. Furthermore, when there is nonlinearity, the forward regression is seriously compromised due to the computational cost and the limited sample size. Thus, variable selection becomes an even more indispensable step in data analysis. Motivated by the multiple linear regression and the profile correlation viewpoint of SIR, Zhong et al. (2012) proposed the correlation pursuit algorithm (COP) selecting the important variables. It is clear that the COP, which selects variables from a forward modeling perspective, could be easily extending to other SDR algorithms (see, e.g., Yu et al. 2016). However, modeling from an inverse perspective can bring us more interesting phenomena. Jiang et al. (2014) proposed the *SIR for variable selection via Inverse modeling* (SIRI) based on the inverse modeling (see, e.g., (12.22) below). It is seen that not only can COP be viewed as an approximation of SIRI (see discussions around

Eq. (12.24)) when conditions required by SDR modeling hold, but also can SIRI be effective in discovering interactions when those conditions are violated.

Last but not least, inverse modeling provides a natural framework for developing nonparametric dependency tests. Since the introduction of the sure independence screening concept by Fan and Lv (2008), looking for effective screening statistics becomes a popular strategy for data analysts. Though screening statistics based on the parametric model have been employed in various applications, a crucial drawback of these methods is that parametric modeling assumptions are often violated in real data and difficult to check. Looking for nonparametric dependency screening statistics is of particular interest from practical consideration. Inverse modeling, especially, the estimation conditioning on discretized response variable values, aka slicing, provides us a large class of nonparametric statistics, which are both theoretically solid and practically effective (see Cui et al. 2014; Jiang et al. 2014; Li et al. 2012; Zhu et al. 2011). In the following sections, we will give a more detailed review of the developments and impacts of the inverse modeling idea.

12.2 SDR and Inverse Modeling

12.2.1 From SIR to PFC

When people try to visualize the data with moderate dimensions (e.g., ≤ 10), a low dimensional projection (e.g., ≤ 2) of predictors $\boldsymbol{X}$ is often desirable, and it would be even better if the projection can explain most of the variation in the response. Li (1991) considered the following model[1]

$$Y = f(\boldsymbol{\beta}_1^\tau \boldsymbol{X}, \ldots, \boldsymbol{\beta}_d^\tau \boldsymbol{X}, \epsilon), \text{ where } \boldsymbol{X} \sim N(0, \mathbf{I}_p), \epsilon \sim N(0, 1) \tag{12.1}$$

and introduced the sliced inverse regression algorithm to estimate the space $\mathcal{S} = \text{span}\{\boldsymbol{\beta}_1, \ldots, \boldsymbol{\beta}_d\}$. We briefly review the SIR procedure here and give an intuitive explanation on why it works.

Suppose we have $n(= Hc)$ observations (x_i, y_i), $i = 1, \ldots, n$ from the model (12.1). We sort the y_i and divide the data along its order statistics $y_{(i)}$ into H equal sized bins. Let $\boldsymbol{x}_{(i)}$ be the concomitant of $y_{(i)}$, $\boldsymbol{x}_{h,i} = \boldsymbol{x}_{((h-1)c+i)}$ and let $\bar{\boldsymbol{x}}_h = \frac{1}{c}\sum_i \boldsymbol{x}_{h,i}$. SIR proceeds by estimating $\text{var}[E(X \mid Y)]$ by

$$\boldsymbol{\Lambda}_{H,c} = \frac{1}{H}\sum_h \bar{\boldsymbol{x}}_h \bar{\boldsymbol{x}}_h^\tau \tag{12.2}$$

[1]To avoid unnecessary technical assumptions, we state the results in this chapter under a stronger condition than their original form in literature.

and the space $\mathcal{S}$ by $\widehat{\mathcal{S}}$ spanned by the first d eigenvector of $\boldsymbol{\Lambda}_{H,c}$. Under mild conditions (e.g., rank(var($\mathbb{E}[\boldsymbol{X}|\,Y]$)) $= d$) the SIR estimate $\widehat{\mathcal{S}}$ and $\boldsymbol{\Lambda}_{H,c}$ were shown to be consistent (Duan and Li 1991; Hsing and Carroll 1992; Zhu et al. 2006). We provide an intuitive explanation here and revisit it later with a more clear vision from the inverse modeling perspective. Suppose we have the following decomposition

$$\boldsymbol{X} = \boldsymbol{Z} + \boldsymbol{W}, \text{ where } \boldsymbol{Z} = \mathbb{P}_{\mathcal{S}}\boldsymbol{X} \text{ and } \boldsymbol{W} = \mathbb{P}_{\mathcal{S}^{\perp}}\boldsymbol{X}. \tag{12.3}$$

We introduce the notations $z_{h,i}$, $\bar{z}_h$, $\boldsymbol{w}_{h,i}$, and $\bar{\boldsymbol{w}}_h$ similarly as $\boldsymbol{x}_{h,i}$, $\bar{\boldsymbol{x}}_h$. Then, we have

$$\begin{aligned}\Lambda_{H,c} = &\frac{1}{H}\sum_h \bar{z}_h\bar{z}_h^{\tau} + \frac{1}{H}\sum_h \bar{\boldsymbol{w}}_h\bar{\boldsymbol{w}}_h^{\tau}\\ &+\frac{1}{H}\sum_h \bar{z}_h\bar{\boldsymbol{w}}_h^{\tau} + \frac{1}{H}\sum_h \bar{z}_h\bar{\boldsymbol{w}}_h^{\tau}\end{aligned}$$

Note that $\boldsymbol{X} \sim N(0, \mathbf{I}_p)$ and $\boldsymbol{W} \perp\!\!\!\perp Y$, it is easily seen that var($\bar{\boldsymbol{w}}_h$) $= \frac{1}{c}$var($\boldsymbol{W}$). As long as $\frac{1}{H}\sum_h \bar{z}_h\bar{z}_h^{\tau}$ is a consistent estimate of var($\mathbb{E}[\boldsymbol{X}|\,Y]$), the cross term is bounded by $\sqrt{\frac{1}{c}\text{var}(\boldsymbol{W})}$, i.e., if the sample size c in each slice $\rightarrow \infty$, the SIR algorithm averages out the noise orthogonal to $\mathcal{S}$ and hence leads to a consistent estimate of $\mathcal{S}$.

SIR, due to its simplicity and computational efficiency, is one of the most popular supervised dimension reduction tools. One of the drawbacks of SIR is that it needs the non-degeneracy of var($\mathbb{E}[\boldsymbol{X}|\,Y]$), i.e., rank (var($\mathbb{E}[\boldsymbol{X}|\,Y]$)) $= d$. Clearly, many functions such as symmetric ones do not satisfy this requirement. When function f does not satisfy this condition, several alternative algorithms have been proposed to rescue (Cook and Weisberg 1991). These alternative algorithms, however, all implicitly imposed some other assumptions on the function f.

The SIR approach of Li (1991) brought us two key innovative ideas. The first one is the *reduction modeling*, i.e., to model y with respect to a low dimension projection of $\boldsymbol{X}$, which motivated the development of *Sufficient Dimension Reduction* (SDR) first proposed by Dennis Cook (Adragni and Cook 2009). The Sufficient Dimension Reduction appears to be a more general framework than Eq. (12.1) and aims at estimating the minimal subspace $\mathcal{S}'$ such that

$$Y \perp\!\!\!\perp \boldsymbol{X}|\mathbb{P}_{\mathcal{S}'}\boldsymbol{X}.$$

In the past two decades, SDR-based algorithms have gained acceptance and from both practitioners and theoreticians. Although the SDR framework brings us a conceptually clear way to treat dimension reduction with a potentially non-linear functional relationship, it is almost equivalent to the model (12.1) when the response is univariate variable. That is, from the forward regression perspective, SDR is a restatement of model (12.1). The second key idea of SIR is *inverse modeling*, which has not been fully taken advantage of until recently.

The inverse modeling view of SDR was first explicitly stated in Cook (2007), where he proposed the Principal Fitted Components (PFC). More precisely, he first proposed the following multivariate inverse regression model:

$$X_Y = \mu + \Gamma \nu_Y + \sigma \varepsilon, \tag{12.4}$$

where $\mu \in \mathbb{R}^p$, $\Gamma \in \mathbb{R}^{p \times d}$, $d < p$, $\Gamma^\tau \Gamma = \mathbf{I}_d$, $\sigma \geq 0$ and d is assumed known. Moreover, ν_Y, the vector value function of Y, is assumed to have a positive definite covariance matrix. Let $\mathcal{S}$ be the column space of Γ, it is easy to see that $Y \perp\!\!\!\perp \boldsymbol{X} | \mathbb{P}_{\mathcal{S}} \boldsymbol{X}$ from the model (12.4). It is clear that the advantage from the inverse modeling is that instead of assuming obscure conditions on the link function f, we can put some more interpretable conditions on the "central curve" ν_Y. With the explicit model (12.4), Cook (2007) proposed the MLE approach to estimate space $\mathcal{S}$, which is not achievable from the forward modeling view without making a parametric assumption on the link function f.

12.2.2 Revisit SDR from an Inverse Modeling Perspective

Due to the rapid technological advances in recent years, it is becoming a common problem for in regression analyses that the sample size is much smaller than the dimension of the data. Although much progress has been made on linear regression in high dimensions, our understanding of behaviors of SIR and SDR in high dimensions is still limited. For instance, it is unclear when the SIR procedure breaks down and if there is any efficient variant of SIR for high dimensional data. Understanding limitations of these SDR/SIR algorithms in high dimension could be the first step towards finding the efficient high dimensional variants.

A few words on the principal component analysis (PCA) are helpful for us to explain what we are pursuing. Following the celebrated work of Johnstone and Lu (2004), where they proved that PCA works if and only if $\rho = \lim \frac{p}{n} = 0$, we expect that a similar structural assumption in high dimensional data is necessary for SDR procedures. By introducing the spiked model with some sparsity constraints on the first several eigenvectors, various sparse PCA algorithms have been proposed (Zou et al. 2006; Johnstone and Lu 2004). Moreover, significant achievements have been made with the sparse PCA procedure. Several research groups (Birnbaum et al. 2013; Cai et al. 2013; Vu and Lei 2012) have established the minimax rate of the sparse PCA problem. Recently, by assuming the hardness of the implied clique problem, Berthet and Rigollet (2013) showed that there is a trade-off between statistical and computational efficiency, which raises the computational complexity issue in the statistical world.

The successes of the PCA and its variants in high dimensional data analyses raise two natural questions for SIR and SDR algorithms: *Can we get similar understandings of the SDR algorithms? Is the computational cost an issue for SDR in high dimensional settings?* These questions seem formidable at the first glance due to the unknown nonlinear link function f. In fact, the minimax rate of the linear regression in high dimensions has only been derived recently (Raskutti et al. 2011). The unknown nonlinear function in index models further increases the difficulty for obtaining the minimax rate for SDR algorithms. A major difficulty is that we do not have explicit and interpretable conditions on the link function f such that $\mathrm{var}(\mathbb{E}[\boldsymbol{X}|f(\boldsymbol{X},\epsilon)])$ is non-degenerate. To prove that the SIR estimate of $\mathcal{S}$ is consistent, Hsing and Carroll (1992) proposed a rather intricate condition on the central curves $\boldsymbol{m}(Y) = \mathbb{E}[\boldsymbol{X}|\, Y]$, as follows:

Definition 1 (Condition A) For $B > 0$ and $n \geq 1$, let $\Pi_n(B)$ be the collection of all the n-point partitions $-B \leq y_1 \leq \cdots \leq y_n \leq B$ of $[-B, B]$. First, assume that the central curve $\boldsymbol{m}(y)$ satisfies the following smooth condition

$$\lim_{n\to\infty} \sup_{y\in\Pi_n(B)} n^{-1/(2+\xi)} \sum_{i=2}^{n} \|\boldsymbol{m}(y_i) - \boldsymbol{m}(y_{i-1})\|_2 = 0, \forall B > 0. \tag{12.5}$$

Second, assume that for $B_0 > 0$, there exists a non-decreasing function $\widetilde{m}(y)$ on (B_0, ∞), such that

$$\widetilde{m}^{2+\xi}(y)P(|\, Y| > y) \to 0 \text{ as } y \to \infty \tag{12.6}$$

$$\|\boldsymbol{m}(y) - \boldsymbol{m}(y')\|_2 \leq |\widetilde{m}(y) - \widetilde{m}(y')| \text{ for } y, y' \in (-\infty, -B_0) \cup (B_0, \infty) \tag{12.7}$$

Condition A has now been well accepted in the SDR community (Li et al. 2007; Zhu et al. 2006). However, it is unclear how to specify the proper class of function f so that the determination of the minimax rate of estimating the central space is possible.

Surprisingly, this difficulty can be overcome from the inverse modeling perspective and Condition A can be replaced by a weaker and more intuitive condition. Recall that in our intuitive explanation about why SIR works (see the discussions around (12.3)), noises orthogonal to the central space are averaged out and the signal along the central space is preserved. The inverse modeling, in a sense, is one more step further than the decomposition (12.3):

$$\begin{aligned} \boldsymbol{X} &= \mathbb{P}_{\mathcal{S}}\boldsymbol{X} + \mathbb{P}_{\mathcal{S}^\perp}\boldsymbol{X} = \mathbb{E}[\boldsymbol{X}|\, Y] + (\mathbb{P}_{\mathcal{S}}\boldsymbol{X} - \mathbb{E}[\boldsymbol{X}|\, Y]) + \mathbb{P}_{\mathcal{S}^\perp}\boldsymbol{X} \\ &:= \boldsymbol{m}(Y) + \boldsymbol{Z}(Y) + \boldsymbol{W}(Y) \end{aligned} \tag{12.8}$$

where $\boldsymbol{m}(Y)$ is the central curve, $\boldsymbol{Z}(Y)$ is the noise lies in the central space, and $\boldsymbol{W}(Y)$ is the noise that lies in the space orthogonal to the central space. Recall the SIR estimator of $\mathrm{var}(\mathbb{E}[\boldsymbol{X}|\, Y])$ in (12.2). If $\boldsymbol{Z}(Y)$ and $\boldsymbol{W}(Y)$ are nearly independent of Y, then one has $\mathrm{var}(\bar{w}_h) = \frac{1}{c}\mathrm{var}(W)$ and $\mathrm{var}(\bar{z}_h) = \frac{1}{c}\mathrm{var}(Z)$. By Cauchy's inequality, the cross terms can be controlled by the principal terms, so we may focus on the other principal term, $\frac{1}{H}\sum_h \bar{\boldsymbol{m}}_h\bar{\boldsymbol{m}}_h^\tau$. In other words, the SIR estimator (12.2) of $\mathrm{var}(\mathbb{E}[\boldsymbol{X}|\, Y])$ is consistent if and only if

$$\frac{1}{H}\sum_h \bar{\boldsymbol{m}}_h\bar{\boldsymbol{m}}_h^\tau \to \mathrm{var}(\boldsymbol{m}(Y)) \tag{12.9}$$

$$\frac{1}{c}\mathrm{var}(\boldsymbol{W}) \to 0 \quad \text{and} \quad \frac{1}{c}\mathrm{var}(\boldsymbol{Z}) \to 0. \tag{12.10}$$

Formula (12.9) is guaranteed by the so-called sliced stable condition introduced in Lin et al. (2018):

Definition 2 (Sliced Stable Condition) For $0 < \mathfrak{a}_1 < 1 < \mathfrak{a}_2$, let $\mathscr{A}_H$ denote all partitions $\{-\infty = a_1 \leq a_2 \leq \cdots \leq a_{H+1} = +\infty\}$ of $\mathbb{R}$, such that

$$\frac{\mathfrak{a}_1}{H} \leq \mathbb{P}(a_h \leq Y \leq a_{h+1}) \leq \frac{\mathfrak{a}_2}{H}.$$

A pair (f, ε) of a k-variate function f and a random variable ε is sliced stable if there exist positive constants $\mathfrak{a}_1 < 1 < \mathfrak{a}_2$ and $M > 0$ such that for any $H \in \mathbb{N},\ H > M$, and all partitions in $\mathscr{A}_H$, there exist two constants $0 \leq \kappa < 1$ and $C > 0$ that only depend on $\mathfrak{a}_1,\ \mathfrak{a}_2$ and M such that for any vector $\boldsymbol{\gamma}$, one has

$$\sum_{h=1}^{H} \gamma^{\tau} \mathrm{var}[\boldsymbol{m}(Y)|a_h < Y \leq a_{h+1}]\gamma \leq CH^{\kappa}\gamma^{\tau}\mathrm{var}[\boldsymbol{m}(Y)]\gamma. \tag{12.11}$$

where $Y = f(\boldsymbol{X}, \varepsilon), \boldsymbol{X} \in N(0, \mathbb{I}_k)$ and $\boldsymbol{m}(Y) = \mathbb{E}[\boldsymbol{X}|\, Y]$.

The sliced stability is almost a necessary condition for (12.9) to hold and can be derived from Condition A (Neykov et al. 2016). Note that the second part of (12.10) holds almost automatically since $\mathrm{var}(\boldsymbol{Z}(Y))$ is bounded. The interesting part is the first term in (12.10). Remember that we have assumed that $\boldsymbol{X}$ is standard Gaussian, hence the $\boldsymbol{W}$ is Gaussian and $\frac{1}{c}\mathrm{var}(W) = O(\frac{p}{c})$. Note that if $\lim \frac{p}{n} = 0$, then one may choose $H = \log \frac{n}{p}$ such that $\lim \frac{p}{c} = 0$ (Recall that we have $n = Hc$). Intuitively, we get the following result (see Lin et al. 2018 for details):

Theorem 1 *The ratio* $\rho = \lim \frac{p}{n}$ *plays the phase transition parameter of the SIR procedure. i.e., the SIR estimate* (12.2) *of* $\mathrm{var}(\mathbb{E}[X|\, Y])$ *is consistent if and only if* $\rho = 0$.

Inspired by the developments of sparse PCA following the work of Johnstone and Lu (2004), Theorem 1 sheds light on the optimality problem of estimating the central space with sparsity constraints. One of our recent results established the optimal sample size for the support recovering problem of the single index model, which provides us a positive evidence regarding the minimax rate problem (see Neykov et al. 2016 and Sect. 12.4 below for details). From the inverse perspective, it would be easy to see that $\rho = \lim \frac{p}{n}$ is the phase transition parameter of the SAVE and PHD procedure. We also speculate that it could be the phase transition parameter of PFC. These problems will be addressed in our future research.

Though it seems quite obvious that the ratio ρ should have played the role of the phase transition parameter for SIR and other SDR algorithms, such a detailed understanding is not achievable from the forward regression perspective. Nevertheless, after determining the phase transition parameter, determining the optimal rate of estimating central subspace in high dimension with sparsity constraints is no longer formidable.

12.3 Variable Selection

12.3.1 Beyond Sufficient Dimension Reduction: The Necessity of Variable Selection

The aforementioned SIR and related methods were developed for dimension reduction via the identification of the SDR space of $\mathbf{X}$. However, the estimation of SDR directions does not automatically lead to variable selection. As pointed out by Theorem 1, in high-dimensional scenarios with $\lim_{n\to\infty} p/n \to 0$, the SIR estimator of $\mathrm{var}\left(\mathbb{E}\left[\mathbf{X} \mid Y\right]\right)$ is inconsistent. Therefore, a direct application of SIR on high-dimensional data can result in a very poor estimation accuracy of SDR directions. To overcome the curse of dimensionality, various methods have been developed to simultaneously perform dimension reduction and variable selection for model in (12.1). For example, Li et al. (2005) designed a backward subset selection method, and Li (2007) developed the sparse SIR (SSIR) algorithm to obtain shrinkage estimates of the SDR directions with sparsity under the L_1 norm.

Motivated by the connection between SIR and multiple linear regression (MLR), Zhong et al. (2012) proposed a forward stepwise variable selection procedure called correlation pursuit (COP) for index models. The aforementioned methods including the original SIR method consider only the information from the first conditional moment, $\mathbb{E}\left(\mathbf{X} \mid Y\right)$, and will miss important variables with interactions or other second-order effects. In order to overcome this difficulty, Jiang et al. (2014) proposed the SIRI method that also utilizes the second conditional moment $\mathrm{var}\left(\mathbf{X} \mid Y\right)$ for selecting variables with higher-order effects. In this section, we review the variable selection methods inspired by SDR algorithms. We first review SIR as transformation-projection pursuit problem, and then review the COP and SIRI methods inspired by this perspective.

12.3.2 SIR as a Transformation-Projection Pursuit Problem

Without loss of generality, we assume that $\mathbf{X}$ is standardized so that $\mathbb{E}\left(\mathbf{X}\right) = 0$ and $\mathrm{var}\left(\mathbf{X}\right) = \mathbf{I}_p$. In multiple linear regression (MLR), the R-squared can be expressed as

$$R^2 = \max_{\eta \in \mathbb{R}^p} \left[\mathrm{Corr}\left(Y, \eta^T \mathbf{X}\right)\right]^2,$$

while in SIR, we can define the profile correlation between Y and $\eta^T\mathbf{X}$ as

$$P\left(\eta\right) = \max_{T} \left(\mathrm{Corr}(T\left(Y\right), \eta^T \mathbf{X})\right), \tag{12.12}$$

where the maximization is taken over any transformation $T(\cdot)$. Using $P(\eta)$ as the projection function, we may look for the first *profile direction* η_1 that maximizes $P(\eta)$. Then we can find the second profile direction η_2, which maximizes (12.12) and is orthogonal to η_1. This process can be continued until all profile directions are found. The following theorem connects $\text{var}(\mathbb{E}(\mathbf{X} \mid Y))$, and thus SIR, with MLR.

Theorem 2 *Let $\eta_1, \ldots, \eta_p$ be the principal directions between Y ad X, and let $\lambda_k = P(\eta_k)$ denote the profile correlation of η_k. Then, λ_k is the kth largest eigenvalue of* $\text{var}(\mathbb{E}(\mathbf{X} \mid Y))$, *for $k = 1, \ldots, p$.*

Given independent observations $\{(y_i, \mathbf{x}_i)\}_{i=1}^n$, SIR first divides the range of the $\{y_i\}_{i=1}^n$ into H disjoint intervals, denoted as $S_1, \ldots, S_H$, and computes for $h = 1, \ldots, n$, $\mathbf{x}_h = n_h^{-1} \sum_{y_i \in S_h} \mathbf{x}_i$, where n_h is the number of y_i's in S_h. Then SIR estimates $\text{var}(\mathbb{E}(\mathbf{X} \mid Y))$ by

$$\hat{\mathbf{M}} = n^{-1} \sum_{h=1}^{H} n_h (\mathbf{x}_h - \mathbf{x})(\mathbf{x}_h - \mathbf{x})^T \tag{12.13}$$

and $\text{var}(\mathbf{X})$ by the sample covariance matrix $\hat{\boldsymbol{\Sigma}}$. Finally, SIR uses the first K eigenvectors of $\hat{\boldsymbol{\Sigma}}^{-1}\hat{\mathbf{M}}$, denoted as $\hat{\eta}_1, \ldots, \hat{\eta}_K$ to estimate $\eta_1, \ldots, \eta_K$. The first K eigenvalues of $\hat{\boldsymbol{\Sigma}}^{-1}\hat{\mathbf{M}}$, denoted as $\hat{\lambda}_1, \ldots, \hat{\lambda}_K$, are used to estimate the first K profile correlations.

12.3.3 COP: Correlation Pursuit

SIR needs to estimate the eigenvalues and eigenvectors of $p \times p$ covariance matrices Σ and M. In high-dimensional scenarios, there are usually a large number of irrelevant variables and the sample size n is relatively small, namely $p \gg n$. Under this setting, $\hat{\Sigma}$ and $\hat{M}$ are usually very unstable, which leads to very inaccurate estimates of principal directions $\hat{\eta}_1, \ldots, \hat{\eta}_K$ and profile correlations $\hat{\lambda}_1, \ldots, \hat{\lambda}_K$.

Zhong et al. (2012) proposed the correlation pursuit (COP), a stepwise procedure for simultaneous dimension reduction and variable selection under the SDR model. Instead of including all covariates, COP starts with a small set of randomly selected predictors and iterates between an addition step, which selects and adds a predictor to the collection, and a deletion step, which selects and deletes a predictor from the collection. The procedure terminates when no new addition or deletion occurs. The addition and deletion steps are briefly described as follows.

Addition Step Let $\mathscr{C}$ denote the set of the indices of the selected predictors. Applying SIR to the data involving only the predictors in $X_{\mathscr{C}}$, we obtain the estimated squared profile correlations $\hat{\lambda}_1^{\mathscr{C}}, \hat{\lambda}_2^{\mathscr{C}}, \ldots, \hat{\lambda}_K^{\mathscr{C}}$. Superscript $\mathscr{C}$ indicates that the estimated squared profile correlations depend on the current subset of selected predictors. Let X_t be an arbitrary predictor outside $\mathscr{C}$ and $\mathscr{C}+t = \mathscr{C} \cup \{t\}$. Applying

SIR to the data involving the predictors in $\mathscr{C} + t$, we obtain the estimated squared profile correlations $\hat{\lambda}_1^{\mathscr{C}+t}, \hat{\lambda}_2^{\mathscr{C}+t}, \ldots, \hat{\lambda}_K^{\mathscr{C}+t}$. Because $\mathscr{C} \subset \mathscr{C} + t$, it is easy to see that $\hat{\lambda}_1^{\mathscr{C}} \leq \hat{\lambda}_1^{\mathscr{C}+t}$. The difference $\hat{\lambda}_1^{\mathscr{C}+t} - \hat{\lambda}_1^{\mathscr{C}}$ reflects the amount of improvement in the first profile correlation due to the incorporation of X_t. COP standardizes this difference and uses the resulting test statistic

$$\mathrm{COP}_i^{\mathscr{C}+t} = \frac{n\left(\hat{\lambda}_i^{\mathscr{C}+t} - \hat{\lambda}_i^{\mathscr{C}}\right)}{1 - \hat{\lambda}_i^{\mathscr{C}+t}},$$

to assess the significance of adding X_t to $\mathscr{C}$ in improving the ith profile correlation, for $2 \leq i \leq K$. The overall contribution of adding X_t to the improvement in all the K profile correlations can be assessed by combining the statistics $\mathrm{COP}_i^{\mathscr{C}+t}$ into one single test statistic,

$$\mathrm{COP}_{1:K}^{\mathscr{C}+t} = \sum_{i=1}^{K} \mathrm{COP}_i^{\mathscr{C}+t}.$$

COP further defines that

$$\overline{\mathrm{COP}}_{1:K}^{\mathscr{C}} = \max_{t \in \mathscr{C}^c} \left(\mathrm{COP}_{1:K}^{\mathscr{C}+t}\right).$$

Let $X_{\bar{t}}$ be a predictor that attains $\overline{\mathrm{COP}}_{1:K}^{\mathscr{C}}$, i.e. $\mathrm{COP}_{1:K}^{\mathscr{C}+\bar{t}} = \overline{\mathrm{COP}}_{1:K}^{\mathscr{C}}$, and let c_e be a prespecified threshold. Then COP adds $\bar{t}$ into $\mathscr{C}$ if $\overline{\mathrm{COP}}_{1:K}^{\mathscr{C}} > c_e$, otherwise no variable will be added.

Deletion Step Let X_t be an arbitrary predictor in $\mathscr{C}$ and define $\mathscr{C} - t = \mathscr{C} \backslash \{t\}$. Let $\hat{\lambda}_1^{\mathscr{C}-t}, \ldots, \hat{\lambda}_K^{\mathscr{C}-t}$ be the estimated squared profile correlations based on the data involving the predictors in $\mathscr{C} - t$ only. The effect of deleting X_t from $\mathscr{C}$ on the ith squared profile correlation can be measured by

$$\mathrm{COP}_i^{\mathscr{C}-t} = \frac{n\left(\hat{\lambda}_i^{\mathscr{C}} - \hat{\lambda}_i^{\mathscr{C}-t}\right)}{1 - \hat{\lambda}_i^{\mathscr{C}}}, \tag{12.14}$$

for $1 \leq i \leq K$. The overall effect of deleting X_t is measured by

$$\mathrm{COP}_{1:K}^{\mathscr{C}-t} = \sum_{i=1}^{K} \mathrm{COP}_i^{\mathscr{C}-t}, \tag{12.15}$$

and the least effect from deleting one predictor from $\mathscr{C}$ is then defined to be

$$\underline{\mathrm{COP}}_{1:K}^{\mathscr{C}} = \min_{t \in \mathscr{A}} \mathrm{COP}_{1:K}^{\mathscr{C}-t}. \tag{12.16}$$

Let $X_{\underline{t}}$ be a predictor that achieves $\underline{\text{COP}}^{\mathscr{C}}_{1:K}$, and let c_d be a pre-specified threshold for deletion. If $\underline{\text{COP}}^{\mathscr{C}}_{1:K} < c_d$, we delete $X_{\underline{t}}$ from $\mathscr{C}$.

Algorithm 1: Summary of the COP algorithm

---

1. **Initialization:** Set the number of principal directions K and randomly select $K+1$ variables as the initial collection of selected variables $\mathscr{C}$.
2. **Addition/deletion:** Iterate until no more addition or deletion of predictors can be performed:

   a. **Addition:**

      i. Find $\bar{t}$ such that $\text{COP}^{\mathscr{C}+\bar{t}}_{1:K} = \overline{\text{COP}}^{\mathscr{C}}_{1:K}$ and
      ii. If $\overline{\text{COP}}^{\mathscr{C}}_{1:K} > c_e$, add $\bar{t}$ to $\mathscr{C}$

   b. **Deletion:**

      i. Find $\underline{t}$ such that $\underline{\text{COP}}^{\mathscr{C}-\underline{t}}_{1:K} = \underline{\text{COP}}^{\mathscr{C}}_{1:K}$ and
      ii. If $\underline{\text{COP}}^{\mathscr{C}}_{1:K} < c_d$, delete $\underline{t}$ from $\mathscr{C}$

The following theorem in Zhong et al. (2012) guarantees that under certain regularity conditions, by appropriately choosing thresholds c_e and c_d, the COP procedure is consistent in variable selection. That is, it will keep adding predictors until all the stepwise detectable predictors have been included, and keep removing predictors until all the redundant variables have been excluded.

Theorem 3 *Let $\mathscr{C}$ be the set of currently selected predictors and let $\mathscr{A}$ be the set of true predictors. Let ϑ be a positive constant determined by technical conditions in Zhong et al. (2012) and C be any positive constant. Then we have*

$$\Pr\left\{\min_{\mathscr{C}:\mathscr{C}^c\cap\mathscr{A}\neq\emptyset}\ \max_{t\in\mathscr{C}^c\cap\mathscr{A}}\left(\text{COP}^{\mathscr{C}+t}_{1:K}\right)\geq\vartheta n^{1-\xi_0}\right\}\to 1, \tag{12.17}$$

and

$$\Pr\left\{\max_{\mathscr{C}:\mathscr{C}^c\cap\mathscr{A}\neq\emptyset}\ \max_{t\in\mathscr{C}^c}\left(\text{COP}^{\mathscr{C}+t}_{1:K}\right)< Cn^{\varrho}\right\}\to 1, \tag{12.18}$$

for any constants $\xi_0 > 0$, $\varrho_0 > 0$, $\varrho > 0$ such that $2\varrho_0 + 2\xi_0 < 1$ and $\varrho > \frac{1}{2} + \varrho_0$.

Because $\max_{t\in\mathscr{C}^c}\left(\text{COP}^{\mathscr{C}+t}_{1:K}\right)\geq\max_{t\in\mathscr{C}^c\cap\mathscr{A}}\left(\text{COP}^{\mathscr{C}+t}_{1:K}\right)$, from (12.17) we have

$$\Pr\left\{\min_{\mathscr{C}:\mathscr{C}^c\cap\mathscr{A}\neq\emptyset}\left(\overline{\text{COP}}^{\mathscr{C}}_{1:K}\right)\geq\vartheta n^{1-\xi_0}\right\}\to 1. \tag{12.19}$$

Consider one set of selected predictors $\tilde{\mathscr{C}} \supset \mathscr{T}$, then

$$\underline{\mathrm{COP}}^{\mathscr{C}}_{1:K} \leq \min_{t\in\tilde{\mathscr{C}}-\mathscr{T}} \left(\mathrm{COP}^{\tilde{\mathscr{C}}-t}_{1:K}\right) \leq \max_{\mathscr{C}:\mathscr{C}^c\cap\mathscr{A}=\emptyset} \max_{t\in\mathscr{C}^c} \left(\sum_{k=1}^{K} \mathrm{COP}^{\mathscr{C}+t}_{k}\right). \tag{12.20}$$

Therefore from (12.18) we have

$$\Pr\left\{\underline{\mathrm{COP}}^{\tilde{\mathscr{C}}}_{1:K} < Cn^{\varrho}\right\} \to 1. \tag{12.21}$$

One possible choice of the thresholds is $\chi^2_e = \vartheta n^{1-\xi_0}$ and $\chi^2_d = \vartheta n^{1-\xi_0}/2$. From (12.19), asymptotically, the COP algorithm will not stop selecting variables until all the true predictors have been included. Moreover, once all the true predictors have been included, according to (12.21), all the redundant variables will be removed from the selected variables.

12.3.4 From COP to SIRI

The original SIR method and COP only consider the information from the first conditional moment, $\mathbb{E}(\mathbf{X} \mid Y)$, and may miss important predictors linked to the response through quadratic functions or interactions. Recently, Jiang et al. (2014) proposed a stepwise variable selection method for general index models based on likelihood ratio test based procedure, named SIRI. SIRI starts with the inverse conditional Gaussian model with equal covariance matrices, whose likelihood ratio test statistic is asymptotically identical to COP, and then extends the first model to allow conditional covariance matrices vary with Y, which leads to a test statistic that can detect interactions and other second-order terms from information in $\mathrm{var}(\mathbf{X} \mid Y)$.

Let $\{(\mathbf{x}_i, y_i)\}_{i=1}^n$ denote n independent observations. SIRI first divides the range of $\{y_i\}_{i=1}^n$ into H disjoint slices, denoted as $S_1, \ldots, S_H$. Let $\mathscr{A}$ denote the set of truly relevant predictors. The first model SIRI utilizes, as shown below, is the one employed in Szretter and Yohai (2009) to show that the maximum likelihood estimate of the subspace $\mathbb{V}^K$ in (12.22) coincides with the subspace spanned by SDR directions estimated from the SIR algorithm:

$$\begin{aligned} \mathbf{X}_{\mathscr{A}} \mid Y \in S_h &\sim \mathscr{N}(\boldsymbol{\mu}_h, \boldsymbol{\Sigma}), \\ \mathbf{X}_{\mathscr{A}^c} \mid \mathbf{X}_{\mathscr{A}}, Y \in S_h &\sim \mathscr{N}\left(\boldsymbol{a} + \boldsymbol{B}^T\mathbf{X}_{\mathscr{A}}, \boldsymbol{\Sigma}_0\right), \end{aligned} \quad h = 1, \ldots, H, \tag{12.22}$$

where $\mu_h \in \mu + \mathbb{V}^K$ belongs to a K-dimensional affine space, $\mathbb{V}^K$ is a K-dimensional space ($K < p$) and $\mu \in \mathbb{R}^p$. It was assumed in (12.22) that the conditional distribution of relevant predictors follows the multivariate normal model and has a common covariance matrix in different slices.

The likelihood ratio test procedure based on (12.22) is as follows. Given the current set of selected predictors indexed by $\mathscr{C}$ and another predictor indexed by $j \notin \mathscr{C}$, SIRI tests the following hypotheses:

$$H_0 : \mathscr{A} = \mathscr{C} \quad \text{v.s.} \quad H_1 : \mathscr{A} = \mathscr{C} \cup \{j\} . \tag{12.23}$$

Let $L_{j|\mathscr{C}}$ denote the likelihood-ratio test statistic. SIRI uses $\hat{D}_{j|\mathscr{C}} = \frac{2}{n} \log\left(L_{j|\mathscr{C}}\right)$ as the test statistic for the hypothesis above. It was shown in Jiang et al. (2014) that

$$\hat{D}_{j|\mathscr{C}} = \sum_{k=1}^{K} \log\left(1 + \frac{\hat{\lambda}_k^{d+1} - \hat{\lambda}_k^d}{1 - \hat{\lambda}_k^{d+1}}\right), \tag{12.24}$$

where $\hat{\lambda}_k^d$ and $\hat{\lambda}_k^{d+1}$ are estimates of the kth profile-R^2 based on $\mathbf{x}_{\mathscr{C}}$ and $\mathbf{x}_{\mathscr{C}\cup\{j\}}$ respectively. Under the null hypothesis, $\frac{\hat{\lambda}_k^{d+1} - \hat{\lambda}_k^d}{1 - \hat{\lambda}_k^{d+1}} \to 0$ as $n \to \infty$, so

$$n\hat{D}_{j|\mathscr{C}} = n \sum_{k=1}^{K} \frac{\hat{\lambda}_k^{d+1} - \hat{\lambda}_k^d}{1 - \hat{\lambda}_k^{d+1}} + o_p(1), \tag{12.25}$$

and this expression coincides with COP statistics,

$$\mathrm{COP}_{1:K}^{d+1} = \sum_{k=1}^{K} \mathrm{COP}_k^{d+1}, \quad \text{where} \quad \mathrm{COP}_k^{d+1} = \frac{n\left(\hat{\lambda}_k^{d+1} - \hat{\lambda}_k^d\right)}{1 - \hat{\lambda}_k^{d+1}}. \tag{12.26}$$

The test statistic in (12.24) based on likelihood in (12.22) captures differences in the mean vectors in different slices (i.e., conditional on different y's). However, it is possible that for a subset $\mathscr{A}_1 \subsetneq \mathscr{A}$ that means $\mathbb{E}\left(X_{\mathscr{A}_1} \mid X_{\mathscr{A}\setminus\mathscr{A}_1}, Y \in S_h\right)$ are the same for $h = 1, \ldots, H$. In such cases neither SIR nor COP can identify $\mathscr{A}_1$ as part of $\mathscr{A}$. The second model of SIRI was designed to address this problem. It utilizes the second conditional moment var $(\mathbf{X} \mid Y)$ to select variables with different (conditional) variances across slices. In particular, SIRI augments model (12.22) to a more general form,

$$\begin{aligned} \mathbf{X}_{\mathscr{A}} \mid Y \in S_h &\sim \mathscr{N}(\boldsymbol{\mu}_h, \boldsymbol{\Sigma}_h), \\ \mathbf{X}_{\mathscr{A}^c} \mid \mathbf{X}_{\mathscr{A}}, Y \in S_h &\sim \mathscr{N}\left(\boldsymbol{a} + \boldsymbol{B}^T\mathbf{X}_{\mathscr{A}}, \boldsymbol{\Sigma}_0\right), \end{aligned} \quad h = 1, \ldots, H, \tag{12.27}$$

which differs from model (12.22) in its allowing for slice-dependent covariance matrices for relevant predictors. Testing the same hypothesis as in (12.23), SIRI's log-likelihood ratio test statistic under the augmented model takes a simpler form than that based on (12.22):

$$\hat{D}_{j|\mathscr{C}}^* = \log \hat{\sigma}_{j|\mathscr{C}}^2 - \sum_{h=1}^{H} \frac{n_h}{n} \log\left[\hat{\sigma}_{j|\mathscr{C}}^{(h)}\right]^2, \tag{12.28}$$

where $\left[\hat{\sigma}^{(h)}_{j|\mathscr{C}}\right]^2$ is the estimated residual variance by regressing X_j on $\mathbf{X}_{\mathscr{C}}$ in slice S_h, and $\hat{\sigma}^2_{j|\mathscr{C}}$ is the estimated residual variance by regressing X_j on $\mathbf{X}_{\mathscr{C}}$ using all observations.

SIRI starts with $\mathscr{C} = \emptyset$ and uses a similar addition/deletion procedure as COP for variable selection. In ultra-high dimensional scenario, SIRI uses $\hat{D}^*_{j|\mathscr{C}}$ as a measure to carry out the sure independence screening to reduce the dimensionality from ultra-high to moderately high. In particular, SIRI ranks predictors according to $\left\{\hat{D}^*_{j|\mathscr{C}}, 1 \le j \le p\right\}$, then SIRI's sure independence screening procedure, named as SIS*, reduces the number of predictors from p to $o(n)$.

Algorithm 2: Summary of the SIRI algorithm

1. **Initialization:** Let $\mathscr{C} = \emptyset$; rank predictors according to $\left\{\hat{D}^*_{j|\emptyset}, 1 \le j \le p\right\}$ and select a subset of predictors, denoted as $\mathscr{S}$. (this step is called SIS*). The proper size of the $\mathscr{S}$ depends on a few constants in technical conditions of Jiang et al. (2014).
2. **Detecting variables with mean effects:** Select predictors from set $\mathscr{S}\backslash\mathscr{C}$ by using the addition and deletion steps based on $\hat{D}_{j|\mathscr{C}}$ in (12.24) and add the selected predictors into $\mathscr{C}$.

   a. **Addition:** Find j_a such that $\hat{D}_{j_a|\mathscr{C}} = \max_{j\in\mathscr{C}^c} \hat{D}_{j|\mathscr{C}}$. Let $\mathscr{C} = \mathscr{C} + \{j_a\}$ if $\hat{D}_{j_a|\mathscr{C}} > \upsilon_a$.
   b. **Deletion:** Find j_d such that $\hat{D}_{j_d|\mathscr{C}\backslash\{j_d\}} = \min_{j\in\mathscr{C}} \hat{D}_{j|\mathscr{C}\backslash\{j\}}$. Let $\mathscr{C} = \mathscr{C} - \{j_d\}$ if $\hat{D}_{j_d|\mathscr{C}\backslash\{j_d\}} < \upsilon_d$.

3. **Detecting variables with second-order effects:** Select predictors from set $\mathscr{S}\backslash\mathscr{C}$ by using the addition and deletion steps based on $\hat{D}^*_{j|\mathscr{C}}$ in (12.28) and add the selected predictors into $\mathscr{C}$.

   a. **Addition:** Find j_a such that $\hat{D}^*_{j_a|\mathscr{C}} = \max_{j\in\mathscr{C}^c} \hat{D}^*_{j|\mathscr{C}}$. Let $\mathscr{C} = \mathscr{C} + \{j_a\}$ if $\hat{D}^*_{j_a|\mathscr{C}} > \upsilon_a$.
   b. **Deletion:** Find j_d such that $\hat{D}^*_{j_d|\mathscr{C}\backslash\{j_d\}} = \min_{j\in\mathscr{C}} \hat{D}^*_{j|\mathscr{C}\backslash\{j\}}$. Let $\mathscr{C} = \mathscr{C} - \{j_d\}$ if $\hat{D}^*_{j_d|\mathscr{C}\backslash\{j_d\}} < \upsilon_d$.

4. **Iterative sure independence screening**: Conditioning on the current selection $\mathscr{C}$, rank the remaining predictors based $\left\{\hat{D}^*_{j|\mathscr{C}}, j \notin \mathscr{C}\right\}$, update set $\mathscr{S}$ using SIS*, and iterate steps 2–4 until no more predictors are selected.

Theorems 1 and 2 in Jiang et al. (2014) establish the variable selection consistency of SIRI under certain technical conditions. It showed that by appropriately choosing thresholds, the SIRI procedure will keep adding predictors until all the stepwise detectable predictors have been included, and keep removing predictors until all the redundant variables have been excluded. Theorem 3 of Jiang et al. (2014) established the independence screening consistency of SIS* under certain technical conditions.

12.3.5 *Simulation Study for Variable Selection and SDR Estimation*

We simulated a few testing examples to compare the performance of original SIR, COP and SIRI methods for variable selection and SDR direction estimation. We considered four simulation settings, indexed as Examples 1–4 and listed below:

$$\begin{aligned}
&\text{Example 1}: \quad && Y = X_1 + X_2 + X_3 + X_4 + X_5 + \sigma\epsilon, \quad p = 20,\\
&\text{Example 2}: \quad && Y = X_1 + X_2 + X_3 + X_4 + X_5 + \sigma\epsilon, \quad p = 200,\\
&\text{Example 3}: \quad && Y = X_1 + X_1 \cdot (X_2 + X_3) + \sigma\epsilon, \quad p = 200,\\
&\text{Example 4}: \quad && Y = X_1 X_2^2 + \sigma\epsilon, \quad p = 200,
\end{aligned}$$

where $\sigma = 0.2$ and ε is independent of $\mathbf{X}$ and follows $N(0, 1)$. In each example, we simulated the predictors $\mathbf{X}$ from multivariate Gaussian with correlation $0.5^{|i-j|}$. Examples 1 and 2 have the same functional relationship between Y and $\mathbf{X}$ but different sample different dimensionality (i.e., $p = 20$ versus $p = 200$), and have only one SDR $X_1 + \cdots + X_5$. We used them to test the methods with both low- and high-dimensional design matrix. Example 3 has two SDRs, X_1 and $X_2 + X_3$, and Example 4 also has two SDRs, X_1 and X_2. In Example 3, X_1, X_2 and X_3 are true predictors but X_2 and X_3 have only second-order (no first-order) effects. In Example 4, X_1 and X_2 are true predictors but X_2 has only second-order (variance/covariance) but no first-order (mean) effects in different slices.

For each simulation setting, we generated one dataset with $n = 250$. For all the three methods, we equally partitioned $\{y_i\}_{i=1}^n$ into $H = 5$ slices. We let COP and SIRI automatically estimate the number of SDRs by cross-validation, but input the true number of SDR for SIR on the four examples. We use COP and SIRI to conduct variable selection, and then run original SIR on the selected variables to estimate the SDR directions. We respectively name them as COP-SIR and SIRI-SIR method.

We first compare the variable selection performance of COP and SIRI on the four examples. As shown in Table 12.1, we observe that COP worked well to select the true predictors for Examples 1 and 2. However, for Examples 3 and 4, as discussed previously, COP was unable to select the predictors without first-order effect. SIRI correctly selected the true predictors for the four examples.

Table 12.1 Variable selection results for COP and SIRI on Examples 1–4

Example	Truth	COP	SIRI
1	X_1, X_2, X_3, X_4, X_5	X_1, X_2, X_3, X_4, X_5	X_1, X_2, X_3, X_4, X_5
2	X_1, X_2, X_3, X_4, X_5	X_1, X_2, X_3, X_4, X_5	X_1, X_2, X_3, X_4, X_5
3	X_1, X_2, X_3	X_1	X_1, X_2, X_3
4	X_1, X_2	X_1	X_1, X_2

We then compared the SDR estimation of original SIR, COP-SIR and SIRI-SIR on Examples 1 and 2. Examples 1 and 2 have only one SDR direction, so it is feasible to plot unitary SDR direction estimations. The estimated SDR directions of Examples 1 and 2 by the three methods are shown in Figs. 12.1 and 12.2. It is observed that, for Example 1, since the data is in low to moderate dimension relative to the sample size, the SDR directions estimated by SIR, COP-SIR and SIRI-SIR are all very close to the truth. SIR-estimated SDR direction has small loadings on redundant predictors $X_6, \ldots, X_{20}$, and COP-SIR and SIRI-SIR have exactly zero loadings on them because of the employment of the variable selection procedures. However, as shown in Fig. 12.2 for Example 2, the estimated SDR direction from the original SIR method is almost random due to the high dimension of the data. On the other hand, COP and SIRI procedures, which first selected the true predictors, and then conducted SIR to estimate the SDR direction only on those selected true predictors, obtained almost perfect answers.

In summary, we demonstrated in these simulations that: (a) for datasets with a large number of predictors, it is essential to first conduct a variable selection procedure, such as COP and SIRI, before estimating the SDR direction since otherwise the curse of dimension can completely overwhelm the SDR direction estimation; and (b) both COP and SIRI can select true predictors with first-order effects, while SIRI can also detect predictors with only second-order effects.

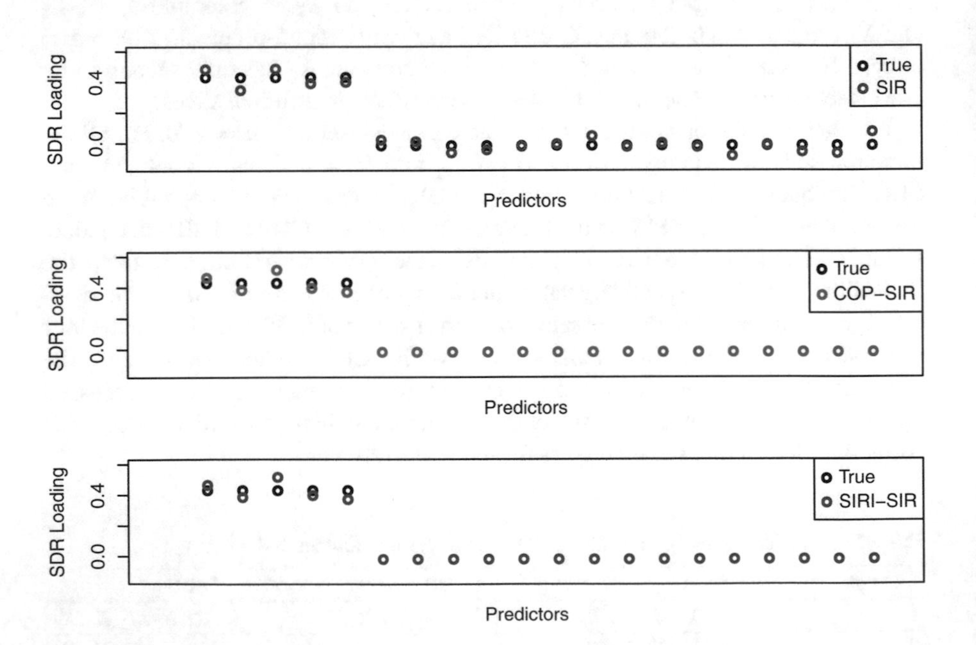

Fig. 12.1 SDR direction estimations for Example 1

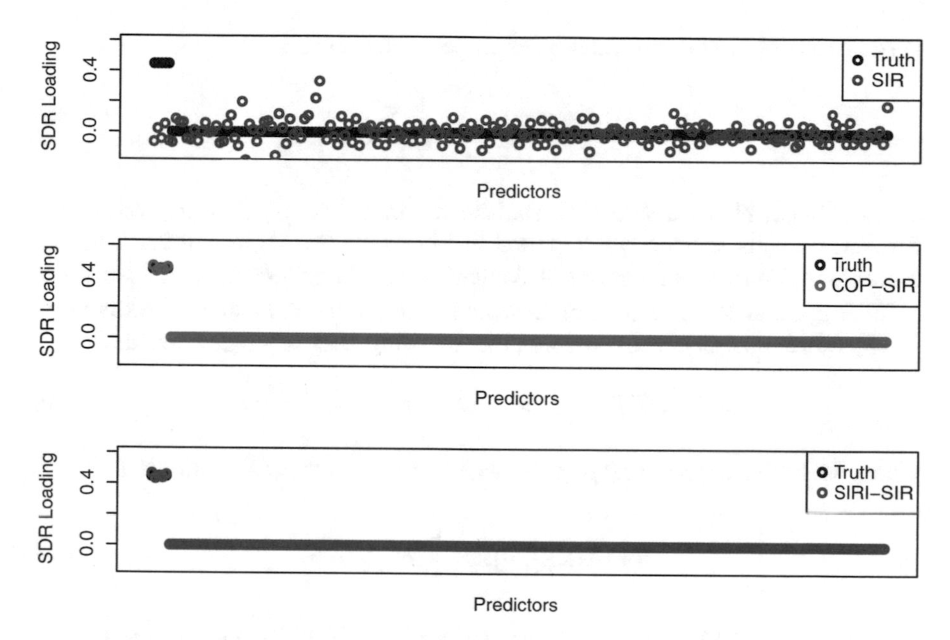

Fig. 12.2 SDR direction estimations for Example 2

12.4 Nonparametric Dependence Screening

When the dimension p is much larger than the sample size n, screening out irrelevant predictors is often crucial for data analysis. Fan and Lv (2008) performed the sure independence screening (SIS) on high-dimensional linear models and proved that, under mild conditions, SIS keeps important features with high probability[2] and reduces the dimension p significantly so that more delicate analyses are possible. Since SIS requires the model to be linear and the error distribution to be normal, which is restrictive for some applications, researchers start to search for other screening statistics for general distributions and nonlinear models. Some early methods (Fan et al. 2009, 2010) were designed for parametric models, which can be difficult to justify in practice. Nonparametric screening statistics are therefore of particular interest in recent years (Zhu et al. 2011; Li et al. 2012; Cui et al. 2014; Fan et al. 2011), some of which will be reviewed in this section.

One of the key ingredients in SIR might stem from a seemingly trivial observation from the total variance decomposition $\text{var}(\boldsymbol{X}) = \text{var}(\mathbb{E}[\boldsymbol{X}|\, Y]) + \mathbb{E}[\text{var}(\boldsymbol{X}|\, Y)]$. In model (12.1), it is clear that, if $X \perp\!\!\!\perp Y$, then all the $\beta_i's$ should be 0, i.e., $\text{var}(\mathbb{E}[X|\, Y]) = 0$. The coverage condition then guarantees that $X \not\!\perp\!\!\!\perp Y$ implies

[2]In the below, we will call this property the SIS property.

$\mathrm{var}(\mathbb{E}[X|\,Y]) \neq 0$. This motivates us to propose the estimate

$$\widehat{\mathrm{VCE}}(\mathbf{X}(i)|\,Y) = \frac{1}{H}\sum_{h} x_h^2. \tag{12.29}$$

of $\mathrm{var}(\mathbb{E}[\mathbf{X}(i)|\,Y])$ as a screening statistic for model (12.1). The numerical performance of this statistic was reported in Lin et al. (2018), in which a two-step algorithm DT-SIR was proposed. A somewhat surprising result is that the statistic $\widehat{\mathrm{VCE}}$ is optimal in terms of sample sizes for single index models (Neykov et al. 2016). More precisely, let us focus on the following class of single index models

$$Y = f(\boldsymbol{\beta}^\tau \mathbf{X}, \varepsilon),\ \mathbf{X} \sim N(0, \mathbf{I}_p),\ \ \varepsilon \sim N(0,1) \tag{12.30}$$

where we further assume that $\boldsymbol{\beta}_i \in \{\pm\frac{1}{\sqrt{s}}, 0\}$. Based on the observation that

$$\mathrm{var}(\mathbb{E}[\mathbf{X}(i)|\,Y]) \propto \frac{1}{s} \ \text{ if } \ \boldsymbol{\beta}_i \neq 0,$$

Neykov et al. (2016) showed that the statistic $\widehat{\mathrm{VCE}}(\mathbf{X}(i)|\,Y)$ achieves the optimal rate:

Theorem 4 *Let $\kappa = \frac{n}{s\log(p)}$, there exist two positive constants $C > c$, such that*

i) if $\kappa > C$, the thresholding algorithm based on the statistics $\widehat{\mathrm{VCE}}$ recovers the support of β with probability 1 as $n \to \infty$,
ii) if $\kappa < c$, any algorithm will fail to recover the support of β with probability at least $\frac{1}{2}$ as $n \to \infty$.

The following simulation study as shown in Fig. 12.3 provides empirical supports to the theorem.[3] The data were generated from the following models (12.31) with $\beta_i \in \{\pm\frac{1}{\sqrt{s}}\}$ for $1 \leq i \leq s$ and $\beta_i = 0$ for $s+1 \leq i \leq p$. We chose the slice number $H = 10$. In general, it should not be expected that the phase transition described in Theorem 4 occurs at the same place for these four models.

$$\begin{aligned}
Y &= \beta^\tau \mathbf{X} + \sin \beta^\tau \mathbf{X} + N(0,1),\\
Y &= 2\,\mathrm{atan}(\beta^\tau \mathbf{X}) + N(0,1),\\
Y &= (\beta^\tau \mathbf{X})^3 + N(0,1),\\
Y &= \sinh(\beta^\tau \mathbf{X}) + N(0,1).
\end{aligned} \tag{12.31}$$

Figure 12.3 presents plots for different p values in the regime $s = \sqrt{p}$. The X axis represents the rescaled sample size $\frac{n}{s\log(p-s)}$ and the Y axis shows the estimated

[3]The examples and figures are borrowed from Neykov et al. (2016).

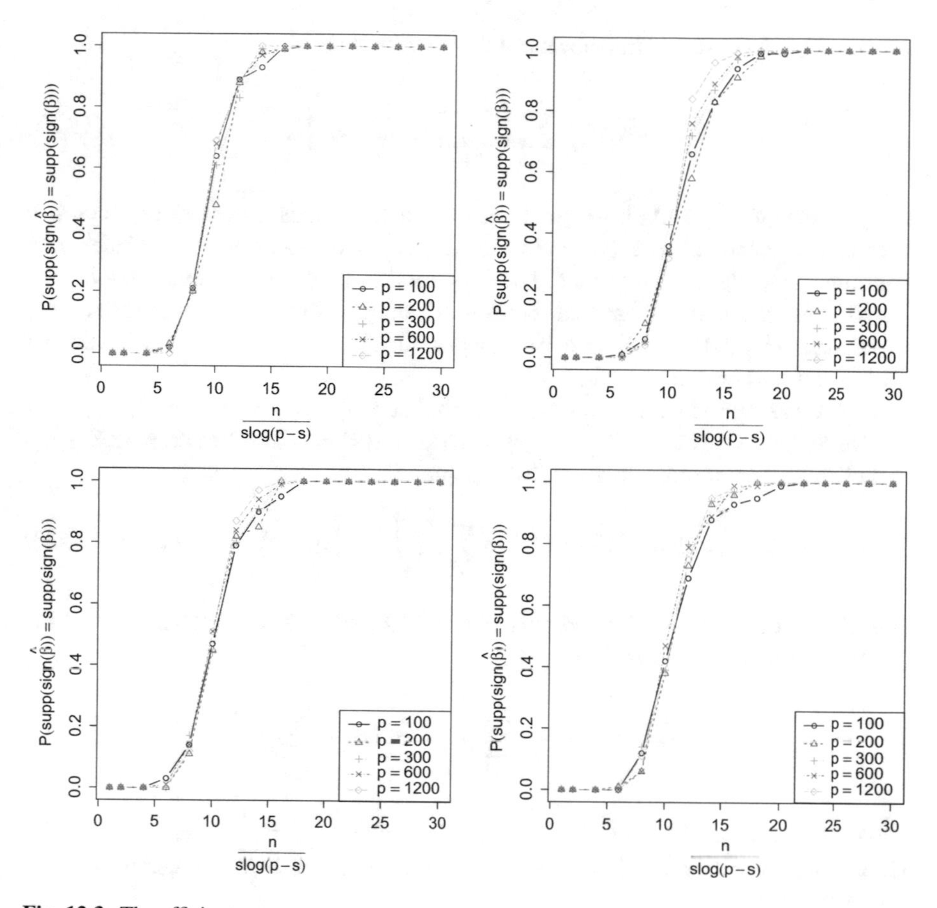

Fig. 12.3 The efficiency curve

probability of successful signed support recovery. We refer to these curves as efficiency curves (EC). It is clear that when $\kappa > 25$ the $\widehat{\text{VCE}}$ separated signals and noises with probability approaching 1 as n grows larger.

Since $\text{var}(\mathbb{E}[X(k)|\, y]) = 0$ does not imply that $X(k) \perp\!\!\!\perp y$ beyond the single index models (12.30), the statistics $\widehat{\text{VCE}}$ is not a satisfactory nonparametric screening statistics for higher-order effects. However, the idea of constructing a screening statistics (dependence test) based on the sliced samples is worth further investigation. We refer to quantities defined based on the information of the order statistics of y_i as *sliced statistics*. It is clear that the screening statistic used in SIRI is a sliced statistic. We will briefly review two other existing sliced statistics and propose a new family of sliced statistics. We believe that these examples have demonstrated the potential of the sliced inverse modeling idea.

To reiterate, we let Y be the response, let $\mathbf{X}$ be a p-dimension vector with $\mathbf{X}(k)$ being its k-th coordinate, and let X be a univariate random variable. To the best of our knowledge, the first statistics sharing the same feature as $\widehat{\text{VCE}}$ is proposed in

Zhu et al. (2011), where they observed that the estimate

$$\widehat{\omega}_k = \frac{1}{n}\sum_{j=1}^{n}\left(\frac{1}{n}\sum_{i=1}^{n}\mathbf{x}_i(k)\mathbf{1}_{y_i<y_j}\right)^2 \tag{12.32}$$

of $\omega_k = \mathbb{E}[\Omega_k(Y)^2]$ could be a candidate of screening statistics where Ω_k is its k-th component of the vector $\Omega(t) = \mathbb{E}\left[\mathbf{X}\mathbb{E}[\mathbf{1}_{y<t}|\mathbf{X}]\right] = \mathrm{cov}(\mathbf{X}, \mathbf{1}(y < t))$. Under mild conditions and the sub-exponential tail assumption on $\mathbf{X}$, they proved that $\widehat{\omega}_k$ has the SIS properties. It is clear that $\widehat{\omega}_k$ takes advantage of the order information of y_i which is similar to the $\widehat{\mathrm{VCE}}$, thus it belongs to the sliced statistics. It is believed that $\mathbf{x}(k) \perp\!\!\!\perp y$ if and only if $\omega_k = 0$.

If Y is a categorical random variable such that $\mathbb{P}(Y = y_r) = p_r, r = 1, \ldots, k$, we denote $F(x|\,Y) = \mathbb{P}(X \le x|\,Y)$, $F_r(x) = \mathbb{P}(X \le x|\,Y = y_r)$ and $F(x) = \mathbb{P}(X \le x)$. Cui et al. (2014) introduced the following quantity

$$\mathrm{MV}(X|\,Y) := E_X[\mathrm{var}_Y(F(X|\,Y))] = \sum_i p_i \int (F_r(x) - F(x))^2 dF(x). \tag{12.33}$$

and observed that $X \perp\!\!\!\perp Y$ if and only if $\mathrm{MV}(X|\,Y) = 0$. Then they introduced the following estimate of $\mathrm{MV}(x|\,y)$

$$\widehat{\mathrm{MV}}(x|\,y) = \frac{1}{n}\sum_{r=1}^{k}\sum_{j=1}^{n}\hat{p}_r[\widehat{F}_r(x_j) - \widehat{F}(x_j)] \tag{12.34}$$

where $\hat{p}_r = \frac{1}{n}\sum_{j=1}^{n}\mathbf{1}_{y_j=y_r}$, $\widehat{F}(x) = \frac{1}{n}\sum_{j=1}^{n}\mathbf{1}_{x_j\le x}$ and $\widehat{F}_r(x) = \frac{1}{n}\sum_{j=1}^{n}\mathbf{1}_{x_j\le x, y_j=y_r}/\hat{p}_r$. Under mild conditions, they proved that the statistics $\widehat{\mathrm{MV}}$ possesses the SIS property.

$\widehat{\omega}_k$ in (12.32) works for both discrete and continuous random variables, however, it requires the sub-exponential tail condition on $\mathbf{X}$. On the other hand, the statistic $\widehat{\mathrm{MV}}(X|\,Y)$ does not require the tail condition on $\mathbf{X}$, but, theoretically, it can only handle the discrete random variables. Besides these deficiencies, one of the drawbacks of these statistics is that they are not computationally efficient. Thus a computationally efficient statistic can handle both continuous and discrete random variables would be of great interest for screening. Recall that $\widehat{\mathrm{VCE}}$ is computationally efficient and can handle both continuous and discrete random variables. However, it requires the sub-Gaussian assumption on the $\mathbf{X}$ and is not a sufficient dependence screening statistic. This motivates us to propose the following family of statistics.

Note that $X \perp\!\!\!\perp Y$ if and only if $\mathrm{var}(\mathbb{E}[\exp(itX)|\,Y]) = 0$ for any $t \in \mathbb{R}$. For any positive weight function $\eta(t)$, we define the Variance of Weighted Conditional Characteristic (VWCC) as:

$$\mathrm{VWCC}_\eta = \int_{\mathbb{R}} \mathrm{var}(\mathbb{E}[\exp^{itX} \mid Y])\eta(t)dt. \tag{12.35}$$

It is easy to see that $\mathrm{VWCC}_\eta = 0$ if and only if $X \perp\!\!\!\perp Y$. Thus, an estimate of VWCC_η can be a candidate for nonparametric independence screening. The integral defined in (12.35) is conceptually clear but is hard to estimate. We resort to a numerical alternative. First, to avoid tedious analysis arguments, we choose a function $\eta(t)$ that is positive on a compact interval $\mathfrak{J}$ and equals 0 outside $\mathfrak{J}$. Without loss of generality, we may fix $\mathfrak{J} = [-1, 1]$. Second, we use the following numerical approximation to the integral in (12.35):

$$V_N(z) = \frac{1}{N}\sum_j \mathrm{var}\left(\mathbb{E}[\exp^{it_jX} \,\middle|\, Y]\right) \eta(t_j) \tag{12.36}$$

where $t_j = -1 + \frac{2j}{N}$ for $j = 0, \ldots, N$. We then let

$$\widehat{V}_{H,c}(z) = \frac{1}{N}\sum_j \mathrm{var}_{H,c}\left(\mathbb{E}[\exp^{it_jX} \,\middle|\, Y]\right) \eta(t_j). \tag{12.37}$$

as an estimate of the approximation formula (12.36). It is well known that, under mild conditions, the approximate error of the numerical approximation (12.36) is of order $O(\frac{1}{N})$. One of the main contributions in Lin et al. (2016) is to prove the SIS properties of $\hat{V}_{H,c}(z)$.

Remark 1 Different choice of η will give us different screening statistics. If we relax the positivity requirement on $\eta(t)$, we obtain more computational efficient screening statistics. For instance, by letting $\eta(t) = \delta_{t_0}$, we have

$$\mathrm{VCC}_{t_0} = \mathrm{var}\left(\mathbb{E}[\exp^{it_0X} \mid Y]\right). \tag{12.38}$$

By letting $\eta(t) = \frac{1}{t^2}\delta_0$ (as a generalized function), we see that VCE is a special case of the VWCC.

To conclude this section, we summarize these screening statistics in Table 12.2. The third column records the model assumptions for each screening statistic: Linear model (LM), Single index model (SIM), and Nonparametric model (NP). The fourth column records if the screening statistics is sufficient for dependence screening. The fifth column records the tail assumptions and (**) stands for very weak assumption on moments.

12.5 Conclusion

In the big data era, modeling nonlinear relationships is a notoriously difficult problem. In this chapter, we summarized some recent developments inspired by the seminal work of Li (1991), the *Sliced Inverse Regression* (SIR), including its high dimensional properties, some variable selection algorithms and a general class of

Table 12.2 Table of some screening statistics

Names	Statistics	Models	Dependence	Tails
SIS	$\text{cov}(Y, X)$	LM	N	Gaussian
$E\Omega$	$\mathbb{E}_Y[(\mathbb{E}[\mathbf{X}\mathbb{E}[\mathbf{1}_{Y<y}\vert\mathbf{X}]])^2]$	NP	N	Sub-Gaussian
SIRI	$\log \frac{\text{var}(X)}{\text{var}(X\vert Y)}$	NP	N	Sub-Gaussian
VCE	$\text{var}(\mathbb{E}[X\vert Y])$	SIM	N	Sub-Gaussian
MV(x)	$\mathbb{E}_X\left[\text{var}_Y\left(F(X\vert Y)\right)\right]$	NP	Y	**
VCC_{t_0}	$\text{var}\left(\mathbb{E}[\exp^{it_0X}\vert Y]\right)$	NP	N	**
VWCC_η	$\int_{\mathbb{R}} \text{var}\left(\mathbb{E}[\exp^{itX}\vert Y]\right)\eta(t)dt$	NP	Y	**

nonparametric screening statistics. Most of these developments benefited from the inverse modeling idea. It is interesting to see that although the *Sufficient Dimension Reduction*, which inherits from and emphasizes on the dimension reduction aspects of SIR, has gained lots of successes, the power of the inverse modeling idea has not been fully recognized. From what we reviewed here, we believe that the inverse modeling can bring us interesting methodologies in the big data era, as well as synergies between Bayesian and frequentist thinking.

References

Adragni KP, Cook RD (2009) Sufficient dimension reduction and prediction in regression. Philos Trans R Soc Lond A Math Phys Eng Sci 367(1906):4385–4405

Berthet Q, Rigollet P (2013) Complexity theoretic lower bounds for sparse principal component detection. In: Conference on learning theory, pp 1046–1066

Birnbaum A, Johnstone IM, Nadler B, Paul D (2013) Minimax bounds for sparse PCA with noisy high-dimensional data. Ann Stat 41(3):1055

Cai TT, Ma Z, Wu Y et al (2013) Sparse PCA: optimal rates and adaptive estimation. Ann Stat 41(6):3074–3110

Candes E, Tao T (2007) The Dantzig selector: statistical estimation when p is much larger than n. Ann Stat 35(6):2313–2351

Chen C-H, Li K-C (1998) Can SIR be as popular as multiple linear regression? Stat Sin 8(2):289–316

Cook RD (2007) Fisher lecture: dimension reduction in regression. Stat Sci 22(1):1–26

Cook RD, Weisberg S (1991) Discussion of a paper by K. C. Li. J Am Stat Assoc 86:328–332

Cui H, Li R, Zhong W (2014) Model-free feature screening for ultrahigh dimensional discriminant analysis. J Am Stat Assoc 110(510):630–641

Duan N, Li K (1991) Slicing regression: a link-free regression method. Ann Stat 19(2):505–530

Fan J, Lv J (2008) Sure independence screening for ultrahigh dimensional feature space. J R Stat Soc Ser B 70(5):849–911

Fan J, Samworth R, Wu Y (2009) Ultrahigh dimensional feature selection: beyond the linear model. J Mach Learn Res 10:2013–2038

Fan J, Song R et al (2010) Sure independence screening in generalized linear models with np-dimensionality. Ann Stat 38(6):3567–3604
Fan J, Feng Y, Song R (2011) Nonparametric independence screening in sparse ultra-high-dimensional additive models. J Am Stat Assoc 106(494):544–557
Hsing T, Carroll R (1992) An asymptotic theory for sliced inverse regression. Ann Stat 20(2):1040–1061
Jiang B, Liu JS et al (2014) Variable selection for general index models via sliced inverse regression. Ann Stat 42(5):1751–1786
Johnstone IM, Lu AY (2004) Sparse principal components analysis. Unpublished manuscript, p 7
Li K-C (1991) Sliced inverse regression for dimension reduction. J Am Stat Assoc 86(414):316–327
Li L (2007) Sparse sufficient dimension reduction. Biometrika 94(3):603–613
Li L, Cook RD, Nachtsheim CJ (2005) Model-free variable selection. J R Stat Soc Ser B (Stat Methodol) 67(2):285–299
Li Y, Zhu L-X et al (2007) Asymptotics for sliced average variance estimation. Ann Stat 35(1):41–69
Li R, Zhong W, Zhu L (2012) Feature screening via distance correlation learning. J Am Stat Assoc 107(499):1129–1139
Lin Q, Zhao Z, Liu JS (2016) Nonparametric screening in ultrahigh dimensions. Technical Report, Harvard University
Lin Q, Zhao Z, Liu JS (2018) On consistency and sparsity for sliced inverse regression in high dimensions. Ann Stat 46(2):580–610
Neykov M, Lin Q, Liu JS (2016) Signed support recovery for single index models in high-dimensions. Ann Math Sci Appl 1(2):379–426
Raskutti G, Wainwright MJ, Yu B (2011) Minimax rates of estimation for high-dimensional linear regression over-balls. IEEE Trans Inf Theory 57(10):6976–6994
Szretter ME, Yohai VJ (2009) The sliced inverse regression algorithm as a maximum likelihood procedure. J Stat Plann Inference 139(10):3570–3578
Tibshirani R (1996) Regression shrinkage and selection via the Lasso. J R Stat Soc Ser B (Methodol) 58(1):267–288
Vu V, Lei J (2012) Minimax rates of estimation for sparse PCA in high dimensions. In: Artificial intelligence and statistics, pp. 1278–1286
Yu Z, Dong Y, Zhu LX (2016) Trace pursuit: a general framework for model-free variable selection. J Am Stat Assoc 111(514):813–821
Zhong W, Zhang T, Zhu Y, Liu JS (2012) Correlation pursuit: forward stepwise variable selection for index models. J R Stat Soc Ser B 74(5):849–870
Zhu L, Miao B, Peng H (2006) On sliced inverse regression with high-dimensional covariates. J Am Stat Assoc 101(474):630–643
Zhu LP, Li L, Li R, Zhu LX (2011) Model-free feature screening for ultrahigh-dimensional data. J Am Stat Assoc 106(496):1464–1475
Zou H, Hastie T, Tibshirani R (2006) Sparse principal component analysis. J Comput Graph Stat 15(2):265–286

Chapter 13
Sufficient Dimension Reduction for Tensor Data

Yiwen Liu, Xin Xing, and Wenxuan Zhong

Abstract With the rapid development of science and technology, a large volume of array data has been collected in areas such as genomics, finance, image processing, and Internet search. How to extract useful information from massive data becomes the key issue nowadays. In spite of the urgent need for statistical tools to deal with such data, there are limited methods that can fully address the high-dimensional problem. In this chapter, we review the general setting of sufficient dimension reduction framework and its generalization to tensor data. Tensor is a multi-way array, and its usage is becoming more and more important with the advancement of social and behavioral science, chemistry, and imaging technology. The vector-based statistical methods can be applied to tensor data by vectorizing a tensor into a vector. However, vectorized tensor usually has a large dimension which may largely exceed the number of samples. To preserve the tensor structure and reduce the dimensionality simultaneously, we revisit the tensor sufficient dimension reduction model and apply it to colorimetric sensor arrays. Tensor sufficient dimension reduction method is simple but powerful and exhibits a competent empirical performance in real data analysis.

Keywords Sufficient dimension reduction · Tensor analysis · Iterative estimation · Colorimetric sensor arrays

13.1 Curse of Dimensionality

With the rapid development of science and technology, more and more massive data sets are routinely generated. For the first time, we have enough data to exploit the structural relationship of a large number of variables and to produce predictions

---

Yiwen Liu and Xin Xing contribute equally to the chapter.

Y. Liu · X. Xing · W. Zhong (✉)
University of Georgia, Athens, GA, USA
e-mail: yiwenliu@uga.edu; xinxing@uga.edu; wenxuan@uga.edu

© Springer International Publishing AG, part of Springer Nature 2018

W. K. Härdle et al. (eds.), *Handbook of Big Data Analytics*, Springer Handbooks of Computational Statistics, https://doi.org/10.1007/978-3-319-18284-1_13

and uncover patterns and anomalies. Regression analysis is probably the most popular statistical tool for data exploration and is commonly used to explore the relationship between a response Y and p predictors $(X_1, \ldots, X_p)$. Various regression models and estimation methods have been developed in the literature, ranging from classic linear regression to nonparametric regression. In general, regression can be considered as an inference about the conditional distribution of Y given X, often with the mean response $E(Y|X)$ of particular interest. When $E(Y|X)$ is a linear combination of X, the corresponding regression model is referred to as the linear model. On the other hand, if there does not exist information to suggest a particular relationship between Y and X, the nonparametric model can be employed. In both linear models and nonparametric models, when the number of predictors is extremely large relative to the sample size, the model may suffer from the curse of dimensionality, which refers to various difficulties a large number of predictors (or dimensions) can cause to function approximation, model fitting, information extraction as well as computation. Therefore, it is necessary to reduce the number of predictors or the dimensions to ensure the success of regression analysis. As a matter of fact, dimension reduction is indeed achievable in many applications where the response Y only depends on a subset of all the present predictors or their lower dimensional projections. It is clearly critical in regression to identify these predictors and their lower dimensional projections.

Dimension reduction in regression analysis has been comprehensively studied under two general frameworks: variable selection and low dimensional projection. Variable selection for linear regression models has been studied over decades, where the original motivation is to find the best predictive subset of variables to improve the prediction accuracy of a model. The early proposals on this direction include forward selection, backward elimination and bidirectional elimination (Akaike 1987; Efroymson 1960; Miller 2002). Although the aforementioned methods are very intuitive, their theoretical properties are difficult to study and establish, simply because we have no unified likelihood function to discuss their statistical properties. There are several studies for linear models in the early literature on stepwise regression (Bendel and Afifi 1977; Wilkinson 1979). However, these proposals have not been developed to their fruition. The first theoretical result along this direction was established in Zhong et al. (2012), where the variable selection consistency was established for both fixed p and growing p under the sufficient dimension reduction regression model that will be introduced later. Compared to the subset selection approach, the shrinkage based methods attract much attention for their theoretical advantages. The basic idea of the shrinkage approach is to introduce small estimation biases to have lower prediction error. Various methods have been proposed along the shrinkage direction, nonnegative garrotte (Breiman 1995; Yuan and Lin 2007), LASSO (Tibshirani 1996), SCAD (Fan and Li 2001; Fan et al. 2004), elastic net (Zou and Hastie 2005; Zou and Zhang 2009), adaptive LASSO (Zhang and Lu 2007; Zou 2006) etc. Their theoretical properties have been extensively studied not only for fixed p but also for varying p that grows with n.

Another line of dimension reduction aims at finding low dimensional projections of the predictors meanwhile preserving full regression information. Popular approaches proposed along this line of thinking include principal component regression, partial least square, projection pursuit (Friedman and Tukey 1974; Huber 1985; Kruskal 1969), and sufficient dimension reduction, where different approaches focus on the reduction of different sources of variations. In this chapter, we will focus on the sufficient dimension reduction approach. The rest of the chapter is organized as follows. In Sect. 13.2, we briefly summarize the sufficient dimension reduction method. The generalization of the sufficient dimension reduction method to tensor data (Zhong et al. 2015) is reviewed in Sect. 13.3, where a formal presentation of the tensor sufficient dimension reduction framework is illustrated in Sect. 13.3.1. A detailed discussion of the estimation and algorithm is provided in Sect. 13.3.2. The performance of tensor sufficient dimension reduction is evaluated in Sect. 13.4. Section 13.5 illustrates the advantages and limitations of the tensor sufficient dimension reduction method using some examples. Some discussions will conclude this chapter in Sect. 13.6.

13.2 Sufficient Dimension Reduction

Let $Y \in \mathbb{R}$ be the response variable and $\mathrm{X} = (\mathrm{x}_1, \ldots, \mathrm{x}_p)^\top \in \mathbb{R}^p$ be the predictors with $E(\mathrm{X}) = \mathbf{0}$ and $\mathrm{cov}(\mathrm{X}) = \Sigma_X$. Throughout this chapter, we considered the sufficient dimension reduction model as introduced in Li (1991) and advocated in Bura and Cook (2001), Chen and Li (1998), Cook et al. (2004), Zhong et al. (2012), Zhu et al. (2006),

$$Y = f(\beta_1^\top \mathrm{X}, \beta_2^\top \mathrm{X}, \ldots, \beta_K^\top \mathrm{X}, \epsilon), \tag{13.1}$$

where $f(\cdot)$ is an unspecified link function on $\mathbb{R}^{K+1}$, $\beta_1, \ldots, \beta_K$ are p-dimensional vectors, and ϵ is the random error independent of X. When model (13.1) holds, p-dimensional variable X is projected onto a K-dimensional subspace $\mathcal{S}$ spanned by $\beta_1, \ldots, \beta_K$, which captures all the information in Y, i.e.

$$Y \perp\!\!\!\perp \mathrm{X} | P_{\mathcal{S}} \mathrm{X}. \tag{13.2}$$

where $P_{(\cdot)}$ is a projection operator in the standard inner product, $\mathcal{S}$ is the subspace of predictor space, and '$\perp\!\!\!\perp$' means "independent of". Cook and Weisberg (2009) formulated the framework of sufficient dimension reduction (SDR) as in model (13.2). The SDR model assumes that Y and X are mutually independent condition on $P_{\mathcal{S}}\mathrm{X}$, which means that the high dimensional predictors can be projected onto a lower dimensional subspace without loss of information.

It has been shown in Zeng and Zhu (2010) that (13.2) is equivalent to (13.1). Thus $\beta_1, \ldots, \beta_K$ are referred to as the SDR directions and the space spanned by these directions as a SDR subspace. The SDR implies that all the information X contains about Y is contained in the K projections $\beta_1^\top \mathrm{X}, \ldots, \beta_K^\top \mathrm{X}$. However, the basis $\boldsymbol{\beta} = (\beta_1, \ldots, \beta_K)$ of SDR space is not unique. Cook and Weisberg (2009) thus introduced the concept of central subspace, which is defined as the intersection of all possible SDR subspaces and is an SDR subspace itself. Under certain conditions, the central subspace is well defined and unique.

Many efforts have been made for estimating the basis of SDR space. Among various methods, one particular family of methods exploits inverse regression. The intuition of inverse regression is to exchange the role of Y and X so that each column of X can be regressed against Y, through which the inverse regression bypasses the high-dimensional problem introduced by large p, and becomes a one-dimension to one-dimension regression problem. Based on the idea of inverse regression, Li (1991) proposed sliced inverse regression (SIR), which became the forerunner of this family of methods. As showed in Li (1991), the centered inverse regression curve $E(\mathrm{X}|\,Y) - E(\mathrm{X})$ is contained in a K-dimensional subspace determined by $\beta_1, \ldots, \beta_K$, assuming the linearity condition. Chen and Li (1998) also illustrated that the solution $\beta_1, \ldots, \beta_K$ to this dimension reduction problem is equivalent to solve the eigenvalue decomposition problem,

$$\begin{aligned} \Sigma_{X|\,Y} u_i &= \lambda_i \Sigma_X u_i, \\ u_i^T \Sigma_X u_i &= 1, i = 1, \ldots, p, \\ \lambda_1 &\geq \lambda_2 \geq \cdots \geq \lambda_p, \end{aligned} \tag{13.3}$$

where $\Sigma_{X|\,Y} = \mathrm{cov}[E(X|\,Y)]$. Intuitively, we may see that $\Sigma_X^{-1}\Sigma_{X|\,Y}$ is degenerate in directions orthogonal to β_ks for $k = 1, \ldots, K$. Therefore, β_ks could be the eigenvectors associated with K largest eigenvalues of $\Sigma_X^{-1}\Sigma_{X|\,Y}$, and estimating the eigenvectors of $\Sigma_X^{-1}\Sigma_{X|\,Y}$ could lead us to find $u_1, \ldots, u_p$, so-called SIR directions.

In the context of projection pursuit, the SIR directions u_is also solve the maximization problem (Chen and Li 1998, Theorem 3.1),

$$\begin{aligned} \mathrm{cov}(\eta_i^T X, \eta_j^T X) &= 0, \text{for } i \neq j, \\ P^2(\eta_i) &= \max_{\eta, T} \mathrm{Corr}^2(T(Y), \eta^T X), \end{aligned} \tag{13.4}$$

where $T(Y)$ represents all the transformations of Y and $\mathrm{Corr}(\cdot, \cdot)$ calculates the correlation coefficient between two vectors. Intuitively, η_1 (or u_1) derived in problem (13.4) (or (13.3)) is a direction in $\mathbb{R}^p$ along which the transformed Y and $\eta_1^T X$ have the largest correlation coefficient. η_2 (or u_2), orthogonal to η_1, is a direction that produces the second largest correlation coefficient between $T(Y)$ and $\eta_2^T X$. Under the assumption of model (13.1) or (13.2), the procedure can be continued until all K directions are found that are orthogonal to each other and have nonzero $P^2(\eta)$, resulting in $\eta_1, \ldots, \eta_K$ that span the K-dimensional subspace $\mathcal{S}$. η_k's are referred to as principal directions.

The procedure above is described in a projection pursuit manner, which builds a connection with SIR such that the SIR directions defined in Eq. (13.3) are the solution for maximization problem (13.4), and the maximum values $P^2(\eta_i), i = 1, \ldots, K$ equal to the eigenvalues defined in Eq. (13.3). Thus the space spanned by principal directions can be identified by obtaining the eigenvectors of $\Sigma_X^{-1}\Sigma_{X|Y}$. The proof was given in Chen and Li (1998) (Appendix A). Basically, observe that the maximization problem (13.4) is a double maximization problem, Chen and Li (1998) derived the following equation by switching the maximizing order,

$$P^2(\eta) = \frac{\eta' \text{cov}[E(X|Y)]\eta}{\eta'\Sigma_X\eta} \triangleq \frac{\eta'\Sigma_{X|Y}\eta}{\eta'\Sigma_X\eta}. \tag{13.5}$$

Therefore, the principal directions $\eta_1, \ldots, \eta_K$ are the solutions of the eigenvalue decomposition of $\Sigma_X^{-1}\Sigma_{X|Y}$, with $\eta_1, \ldots, \eta_K$ as the first K eigenvectors, and the corresponding eigenvalues as the squared profile correlations, i.e. $\lambda_i = P^2(\eta_i)$ for $i = 1, \ldots, K$.

Regarding the implementation of SIR, Li (1991) formed the weighted sample covariance matrix estimation of $\Sigma_{X|Y}$ with the idea of slicing. Basically, the range of Y is divided into H slices. Observe $(\mathbf{x}_i^\top, y_i)|_{i=1}^n$, the probability that y_i falls into slice S_h is p_h for $h = 1, \ldots, H$. Within each slice, mean of $\mathbf{x}_i, i = 1, \ldots, n$ is calculated and denoted as $\bar{\mathbf{x}}_h, h = 1, \ldots, H$. Thus $\Sigma_{X|Y}$ is estimated by

$$\hat{\Sigma}_{X|Y} = \sum_{h=1}^{H} \hat{p}_h(\bar{\mathbf{x}}_h - \bar{\mathbf{x}})(\bar{\mathbf{x}}_h - \bar{\mathbf{x}})^\top, \tag{13.6}$$

where $\bar{\mathbf{x}}_h = (n\hat{p}_h)^{-1}\sum_{i=1}^n \mathbf{x}_i I(y_i \in S_h)$, and $\bar{\mathbf{x}}$ is the sample mean. The covariance matrix is estimated by sample covariance matrix

$$\hat{\Sigma}_X - \frac{1}{n-1}\sum_{i=1}^{n}(\mathbf{x}_i - \bar{\mathbf{x}})(\mathbf{x}_i - \bar{\mathbf{x}})^\top. \tag{13.7}$$

The first K eigenvalues of $\hat{\Sigma}_X^{-1}\hat{\Sigma}_{X|Y}$ are used to estimate the first K squared profile correlations, and the K eigenvectors are used to estimate the first K principal directions $\hat{\eta}_1, \ldots, \hat{\eta}_K$.

To determine the dimensionality K, statistical tests are also available. For example, given that X is normally distributed, Li (1991) developed a test statistic

$$\hat{\Lambda} = n\sum_{j=K+1}^{p} \hat{\lambda}_j, \tag{13.8}$$

which has an asymptotic χ^2-distribution with degree of freedom $(p-K)(H-K-1)$, where $\hat{\lambda}_{K+1}, \hat{\lambda}_{K+2} \ldots, \hat{\lambda}_p$ are the smallest $p-K$ eigenvalues of $\hat{\Sigma}_X^{-1}\hat{\Sigma}_{X|Y}$ (see Li 1991, Theorem 5.1). Thus if the test statistic $\hat{\Lambda}$ is larger than $\chi^2_\alpha[(p-K)(H-K-1)]$ given a certain significance level α, there must exist $(K+1)$th component. The result provides a simple but effective way to help determine the number of components in the model.

13.3 Tensor Sufficient Dimension Reduction

A tensor is a multidimensional array. More precisely, an mth-order tensor is an element of the tensor product of m vector spaces, each of which has its own coordinate system termed as mode. For example, a first-order tensor is a vector with one mode, and a second-order tensor is a matrix with two modes (row and column). Tensor data has been extensively used in social and behavioral sciences studies. With the rapid development of science and technology in the past decades, it is now becoming more and more popular in numerous other scientific disciplines, including life science, econometrics, chemical engineering as well as imaging and signal processing (Kroonenberg 2008; Smilde et al. 2005).

Existing methods for the aforementioned applications largely ignore the tensor structures by simply vectorizing each tensor observation into a vector, and offering solutions using the vector-based statistical methods. These solutions, however, are far from satisfactory. First, the simple vectorization destroys the original design information and leads to interpretation difficulties. Second, the simple vectorization significantly aggregates the dimensionality of the parameter space. For example, assume there are $p \times q$ parameters for a vectorized X, where $X \in \mathbb{R}^{p \times q}$, the number of parameters can be $p + q$ if $X = \alpha \circ \beta$, where $\alpha \in \mathbb{R}^p$, $\beta \in \mathbb{R}^q$, and $\circ$ denotes the outer product of vectors. Thus the simple vectorization renders many vector-based approaches infeasible if the sample size is smaller than $p \times q$ observations. Moreover, even if we have a fairly large sample size, both the computational efficiency and the estimation accuracy of the classical vector-based analysis will be compromised by simple vectorization approach.

In this section, we review a significant different procedure named tensor sufficient dimension reduction (TSDR) (Zhong et al. 2015), which generalizes the classic SDR approach to tensor predictors. The model used by the TSDR assumes that the response depends on a few linear combinations of some tensors through an unknown link function. A sequential-iterative dimension reduction approach is proposed to estimate the TSDR model. Consequently, the TSDR procedure can be automated to not only estimate the dimension reduction subspace but also keep the tensor structure. The advantages of this new model are its ability to reduce the number of parameters by assuming tensor structure on the regression parameters, to recover a subspace that includes all important regression information of the predictors, and to increase the model flexibility by assuming an unknown dependency between the response and a projection of the predictors onto the subspace.

13.3.1 Tensor Sufficient Dimension Reduction Model

We first review some basic notations for the simplicity of our description. In the rest of the chapter, we assume $\mathbf{X} \in \mathbb{R}^{p_1 \times p_2 \times \cdots \times p_m}$ to be an mth-order tensor and vec($\mathbf{X}$) to be the vectorized $\mathbf{X}$. Let $\gamma_j = \beta_j^{(1)} \otimes \beta_j^{(2)} \otimes \cdots \otimes \beta_j^{(m)}$ be the Kronecker product of vectors $\beta_j^{(1)}, \ldots, \beta_j^{(m)}$ where $\beta_j^{(i)} \in \mathbb{R}^{p_i}$. Then, given the tensor predictor $\mathbf{X} \in \mathbb{R}^{p_1 \times \ldots, p_m}$ and the response $Y \in \mathbb{R}$, the TSDR model is of the form

$$Y = f(\gamma_1^\top \text{vec}(\mathbf{X}), \ldots, \gamma_D^\top \text{vec}(\mathbf{X}), \varepsilon), \quad (13.9)$$

where $f(\cdot)$ is an unknown function, γ_j for $j = 1, \ldots, D$ are regression indexes, and ε is a random error that is independent of vec($\mathbf{X}$).

Clearly, TSDR model aims at finding $\{\gamma_1, \ldots, \gamma_D\}$ such that:

$$Y \perp \text{vec}(\mathbf{X}) | (\gamma_1^\top \text{vec}(\mathbf{X}), \ldots, \gamma_D^\top \text{vec}(\mathbf{X})). \quad (13.10)$$

Let $\mathcal{S}$ be the linear space spanned by $\{\gamma_1, \ldots, \gamma_D\}$, i.e., for any $\gamma \in \mathcal{S}$ we have $\gamma = \sum_{j=1}^D c_j \gamma_j$. $\mathcal{S}$ is referred to as the TSDR subspace. It is worth noting that not all the elements in $\mathcal{S}$ can be written in the form of a Kronecker product of vectors. For example, $\beta_1^{(1)} \otimes \beta_1^{(2)} + \beta_2^{(1)} \otimes \beta_2^{(2)} \in \mathcal{S}$ but cannot be expressed in a Kronecker product of any vectors in $\mathcal{S}_1$ and $\mathcal{S}_2$, unless $\beta_1^{(1)} \propto \beta_2^{(1)}$ or $\beta_1^{(2)} \propto \beta_2^{(2)}$.

We want to emphasize that model (13.9) is substantially different from model (13.1), although they look similar in expression. The TSDR model can naturally alleviate the curse of dimensionality without increasing the estimation bias. For example, assuming $d = 1$, the index β_1 in model (13.1) with vec($\mathbf{X}$) has $p_1 p_2$ parameters, while γ_1 in model (13.9) has only $p_1 + p_2$ parameters. It is also important to know that model (13.9) is significantly different from the dimension folding model (Li et al. 2010), which assumes

$$Y \perp \text{vec}(\mathbf{X}) | P_{\mathcal{S}_1} \otimes P_{\mathcal{S}_2} \otimes \cdots \otimes P_{\mathcal{S}_m} \text{vec}(\mathbf{X}), \quad (13.11)$$

where $\mathcal{S}_i$ is the d_i dimensional SDR space of $\mathbb{R}^{p_i}$ and $P_{\mathcal{S}_i}$ for $i = 1, \ldots, m$ are projection operator from $\mathbb{R}^{p_i}$ to $\mathcal{S}_i$ in a standard inner product. Let $\mathcal{S}_1 \otimes \cdots \otimes \mathcal{S}_m$ be the space that spanned by $\{v_1 \otimes \cdots \otimes v_m \mid \forall v_j \in \mathcal{S}_j\}$. Then $\mathcal{S}_1 \otimes \cdots \otimes \mathcal{S}_m$ is referred to as the dimension folding subspace in Li et al. (2010) and central tensor subspace in Ding and Cook (2015). It is easy to see that $\mathcal{S}$ in model (13.9) is in general much smaller than $\mathcal{S}_1 \otimes \cdots \otimes \mathcal{S}_m$, i.e., $\mathcal{S} \subseteq \mathcal{S}_1 \otimes \cdots \otimes \mathcal{S}_m$. For example, for model $Y = f(\gamma_1^\top \text{vec}(\mathbf{X}), \gamma_2^\top \text{vec}(\mathbf{X}), \varepsilon)$, where $\gamma_1 = \beta_1^{(1)} \otimes \beta_1^{(2)}$ and $\gamma_2 = \beta_2^{(1)} \otimes \beta_2^{(2)}$, $\mathcal{S}$ is the space spanned by γ_1 and γ_2, while the central dimension folding subspace is spanned by $(\beta_1^{(1)} \otimes \beta_1^{(2)}, \beta_1^{(1)} \otimes \beta_2^{(2)}, \beta_2^{(1)} \otimes \beta_1^{(2)}, \beta_2^{(1)} \otimes \beta_2^{(2)})$.

13.3.2 *Estimate a Single Direction*

A sequential-iterative algorithm can be used to estimate γ_j, $1 \le j \le D$ while leaving $f(\cdot)$ unspecified. Let $\mathcal{R}$ be the set of all rank one tensors in $\mathbb{R}^{p_1} \otimes \cdots \otimes \mathbb{R}^{p_m}$ ($X \in \mathcal{R}^{p_1 \times p_2 \times \cdots \times p_n}$ is rank one if it can be written as the outer product of n vectors). Let $P^2(\eta) = \max_T \text{Corr}^2(T(Y), \eta^\top \text{vec}(\mathbf{X}))$, and it is also equivalent to

$$P^2(\eta) = \frac{\eta^\top \text{cov}[E(\text{vec}(\mathbf{X})| Y)]\eta}{\eta^\top \text{cov}(\text{vec}(\mathbf{X}))\eta}. \quad (13.12)$$

When vec(**X**) is a vector, we may naturally attempt the optimization of (13.12) using the matrix spectral decomposition in a projection pursuit manner. However, unlike its matrix sibling, spectral decomposition, such as the PARAFAC (Harshman and Lundy 1984) or Tucker model (Tucker 1951, 1966), cannot provide the maximizer of $P^2(\eta)$ for higher order tensor. Beyond this, the definition and algorithm of the tensor spectral decomposition are far from mature and have many intrinsic problems. For example, the orthogonality on each mode is not assumed and the decomposition on the same mode is not unique using different algorithms (Kolda and Bader 2009; Smilde et al. 2005). Here, we are aiming to estimate a series of γ_is which maximize (13.12). An algorithm is introduced in Zhong et al. (2015) to find γ_is iteratively.

We assume $m = 2$ and briefly summarize the algorithm in this section. Recall that vec(**X**) is the vectorization of **X**, where $\mathbf{X} \in \mathbb{R}^{p_2 \times p_1}$, and for any $\eta \in \mathcal{R}$, we have $\eta^\top \text{vec}(\mathbf{X}) = ({\beta_1^{(2)}}^\top \mathbf{X} \beta_1^{(1)})$, where $\beta_1^{(j)} \in \mathbb{R}^{p_j}$ for $j = 1$ and 2. Maximizing (13.12) is equivalent to maximizing

$$P^2(\beta_1^{(1)}, \beta_1^{(2)}) = \frac{\text{cov}[E({\beta_1^{(2)}}^\top \mathbf{X} \beta_1^{(1)} \mid Y)]}{\text{cov}({\beta_1^{(2)}}^\top \mathbf{X} \beta_1^{(1)})} \tag{13.13}$$

with respect to $\beta_1^{(1)}$ and $\beta_1^{(2)}$. This sequential-iterative algorithm is summarized as follows.

Algorithm 1: A sequential-iterative algorithm for TSDR

Observing $\{(\mathbf{X}_i, y_i)\}_{i=1,\ldots,n}$, where $\mathbf{X}_i \in \mathbb{R}^{p_1 \times p_2}$ and $y_i \in \mathbb{R}$

1. Estimate $\text{cov}(\mathbf{X}^\top \beta_1^{(2)})$, $\text{cov}(\mathbf{X} \beta_1^{(1)})$, $\text{cov}[E(\mathbf{X}^\top \beta_1^{(2)} \mid Y)]$ and $\text{cov}[E(\mathbf{X} \beta_1^{(1)} \mid Y)]$ by their sample version.
2. For given $\beta_1^{(2)}$ and $\beta_1^{(1)}$, maximize $P_1^2(\eta)$ and $P_2^2(\eta)$ respectively, where

$$P_1^2(\eta) = \frac{\eta^\top \text{cov}[E(\mathbf{X}^\top \beta_1^{(2)} \mid Y)]\eta}{\eta^\top \text{cov}(\mathbf{X}^\top \beta_1^{(2)})\eta}, \quad P_2^2(\eta) = \frac{\eta^\top \text{cov}[E(\mathbf{X} \beta_1^{(1)} \mid Y)]\eta}{\eta^\top \text{cov}(\mathbf{X} \beta_1^{(1)})\eta}. \tag{13.14}$$

3. Repeat Step 2 until convergence.

Notice that $P_2^2(\eta)$ is a quadratic form on the vector space $\mathbb{R}^{p_2}$. Thus, if $\beta_1^{(1)}$ is given, $\beta_1^{(2)}$ can be obtained by SIR, and vice versa. Thus γ_1 can be obtained by iteratively maximizing $P_1^2(\eta)$ and $P_2^2(\eta)$, respectively. Notice that $P_j^2(\cdot)$ for $j = 1, 2$ are bounded above by one, and the iterative maximization approach ensures the non-decreasing of $P^2(\cdot)$ in each iteration. Thus the convergence of iteration is guaranteed.

More generally, given $B_k = (\gamma_1, \ldots, \gamma_k)$, we can estimate γ_{k+1} in a sequential way. Let $\mathcal{S}_{B_k}$ be the space that is spanned by B_k, and $\mathcal{R}_k$ be the set of all decomposable tensors of $\mathbb{R}^{p_2} \otimes \mathbb{R}^{p_1}$ that are orthogonal to $\mathcal{S}_{B_k}$. We have

$$\gamma_{k+1} = \text{argmax}_{\eta \in \mathcal{R}_k} \frac{\eta^\top \text{cov}[E(\text{vec}(\mathbf{X}^{(k)})|\, Y)]\eta}{\eta^\top \text{cov}(\text{vec}(\mathbf{X}^{(k)}))\eta}, \tag{13.15}$$

where $\text{vec}(\mathbf{X}^{(k)})$ is the projection of $\text{vec}(\mathbf{X})$ in the complementary space of $\mathcal{S}_{B_k}$, i.e., $\text{vec}(\mathbf{X}^{(k)}) = (I - P_k)\text{vec}(\mathbf{X})$, and

$$P_k = \Sigma_{\text{vec}(\mathbf{X})}^{\frac{1}{2}} B_k (B_k^\top \Sigma_{\text{vec}(\mathbf{X})} B_k)^{-1} B_k^\top \Sigma_{\text{vec}(\mathbf{X})}^{\frac{1}{2}}$$

is the projection matrix from $\mathbb{R}^{p_2} \otimes \mathbb{R}^{p_1}$ onto $\mathcal{S}_{B_k}$ with respect to $\Sigma_{\text{vec}(\mathbf{X})} \triangleq \text{cov}(\text{vec}(\mathbf{X}))$. Clearly, γ_{k+1} can be estimated in the same fashion as γ_1.

13.4 Simulation Studies

Comprehensive empirical studies were carried out to evaluate the performance of TSDR. In this section, we report two simulation studies. To assess the accuracy of the estimated subspace, we used the following distance

$$d = 1 - \mathcal{M}(\mathcal{S}, \hat{\mathcal{S}}), \tag{13.16}$$

where $\mathcal{M}(\cdot)$ is the similarity score between the subspaces $\mathcal{S}$ and $\hat{\mathcal{S}}$, and

$$\mathcal{M}(\mathcal{S}, \hat{\mathcal{S}}) = \sqrt{\frac{1}{r} \sum_{i=1}^{r} \lambda_i(\hat{P}^T P_0 P_0^T \hat{P})}, \tag{13.17}$$

where P_0 and $\hat{P}$ are the projection matrices corresponding to $\mathcal{S}$ and $\hat{\mathcal{S}}$, respectively, and $\lambda_i(\cdot)$ is the ith eigenvalue of a matrix.

Simulation Study 1 Let $\mathbf{X} \in \mathbb{R}^{p_1 \times p_2}$ be a second-order tensor and $Y \in \mathbb{R}$ be the response variable. In this study, we assume the following TSDR model,

$$Y = \frac{\gamma_1^T \text{vec}(\mathbf{X})}{2 + \left(3 + \gamma_2^T \text{vec}(\mathbf{X})\right)^2} + \epsilon, \tag{13.18}$$

where $\gamma_1 = \beta_1 \otimes \alpha_1$, $\gamma_2 = \beta_2 \otimes \alpha_2$ with $\alpha_1, \alpha_2 \in \mathbb{R}^{p_1}$ and $\beta_1, \beta_2 \in \mathbb{R}^{p_2}$, and ϵ is the stochastic error with mean 0 and variance 0.5.

We generated n pairs of observations $\{\mathbf{x}_i, y_i\}_{i=1}^n$ from this model, where $\mathbf{x}_i \in \mathbb{R}^{p_1 \times p_2}$ and $y_i \in \mathbb{R}$. Let $n = 200, 400, 600, 800, p_1 = p_2 = 5, 8$ and

$$\beta_1 = (1, 0, 1, 0, 0), \ \ \alpha_1 = (1, 1, 0, 0, 0)$$
$$\beta_2 = (0, 1, 0, 0, 1), \ \ \alpha_2 = (0, 0, 0, 1, 1)$$

when $p_1 = p_2 = 5$,

$$\beta_1 = (1, 1, 1, 1, 0, 0, 0, 0), \ \ \alpha_1 = (1, 1, 0, 1, 1, 0, 0, 0)$$
$$\beta_2 = (0, 0, 0, 0, 1, 1, 1, 1), \ \ \alpha_2 = (0, 0, 0, 0, 0, 0, 1, 1)$$

when $p_1 = p_2 = 8$. For each combination of n and p, we generated 200 datasets and applied SIR, folded-SIR and TSDR to those datasets. For SIR, the central subspace is spanned by $\beta_1 \otimes \alpha_1$ and $\beta_2 \otimes \alpha_2$. For folded-SIR, the central left-folding and right-folding subspace is spanned by (α_1, α_2) and (β_1, β_2), respectively. Thus, we use folded-SIR(1,2) and folded-SIR(2,1) in this simulation study to find a subspace of two dimensions.

Figure 13.1 displays means and standard deviations of distances evaluated by criterion (13.16) using the above-mentioned methods. In all scenarios, TSDR outperforms other methods. Both TSDR and Folded-SIR(2,1) have improvements over SIR, especially in the cases where the sample sizes are small and the dimensions are relatively large. This is due to the fact that the subspaces estimated by TSDR and Folded-SIR contain fewer parameters than the central subspace estimated by SIR. When sample size gets larger, the performance of SIR is substantially improved.

Simulation Study 2 Let $\mathbf{X} \in \mathbb{R}^{p_1 \times p_2}$ be a second-order tensor and $Y \in \mathbb{R}$ be the response variable. In this study, we assume the following TSDR model,

$$Y = \gamma_1^T \text{vec}(\mathbf{X})\Big(1 + \gamma_1^T \text{vec}(\mathbf{X}) + \gamma_2^T \text{vec}(\mathbf{X})\Big) + \epsilon, \tag{13.19}$$

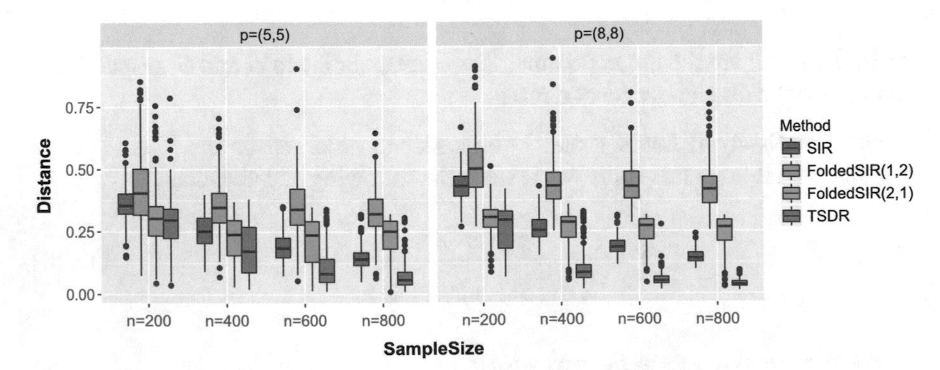

Fig. 13.1 Left panel: the boxplots of the distance (13.16) of SIR(2), Folded-SIR(1,2), Folded-SIR(2,1) and TSDR(2) when $p = (5, 5)$ and $n = 200, 400, 600, 800$ under model (13.18). Right panel: the boxplots of the distance (13.16) of SIR(2), Folded-SIR(1,2), Folded-SIR(2,1) and TSDR(2) when $p = (8, 8)$ and $n = 200, 400, 600, 800$ under model (13.18)

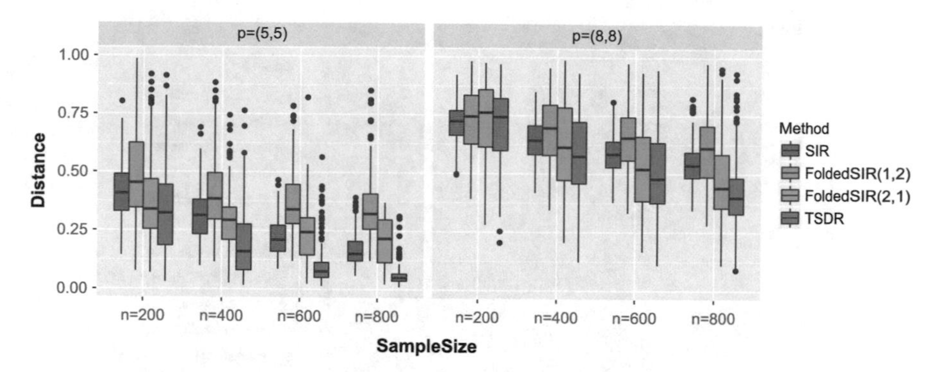

Fig. 13.2 Left panel: the boxplots of the distance (13.16) of SIR(2), Folded-SIR(1,2), Folded-SIR(2,1), and TSDR(2) when $p = (5, 5)$ and $n = 200, 400, 600, 800$ under model (13.19). Right panel: the boxplots of the distance (13.16) of SIR(2), Folded-SIR(1,2), Folded-SIR(2,1), and TSDR(2) when $p = (8, 8)$ and $n = 200, 400, 600, 800$ under model (13.19)

where $\gamma_1 = \beta_1 \otimes \alpha_1$, $\gamma_2 = \beta_2 \otimes \alpha_2$ with $\alpha_1, \alpha_2 \in \mathbb{R}^{p_1}$ and $\beta_1, \beta_2 \in \mathbb{R}^{p_2}$, and ϵ is the stochastic error with mean 0 and variance 0.5. We generated n pairs of observations $\{\mathbf{x}_i, y_i\}_{i=1}^n$ from this model. The parameters $n, p, \alpha_1, \alpha_2, \beta_1$, and β_2 were defined the same as in simulation study 1. For each combination of n and p, we simulated 200 datasets and applied SIR, folded-SIR and TSDR to those datasets.

Figure 13.2 displays means and standard deviations of distances evaluated by criterion (13.16) using the above-mentioned methods. In Fig. 13.2, we also observe improvements by TSDR. Left panel of Fig. 13.2 shows that in the scenarios of low sample sizes ($n = 200, 400$), both folded-SIR(2,1) and TSDR perform better than SIR. In high-dimensional scenarios where $p = (8, 8)$ (right panel of Fig. 13.2), the same pattern appears when sample size gets larger, for the same reason explained in study 1. While in this scenario when $n = 200$ under model (13.19), the signal-to-noise ratio is very small such that all three methods fail to identify their corresponding dimension reduction subspaces. As n gets larger, TSDR outperforms other methods.

13.5 Example

Colorimetric sensor array (CSA), a.k.a. the "optical electronic nose," uses a number of chemical dyes to turn smell into digital signals. The CSA is simply a digitally-imaged, two-dimensional extension of litmus paper (Rakow and Suslick 2000). Thirty-six chemo-responsive dyes are assigned to 36 spots scattered as a 6×6 array on a chip. For any smell generated from fungus, a CSA response is generated by digital subtraction of the color of 36 pre-print chemo-responsive dyes before and after exposure: red value after exposure minus red value before, green minus green, blue minus blue (Zhong and Suslick 2014). The resulting "color difference map"

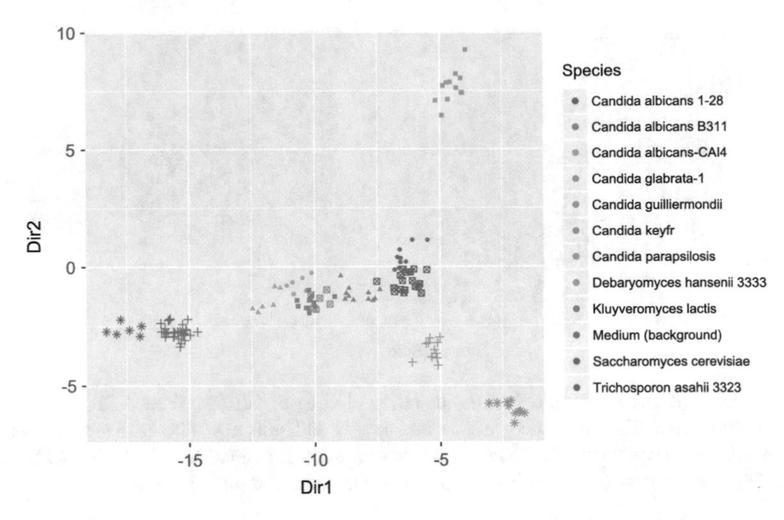

Fig. 13.3 Plotted here is the projection of 140 third-order tensor data points on the first two dimension reduction directions obtained by TSDR

is a 36×3 matrix, where each row represents the color change of a dye, and each column represents one of the three spectrum coordinates (red, green, and blue) of a color cube. When repeated CSA measurements were taken over time, the data can be cast as third-order tensor data (dyes$\times$ color spectrum $\times$ time points).

We use a dataset with the color difference maps of CSA measurements of 140 fungi coming from 12 fungal species. The dataset includes color difference maps over 17 time points with 30 min time intervals after the CSA was exposed to each fungus. That is, for each fungus, we have a $36 \times 3 \times 17$ tensor. We aim to classify these 140 fungi into 12 species based on the CSA data. We applied our TSDR to the data.

Plotted in Fig. 13.3 is the projection of the 140 tensor data points onto the first and second directions obtained using the sequential-iterative algorithm. We observe that most of the fungi are well separated. The exceptions are the data of *Kluyveromyces lactis* and *Candida parapsilosis*, which cannot be clearly separated using two directions. It is worth noting that existing vector-based discriminant analysis is inapplicable for this data because of the "small n, large p" problem. That is, the data has 36 dyes $\times$ 3 colors $\times$17 time points$=$ 1836 parameters whereas the sample size is only 140. In contrast, TSDR only needs to estimate $(36+3+17)\times d$ parameters where $d \leq 5$ is enough for many practical examples.

13.6 Discussion

The TSDR model provides an effective and flexible framework for dimension reduction in regression with tensor predictors. It can effectively address the curse of dimensionality in regression and maintain the data structure and the model interpretability. It does not impose any assumption on the relationship between the response variable and the predictors, and the TSDR model includes fully non-parametric models as special cases. Because the aforementioned sequential-iterative algorithm requires the least amount of prior information of the model, it becomes an indispensable member of the repository of dimension reduction tools in broad applications, such as in chemical sensing and in brain image analysis.

However, because TSDR is a generalization of SIR, it inherits all the limitations that SIR has. For example, TSDR imposes various assumptions on the distribution of the predictors, of which the linearity assumption is the most fundamental and crucial. However, violation of the assumptions can be remedied using some data reweighting schemes.

Like all the iterative estimation approaches, the sequential-iterative algorithm procedure may also encounter issues typical to all the other iterative estimation approaches. One major limitation of the iterative estimation approach is that the estimate may practically be attracted to a local optimal estimate and fail to reach the global optimal. To solve this issue, we advocate trying multiple starting points for iteration and choosing the best estimate. This approach is currently under intensive investigation and will be reported in the future publication.

Acknowledgements This work was supported by National Science Foundation grant DMS-1222718 and DMS-1055815, National Institutes of Health grant R01 GM113242, and National Institute of General Medical Sciences grant R01GM122080.

References

Akaike H (1987) Factor analysis and AIC. Psychometrika 52(3):317–332
Bendel RB, Afifi AA (1977) Comparison of stopping rules in forward "stepwise" regression. J Am Stat Assoc 72(357):46–53
Breiman L (1995) Better subset regression using the nonnegative garrote. Technometrics 37(4):373–384
Bura E, Cook RD (2001) Extending sliced inverse regression: the weighted chi-squared test. J Am Stat Assoc 96(455):996–1003
Chen C-H, Li K-C (1998) Can SIR be as popular as multiple linear regression? Stat Sin 8(2):289–316
Cook RD, Weisberg S (2009) An introduction to regression graphics, vol 405. Wiley, New York
Cook RD et al (2004) Testing predictor contributions in sufficient dimension reduction. Ann Stat 32(3):1062–1092
Ding S, Cook RD (2015) Tensor sliced inverse regression. J Multivar Anal 133:216–231
Efroymson M (1960) Multiple regression analysis. Math Methods Digit Comput 1:191–203
Fan J, Li R (2001) Variable selection via nonconcave penalized likelihood and its oracle properties. J Am Stat Assoc 96(456):1348–1360
Fan J, Peng H, et al (2004) Nonconcave penalized likelihood with a diverging number of parameters. Ann Stat 32(3):928–961
Friedman JH, Tukey JW (1974) A projection pursuit algorithm for exploratory data analysis. IEEE Trans Comput 23(9):881–890

Harshman RA, Lundy ME (1984) The PARAFAC model for three-way factor analysis and multidimensional scaling. In: Research methods for multimode data analysis. Praeger, New York, pp 122–215

Huber PJ (1985) Projection pursuit. Ann Stat 13(2):435–475

Kolda TG, Bader BW (2009) Tensor decompositions and applications. SIAM Rev 51(3):455–500

Kroonenberg PM (2008) Applied multiway data analysis, vol 702. Wiley, New York

Kruskal JB (1969) Toward a practical method which helps uncover the structure of a set of multivariate observations by finding the linear transformation which optimizes a new 'index of condensation'. In: Statistical computation. Academic, New York, pp 427–440

Li K-C (1991) Sliced inverse regression for dimension reduction. J Am Stat Assoc 86(414):316–327

Li B, Kim MK, Altman N (2010) On dimension folding of matrix-or array-valued statistical objects. Ann Stat 38(2):1094–1121

Miller A (2002) Subset selection in regression. CRC Press, Boca Raton

Rakow NA, Suslick KS (2000) A colorimetric sensor array for odour visualization. Nature 406(6797):710–713

Smilde A, Bro R, Geladi P (2005) Multi-way analysis: applications in the chemical sciences. Wiley, New York

Tibshirani R (1996) Regression shrinkage and selection via the lasso. J R Stat Soc Ser B Methodol 58(1):267–288

Tucker LR (1951) A method for synthesis of factor analysis studies. Technical report, DTIC Document

Tucker LR (1966) Some mathematical notes on three-mode factor analysis. Psychometrika 31(3):279–311

Wilkinson L (1979) Tests of significance in stepwise regression. Psychol Bull 86(1):168

Yuan M, Lin Y (2007) On the non-negative garrotte estimator. J R Stat Soc Ser B Stat Methodol 69(2):143–161

Zeng P, Zhu Y (2010) An integral transform method for estimating the central mean and central subspaces. J Multivar Anal 101(1):271–290

Zhang HH, Lu W (2007) Adaptive lasso for cox's proportional hazards model. Biometrika 94(3):691–703

Zhong W, Suslick K (2014) Matrix discriminant analysis with application to colorimetric sensor array data. Technometrics 57(4):524–534

Zhong W, Zhang T, Zhu Y, Liu JS (2012) Correlation pursuit: forward stepwise variable selection for index models. J R Stat Soc Ser B Stat Methodol 74(5):849–870

Zhong W, Xing X, Suslick K (2015) Tensor sufficient dimension reduction. Wiley Interdiscip Rev Comput Stat 7(3):178–184

Zhu L, Miao B, Peng H (2006) On sliced inverse regression with high-dimensional covariates. J Am Stat Assoc 101(474):630–643

Zou H (2006) The adaptive lasso and its oracle properties. J Am Stat Assoc 101(476):1418–1429

Zou H, Hastie T (2005) Regularization and variable selection via the elastic net. J R Stat Soc Ser B Stat Methodol 67(2):301–320

Zou H, Zhang HH (2009) On the adaptive elastic-net with a diverging number of parameters. Ann Stat 37(4):1733

Chapter 14
Compressive Sensing and Sparse Coding

Kevin Chen and H. T. Kung

Abstract Compressive sensing is a technique to acquire signals at rates proportional to the amount of information in the signal, and it does so by exploiting the sparsity of signals. This section discusses the fundamentals of compressive sensing, and how it is related to sparse coding.

Keywords Compressive sensing · Sparse coding

Compressive sensing is a way to sample signals at rates proportional to the amount of information in the signal by exploiting the sparsity of signals. The measurements $y \in \mathbb{R}^m$ of a signal $x \in \mathbb{R}^n$ are obtained through linear projection Φ:

$$y = \Phi x \tag{14.1}$$

where $\Phi \in \mathbb{R}^{m \times n}$ (with $m \ll n$) is called the sensing matrix or measurement matrix. The signal x is then recovered by finding a sparse solution to 14.1.

14.1 Leveraging the Sparsity Assumption for Signal Recovery

Normally, it would be impossible to recover x from this underdetermined linear system, because given y and Φ there are infinite solutions to x. The first insight is to reduce the solution space by leveraging the signals' sparsity. With the sparse assumptions, it becomes possible to have a unique solution even when the system is underdetermined. For example, suppose $\|x\|_0$ is very small, then the idea is to

K. Chen (✉) · H. T. Kung
School of Engineering and Applied Sciences, Harvard University, Cambridge, MA, USA
e-mail: kevin@seas.harvard.edu; kung@harvard.edu

© Springer International Publishing AG, part of Springer Nature 2018

W. K. Härdle et al. (eds.), *Handbook of Big Data Analytics*, Springer Handbooks of Computational Statistics, https://doi.org/10.1007/978-3-319-18284-1_14

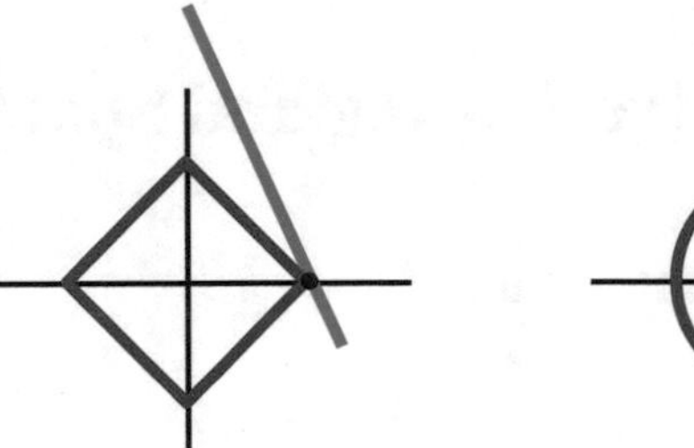
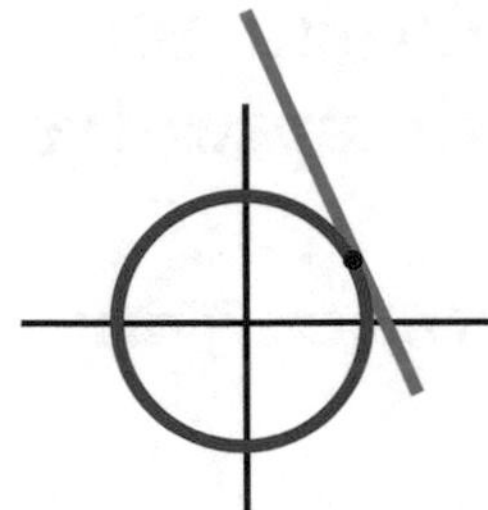

Fig. 14.1 Illustration of why ℓ_1 minimization finds sparse solutions. (Left) When we expand the ℓ_1 ball, it tends to intersect with the solution space on sparse solutions. (Right) In contrast, the ℓ_2 ball does not favor sparse solutions at all

reconstruct the signal with[1]:

$$\hat{x} \in \arg\min_x \|x\|_0 \text{subjectto} y = \Phi x \tag{14.2}$$

A potential issue is that (14.2) is a combinatorial optimization problem, and cannot be solved efficiently in practice. This leads us to the second insight of compressive sensing.

14.2 From Combinatorial to Convex Optimization

The most straightforward way to represent the sparsity of a vector is the ℓ_0 norm, which simply counts the number of non-zero elements. However, introducing ℓ_0 norm in the objective function leads to combinatorial optimization problems. One of the surprising key results in compressive sensing is that minimizing the ℓ_1 norm can lead to the same solution. This makes the signal recovery problem a convex problem, which is a lot easier to solve. Specifically, instead of solving (14.2), one solves the following:

$$\hat{x} \in \arg\min_x \|x\|_1 \text{subjectto} y = \Phi x \tag{14.3}$$

The main intuition of why minimizing ℓ_1 norm would result in sparse solutions is illustrated in Fig. 14.1. Namely, the ℓ_1 ball has pointy ends that tend to intersect with solution planes on the points with more zero-coefficients.

14.3 Dealing with Noisy Measurements

Up to now we have assumed that the signal is measured perfectly (that is, $y = \Phi x$). In practice, we expect certain noise ϵ in the measurements:

$$y = \Phi x + \epsilon \tag{14.4}$$

[1]The details about *when* this would work is presented in later sections.

which leads to the noisy signal recovery problem:

$$\hat{x} \in \arg\min_{x} \|x\|_1 \text{subjectto} \| y = \Phi x\|_2^2 \leq \eta \tag{14.5}$$

where $\eta \sim \|\epsilon\|_2^2$ is an estimation of the noise level. Surprisingly, compressive sensing also has strong guarantees under the presence of noise.

14.4 Other Common Forms and Variations

We have introduced the canonical forms for cleaner presentation. However, note that there are variations that are essentially equivalent:

$$\hat{x} \subset \arg\min_{x} s(x) \text{subjcctto} \| y - \Phi x\|_2^2 \leq \eta \tag{14.6}$$

$$\hat{x} \in \arg\min_{x} \| y - \Phi x\|_2^2 + \lambda s(x)$$

$$\hat{x} \in \arg\min_{x} \| y - \Phi x\|_2^2 \text{subjectto} \lambda s(x) \leq k$$

Depending on the situation, one might prefer one form to the others based on which parameter is easier to estimate.

14.5 The Theory Behind

This section presents three key results in compressive sensing without going into the details.

14.5.1 The Restricted Isometry Property

A matrix Φ satisfies the Restricted Isometry Property (RIP) with parameter k if there exist $\delta_k \in (0, 1)$ such that

$$(1 - \delta_k)\|x\|_2^2 \leq \|\Phi x\|_2^2 \leq (1 - \delta_k)\|x\|_2^2 \tag{14.7}$$

holds for all $x \in \mathbb{R}^n$ with $\|x\|_0 \leq k$. The intuition is that the projection by Φ roughly preserves the distance between sparse vectors, so that these sparse vectors are still separable after the projection.

14.5.2 Guaranteed Signal Recovery

Assume that $y = \Phi x$ with Φ that satisfies RIP($3k$) with $\delta_{3k} < 1$. Candes and Tao (2005) showed that the solution found with

$$\hat{x} \in \arg\min_{x} \|x\|_1 \text{subjectto} y = \Phi x \quad (14.8)$$

will satisfy

$$\|\hat{x} - x\|_2 \leq \frac{C \cdot \|x - A_k(x)\|_1}{k} \quad (14.9)$$

where $A_k(x)$ represents the best k-term approximation in terms of ℓ_1 norm, and C is a constant that only depends on δ_{3k}. This is a very strong result, as it shows that if a signal x can be approximated with k terms (i.e., that x is sparse), then the recovered signal would be close to x. More strikingly, the recovery is exact if $\|x\|_0 \leq k$.

14.5.3 Random Matrix is Good Enough

If we generate a random matrix Φ of size $m \times n$ with entries drawn from the Gaussian distribution,[2] then there exist constants $C_1, C_2 > 0$ depending only on δ such that the RIP holds for Φ with the prescribed δ and any

$$k \leq \frac{C_1 m}{\log(n/k)} \quad (14.10)$$

with probability greater than $1 - 2^{-C_2 n}$. This is to say that if we want to apply compressive sensing on signals that are expected to be k sparse, then randomly generating sensing matrix with

$$m \geq \frac{k \log(n/k)}{C_1} \quad (14.11)$$

would give us good recovery result using compressive sensing! Note that the required number of measurements (m) mostly depends on k, and scales logarithmically with the signal dimension (n).

[2]In fact, any distribution satisfying a specific concentration inequality would do (Baraniuk et al. 2008).

14.6 Compressive Sensing in Practice

14.6.1 Solving the Compressive Sensing Problem

With the ℓ_1 regularization ($s(\cdot) := \|\cdot\|_1$), the problem in (14.13) becomes a convex optimization problem known as the LASSO (Tibshirani 1996). There are many efficient solvers for the LASSO, including least-angle regression (LARS) (Efron et al. 2004), feature-sign search (Lee et al. 2006), FISTA (Beck and Teboulle 2009), and alternating direction method of multipliers (ADMM) (Boyd et al. 2011). Candes and Tao (2006) and Donoho (2006) first derived the signal recovery guarantees and showed that minimizing ℓ_1 is the same as minimizing ℓ_0 under RIP assumptions.

The optimization problem in (14.13) is combinatorial with the ℓ_0 regularization ($s(\cdot) := \|\cdot\|_0$), but we can use greedy algorithms such as orthogonal matching pursuit (OMP) (Pati et al. 1993), CoSamp (Needell and Tropp 2009), and iterative hard thresholding (IHT) (Blumensath and Davies 2009), etc. Some methods have stronger guarantees on signal recovery rate, but in practice the performance of these solvers is often much better than the lower bounds given by theoretical analysis.

14.6.2 Sparsifying Basis

It has been shown that for a randomized sensing matrix Φ, we can select an orthonormal basis Ψ, and $\Phi\Psi$ would also satisfy RIP with high probability. This means compressive sensing works for any x that can be written as $x = \Psi z$ for some sparse z. For example, many time series signals are sparse in the frequency domain, and one could use discrete cosine transform (DCT) basis as the sparsifying basis.

Empirically, we often find that compressive sensing also works under a more general setting: Compressive sensing is applicable to signal x which can be represented as Dz, where D is an overcomplete dictionary learned from data.[3] The orthonormal basis Ψ mentioned above can be considered as a special case of the dictionary D.

For signals that are sparse with respect to a basis D, the compressive sensing reconstruction solves:

$$\hat{z} \in \arg\min_{z} \| y - \Phi D z\|_2^2 + \lambda s(z) \tag{14.12}$$

While the matrix ΦD may not satisfy the properties of the sensing matrix normally assumed in conventional CS literature, this formulation still works as long as ΦD is sufficiently incoherent (atoms are sufficiently dis-similar). By being overcomplete and data-dependent, the use of dictionary learned from sparse

[3]More information about dictionary training can be found in Sect. 14.7.

coding can improve reconstruction substantially, and thus allow an extended set of compressive sensing applications.

14.6.3 Sensing Matrix

Many natural time series are sparse in frequency domain, and can be effectively compressed by a random subsampling matrix Φ_S. The reason is that Φ_S is incoherent with the Fourier basis. The work by Candès et al. (2006) shows that signal is recoverable when Ψ is identity and Φ is a randomly subsampled Fourier matrix. Moreover, if the dictionary atoms are sufficiently incoherent in the low frequencies, then a uniform subsampling matrix can be used for high resolution recovery. Using these subsampling matrices as the sensing matrix is desirable because it is easy to implement on a sensor.

For radar signal acquisition, a sensing matrix corresponding to random convolution is more suitable since the way radar signals bounce back is naturally convolutional (see Fig. 14.2). The random convolution sensing matrix can be implemented by randomly subsampling the received signal in time domain. This allows us to recover high resolution signals from low number of measurements (Romberg 2009).

If the basis D is known ahead of time, then using PCA on D to form the sensing matrix Φ is provably optimal for reconstruction. More specifically, signal recovery with the projection $y = V^T x$ where V contains the m largest eigenvectors of $D^T D$ requires least number of measurement (Gkioulekas and Zickler 2011).

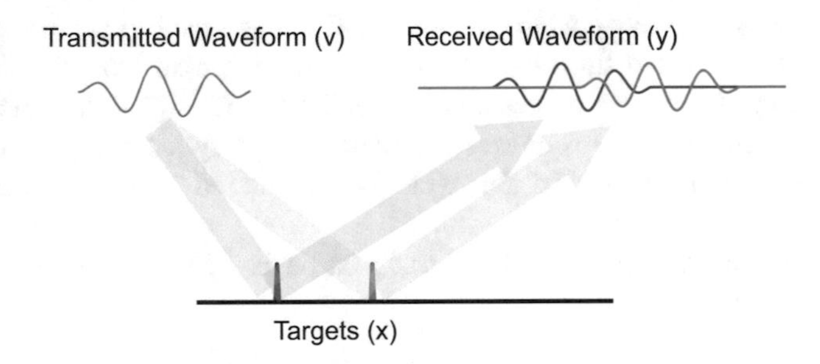

Fig. 14.2 Illustration of radar signal acquisition. The received waveform y is a convolution between the transmitted waveform v and the targets x. We can therefore write $y = Vx$ where each row in V contains a shifted v

14.7 Sparse Coding Overview

In sparse coding, a signal x is transformed into a sparse representation z by solving a linear system with a sparsity-inducing regularizer $s(\cdot)$

$$z \in \arg\min_{z} \|x - Dz\|_2^2 + \lambda s(z) \tag{14.13}$$

where D is a given dictionary (or basis). The first term requires that z captures the information about x (since $x \sim Dz$), and the second term encourages z to have small number of nonzeros.

A dictionary D can be learned from a set of samples $X = \{x_i\}$ by solving

$$\{D, Z\} \in \arg\min_{D,Z} \sum_i \{\|x_i - Dz_i\|_2^2 + \lambda s(z_i)\} \tag{14.14}$$

where the second term regularizes z on its ℓ_0 or ℓ_1-norm (that is, $s(\cdot) := \|\cdot\|_0$ or $s(\cdot) := \|\cdot\|_1$.) with certain parameter $\lambda > 0$. This is often referred to as dictionary learning, and each column in D is called an atom. It is hard to solve (14.14) directly, so in practice people often use alternating methods (e.g., Mairal et al. 2009): solve Z while holding D fixed, and then solve D while holding Z fixed. Note that both subproblems can be solved efficiently: the former is exactly the same as the compressive sensing problem mentioned in Sect. 14.6.1, and can be solved with the same tools; the latter problem

$$\{D\} \in \arg\min_{D} \sum_i \{\|x_i - Dz_i\|_2^2\} \tag{14.15}$$

is a standard least squares problem.

The sparse representation of signals has been shown to be useful for classification tasks in various models, and is therefore often referred to as *feature vectors* (e.g., in spatial pyramid matching (Yang et al. 2009), hierarchical sparse coding (Lin and Kung 2014), and single layer networks (Coates et al. 2011)).

There are many reasons to use sparse coding for feature extraction (Fig. 14.3). To better understand what it does, we can look at the data model corresponding to the optimization problem in (14.13). When $s(\cdot) := \|\cdot\|_0$, it solves a subspace clustering problem, which is like a generalization of gain shape (spherical) k-means. When $s(\cdot) := \|\cdot\|_1$, there is a Bayesian interpretation: Suppose the data x and latent variable z are modeled by Gaussian and Laplace distribution, respectively:

$$x \sim \mathcal{N}(Dz, \sigma) = \frac{1}{\sqrt{2\pi}\sigma} \exp\left(-\frac{\|x - Dz\|_2^2}{2\sigma^2}\right) \tag{14.16}$$

$$z \sim \mathcal{L}(z, \beta) = \frac{1}{2\beta} \exp\left(-\frac{|z|}{\beta}\right) \tag{14.17}$$

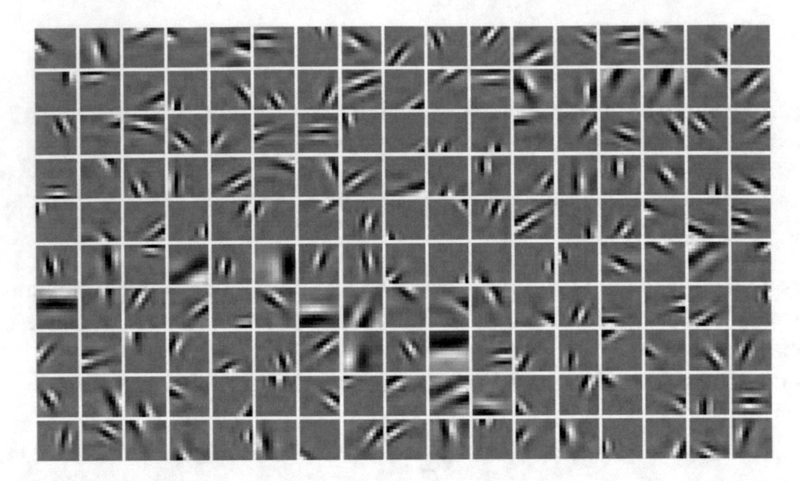

Fig. 14.3 Dictionary atoms learned from patches of natural images. As pointed out by Olshausen and Field, these patterns resemble the receptive fields found in primary visual cortex (Olshausen and Field 1997)

Then the MAP estimation is the solution from sparse coding:

$$z^* = \arg\max_z p(z|x) = \arg\min_z \{-\log p(z|x)\} \tag{14.18}$$

$$= \arg\min_z \left\{ \frac{1}{2\sigma^2} \cdot \|x - Dz\|_2^2 + \frac{1}{\beta} \cdot |z| - \log(2\sqrt{2\pi}\sigma\beta) \right\} \tag{14.19}$$

$$= \arg\min_z \|x - Dz\|_2^2 + \frac{2\sigma^2}{\beta}|z| \tag{14.20}$$

Note that while this simple model may be applicable to many natural signals such as images, it is not a universal principle for finding feature representations (this is unlike PCA, which can serve as a general principle for dimensionality reduction). Sparse coding should only be applied when the data actually has sparse structures.

14.7.1 Compressive Sensing and Sparse Coding

In sparse coding, one is interested in finding a sparse representation z that parsimoniously describes a given data x. It is often viewed as an unsupervised learning method that extracts or finds structures in data. In contrast, compressive sensing assumes such structures in the data, and exploits them to narrow down the solution

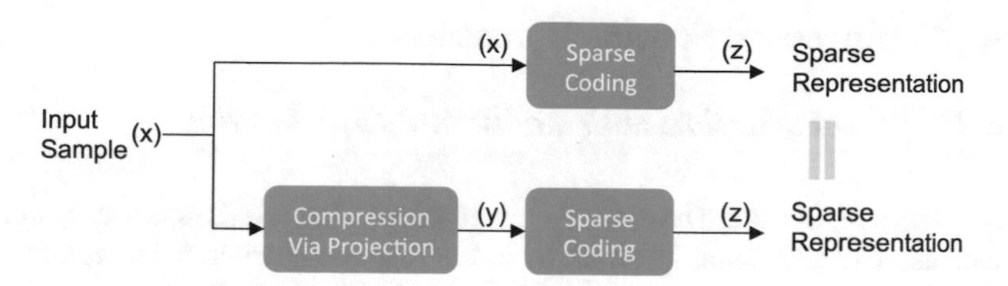

Fig. 14.4 Using compressive sensing to speed up sparse coding pipeline. Note that both approaches generate the same feature representation, but the one with compression does so with much lower cost due to reduced input dimension

space in the signal recovery process. While sparse coding and compressive sensing are designed for very different applications, they are clearly deeply connected through the sparse assumption about data. Here we explore some interesting interactions when we apply both in signal processing.

14.7.1.1 Compressed Domain Feature Extraction

In standard feature extraction, the feature representation is extracted from input x using (14.13). However, we can compress x first to reduce computation cost. Assuming that x is sparse and ΦD is sufficiently incoherent, the feature representation z extracted from x or $y = \Phi x$ would be the same (Fig. 14.4). This is particularly useful if the signal was obtained through compressive sensing in the first place.

14.7.1.2 Compressed Domain Classification

In compressive sensing, Φ is a distance preserving projection such that

$$(1-\delta)\|x_i - x_j\| < \|\Phi(x_i) - \Phi(x_j)\| < (1+\delta)\|x_i - x_j\|$$

for a small constant δ. This means distance-based classification methods such as KNN can work directly on compressed measurement, and enjoys graceful degradation as the number of measurement decreases. Another example is signal detection using likelihood ratio tests. If the tests are based on radial distributions such as Gaussian, then the likelihood ratio test can be easily modified to work with $\Phi(x)$ instead, since the distance $\|x - \mu\|$ is preserved.

14.8 Compressive Sensing Extensions

14.8.1 Reconstruction with Additional Information

In sparse coding, one often has stronger assumptions than just the sparsity. Similarly, we can also use additional information to improve reconstruction for compressive sensing. Sparse coding can be used jointly with these other constraints:

$$z \in \arg\min_{z} \| y - \Phi Dz\|_2^2 + g(z) + \lambda s(z) \tag{14.21}$$

where $g(\cdot)$ incorporates the additional information.

A notable example is super-resolution signal recovery. In this setting, we have $g(z) = \|x^* - Dz\|_2^2$ where x^* is estimated based on self-similarities in the image (epitomic matching). Epitomic matching searches for similar example patches from the input image itself, based on the fact that patches often tend to recur within the image across different scales (Mairal et al. 2009; Glasner et al. 2009). In this approach one estimates z and x^* alternatingly to improve the reconstruction quality over iterations.

14.8.2 Compressive Sensing with Distorted Measurements

In compressive sensing, measurements are assumed to be the linear projection of signals. However, in practice non-linearity could easily arise from the physical devices that perform the measuring. For example, amplifiers and ADCs only have a small region of operation where the gain is constant (Fig. 14.5). As the input power exceeds the linear region, the gain gradually decreases and the relation between input and output is no longer linear. One would need to either have better hardware to handle large linear regions or model the nonlinearity for signal reconstruction.

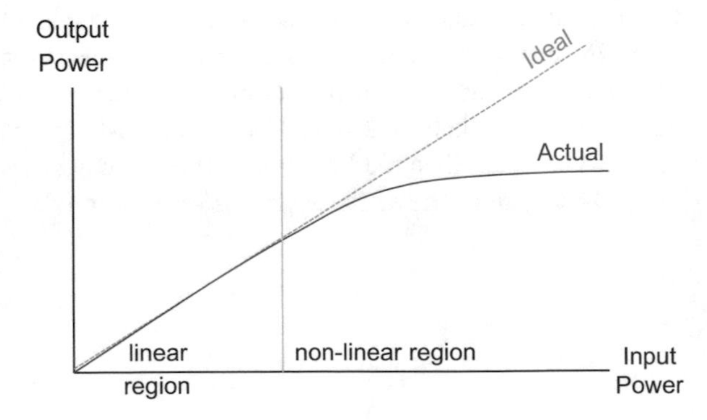

Fig. 14.5 Illustration of amplifier nonlinearity. Amplifiers have a region of operation where the gain is constant. As the input power exceeds the linear region, the gain gradually decreases and the relation between input and output is no longer linear

Assuming that the measuring process is distorted by some nonlinear function F, and we obtain $y = F(\Phi x)$ instead of $y = \Phi x$ in the standard compressive sensing setting. In nonlinear compressive sensing, we need to solve the nonlinear system with sparse regularization:

$$\hat{z} = \arg\min_{z} \| y - F(\Phi Dz)\|_2^2 + \lambda s(z) \tag{14.22}$$

For differentiable F, this formulation can be solved with proximal methods (e.g., FISTA). It has been shown that modeling the non-linearity directly could lead to significant power savings (Comiter et al. 2017).

Another example is the non-linearity due to quantization. It has been shown that even under the most extreme case of binary quantization, we could still recover the signal[4] using a modified compressive sensing (Jacques et al. 2013). Similar to nonlinear compressive sensing, one can modify the objective function in single-bit compressive sensing:

$$\hat{z} = \arg\min_{z} \|\mathbf{sign}(y) - \mathbf{sign}(\Phi Dz)\|_1 + \lambda s(z) \tag{14.23}$$

which can then be solved with a variety of dedicated solvers, including binary iterative hard-thresholding (Jacques et al. 2013), adaptive outlier pursuit (Yan et al. 2012), and even convex programming (Plan and Vershynin 2013).

References

Baraniuk R, Davenport M, DeVore R, Wakin M (2008) A simple proof of the restricted isometry property for random matrices. Constr Approx 28(3):253–263

Beck A, Teboulle M (2009) A fast iterative shrinkage-thresholding algorithm for linear inverse problems. SIAM J Imag Sci 2(1):183–202

Blumensath T, Davies ME (2009) Iterative hard thresholding for compressed sensing. Appl Comput Harmon Anal 27(3):265–274

Boyd S, Parikh N, Chu E, Peleato B, Eckstein J (2011) Distributed optimization and statistical learning via the alternating direction method of multipliers. Found Trends Mach Learn 3(1):1–122

Candes EJ, Tao T (2005) Decoding by linear programming. IEEE Trans Inf Theory 51(12):4203–4215

Candes EJ, Tao T (2006) Near-optimal signal recovery from random projections: universal encoding strategies? IEEE Trans Inf Theory 52(12):5406–5425

Candès EJ, Romberg J, Tao T (2006) Robust uncertainty principles: exact signal reconstruction from highly incomplete frequency information. IEEE Trans Inf Theory 52(2):489–509

Coates A, Ng AY, Lee H (2011) An analysis of single-layer networks in unsupervised feature learning. In: International conference on artificial intelligence and statistics, pp 215–223

[4]The magnitude of the signal is lost in this setting.

Comiter M, Chen H-C, Kung HT (2017) Nonlinear compressive sensing for distorted measurements and application to improving efficiency of power amplifiers. In: IEEE international conference on communications

Donoho DL (2006) Compressed sensing. IEEE Trans Inf Theory 52(4):1289–1306

Efron B, Hastie T, Johnstone I, Tibshirani R, et al (2004) Least angle regression. Ann Stat 32(2):407–499

Gkioulekas IA, Zickler T (2011) Dimensionality reduction using the sparse linear model. In: Advances in neural information processing systems, pp 271–279

Glasner D, Bagon S, Irani M (2009) Super-resolution from a single image. In: 2009 IEEE 12th international conference on computer vision. IEEE, New York, pp 349–356

Jacques L, Laska JN, Boufounos PT, Baraniuk RG (2013) Robust 1-bit compressive sensing via binary stable embeddings of sparse vectors. IEEE Trans Inf Theory 59(4):2082–2102

Lee H, Battle A, Raina R, Ng AY (2006) Efficient sparse coding algorithms. In: Advances in neural information processing systems, pp 801–808

Lin T-H, Kung HT (2014) Stable and efficient representation learning with nonnegativity constraints. In: Proceedings of the 31st international conference on machine learning (ICML-14), pp 1323–1331

Mairal J, Bach F, Ponce J, Sapiro G (2009) Online dictionary learning for sparse coding. In: Proceedings of the 26th annual international conference on machine learning. ACM, New York, pp 689–696

Mairal J, Bach F, Ponce J, Sapiro G, Zisserman A (2009) Non-local sparse models for image restoration. In: 2009 IEEE 12th international conference on computer vision. IEEE, New York, pp 2272–2279

Needell D, Tropp JA (2009) Cosamp: iterative signal recovery from incomplete and inaccurate samples. Appl Comput Harmon Anal 26(3):301–321

Olshausen BA, Field DJ (1997) Sparse coding with an overcomplete basis set: a strategy employed by v1? Vision Res 37(23):3311–3325

Pati YC, Rezaiifar R, Krishnaprasad PS (1993) Orthogonal matching pursuit: recursive function approximation with applications to wavelet decomposition. In: 1993 conference record of the twenty-seventh Asilomar conference on signals, systems and computers. IEEE, New York, pp 40–44

Plan Y, Vershynin R (2013) Robust 1-bit compressed sensing and sparse logistic regression: a convex programming approach. IEEE Trans Inf Theory 59(1):482–494

Romberg J (2009) Compressive sensing by random convolution. SIAM J Imag Sci 2(4):1098–1128

Tibshirani R (1996) Regression shrinkage and selection via the lasso. J R Stat Soc Ser B Methodol 58(1):267–288

Yan M, Yang Y, Osher S (2012) Robust 1-bit compressive sensing using adaptive outlier pursuit. IEEE Trans Signal Process 60(7):3868–3875

Yang J, Yu K, Gong Y, Huang T (2009) Linear spatial pyramid matching using sparse coding for image classification. In: IEEE conference on computer vision and pattern recognition, CVPR 2009. IEEE, New York, pp 1794–1801

Chapter 15
Bridging Density Functional Theory and Big Data Analytics with Applications

Chien-Chang Chen, Hung-Hui Juan, Meng-Yuan Tsai, and Henry Horng-Shing Lu

Abstract The framework of the density functional theory (DFT) reveals both strong suitability and compatibility for investigating large-scale systems in the *Big Data* regime. By technically mapping the data space into physically meaningful bases, the chapter provides a simple procedure to formulate global Lagrangian and Hamiltonian density functionals to circumvent the emerging challenges on large-scale data analyses. Then, the informative features of mixed datasets and the corresponding clustering morphologies can be visually elucidated by means of the evaluations of global density functionals. Simulation results of data clustering illustrated that the proposed methodology provides an alternative route for analyzing the data characteristics with abundant physical insights. For the comprehensive demonstration in a high dimensional problem without prior ground truth, the developed density functionals were also applied on the post-process of magnetic resonance imaging (MRI) and better tumor recognitions can be achieved on the T1 post-contrast and T2 modes. It is appealing that the post-processing MRI using the proposed DFT-based algorithm would benefit the scientists in the judgment of clinical pathology. Eventually, successful high dimensional data analyses revealed that

C.-C. Chen
Bio-MicroSystems Integration Laboratory, Department of Biomedical Sciences and Engineering, National Central University, Taoyuan City, Taiwan

Shing-Tung Yau Center, National Chiao Tung University, Hsinchu City, Taiwan

H.-H. Juan
Shing-Tung Yau Center, National Chiao Tung University, Hsinchu City, Taiwan

M.-Y. Tsai
Institute of Statistics, National Chiao Tung University, Hsinchu City, Taiwan
e-mail: u9826804@stat.nctu.edu.tw

H. H.-S. Lu (✉)
Institute of Statistics, National Chiao Tung University, Hsinchu City, Taiwan

Big Data Research Center, National Chiao Tung University, Hsinchu City, Taiwan

Shing-Tung Yau Center, National Chiao Tung University, Hsinchu City, Taiwan
e-mail: hslu@stat.nctu.edu.tw

© Springer International Publishing AG, part of Springer Nature 2018

W. K. Härdle et al. (eds.), *Handbook of Big Data Analytics*, Springer Handbooks of Computational Statistics, https://doi.org/10.1007/978-3-319-18284-1_15

the proposed DFT-based algorithm has the potential to be used as a framework for investigations of large-scale complex systems and applications of high dimensional biomedical image processing.

Keywords Density functional theory (DFT) · Big data · High dimensional data analysis · Image analysis

15.1 Introduction

Since the mid-twentieth century, scientists have been recognizing the fact that the exploitation of N-particle Schrödinger wave function would eventually suffer a provocative problem in the judgment of legitimate significance while N is larger than 10^3 orders which is even an over-optimistic value on some occasions (Kohn 1999). In spite of the rapid progress of computer science nowadays, validity of contemporary physical theories, from the Heitler-London treatment of chemical bonding to the Hartree-Fock approximation, and the associated computational complexities are still going through a terrible ordeal while dealing with enormous electron interaction within diverse materials or large-scale physical systems.

To escape from these muddles, the density functional theory (DFT), founded on the Hohenberg–Kohn theorem and Kohn–Sham theorem (Hohenberg and Kohn 1964; Kohn and Sham 1965), provides an elegant framework to handle those aggravating situations and has been employed in many interdisciplinary applications, such as quantum chemistry, solid state theory, material science, bioscience, molecular dynamics, and so forth (Lebègue et al. 2013; Grimme et al. 2007; Riley et al. 2010; Neese 2009; Cramer and Truhlar 2009; Wu 2006). Physical configuration and characteristics within a system of interest can be sufficiently elucidated using a three-dimensional electronic probability density function (PDF) (Cramer and Truhlar 2009; Wu 2006; Daw and Baskes 1983) rather than processing $3N$-dimensional many-particle wave functions. Once a PDF of an investigating system is given, demanding characteristics, including cluster distribution, intensity, similarity, affinity, morphology, clustering tendency, spatial configuration, and so forth, can be further analyzed with the convenience of immensely computational complexity reduction. Thus, the mathematical framework of DFT in these aspects reveals highly beneficial suitability and compatibility for investigating large-scale systems such as the applications in the *Big Data* regime. To be specific, the data configuration as well as the derived dimension-reduced data PDF might be analogically treated as in a many-particle scheme embedded in the Hilbert space. Transforming the implicit or explicit data features to useful knowledge would be obviously essential in the big data system.

Additionally, due to multifarious data expansions inevitably and rapidly arise from user generated contents and log-data formed in Internet surfing, community of networks (Newman and Girvan 2004; Girvan and Newman 2002; Clauset et al. 2004; Sporns 2011; Esquivel and Rosvall 2011), investigations on pathology and

DNA sequence (Jacobs 2009; Rozas et al. 2003), bioscience (Girvan and Newman 2002; Bullmore and Bassett 2011; Lichtman et al. 2008; Hampel et al. 2011; Kobiler et al. 2010), cloud and heterogeneous computing (Schadt et al. 2011), explosive growth of global financial information (McAfee and Brynjolfsson 2012; Chen et al. 2012; Tóth et al. 2011), and so forth, relevant problems implicated in the *Big Data* regime are confronting with technical challenges on demands of data recognition, reconstruction, connection and correlation, and visualization in the considerations of commercial strategy and scientific investigations in this era. For instance, the methodologies on pattern recognition of neural networks in human brains especially attract wide attentions due to the desires of seeking correspondences among brain functionalities with physiological and psychological modulations, pathological diagnoses, perceptional characteristics, and so forth. A wiring diagram in a brain delineated by a comprehensive map of neural circuitry, i.e., the connectome, is then pursued in order to thoroughly understand the basic elements within a brain. A well-investigated morphology of cerebral neurons can benefit clinical diagnoses to detect regions of neural mis-wiring connection that may result in Alzheimer's and Parkinson's diseases (Bas et al. 2012).

However, the progress on pattern recognition of biomedical applications staggered (Bas et al. 2012; Hsu and Lu 2013; Kreshuk et al. 2011; Bas and Erdogmus 2010; Livet et al. 2007). The reason can be attributed to the scarcity of robust and reliable automatic artifices for segmentation, thus immense and tedious manual interventions became inevitable when dealing with humongous and intricate biomedical imagery. To circumvent these deficiencies, state-of-the-art techniques based on machine learning in probabilistic perspectives have brought in fruitful achievements (Hsu and Lu 2013; Kreshuk et al. 2011; Shao et al. 2012; Wu et al. 2011; Gala et al. 2014; Chothani et al. 2011; Türetken et al. 2011). By combining graphic theory with geometric features (Bas et al. 2012; Bas and Erdogmus 2010; Vasilkoski and Stepanyants 2009; Wang et al. 2011; Türetken et al. 2013; Gala et al. 2014; Chothani et al. 2011; Türetken et al. 2011; Zhang et al. 2010; Peng et al. 2011; Rodriguez et al. 2009) and/or topological priors (Hsu and Lu 2013; Shao et al. 2012; Wu et al. 2011; Rodriguez et al. 2009), morphologies of human physical structures could be delineated thoroughly and visually. However, several technical limitations still obstruct on the way to mapping. Machine learning-based algorithms inseparably rely on certain requirements of training sets (Gala et al. 2014) and seeding voxels (Hsu and Lu 2013; Shao et al. 2012; Wu et al. 2011; Rodriguez et al. 2009), specific regular curve or shape (Bas et al. 2012; Bas and Erdogmus 2010; Wang et al. 2011; Türetken et al. 2013, 2011; Zhang et al. 2010; Peng et al. 2011; Rodriguez et al. 2009), designated size (Zhang et al. 2010), and so forth. Once these prerequisites eventually become indispensable, user-supervised interventions and complex filtering/classifying mechanisms might make the algorithms more unstable for irregular or unanticipated circumstances.

Therefore, to convey the advantages and the bonanza from the DFT to the applications of *Big Data* analytics, an intuitive and fully energy-based methodology for unsupervised automated pattern recognitions, embedded in the framework of DFT in physical perspectives, is introduced in the chapter. Evolution of mixed data migration under diverse circumstances was first visually studied. The proposed methodologies provided abundant physical insights when analyzing the characteristics of mixed data groups. Furthermore, as tentative demonstrations, typical clustering problem on the dataset of Fisher's iris was then studied. Meanwhile, the large-scale MRI morphology and the corresponding tumor detections were also visually exhibited by means of a derived dimension-reduced pixel intensity function based on the proposed DFT-based algorithm.

15.2 Structure of Data Functionals Defined in the DFT Perspectives

Purpose of the investigation is intended to create a connection between the *Big Data* analysis and the mathematical framework of the DFT. That is, we are extending a well-developed technique in pure physics to interdisciplinary scientific researches and engineering applications. First of all, each specific feature of data points should be smoothly mapped to a corresponding physical basis, i.e., the oriented Riemannian manifold. However, the intrinsic properties of the original data configuration would result in difficulties on the basis transformation mathematically. For instance, labeled or non-numerical features could not construct the corresponding Euclidean distances, so that the existence of Cauchy sequences cannot be definitely guaranteed as well as a complete metric space. To conquer the predicament, those features can be technically transferred to a high dimensional pseudo-numerical or a spatial eigen-space artificially, in which each norm of pseudo-vectors would be finite to meet the formal definition of the Hilbert space. Many contemporary technical articles have verified the feasibility mathematically (Levin et al. 2008; Wu et al. 2012).

Under the framework of DFT, the information of specific data features, such as adjacent matrix in social networks, density of a statistical distribution, proliferation rate of animal populations, or color intensity in an image set, can be extracted from the studied objectives and directly mapped into a high-dimensional energy-space (k-space) in physical views to form a hyper-Fermi surface composed of vectors $\mathbf{k}_F$ and provides a significant enclosure of the dataset. Finding an enclosed hyper-surface of a dataset is the first step for the studies on data clustering and threshold of data mixtures. For further comprehension, properties of N data features confined to a D-dimensional volume V with specific boundary conditions (Lebowitz and Lieb 1969) were analyzed in the high-dimensional Hilbert space. The D-dimensional volume of k-space per allowed $\mathbf{k}$ value, for instance, has generally the value of $\frac{(2\pi)^D}{\prod_i^D L_i}$ under the Born-von Karman boundary condition, and then the quantity of allowed data features N becomes $\frac{\int_0^{\mathbf{k}_F} d\Omega(\mathbf{k}_1,\mathbf{k}_2,\ldots,\mathbf{k}_D) f_{FD}(\mathbf{k}_1,\mathbf{k}_2,\ldots,\mathbf{k}_D)}{(2\pi)^D/\prod_i^D L_i} \equiv \frac{\Omega(\mathbf{k}_F)V}{(2\pi)^D}$, where the volume V is the multiplication of characteristic length L_i spanned in the D-dimensional space and f_{FD} is the Fermi-Dirac distribution. Once the hyper-volume $\Omega(\mathbf{k}_F)$ and the dimensionality D are clarified, the average density of data feature, i.e., $n = \frac{N}{V}$, could be estimated as expected in viewpoints of solid state theory. However, we typically need a definitely localized data feature PDF for data analysis rather than their average density, due to the reason that the latter provides less information about localized variations. In several applications, such as the community of networks, Internet and website searching, image processing with large sizes, object tracking of video stream, and so forth, the localized features are necessary to illustrate the morphological varieties in detail (Newman and Girvan 2004; Girvan and Newman 2002; Clauset et al. 2004; Hampel et al. 2011; McAfee and Brynjolfsson 2012; Kalal et al. 2012; Casagrande et al. 2012).

Thus, to extend the feasibility of DFT into the interdisciplinary applications, a localized functional N' within a D-dimensional localized infinitesimal element $d^D\mathbf{r}'$ was constructed as:

$$N'[n_{\text{local}}] = \int n_{\text{local}}[k_F]d^D\mathbf{r}' = v_D \cdot n_{\text{local}}[k_F], \tag{15.1}$$

where physically

$$n_{\text{local}}[k_F] = \frac{1}{(2\pi)^D}\int_0^{k_F} dk \cdot \alpha_D k^{D-1} = \frac{\alpha_D k_F^D}{D(2\pi)^D}. \tag{15.2}$$

The parameters, v_D and α_D, are D-dimensional hyper-volume and dimension-dependent integral constant, respectively. In the following formula derivation, the α_D can be directly merged into the scaling factor as indicated in Eq. (15.11), thus it can be safely assigned as 1 for convenience. In order to construct the data PDF that can be used to specify the local variations in a system of interest, the local density approximation (LDA) (Wu 2006) was adopted to study the corresponding circumstances. Thus, the functional relation between the data PDF I and the hyper Fermi level k_F can be obtained using Eqs. (15.1) and (15.2):

$$I[k_F] = n_{\text{local}}[k_F] = \frac{N}{v_D} = \frac{k_F^D}{D(2\pi)^D} \qquad \text{or} \qquad k_F[I] = 2\pi[D\cdot I]^{\frac{1}{D}}. \tag{15.3}$$

Furthermore, by means of the LDA and Eq. (15.3), the D-dimensional kinetic energy density functional (KEDF) can be derived as:

$$t_s[I] = \frac{v_D}{N(2\pi)^D}\int_0^{k_F} dk \cdot \alpha_D k^{D-1} \cdot \frac{k^2}{2} = \frac{2\pi^2}{D+2} \cdot D^{\frac{(D+2)}{D}} \cdot I^{\frac{2}{D}} \tag{15.4}$$

It should be noted that the KEDF $t_s[I] = \delta(\text{kinetic energy}))/\delta I$ and is directly proportional to $I^{\frac{2}{D}}$ as indicated in Eq. (15.4).

In general, the ground-state energy of an electron-gas system with an external potential functional $V[n_e]$ and under the atomic units (a.u.) is (Hohenberg and Kohn 1964; Kohn and Sham 1965):

$$E[n_e] = V[n_e] + E_H[n_e](\mathbf{r}_1, \mathbf{r}_2) + G[n_e], \tag{15.5}$$

where $\mathbf{r}_1$ and $\mathbf{r}_2$, n_e, $E_H[n_e]$, and $G[n_e]$ are three-dimensional position vectors of electrons, the electronic PDF, the Hartree energy functional, and the universal functional, respectively. The $G[n_e]$ includes $T_s[n_e]$ and $E_{xc}[n_e]$, where the former is the global kinetic energy functional of non-interacting electrons and the latter the exchange-correlation (xc) energy functional. Based on the Hohenberg–Kohn theorem, it can safely deduce that once the data configuration can be guaranteed to be entirely embedded in oriented Riemannian manifold, the data feature functional in energy perspectives could be likewise elaborated while all of the corresponding functionals mentioned in Eq. (15.5) are clearly specified. Under the circumstances, the data weighting (or the data significance) might be characterized by the kinetic energy functional, and the external potential could denote external influences from the outside environment. Morphology of the data similarity can be mathematically clarified by means of the form of Hartree potential functional. Then, eventually, the data exchangeability might be symbolized by the xc energy functional. Therefore, the localized data feature energy functional in a D-dimensional space could be expressed by linking up $I(\mathbf{r}')$ and $n_e(\mathbf{r})$:

$$E[I] = V[I](\mathbf{r}') + E_H[I](\mathbf{r}') + G[I](\mathbf{r}'). \tag{15.6}$$

Detailed transform relation is shown in Fig. 15.1. Theoretically, the developed methodology can be mathematically extended to arbitrary high-dimensions in the Hilbert space (Finken et al. 2001). Since each data feature can carry $\log_2 D^2$ bits of information in a two-group communication (Wang et al. 2005), the high dimensional data with huge sizes ideally provides the convenience of information storage. Additionally, since the input data PDF $I(\mathbf{r}')$ is a nonnegative function, the considered system can be released from the N-representation problem. Meanwhile, from the

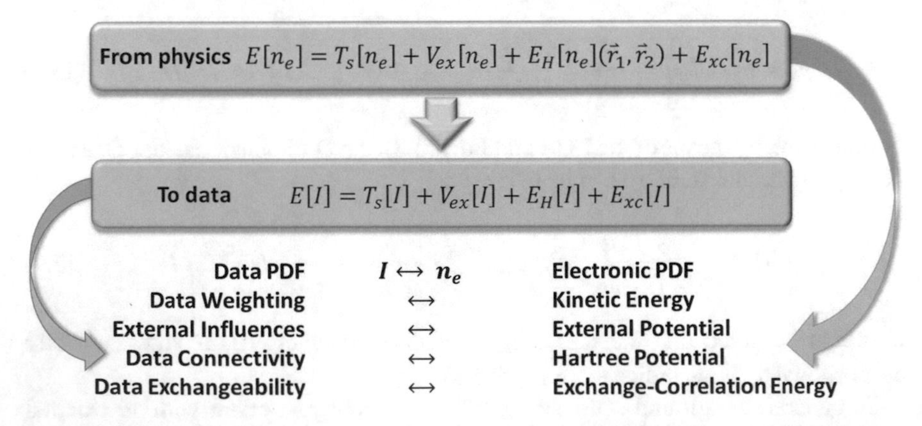

Fig. 15.1 Transform relation of energy functionals between the DFT scheme and the data space. The Hohenberg–Kohn theorem can safely guarantee that once the data configuration can be entirely embedded in the oriented Riemannian manifold, the data feature functional in energy perspectives could be likewise elaborated while all of the corresponding functionals mentioned in Eq. (15.5) are clearly specified

constrained search algorithm (Levy 1982, 2010), issues of the v-representation of a system can also be avoided for the Hohenberg–Kohn theorem (Capelle 2006). Thus, the developed methodology can escape from these non-uniqueness problems.

Furthermore, the Hartree energy with a coulomb form was then adopted in the following tentative investigation to illustrate the data morphology and similarity assessed using the Euclidean distance so that the pure relationship and regularization between any two data features can be definitely clarified. The coulomb form indicates that physical particles must obey the Stokes' theorem and the Gauss' theorem. Thus, in the first stage, the form of coulomb interaction was introduced in the methodology due to its pure and simple physical meanings. However, it does not mean that the methodology is limited to using the coulomb form in the DFT-based algorithm. The selection of a potential form relies on the configuration of the investigating system. To describe the physical phenomena between pair-data-features, the coulomb form might be a good candidate. Meanwhile, for simplicity, the external potential $V[I](\mathbf{r}')$ was set only to be as a uniform background and all other external influences were also ignored. Additionally, as usual the xc energy functional is physically decomposed as the exchange energy derived from the Pauli principle and the correlation energy attributed to the particle correlations. In general, the exchange energy is simply an explicit functional of Kohn–Sham orbitals and can be derived directly using the anti-symmetrized product. On the other hand, there are several approaches to deal with the correlation energy, such as the additional energy lowering for avoiding the mutual interaction between electrons, mean-field approximation without considering the quantum fluctuations, the method of xc hole, and so forth (Capelle 2006). Therefore, according to their physical properties, the xc energy can amend the energy fluctuation while the data points are interchanged among data groups for further problems from social networks or animal populations clustering. In our tentative case, the data interchanges between data groups were temporarily forbidden so that the $E_{xc}[I]$ was only used to tackle the issue of self-interacting in terms of the Hartree energy and the rest long-range part was safely ignored (Capelle 2006) as the conception of range-separated hybrids (Baer et al. 2010) in the methodology derivation so far.

Thus, the rest localized kinetic energy functional in $G[I]$ can be simply estimated using LDA (Wu 2006; Zupan et al. 1997a,b) and Eq. (15.4):

$$T_s[I] = \int d^D\mathbf{r}' t_s[I] I(\mathbf{r}'). \tag{15.7}$$

Then, the localized Hamiltonian functional can be summarized as:

$$\begin{aligned} H[I] &= V + \int I(\mathbf{r}') d^D\mathbf{r}' \left\{ \frac{1}{2} \int \frac{I(\mathbf{r}'')}{|\mathbf{r}' - \mathbf{r}''|} d^D\mathbf{r}'' + t_s[I] \right\} \\ &\equiv V + \int I(\mathbf{r}') d^D\mathbf{r}' \left\{ u[I](\mathbf{r}'') + t_s[I] \right\}, \end{aligned} \tag{15.8}$$

where $u[I]$ is the potential energy density functional (PEDF) in the Hartree's from. According to the Jacob's ladder of DFT, the accuracy of Eq. (15.8) will be limited due to the fact that the formulations are constrained to the Hartree scheme and the employment of LDA (Ullrich 2012). However, the errors produced from the trick of dealing with the xc energy in the chapter could be vastly restrained because of the myriad of data dimensionality and sizes (Voorhis and Scuseria 1998; Langreth and Perdew 1977).

In order to seek the data morphology and global distribution in the development stage, the method of Lagrange multipliers with a constrained condition of constant $\int I(\mathbf{r}')d^D\mathbf{r}'$ was utilized to elaborate the stationary information. The constrained condition indicated the fraction of data features change smoothly and slowly in the considered localized region (Wu 2006; Zupan et al. 1997a,b). Thus, by taking a variational calculation on the localized auxiliary function, $A[I] = H[I] - \mu(\mathbf{r}', \mathbf{r}'') \int I(\mathbf{r}')d^D\mathbf{r}'$, with a Lagrangian multiplier μ, the D-dimensional global Hamiltonian density distribution, or the Hamiltonian density functional (HDF), can be expressed as:

$$\mu[I](\mathbf{r}'') = \frac{\delta H[I]}{\delta I(\mathbf{r}')} = u[I] + t_s[I] + \frac{\delta V}{\delta I(\mathbf{r}')}, \tag{15.9}$$

where the first and second terms correspond to the PEDF and the KEDF, respectively. Since the external potential was treated as a uniform background, the last term can be eliminated. It is noted that Eq. (15.9) also yields a D-dimensional Thomas-Fermi equation and has a similar form as the chemical potential as expected. The advantage of employing the HDF here is to provide a theoretical foundation for the data clustering/segmentation that can be easily and physically elaborated by finding the iso-surfaces of minimum energy. Additionally, once the data space can be mapped into a configuration space of smooth functions and then being constructed to an oriented Riemannian manifold using the mentioned contemporary techniques, boundaries of data groups, according to the Noether's theorem, can also be explicitly defined using $\frac{\delta L[I]}{\delta I(\mathbf{r}')} = 0$. Symbol L here is the corresponding Lagrangian that can be definitely determined from the Hamiltonian with Legendre transform (Salasnich and Toigo 2008). Then the corresponding Lagrangian density functional (LDF) can be expressed as:

$$l[I](\mathbf{r}'') = \frac{\delta L[I]}{\delta I(\mathbf{r}')} = -u[I] + t_s[I] - \frac{\delta V}{\delta I(\mathbf{r}')}, \tag{15.10}$$

Thus, the localized data feature PDFs that satisfy those boundary conditions will construct a subspace of on-shell solutions. Meanwhile, locations of the cusps of HDF and LDF morphologies as well as the amount of data groups can be theoretically determined by adopting the Kato's theorem (March 1986), that is, finding the global maximums of gradient of HDF. Therefore, the enclosure of a data group from a mixed dataset can be uniquely determined by removing the background potential in Eqs. (15.9) and (15.10) and combining the sequence of theorems mentioned above. It should be emphasized that based on machine learning-based algorithms, there are several well-defined supervised clustering or classification methods, such as k-means algorithm (Wu et al. 2012), support vector machine (SVM) (Türetken et al. 2011), and so forth, can be also adopted to achieve similar tasks. The advantage of the proposed DFT-based algorithm is introducing an unsupervised learning method to the problems of interest.

However, the incomparable effect of uniform coordinate scaling between PEDF and KEDF (Levy and Perdew 1985) will result in an incorrect HDF of an arbitrary dimensional system. A solution is to transfer the functional spaces into a normal coordinates using an adaptive scaling factor γ. By employing the Hellmann-Feynman theorem to minimize the average global Lagrangian of the high-dimensional system, the non-trivial solution was given as:

$$\gamma = \frac{1}{2}\frac{\int u[I](\gamma\mathbf{r}')\cdot I(\gamma\mathbf{r}')d\mathbf{r}'}{\int t_s[I](\gamma\mathbf{r}')\cdot I(\gamma\mathbf{r}')d\mathbf{r}'} = \frac{1}{2}\frac{\langle u[I](\gamma\mathbf{r}')\rangle}{\langle t_s[I](\gamma\mathbf{r}')\rangle}. \tag{15.11}$$

Consequently, the adaptive scaling factor is simply the ratio of expected values between the global high-dimensional PEDF and KEDF. Eventually, the scaled HDF and LDF can be, respectively, likewise read as $\gamma^2 t_s[I](\gamma\mathbf{r}') + \gamma u[I](\gamma\mathbf{r}')$ and $\gamma^2 t_s[I](\gamma\mathbf{r}') - \gamma u[I](\gamma\mathbf{r}')$.

15.3 Determinations of Number of Data Groups and the Corresponding Data Boundaries

For realistic clustering problems without any prior information, the number of data groups and the corresponding data boundaries are actually unknown. Trial-and-error methods can be used for typical clustering methods, for instance, the Gaussian mixture model (GMM) or k-means clustering, but they are time-consuming with low efficiency. In the proposed DFT-based algorithm, the mentioned problems can be simply resolved by directly adopting the Kato's theorem to sequentially search the stable points of averaged global high-dimensional PEDF $\langle u[I](\gamma\mathbf{r}')\rangle$ and KEDF $\langle t_s[I](\gamma\mathbf{r}')\rangle$ as listed in Eq. (15.11) from single data group to infinite amount of groups.

A two-dimensional GMM was first employed to generate true data PDFs for the system of interest. The KEDF can be specifically estimated from Eq. (15.4), and the value was given as $t_s[I] = 2\pi^2 I$ for the two-dimensional case. The background potential was kept as a constant so that it can be removed from the estimations. To embody the presented methodology, two identical data groups with huge data size were constructed using the GMM technique. Data groups totally contained 2×10^5 indistinguishable data points with explicit spatial features in two-dimensional coordinates. In order to strongly blend the techniques provided by the DFT with the field of statistical learning, the expectation-maximization (EM) technique (Dempster et al. 1977) was employed in the assessment of the data distribution, and then the estimated normalized PDFs became the input data in the following analysis. It should be emphasized that for the problem of extreme high dimensional PDF estimation, the Bayesian Sequential Partitioning (BSP) method can provide an elegant solution rather than the conventional machine learning methods (Lu et al. 2013). In Fig. 15.2a, each true data group had the same standard deviation σ and their peak-to-peak distance (PtPD) was set to be 10σ to precisely describe the group independence. Monte Carlo method with importance sampling was also technically introduced to reduce the computational complexity of the integral in PEDF, where the computational complexity had an order of $O(N)$ (Foulkes et al. 2001; Needs et al. 2010).

To validate the theory, several different data mixtures were studied as shown in Fig. 15.2. In Fig. 15.2a, two independent data groups were modeled using the GMM method. It is obvious that $\langle t_s[I](\gamma \mathbf{r}')\rangle$ and $\langle u[I](\gamma \mathbf{r}')\rangle$ had obvious gradient changes at the group index with a value of 2 as well as the averaged HDF $\langle HDF\rangle$ estimated using Eqs. (15.9) and (15.11). Thus, it indicates that there are two groups in the studying system. Each group index was estimated by averaging 30 times of trial simulations of corresponding $\langle t_s[I](\gamma \mathbf{r}')\rangle$, $\langle u[I](\gamma \mathbf{r}')\rangle$, and $\langle HDF\rangle$, respectively. Small fluctuations occurred after group index of 2 exhibited the energy gradually became more stable (with small noises) as the core spirit of the Kato's theorem. Of course, an obvious gradient change would emerge again once the group index approached the total number of data points. When the data groups were physically approaching each other accordingly, the corresponding group indexes still exhibited correct estimations as expected and as shown in Fig. 15.2b–d, respectively. The fluctuations behind the group indexes, however, were gradually and sequentially enlarged as also obviously shown from Fig. 15.2c, d. To simply identify the group index in Fig. 15.2d, the averaged curves were colored in black and depicted in each corresponding spectra, respectively. These phenomena exhibited in the group index estimations can be attributed to the reinforced strong similarity between pair-data points with shorter distances. Figure 15.2e shows the morphology of 3 data group mixture with different PDF relative intensities. The stable points on the group index with a value of 3 validated again the feasibility of the DFT-based algorithm in searching the group numbers. The quantum clustering can provide the number of groups as well, but it needs a further searching algorithm to identify the demanding factor (Horn and Gottlieb 2001).

The obtained group numbers were then used to find the corresponding data boundaries of each case. By recalling the Noether's theorem, the optimized data boundary can be defined by searching the locations with the features of $\frac{\delta L[I]}{\delta I(\mathbf{r}')} = 0$. In the practical programming processes, the mathematical approach can be simply realized and should be equivalent to searching the iso-surface of minimum energy that only contains one data group. Figure 15.3 shows the searching results of each mentioned case. The arrows in Fig. 15.3a indicate the modeled outmost enclosure of each group, i.e., the boundary of each iso-surface of minimum energy. Then each recognized data group was respectively colored in red and blue. Additionally, because the modeled boundaries were definitely defined by estimating the corresponding iso-surface families, the data points that exactly enclosed in the

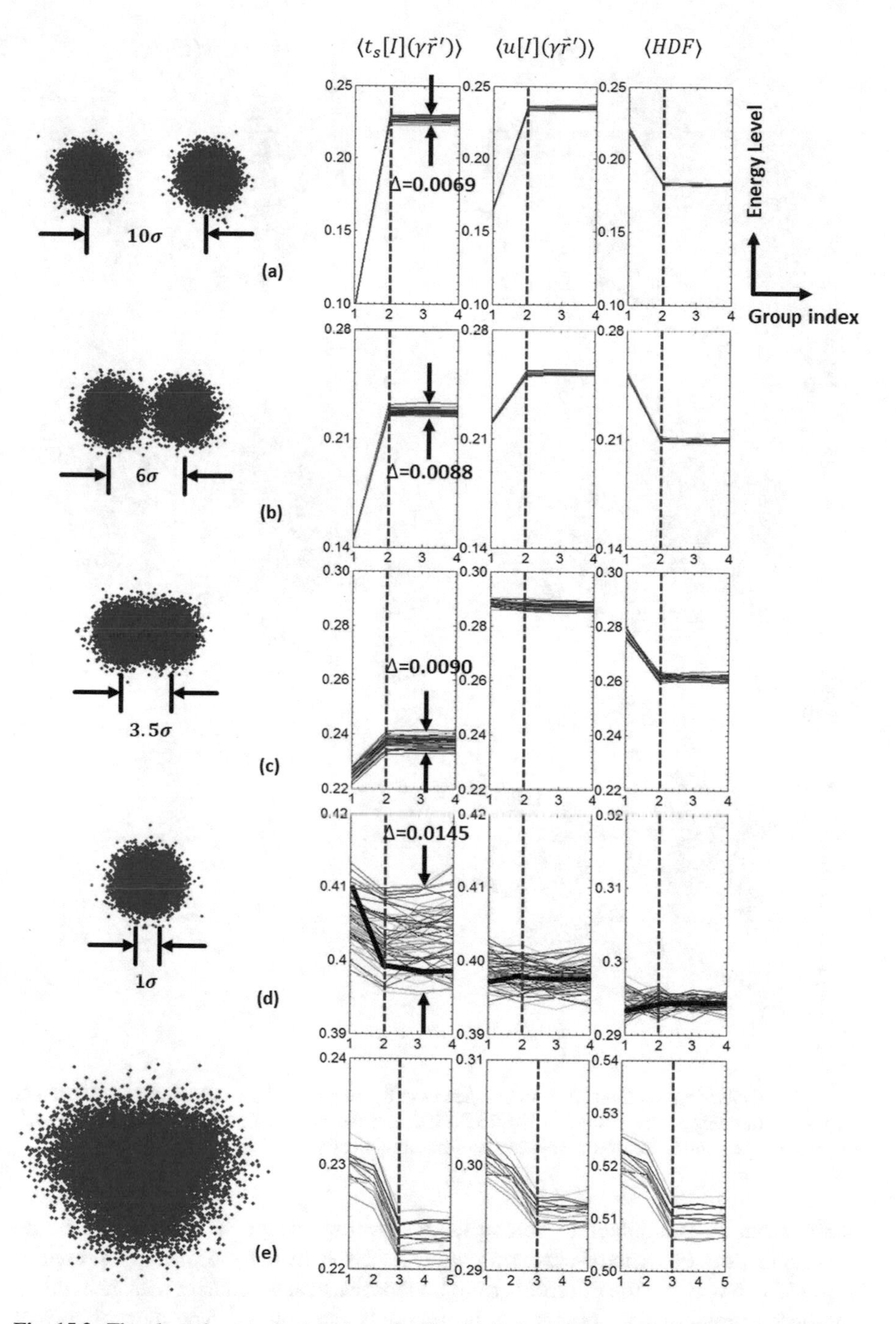

Fig. 15.2 The clustering results based on the Kato's theorem by means of searching the stable points

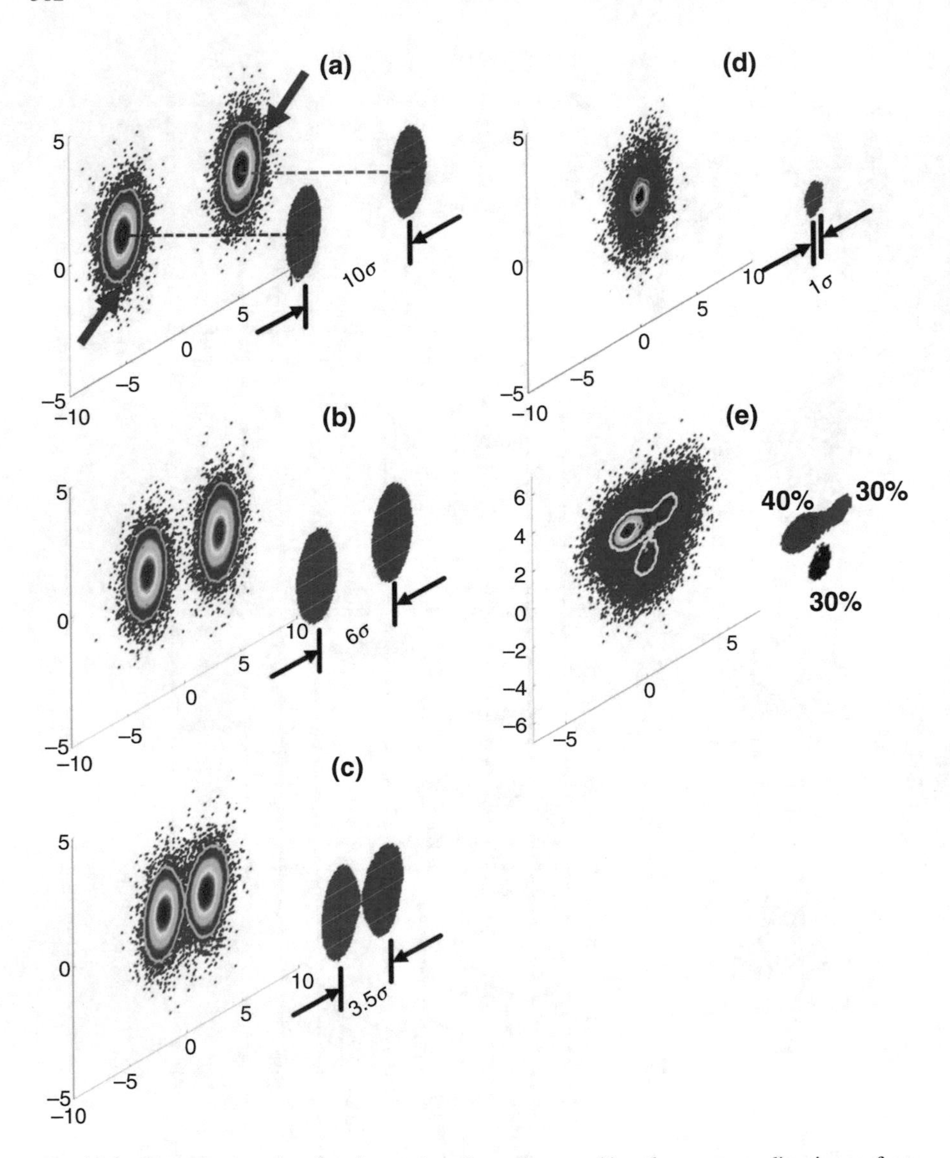

Fig. 15.3 Clustering results of each mentioned case by searching the corresponding iso-surfaces of minimum energy. The arrows in Fig. 15.3a indicate the modeled outmost enclosure of each group, i.e., the boundaries of the iso-surface of minimum energy

surface can be guaranteed to belong to the owning groups. Thus, the DFT-based algorithm provides a hard-clustering strategy. Even in the morphology shown in Fig. 15.3c, e wherein the different group boundaries exactly contact with each other, the adscriptions of data points can be uniquely determined. The morphology of Fig. 15.3d, however, could not show distinguishable boundaries due to its severe overlapping.

15.4 Physical Phenomena of the Mixed Data Groups

15.4.1 Physical Structure of the DFT-Based Algorithm

For the purpose of visualized demonstrations, the original data space would be projected to high-dimensional physical spaces. Figure 15.4 exhibits a three-dimensional morphology of a single dataset (see Fig. 15.4b) in energy space in detail. Relationships between the principal axes and energy level of PDF are illustrated in Fig. 15.4a, c in specific directions. The KEDF and the PEDF (also called the Hartree potential density functional) were estimated as well as the LDF. It is noted that there was an obvious minimum LDF set (the white circle region) surrounding the main part of LDF as shown in Fig. 15.4d. Comparing Fig. 15.4a and c to d, the minimum LDF set just met the data boundary (see the edge of KEDF depicted by red line) and therefore obeyed the boundary features indicated by the Noether's theorem as expected. The cross symbol depicted on the peak of LDF as shown in Fig. 15.4d was estimated by associating the GMM method with the *k*-means, where a prior number of data group is used for a further check.

In Fig. 15.4h, as the case shown in Fig. 15.2a, the PtPD was set to be 10σ. The morphological illustrations of LDF visually revealed two distinct trenches at C and C' which were entirely surrounding each corresponding LDF peak associated with D and D', respectively. Since the steeper distribution of KEDF was proportional to the data PDF, it can represent the data intensity with weighting information as mentioned above. Then the wider shape of PEDF depicted there existed a significant data similarity due to the employed form of long-range coulomb potential by estimating the Euclidean distance of pair-data. Thus, the formation of trenches can be attributed to that the internal intensity information inhibited the external data similarity which was carried by the PEDF. In other words, these two data groups were independent. This inference can be evidenced by comparing the cross sections of principal axis 1 and 2 of Fig. 15.4e and g respectively. Apparently, the KEDF at the common boundary between those two groups were almost vanished, whereas that in PEDF had significant values due to data similarity caused by the coulomb form. Meanwhile, it can be found that the original point of LDF, O, was a local maxima in AA' direction but a local minima in BB' direction. In other words, the O point was also a saddle point and can represent an energy barrier between the data groups. Therefore, in the AA' direction, if these two data groups would like to move towards each other, they should first conquer the obstacle of energy barrier at the O point.

In a nutshell, the morphology of LDF trenches qualitatively revealed the oriented tendency of data similarity and simultaneously indicated the enclosures of each corresponding data groups. The LDF barrier between those groups carried different information regarding the data affinity. Physically, the configuration of the LDF barrier definitely revealed the strength of data affinity and the corresponding localized values can be quantitatively estimated by calculating the difference between the local minimum (about −0.10 a.u./PDF) and the local maximum (about −0.08

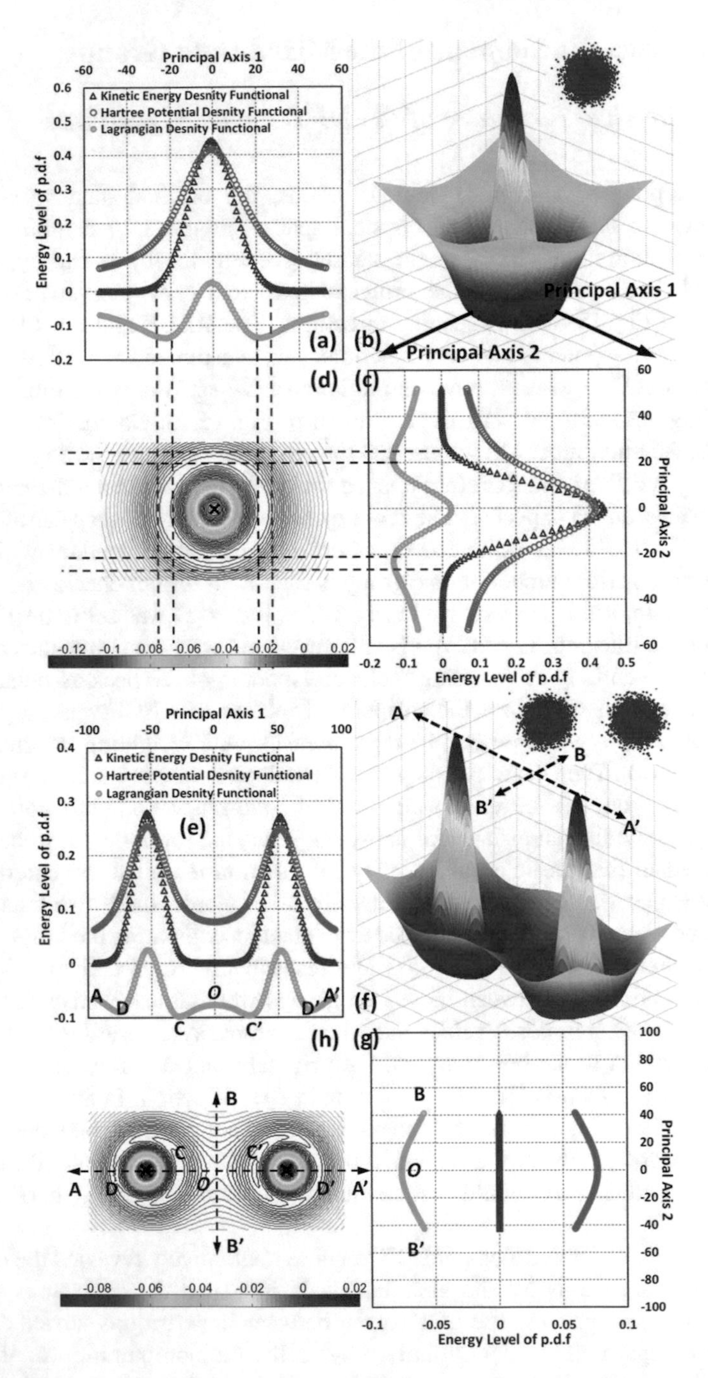

Fig. 15.4 LDF morphologies of single and two-component datasets. Two distinct trenches at C and C' in (h) were entirely surrounding each corresponding LDF peak associated with D and D', respectively. The original point of LDF, *O*, was a saddle point and can represent an energy barrier between the data groups

a.u./PDF) in the AA' direction. In this case, the estimated barrier was 0.02 a.u./PDF and was almost purely contributed by PEDF. Once the LDF barriers were overcome, the data affinity raised then the circumstances of data mixture emerged as well. Mathematically, the localized values of LDF trench and barrier can be, respectively, determined by estimating the gradient and the Laplacian of LDF in practice.

Evolution of data migration is demonstrated in Fig. 15.5. In Fig. 15.5a, the groups were just under the contact mode with 6σ PtPD. The LDF barrier, as mentioned in Fig. 15.4f, disappeared due to the trenches merged with each other as shown in the *O* point of insertion. While PtPD approached to 3.5σ, as shown in Fig. 15.5b, the mode of weak data overlap occurred. Visually, the original merged trench was split by the approaching groups into two ripples (E and E') to indicate the residual information of data similarity that stood aside the groups as shown in insertion. Eventually, while the PtPD approached to 2σ and even 1σ, the groups would suffer severe overlapping and even mixture situations, respectively, as shown in Fig. 15.5c. Significant level decrease of LDF ripples under this mode generally revealed that the enclosures and the main bodies of these groups are gradually becoming indistinguishable. Fortunately, those demanding information for pattern recognition under these circumstances still can be extracted from the morphologies of LDF ripples, i.e., the residual information of data similarity (see E and E'). In Fig. 15.5c, two symmetrical LDF ripples (see also E and E') beside the dataset showed there used to be two groups even though the groups were severely mixed with each other.

15.4.2 *Typical Problem of the Data Clustering: The Fisher's Iris*

To further illustrate the advantages using DFT-based algorithm, Fig. 15.6 also shows the clustering results of a typical problem, the Fisher's iris, which was extracted from the UC Irvine Machine Learning Repository (UC Irvine Machine Learning Repository Database 1988). In this dataset, there are three data groups (*Iris setosa*, *Iris versicolour*, and *Iris virginica*) with four outward appearance features (the length and the width of sepal and petal, respectively), and each feature had 50 observed values. In this case, the Kato's theorem was first used to estimate the group number, and then the values of group number were set to be as start points for searching each local enclosure of groups sequentially. The boundary conditions were then numerically provided by searching the iso-surfaces of minimum energy based on the Noether's theorem for the practically programming. Obviously, the group number and the data boundaries can be well evaluated and the clustering accuracy was limited by the hard-clustering method.

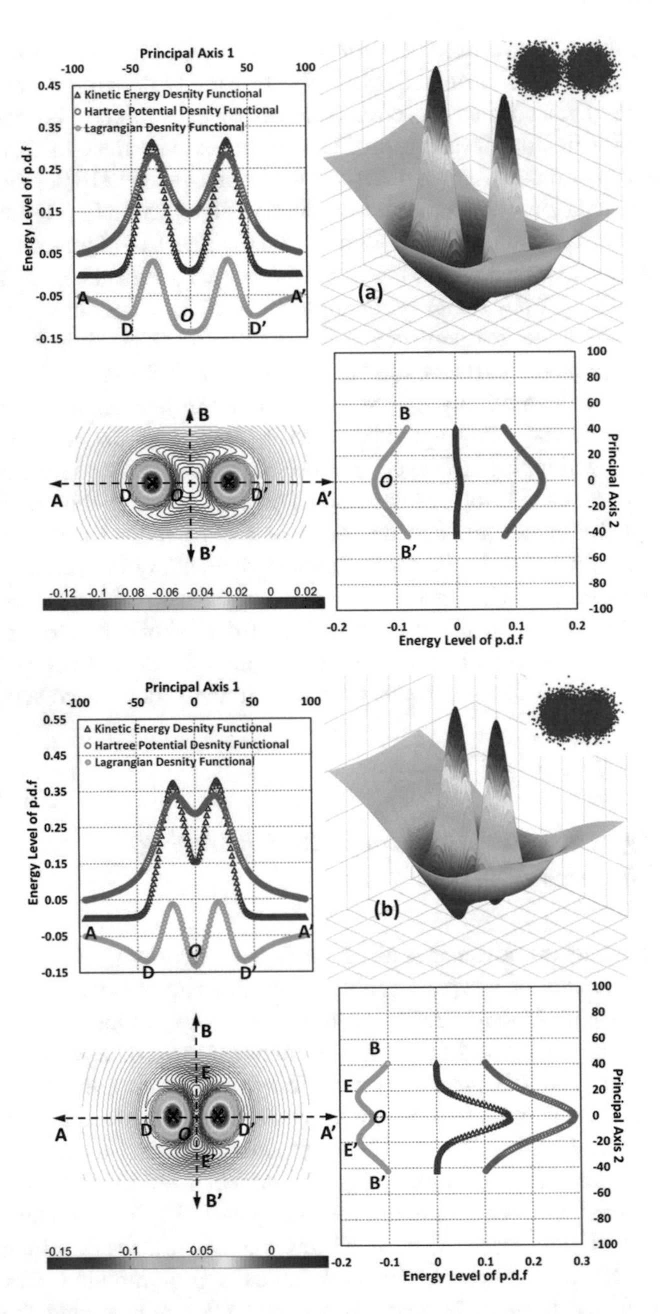

Fig. 15.5 Evolution of data migration. (**a**) The two groups were just under the contact mode with 6σ PtPD. The LDF barrier disappeared due to the trenches merge with each other in the O. (**b**) The mode of weak data mixture with PtPD approaching to 3.5σ. The original merged trench was divided into two ripples (E and E′) to indicate the residual information of data similarity. (**c**) The mixture mode with PtPD approaching to 1σ. The demanding information for data clustering under these circumstance still can be extracted from the morphologies of LDF ripples, i.e., the residual information of data similarity (see E and E′), as well

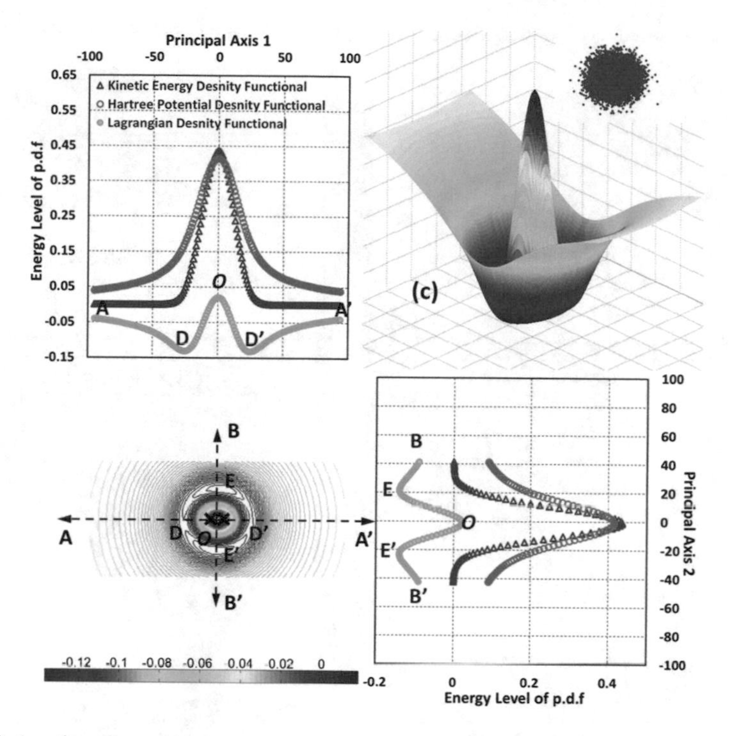

Fig. 15.5 (continued)

15.4.3 Tentative Experiments on Dataset of MRI with Brain Tumors

Unsupervised learning for brain tumor segmentation of MRI relies on the strategies of feature selection and is also an important topic both in the fields of computer vision application and especially the recognition of pathological morphologies. Several state-of-the-art methods have provided alternative ways for separating different tumor tissues (Gordillo et al. 2013), and most of them are based on statistical classifications: probabilistic geometric model (Moon et al. 2002), outlier detection (Prastawa et al. 2004), Knowledge-Based Techniques (Clark et al. 1998), SVM (Zhang et al. 2004), and so forth. In the investigation, a physical-based classification was achieved using the developed DFT-based algorithm. A pixel classification technique was then proposed by taking the boundary conditions according to the Noether's theorem.

Locations of each pixel of the MRI are a set of prior information and thus can be used to construct a pixel intensity PDF. Thus the corresponding normalized data PDF was modified as $I(\mathbf{r}') = \sum_{n=1}^{W\times H} M_n \times \delta(\mathbf{r}' - \mathbf{r}'_n)$, where W and H are, respectively, width and height of MRI images, $\mathbf{r}'_n$ is the position of the n-th pixel, and M_n is the corresponding normalized local intensity. Eventually, the PEDF can

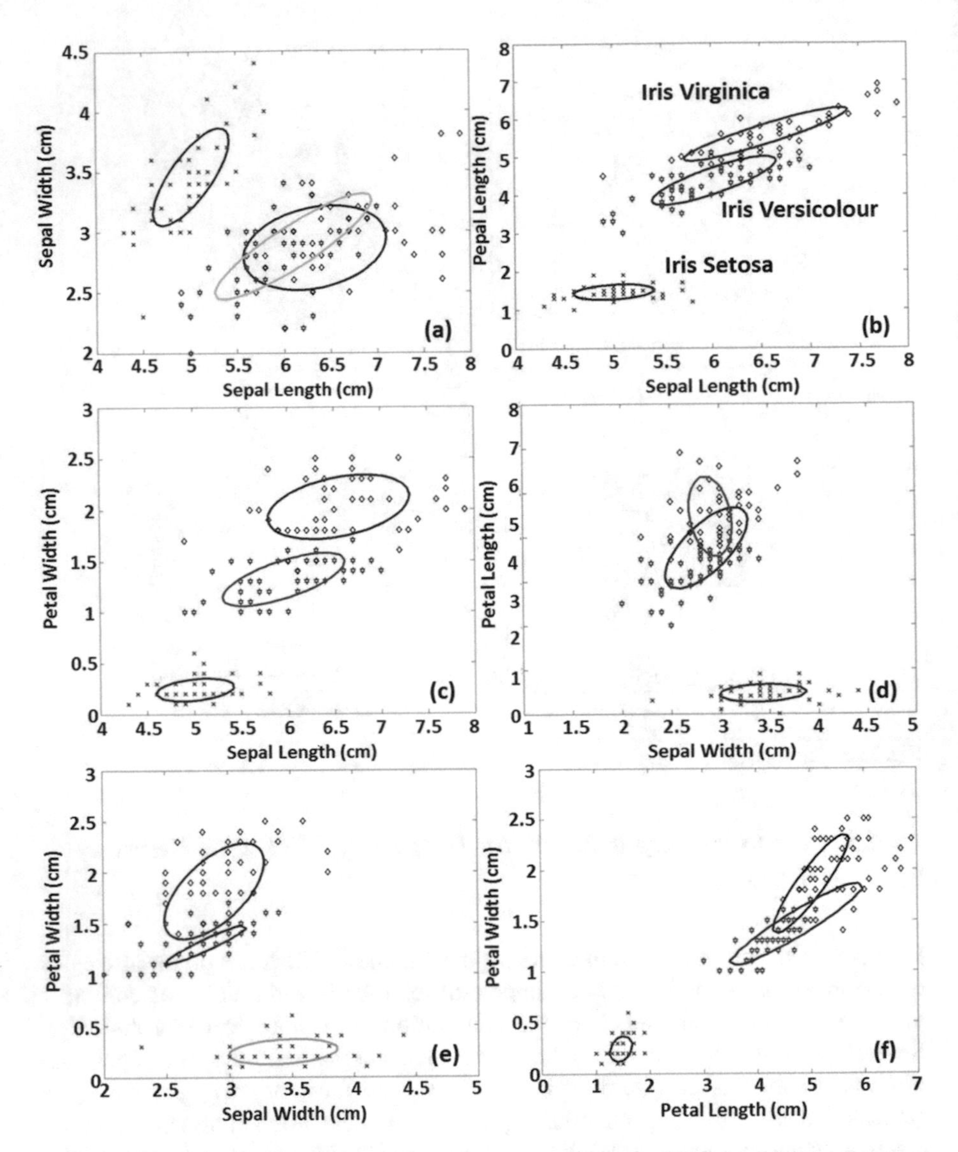

Fig. 15.6 Clustering results of the Fisher's iris, extracted from the UC Irvine Machine Learning Repository (UC Irvine Machine Learning Repository Database 1988). In this dataset, there are three data groups (*Iris setosa*, *Iris versicolour*, and *Iris virginica*) with four outward appearance features (the length and the width of sepal and petal, respectively), and each feature has 50 observed values

be simplified as:

$$u[I](\mathbf{r}'_n) = \int \frac{I(\mathbf{r}'')}{||\mathbf{r}'' - \mathbf{r}'_n||} d^D\mathbf{r}'' \cong \sum_{n=1}^{W\times H} \frac{M_n}{||\mathbf{r}'' - \mathbf{r}'_n||}, \tag{15.12}$$

or in matrix form:

$$\sum_{n=1}^{W\times H}\frac{M_n}{||\mathbf{r}''-\mathbf{r}'_n||}=\left[\frac{1}{||\mathbf{r}''-\mathbf{r}'_1||}\ \frac{1}{||\mathbf{r}''-\mathbf{r}'_2||}\cdots\frac{1}{||\mathbf{r}''-\mathbf{r}'_N||}\right]\begin{bmatrix}M_1\\M_2\\\vdots\\M_N\end{bmatrix}\equiv\mathbb{DM}.\quad(15.13)$$

The factor $N = W \times H$ is the total number of pixels of the studied MRI image. Therefore, the problem was simplified to estimate the distance matrix $\mathbb{D}$ since it contained the information of similarity in terms of $\frac{1}{||\mathbf{r}''-\mathbf{r}'_n||}$ $\forall n$. It is also noted that the distance matrix should be artificially defined as follows to avoid the issue of singularity (Capelle 2006):

$$D_{ij}=\begin{cases}\frac{1}{||\mathbf{r}_i-\mathbf{r}_j||} & \forall i\neq j\\ 0 & \forall i=j\end{cases}.\quad(15.14)$$

Thus, by combining from Eqs. (15.9) to (15.14), the weighting of pixel intensities and the similarity among the pixels within the MRI sets can be clearly delineated by the LDF morphologies in an energy representation. Figure 15.7 sequentially shows sets of MRI image using T1 pre-contrast, T1 post-contrast, and T2 processes, and their corresponding treatments using DFT-based algorithm, respectively. The MRI datasets were extracted from Moon et al. (2002). For pure demonstrations of the subsets defined only by the developed methodology, all of the post-processing MRI datasets were not treated using any contemporary technique of segmentation or pattern screening. The data boundary defined purely by the Noether's theorem can reflect the steep boundaries at the interfaces between normal and abnormal tissues. All of the marked errors from the bone and parts of undesired tissues defined by the subsets were left in all of DFT-treated results of the post-processing MRI. It is obviously that the DFT treatments on the cases using T1 post-contrast and T2 processes revealed better outcomes of recognitions of brain tumors than that using T1 pre-contrast. The result of tumor recognition in the T1 pre-contrast image can be attributed to that the interface changes between normal tissues and the tumor in this case were smoother than that in the others studied case. It is appealing that the post-processing MRI using the proposed DFT-based algorithm associated with other methods of pattern recognition or clustering could benefit the scientists in the judgment of clinical pathology and/or the applications of high dimensional biomedical image processing.

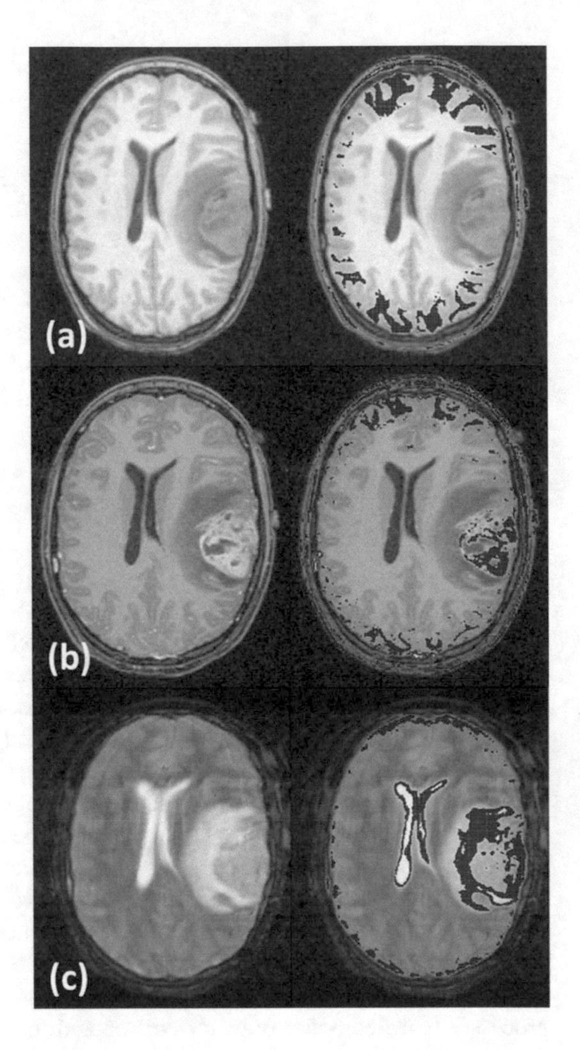

Fig. 15.7 The set of MRI (**a**) T1 pre-contrast, (**b**) T1 post-contrast, and (**c**) T2, and their corresponding treatments using DFT-based algorithm, respectively. The DFT treatments on the cases of T1 post-contrast and T2 revealed better morphologies of recognitions of brain tumors than that on the T1 pre-contrast. This circumstance can be attributed to the interface changes between normal tissue and tumor in the case of T1 pre-contrast were smoother than others in the studied case

15.5 Conclusion

The mathematical framework of the density functional theory potentially provides an elegant configuration for analyzing the high-dimensional dataset, the *Big Data*. Once the dataset can be technically mapped into the oriented Riemannian manifold, the transformations in energy representations from the pure physics to the data space can be derived. Then the goal has been achieved in the chapter in a high-dimensional data perspective. In the presented methodology, scaling effects have been considered so that the high-dimensional data LDF and EDF can be rebuilt in a corresponding high-dimensional uniformed coordinates. LDF can be realized as a data distribution subtracts its local mean, so that the localized features can be significantly reinforced. Physical phenomena of data migration and the information of data similarity were visually clarified by employing LDF trenches and ripples. The LDF barriers were also used to indicate the threshold of data mixtures and the strength of data affinity. Meanwhile, the presented clustering methods using LDF provide routes of group segmentation. The group number can be estimated by the HDF curve.

For a comprehensive demonstration, the proposed DFT-based algorithm constitutes a potential methodology for automatically pattern recognition of the biomedical imageries. By simultaneously considering the weighting of intensities and the similarity among the pixels in an energy representation, the imaging topologies can be visually illustrated in LDF morphologies. Hopefully, the presented DFT-based algorithm can be likewise implemented in the major topics in science, such as the connectomics, DIADEM (digital reconstruction of axial and dendritic morphology) challenges, and so forth.

Acknowledgements We would like to acknowledge the supports from National Science Council, National Center for Theoretical Sciences, Shing-Tung Yau Center, Center of Mathematical Modeling and Scientific Computing at National Chiao Tung University in Taiwan.

References

Baer R, Livshits E, Salzner U (2010) Tuned range-separated hybrids in density functional theory. Annu Rev Phys Chem 61:85–109

Bas E, Erdogmus D (2010) Piecewise linear cylinder models for 3-dimensional axon segmentation in Brainbow imagery. In: International symposium on biomedical imaging (ISBI), pp 1297–1300

Bas E, Erdogmus D, Draft RW, Lichtman JW (2012) Local tracing of curvilinear structures in volumetric color images: application to the Brainbow analysis. J Vis Commun Image R 23:1260–1271

Bullmore ET, Bassett DS (2011) Brain graphs: graphical models of the human brain connectome. Annu Rev Clin Psychol 7:113–140

Capelle K (2006) A bird's-eye view of density-functional theory. Braz J Phys 36:1318–1343

Casagrande D, Sassano M, Astolfi A (2012) Hamiltonian-based clustering: algorithms for static and dynamic clustering in data mining and image processing. IEEE Control Syst 32:74–91

Chen H, Chiang RHL, Storey VC (2012) Business intelligence and analytics: from big data to big impact. MIS Q 36:1165–1188

Chothani P, Mehta V, Stepanyants A (2011) Automated tracing of neurites from light microscopy stacks of images. Neuroinformatics 9:263–278

Clark MC, Hall LO, Goldgof DB, Velthuizen R, Murtagh FR, Silbiger MS (1998) Automatic tumor segmentation using knowledge-based techniques. IEEE Trans Med Imag 17:187–201

Clauset A, Newman MEJ, Moore C (2004) Finding community structure in very large networks. Phys Rev E 70:066111

Cramer CJ, Truhlar DG (2009) Density functional theory for transition metals and transition metal chemistry. Phys Chem Chem Phys 11:10757–10816

Daw MS, Baskes MI (1983) Semiempirical, quantum mechanical calculation of hydrogen embrittlement in metals. Phys Rev Lett 50:1285–1288

Dempster AP, Laird NM, Rubin DB (1977) Maximum likelihood from incomplete data via the EM algorithm. J R Stat Soc B 39:1–38

Esquivel AV, Rosvall M (2011) Compression of flow can reveal overlapping-module organization in networks. Phys Rev X 1:021025

Finken R, Schmidt M, Löwen H (2001) Freezing transition of hard hyperspheres. Phys Rev E 65:016108

Foulkes WMC, Mitas L, Needs RJ, Rajagopal G (2001) Quantum Monte Carlo simulations of solids. Rev Mod Phys 73:33–83

Gala R, Chapeton J, Jitesh J, Bhavsar C, Stepanyants A (2014) Active learning of neuron morphology for accurate automated tracing of neurites. Front Neuroanat 8:1–14

Girvan M, Newman MEJ (2002) Community structure in social and biological networks. Proc Natl Acad Sci USA 99:7821–7826

Gordillo N, Montseny E, Sobrevilla, P (2013) State of the art survey on MRI brain tumor segmentation. Magn Reson Imaging 31:1426–1438

Grimme S, Antony J, Schwabe T, Mück-Lichtenfeld C (2007) Density functional theory with dispersion corrections for supramolecular structures, aggregates, and complexes of (bio)organic molecules. Org Biomol Chem 5:741–758

Hampel S, Chung P, McKellar CE, Hall D, Looger LL, Simpson JH (2011) Drosophila Brainbow: a recombinase-based fluorescence labeling technique to subdivide neural expression patterns. Nat Methods 8:253–259

Hohenberg P, Kohn W (1964) Inhomogeneous electron gas. Phys Rev 136:B864–B871

Horn D, Gottlieb A (2001) Algorithm for data clustering in pattern recognition problems based on quantum mechanics. Phys Rev Lett 88:018702

Hsu Y, Lu HH-S (2013) Brainbow image segmentation using Bayesian sequential partitioning. Int J Comput Electr Autom Control Inf Eng 7:897–902

Jacobs A (2009) The pathologies of big data. Commun ACM 52:36–44

Kalal Z, Mikolajczyk K, Matas J (2012) Tracking-learning-detection. IEEE Trans Pattern Anal Mach Intell 34:1409–1422

Kobiler O, Lipman Y, Therkelsen K, Daubechies I, Enquist LW (2010) Herpesviruses carrying a Brainbow cassette reveal replication and expression of limited numbers of incoming genomes. Nat Commun 1:1–8

Kohn W (1999) Nobel lecture: electronic structure of matter-wave functions and density functionals. Rev Mod Phys 71:1253–1266

Kohn W, Sham LJ (1965) Self-consistent equations including exchange and correlation effects. Phys Rev 140:A1133–A1138

Kreshuk A, Straehle CN, Sommer C, Korthe U, Knott G, Hamprecht FA (2011) Automated segmentation of synapses in 3D EM data. In: International symposium on biomedical imaging (ISBI), pp 220–223

Langreth DC, Perdew JP (1997) Exchange-correlation energy of a metallic surface: wave-vector analysis. Phys Rev B 15:2884–2901

Lebègue S, Björkman T, Klintenberg M, Nieminen RM, Eriksson O (2013) Two-dimensional materials from data filtering and Ab initio calculations. Phys Rev X 3:031002

Lebowitz JL, Lieb EH (1969) Existence of thermodynamics for real matter with Coulomb forces. Phys Rev Lett 22:631–634

Lichtman JW, Livet J, Sanes JR (2008) A technicolour approach to the connectome. Nat Rev Neurosci 9:417–422

Livet J, Weissman TA, Kang H, Draft RW, Lu J, Bennis RA, Sanes JR, Kichtman JW (2007) Transgenic strategies for combinatorial expression of fluorescent proteins in the nervous system. Nature 450:56–62

Levin A, Rav-Acha A, Lischinski D (2008) Spectral matting. IEEE Trans Pattern Anal Mach Intell 30:1699–1712

Levy M (1982) Electron densities in search of Hamiltonians. Phys Rev A 26:1200–1208

Levy M (2010) On the simple constrained-search reformulation of the Hohenberg-Kohn theorem to include degeneracies and more (1964–1979). Int J Quant Chem 110:3140–3144

Levy M, Perdew JP (1985) Hellmann-Feynman, virial, and scaling requisites for the exact universal density functionals. Shape of the correlation potential and diamagnetic susceptibility for atoms. Phys Rev A 32:2010–2021

Lu L, Jiang H, Wong WH (2013) Multivariate density estimation by Bayesian Sequential Partitioning. J Am Stat Assoc Theory Methods 108:1402–1410

March NH (1986) Spatially dependent generalization of Kato's theorem for atomic closed shells in a bare Coulomb field. Phys Rev A 33:88–89

McAfee A, Brynjolfsson E (2012) Big data: the management revolution. Harv Bus Rev 90:59–68

Moon N, Bullitt E, van Leemput K, Gerig G (2002) Automatic brain and tumor segmentation. In: MICCAI proceedings. Lecture notes in computer science, vol 2488, pp 372–379

Needs RJ, Towler MD, Drummond ND, Ríos PL (2010) Continuum variational and diffusion quantum Monte Carlo calculations. J Phys Condens Matter 22:023201

Neese F (2009) Prediction of molecular properties and molecular spectroscopy with density functional theory: from fundamental theory to exchange-coupling. Coord Chem Rev 253:526–563

Newman MEJ, Girvan M (2004) Finding and evaluating community structure in networks. Phys Rev E 69:026113

Peng H, Long F, Myers G (2011) Automatic 3D neuron tracing using all-path pruning. Bioinformatics 27:i239–i247

Prastawa M, Bullitt E, Ho S, Gerig G (2004) A brain tumor segmentation framework based on outlier detection. Med Image Anal 8:275–283

Riley KE, Pitoňák M, Jurečka P, Hobza P (2010) Stabilization and structure calculations for noncovalent interactions in extended molecular systems based on wave function and density functional theories. Chem Rev 110:5023–5063

Rodriguez A, Ehlenberger DB, Hof PR, Wearne SL (2009) Three-dimensional neuron tracing by voxel scooping. J Neurosci Methods 184:169–175

Rozas J, Sánchez-DelBarrio JC, Messeguer X, Rozas R (2003) DnaSP, DNA polymorphism analyses by the coalescent and other methods. Bioinformatics 19:2496–2497

Salasnich L, Toigo F (2008) Extended Thomas-Fermi density functional for the unitary Fermi gas. Phys Rev A 78:053626

Schadt EE, Linderman MD, Sorenson J, Lee L, Nolan GP (2011) Cloud and heterogeneous computing solutions exist today for the emerging big data problems in biology. Nat Rev Genet 12:224–224

Shao H-C, Cheng W-Y, Chen Y-C, Hwang W-L (2012) Colored multi-neuron image processing for segmenting and tracing neural circuits. In: International conference on image processing (ICIP), pp 2025–2028

Sporns O (2011) The human connectome: a complex network. Ann NY Acad Sci 1224:109–125

Tóth B, Lempérière Y, Deremble C, de Lataillade J, Kockelkoren J, Bouchaud, J-P (2011) Anomalous price impact and the critical nature of liquidity in financial markets. Phys Rev X 1:021006

Türetken E, González G, Blum C, Fua P (2011) Automated reconstruction of dendritic and axonal trees by global optimization with geometric priors. Neuroinformatics 9:279–302

Türetken E, Benmansour F, Andres B, Pfister H, Fua P (2013) Reconstructing loopy curvilinear structures using integer programming. In: IEEE conference on computer vision and pattern recognition (CVPR), pp 1822–1829

UCI Machine Learning Repository Iris Dataset (1988) Available at https://archive.ics.uci.edu/ml/datasets/Iris

Ullrich CA (2012) Time-dependent density-functional theory. Oxford University Press, New York, pp 21–41

Vasilkoski Z, Stepanyants A (2009) Detection of the optimal neuron traces in confocal microscopy images. J Neurosci Methods 178:197–204

Voorhis TV, Scuseria GE (1998) A novel form for the exchange-correlation energy functional. J Chem Phys 109:400–410

Wang C, Deng F-G, Li Y S, Liu X-S, Long GL (2005) Quantum secure direct communication with high-dimension quantum superdense coding. Phys Rev A 71:044305

Wang Y, Narayanaswamy A, Tsai C-L, Roysam B (2011) A broadly applicable 3-D neuron tracing method based on open-curve snake. Neuroinformatics 9:193–217

Wu J (2006) Density functional theory for chemical engineering: from capillarity to soft materials. AIChE J 52:1169–1193

Wu T-Y, Juan H-H, Lu HH-S, Chiang A-S (2011) A crosstalk tolerated neural segmentation methodology for Brainbow images. In: Proceedings of the 4th international symposium on applied sciences in biomedical and communication technologies (ISABEL)

Wu T-Y, Juan H-H, Lu, HH-S (2012) Improved spectral matting by iterative K-means clustering and the modularity measure. In: IEEE international conference on acoustics, speech, and signal processing (IEEE ICASSP), pp 1165–1168

Zhang J, Ma K-K, Er M-H, Chong V (2004) Tumor segmentation from magnetic resonance imaging by learning via one-class support vector machine. In: International Workshop on Advanced Image Technology, IWAIT, pp 207–211

Zhang Y, Chen K, Baron M, Teylan MA, Kim Y, Song Z, Greengard P, Wong STC (2010) A neurocomputational method for fully automated 3D dendritic spine detection and segmentation of medium-sized spiny neurons. Neuroimage 50:1472–1484

Zupan A, Burke K, Ernzerhof M, Perdew JP (1997) Distributions and averages of electron density parameters: explaining the effects of gradient corrections. J Chem Phys 106:10184–10193

Zupan A, Perdew JP, Burke K, Causá M (1997) Density-gradient analysis for density functional theory: application to atoms. Int J Quant Chem 61:835–845

Part III
Software

Chapter 16
Q3-D3-LSA: D3.js and Generalized Vector Space Models for Statistical Computing

Lukas Borke and Wolfgang K. Härdle

Abstract QuantNet is an integrated web-based environment consisting of different types of statistics-related documents and program codes. Its goal is creating reproducibility and offering a platform for sharing validated knowledge native to the social web. To increase the information retrieval (IR) efficiency there is a need for incorporating semantic information. Three text mining models will be examined: vector space model (VSM), generalized VSM (GVSM), and latent semantic analysis (LSA). The LSA has been successfully used for IR purposes as a technique for capturing semantic relations between terms and inserting them into the similarity measure between documents. Our results show that different model configurations allow adapted similarity-based document clustering and knowledge discovery. In particular, different LSA configurations together with hierarchical clustering reveal good results under M^3 evaluation. QuantNet and the corresponding Data-Driven Documents (D3) based visualization can be found and applied under http://quantlet.de. The driving technology behind it is Q3-D3-LSA, which is the combination of "GitHub API based QuantNet Mining infrastructure in R", LSA and D3 implementation.

Keywords Computational statistics · Transparency · Dissemination or quantlets · Quantlets

---

L. Borke
Humboldt-Universität zu Berlin, R.D.C - Research Data Center, SFB 649 "Economic Risk", Berlin, Germany
e-mail: lukas.borke@hu-berlin.de; https://github.com/lborke

W. K. Härdle (✉)
Humboldt-Universität zu Berlin, C.A.S.E. - Center for Applied Statistics and Economics, Berlin, Germany

School of Business, Singapore Management University, Singapore, Singapore
e-mail: haerdle@hu-berlin.de

© Springer International Publishing AG, part of Springer Nature 2018

W. K. Härdle et al. (eds.), *Handbook of Big Data Analytics*, Springer Handbooks of Computational Statistics, https://doi.org/10.1007/978-3-319-18284-1_16

16.1 Introduction: From Data to Information

The "QuantNet" concept is the effort to collect, interlink, retrieve, and visualize all the information in the scientific community with the particular emphasis on statistics. The richness and diversity of various and heterogeneous data types, descriptions and data sets submitted by numerous authors require an appropriate text mining model to be established and tuned. The big collection of data has now to be distilled to human-readable and applicable information and at the same time a modern and robust visualization framework is crucial.

QuantNet was originally designed as a platform to freely exchange empirical as well as quantitative-theoretical methods, called Quantlets. It supported the deployment of computer codes (R, Matlab, SAS, and Python), thus helping to establish collaborative reproducible research (CRR) in the field of applied statistics and econometrics at the Collaborative Research Center 649 (http://sfb649.wiwi.hu-berlin.de/), operated at the Humboldt University of Berlin. The former PHP-based QuantNet provided users a series of basic functions including registration, Quantlet uploading, searching, demonstrating, and downloading. Heterogeneous resources submitted by diverse contributors were stored on a proprietary Linux server having its own Oracle database. Hence, this IT-infrastructure was quite restrictive, maintenance-intensive, and also relatively susceptible to errors due to strict data type requirements, complexity and constraints of the Oracle database.

With the time, some problems and drawbacks became increasingly apparent:

1. lack of version control (VC) and source code management (SCM)
2. lack of distinct abilities of collaboration and project management between teams and heterogeneous groups of people
3. high personal maintenance costs of the infrastructure
4. database-restrictions and inflexibility of data handling
5. lack of a clear abstraction barrier between the data storage and the text mining (TM) and visualization layer of the system architecture

The points 1, 2, and 3 could be easily solved by the immanent features of the "GitHub's philosophy." As Marcio von Muhlen (Product Manager at Dropbox) eloquently expresses (http://marciovm.com/i-want-a-github-of-science/):

> GitHub is a social network of code, the first platform for sharing validated knowledge native to the social web. Open Science efforts like arXiv and PLoS ONE should follow GitHub's lead and embrace the social web.

Point 4 could be tackled by using the YAML standard (http://yaml.org/) for meta information of the resources, thus replacing the necessity of a database system. More about this human-readable data serialization language can be found on https://github.com/yaml/yaml-spec. Point 5 could be realized via the GitHub API (Cosentino et al. 2016). After the challenge of the abstraction barrier was solved, it was a straightforward procedure to connect the newly created Quantlet organization (https://github.com/Quantlet) on GitHub with the rest of the existing system architecture, comprising the TM and D3.js visualization layer.

QuantNet (http://quantlet.de) is now an online GitHub based organization with diverse repositories of scientific information consisting of statistics related documents and program codes. The advantages of QuantNet are:

- Full integration with GitHub
- Proprietary GitHub-R-API implementation developed from the core R package **github** (Scheidegger 2016) available as GitHub repository “R Bindings for the Github v3 API” (https://github.com/cscheid/rgithub) from Carlos Scheidegger, professor in the Department of Computer Science at the University of Arizona
- TM Pipeline providing IR, document clustering and D3 visualizations realized via QuantMining, a “GitHub API based QuantNet Mining infrastructure in R”
- Tuned and integrated search engine within the main D3 Visu based on validated meta information in Quantlets
- Ease of discovery and use of your technology and research results, everything in a single GitHub Markdown page
- Standardized audit and validation of your technology by means of the Style Guide (https://github.com/Quantlet/Styleguide-and-FAQ) and Yamldebugger (https://github.com/Quantlet/yamldebugger) (Borke 2017b)

16.1.1 Transparency, Collaboration, and Reproducibility

QuantNet: Open Access Code-Sharing Platform

- Quantlets: R, Matlab, SAS, and Python programs, various authors and topics
- QuantNetXploRer: Q3-D3-LSA driven and GitHub based search engine
- Knowledge discovery of brand-new research topics but also of dormant and archived research materials as required by good scientific practice

The Q3-D3-LSA Technology Comprises the Following Main Components

- Q3 (Quantlets, QuantNet, QuantMining): Scientific data pool and data mining infrastructure for CRR
- D3 (Data-Driven Documents): Knowledge discovery via information visualization by use of the D3 JavaScript library combining powerful visualization components and a data-driven approach

- LSA (Latent Semantic Analysis): Semantic embedding for higher clustering performance and automatic document classification by topic labeling

16.2 Related Work

Feinerer and Wild (2007) applied LSA based algorithms in a fully automated way on transcripts of interviews. The machine results were compared against marketing expert judgments with the outcome that the proposed algorithms provided perfect reliability with appropriate validity in automated coding and textual analysis. Feinerer and Wild (2007) could guarantee reliability on a very high level, while at the same time avoiding the main disadvantages of qualitative methods performed by humans like their inherent subjectivity and their high costs.

Linstead et al. (2008) pointed out that while there has been progress in developing sourcecode-specific search engines in recent years (e.g., Koders, Krugle, and Google's CodeSearch), these systems continue to focus strictly on text information retrieval, and do not appear to leverage the copious relations that can be extracted and analyzed from code. By combining software textual content with structural information captured by their CodeRank approach, they were able to significantly improve software retrieval performance. Developing and applying probabilistic models to automatically discover the topics embedded in the code and extracting topic-word and author-topic distributions, the authors provided a statistical and information-theoretic basis for quantifying and analyzing developer similarity and competence, topic scattering, and document tangling, with direct applications to software engineering.

Encouraged by the presented studies, we propose in this paper to use the latent semantic analysis (LSA) (Deerwester et al. 1990) as a technique capturing semantic relations between terms and inserting them into the similarity measure between documents. In this approach, the documents are implicitly mapped into a "semantic space," where documents that do not share any terms can still be close to each other if their terms are semantically related. The semantic similarity between two terms is inferred by an analysis of their co-occurrence patterns: terms that co-occur often in the same documents are considered as related. This statistical co-occurrence information is extracted by means of a singular value decomposition (SVD) of the "term by document" matrix in the way described in Sect. 16.4.

16.3 Q3-D3 Genesis

D3 (https://d3js.org/) is a rather new and not traditional visualization framework introduced by Bostock et al. (2011). D3.js (or D3 for Data-Driven Documents) is a JavaScript library for producing dynamic, interactive data visualizations in web browsers. It makes use of the widely implemented SVG, HTML5, and CSS standards. Instead of establishing a novel graphical grammar, D3 solves a different, smaller problem: efficient manipulation of documents based on data. The software design is heavily influenced by prior visualization systems, including Protovis.

The D3 gallery (available at http://bl.ocks.org/mbostock) demonstrates diverse capabilities and performance of the D3 technology, providing a huge collection of D3 visualization examples. Moreover, various applications and frameworks for data visualization have been built using D3, combining its methods with other modern technologies. Examples of these include, among many others, a data visualization library Plotly (see https://plot.ly) and a Force-directed Network Visualization developed by Jim Vallandingham (see https://flowingdata.com/2012/08/02/how-to-make-an-interactive-network-visualization/).

Impressed by the performance and universal applicability of the D3 framework, we decided to build the new QuantNet visualization upon the D3 architecture. The first steps are summarized in chapter "I. Genesis (Nov 2013–Aug 2014)." Basically, all main data objects from QuantNet could be exported to and visualized in the D3 framework templates, amongst them the whole "QuantNet universe" and "galaxies" representing individual subsets like books and projects. Further, co-occurrence information about authors and keywords as well as further details like creation times, etc., could be exploited. Not only all source code files from Q3-D3 Genesis are available for free use and reproducibility but also live examples on GitHub pages (Borke and Bykovskaya 2017b).

QuantNet contains also all Quantlets (which serve as supplementary examples and exercises) from the following books: MVA (Härdle and Simar 2015), SFE (Franke et al. 2015), SFS (Borak et al. 2013), XFG (Härdle et al. 2008). These book abbreviations are used in some figures in this section and in Sect. 16.5.

One of the most popular D3 layouts is the "Force-Directed Graph," which was extensively deployed in the "Genesis" chapter and which is a fundamental part of the final QuantNet visualization (https://bl.ocks.org/mbostock/4062045). The layout is based on special graph-drawing methods called force-directed techniques. These techniques represent a graph as a system of physical objects with diverse physical forces (e.g., electric) pulling and pushing the objects apart. The optimal visualization layout implies that all these forces are in equilibrium, see for more details Michailidis (2008) (Figs. 16.1, 16.2, 16.3, 16.4, 16.5, 16.6, and 16.7).

Subsequently, other D3 layouts were examined, which is documented in the chapters from "II. Shakespeare works" to "VI. QuantNet 2.0 @ GitHub." Figure 16.27 shows two visualization examples based on two different D3 layouts: Circle Packing and Expandable Tree. They are realized via the following D3 classes: `d3.layout.pack` and `d3.layout.tree`.

Chapter "II. Shakespeare works" served as a simple and impressive example. Further, diverse subsets of QuantNet documents and code files in different stages of development were visualized in five D3 layouts, which are mainly designed for the graphical representation of hierarchically structured data. Specially for this purpose, a dendrogram parser was constructed. Starting with the "document term matrix" of the Quantlets, the R code generated the tree structure and cluster labels based on the dendrogram which was created by the R function `hclust`. Finally, the recursively

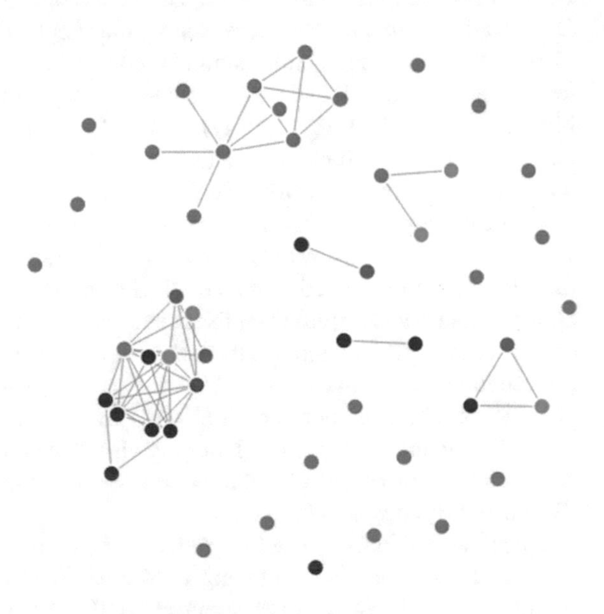

Fig. 16.1 Q3-D3 Genesis—Chapters

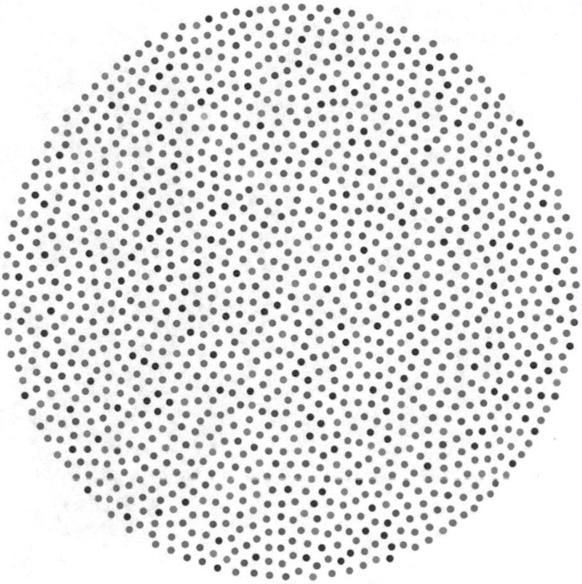

Fig. 16.2 The entire QNet-Universe

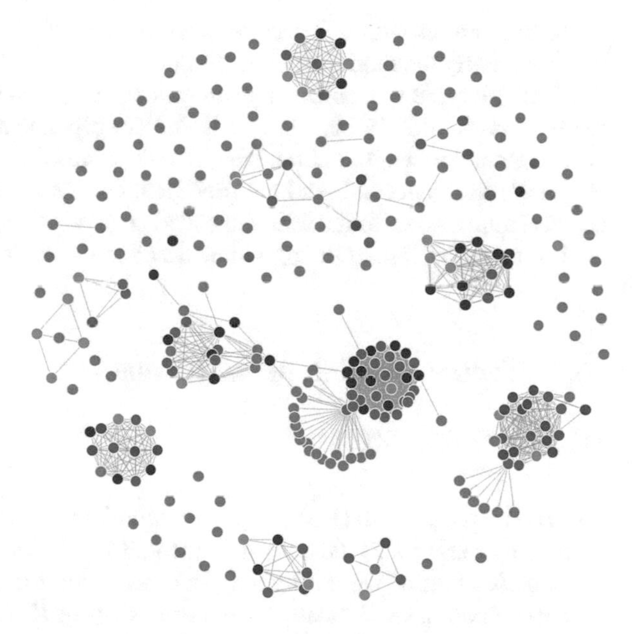

Fig. 16.3 Galaxy MVA with clusters

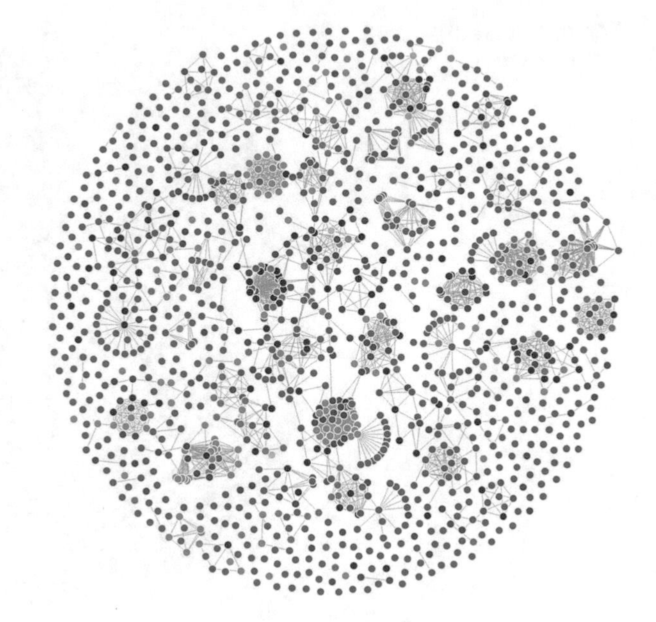

Fig. 16.4 The entire QNet-Universe with clusters

structured tree list within R was transformed to a JSON (http://json.org) file, which is subsequently required by the D3 designs.

Finally, we see four different examples of the QuantNet Visu from quantlet.de, see Figs. 16.8, 16.9, 16.10, and 16.11. The TM pipeline retrieves the meta information of Quantlets via the GitHub-R-API, then the LSA model is applied, clusters and labels are generated, and the processed data is transferred via JSON into the D3 Visu application. In the following section the vector space representations (with LSA as a special case of them) will be described in more detail (Fig. 16.12).

16.4 Vector Space Representations

16.4.1 Text to Vector

The vector space model (VSM) representation for a document d has been introduced by Salton et al. (1975). Given a document, it is possible to associate with it a bag of terms (or bag of words) by simply considering the number of occurrences of all terms contained. Typically words are “stemmed,” meaning that the inflection information contained in the last few letters is removed.

A bag of words has its natural representation as a vector in the following way. The number of dimensions is the same as the number of different terms in the corpus, each entry of the vector is indexed by a specific term, and the components of the vector are formed by integer numbers representing the frequency of the term in the given document. Typically such a vector is then mapped/transformed into some

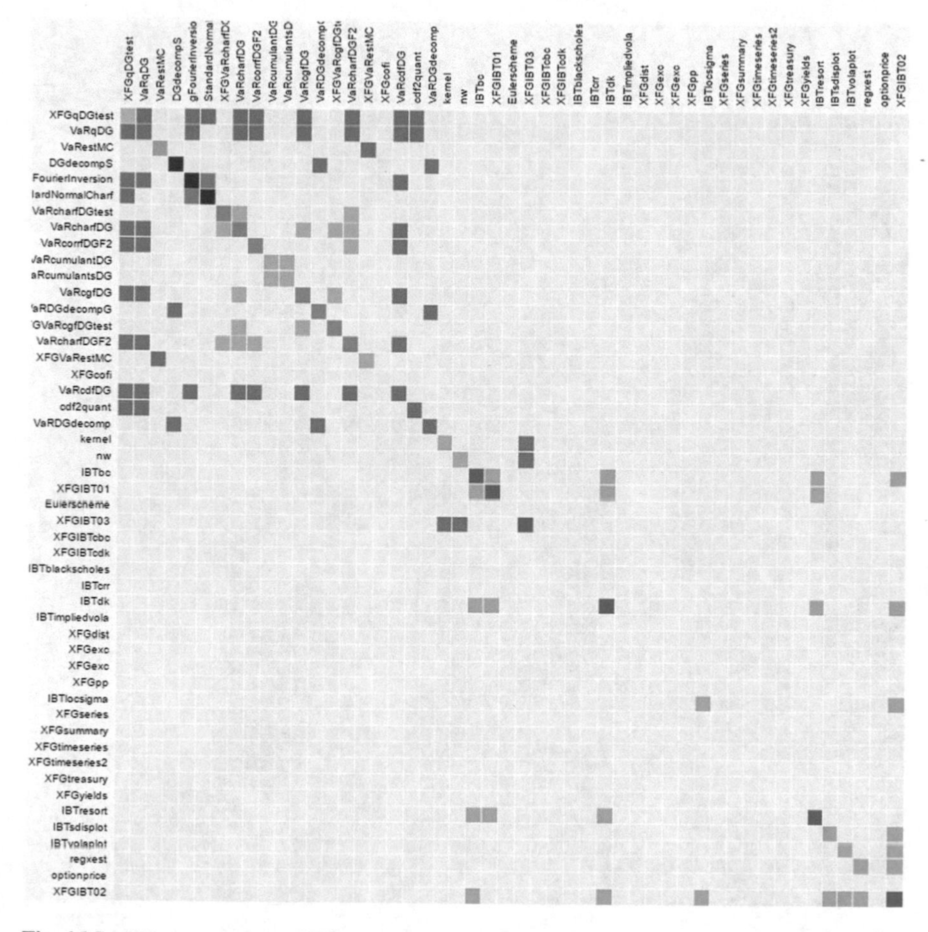

Fig. 16.5 Adjacency matrix of XFG

other space, where the word frequency information is merged/rescaled considering other information like word importance, relevance and semantic, assigning to uninformative words lower or no weight.

Suppose we have a set of documents Q and a set of terms T. Define $tf(d,t)$ as the absolute frequency of term $t \in T$ in $d \in Q$ and $idf(t) = \log(|Q|/n_t)$ as the inverse document frequency, with $n_t = |\{d \in Q | t \in d\}|$. Let $w(d) = \{w(d,t_1), \ldots, w(d,t_m)\}^\top$, $d \in Q$, be the weighting vector of the given document. Each $w(d,t_i)$ is calculated by a weighting scheme, see next Sect. 16.4.2. Then $D = [w(d_1), \ldots, w(d_n)]$ is the "term by document" matrix, or in abbreviated form TDM.

In this way a document is represented by a (column) vector $w(d)$, in which each entry reflects the relevance/importance of a particular word stem used in the document. Typically d can have tens of thousands of entries, often more than the number of documents. Furthermore, for a particular document the representation is

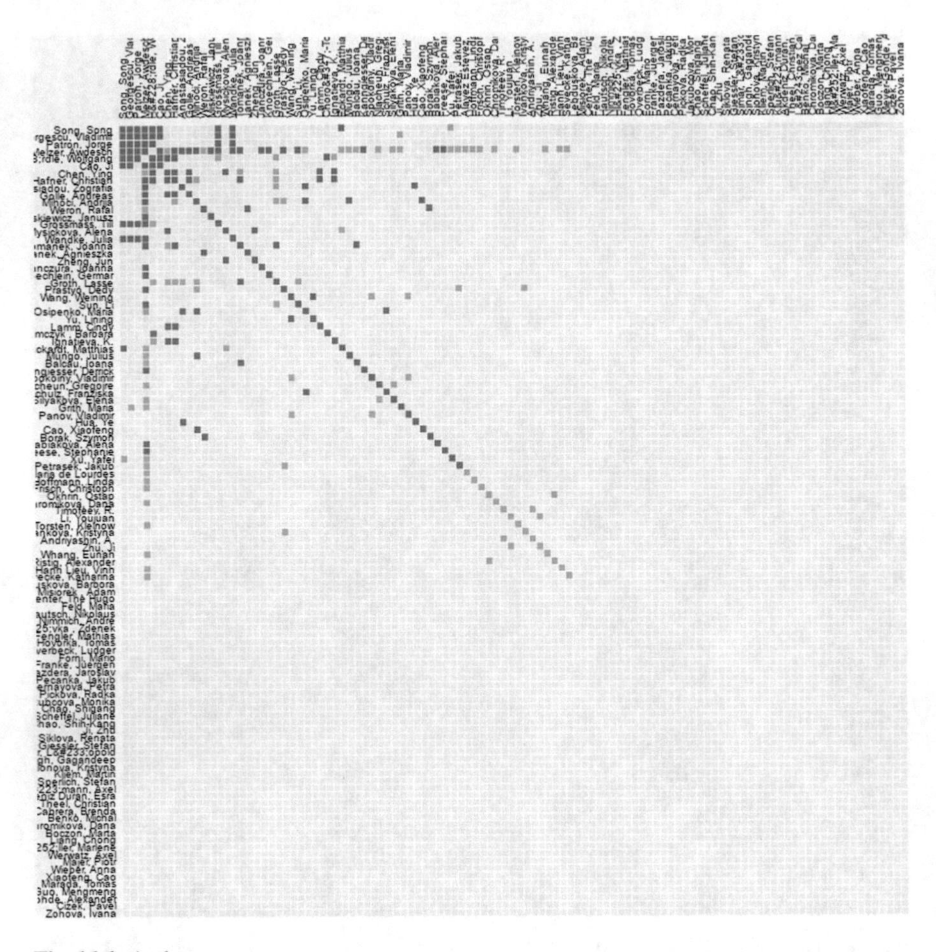

Fig. 16.6 Authors: co-occurrence

typically extremely sparse, having only relatively few non-zero entries, more details in Sect. 16.6.2.

16.4.2 Weighting Scheme, Similarity, Distance

A widely used weighting scheme in IR and TM is the *tf-idf*, short for *term frequency–inverse document frequency*. The concept of *idf* was introduced as "term specificity" by Jones (1972). Although it has worked well as a heuristic, its theoretical foundations have been troublesome for at least three decades afterward, with many researchers trying to find information theoretic justifications for it. Robertson (2004) (who worked from 1998 to 2013 in the Cambridge laboratory of

Fig. 16.7 Keywords: co-occurrence

Microsoft Research and contributed to the Microsoft search engine Bing) concludes 32 years later in the same journal "Journal of Documentation":

> However, there is a relatively simple explanation and justification of IDF in the relevance weighting theory of 1976. This extends to a justification of TF*IDF in the Okapi BM25 model of 1994. IDF is simply neither a pure heuristic, nor the theoretical mystery many have made it out to be. We have a pretty good idea why it works as well as it does.

The (normalized) *tf-idf* weighting scheme is defined as

$$w(d,t) = \frac{tf(d,t)idf(t)}{\sqrt{\sum_{j=1}^{m} tf(d,t_j)^2 idf(t_j)^2}}, m = |T|. \tag{16.1}$$

Hence, the similarity of two documents d_1 and d_2 (or the similarity of a document and a query vector q) can be computed based on the inner product of the vectors.

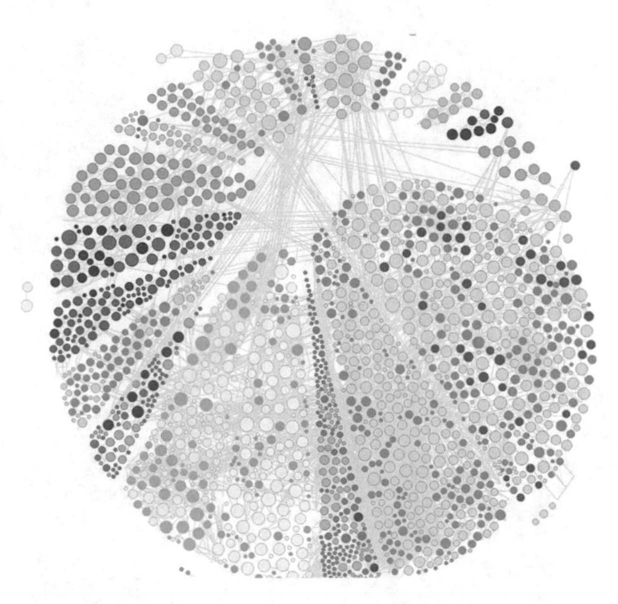

Fig. 16.8 Orbit clustering of QuantNet, grouped by books and projects

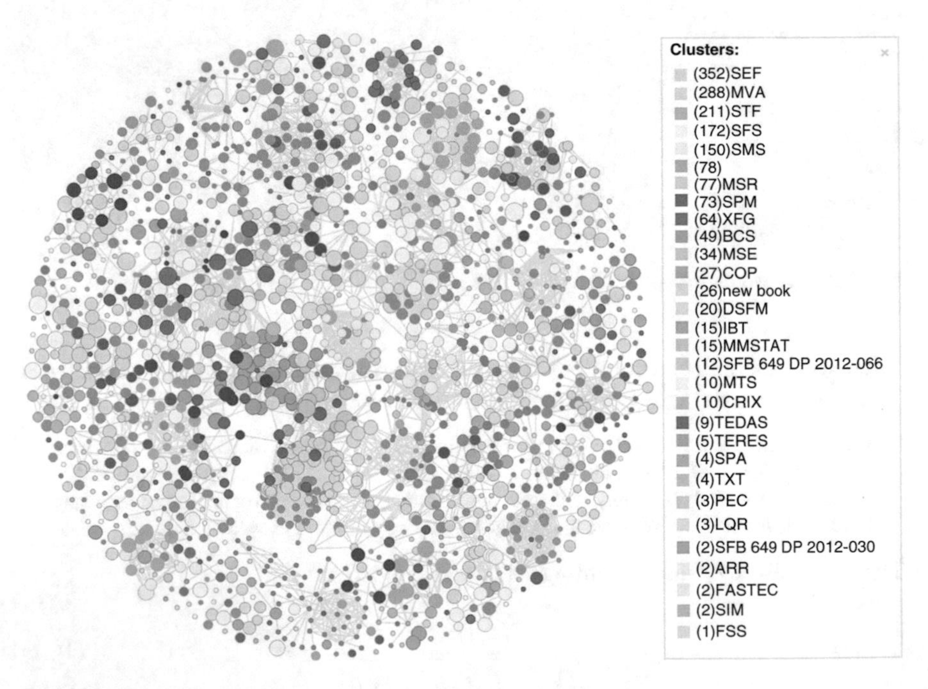

Fig. 16.9 Force-Directed Graph of QuantNet, linked by "see also" connections

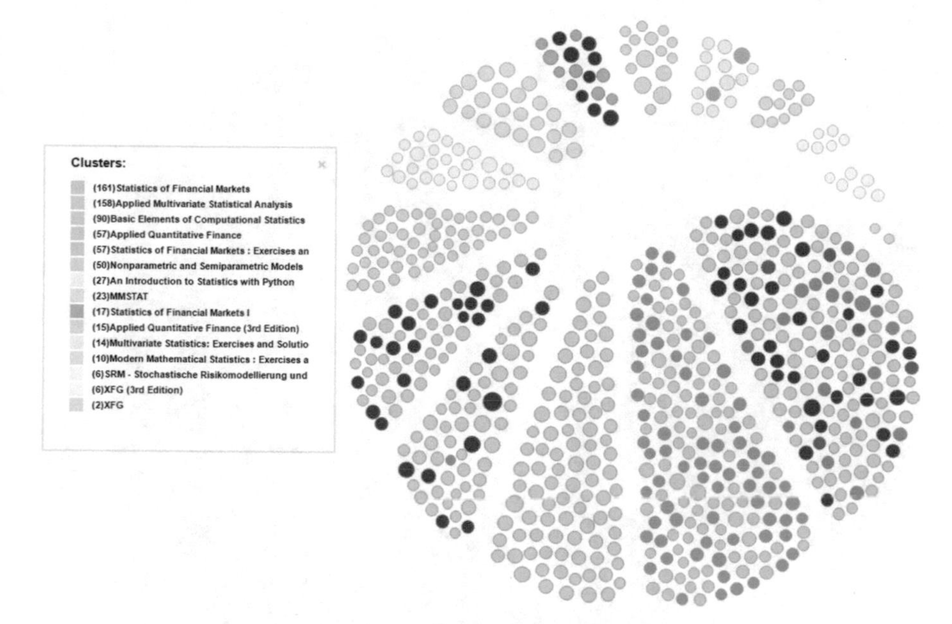

Fig. 16.10 Orbit clustering of QuantNet, subset grouped by Springer books. Quantlets containing the search query "black scholes" are highlighted in red

The (normalized *tf-idf*) similarity S of two documents d_1 and d_2 is given by

$$S(d_1, d_2) = \sum_{k=1}^{m} w(d_1, t_k) \cdot w(d_2, t_k) = w(d_1)^\top w(d_2). \tag{16.2}$$

A frequently used distance measure is the Euclidean distance:

$$dist_2(d_1, d_2) = \sqrt{\sum_{k=1}^{m} \{w(d_1, t_k) - w(d_2, t_k)\}^2}. \tag{16.3}$$

It holds the general relationship:

$$\cos\phi = \frac{x^\top y}{|x| \cdot |y|} = 1 - \frac{1}{2} dist^2 \left(\frac{x}{|x|}, \frac{y}{|y|} \right), \tag{16.4}$$

with ϕ as the angle between x and y. Substituting $\frac{x}{|x|}$ by $w(d_1)$ and $\frac{y}{|y|}$ by $w(d_2)$, we have an easily computable transformation between the *tf-idf* similarity and the Euclidean distance. In particular when dealing with big data this fact can be exploited, since many standard clustering methods expect a distance matrix in advance. Usually, it is more efficient to first calculate the similarity matrix, exploiting the strong sparsity in text documents, and then apply the transformation

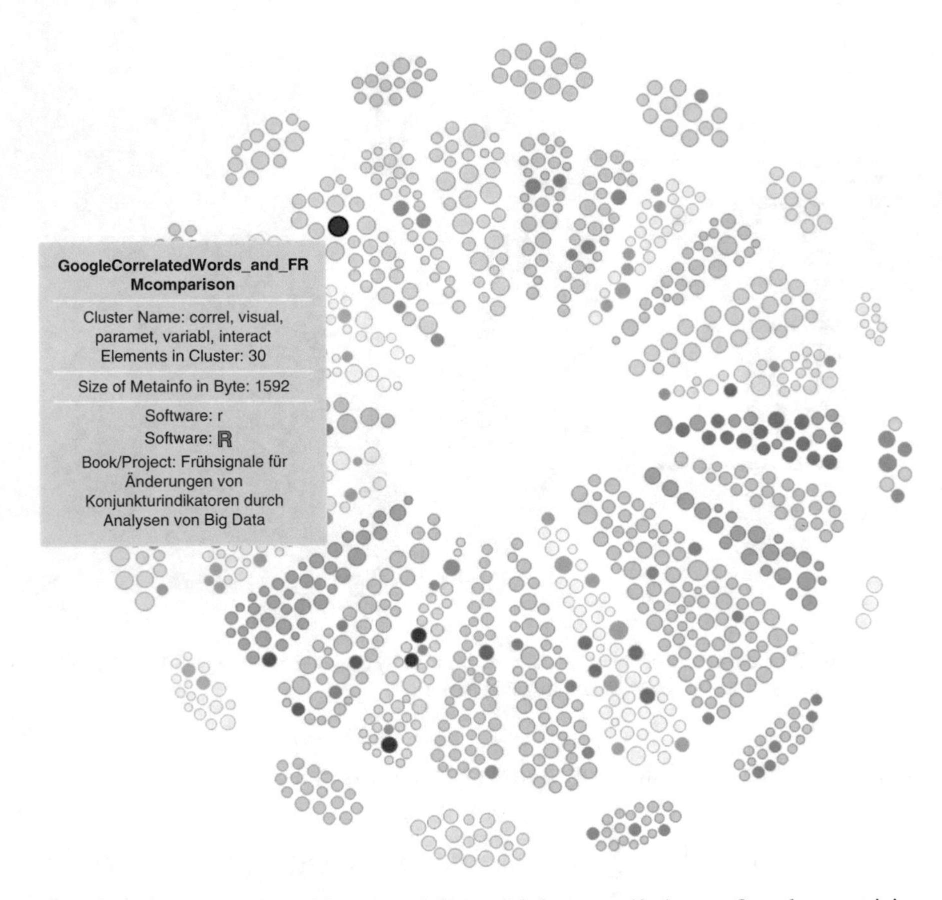

Fig. 16.11 Orbit clustering of QuantNet, LSA model, k-means, 40 clusters. Quantlets containing the search query "big data" are highlighted in red

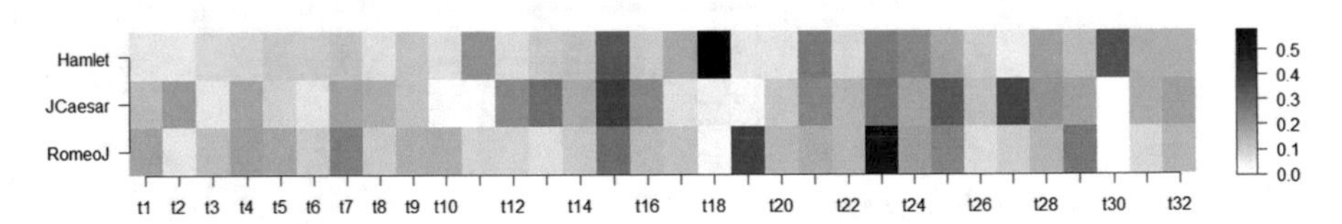

Fig. 16.12 Heatmap of T_s in three Shakespeare's tragedies

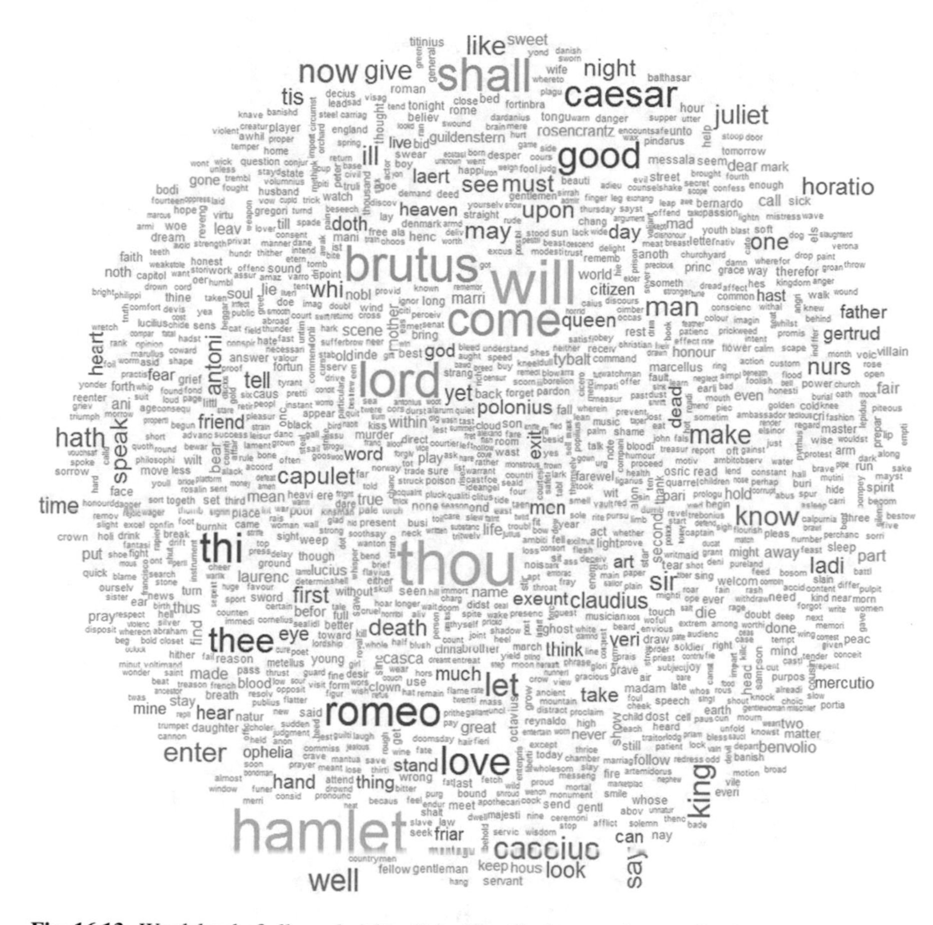

Fig. 16.13 Wordcloud of all words ($tf \geq 5$) in three Shakespeare's tragedies in corpus Q

in Formula (16.4) to obtain the distance matrix. Figure 16.13 displays a word cloud of the 3 Shakespeare tragedies (Hamlet, Romeo and Juliet, Julius Caesar) that constituted the corpus Q. For clarity only word with term frequency $> = 5$ are shown.

16.4.3 Shakespeare's Tragedies

The basic concepts of the introduced vector space representations will be illustrated by the example of Shakespeare's works, available under http://shakespeare.mit.edu. Let $Q = \{d_1, d_2, d_3\}$ be the document corpus containing the following Shakespeare's tragedies: d_1 = "Hamlet" (total word number: 16,769); d_2 = "Julius Caesar" (total word number: 11,003); d_3 = "Romeo and Juliet" (total word number: 14,237). After some text preprocessing as in Sect. 16.6.1, the TDM is a 5521×3

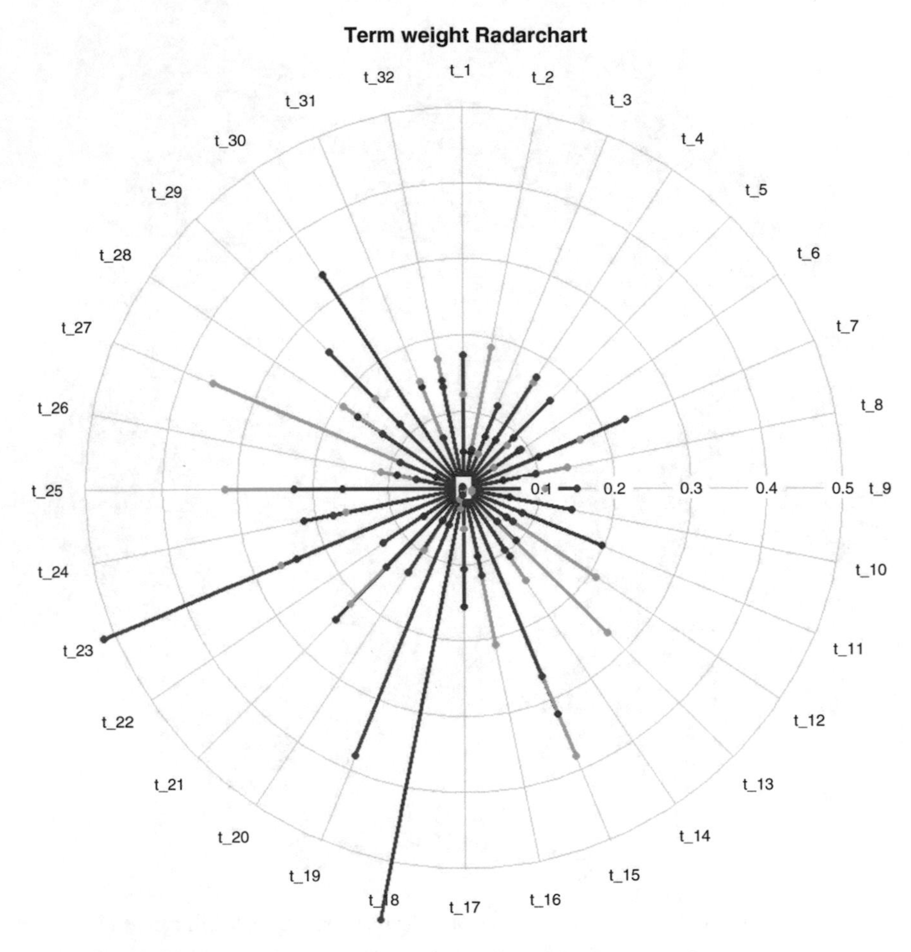

Fig. 16.14 Radar chart: weightings of terms in T_s of tragedies in corpus Q

matrix. Consider the special vocabulary T_s, selected amongst 100 most frequent words:

$$\begin{aligned} T_s = \{&\text{art, bear, call, day, dead, dear, death, die, eye, fair, father, fear,}\\ &\text{friend, god, good, heart, heaven, king, ladi, lie, like, live, love,}\\ &\text{make, man, mean, men, must, night, queen, think, time}\}\\ = \{&t_1, \dots, t_{32}\} \end{aligned}$$

Figure 16.14 shows the weighting vectors $w(d)$ of the tragedies in Q (Hamlet, Julius Caesar, Romeo and Juliet) wrt. to the special vocabulary T_s in a radar chart. The highest term weightings $w(d, t)$ are distributed as follows: $w(d_1, t_{18})$, $t_{18} \hat{=}$

"king"; $w(d_1, t_{30})$, $t_{30}\hat{=}$ "queen"; $w(d_2, t_{15})$, $t_{15}\hat{=}$ "good"; $w(d_2, t_{27})$, $t_{27}\hat{=}$ "men"; $w(d_3, t_{19})$, $t_{19}\hat{=}$ "ladi"; $w(d_3, t_{23})$, $t_{23}\hat{=}$ "love." The heatmap in Fig. 16.12 displays the same information in another representation.

M_S and M_D for 32 special terms in T_s:

$$M_S = \begin{pmatrix} 1 & 0.64 & 0.63 \\ 0.64 & 1 & 0.77 \\ 0.63 & 0.77 & 1 \end{pmatrix} \quad M_D = \begin{pmatrix} 0 & 0.85 & 0.87 \\ 0.85 & 0 & 0.68 \\ 0.87 & 0.68 & 0 \end{pmatrix}$$

M_S and M_D for all 5521 terms:

$$M_S = \begin{pmatrix} 1 & 0.39 & 0.46 \\ 0.39 & 1 & 0.42 \\ 0.46 & 0.42 & 1 \end{pmatrix} \quad M_D = \begin{pmatrix} 0 & 1.10 & 1.04 \\ 1.10 & 0 & 1.07 \\ 1.04 & 1.07 & 0 \end{pmatrix}$$

Finally, we present the similarity matrices M_S and distance matrices M_D for the selected tragedies in Q. On the one hand, wrt. to the special vocabulary T_s, on the other hand, wrt. to the full vocabulary containing 5521 terms. Every entry in M_S and M_D corresponds to the value calculated by Formula (16.2) and (16.3), respectively, for any given document pair $d_i, d_j \in Q$. The weighting scheme was calculated via the normalized *tf* weight. In the case of a few documents in the corpus the document frequency *idf* is inappropriate as many frequent terms have a high probability to be present in all documents, in this case only three. Therefore, the *idf* weighting share would make many terms vanish, which would considerably decrease the overall similarity between two documents which is calculated by the scalar product of their term weights.

16.4.4 Generalized VSM (GVSM)

One of the problems with basic VSM representations as presented in Sect. 16.4.1 is that they treat terms as uncorrelated, assigning them into orthogonal directions in the feature space. A classical example is synonymous words which contain the same information, but are assigned distinct components (Srivastava and Sahami 2009). As a consequence, only documents that share many terms (which serve as vector components) can be clustered into common topics and clusters. But in reality words are correlated, and sometimes even synonymous, so that documents with very few common terms can potentially be on closely related topics. Such similarities cannot be detected by the basic vector space model (BVSM) (Salton et al. 1975). This raises the question of how to incorporate information about semantics into the feature map, so as to link documents that share "related" terms?

So far, we have identified the following drawbacks of the classical *tf-idf* approach and of the BVSM in general: (1) uncorrelated/orthogonal terms in the feature space, (2) documents must have common terms to be similar, (3) sparsity of document vectors and similarity matrices.

Over the time many solutions were proposed by various researchers, first of them Wong et al. (1985) and Deerwester et al. (1990). We will treat them later in this section. Other noteworthy books giving a general survey of the big topic "Text mining and different models" are Berry (2003) and Srivastava and Sahami (2009). The most popular solutions are: (1) using statistical information about term–term correlations (GVSM in Sect. 16.4.4.2); (2) incorporating information about semantics (semantic smoothing, LSA in Sect. 16.4.4.3).

More generally, we can consider transformations of the document vectors by some mapping P. The simplest case involves linear transformations, where P is any appropriately shaped matrix. In this case the generalized similarity S has the form:

$$S_P(d_1, d_2) = (Pd_1)^\top (Pd_2) = d_1^\top P^\top P d_2, \quad d_1, d_2 \in Q. \tag{16.5}$$

Every P defines another generalized vector space model (GVSM), resulting in the similarity matrix:

$$M_S^{(P)} = D^\top \left(P^\top P\right) D,$$

with D being the "term by document" matrix as defined in Sect. 16.4.1.

16.4.4.1 Basic VSM (BVSM)

The BVSM was introduced by Salton et al. (1975) and uses the vector representation with no further mapping, the VSM shows $P = I$ in this case. Even in this simple case the "matrix nature" of VSM allows different embeddings of *tf-idf* weightings into the matrix representations.

- $P = I_m$ and $w(d) = \{tf(d, t_1), \ldots, tf(d, t_m)\}^\top$ lead to the classical *tf*-similarity $M_S^{tf} = D^\top D$
- diagonal $P(i, i)^{idf} = idf(t_i)$ and $w(d) = \{tf(d, t_1), \ldots, tf(d, t_m)\}^\top$ lead to the classical *tf-idf*-similarity $M_S^{tfidf} = D^\top (P^{idf})^\top P^{idf} D$
- starting with $w(d) = \{tf(d, t_1) idf(t_1), \ldots, tf(d, t_m) idf(t_m)\}^\top$ and $P = I_m$ results in the classical *tf-idf*-similarity $M_S^{tfidf} = D^\top I_m D = D^\top D$ as well

16.4.4.2 GVSM: Term–Term Correlations

An early attempt to overcome the limitations of the BVSM was proposed by Wong et al. (1985) under the name of generalized VSM, or GVSM. A document is characterized by its relation to other documents in the corpus as measured by the BVSM. The mapping P and the resulting model specifications are as follows:

- $P = D^\top$ is the linear mapping
- $S(d_1, d_2) = \left(D^\top d_1\right)^\top \left(D^\top d_2\right) = d_1^\top D D^\top d_2$ is the document similarity
- $M_S^{TT} = D^\top \left(DD^\top\right) D$ is the similarity matrix

$DD^\top$ is called a "term by term" matrix, having a nonzero ij entry if and only if there is a document containing both the i-th and the j-th term. Thus, terms become semantically related if they co-occur often in the same documents. The documents are mapped into a feature space indexed by the documents in the corpus, as each document is represented by its relation to the other documents in the corpus. If the BVSM represents a document as bag of words, the GSVM represents a document as a vector of its similarities relative to the different documents in the corpus. If there are less documents than terms, then we additionally achieve a dimensionality reduction effect. In order to avoid misleading we will refer to this model as the GVSM(TT) for the rest of our article, hence distinguishing it from other possible GVSM representations which are induced by another mapping P.

16.4.4.3 GVSM: Latent Semantic Analysis (LSA)

Latent semantic analysis (LSA) is a technique to incorporate semantic information in the measure of similarity between two documents (Deerwester et al. 1990). LSA measures semantic information through co-occurrence analysis in the corpus. The document feature vectors are projected into the subspace spanned by the first k singular vectors of the feature space. The projection is performed by computing the singular value decomposition (SVD) of the matrix $D = U\Sigma V^\top$. Hence, the dimension of the feature space is reduced to k and we can control this dimension by varying k. This is achieved by constructing a modified (or truncated) matrix D_k from the k-largest singular values σ_i, $i = 1, 2, 3, \ldots, k$, and their corresponding vectors: $D_k = U_k \Sigma_k V_k^\top$. Based on the SVD factors, the resulting model specifications are as follows:

- $P = U_k^\top := I_k U^\top$ is the projection operator onto the first k dimensions, I_k is a $m \times m$ identity matrix having ones only in the first k diagonal entries, $k < m$
- $M_S^{\mathrm{LSA}} = D^\top \left(U I_k U^\top\right) D$ is the similarity matrix
- $D_k = UPD = U_k \Sigma_k V_k^\top = U \Sigma_k V^\top$ is the truncated TDM which is re-embedded into the original feature space, $PD = \Sigma_k V^\top$ is the corresponding counterpart in the semantic space
- $D_{\mathrm{err}} = D - D_k = U(\Sigma - \Sigma_k) V^\top$ is the approximation error of the SVD truncation

The k dimensions can be interpreted as the main semantic components/concepts and $U_k U_k^\top = U I_k U^\top$ as their correlation. Some authors refer to $U I_k U^\top$ as a "semantic kernel" or "latent semantic kernel." It can be shown that $M_S^{\mathrm{LSA}} =$

$V\Lambda_k V^\top$. Starting with $V\Lambda V^\top = V\Sigma^\top \Sigma V^\top = V\Sigma^\top U^\top U\Sigma V^\top = D^\top D$ and diagonal $\Lambda_{ii} = \lambda_i = \sigma_i^2$ with eigenvalues of $D^\top D$, the truncated diagonal Λ_k consists of the first k eigenvalues and zero-values else. It should be noted that $D^\top D$ is the BVSM similarity matrix. For more technical and scientific proofs and interpretations of this paragraph, we recommend the following publications: Cristianini et al. (2002), Berry (2003) and Srivastava and Sahami (2009). The visualization of the "LSA anatomy" in Sect. 16.6.4 may also be helpful.

16.4.4.4 Closer Look at the LSA Implementation

Several classes of adjustment parameters can be functionally differentiated in the LSA process. Every class introduces new parameter settings that drive the effectiveness of the algorithm. The following classes have been identified so far by Wild and Stahl (2007):

1. Textbase compilation and selection
2. Preprocessing: stemming, stopword filtering, special vocabulary, etc.
3. Weighting schemes: local weights (none (i.e., *tf*), binary *tf*, log *tf*, etc.); global weights (normalization, *idf*, entropy, etc.)
4. Dimensionality: singular values k (coverage of total weight = 0.3, 0.4, 0.5, etc.)
5. Similarity measurement: cosine, best hit, mean of best, pearson, spearman, etc.

The latent semantic space can be either created directly by using the documents, in our case Quantlets, letting the matrix D be the weighting vectors of the Quantlets or it can be first trained by domain-specific and generic background documents. Generic texts add thereby a reasonably heterogeneous amount of general vocabulary, whereas the domain-specific texts provide the professional vocabulary. The Quantlets would be then folded into the semantic space which was created in the previous SVD process. By doing so, one gains in general a higher retrieval performance as the vocabulary set is bigger and more semantic structure is embedded.

Bradford (2009) presented an overview of 30 sets of studies in which the LSA performance in text processing tasks could be compared directly to human performance on the same tasks. In half of the studies, performance of LSA was equal to or better than that of humans.

Miller et al. (2009) proposed a family of LSA-based search algorithms which is designed to take advantage of the semantic properties of well-styled hyperlinked texts such as wikis. Performance was measured by having human judges rating the relevance of the top four search results returned by the system. When given single-term queries, the highest-performing search algorithm performed as well as the proprietary PageRank-based Google search engine. The comparison with respect to Google is especially promising, given that the presented system operated on less than 1% of the original corpus text, whereas Google uses not only the entire corpus text but also meta data internal and external to the corpus.

Fernández-Luna et al. (2011) proposed a recommender agent based on LSA formalism to assist the users that search alone to find and join to groups with similar information needs. With this mechanism, a user can easily change her solo search intent to explicit collaborative search.

A comparison of three WordNet related methods for taxonomic-based sentence semantic relatedness was examined in Mohamed and Oussalah (2014). Using a human annotated benchmark data set, all three approaches achieved a high positive correlation, reaching up to $r = 0.88$ with comparison to human ratings. In parallel, two other baseline methods (LSA as part of it) evaluated on the same benchmark data set. LSA showed comparable correlation as the more sophisticated WordNet based methods, (https://wordnet.princeton.edu).

16.4.4.5 GVSM Applicability for Big Data

Having n documents with a vocabulary of m terms and LSA truncation to k dimensions, there are the following memory space requirements for the TDM representations: $m \times n$ matrix cells in BVSM ($\mathcal{O}(mn)$); n^2 matrix cells in GVSM(TT) ($\mathcal{O}(n^2)$); $k \times (k + m + n)$ matrix cells in LSA(k) ($\mathcal{O}(kn)$). In the context of big data the n will usually dominate the other quantities m and k, furthermore k is fixed, see for comparison Table 16.1. Clearly, the TDM D in the BVSM is the first step for all three models. Hence, the basic calculation and storage demand is dictated by D. Concerning the memory demands, the GVSM(TT)-TDM would be maximal. For a fixed k the memory demand for a TDM in LSA would be less than in BVSM: $\mathcal{O}(kn)$ versus $\mathcal{O}(mn)$. The calculation of the GVSM(TT)-TDM would involve a matrix multiplication $D^{\top}D$, see Sect. 16.4.4.2, implying $n^2 \times m$ multiplications. Concerning the LSA, which is performed by SVD, the situation is more complex.

There are numerous theoretical approaches and software implementations with respect to the SVD topic. Several state-of-the-art algorithms including the Lanczos-based truncated SVD and the corresponding implementations are outlined in Korobeynikov (2010) and Golyandina and Korobeynikov (2014). The R package **svd** (Korobeynikov et al. 2016) provides "Interfaces to Various State-of-Art SVD and Eigensolvers" (https://github.com/asl/svd). This package is basically an R interface to the software package PROPACK containing a set of functions for computing the singular value decomposition of large and sparse or structured

Table 16.1 Benchmark for TDM matrix creation in BVSM (package **tm**) and LSA(k) (propack.svd from package **svd**), $k = 100$, elapsed time in seconds

Time in seconds for	BVSM	LSA(k)	BVSM + LSA(k)	Size of TDM (BVSM)
10.570 Org's	39	149	188	14238 × 10570
16.803 Org's	51	264	315	16029 × 16803
30.437 Org's	69	637	706	18501 × 30437
45.669 Org's	93	990	1083	20368 × 45669
97.444 Org's	159	2673	2832	23667 × 97444

matrices, which are written in Fortran and C (http://sun.stanford.edu/%7Ermunk/PROPACK/). Although the R package **lsa** (Wild 2015), which performs a full SVD, is sufficient for the QuantNet data, Lukas Borke has run some benchmarks applying the function `propack.svd` from the R package **svd** to examine its performance. The main advantages are the time saving partial SVD calculation (depending on k) and the fast C optimized implementation. For this purpose he has extracted several data sets from GitHub by means of the "GitHub Mining Infrastructure in R" (see Sect. 16.8.1). The collected data are meta information describing samples of GitHub organizations.

As can be inferred from Table 16.1, the time complexity both for BVSM and LSA TDM matrix creation is feasible. 10,570 data sets from GitHub organizations require less than 1 min for BVSM and two and a half minutes for LSA. Increasing the number of data up to roughly 100,000 samples leads to less than 3 min calculation time for BVSM and 45 min for LSA. In simpler terms, one can create a TDM for 100,000 documents both in BVSM and LSA in less than 1 h on a single CPU core without any parallelization expense. A smaller data set like 10,000 documents can be handled on a usual PC with 8 GByte RAM. For larger data sets a Linux server (Research Data Center) with an available memory of 256 GiB was used. Since this benchmark was focused on the time complexity, no deeper analysis was undertaken concerning the memory demand. At any time point of the benchmark process the available RAM of 256 GiB was far away from being exhausted. Some of the above listed data sets are available for visual data mining under http://bitquery.de/dp (Borke and Bykovskaya 2017a).

Concluding we can say that a Linux server with 256 GiB RAM has sufficient performance reserves for BVSM and LSA processing of big data, having 100,000 documents and an hour processing time as a "lower boundary." As software one needs only an R installation and some freely available R packages (**tm**, **svd** as the most crucial ones). All tests were conducted on a single core, hence there is additional potential to speed up the calculation time. In Theußl et al. (2012) a **tm** plug-in called **tm.plugin.dc** is presented implementing a distributed corpus class which can take advantage of the Hadoop MapReduce library for large-scale text mining tasks. With a quadratic space complexity (memory demand) of $\mathcal{O}(n^2)$ and a cubic time complexity of $n^2 \times m$ multiplications, the GVSM(TT) model is the worst choice among the considered TM models, unless some optimization (like parallelization, exploiting theoretical properties like sparsity, etc.) is done.

16.5 Methods

16.5.1 Cluster Analysis

If the data can validly be summarized by a small number of groups of objects, then the group labels may provide a very concise description of patterns of similarities and differences in the data. The need to summarize data sets in this way is increasingly important because of the growing volumes of data now available in many areas of science, and the exploration process of such data sets using cluster analysis and other multivariate analysis techniques is now often called data mining. In the twenty-first century, data mining has become of particular interest for investigating material on the World Wide Web, where the aim is to gather and analyze useful information or knowledge from web page contents (Everitt et al. 2011).

Our objectives are to determine topic labels and assign them to (text) documents. A confident and reliable automatic process would completely bypass the expense of having humans, whose task is to provide labels. But the process known as document clustering is less than perfect. The labels and their assignment may vary depending on humans or different objective processes that incorporate external information such as stock price change. Document clustering assigns each of the documents in a collection to one or more smaller groups called clusters (Weiss et al. 2010).

The result of clustering is typically a partition (also called clustering) $\mathscr{C}$, a set of clusters C. Each cluster/group consists of a number of documents d. Objects - in our case documents - of a cluster should be similar within the same group and dissimilar to documents of other groups. The code for the reproducibility of the clustering in Fig. 16.15 is available as interactive Q in http://quantlet.de/

16.5.1.1 Partitional Clustering

k-Means is a classical clustering method that has been adapted to documents. It is very widely used for document clustering and is relatively efficient. The k-means algorithm aims to partition n observations/objects into k clusters in which each observation is assigned to the cluster with the nearest mean, serving as a prototype of the cluster. k-Means typically converges to its minimum after relatively few iterations.

k-Medoids clustering is related to the k-means. It is also referred to as partitioning around medoids or PAM. Both variants attempt to minimize the distance between points labeled to be in a cluster and a point designated as the center/medoid of that cluster. In contrast to the k-means, k-medoids chooses datapoints as centers and works with an arbitrary matrix of distances. Concerning their R implementations `kmeans` and `pam`, the function `pam` is more robust because it minimizes a sum of unsquared dissimilarities. Moreover, `pam` does not need initial guesses for the cluster centers, contrary to `kmeans` (Kaufman and Rousseeuw 2008).

In Fig. 16.16 `kmeans` produced eight clusters with the following topic assignments: (1) "distribut copula normal gumbel pdf"; (2) "call option blackschol put price"; (3) "return timeseri dax stock financi"; (4) "portfolio var pareto return risk"; (5) "interestr filter likelihood cir term"; (6) "visual dsfm requir kernel test"; (7) "regress nonparametr linear logit lasso"; (8) "cluster analysi pca principalcompon dendrogram." The cluster topics were created based on the most frequent terms of cluster centroids. A multidimensional scaling (MDS) output of the `pam` function with cluster labeling can be reproduced by Q YAMLcentroids.

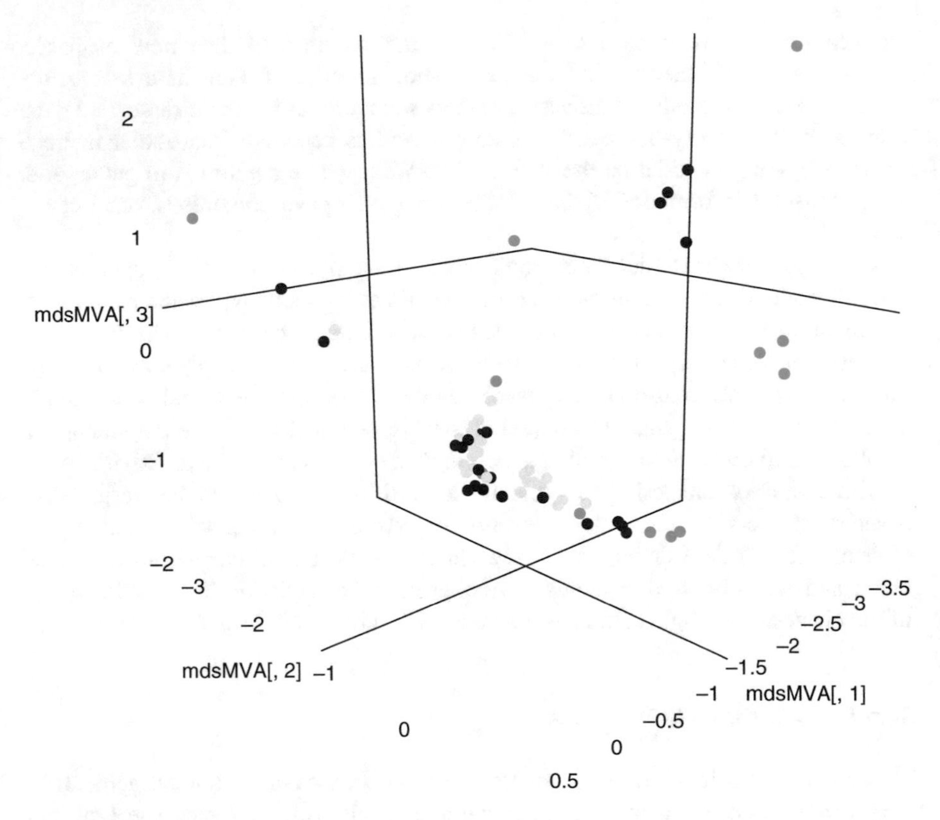

Fig. 16.15 *k*-Means clustering and metric MDS for MVA quantlets via Plotly

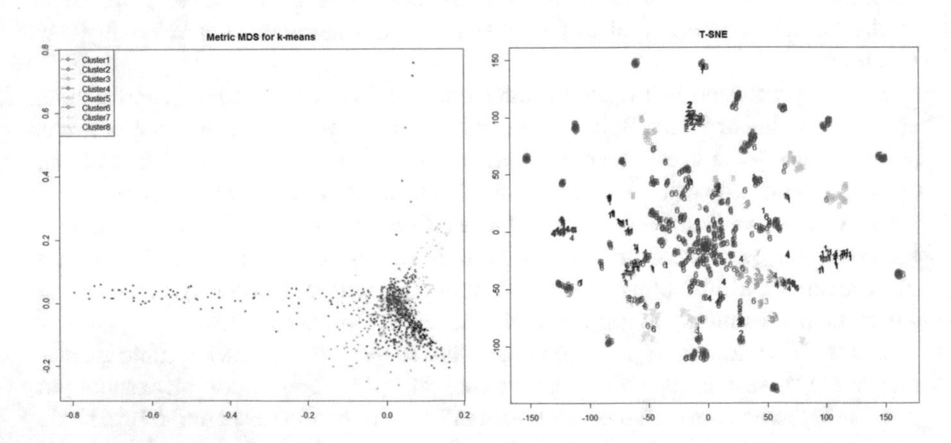

Fig. 16.16 LSA:50 geometry of Quantlets via MDS (left) and t-SNE (right), clustered by *k*-means with generated topics

16.5.1.2 Hierarchical Clustering

Hierarchical clustering algorithms got their name since they form a sequence of groupings or clusters that can be represented in a hierarchy of nested clusters (Steinbach et al. 2000). This hierarchy can be obtained either in a top-down or bottom-up fashion. Top-down means that we start with one cluster that contains all documents. This cluster is stepwise refined by splitting it iteratively into sub-clusters. One speaks in this case also of the so-called "divisive" algorithm. The bottom-up or "agglomerative" procedures start by considering every document as an individual cluster. Then the most similar clusters are iteratively merged, until all documents are contained in one single cluster. In practice the divisive procedure is almost of no importance due to its generally bad results. Therefore, only the agglomerative variants are outlined in the following. Typical agglomeration methods are "ward.D", "ward.D2", "single", "complete" and "average". This family of agglomeration methods will be abbreviated as HC in the following, all of them are available by means of the R function `hclust`.

Hierarchical (agglomerative) clustering is a popular alternative to k-means clustering of documents. As explained above, the method produces clusters, but they are organized in a hierarchy comparable with a table of contents for a book. The binary tree produced by HC is a map of many potential groupings of clusters. One can process this map to get an appropriate number of clusters. That is more difficult with k-means, where the procedure usually must be restarted when we specify a new value of k.

Hierarchical classifications produced by either the agglomerative or divisive route may be represented by a two-dimensional diagram known as a *dendrogram*, which illustrates the fusions or divisions made at each stage of the analysis. Two examples of such a dendrogram are given in Figs. 16.17 and 16.29.

16.5.2 Cluster Validation Measures

Internal validation measures take only the data set and the clustering partition as input and use intrinsic information in the data to assess the quality of the clustering. For internal validation, we decided for measures that reflect the *compactness*, *connectedness* and *separation* of the cluster partitions. Connectedness relates to what extent observations are placed in the same cluster as their nearest neighbors in the data space, and is measured by the *connectivity* method as suggested by Handl et al. (2005). Compactness assesses cluster homogeneity, usually by looking at the intra-cluster variance, while separation quantifies the degree of separation between clusters, usually by measuring the distance between cluster centroids. Since compactness and separation demonstrate opposing trends (compactness increases with the number of clusters but separation decreases), popular methods combine the two measures into a single score. The *Dunn Index* (Dunn 1974) and *Silhouette Width* (Rousseeuw 1987) are both examples of non-linear combinations of the

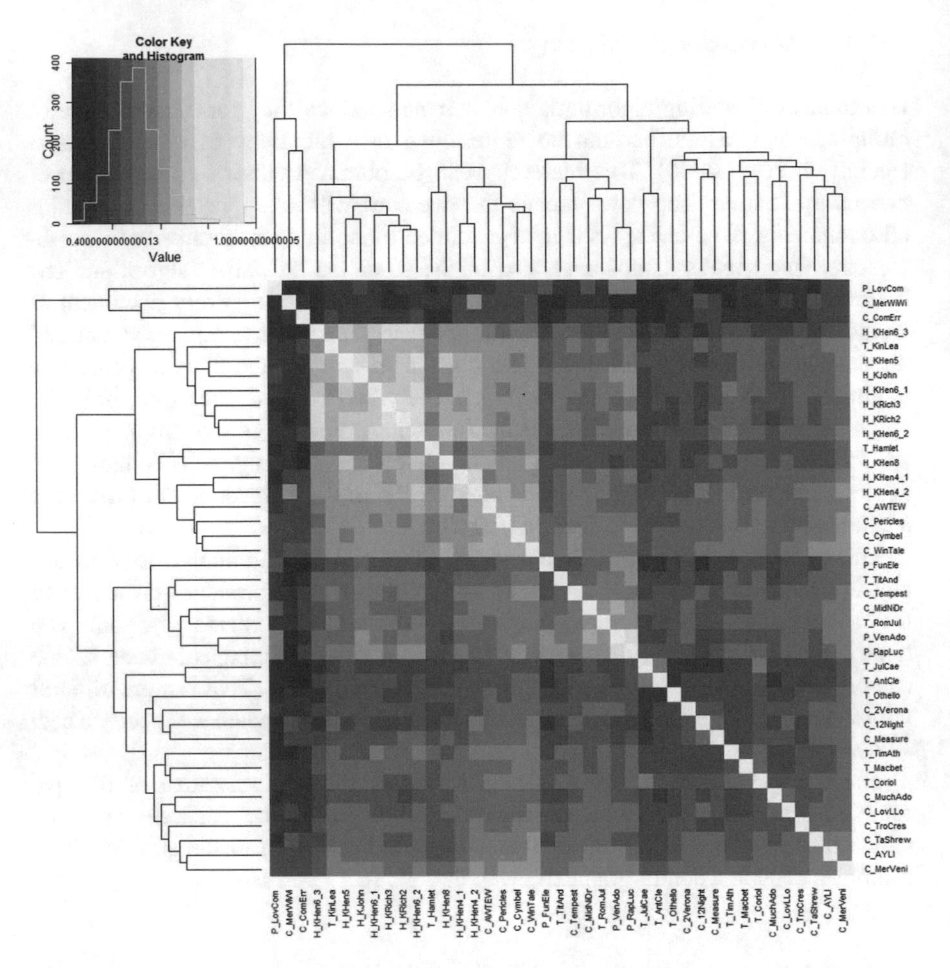

Fig. 16.17 Combined representation of Shakespeare's works: their similarity matrix via heat map, histogram of the matrix values and dendrograms of the row and column values (created via heatmap.2 function from the R package **gplots**)

compactness and separation. Together with the connectivity method they constitute the three internal measures available in the R package **clValid** (Brock et al. 2008). The details of each measure are given below, and for a good overview of internal measures in general, see Handl et al. (2005).

16.5.2.1 Connectivity

The *connectivity* indicates the degree of connectedness of the clusters, as determined by the k-nearest neighbors. Let N denote the total number of observations (documents) in a data set. Define $nn_{i(j)}$ as the jth nearest neighbor of observation

i, and let $x_{i,nn_i(j)}$ be zero if i and j are in the same cluster and $1/j$ otherwise. Then, for a particular clustering partition $\mathscr{C} = \{C_1, \ldots, C_K\}$ of the N observations into K disjoint clusters, the connectivity is defined as

$$\mathrm{Conn}(\mathscr{C}) = \sum_{i=1}^{N} \sum_{j=1}^{L} x_{i,nn_i(j)} , \tag{16.6}$$

where L is a parameter giving the number of nearest neighbors to use. The connectivity has a value between zero and ∞ and should be minimized.

16.5.2.2 Silhouette

The *Silhouette* of a datum is a measure of how closely it is matched to data within its cluster and how loosely it is matched to data of the neighboring cluster, i.e. the cluster whose average distance from the datum is lowest. A Silhouette close to 1 implies the datum is in an appropriate cluster, while a Silhouette close to -1 implies the datum is in the wrong cluster. For observation i, it is defined as

$$S(i) = \frac{b_i - a_i}{\max(b_i, a_i)} , \tag{16.7}$$

where a_i is the average distance between i and all other observations in the same cluster, and b_i is the average distance between i and the observations in the "nearest neighbouring cluster," i.e.

$$b_i = \min_{C_k \in \mathscr{C} \setminus C(i)} \sum_{j \in C_k} \frac{dist(i,j)}{n(C_k)} , \tag{16.8}$$

where $C(i)$ is the cluster containing observation i, $dist(i,j)$ is the distance (e.g., Euclidean, Manhattan) between observations i and j, and $n(C)$ is the cardinality of cluster C. The Silhouette Width is the average of each observation's Silhouette value:

$$\mathrm{Silh}(\mathscr{C}) = \frac{1}{N} \sum_{i=1}^{N} S(i) . \tag{16.9}$$

The Silhouette Width thus lies in the interval $[-1, 1]$, and should be maximized. For more information, see the help page for the `silhouette` function in the package **cluster** (Maechler et al. 2016).

16.5.2.3 Dunn Index

The *Dunn Index* is the ratio of the smallest distance between observations not in the same cluster to the largest intra-cluster distance. It is computed as

$$\mathrm{Dunn}(\mathscr{C}) = \frac{\min\limits_{C_k, C_l \in \mathscr{C}, C_k \neq C_l} \left(\min\limits_{i \in C_k, j \in C_l} dist(i,j) \right)}{\max\limits_{C_m \in \mathscr{C}} diam(C_m)}, \tag{16.10}$$

where $diam(C_m)$ is the maximum distance between observations in cluster C_m. The Dunn Index has a value between zero and ∞, and should be maximized.

16.5.3 Visual Cluster Validation

As long as the data set remains limited and the topic number is of modest size, cluster validation can be easily conducted using visual inspection of the generated topics and the resulting cluster content, comparing them with prior domain specific knowledge. Figure 16.18 demonstrates that through the example of the Quantlets belonging to the book SFE : "Statistics of Financial Markets" (Franke et al. 2015). Incorporating the domain knowledge of the SFE book, the dominating first 8 clusters/topics (corresponding to 96% of the data set) deal with "stochastic process simulation", "returns", "dax", "financial stocks", "call option prices", "assets", "black scholes", "normal distribution density", "probability", "parameter com-

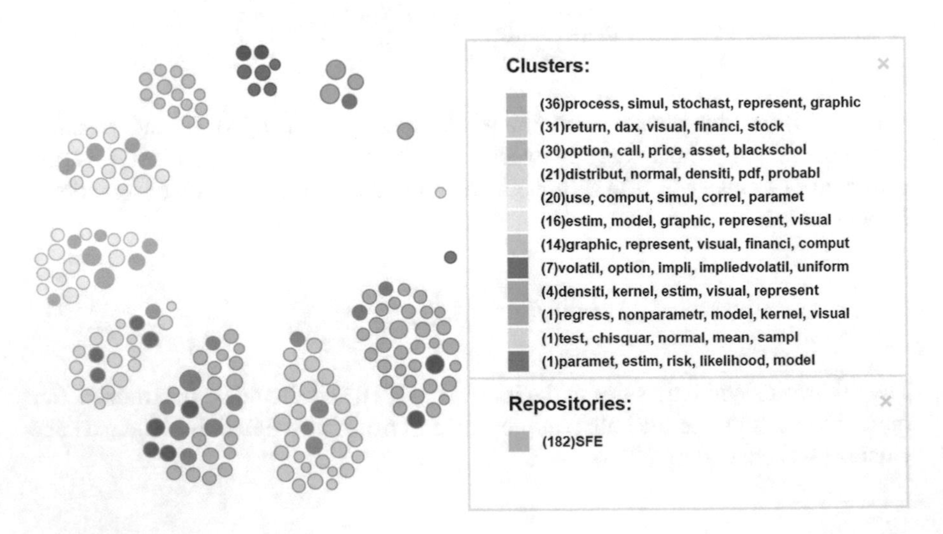

Fig. 16.18 SFE Quantlets clustered by *k*-means into 12 clusters, the tooltip on the right shows their topics

putation", "simulation", "correlation", "model estimation", "finance", "options", "implied volatility". The cluster topics are displayed in the cluster legend on the right in Fig. 16.18. The remaining four topics (corresponding to 4% of the data set) also show good concordance with the appropriate cluster content like "kernel density estimation", "nonparametric regression", "risk", etc. Since the automatically generated topic labels consist of stemmed words, the above listed "human readable versions" were syntactically improved for illustration purpose by the authors, see also Sect. 16.8.2.

16.6 Results

As data set for the following examination and analysis the whole QuantNet data base was taken. At the time of the big data analysis the documents structure was as follows: 1170 Gestalten (from 1826 individual Quantlets). That means that the meta information was extracted from Quantlets, in the case that several Quantlet versions in different programming languages were available, their meta information was merged to a single and unique representation, called "Gestalt." Q SFEGBMProcess is such an example, see Fig. 16.30.

- $Q = \{d_1, \dots, d_n\}$: set of documents (Quantlets/Gestalten)
- $T = \{t_1, \dots, t_m\}$: dictionary (set of all terms)
- $tf(d, t)$: absolute frequency of term $t \in T$ in $d \in Q$
- $D = [w(d_1), \dots, w(d_n)]$: "term by document" matrix TDM

Throughout the whole Sect. 16.6 we will use the definitions and notations from Sects. 16.4 and 16.5. The first step is to transform the text documents into the quantities listed above. This will be demonstrated in Sect. 16.6.1.

16.6.1 Text Preprocessing Results

For the basic text preprocessing and calculation of the TDM the R package **tm** (Feinerer and Hornik 2015) was applied, see Listing 16.1. It provides a framework for text mining applications within R (Feinerer et al. 2008). According to Table 16.2 we selected the preprocessing configuration "discarding $tf \leq 2$", resulting in a TDM with 1039×1170 entries.

Listing 16.1 Text preprocessing via R package **tm**

```
# preprocessing text with this function
cleanCorpus = function(corpus) {
        corpus.tmp <- tm_map(corpus, removePunctuation)
        corpus.tmp <- tm_map(corpus.tmp, stripWhitespace)
        corpus.tmp <- tm_map(corpus.tmp, removeNumbers)
        corpus.tmp <- tm_map(corpus.tmp, content_transformer(tolower))
        corpus.tmp <- tm_map(corpus.tmp, stemDocument)
        corpus.tmp <- tm_map(corpus.tmp, removeWords, stopwords("english"))
        corpus.tmp <- tm_map(corpus.tmp, removeWords, qn_stopwords)
        return(corpus.tmp)
}

doc_corpus <- VCorpus(DirSource(dir.name, encoding = "UTF-8"),
                      readerControl = list(language = "en"))
corpus.cleaned <- cleanCorpus(doc_corpus)

# TDM with all terms
tdm_cleaned <- TermDocumentMatrix(corpus.cleaned)

# trimmed TDM, discarding tf <= 2
tdm_cleaned_tf2 <- TermDocumentMatrix(corpus.cleaned,
                                      list(bounds = list(global = c(3, Inf))))
```

Table 16.2 Total number of documents in QuantNet: 1170 Gestalten/1826 Quantlets; term sparsity: 98–99%

	Terms	Non-/sparse entries
All terms (after preprocessing)	2223	17,878/2,583,032
Discarding $tf = 1$	1416	17,071/1,639,649
Discarding $tf \leq 2$	1039	16,317/1,199,313
Discarding $tf \leq 3$	846	15,738/974,082

Table 16.3 Model performance regarding the sparsity of the "term by document" matrix TDM and the similarity matrix M_S in the appropriate models (weighting scheme: tf-idf normalized)

	BVSM	GVSM(TT)	LSA:300	LSA:171(50%)	LSA:50
TDM	0.99	0.65	0.51	0.51	0.47
M_S	0.65	0.07	0.35	0.36	0.35

16.6.2 Sparsity Results

The BVSM, GVSM(TT), and three LSA configurations with the dimension parameter k equal to 300, 171 (50% of the weight of all singular values) and 50 were considered, see Table 16.3. *Sparsity* and *density* are terms used to describe the percentage of cells in a database table (or a matrix) that are not populated and populated, respectively. The sum of the sparsity and density should equal 100%. Sparsity is the ratio of the number of zero entries to the total number of entries of a matrix. In general, the lower the sparsity, the better, see also "drawbacks of the BVSM" in Sect. 16.4.4.

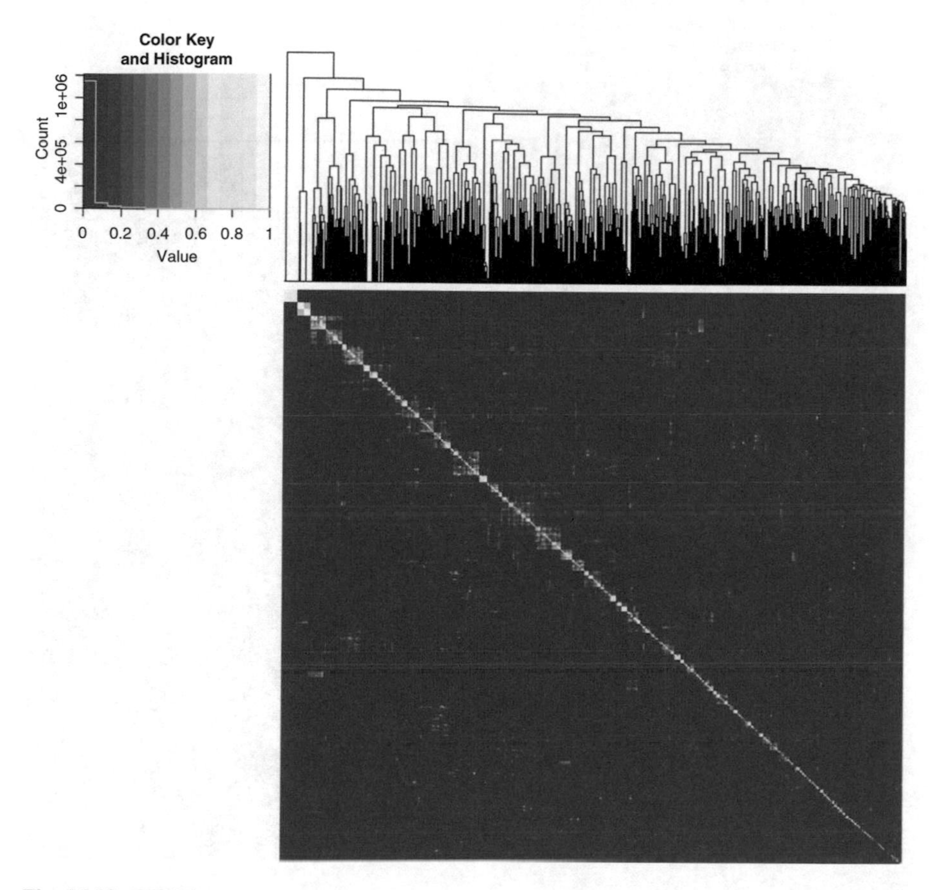

Fig. 16.19 BVSM

Heat maps with dendrograms of the similarity matrices in the appropriate model configurations are displayed in Figs. 16.19, 16.20, 16.21, 16.22, and 16.23. They allow an extensive visual interpretation and characterization of the inherent cluster structure of the included text documents. The method `heatmap.2` from the R package **gplots** was used for creating the heat maps (Warnes et al. 2016). This method simultaneously performs reordering of the matrix rows and/or columns according to the row and/or column means within the restrictions imposed by the dendrogram. Hence, an easier identification of "similarity clusters" within the matrix is provided. The color map on the left displays the meaning of the color keys: yellow values show the similarity values close to 1, red values those close to zero, see also Formula (16.2).

Two interesting effects can be stated. (1) GVSM(TT) and LSA similarity matrices pronounce a higher concentration of "similarity clusters" around the diagonal than those in the BVSM, thereby indicating subsets of documents allowing good clusterization into one particular group. (2) LSA allows an adjusted sparsity

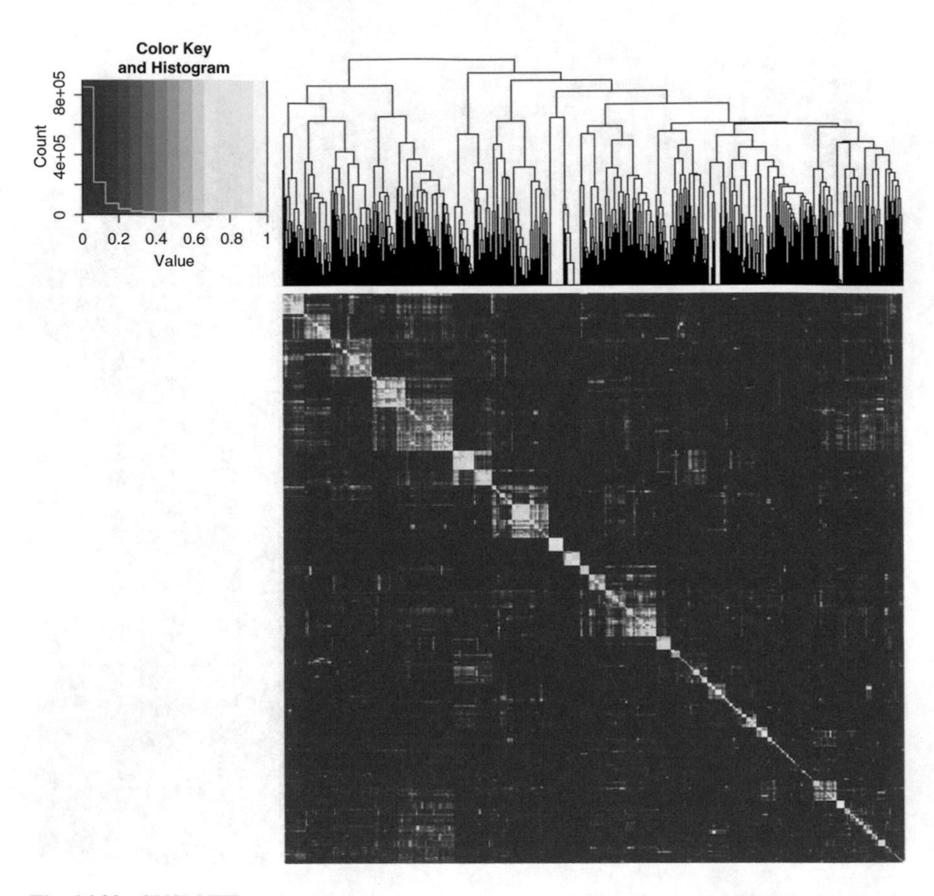

Fig. 16.20 GVSM(TT)

reduction and similarity enhancement, respectively, by varying the k parameter. We can see the apparent relationship that lower k values imply clearer "similarity clusters" within the matrix, compare Figs. 16.21, 16.22, and 16.23.

We can conclude that the more sophisticated models GVSM(TT) and LSA clearly outperform the BVSM, concerning both the TDM and similarity matrices. Given the pure numbers in Table 16.3, we observe that the LSA configurations reduce the TDM sparsity to the greatest extent. In the case of similarity matrices GVSM(TT) achieves the greatest sparsity reduction.

16.6.3 Three Models, Three Methods, Three Measures

For evaluation and benchmark purpose we have introduced the so-called M^3 evaluation. All TM **models**, clustering **methods** and validation **measures** as presented

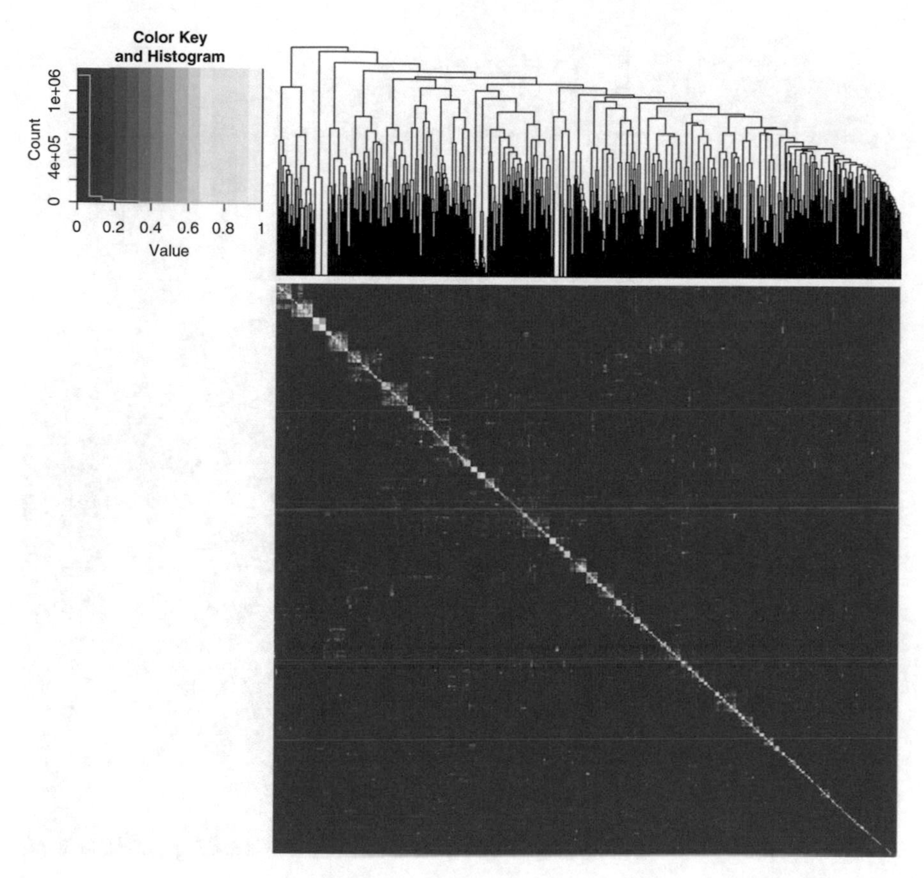

Fig. 16.21 LSA:300

in the previous sections are combined in a $3 \times 3 \times 3$ benchmark setup, hence the name M^3 evaluation. Every M stands for one of the dimensions: models, methods, and measures.

- 3 models: BVSM, GVSM(TT), and LSA
- 3 clustering methods: *k*-means, *k*-medoids, HC
- 3 cluster validation measures: connectivity, Silhouette width, Dunn index

More precisely, the current experimental design should be named as $M^3_{3,3,3,250}$ (250 as the maximal cluster size to be evaluated). Later we will explain how it can be extended to $M^3_{d_1,d_2,d_3,\max}$, with d_i encompassing more settings in the appropriate dimension.

Concerning the LSA, two configurations were taken: k equal to 171 (50% of the weight of all singular values) and 50. There is another implicit dimension in the experimental design, namely the number of possible clusters, let's call it i, which is captured on the x-axis in the plot matrix in Fig. 16.24. We have decided

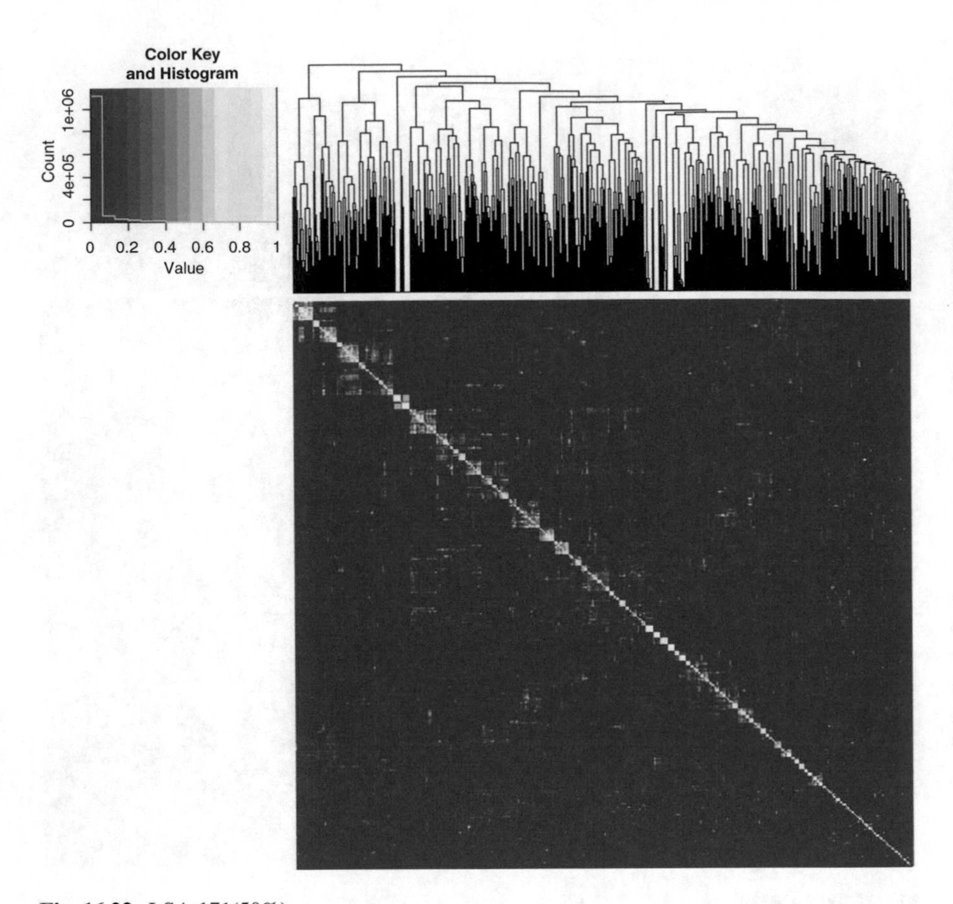

Fig. 16.22 LSA:171(50%)

to run the validation for the first $2, \ldots, 250$ i-values. Since our TDM has 1170 documents/columns, we regard the choice of 250 as the maximal cluster size as appropriate. On the one hand, 250 is more than enough for the practical needs. On the other hand, in the case of 250 clusters amongst 1170 objects one would obtain around 5 objects in one cluster at average. This is quite close to the extreme case, one object in one cluster, what is trivial and honored by the most validation measures with the "highest score." All things considered, our choice of the maximal cluster size was a good compromise between the practical needs, the theoretical limits and computational expense.

Listing 16.2 demonstrates the main idea of the M^3 experimental design. For any given TDM the main function `clValid` is executed. Afterwards, the evaluation results for all considered TM models are aggregated with respect to any considered validation measure and clustering method, in our example, Silhouette and HC. Apparently, the experimental design can be extended in any dimension: more TM models, more clustering methods, more validation measures and, if necessary, more

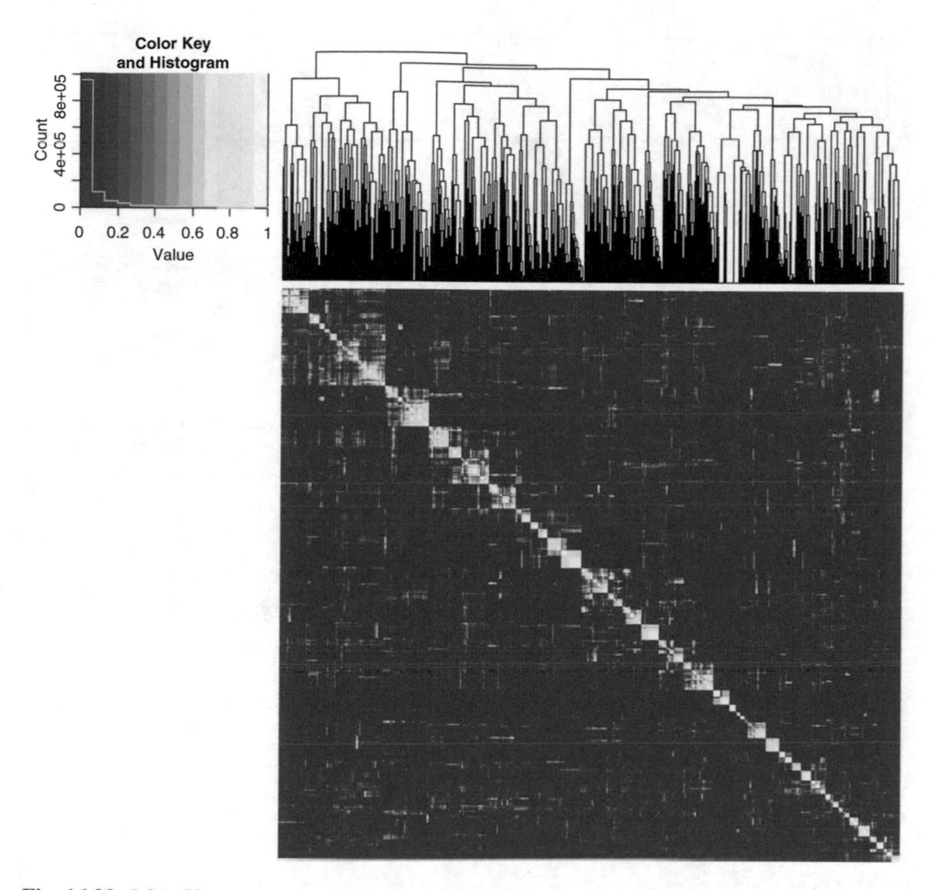

Fig. 16.23 LSA:50

cluster sizes. The increasing calculation time of the overall experiment should be considered. A contemporary Intel Core i5 CPU needed one night to finish all calculations.

For any given M^3-combination (fixed measure, method and model) the shape of the function graph in the appropriate M^3 plot matrix cell (a particular row, column and color) can exhibit an individual behavior, see Fig. 16.24. Characteristic for validation measures in our setup is the monotonous growth. In some cases there are some fluctuations and oscillations for lower i values. After an initial period of some i's all function graphs start to consolidate their growth trend. Remarkable is the unstable and noisy behavior of the k-means method, in particular in the BVSM. Another interesting observation is the combination Silhouette and LSA50. First the graph has a strong oscillation with a decreasing trend, then a relatively steep ascent and finally, after around a quarter of the interval length of i-values, the graph shows a stable sideways movement.

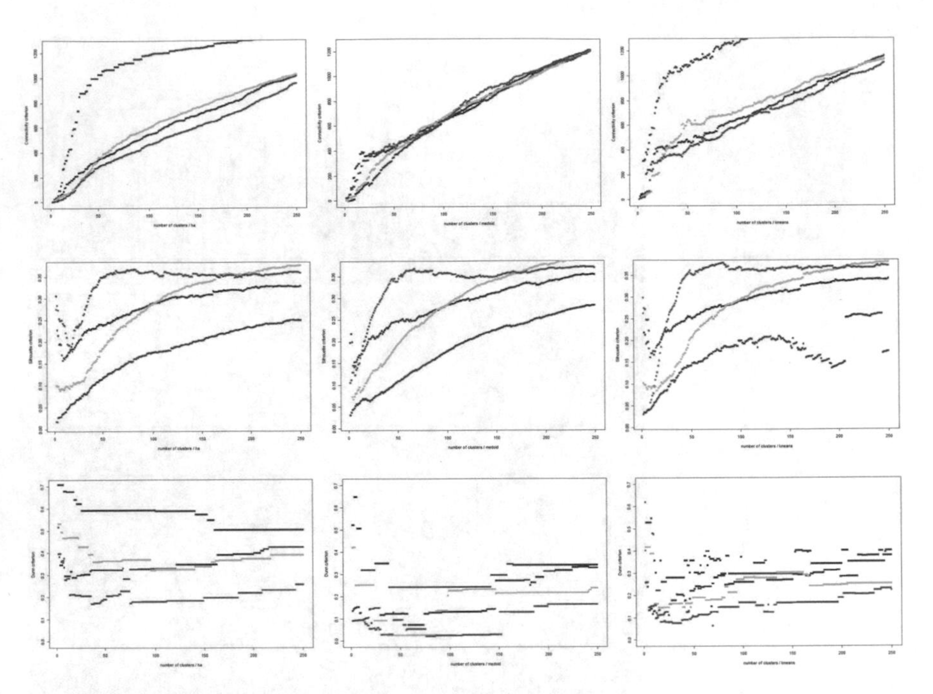

Fig. 16.24 M^3 plot matrix. Rows: connectivity, Silhouette, Dunn. Columns: HC, *k*-medoids, *k*-means. Colors: BVSM, GVSM(TT), LSA, LSA50

Listing 16.2 Cluster validation via R package **clValid**

```
# load the R package
library(clValid)
# transpose the TDM in the LSA model
A = t(m_lsa_mat)
# run the main evaluation function
intern <- clValid(A, 2:250, clMethods=c("hierarchical","kmeans","pam"),
                  validation="internal")
# basic inspection methods
summary(intern)
plot(intern)
m_lsa = measures(intern)

# aggregate evaluation results for 4 different TM models; Silhouette / HC
x_l = 250
plot(2:x_l, m_b[3,,1], pch=15, ylim=c(0.01,0.7), col="blue",
     xlab="number of clusters / hc", ylab="Silhouette criterion")
lines(2:x_l, m_tt[3,,1], type = "p", pch=15, col="red")
lines(2:x_l, m_lsa[3,,1], type = "p", pch=15, col="green")
lines(2:x_l, m_lsa50[3,,1], type = "p", pch=15, col="magenta")
legend("topright", col= c("blue", "red", "green", "magenta"), pch=15,
       legend = c("BVSM","GVSM(TT)","LSA", "LSA50"), lty=3)
```

The results of our M^3 evaluation are summarized in Table 16.4. The most important observations and conclusions are:

- HC better or comparable to other methods under all measures and in all models
- LSA50 superior with respect to the connectivity and Silhouette measures

Table 16.4 M^3 evaluation results

Measure	Model	Method
Connectivity	LSA50	HC
Silhouette	LSA50	HC
Dunn	BVSM/LSA	HC

- BVSM/LSA slightly better than LSA50 with respect to the Dunn measure, but still comparable (small range of values in all models)
- Conclusion: LSA/LSA50 and HC is the optimal model/method combination under M^3 evaluation

16.6.4 LSA Anatomy

Since the SVD truncation as performed in Sect. 16.4.4.3 results in the following decomposition:

$$D = D_k + D_{\text{err}} \tag{16.11}$$

and

$$D_k = U_k \Sigma_k V_k^\top, \tag{16.12}$$

the question arises how these six matrices, namely $D, D_k, D_{\text{err}}, U_k, \Sigma_k$ and $V_k^\top$, look like?

All results, in particular all plots and figures, concerning the LSA anatomy can be examined and reproduced by the corresponding Quantlets, available under https://github.com/Quantlet/Q3D3LSA. The reader can also "just browse" through the GitHub repository and study the plots in a higher resolution, in particular the high dimensional matrix representations. The most important incorporated R packages are **lsa** (Wild 2015), **gplots** (Warnes et al. 2016), and **ggplot2** (Wickham 2009). In the beginning of every Quantlet the LSA space is created from the term document matrix TDM of the Quantlets, which was created as described in Sect. 16.6.1.

16.7 Application

The current implementation of the self-developed visualization framework for knowledge discovery in QuantNet is displayed in Fig. 16.25. The so-called D3 Visu application is available as web page at http://quantlet.de. Driven by the Q3-D3-LSA technology, which is the combination of our research findings, the integrated search engine facilitates easier discovery of shared validated knowledge and collaborative

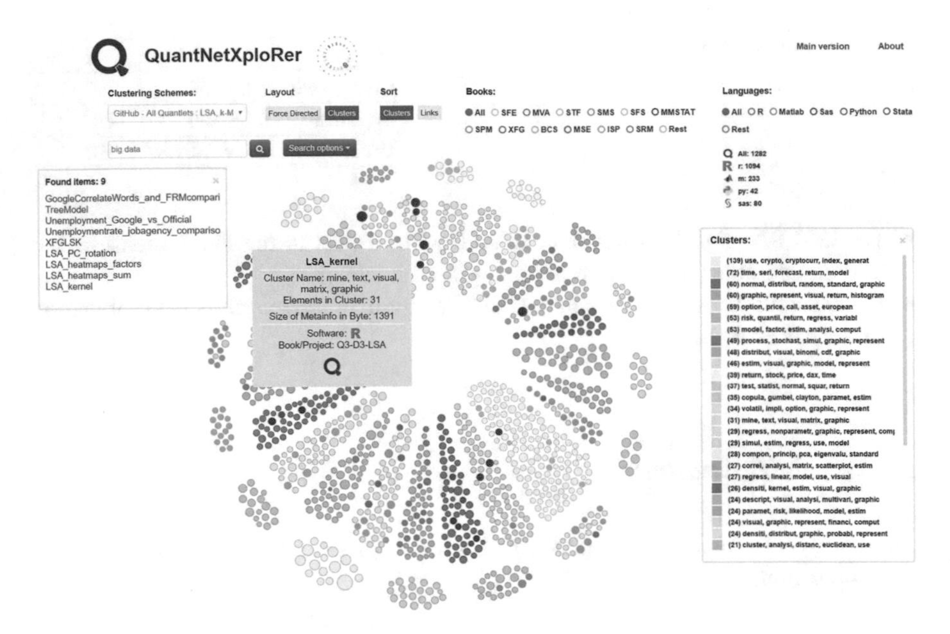

Fig. 16.25 Front end view: all Quantlets in QuantNetXploRer, search term "big data"

reproducible research (CRR). While the D3 based application provides an interactive front end of IR, document clustering and visualization elements, one can rely on the robust data storage infrastructure of GitHub in the background, comprising the distinct abilities of version control (VC) and source code management (SCM). A start page screenshot of the Quantlet GitHub organization is given in Fig. 16.31.

The GitHub platform, having more than 14 million users and more than 35 million repositories, is currently the largest host of source code in the world. It provides access control, task management, and collaboration features for all project types. Thanks to the Style Guide (https://github.com/Quantlet/Styleguide-and-FAQ), Yamldebugger R package (https://github.com/Quantlet/yamldebugger) and introductory Quantlets Q https://github.com/Quantlet/yamldebugger_intro, the Quantlet members have all necessary tools for a fast, transparent and iterative code development and documentation process. Once a member or outside collaborator has contributed valid Quantlets, the TM pipeline retrieves the meta information of Quantlets via the GitHub-R-API and distills them to human-readable and applicable information by means of the Q3-D3-LSA technology.

Quantlets, which have been processed in that manner, are finally extracted into the D3 Visu application layer, called QuantNetXploRer. Figure 16.25 demonstrates a typical application. The hits of the entered search query, in this case "big data", are displayed both in textual form and in graphical form. Quantlets (represented by nodes) containing the expression "big data" are highlighted in red color. The application screen is divided into the central main visualization ("orbit clustering" scheme), and auxiliary components like buttons, tool tips, and legends. The upper control panel allows the choice of different clustering schemes, D3 layouts, color palettes and allows the configuration of the dynamic and draggable legends. Two legends allow to filter the nodes by programming languages or books and projects. Other two legends display the cluster topics or GitHub repositories of the visualized Quantlets. All relevant auxiliary components are draggable, can be deactivated and are responsive, what means that the action performed in one element is reflected in all other visualization components. For instance, if the user filters the nodes by the programming language R, the contents of the main D3 Visu, the cluster topics legend and the GitHub repositories legend are updated. The statistic of the remaining language combinations in the programming languages legend is recalculated, too. All updates of the main D3 Visu are realized via dynamic transition effects.

The Q3-D3-LSA engine of the QuantNetXploRer has many other characteristics and features which are best explored by "learning by doing":

Build Quantlets better, together, now (QuantNet @ GitHub).

16.8 Outlook

The benchmarks in Sect. 16.6 have shown that different GVSM configurations allow adapted similarity based document clustering. Concerning sparsity and higher concentration of "similarity clusters" (as shown in Sect. 16.6.2) both the GVSM(TT) and LSA configurations clearly outperform the classical BVSM. Incorporating term–term correlations and semantics, GVSM(TT) and LSA provide considerable sparsity reduction, thereby achieving higher clustering performance. The main advantage of LSA is the flexible dimension reduction property which is controlled by the truncation parameter k within the SVD process. Additionally, the M^3 evaluation identifies the LSA/LSA50 and HC as the optimal model/method combination. The benefits of the dimension reduction effect with smaller k values can also be observed in the M^3 plot matrix (see Fig. 16.24).

First benchmark results in Sect. 16.4.4.5 show that the LSA model seems to be applicable for big data and has a modest time complexity. Thus, samples of 100,000 GitHub organizations could be processed within an hour. Potential bottlenecks are the GitHub API extraction process or the calculation of big distance matrices for some clustering methods. Both issues could be tackled by massive parallelization and are beyond the actual subject "TM models."

16.8.1 *GitHub Mining Infrastructure in R*

Our TM pipeline together with the GitHub-R-API implementation relies on several sophisticated R packages like **tm**, **lsa**, **svd**, **cluster**, **yaml**, **jsonlite** and some more. An essential element is the R package **github** (Scheidegger 2016) "R Bindings

Listing 16.3 GitHub API method **Get contents** returns the contents of a file or directory in any repository on GitHub which is publicly available

```
get.repository.path <- function(owner, repo, path,
                        ..., ctx = get.github.context())

# Browser as function for
# https://github.com/thomas-haslwanter/statsintro_python
QBrowser_2Dir_Offset = function(gh_user = "thomas-haslwanter",
                        reponame = "statsintro_python",
                        path_offset = "ISP/Code_Quantlets", showSummary = TRUE)
  ### Start
  rep_c = get.repository.path(gh_user, reponame, path_offset, ctx = ctx)
```

for the Github v3 API". Taken as a whole, we have a powerful "GitHub Mining infrastructure in R" which allows to incorporate any GitHub organization with its content for further analysis and possible data mining thanks to the official GitHub API v3 (https://developer.github.com/v3/). Currently, there are more than one million organizations on GitHub, among them Google, Facebook, Twitter, Yahoo, CRAN, RStudio, D3, Plotly, and many more. Borke and Bykovskaya (2017a) show how the "GitHub Mining Infrastructure in R" can be applied to mine some popular GitHub organizations containing several ten thousand repositories.

Listing 16.3 shows how the content of any publicly available repository on GitHub can be retrieved within R. The first parameter `owner` can be substituted by any organization or user name. Basically, the operating and mining scope of QuantNet can be extended to any subset of GitHub. One challenge is to implement the appropriate parsers for the specific repository structures and contents of new organizations. The other is to adjust and calibrate the TM models to the new kind of information. Actually, QuantNet has already several parsers implemented. In addition to the Quantlet organization, the repository "Introduction to Statistics with Python" (Haslwanter 2016) (https://github.com/thomas-haslwanter/statsintro_python) is also incorporated via the Q3-D3-LSA engine.

16.8.2 Future Developments

In the near future, we are going to publish three R packages under the overall heading "GitHub API based QuantNet Mining infrastructure in R" (Q3). At this stage it seems reasonable to organize this R infrastructure in the following packages:

- **rgithubQ** (Scheidegger and Borke 2017): an extension of the R package **github**, first of all, enabling file operations like *Create a file*, *Update a file* and providing a series of low level API helping functions (see also https://github.com/cscheid/rgithub/blob/master/todo.org).

- **TManalyzerQ** (Borke 2017a): comprising the parser layer, TM models layer, clustering layer, and D3 export layer. This is the main component of the Q3-D3-LSA engine.
- **mdGeneratorQ** (Borke and Bykovskaya 2017c): GitHub Markdown generator, a special parser runs through the QuantNet repository structure, extracts resources like meta information, source code, pictures, etc., reformats, integrates, and exports them via the GitHub API into a single Markdown file for every Quantlet, see, e.g., Fig. 16.30.

The prototypes of the aforementioned three packages are already in operational and working state and are continuously tested and improved. The **TManalyzerQ** and **mdGeneratorQ** prototypes operate independently from each other. Both components require the **rgithubQ** functionality. The final design and structure of the Q3 packages is subject of current research and will be presented in Borke and Härdle (2017).

Furthermore, more TM models, clustering methods and validation measures could be considered and studied for performance validation: from M^3 to $M^3_{d_1,d_2,d_3,\max}$, see Sect. 16.6.3 and (Borke 2017). Optimization of the automatically generated cluster labels for easier human readability and implementation of new "upgrades" into the D3 Visu could contribute to a better usability of the Q3-D3-LSA technology.

Acknowledgements Financial support from the Deutsche Forschungsgemeinschaft via CRC "Economic Risk" and IRTG 1792 "High Dimensional Non Stationary Time Series," Humboldt-Universität zu Berlin, is gratefully acknowledged.

Appendix

See Figs. 16.26, 16.27, 16.28, 16.29, 16.30, and 16.31.

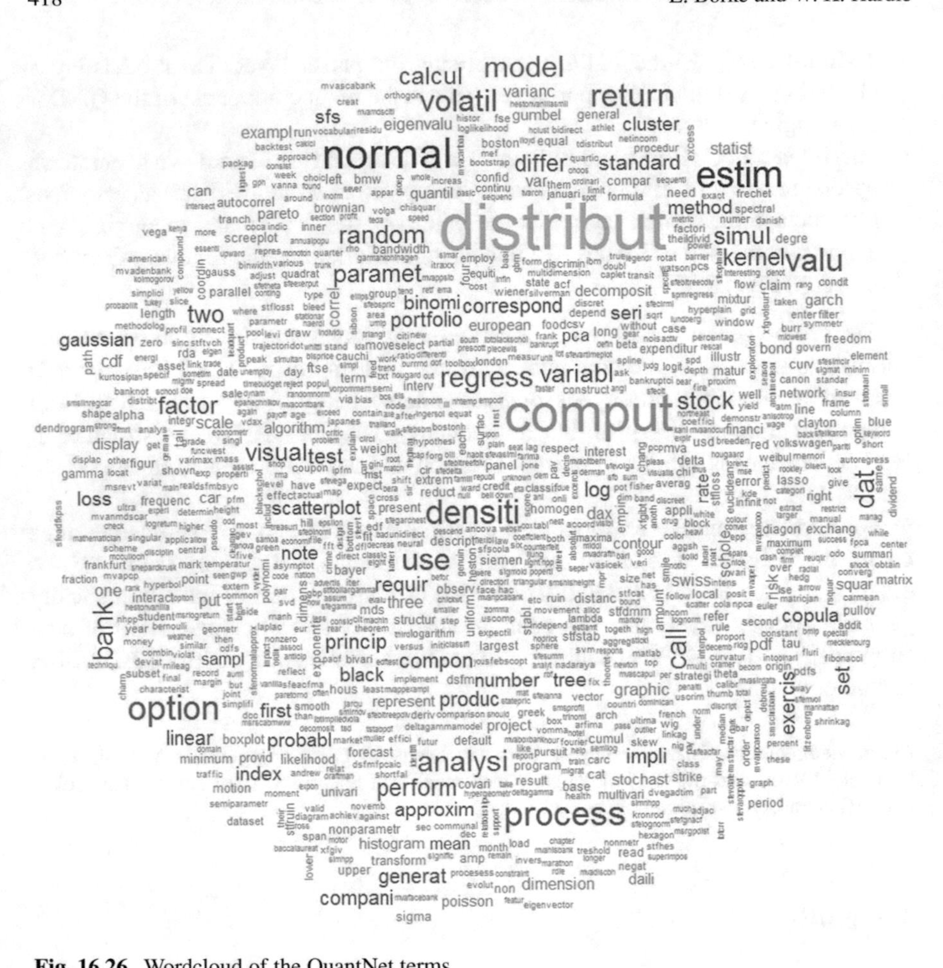

Fig. 16.26 Wordcloud of the QuantNet terms

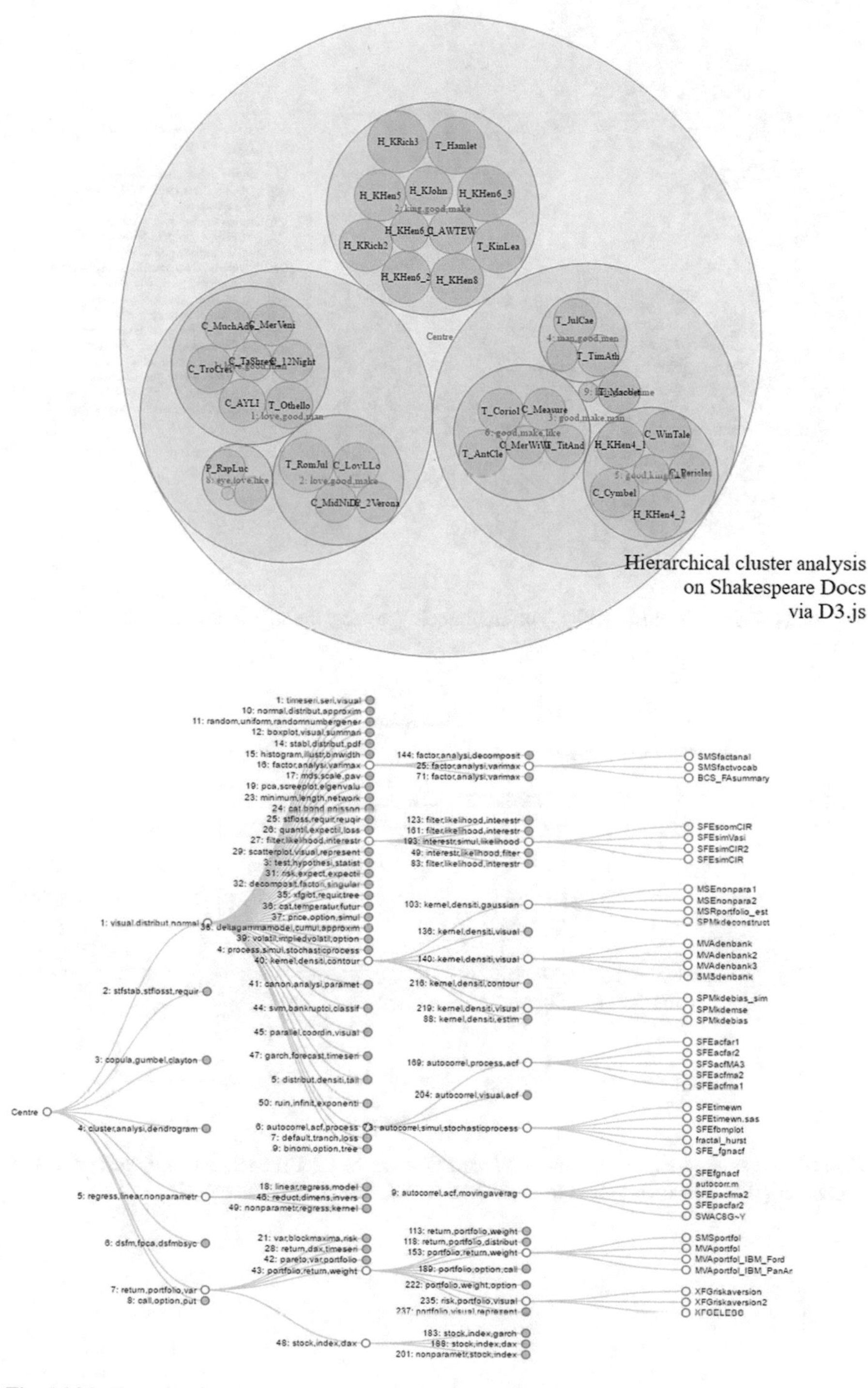

Fig. 16.27 Two visualization examples from Q3-D3 Genesis Chapters II–VI

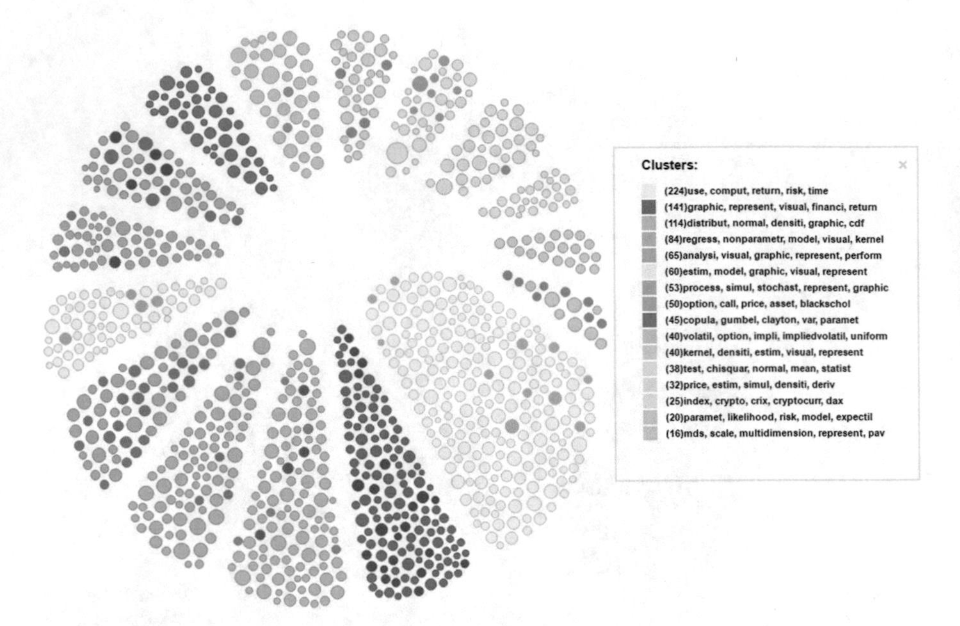

Fig. 16.28 Quantlets clustered by *k*-means into 16 clusters, the tooltip on the right shows their topics

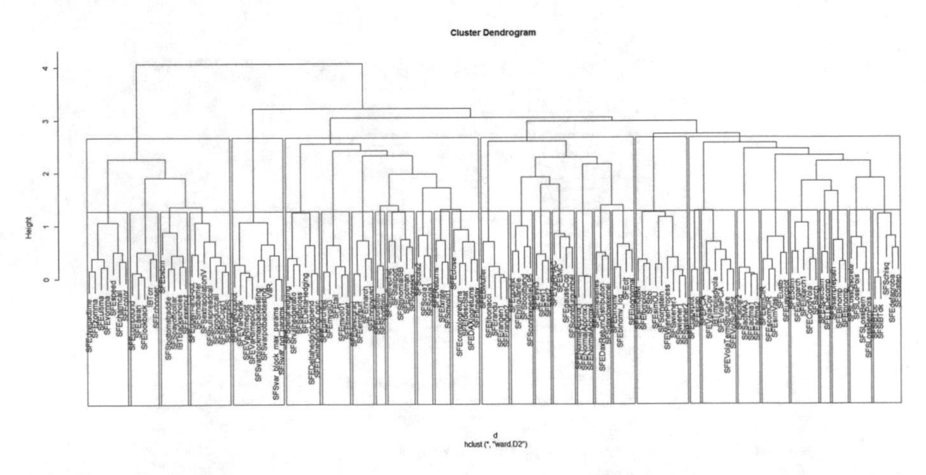

Fig. 16.29 Dendrogram created by HC (ward-method) in LSA model, cut in 6 clusters and 30 subclusters, 137 Gestalten, subset from the books SFE, SFS, and the project IBT

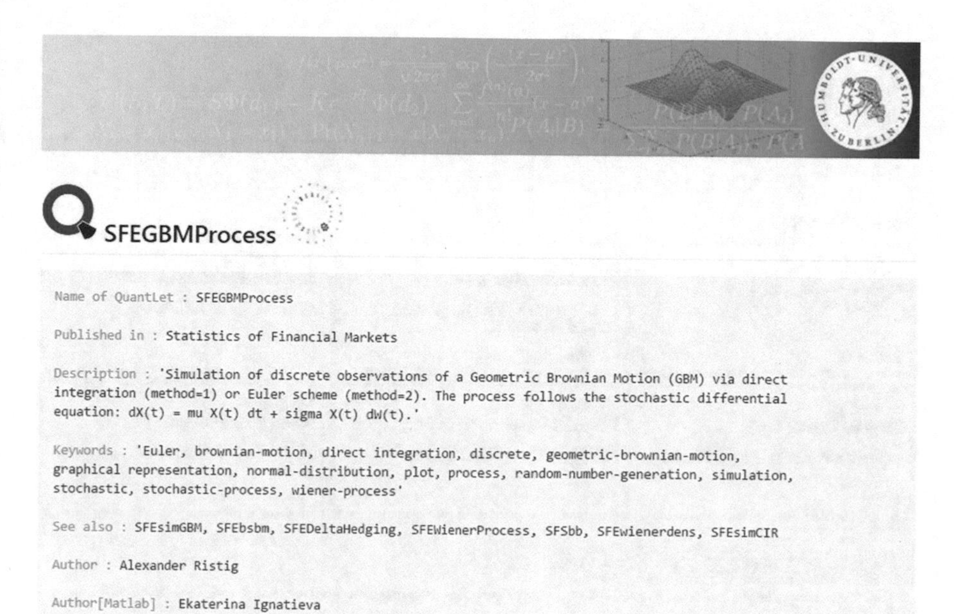

SFEGBMProcess

Name of QuantLet : SFEGBMProcess

Published in : Statistics of Financial Markets

Description : 'Simulation of discrete observations of a Geometric Brownian Motion (GBM) via direct integration (method=1) or Euler scheme (method=2). The process follows the stochastic differential equation: dX(t) = mu X(t) dt + sigma X(t) dW(t).'

Keywords : 'Euler, brownian-motion, direct integration, discrete, geometric-brownian-motion, graphical representation, normal-distribution, plot, process, random-number-generation, simulation, stochastic, stochastic-process, wiener-process'

See also : SFEsimGBM, SFEbsbm, SFEDeltaHedging, SFEWienerProcess, SFSbb, SFEwienerdens, SFEsimCIR

Author : Alexander Ristig

Author[Matlab] : Ekaterina Ignatieva

Author[SAS] : Daniel T. Pele

Submitted : Tue, June 17 2014 by Franziska Schulz

Input:
- method: 'type of method used: 1 - direct integration, 2 - Euler scheme'

Example : 'A plot of typical path of a geometric brownian motion is provided for the case n=1, x0=0.084, mu=0.02, sigma=sqrt(0.1) delta=1/1000.'

Fig. 16.30 Gestalt "SFEGBMProcess" simulating the geometric Brownian motion comprises three Quantlets in three programming languages: R, Matlab, and SAS

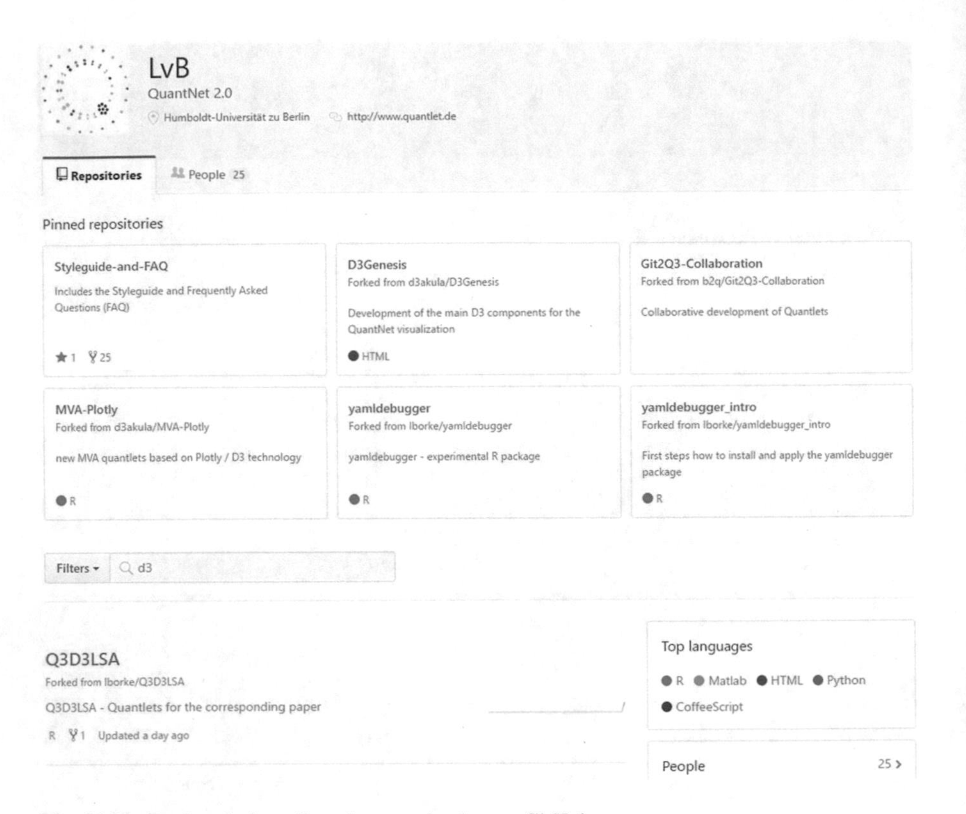

Fig. 16.31 Back end view: Quantlet organization on GitHub

References

Berry M (2003) Survey of text mining: clustering, classification, and retrieval, 1st edn. Springer, New York

Borak S, Härdle W, López-Cabrera B (2013) Statistics of financial markets: exercises and solutions, 2nd edn. Springer, Berlin

Borke L (2017) Dynamic clustering and visualization of smart data via D3-3D-LSA. Humboldt-Universität zu Berlin, Wirtschaftswissenschaftliche Fakultät. http://dx.doi.org/10.18452/18307

Borke L (2017a) TManalyzerQ: provides IR tools in 3 text mining models: BVSM, GVSM(TT) and LSA - QuantNet edition. R package version 0.5.0

Borke L (2017b) yamldebugger: YAML parser debugger according to the QuantNet style guide. R package version 1.0

Borke L, Bykovskaya S (2017a) BitQuery: a GitHub API driven and D3 based search engine for open source repositories. http://bitquery.de

Borke L, Bykovskaya S (2017b) D3 for visual analytics. https://github.com/d3VA

Borke L, Bykovskaya S (2017c) mdGeneratorQ: GitHub Markdown generator according to the QuantNet style guide. R package version 0.4.0

Borke L, Härdle WK (2017) GitHub API based QuantNet mining infrastructure in R. SFB 649 discussion paper. Humboldt Universität zu Berlin

Bostock M, Ogievetsky V, Heer J (2011) D3 data-driven documents. IEEE Trans Vis Comput Graph 17(12):2301–2309

Bradford RB (2009) Comparability of LSI and human judgment in text analysis tasks. In: Proceedings of the 11th WSEAS international conference on mathematical methods and computational techniques in electrical engineering, MMACTEE'09. World Scientific and Engineering Academy and Society (WSEAS), Stevens Point, pp 359–366

Brock G, Pihur V, Datta S, Datta S (2008) clValid: an R package for cluster validation. J Stat Softw 25(1):1–22

Cosentino V, Luis J, Cabot J (2016) Findings from GitHub: methods, datasets and limitations. In: Proceedings of the 13th international conference on mining software repositories, MSR '16. ACM, New York, pp 137–141

Cristianini N, Shawe-Taylor J, Lodhi H (2002) Latent semantic kernels. J Intell Inf Syst 18(2):127–152

Deerwester S, Dumais ST, Furnas GW, Landauer TK, Harshman R (1990) Indexing by latent semantic analysis. J Am Soc Inf Sci 41(6):391–407

Dunn JC (1974) Well separated clusters and fuzzy partitions. J Cybern 4:95–104

Everitt BS, Landau S, Leese M, Stahl D (2011) Cluster analysis, 5th edn. Wiley, Hoboken

Feinerer I, Hornik K (2015) tm: text mining package. R package version 0.6-2

Feinerer I, Wild F (2007) Automated coding of qualitative interviews with latent semantic analysis. In: Mayr HC, Karagiannis D (eds) Information systems technology and its applications, 6th international conference ISTA. Gesellschaft für Informatik, Bonn, pp 66–77

Feinerer I, Hornik K, Meyer D (2008) Text mining infrastructure in R. J Stat Softw 25(5):1–54

Fernández-Luna JM, Huete JF, Rodríguez-Cano JC (2011) User intent transition for explicit collaborative search through groups recommendation. In: Proceedings of the 3rd international workshop on collaborative information retrieval, CIR '11. ACM, New York, pp 23–28

Franke J, Härdle W, Hafner C (2015) Statistics of financial markets: an introduction, 4th edn. Springer, Berlin

Golyandina N, Korobeynikov A (2014) Basic singular spectrum analysis and forecasting with R. Comput Stat Data Anal 71:934–954. R package version 0.14

Handl J, Knowles J, Kell DB (2005) Computational cluster validation in post-genomic data analysis. Bioinformatics 21(15):3201–3212

Härdle W, Simar L (2015) Applied multivariate statistical analysis, 4th edn. Springer, Berlin

Härdle W, Hautsch N, Overbeck L (2008) Applied quantitative finance, 2nd edn. Springer, Berlin

Haslwanter T (2016) An introduction to statistics with Python: with applications in the life sciences, 1st edn. Springer International Publishing, Berlin

Jones KS (1972) A statistical interpretation of term specificity and its application in retrieval. J Doc 28(1):11–21

Kaufman L, Rousseeuw PJ (2008) Partitioning around medoids (program PAM). In: Finding groups in data. Wiley, Hoboken, pp 68–125

Korobeynikov A (2010) Computation- and space-efficient implementation of SSA. Stat Interface 3(3):357–368. R package version 0.14.

Korobeynikov A, Larsen RM, Laboratory LBN (2016) svd: interfaces to various state-of-art SVD and eigensolvers. R package version 0.4

Linstead E, Rigor P, Bajracharya S, Lopes C, Baldi PF (2008) Mining internet-scale software repositories. In: Platt J, Koller D, Singer Y, Roweis S (eds) Advances in neural information processing systems 20. Curran Associates, Red Hook, pp 929–936

Maechler M, Rousseeuw P, Struyf A, Hubert M, Hornik, K (2016) cluster: cluster analysis basics and extensions. R package version 2.0.5

Michailidis G (2008) Data visualization through their graph representations. In: Handbook of data visualization. Springer handbooks of computational statistics. Springer, Berlin, pp 103–120

Miller T, Klein B, Wolf E (2009) Exploiting latent semantic relations in highly linked hypertext for information retrieval in wikis. In: Proceedings of the international conference RANLP-2009. Association for Computational Linguistics, Borovets, pp 241–245

Mohamed M, Oussalah M (2014) A comparative study of conversion aided methods for WordNet sentence textual similarity. In: Proceedings of the first AHA!-workshop on information discovery in text. Association for Computational Linguistics, Borovets and Dublin City University, Dublin, pp 37–42

Robertson S (2004) Understanding inverse document frequency: on theoretical arguments for IDF. J Doc 60(5):503–520

Rousseeuw PJ (1987) Silhouettes: a graphical aid to the interpretation and validation of cluster analysis. J Comput Appl Math 20:53–65

Salton G, Wong A, Yang CS (1975) A vector space model for automatic indexing. Commun ACM 18(11):613–620

Scheidegger C (2016) github: github API. R package version 0.9.8

Scheidegger C, Borke L (2017) rgithubQ: GitHub API bindings for R - QuantNet edition. R package version 0.5.0

Srivastava A, Sahami M (2009) Text mining: classification, clustering, and applications, 1st edn. Chapman & Hall/CRC, Boca Raton

Steinbach M, Karypis G, Kumar V (2000) A comparison of document clustering techniques. In: KDD workshop on text mining

Theußl S, Feinerer I, Hornik K (2012) A tm plug-in for distributed text mining in R. J Stat Softw 51(5):1–31

Warnes GR, Bolker B, Bonebakker L, Gentleman R, Liaw WHA, Lumley T, Maechler M, Magnusson A, Moeller S, Schwartz M, Venables B (2016) gplots: various R programming tools for plotting data. R package version 3.0.1

Weiss SM, Indurkhya N, Zhang T (2010) Fundamentals of predictive text mining. Springer, London

Wickham H (2009) ggplot2: elegant graphics for data analysis. Springer, New York

Wild F (2015) lsa: latent semantic analysis. R package version 0.73.1

Wild F, Stahl C (2007) Investigating unstructured texts with latent semantic analysis. In: Decker R, Lenz HJ (eds) Advances in data analysis. Proceedings of the 30th annual conference of the Gesellschaft für Klassifikation e.V., Freie Universität Berlin, 8–10 March 2006. Springer, Berlin, pp 383–390

Wong SKM, Ziarko W, Wong PCN (1985) Generalized vector spaces model in information retrieval. In: Proceedings of the 8th annual international ACM SIGIR conference on research and development in information retrieval, SIGIR '85. ACM, New York, pp 18–25

Chapter 17
A Tutorial on `Libra`: R Package for the Linearized Bregman Algorithm in High-Dimensional Statistics

Jiechao Xiong, Feng Ruan, and Yuan Yao

Abstract The R package, `Libra`, stands for the LInearized BRegman Algorithm in high-dimensional statistics. The Linearized Bregman Algorithm is a simple iterative procedure which generates sparse regularization paths of model estimation. This algorithm was firstly proposed in applied mathematics for image restoration, and is particularly suitable for parallel implementation in large-scale problems. The limit of such an algorithm is a sparsity-restricted gradient descent flow, called the Inverse Scale Space, evolving along a parsimonious path of sparse models from the null model to overfitting ones. In sparse linear regression, the dynamics with early stopping regularization can provably meet the unbiased oracle estimator under nearly the same condition as LASSO, while the latter is biased. Despite its successful applications, proving the consistency of such dynamical algorithms remains largely open except for some recent progress on linear regression. In this tutorial, algorithmic implementations in the package are discussed for several widely used sparse models in statistics, including linear regression, logistic regression, and several graphical models (Gaussian, Ising, and Potts). Besides the simulation examples, various applications are demonstrated, with real-world datasets such as diabetes, publications of COPSS award winners, as well as social networks of two Chinese classic novels, Journey to the West and Dream of the Red Chamber.

J. Xiong
Peking University, School of Mathematical Sciences, Beijing, China
e-mail: xiongjiechao@pku.edu.cn

F. Ruan
Stanford University, Department of Statistics, Stanford, CA, USA
e-mail: fengruan@stanford.edu

Y. Yao (✉)
Hong Kong University of Science & Technology, Clear Water Bay, Hong Kong

Peking University, Beijing, China
e-mail: yuany@ust.hk

© Springer International Publishing AG, part of Springer Nature 2018

W. K. Härdle et al. (eds.), *Handbook of Big Data Analytics*, Springer Handbooks of Computational Statistics, https://doi.org/10.1007/978-3-319-18284-1_17

Keywords Linearized Bregman iteration · LASSO · Variable selection · Regularization path

17.1 Introduction to `Libra`

The free R package, `Libra`, is named as an acronym of the LInearized BRegman Algorithm (also known as Linearized Bregman Iteration in literature). It can be downloaded at

https://cran.r-project.org/web/packages/Libra/index.html

A parsimonious model selection with sparse parameter estimation has been a central topic in high-dimensional statistics in the past two decades. For example, the following models are included in the package:

- sparse linear regression,
- sparse logistic regression (binomial, multinomial),
- sparse graphical models (Gaussian, Ising, Potts).

A widespread traditional approach is based on penalized M-estimators, i.e.

$$\min_{\theta} L(\theta) + \lambda P(\theta), \quad L(\theta) := \frac{1}{n}\sum_{i=1}^{n} l((x_i, y_i), \theta), \tag{17.1}$$

where $l((x_i, y_i), \theta)$ measures the loss of θ at sample (x_i, y_i) and $P(\theta)$ is a sparsity-enforced penalty function on θ such as the l_1-penalty in LASSO (Tibshirani 1996) and the nonconvex SCAD (Fan and Li 2001), etc. However, there are several known shortcomings of this approach: a convex penalty function will introduce bias to the estimators, while a nonconvex penalty, which may reduce the bias, yet suffers the computational hurdle to locate the global optimizer. Moreover, in practice a regularization path is desired which needs to search many optimizers θ_λ over a grid of regularization parameters $\{\lambda_j \geq 0 : j \in \mathbb{N}\}$.

In contrast, the Linearized Bregman (Iteration) Algorithm implemented in `Libra` is based on the following iterative dynamics:

$$\rho^{k+1} + \frac{1}{\kappa}\theta^{k+1} - \rho^k - \frac{1}{\kappa}\theta^k = -\alpha_k \nabla_\theta L(\theta^k), \tag{17.2a}$$

$$\rho^k \in \partial P(\theta^k), \tag{17.2b}$$

with parameters $\alpha_k, \kappa > 0$, and initial choice $\theta^0 = \rho^0 = 0$. The second constraint requires that ρ^k must be a subgradient of the penalty function P at θ^k. The iteration above can be restated in the following equivalent form with the aid of proximal map,

$$z^{k+1} = z^k - \alpha_t \nabla_\theta L(\theta^k), \tag{17.3a}$$

$$\theta^{k+1} = \kappa \cdot \mathrm{prox}_P(z^{k+1}), \tag{17.3b}$$

where the proximal map associated with the penalty function P is given by

$$\mathrm{prox}_P(z) = \arg\min_u \left(\frac{1}{2}\|u - z\|^2 + P(u) \right).$$

The Linearized Bregman Iteration (17.2) generates a parsimonious path of sparse estimators, θ^t, starting from a null model and evolving into dense models with different levels of sparsity until reaching overfitting ones. Therefore the dynamics can be viewed as regularization paths. Such an iterative algorithm was first introduced in Yin et al. (2008) (Section 5.3, Equations (5.19) and (5.20)) as a scalable algorithm for large-scale problems of image restoration with TV-regularization and compressed sensing, etc. As $\kappa \to \infty$ and $\alpha_t \to 0$, the iteration has a limit dynamics, known as Inverse Scale Space (ISS) (Burger et al. 2005) describing its evolution direction from the null model to full ones,

$$\frac{d\rho(t)}{dt} = -\nabla_\theta L(\theta(t)), \tag{17.4a}$$

$$\rho(t) \in \partial P(\theta(t)). \tag{17.4b}$$

The computation of such ISS dynamics is discussed in Burger et al. (2013). With the aid of ISS dynamics, recently Osher et al. (2016) establish the model selection consistency for early stopping regularization in both ISS and Linearized Bregman Iterations for the basic linear regression models. In particular, under nearly the same conditions as LASSO, ISS finds the oracle estimator which is bias-free while the LASSO is biased. However, it remains largely open to explore the statistical consistency for general loss and penalty functions, despite successful applications of (17.2) in a variety of fields such as image processing and statistical modeling that will be illustrated below. One purpose of writing this tutorial is the hope that more statisticians will benefit from the usage of this simple algorithm with the aid of this R package, Libra, and eventually reach a deep understanding of its statistical nature.

In the sequel we shall consider two types of parameters, (θ_0, θ), where θ_0 denotes the unpenalized parameters (usually intercept in the model) and θ represents all the penalized sparse parameters. Correspondingly, $L(\theta_0, \theta)$ denotes the Loss function. In most cases, $L(\theta_0, \theta)$ is the same as the negative log-likelihood function of the model.

Two types of sparsity-enforcement penalty functions will be studied here:

- LASSO (l_1) penalty for entry-wise sparsity:

$$P(\theta) = \|\theta\|_1 := \sum_j |\theta_j|;$$

- Group LASSO (l_1-l_2) penalty for group-wise sparsity:

$$P(\theta) = \|\theta\|_{1,2} = \sum_g \|\theta_g\|_2 := \sum_g \sqrt{\sum_{j:g_j=g} \theta_j^2},$$

where we use $\mathcal{G} = \{g_j : g_j \text{ is the group of } \theta_j, j = 1, 2, \ldots, p\}$ to denote a disjoint partition of the index set $\{1, 2, \ldots, p\}$—that is, each group g_j is a subset of the index set. When $\mathcal{G}$ is degenerate, i.e., $g_j = j, j = 1, 2, \ldots, p$, the Group Lasso penalty is the same as the LASSO penalty. The proximal map for Group LASSO penalty is given by

$$\mathrm{prox}_{\|\theta\|_{1,2}}(z)_j := \begin{cases} \left(1 - \frac{1}{\sqrt{\sum_{i:g_i=g_j} z_i^2}}\right) z_j, & \|z_{g_j}\|_2 \geq 1, \\ 0, & \text{otherwise}, \end{cases} \tag{17.5}$$

which is also called the **Shrinkage** operator in literature.

When entry-wise sparsity is enforced, the parameters to be estimated in the models are encouraged to be "sparse" and treated independently. On the other hand, when group-wise sparsity is enforced, it not only encourages the estimated parameters to be sparse, but also expects variables within the same group to be either selected or not selected simultaneously. Hence, the group-wise sparsity requires prior knowledge of the group information of the correlated variables.

Once the parameters (θ_0, θ), the loss function and group vectors are specified, the Linearized Bregman Iteration algorithm in (17.2) or (17.3) can be adapted to the new setting with partial sparsity-enforcement on θ, as shown in Algorithm 1. The iterative dynamics compute a regularization path for the parameters at different levels of sparsity—starting from the null model with $(\theta_0, 0)$, the solution evolves along a path of sparse models into the dense ones minimizing the loss.

Algorithm 1: Linearized Bregman Algorithm

1 **Input:** Loss function $L(\theta_0, \theta)$, group vector $\mathcal{G}$, damping factor κ, step size α.
2 **Initialize:** $k = 0, t^k = 0, \theta^k = 0, z^k = 0, \theta_0^k = \arg\min_{\theta_0} L(\theta_0, 0)$.
3 **for** $k = 1, \ldots, K$ **do**

- $z^{k+1} = z^k - \alpha \nabla_\theta L(\theta_0^k, \theta^k)$.
- $\theta^{k+1} = \kappa \cdot \mathbf{Shrinkage}(z^{k+1}, \mathcal{G})$.
- $\theta_0^{k+1} = \theta_0^k - \kappa\alpha \nabla_{\theta_0} L(\theta_0^k, \theta^k)$.
- $t^{k+1} = (k+1)\alpha$.

end for
4 **Output:** Solution path $\{t^k, \theta_0^k, \theta^k\}_{k=0,1,\ldots,K}$.

where $\theta = \mathbf{Shrinkage}(z, \mathcal{G})$ is defined as: $\theta_j = \mathbf{max}\left(0, 1 - \frac{1}{\sqrt{\sum_{i:g_i=g_j} z_i^2}}\right) z_j$.

In the following Sects. 17.2–17.4, we shall specialize such a general algorithm in linear regression, logistic regression, and graphical models, respectively. Section 17.5 includes a discussion on some universal parameter choices. Application examples will be demonstrated along with code snippets.

17.2 Linear Model

In this section, we are going to show how the Linearized Bregman (LB) algorithm and the Inverse Scale Space (ISS) fit sparse linear regression model. Suppose we have some covariates $x_i \in \mathbb{R}^p$ for $i = 1, 2, \ldots, n$. The responses y_i with respect to x_i, where $i = 1, 2, \ldots, n$, are assumed to follow the linear model below:

$$y_i = \theta_0 + x_i^T \theta + \epsilon, \quad \epsilon \sim \mathcal{N}(0, \sigma^2).$$

Here, we allow the dimensionality of the covariates p to be either smaller or greater than the sample size n. Note that, in the latter case, we need to make additional sparsity assumptions on θ in order to make the model identifiable (and also, make recovery of θ possible). Both the Linearized Bregman Algorithm and ISS compute their own "regularization paths" for the (sparse) linear model. The statistical properties for the two regularization paths for linear models are established in Osher et al. (2016) where the authors show that under some natural conditions, some points on the paths determined by a data-dependent early-stopping rule can be nearly unbiased and exactly recover the support of signal θ. Note that the latter exact recovery of signal support can have a significant meaning in the regime where $p \gg n$, in which case, an exact variable selection work is done simultaneously with the model fitting process. In addition, the computational cost for regularization path generated by LB algorithm is relatively cheap in linear regression model case, compared to many other existing methods. We refer the readers to Osher et al. (2016) for more details. Own to both statistical and computational advantages over other methods, the Linearized Bregman Algorithm is strongly recommended for practitioners, especially for those who are dealing with computationally heavy tasks.

Here, we give a more detailed illustration on how the Linearized Bregman Algorithm computes the solution path for the linear model. We use negative log-likelihood as our loss function,

$$L(\theta_0, \theta) = \frac{1}{2n} \sum_{i=1}^{n} (y_i - \theta_0 - x_i^T \theta)^2.$$

To compute the regularization path, we need to compute the gradient of loss with respect to its parameters θ_0 and θ, as is shown in Algorithm 1,

$$\nabla_{\theta_0} L(\theta_0, \theta) = \frac{1}{n}\sum_{i=1}^{n} -(y_i - \theta_0 - x_i^T\theta),$$

$$\nabla_{\theta} L(\theta_0, \theta) = \frac{1}{n}\sum_{i=1}^{n} -x_i(y_i - \theta_0 - x_i^T\theta).$$

In the linear model, each iteration of the Linearized Bregman Algorithm requires $O(np)$ FLOPs in general (and the cost can be cheaper if additional sparsity structure on parameters is known), and the overall time complexity for the entire regularization path is $O(npk)$, where k is the number of iterations. The number of iterations in the Linearized Bregman Algorithm is dependent on the underlying step-size α, which can be understood as the counterpart of the learning rate which appears in the standard gradient descent algorithms. In general, choosing the parameter α needs a trade off between statistical and computational concerns here. For example, with a large learning rate α, the Linearized Bregman Algorithm can generate a "coarse" regularization path in only a few iterations. But such a "coarse" solution path might be highly biased since it cannot approximate well the continuous solution path of ISS; hence with only a few points on the path, users may not be able to recover the true support of the unknown signal θ from these coarse estimates. On the other hand, a "denser" solution path generated by low learning rate α provides more information about the true signal θ, yet it might lose some computational efficiency of the algorithm itself.

In addition to the parameter α, another parameter κ is needed in the algorithm. As $\kappa \to \infty$ and $\alpha \to 0$, the Linearized Bregman Algorithm (17.2) will converge to its limit ISS (17.4). Therefore, with a higher value of κ, the Linearized Bregman Algorithm will have a stronger effect on "debiasing" the path, and hence will give a better estimate of the underlying signal at a cost of possible high variance. Moreover, the parameters α and κ need to satisfy

$$\alpha\kappa\|S_n\| \leq 2, \quad S_n = \frac{1}{n}\sum_{i=1}^{n} x_i x_i^T, \tag{17.6}$$

otherwise the Linearized Bregman iterations might oscillate and suffer numerical convergence issues (Osher et al. 2016). Therefore in practice, one typically first chooses κ which might be large enough, then follows a large enough α according to (17.6). In this sense, κ is the essential free parameter.

Knowing how the Linearized Bregman Algorithm works in the linear model, we are ready to introduce the command in `Libra` which can be used to generate the path,

```
lb(X, y, kappa, alpha, tlist, family = "gaussian", group, index)
```

In using the command above, the user *must* give inputs for the design matrix $X \in \mathbb{R}^{n\times p}$, the response vector $y \in \mathbb{R}^n$, and the parameter kappa. Notably, the parameter alpha is not required to be given in the use of this command, and in the case when it's missing, an internal value for alpha satisfying (17.6) would be used. This internally-generated alpha would guarantee the convergence of the algorithm. The tlist is a group of parameters t that determine the output of the above command. When the tlist is given, only points at the pre-decided set of tlist on the regularization path will be returned. When it is missing, then a data dependent tlist will be calculated. See Sect. 17.5 for more details on the tlist. Finally, when group sparsity is considered, the user needs to input an additional argument index to the algorithm so that it knows the group information on the covariates.

As the limit of Linearized Bregman iterations when $\kappa \to \infty, \alpha \to 0$, the Inverse Scale Space for linear model with l_1-penalty is also available in our Libra package:

$$\texttt{iss}(\texttt{X}, \texttt{y}, \texttt{intercept} = \texttt{TRUE}, \texttt{normalize} = \texttt{TRUE}).$$

As is suggested by the previous discussion on the effect of κ on the regularization path, the ISS has the strongest power of "debiasing" the path; once the model selection consistency is reached, it can return the "oracle" unbiased estimator! Yet one disadvantage of ISS solution path is its relative computational inefficiency compared to the Linearized Bregman Algorithm.

17.2.1 Example: Simulation Data

Here is the example in Osher et al. (2016). A comparison of regularization paths generated by LASSO, ISS, and the Linearized Bregman iterations is shown in Fig. 17.1.

```
library(MASS)
library(lars)
library(Libra)
n = 80;p = 100;k = 30;sigma = 1
Sigma = 1/(3*p)*matrix(rep(1,p^2),p,p)
diag(Sigma) = 1
A = mvrnorm(n, rep(0, p), Sigma)
u_ref = rep(0,p)
supp_ref = 1:k
u_ref[supp_ref] = rnorm(k)
u_ref[supp_ref] = u_ref[supp_ref]+sign(u_ref[supp_ref])
b = as.vector(A%*%u_ref + sigma*rnorm(n))
lasso = lars(A,b,normalize=FALSE,intercept=FALSE,max.steps=100)
par(mfrow=c(3,2))
matplot(n/lasso$lambda, lasso$beta[1:100,], xlab = bquote(n/
    lambda),
```

```
  ylab = "Coefficients", xlim=c(0,3),ylim=c(range(lasso$beta)),
    type='l', main="Lasso")
object = iss(A,b,intercept=FALSE,normalize=FALSE)
plot(object,xlim=c(0,3),main=bquote("ISS"))
kappa_list = c(4,16,64,256)
alpha_list = 1/10/kappa_list
for (i in 1:4){
  object <- lb(A,b,kappa_list[i],alpha_list[i],family="gaussian",
    group=FALSE,
           trate=20,intercept=FALSE,normalize=FALSE)
  plot(object,xlim=c(0,3),main=bquote(paste("LB ",kappa,"=",.(
    kappa_list[i]))))
}
```

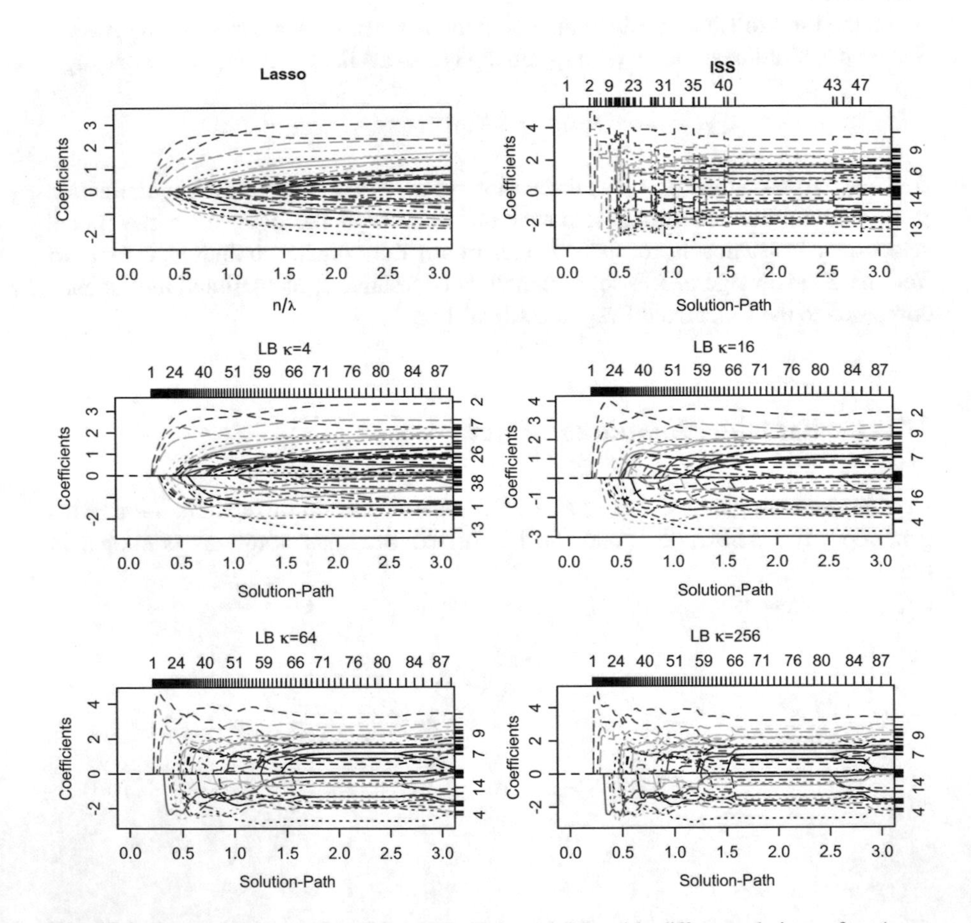

Fig. 17.1 Regularization paths of LASSO, ISS, and LB with different choices of κ ($\kappa = 2^2, 2^4, 2^6, 2^8$, and $\alpha\kappa = 1/10$). As κ grows, the paths of Linearized Bregman iterations approach that of ISS. The x-axis is t

17.2.2 Example: Diabetes Data

A diabetes dataset is used as an example in Efron et al. (2004) to illustrate the `lars` algorithm. The dataset contains 442 samples (diabetes patients) with 10 baseline variables. Here, we show the solution paths of both the Linearized Bregman Algorithm and ISS on the data, assuming a sparse linear regression model between the baseline variables and the response. The LASSO regularization path is computed by R-package `lars`. Figure 17.2 shows the comparison of different paths. It can be seen that the LASSO path is continuous, while the ISS path is piece-wise constant exhibiting the strong "debiasing" effect. The paths generated by discrete Linearized Bregman iterations somehow lie between them. It is easy to see the sudden "shocks" in the figure when the variables are picked up in the regularization path of the ISS or in the paths of Linearized Bregman iterations with large κ. These "shocks" correspond to the stronger debiasing effect of the Linearized Bregman Algorithm and ISS compared to LASSO. Hence our algorithm can fit the signals more "aggressively" compared to the LASSO when we use strong regularization. Although the curve shapes of these paths are different, it is noticeable that the order of those paths entering into nonzero regimes bears great similarity, which implies that the model selection effects of these algorithms are similar on this dataset.

```
library(lars)
library(Libra)
data(diabetes)
attach(diabetes)

lasso <- lars(x,y)
par(mfrow=c(2,2))
plot(lasso)

issobject <- iss(x,y)
plot(issobject,xtype="norm")  #plot.lb
title("ISS",line = 2.5)

kappa <- c(100,500)
for (i in 1:2){
    object <- lb(x,y,kappa[i],family="gaussian",trate=1000)
    plot(object,xtype="norm")
    title(paste("LBI:kappa =",kappa[i]),line = 2.5)
}
detach(lasso)
```

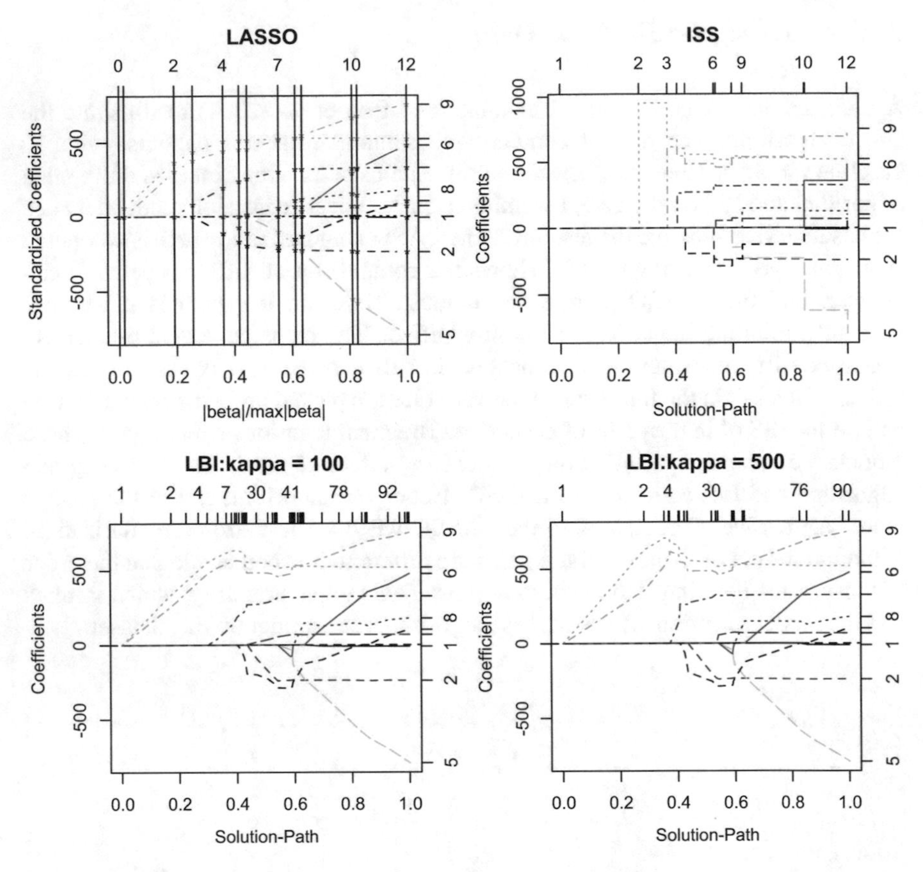

Fig. 17.2 Regularization paths of LASSO, ISS, and Linearized Bregman Iterations on diabetes data. The piecewise constant regularization path of ISS exhibits strong debiasing effect. The ordering of how the variables enter in nonzero regimes is similar across different paths. The x-axis is $\|\theta\|_1$

17.3 Logistic Model

17.3.1 Binomial Logistic Model

One of the mostly widely used model in binary classification is the binomial logistic model, see Hastie et al. (2009). Given i.i.d data $(x_i, y_i) \in \mathbb{R}^p \times \{\pm 1\}$, the standard binomial logistic model assumes the following predictive relationship between the covariates $x_i \in \mathbb{R}^p$ and their response $y_i \in \{\pm 1\}$ for $i = 1, 2, \ldots, n$:

$$\frac{P(y_i = 1|x_i)}{P(y_i = -1|x_i)} = \exp(\theta_0 + x_i^T \theta),$$

where, in the above equation, $\theta \in \mathbb{R}^p$ represents the regression coefficients and $\theta_0 \in \mathbb{R}$ represents the intercept in the regression model. Here, we allow the dimensionality p to be greater than or equal to the sample size n. As is discussed in the linear regression case, when $p > n$, additional sparsity assumptions on the regression coefficient θ should be enforced to make the logistic model identifiable from the data (and also, recovery of the parameters θ possible). The goal of this section is to show how the Linearized Bregman Algorithm fits the sparse binomial logistic regression model in high dimensions. An early version of the Linearized Bregman iterations was implemented in Shi et al. (2013), which differs from Algorithm 1 mainly in its zero initialization. In contrast to Algorithm 1 where we exploit an optimal choice of θ_0, Shi et al. (2013) initialize $\theta_0 = 0$. See more discussions on initializations in Sect. 17.5.

As is discussed similarly in the linear regression case, a regularization path is returned by the Linearized Bregman Algorithm, where practitioners can find different estimates of the same parameters under different levels of sparsity assumptions on the true parameter θ. To give a more detailed illustration on how the Linearized Bregman Algorithm computes the regularization path, we first introduce the loss function of the algorithm, which is given by the negative log-likelihood of the binomial model:

$$L(\theta_0, \theta) = \frac{1}{n}\sum_{i=1}^{n} \log(1 + \exp(-y_i(\theta_0 + x_i^T\theta))).$$

To compute the regularization path, the Linearized Bregman Algorithm 1 needs to evaluate the derivatives of the loss function with respect to θ and θ_0 for each of the iteration point in the path,

$$\nabla_{\theta_0} L(\theta_0, \theta) = \frac{1}{n}\sum_{i=1}^{n} \frac{-y_i}{1 + \exp(y_i(\theta_0 + x_i^T\theta))},$$

$$\nabla_{\theta} L(\theta_0, \theta) = \frac{1}{n}\sum_{i=1}^{n} \frac{-y_i x_i}{1 + \exp(y_i(\theta_0 + x_i^T\theta))}.$$

In binomial logistic model, each iteration of the Linearized Bregman Algorithm requires $O(np)$ FLOPS in general, and the overall time complexity for the entire solution path is $O(npk)$, where k is the number of iterations.

Here, we give the command in `Libra` that can be used to generate the path for the logistic model,

```
lb(X, y, kappa, alpha, tlist, family = "binomial", group, index).
```

As shown in the above command, the user is required to provide data `X`, `y`, as well as the parameters `alpha`, `kappa`, and `tlist`. The effects of these parameters on the resulting regularization paths for binomial logistic model parallel that of the linear model. Hence, we refer the reader to Sect. 17.2 for a detailed explanation on how the parameters affect the regularization paths. Finally, similarly to the case of linear regression, if one needs to enforce a particular group sparsity structure on the output parameters θ, he/she has to input the `index` argument so that the algorithm can know the group information assumption on the covariates.

17.3.1.1 Example: Publications of COPSS Award Winners

The following example explores a statistician publication dataset provided by Professor Jiashun Jin at Carnegie Mellon University (Ji and Jin 2016). The dataset consists of 3248 papers by 3607 authors between 2003 and the first quarter of 2012 from the following four journals: the Annals of Statistics, Journal of the American Statistical Association, Biometrika and Journal of the Royal Statistical Society Series B. Here we extract a subset of 382 papers co-authored by 35 COPSS award winners. Peter Gavin Hall (20 November 1951–9 January 2016) is known as one of the most productive statisticians in history and contributed 82 papers in this dataset. Can we predict the probability of his collaborations with other COPSS award winners? A logistic regression model will be used for this exploration. For a better visualization, we only choose nine other COPSS winners who have no less than 10 papers in this dataset. The following codes compute regularization paths of the Linearized Bregman iterations for logistic regression model to predict the probability of Peter Hall's collaborations with them. From the regularization paths shown in Fig. 17.3, it can be seen that the probability of collaborations between Peter Hall and other COPSS winners are all reduced below the average indicated by the negative coefficients, which suggests that these COPSS winners usually work independently even occasionally coauthor some papers. The three paths which level off as iterations go correspond to Jianqing Fan, Tony Cai, and Raymond J Carroll, who are the only collaborators of Peter Hall in this dataset.

```
library(Libra)
data<-read.table("copss.txt")
s0<-colSums(data)
data1<-data[,s0>=10]    # choose the authors whose publications
    are of no less than 10

y<-as.vector(2*as.matrix(data1[,5])-1); # Peter.Hall as response
X<-as.matrix(2*as.matrix(data1[,-5])-1); # Other COPSS winners as
     predictors
path <- lb(X,y,kappa = 1,family="binomial",trate=100,normalize =
    FALSE)

plot(path,xtype="norm",omit.zeros=FALSE)
title(main=paste("Logistic: ",attributes(data1)$names[5],"~."),
    line=3)
legend("bottomleft", legend=attributes(data1)$names[-5], col=c
    (1:6,1:3),lty=c(1:5,1:4))
```

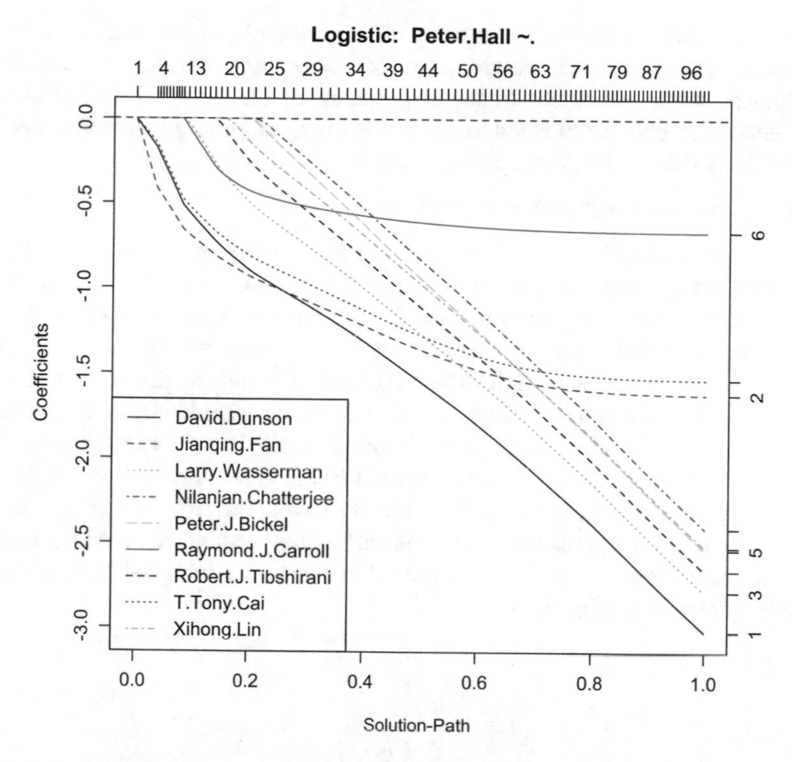

Fig. 17.3 Regularization path of logistic regression by LB on COPSS data. The x-axis is normalized $\|\theta\|_1$. As all the coefficients on the paths appear to be negative, it suggests that the probability of these COPSS award winners collaborating with Peter Hall is below the average in contrast to his fruitful publications. The three paths which level off as $\|\theta\|_1$ grows correspond to Jianqing Fan, Tony Cai, and Raymond J. Carroll, who are the only collaborators of Peter Hall in this dataset

17.3.1.2 Example: Journey to the West

Journey to the West is one of the Four Great Classical Novels of Chinese novel. The literature describes an adventure story about `Tangseng` who travelled to the "West Regions" for Sacred Texts. The novel contains more than a hundred chapters and involves more than a thousand of characters. One interesting study on the novel would be to understand the social relationships between the main characters, i.e., to understand how those with different personalities and power can come along with each other.

Here, we give a simple example showing how the Linearized Bregman Algorithm can be used to analyze the relationship between one main character, `MonkeyKing (Sunwukong)`, to the other main characters. We collect some data that documents the appearance/disappearance of the top 10 main characters under the pre-specified 408 different scenes in the novel. To analyze the relationship between `MonkeyKing` to the other nine main characters, we build up a logistic regression model, where the response Y corresponds to the indicator of the appearance of the MonkeyKing in these scenes and the other covariates X correspond to the indicators of the appearance of the other nine characters in the scenes. The data is collected via crowdsourcing at Peking University, and can be downloaded at the following course website

https://github.com/yuany-pku/journey-to-the-west.

Below we analyze the result of the logistic regression model fitted by the Linearized Bregman Algorithm. Notice that, Tangseng, Pig (Zhubajie), and FriarSand (Shaseng) are the first three main characters that are picked up in the regularization path. In addition, the coefficients of their corresponding covariates are all positive, meaning that they probably show up the same time as the MonkeyKing in the story. A combination of the above two phenomena is explained by the fact that in the novel they together with MonkeyKing (Sunwukong) form the fellowship of the journey to the west. On the other hand, Yuhuangdadi, Guanyinpusa, and Muzha are less involved with the MonkeyKing, as they didn't show up in the paths until very late stages, with estimated coefficients being negative, indicating that they just appeared occasionally with the MonkeyKing when he got troubles (Fig. 17.4).

```
library(Libra)
data(west10)
y<-2*west10[,1]-1;
X<-as.matrix(2*west10[,2:10]-1);

path <- lb(X,y,kappa = 1,family="binomial",trate=100,normalize =
    FALSE)
plot(path,xtype="norm",omit.zeros=FALSE)
title(main=paste("Logistic",attributes(west10)$names[1],"~."),
    line=3)
legend("bottomleft", legend=attributes(west10)$names[-1], col=c
    (1:6,1:3),lty=c(1:5,1:4))
```

17.3.2 Multinomial Logistic Model

Multinomial logistic regression is a method which generalizes the binary logistic model to multi-class classification problems, where the response y has $K(\geq 2)$ different outcomes (Hastie et al. 2009). The model assumes the following relationship between the response $y \in \{1, 2, \ldots, K\}$ and its covariate $x \in \mathbb{R}^p$:

$$P(y = k|x) = \frac{\exp(\theta_{k0} + x^T\theta_k)}{\sum_{k=1}^{K} \exp(\theta_{k0} + x^T\theta_k)}$$

As discussed in the previous sections, often additional sparsity assumptions on the coefficients θ_k for $k = 1, 2, \ldots, K$ are added by researchers to make the model

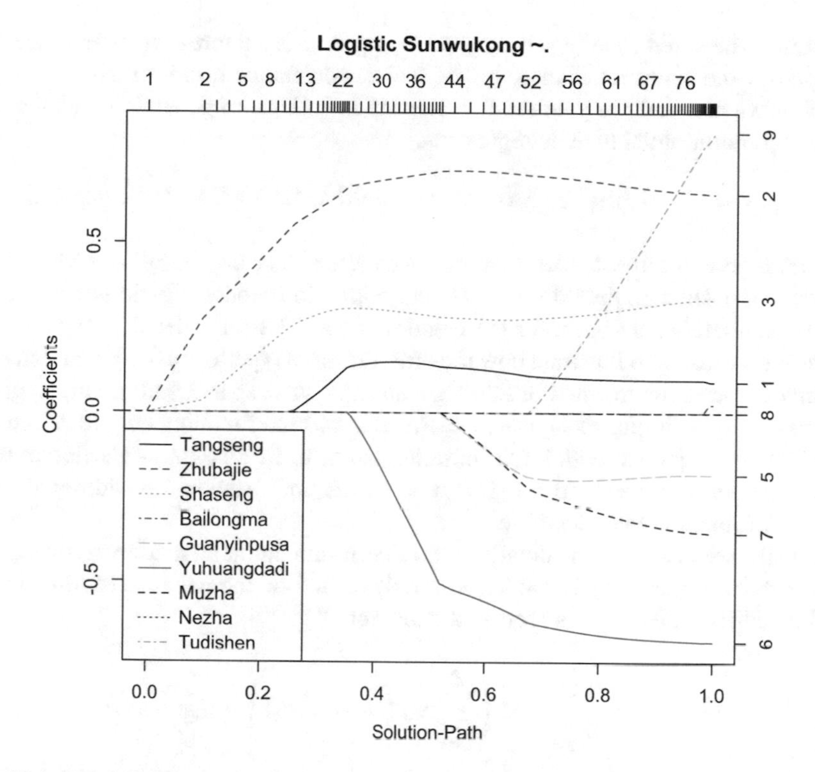

Fig. 17.4 Regularization path of `lb` on `west10` data using `family` = `"binomial"`. The fellowship of the journey to the west is formed by `Sunwukong` (`MonkeyKing`) and his three peers: `Tangseng`, `Zhubajie`, and `Shaseng`, corresponding to the first three paths

more identifiable/more interpretable in high dimensions. Usually, researchers can have different prior beliefs on the underlying sparsity structure of the models, and these different types of sparsity structures correspond to different types of sparse multinomial logistic regression model. In our package, we consider three major variants of the original multinomial logistic model, i.e., the *entry-wise* sparse, the *column-wise* sparse, and the *block-wise* sparse multinomial logistic model. The entry-wise sparse model corresponds to adding an LASSO (l_1) penalty on all the parameters θ_k for $k = 1, 2, \ldots, K$. The column-wise sparsity corresponds to adding a more complicated group LASSO penalty on each column group of parameters θ_k, $\sum_{j=1}^{p} \sqrt{\sum_{k=1}^{K} \theta_{kj}^2}$. Since each column of θ corresponds to a feature x_i for some $1 \leq i \leq p$, getting column-wise sparse estimates will select the same set of features for different response classes simultaneously. Finally, as a generalization of the previous group sparse model, the block-wise sparse model assumes an additional group structure on the coefficients θ, and penalizes our model through the following block-wise penalty $\sum_g \sqrt{\sum_{k=1}^{K} \sum_{j:g_j=g} \theta_{kj}^2}$. Similar to the column-wise sparse model, the block-wise sparse model does feature selection for all response classes at the same time, yet it may select a group of features together instead of singletons and hence relies more on the feature correlation group structure.

Now we are ready to give the R command in `Libra` to generate regularization paths for multinomial logistic regression.

```
lb(X, y, kappa, alpha, tlist, family="multinomial", group, index)
```

We note here for the reader that the parameters `alpha`, `kappa`, and `tlist` function the same as they do in the linear regression model. We do not introduce these parameters here but refer the reader to Sect. 17.2 for a detailed explanation. Now, we are going to illustrate how the three different types of sparsity structures on parameters are implemented in R. To get an entry-wise sparse multinomial logistic regression, one simply sets `group` = `FALSE`, and the function `lb` will return the solution path for this model. On the other hand, to fit a column-wise/block-wise sparse model, one needs to set `group` = `TRUE` and provide the additional prior group information when possible.

Finally, we discuss some details of the algorithmic implementation in solving the sparse multinomial logistic model. Similarly to before, the negative log-likelihood of the multinomial model is used as the loss function:

$$L(\theta_0, \theta) = \frac{1}{n}\sum_{i=1}^{n}\log\left(\sum_{k=1}^{K}\exp(\theta_{k0} + x_i^T\theta_k)\right) - \theta_{y_i0} - x_i^T\theta_{y_i}$$

One can compute the derivatives of the above loss function with respect to its parameters:

$$\nabla_{\theta_{j0}}L(\theta_0, \theta) = \frac{1}{n}\sum_{i=1}^{n}\frac{\exp(\theta_{j0} + x_i^T\theta_j)}{\sum_{k=1}^{K}\exp(\theta_{k0} + x_i^T\theta_k)} - 1(y_i = j),$$

$$\nabla_{\theta_j}L(\theta_0, \theta) = \frac{1}{n}\sum_{i=1}^{n}\frac{\exp(\theta_{j0} + x_i^T\theta_j)x_i}{\sum_{k=1}^{K}\exp(\theta_{k0} + x_i^T\theta_k)} - x_i 1(y_i = j).$$

Therefore, the computational complexity for each iteration of the Linearized Bregman Algorithm is of $O(npK)$ FLOPs.

17.4 Graphical Model

Undirected graphical models, also known as Markov random fields, have many applications in different fields including statistical physics (Ising 1925), nature language processing (Manning and Schütze 1999), image analysis (Hassner and Sklansky 1980), etc. A Markov random field models the joint probability distribution of set random variables $\{X_v\}$, where the subscript v belongs to some set V, by some undirected graph $G = (V, E)$, where $E \in \{0,1\}^{V\times V}$ denotes the edges among V that determine the (conditional) independence between subsets of random variables of $\{X_v\}_{v\in V}$. In this section, we introduce three types of undirected graphical models implemented in `Libra`: Gaussian Graphical Models, Ising Models, and Potts Models.

17.4.1 Gaussian Graphical Model

The Gaussian graphical model assumes the data $x \in \mathbb{R}^p$ follow a joint normal distribution $\mathcal{N}(\mu, \Theta^{-1})$, where Θ is a sparse p-by-p inverse covariance (precision) matrix which encodes the conditional independence relations between variables, i.e. $\{x_i \perp x_j : x_{\{-i,-j\}}\} \Leftrightarrow \Theta_{ij} = 0$. Note that θ_0 here is the diagonal of Θ which is not penalized and the sparse parameter θ contains the off-diagonal elements.

Graphical LASSO (Friedman et al. 2008) exploits the maximum likelihood estimate with l_1 regularization on θ. However the gradient of Gaussian likelihood with respect to θ involves matrix inverse and is thus not good for implementing the Linearized Bregman Algorithm. To avoid this issue, here we exploit the composite conditional likelihood as the loss function.

It is easy to calculate the distribution of x_j conditional on x_{-j} is also a normal distribution:

$$x_j | x_{-j} \sim \mathcal{N}\left(\mu_j - \sum_{k\neq j} \frac{\Theta_{jk}}{\Theta_{jj}}(x_k - \mu_k), \frac{1}{\Theta_{jj}}\right)$$

For simplicity assume that the data is centralized, then the composite conditional likelihood becomes

$$L(\Theta) = \sum_j^p \frac{1}{n} \sum_{i=1}^n \frac{\Theta_{jj}}{2} \left(x_{i,j} + \sum_{k\neq j} \frac{\Theta_{jk}}{\Theta_{jj}} x_{i,k}\right)^2 - \frac{1}{2}\log(\Theta_{jj}).$$

or equivalently,

$$L(\Theta) = \sum_j \frac{1}{2\Theta_{jj}} \Theta_{\cdot j}^T S \Theta_{\cdot j} - \frac{1}{2}\log(\Theta_{jj})$$

where $S = \frac{1}{n}\sum_{i=1}^n x_i x_i^T$ is the covariance matrix of data. Such a loss function is convex.

The corresponding gradient is defined by

$$\nabla_{\theta_{jj}} L(\Theta) = \frac{1}{\Theta_{jj}} S_{j\cdot}\Theta_{\cdot j} - \frac{1}{2\Theta_{jj}^2}\Theta_{\cdot j}^T S \Theta_{\cdot j} - \frac{1}{2\Theta_{jj}}$$

$$\nabla_{\theta_{jk}} L(\Theta) = \frac{1}{\Theta_{jj}} S_{k\cdot}\Theta_{\cdot j} + \frac{1}{\Theta_{kk}} S_{j\cdot}\Theta_{\cdot k},$$

and the computation of gradient is $O(\min(p^3, np^2))$.

The Libra command to estimate the Gaussian Graphical Model is

$$\texttt{ggm(X, kappa, alpha, S, tlist, nt = 100, trate = 100)}$$

where X is the data matrix. If X is missing, the covariance matrix S should be provided. Moreover nt is the number of models on solution path which decides the length of tlist and trate $:= t_{\max}/t_{\min}$ as the scale span of t. Their choices are further discussed in Sect. 17.5.

17.4.1.1 Example: Journey to the West

Here we demonstrate the application of function ggm to the same dataset west10 which was introduced before. We choose a particular model at sparsity level 51% and plot it in Fig. 17.5 against the outcome of Graphical LASSO implemented by R package huge (Zhao and Liu 2012). It can be seen that the resulting graphs bear a globally similar sparsity pattern with several distinct edges.

```
library(Libra)
library(igraph)
library(huge)
library(clime)
data(west10)

X <- as.matrix(2*west10-1);
obj = ggm(X,1,alpha = 0.01,nt=1000,trate=100)
g<-graph.adjacency(obj$path[,,720],mode="undirected",weighted=
    TRUE,diag=FALSE)
E(g)[E(g)$weight<0]$color<-"red"
E(g)[E(g)$weight>0]$color<-"green"
V(g)$name<-attributes(west10)$names
plot(g,vertex.shape="rectangle",vertex.size=35,vertex.label=V(g)
    $name,
edge.width=2*abs(E(g)$weight),main="GGM (LB): sparsity=0.51")

obj2<- huge(as.matrix(west10), method = "glasso")
obj2.select = huge.select(obj2,criterion = "ebic")
g2<-graph.adjacency(as.matrix(obj2.select$opt.icov),mode="plus",
    weighted=TRUE,diag=FALSE)
E(g2)[E(g2)$weight<0]$color<-"red"
```

```
E(g2)[E(g2)$weight>0]$color<-"green"
V(g2)$name<-attributes(west10)$names
plot(g2,vertex.shape="rectangle",vertex.size=35,edge.width=2*abs(
    E(g2)$weight),vertex.label=V(g2)$name,main="Graphical LASSO:
    sparsity=0.51")

obj3<- clime(as.matrix(west10),linsolver = "simplex")
g3<-graph.adjacency(as.matrix(obj3$Omegalist[[70]]),mode="plus",
    weighted=TRUE,diag=FALSE)
E(g3)[E(g3)$weight<0]$color<-"red"
E(g3)[E(g3)$weight>0]$color<-"green"
V(g3)$name<-attributes(west10)$names
plot(g3,vertex.shape="rectangle",vertex.size=35,edge.width=2*abs(
    E(g3)$weight),vertex.label=V(g3)$name,main="CLIME: sparsity
    =0.51")
```

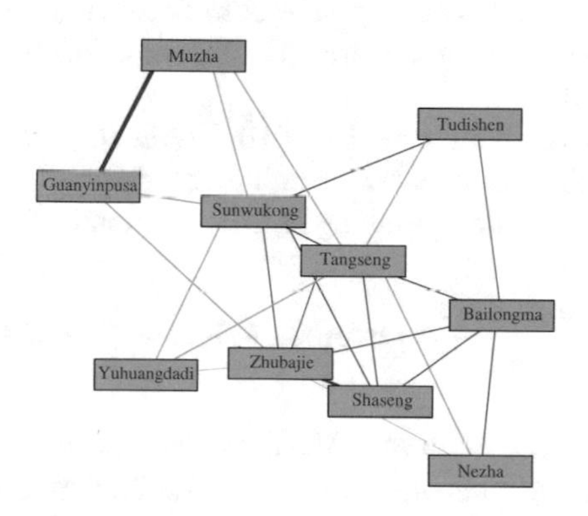

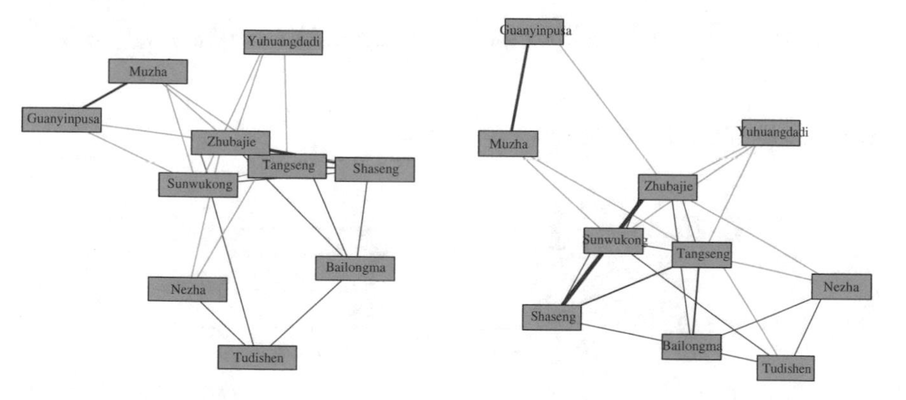

Fig. 17.5 A comparison of sparse Gaussian Graphical models returned by LB for composite conditional likelihood (upper), Graphical LASSO (left) and CLIME (right). Green for the positive coefficients and Red for the negative in the inverse covariance matrix Θ. The width of edge represents the magnitude of coefficients

17.4.2 Ising Model

One important graphical model for *binary* random variables (i.e., $X_v \in \{0, 1\}$ for any $v \in V$) is the Ising model, which specifies the underlying distribution on $\{X_v\}$ by the following Boltzmann distribution:

$$\mathbb{P}(x) = \frac{1}{Z(\theta_0, \theta)} \exp\left(x^T \theta_0 + \frac{1}{2} x^T \theta x\right),$$

In the above equation, $\theta_0 \in \mathbb{R}^{|V|}$ and $\theta \in \mathbb{R}^{|V| \times |V|}$ are the parameters of the Ising model and Z is the normalizing function. (Z is also called the partition function in the literature.) Notably, the nonzero entries of $|V|$ by $|V|$ symmetric matrix $\theta \in \mathbb{R}^{|V| \times |V|}$ correspond to the edge-set E, which determines the dependence structure (conditional independence) between $\{X_v\}$. Therefore, given the data $\{x_i\}_{i=1}^n$, where $x_i \in \{0, 1\}^{|V|}$, the objective of learning here is to determine the support of θ (i.e., the graph structure) and estimate the strength of θ simultaneously (strength of the dependency relationship).

To solve this model, Ravikumar et al. (2010), Xue et al. (2012) etc. suggest using logistic regression by observing that the conditional distribution of X_v given all the other variables X_{-v} satisfies the following logistic distribution,

$$\frac{P(X_v = 1|X_{-v})}{P(X_v = 0|X_{-v})} = \exp(\theta_{v0} + \theta_{v,-v} X_{-v}) \;\; v \in V.$$

To fully utilize all the information from the data while preserving the symmetry of the parameters, we use the following composite conditional likelihood (Xue et al. 2012) as our loss function in `Libra`,

$$L(\theta_0, \theta) = \sum_{v=1}^{|V|} \frac{1}{n} \sum_{i=1}^{n} \log(1 + \exp(\theta_{v0} + \theta_{v,-v} x_{i,-v})) - x_{iv}(\theta_{v0} + \theta_{v,-v} x_{i,-v}).$$

The gradient of the above loss is shown below

$$\nabla_{\theta_{v0}} L(\theta_0, \theta) = \frac{1}{n} \sum_{i=1}^{n} \frac{1}{1 + \exp(-\theta_{v0} - \theta_{v,-v} x_{i,-v})} - x_{iv}$$

$$\begin{aligned} \nabla_{\theta_{v_1 v_2}} L(\theta_0, \theta) = \frac{1}{n} \sum_{i=1}^{n} & \frac{x_{iv_2}}{1 + \exp(-\theta_{v_1 0} - \theta_{v_1,-v_1} x_{i,-v_1})} \\ & + \frac{x_{iv_1}}{1 + \exp(-\theta_{v_2 0} - \theta_{v_2,-v_2} x_{i,-v_2})} - 2 x_{iv_1} x_{iv_2}. \end{aligned}$$

When fitting the Ising model, each iteration of the Linearized Bregman Algorithm requires $O(n|V|^2)$ FLOPS in general, and the overall time complexity for the entire solution path is $O(n|V|^2k)$, where k is the number of iterations.

The command in Libra that can be used to generate the path for the Ising model is

```
ising(X, kappa, alpha, tlist, responses = c(0, 1), nt = 100, trate = 100)
```

The arguments kappa, alpha, and tilst have similar functions to the corresponding arguments in the linear, binomial logistic and multinomial logistic model. Hence, we refer the reader to Sect. 17.2 for detailed explanations of these arguments. There are several arguments specific to the Ising model, e.g. nt is the number of models on the solution path which decides the length of tlist and trate $= t_{\max}/t_{\min}$ is the scale span of t. See Sect. 17.5 for more details on these two arguments. The choice of the argument responses can be either c(0, 1) or c(−1, 1). The choice c(−1, 1) corresponds to the following model formulation, where we instead assume our data x coming from $\{-1, 1\}$ and our distribution on data x having the following specification:

$$P(x) = \frac{1}{Z}\exp\left(\frac{1}{2}x^T h + \frac{1}{4}x^T Jx\right),$$

where $h \in \mathbb{R}^{|V|}$ and $J \in \mathbb{R}^{|V|\times|V|}$. Since such model formulations appear quite often in some scientific fields including computational physics, for convenience, we include Linearized Bregman Algorithm solvers for this type of model in our package. For clarity, we also give the one-to-one correspondence between the two model formulations:

$$\begin{aligned} x_{-1/1} &= 2x_{0/1} - 1, \\ J &= \theta/2, \\ h &= \theta_0 + J\mathbf{1}. \end{aligned}$$

17.4.2.1 Example: Simulation Data

In this section, we give some simulation results that illustrate the performance of the Linearized Bregman Algorithm in solving the Ising model. In our simulation setting, we choose our sample size n to be 5000 and choose our underlying graph G to be the standard 10-by-10 grid (see Fig. 17.6). We set the intercept coefficients h to be 0 for all nodes. Each entry in the interaction matrix J_{jk} is set to be $2/2.3$

whenever j and k are neighbors on the 10-by-10 grid or set to 0 otherwise. The code to reproduce this simulation is shown below.

```
library(Libra)
data(isingdata)
obj = ising(isingdata$X,10,alpha=0.1,trate=30)

TPrate <- rep(0,100)
FPrate <- rep(0,100)
for (i in 1:100){
      TPrate[i] = sum((obj$path[,,i]!=0)&(isingdata$J!=0))
      FPrate[i] = sum((obj$path[,,i]!=0)&(isingdata$J==0))
}
TPrate <- TPrate/sum(isingdata$J!=0)
FPrate <- FPrate/sum(isingdata$J==0)
tmin <- log(obj$t[min(which(TPrate==1))])
tmax <- log(obj$t[max(which(FPrate==0))])

coord = matrix(c(rep(1:10,each=10),rep(1:10,10)),ncol=2)
g<-graph.adjacency(as.matrix(isingdata$J),mode="plus",weighted=
   TRUE,diag=FALSE)
png(file="Grid_true.png", bg="transparent")
plot(g,vertex.shape="circle",vertex.size=10,edge.width=2*abs(E(g)
   $weight),layout=coord)
dev.off()
png(file="Ising_TPFP.png", bg="transparent")
plot(log(obj$t),TPrate,col='red',type='l',lty=1,xlab=expression(
   log(t)),ylab='TPrate & FPrate')
lines(log(obj$t),FPrate,col='blue',type='l',lty=2)
abline(v = c(tmin,tmax),lty=3)
axis(1,at = c(tmin,tmax),labels = c(expression(t[1]),expression(t
   [2])))
legend(x = 3, y = 0.58, lty=1:2,col=c('red','blue'), legend=c('
   TPrate','FPrate'))
dev.off()
```

Figure 17.6 shows the True-Positive-Rate curve and False-Positive-Rate curve along the model path computed by `ising`. There is a segment in the LB path which gives the same sparsity pattern as the ground truth.

17.4.2.2 Example: Journey to the West

In this section, we revisit our example of Sect. 17.3.1.2. In Sect. 17.3.1.2, we analyze the social relationship between the main character `MonkeyKing` and the other nine characters for the classic novel Journey to the West via a single logistic regression. However, such an analysis doesn't take into account the pairwise relationships between the other top nine main characters, and hence without using the joint information among the other nine characters, our estimate of social networking structure may be statistically inefficient. In this section, we are going to jointly estimate the social networking among all ten main characters simultaneously

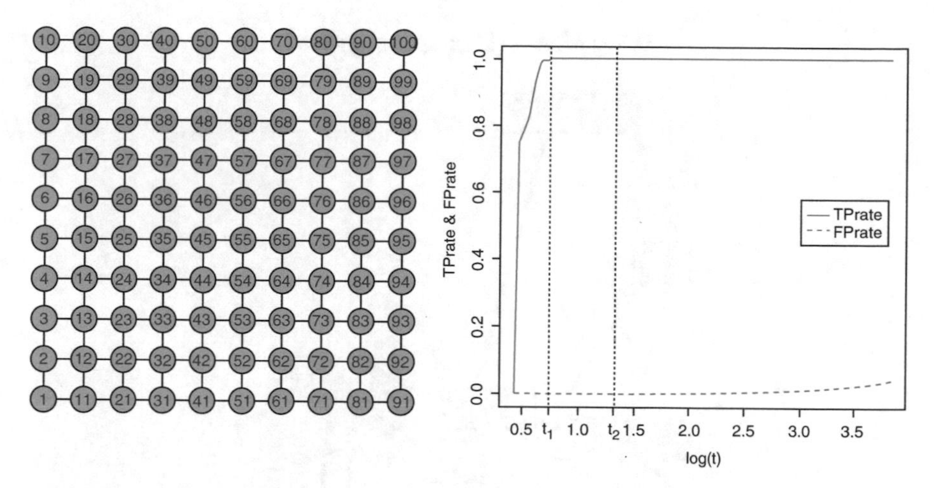

Fig. 17.6 Left: True Grid. Right: TPrate and FPrate vs. $\log(t)$. The path between t_1 and t_2 gives the correct sparsity pattern of models

by applying techniques from graphical models. Note that, this can result in a statistically more efficient estimate of the social networking, compared to the result coming from multiple times of single logistic regressions.

We first consider using an Ising model to model the interaction relationships between the top ten main characters in the novel Journey to the West. Figure 17.7 shows an Ising model estimate at sparsity level of 51%. The Ising model was fit with the command `ising`. Comparing it with the Gaussian graphical model in Fig. 17.5, note that the color of these two types of graphs is almost opposite. This is because there is a negative sign on the exponential term in Gaussian likelihood function, which means a negative interaction coefficient actually increases the probability of co-presence in Gaussian graphical models. Up to the sign difference, the sparsity patterns in all these models are qualitatively similar.

```
library(Libra)
library(igraph)
data(west10)
X <- as.matrix(2*west10-1);
obj = ising(X,10,0.1,nt=1000,trate=100)

g<-graph.adjacency(obj$path[,,770],mode="undirected",weighted=
    TRUE)
E(g)[E(g)$weight<0]$color<-"red"
E(g)[E(g)$weight>0]$color<-"green"
V(g)$name<-attributes(west10)$names
plot(g,vertex.shape="rectangle",vertex.size=35,vertex.label=V(g)
    $name,edge.width=2*abs(E(g)$weight),main="Ising Model (LB):
    sparsity=0.51")
```

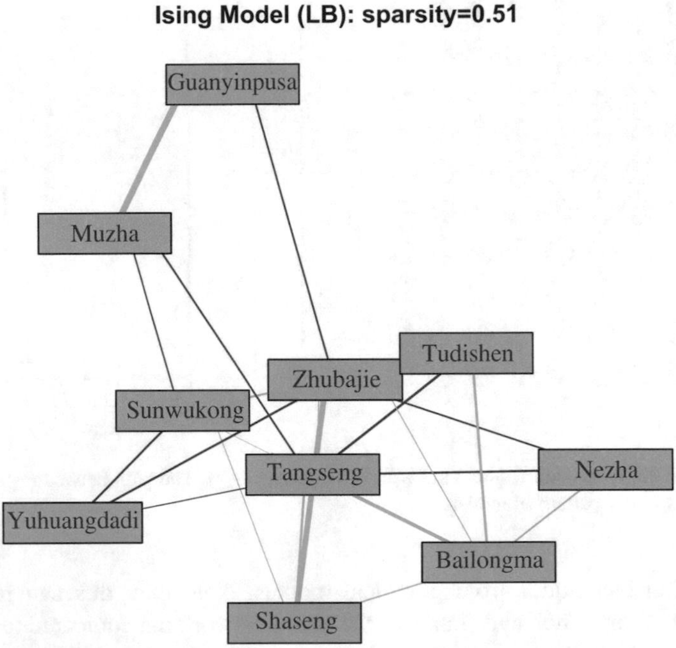

Ising Model (LB): sparsity=0.51

Guanyinpusa
Muzha
Tudishen
Zhubajie
Sunwukong
Tangseng
Nezha
Yuhuangdadi
Bailongma
Shaseng

Fig. 17.7 An Ising model of sparsity level 51% on LB path. Green edges are used for positive coefficients which increases the probability of co-appearance, while red edges are for negative which decreases this probability. The width of edge represents the magnitude of coefficients. Despite that the signs of coefficients are almost opposite compared with Gaussian graphical models, the sparsity patterns in these models are qualitatively similar

17.4.2.3 Example: Dream of the Red Chamber

Dream of the Red Chamber, often regarded as the pinnacle of Chinese fiction, is another one of the Four Great Classical Novels of Chinese Literature, composed by Cao, Xueqin for the first 80 chapters and Gao, E for the remaining 40 chapters. With a precise and detailed observation of the life and social structures typical of eighteenth-century society in Qing Dynasty, the novel describes a tragic romance between `Jia, Baoyu` and `Lin, Daiyu` among other conflicts. Our interest is to study the social network of interactions among the main characters. Our dataset records 375 characters who appear ("1") or do not show up ("0") in 475 events extracted from the 120 chapters. The data is collected via crowdsourcing at Peking University, and can be downloaded at the following course website:

https://github.com/yuany-pku/dream-of-the-red-chamber.

The following R codes give a simple example showing how the Linearized Bregman Algorithm can be used to build up sparse Ising models from the data, focusing on the most frequently appeared 18 characters. To compare the structural difference of the first 80 chapters by Cao, Xueqin and the latter 40 chapters by Gao, E, we run `ising` on two subsets of data to extract two Ising models shown in Fig. 17.8. The links shed light on conditional independence relations among characters learned from data. It is clear that in the first part of the novel, `Jia, Baoyu` has a strong connection with `Lin, Daiyu` and is conditionally independent to another main character `Xue, Baochai` as Cao, Xueqin depicts; while in the second part `Jia, Baoyu` connects to `Xue, Baochai` directly and becomes conditional independent to `Lin, Daiyu` as Gao, E implies. Such a transition is consistent with the split of the novel.

```
library(Libra)
library(igraph)

load("dream.RData")
# Choose the first 80 chapters authored by Cao, Xueqin
data<-dream[dream[,1]>0,]
dim(data)
s0<-colSums(data)
# restrict to the most important characters
data1<-data[,s0>=30]
#Eng_names <- c('Jia, Zheng','Jia, Zhen','Jia, Lian','Jia, Baoyu
    ','Jia, Tanchun','Jia, Rong','Lady Dowager','Shi, Xiangyun','
    Lady Wang','Wang, Xifeng','Aunt Xue','Xue, Baochai','Lin,
    Daiyu','Lady Xing','Madam You','Li, Wan','Xiren','Ping\'er')
p = dim(data1)[2];
X<-as.matrix(2*as.matrix(data1[,2:p])-1);
```

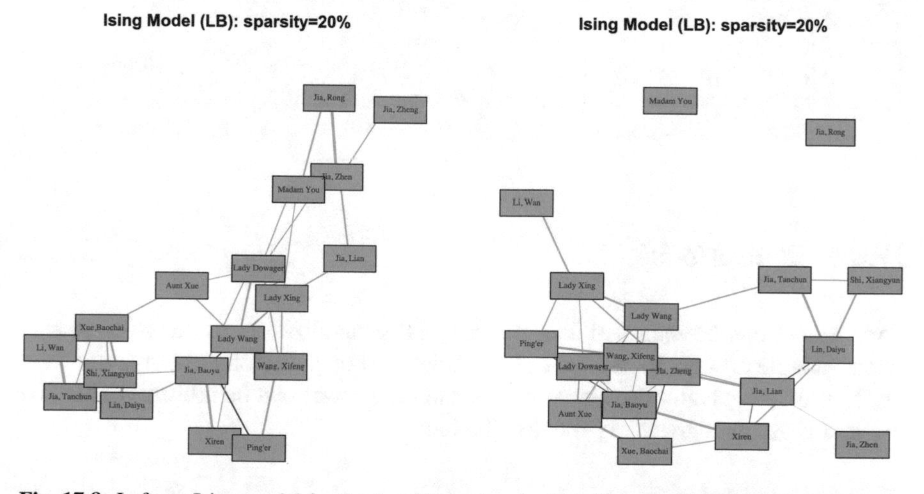

Fig. 17.8 Left: an Ising model for the first 80 chapters by Cao, Xueqin. Right: an Ising model for the remaining chapters by Gao, E. Sparsity levels are all chosen as 20% on LB path. Green edges are used for positive coefficients which increases the probability of co-appearance, while red edges are for negative which decreases this probability. The width of edge represents the magnitude of coefficients. Comparing the two models, one can see that `Jia, Baoyu` has a strong link with `Lin, Daiyu` in the first part, and changes the link to `Xue, Baochai` who becomes his wife in the second part of the novel

```
obj = ising(X,10,0.1,nt=1000,trate=100)
sparsity=NULL
for (i in 1:1000) {sparsity[i]<-(sum(abs(obj$path[,,i])>1e-10))/(
    p^2-p) }

# Choose sparsity=20% at point 373
g<-graph.adjacency(obj$path[,,373],mode="undirected",weighted=
    TRUE)
E(g)[E(g)$weight<0]$color<-"red"
E(g)[E(g)$weight>0]$color<-"green"
V(g)$name<-attributes(data1)$names[2:p]
plot(g,vertex.shape="rectangle",vertex.size=25,vertex.label=V(g)
    $name,edge.width=2*abs(E(g)$weight),vertex.label.family='
    STKaiti',main="Ising Model (LB): sparsity=20%")

# Choose the later 40 chapters authored by Gao, E
data<-dream[dream[,1]<1,]
data2<-data[,s0>=30]
X<-as.matrix(2*as.matrix(data2[,2:p])-1);
obj = ising(X,10,0.1,nt=1000,trate=100)
sparsity=NULL
for (i in 1:1000) {sparsity[i]<-(sum(abs(obj$path[,,i])>1e-10))/(
    p^2-p) }

# Choose sparsity=20% at point 344.
g<-graph.adjacency(obj$path[,,344],mode="undirected",weighted=
    TRUE)
E(g)[E(g)$weight<0]$color<-"red"
E(g)[E(g)$weight>0]$color<-"green"
V(g)$name<-attributes(data2)$names[2:p]
plot(g,vertex.shape="rectangle",vertex.size=25,vertex.label=V(g)
    $name,edge.width=2*abs(E(g)$weight),vertex.label.family='
    STKaiti',main="Ising Model (LB): sparsity=20%")
```

17.4.3 Potts Model

Potts Model can be regarded as a multinomial generalization of the Ising model. Each variable x_j can be a multi-class variable. For simplicity we assume $x \in \{1, 2, \ldots, K\}^p$, actually the class number and class name can be arbitrary. Then the model x is assumed to satisfy the distribution:

$$P(x) = \frac{1}{Z} \exp\left(\sum_{\substack{j=1,\ldots,p \\ s=1,\ldots,K}} \theta_{js,0} 1(x_j = s) + \frac{1}{2} \sum_{\substack{j=1,\ldots,p; s=1,\ldots,K \\ k=1,\ldots,p; t=1,\ldots,K}} \theta_{js,kt} 1(x_j = s) 1(x_k = t) \right)$$

where Z is the normalization factor. The vector of intercept coefficients θ_0 is of length pK and the vector of interaction coefficients θ is a pk-by-pk symmetric matrix with zero diagonal block. So the distribution of x_j conditional on the rest variables x_{-j} satisfies

$$P(x_j = s|x_{-j}) = \frac{\exp(\theta_{js,0} + \sum_{k=1,\ldots,p;t=1,\ldots,K} \theta_{js,kt} 1(x_k = t))}{\sum_{s=1,\ldots,K} \exp(\theta_{js,0} + \sum_{k=1,\ldots,p;t=1,\ldots,K} \theta_{js,kt} 1(x_k = t))}$$

which is actually a multinomial logistic distribution.

So the loss function is defined as the composite conditional likelihood:

$$L(\theta_0, \theta) = \sum_{j=1}^{p} \frac{1}{n} \sum_{i=1}^{n} \log\left(\sum_{s=1,\ldots,K} \exp\left(\theta_{js,0} + \sum_{\substack{k=1,\ldots,p \\ t=1,\ldots,K}} \theta_{js,kt} 1(x_{i,k} = t) \right)\right) + \cdots$$

$$\cdots - \theta_{jx_{i,j},0} - \sum_{\substack{k=1,\ldots,p \\ t=1,\ldots,K}} \theta_{jx_{i,j},kt} 1(x_{i,k} = t)$$

The corresponding gradient is

$$\nabla_{\theta_{js,0}} L(\theta_0, \theta) = \frac{1}{n} \sum_{i=1}^{n} \frac{\exp\left(\theta_{js,0} + \sum_{\substack{k=1,\ldots,p \\ t=1,\ldots,K}} \theta_{js,kt} 1(x_{i,k} = t) \right)}{\sum_{s=1,\ldots,K} \exp\left(\theta_{js,0} + \sum_{\substack{k=1,\ldots,p \\ t=1,\ldots,K}} \theta_{js,kt} 1(x_{i,k} = t) \right)}$$
$$-1(x_{ij} = s)$$

$$\nabla_{\theta_{js,kt}} L(\theta_0, \theta) = \frac{1}{n} \sum_{i=1}^{n} \frac{1(x_{i,k} = t) \exp\left(\theta_{js,0} + \sum_{\substack{k=1,\ldots,p \\ t=1,\ldots,K}} \theta_{js,kt} 1(x_{i,k} = t) \right)}{\sum_{s=1,\ldots,K} \exp\left(\theta_{js,0} + \sum_{\substack{k=1,\ldots,p \\ t=1,\ldots,K}} \theta_{js,kt} 1(x_{i,k} = t) \right)}$$
$$-1(x_{ij} = s, x_{ik} = t)$$
$$+ \frac{1(x_{i,j} = s) \exp\left(\theta_{kt,0} + \sum_{\substack{j=1,\ldots,p \\ s=1,\ldots,K}} \theta_{kt,js} 1(x_{i,j} = s) \right)}{\sum_{t=1,\ldots,K} \exp\left(\theta_{kt,0} + \sum_{\substack{j=1,\ldots,p \\ s=1,\ldots,K}} \theta_{kt,js} 1(x_{i,j} = s) \right)}$$
$$-1(x_{ik} = t, x_{ij} = s)$$

and the computation cost of gradient tanks $O(np^2K^2)$ computations(or $O(np^2)$ if using sparse encoding to represent x).

The function to estimate the Potts model in Libra is

```
potts(X, kappa, alpha, tlist, nt = 100, trate = 100, group = FALSE)
```

The data matrix X is expected to be of size n-by-p, and each column is a class vector (the number of class for each variable can be different). If `group = TRUE`, then the group penalty is used;

$$\sum_{\substack{k=1,\ldots,p \\ k=1,\ldots,p}} \sqrt{\sum_{\substack{s=1,\ldots,K \\ t=1,\ldots,K}} \theta_{js,kt}^2}.$$

17.5 Discussion

In this section, we include some comments on the choice of some universal parameters that are used throughout the `Libra` package.

- *Initialization of intercept parameter* θ_0 *:* The initialization of θ_0 in the Linearized Bregman Algorithm is $\theta_0^0 = \arg\min_{\theta_0} L(\theta_0, 0)$, not $\theta_0 = 0$. The reason for this is to avoid picking up the variables whose effects are very close to the intercept term. If $\theta_0 = 0$ at first, then the gradient of those spurious variables close to the intercept may become very large due to the influence of intercept, such that they are much easier to be picked out. This issue is highlighted when the Ising model suffers from imbalanced sampling such as in low-temperature effects. For example, when 1 or -1 dominates a sample, the corresponding variable is very close to the intercept term by exhibiting nearly a constant in sample and thus becomes a spurious variable being selected early. In this case, initialization using $\arg\min_{\theta_0} L(\theta_0, 0)$ can avoid this issue.
- *Initialization of t:* Because the initial value of θ_0 is a minimal point, the gradient of the loss is always zero unless a new variable is added in. So in the package, the iteration actually begins from the first entry time

$$t_0 = \inf\{t : \theta_j(t) \neq 0, \text{ forsome } j\}$$

and $z(t_0)$ can be calculated easily because $\nabla_\theta L(\theta_0^0, 0)$ is constant.
- *Parameter* `tlist`*:* Instead of returning all the results of iteration steps, we need to return the results at a pre-decided set of t, `tlist`, along the path. However the Linearized Bregman Iterations only compute the value at a regular grid of time $t_0 + k\alpha, k = 0, 1, \ldots$, which may not consist a particular t in `tlist`. To solve this issue, for a point t in `tlist` but not on the computed time grid, a linear interpolation of $z^k(\theta_0^k)$ and $z^{k+1}(\theta_0^{k+1})$ is used to compute $z(t)$ or $\theta_0(t)$, $\theta(t)$ is further obtained by using **Shrinkage** on $z(t)$. Finally if `tlist` is not specified by the user, a geometric sequence from t_0 to $t_0 \cdot$ `trate` (`trate` $= t_{\max}/t_{\min}$)

with length `nt` (number of models on path to show) is used as the default choice `tlist`.

Acknowledgements The authors would like to thank Chendi Huang, Stanley J. Osher, Ming Yan, and Wotao Yin for helpful discussions. The research of Jiechao Xiong and Yuan Yao was supported in part by National Basic Research Program of China: 2015CB85600 and 2012CB825501, National Natural Science Foundation of China: 61370004 and 11421110001 (A3 project), as well as grants from Baidu and Microsoft Research Asia. The research of Feng Ruan was partially supported by the E.K. Potter Stanford Graduate Fellowship.

References

Burger M, Osher S, Xu J, Gilboa G (2005) Nonlinear inverse scale space methods for image restoration. In: Variational, geometric, and level set methods in computer vision. Springer, Berlin, pp 25–36

Burger M, Möller M, Benning M, Osher S (2013) An adaptive inverse scale space method for compressed sensing. Math Comput 82(281):269–299

Efron B, Hastie T, Johnstone I, Tibshirani R (2004) Least angle regression. Ann Stat 32(2):407–499

Fan J, Li R (2001) Variable selection via nonconcave penalized likelihood and its oracle properties. J Am Stat Assoc 96:1348–1360

Friedman J, Hastie T, Tibshirani R (2008) Sparse inverse covariance estimation with the graphical lasso. Biostatistics 9(3):432–441

Hassner M, Sklansky J (1980) The use of Markov random fields as models of texture. Comput Graph Image Process 12(4):357–370

Hastie TJ, Tibshirani RJ, Friedman JH (2009) The elements of statistical learning: data mining, inference, and prediction, 2nd edn. Springer, New York. http://opac.inria.fr/record=b1127878. Autres impressions: 2011 (corr.), 2013 (7e corr.)

Ising E (1925) Beitrag zur theorie des ferromagnetismus. Z Phys A Hadrons Nucl 31(1):253–258

Ji P, Jin J (2016) Coauthorship and citation networks for statisticians. Ann Appl Stat 10(4):1779–1812. http://dx.doi.org/10.1214/15-AOAS896

Manning CD, Schütze H (1999) Foundations of statistical natural language processing, vol 999. MIT, Cambridge

Osher S, Ruan F, Xiong J, Yao Y, Yin W (2016) Sparse recovery via differential inclusions. Appl Comput Harmon Anal. https://doi.org/10.1016/j.acha.2016.01.002

Ravikumar P, Wainwright MJ, Lafferty JD et al (2010) High-dimensional Ising model selection using l_1-regularized logistic regression. Ann Stat 38(3):1287–1319

Shi JV, Yin W, Osher SJ (2013) Linearized Bregman for l_1-regularized logistic regression. In: Proceedings of the 30th international conference on machine learning (ICML)

Tibshirani R (1996) Regression shrinkage and selection via the lasso. J R Stat Soc Ser B 58:267–288

Xue L, Zou H, Cai T (2012) Nonconcave penalized composite conditional likelihood estimation of sparse Ising models. Ann Stat 40(3):1403–1429. https://doi.org/10.1214/12-AOS1017

Yin W, Osher S, Darbon J, Goldfarb D (2008) Bregman iterative algorithms for compressed sensing and related problems. SIAM J Imag Sci 1(1):143–168

Zhao T, Liu H (2012) The `huge` package for high-dimensional undirected graph estimation in R. J Mach Learn Res 13:1059–1062

Part IV
Application

Chapter 18
Functional Data Analysis for Big Data: A Case Study on California Temperature Trends

Pantelis Zenon Hadjipantelis and Hans-Georg Müller

Abstract In recent years, detailed historical records, remote sensing, genomics and medical imaging applications as well as the rise of the Internet-of-Things present novel data streams. Many of these data are instances where functions are more suitable data atoms than traditional multivariate vectors. Applied functional data analysis (FDA) presents a potentially fruitful but largely unexplored alternative analytics framework that can be incorporated directly into a general Big Data analytics suite. As an example, we present a modeling approach for the dynamics of a functional data set of climatic data. By decomposing functions via a functional principal component analysis and functional variance process analysis, a robust and informative characterization of the data can be derived; this provides insights into the relationship between the different modes of variation, their inherent variance process as well as their dependencies over time. The model is applied to historical data from the Global Historical Climatology Network in California, USA. The analysis reveals that climatic time-dependent information is jointly carried by the original processes as well as their noise/variance decomposition.

Keywords Functional principal components · Functional variance process · Temperature curves

18.1 Introduction

Functional Data Analysis (FDA) is the statistical framework for the analysis of function-valued data. Under this framework each sample instance is considered to be an individual realization of an (often smooth) underlying stochastic process (Rao 1965). Common examples are curves (one-dimensional functions) or images (two-dimensional functions).

P. Z. Hadjipantelis · H.-G. Müller (✉)
University of California, Davis, Department of Statistics, Davis, CA, USA
e-mail: pantelis@ucdavis.edu; hgmueller@ucdavis.edu

© Springer International Publishing AG, part of Springer Nature 2018

W. K. Härdle et al. (eds.), *Handbook of Big Data Analytics*, Springer Handbooks of Computational Statistics, https://doi.org/10.1007/978-3-319-18284-1_18

A typical real-life dataset is a set of growth curves; a size/development-related process Y is measured in a sample of N individuals along a time domain T (Müller et al. 2009). The observed process Y is therefore a *height process* of which a sample of individual growth trajectories $y_i(t)$, $i = 1, \ldots, n$, are observed for n individuals. Another example is auction data (Liu and Müller 2009); the bid amounts from N individuals from the time an auction starts until it ends. In this case the observed process Y is a *price process*. The continuum over which a process is observed is not exclusively time; for example, spatial coordinates (Delicado et al. 2010) or mass (Harezlak et al. 2008) and mass-charge spectra (Knight et al. 2004) are measured over various continua that occur in FDA applications. The fact that one has repeated observations from the process at hand distinguishes FDA from standard time-series analysis techniques for which certain stationarity assumptions are made. Similarly the continuum needs not be one-dimensional; in shape analysis (Dryden 2005) and medical imaging (Dryden et al. 2009; Jiang et al. 2009) functional data are often two- (Kenobi et al. 2010; Sangalli et al. 2013) or even three-dimensional (Kurtek et al. 2010; Pigoli and Sangalli 2012).

As expressed by Valderrama: "[extensions towards] FDA have been done [in] two main ways: by extending multivariate techniques from vectors to curves and by descending from stochastic processes to [the] real world" (Valderrama 2007). FDA has been developed under diverse frameworks: e.g., Dirichlet (Petrone et al. 2009), Gaussian (Hall et al. 2008) or Poisson (Illian et al. 2006) data. Furthermore the data themselves can be represented using splines (Ramsay and Silverman 2005), wavelets (Guo 2002) or principal components as a basis; any basis (e.g., Fourier Ramsay and Silverman (2005) or Legendre Grabe et al. (2007) polynomials) could be used depending on the suitability of the basis in question to represent the data.

A major distinction from multivariate data is the smoothness and differentiability of functional data. If the functional nature of a functional dataset is ignored, results typically are sub-optimal as valuable information related to the functional nature of the dataset is not properly utilized.

This latter point emphasizes the connection of FDA with big data. Big data are assumed to be primarily characterized in terms of *v*olume, *v*ariety, *v*elocity (Gartner 2011) and complexity. This can be true for multivariate and functional data alike; what is specific for functional data is that they are infinite-dimensional objects and therefore even highly complex multivariate analysis techniques cannot account for the complexity reflected in a stochastic process. In addition traditional multivariate techniques will suffer as the number of components gets bigger. Their direct applicability diminishes due to increased computational demands as well as theoretical shortcomings as the dimensions increase. Standard principal component analysis (PCA) does not account for smoothness or continuity. One can consider penalization/sparsity-inducing (Tibshirani 1996; Bühlmann and Van De Geer 2011) or re-sampling techniques (Breiman 1996; Efron and Tibshirani 1994) to intelligently reduce the number of variables required to work with and then potentially post-process the data in a heuristic way to enforce certain modeling assumptions [e.g., the non-negativeness of the generated factors in the case of non-negative matrix factorization (Lee and Seung 2001)].

These are valid approaches that are however inapplicable when the data have a smooth functional structure and are in this sense infinite dimensional. For example differentiation and convolution operations are natural steps when working with functions. They allow practitioners to quantify changes in the underlying dynamics of time evolving patterns (e.g., Ramsay et al. 1996; Wang et al. 2008) or to apply appropriate transformations (e.g., Hadjipantelis et al. 2015). Finally, it should be noted that FDA can tackle cases where the data are recorded over an irregular or even sparse design of points where measurements are taken, while multivariate techniques assume that the dimensions of all data vectors are the same.

18.2 Basic Statistics for Functional Data

When working with functional data the unit of observation is a smooth function $Y(t)$, $t \in T$. For simplicity we assume that $T \subset R$. One then observes a sample of size N realizations of Y; we will use Y_{ij} to describe the observed value of i-th realization at the j-th time-point. Importantly, we need to also assume that $Y \in L^2(T)$. For the smooth random processes Y we define the mean function:

$$\mu_Y(t) = E[Y(t)] \tag{18.1}$$

and the symmetric positive definite auto-covariance function:

$$C_{YY}(s,t) = \mathrm{Cov}[Y(s), Y(t)] \tag{18.2}$$

$$= E[(Y(t) - \mu_Y(t))(Y(s) - \mu_Y(s))]. \tag{18.3}$$

Noting that the auto-covariance function $C_{YY}(s,t)$ is by definition symmetric and positive semi-definite, its spectral decomposition is given by Mercer's theorem (Mercer 1909):

$$C_{YY}(s,t) = \sum_{k=1}^{\infty} \lambda_k \phi_k(s)\phi_k(t), \tag{18.4}$$

where $\lambda_1 \geq \lambda_2 \geq \cdots \geq 0$ are the ordered eigenvalues of the Hilbert–Schmidt operator with kernel C_{YY} with $\sum_{k=1}^{\infty} \lambda_k < \infty$. The ϕ_k are the orthonormal eigenfunctions of this operator. The Karhunen–Loève expansion of the observations Y (Hall et al. 2006) is given as:

$$Y_i(t) = \mu_Y(t) + \sum_{k=1}^{\infty} \xi_{ik}\phi_k(t), \tag{18.5}$$

where the ξ_{ik} are the zero-mean functional principal component scores with variance equal to the corresponding eigenvalues λ_k and the eigenfunction ϕ_k act as the orthonormal basis for the space spanned by the process Y.

Assuming a second process X similar to Y, in a similar manner we define the kernel C_{YX} of the cross-covariance operator $\mathcal{C}_{YX}$ as:

$$C_{YX}(s,t) = E\{(Y_i(s) - \mu_Y(s))\,(X_i(t) - \mu_X(t))\} \tag{18.6}$$

which has the following functional singular decomposition (Yang et al. 2011):

$$C_{YX}(s,t) = \sum_{p=1}^{\infty} \sigma_p \phi_{Y,p}(s)\phi_{X,p}(t), \tag{18.7}$$

where now the σ_p denote the singular values associated with the decomposition of the covariance between the processes Y and X (Yang et al. 2011) and $\phi_{Y,p}$ are the "left" and $\phi_{X,p}$ the "right" eigenfunctions. Notably the cross-covariance surface, while adjoint, is not symmetric and not self-adjoint.

In real-life applications statistical analysts are increasingly presented with massive datasets stemming from continuous multi-channel recordings: from anatomical brain regions (Worsley et al. 2002) to climatic variables across different geographic regions (Zhang et al. 2011) and multiple sensors tracking humans or animals (Coffey et al. 2011). The functional data analysis framework is essential to analyze such data.

18.3 Dimension Reduction for Functional Data

As stated by Delicado (2007), functional data can be viewed as augmenting multivariate techniques to a functional domain. One is therefore presented with infinite dimensional objects that require extension of traditional inferential procedures. Additionally, an infinite dimensional object cannot be fully represented when using a finite amount of memory.

Aside the obvious size-constraints, it is possible that we are encoding redundant, unrelated or even misleading information in a high-dimensional dataset. It is therefore to a modeler's benefit to extract features or modes of variation that are informative representations. In that regard, even higher dimensional datasets might have an adequate, in terms of variation explained, representation in two or three dimensions enabling their visualization. Classical methods to obtain such representations are PCA and Multi-Dimensional Scaling (MDS). Ideally, the reduced dimension representation of the dataset is useful as a surrogate dataset for the original high-dimensional data analyzed, both for dimension reduction and to filter unstructured information out of the original dataset.

Dimension reduction is based on the notion that we can produce a compact low-dimensional encoding of a given high-dimensional dataset. The currently main methodology to achieve this is Functional Principal Components Analysis (FPCA) (Hall et al. 2006; Kleffe 1973).

FPCA is an inherently linear and unsupervised method for dimension reduction (Ghodsi 2006) that has been implemented for diverse applications of FDA. Linear refers to the fact that the resulting representation is linear in the random functions that are represented. In the case of FPCA the original zero-mean process Y is approximated by forming projections of the form:

$$\xi_{\nu,n} = \int_0^T \phi_\nu(t)(Y(t) - \mu_y(t))dt \tag{18.8}$$

where $E\{Y(t)\} = 0$, ϕ_ν is the ν-th eigenfunction and ξ_ν is the corresponding FPC score; $\text{Var}(\xi_\nu) = \lambda_\nu$ as in Eq. (18.5). Alternative non-linear (but still unsupervised) dimension reduction algorithms include kernel PCA (Schölkopf et al. 1998), locally linear embedding (LLE) (Roweis and Saul 2000), and semi-definite embedding (SDE) (Weinberger et al. 2004); the latter can also be casted as kernel PCA (Ghodsi 2006). Chen and Müller demonstrate that some functional data can be successfully represented as non-linear manifolds (Chen and Müller 2012a) using Isomap (Tenenbaum et al. 2000). Fisher's Linear Discriminant analysis is a typical example of an alternative supervised linear dimensionality reduction algorithm (Barber 2012) that also has been successfully applied to functional data (James and Hastie 2001).

As a final note, a problem that is common to FPCA applications is the selection of the number of components to retain. This is still an open problem that is effectively a model selection problem (Hoyle 2008; Minka 2001). One can formulate the selection problem as an optimization problem of an equivalent information criterion; for example, Yao et al. (2005) propose pseudo-AIC and BIC criteria for the determination of the relevant k; other heuristic methods (e.g., the *broken stick* model) (Cangelosi and Goriely 2007) are popular. A final important note on dimension reduction is that, as Ma and Zhu emphasize: "because dimension reduction is generally viewed as a problem of estimating a space, inference is strikingly left out of the main-stream research" (Ma and Zhu 2013). Sliced inverse regression can be a fruitful alternative approach (Li 1991) for dimension reduction. Ferré and Yao (2003) have formulated a relevant application framework for applying sliced inverse regression to functional data; Jiang et al. (2014) have moved this framework even further with results on irregularly sampled data.

18.4 Functional Principal Component Analysis

Tucker (1958) introduced the idea of a function as the fundamental unit of statistical analysis in the context of Factor Analysis (FA). Nevertheless it was the later work of Castro et al. (1986) that popularized the concept of dimensionality reduction via covariance function eigendecomposition for functional data as in Eq. (18.4). Returning to the original notion of a stochastic process Y, the "best" K-dimensional linear approximation in the sense of minimizing the variance of the residuals is:

$$Y_i(t) = \mu(t) + \sum_{k=1}^{K} \xi_{i,k}\phi_k(t), \tag{18.9}$$

where the $\xi_{i,k}$ are the k-th principal component scores and ϕ_k ($k \geq 1$) are the eigenfunctions that form a basis of orthonormal functions. In this functional setting the usual mean squared error optimality condition found in standard multivariate PCA minimizing $\sum_{i=1}^{K} || y_i - \hat{y} ||^2$ can now be restated as the integrated square error $\int [y(t) - \hat{y}(t)]^2 dt$. The fraction of the sample variance explained is maximized along the modes of variation represented by the eigenfunctions ϕ_k.

Empirically finding the eigenfunctions ϕ_k requires first the estimation of the sample auto-covariance function $\hat{C}_Y(s,t)$; see Eq. (18.3):

$$\hat{C}_{YY}(s,t) = \frac{1}{N} \sum_{i=1}^{N} (Y_i(s) - \hat{\mu}_Y(s))(Y_i(t) - \hat{\mu}_Y(t)), \tag{18.10}$$

where $\hat{\mu}_Y(t) = \frac{1}{N} \sum_{i=1}^{N} Y_i(t)$. Then the subsequent eigendecomposition of $\hat{C}_{YY}(s,t)$ for the zero-meaned sample Y is approximated by the first K components as:

$$\hat{C}_{YY}(s,t) = \sum_{\nu=1}^{K} \hat{\lambda}_\nu \hat{\phi}_\nu(s) \hat{\phi}_\nu(t). \tag{18.11}$$

This is in direct analogy with the *principal axis theorem* (Jolliffe 2005) for multivariate data where the sample covariance matrix $\hat{C}_{YY}$ is decomposed as $\hat{C}_{YY} = \hat{\Phi}\hat{\Lambda}\hat{\Phi}^T$, where $\hat{\Phi}$ is the orthogonal matrix composed of the eigenvectors for the multivariate sample Y and $\hat{\Lambda}$ is the diagonal matrix composed of estimated eigenvalues. Ultimately, because of the optimality of the FPCs in terms of variance explained, these new axes explain the maximum amount of variance in the original sample. The estimated eigenfunctions $\hat{\phi}_k$ are restricted to be orthonormal. The corresponding mode of variation has the form $\mu(t) + \zeta\phi_k(t)$ for $\zeta \in R$. For a more detailed introduction to FPCA, Horváth and Kokoszka provide an excellent resource (Horváth and Kokoszka 2012).

FPCA serves not only as a convenient method for dimensionality reduction but also provides a way to build characterizations of the sample trajectories around an overall mean trend function (Yao et al. 2005). An additional geometric perspective is that the eigenfunctions represent a rotation of the original functional data that diagonalizes the covariance matrix of the data. The resulting scores ξ_k are uncorrelated (Barber 2012), as in the case of multivariate PCA. Table 18.1 lists the analogies between standard multivariate and functional PCA.

Common assumptions are that the mean curve and the first few eigenfunctions are smooth and that the eigenvalues λ_k tend to zero rapidly; the faster this convergence happens the smoother are the trajectories. The smoothness of the

Table 18.1 The analogies between multivariate PCA and functional PCA

Concept	MVPCA	FPCA
Data	$Y \in R^p$	$Y(t) \in L^2(T)$
Dimension	$p < \infty$	∞
Mean	$\mu_Y = E(Y)$	$\mu_Y(t) = E(Y(t))$
Inner product $\langle x,y \rangle$	$\sum x_j y_j$	$\int x(t)y(t)dt$
Covariance	$\Sigma_{Y,Y}(j,l)$ (matrix)	$C_{Y,Y}(s,t)$ (kernel)
Eigenvalues	$\lambda_1 \geq \lambda_2 \geq \cdots \geq \lambda_p \geq 0$	$\lambda_1 \geq \lambda_2 \geq \cdots \geq 0$
Eigen-vectors/functions	$\phi_1 \perp \phi_2 \perp \cdots \perp \phi_p$	$\phi_1(t) \perp \phi_2(t) \perp \ldots$
Principal Component Scores	$\xi_k = \sum(y_j - \mu_y)\phi_k$	$\xi = \int(Y(t) - \mu_Y(t)\phi(t)dt$

underlying trajectories is critical so that the discrete sample data can be considered functional (Horváth and Kokoszka 2012), even when there is usually additive measurement noise. As seen in Chen and Müller (2012b) for the case of two-dimensional functional data, the discretization and the subsequent interpolation can have significant implications.

FPCA can be implemented in many different ways. A number of regularized or smoothed functional PCA approaches have been developed over the years. Smoothness is imposed in two main ways. The first approach is by penalizing the roughness of the candidate eigenfunctions ϕ_k directly, where their roughness is measured by the integrated squared second derivative over the interval of interest (Ramsay 2002) or by approximating ϕ_k with a family of smooth functions [e.g., B-splines (James et al. 2000) or Fourier polynomials (Graves et al. 2009)]. The second approach is by smoothing the data or their equivalent representation (their autocovariance) directly and then projecting the dataset to a lower dimensional domain where smoothness is ensured. The actual smoothing in these cases can be done using usual smoothing procedures (e.g., Yao et al. 2005). The basic qualitative difference between these two smoothing approaches is that in the first case smoothing occurs directly on the FPCA eigen decomposition step, while in the second case we smooth the raw data or their higher order moments. Both approaches have been used extensively in the literature.

18.4.1 Smoothing and Interpolation

Common design assumptions in FDA are that the sample consists either of sparse (and possible irregularly sampled) observations (Yao et al. 2005) or of densely/perfectly observed discretized instances of smooth varying functions (Aston et al. 2010). Situations in between these two have also been considered. In all cases, we assume there is an underlying smooth generating process such that the observed data result from possibly noisy measurements of this process. Due to the assumed smoothness of the underlying process the following smoothing techniques can be employed: localized kernel smoothing (Chiou et al. 2003), smoothing splines (Guo 2002), or wavelet smoothing (Antoniadis et al. 1994).

A local smoother based on kernel weights is analogous to the *windowing* of a digital signal to conduct short-time analysis. It corresponds roughly to the windowing idea in Digital Signal Processing that the signal/sample within that window is stationary and the model's inferred parameters are reasonable estimates for the overall behavior within the window. In this setting, a smoothing kernel is a non-negative, symmetric, continuous (on its support), real-valued integrable function K satisfying: $\int_{-\infty}^{+\infty} K(t)dt = 1$. The bandwidth (or tuning parameter) b plays a key role in kernel smoothing; it can be viewed as analogous to the characteristic length scale of a Gaussian process (Rasmussen 2004).

A basic kernel smoother is the Nadaraya–Watson estimator which is effectively a weighted mean $\hat{\mu}_{\mathrm{NW}}$, where the weights are determined by the kernel function K and bandwidth b used. For data $(t_i, y_i), i = 1, \ldots, s$, generated by $y_i = m(t_i) + \epsilon_i$, assuming a smooth function m, the Nadaraya–Watson estimator is given by $\hat{\mu}_{\mathrm{NW}}(t) = \frac{\sum_{i=1}^{s} K(\frac{t-t_i}{b}) y_i(t)}{\sum_{j=1}^{s} K\left(\frac{t-t_j}{b}\right)}$ (Davison 2003); alternative kernel smoothers have been proposed (Gasser and Müller 1984). A closely related method is weighted least squares smoothing where one fits local linear lines to the data within the windows $[t-b, t+b]$ and produces smooth curves. Using the notation S_L for the smoother, evaluating the non-parametric regression function at point t and considering a sample $(t_i, y_i), i = 1, \ldots, s$, as inputs, the smoother S_L takes the form:

$$S_L\{t; b, (t_i, y_i)_{i=1,\ldots,s}\} \tag{18.12}$$
$$= \underset{\alpha_0}{\operatorname{argmin}} \left\{ \min_{\alpha_1} \left(\sum_{i=1}^{s} K\left(\frac{t-t_i}{b}\right) [y_i - \{\alpha_0 + \alpha_1(t-t_i)\}]^2 \right) \right\}.$$

Kernels commonly used include the uniform (or rectangular) kernel function: $K(x) = 0.5 I_{|x| \leq 1}$ and the Epanechnikov kernel function: $K(x) = \frac{3}{4}(1-x^2) I_{|x| \leq 1}$, where $I_{|x| \leq 1}$ is an indicator function; the Gaussian kernel function: $K(x) = \frac{1}{\sqrt{2\pi}} e^{-\frac{x^2}{2}}$ and the triangular kernel function: $K(x) = (1 - |x|) I_{|x| \leq 1}$; see Izenman (2008). The bandwidth b is commonly estimated using cross-validation (Izenman 2008) or generalized cross-validation (Silverman 1985). The uniform, Epanechnikov, and the triangular kernels have a finite support which ensures that the size of the linear system solved by Eq. (18.12) always remains finite. Qualitatively, smaller values of b are associated with large variability while larger values of b are associated with larger bias that results from broader smoothing windows.

It is important to note that by employing a locally weighted least squares smoother we can adjust for irregular sampling and/or data to some extent. While ideally all curves are sampled over the same dense grid of points, in practice this is often not the case. Smoothing can be used to transform the original irregularly spaced observations to smooth curves that can then be sampled on a regular output grid. The number of grid points for the output grid is usually determined empirically;

in practice L is set equal to the expected number of readings per case for the sample at hand.

For example, when weekly measurements of a reasonably smooth varying process are available, 53 equidistant points are an obvious choice for the output grid; thus ensuring that there are smooth values on weekly support points. As Levitin et al. mention: "this [the choice] is largely a matter of experience, experimenter intuition and trial-and-error" (Levitin et al. 2007). The choice of L strongly relates to the binning of the data and has a bearing on the computational load. When analyzing datasets with few missing values (i.e., not significantly sparse datasets) often the use of the Epanechnikov kernel function is a good choice. As an alternative to the finite support kernels, the Gaussian kernel, which enjoys an infinite support, can always "interpolate over holes" in situations with extremely sparse designs. One must be careful though to avoid situations where the dimensions of the actual systems solved [Eq. (18.12)] become prohibitively large; explicit solutions are available (Fan and Gijbels 1996). Implementation-wise, based on simulations we recommend truncating a Gaussian kernel window at three bandwidth lengths from each side to ensure a reasonable maximal length-scale.

As an alternative to kernel smoothing, smoothing splines are a popular choice (Eubank 1999). For FDA applications Guo presented a test-case where the smoothing framework employed a generalization of smoothing splines (Guo 2002). For standard spline smoothing, the number of knots, the number of basis functions, and the order of the spline employed affect the final result. A third alternative is based on the notion of multi-resolution signal decomposition as implemented by wavelets (Mallat 1989). Wavelet approximations have been widely employed; for example, the popularly used JPEG standard is based on wavelets. Wavelet estimators use projections onto subspaces of $L^2[0, 1]$ to represent successive approximations. In contrast to a standard orthogonal basis that is localized only in a single domain (e.g. Fourier polynomials are localized in frequency), wavelets can be localized both in time and frequency. Wavelets allow for fast implementations based on the Discrete Wavelet Transformation (DWT) and can be used to smooth original noisy functional data (Morris and Carroll 2006). Under this paradigm, smoothing is achieved by thresholding; certain wavelet coefficients are "thresholded" in order to exclude parts considered to be noise. Often however, the smoothed curves resulting from wavelets are not as smooth as desired.

All techniques require a tuning parameter which in the case of wavelets is the original mother wavelet and the threshold used, in the case of spline smoothing, the type of splines and the penalty parameter used, and in the case of kernel smoothing the choice of kernel type and bandwidth.

Kernel, splines or wavelets estimators are linear in the data y:

$$y_{\text{smooth}} = Hy, \tag{18.13}$$

where H is a projection or *hat* matrix. On a conceptual level one might argue that smoothing leads to *dimension enhancement* as it generates readings across a whole continuum that was previously unpopulated as this is obvious in the case of irregularly sampled data. Numerically, both spline and wavelet smoothing have been reported to be more computationally efficient than kernel smoothing (Wink and Roerdink 2004; Bruns 2004; Theis 2005) in its naive form. Binning and computational parallelizing overcome this issue and offer significant speed-ups compared to naive kernel smoothing, while providing a straightforward implementation that is easily adaptable to specific restrictions and modifications one may wish to adopt. An additional advantage of this approach is that it is easily amenable to mathematical analysis.

18.4.2 Sample Size Considerations

When working with a large sample of dense functional data, sample size considerations come into play not only because the number of design points may increase. To address the computational challenges we propose two basic remedies, binning and the use of the Gram matrix.

Binning of data is a basic data pre-processing procedure. For the analysis of large data, binning has significant utilities; it has an important place in neuroscience applications (Cunningham et al. 2009).

When sampling a signal, the notion of the Nyquist–Shannon sampling theorem is relevant when choosing the sampling frequency, as this corresponds to the bin width. There are a number of heuristics available to choose the number of bins, e.g., Sturges (1926), Scott (1979) and Freedman and Diaconis (1981) as well information-theoretic approaches (e.g., Kohavi and Sahami 1996); in any case using some prior knowledge as well as visualizing the data is essential. The number of bins must be large enough to allow the process dynamics to be expressed without aliasing. This is clearly a common problem in situations where densely sampled signals are presented. When implementing FDA, one may face situations with "*small n, large p*," sometimes abbreviated as "$n \ll p$".

For *small n, large p*, the eigendecompositions become non-trivial in terms of space as well computations required. A common computational trick that supports binning is to recognize that for the zero-meaned sample matrix Y_0 the covariance matrix is:

$$C = \frac{1}{n-1} Y_0^T Y_0, \tag{18.14}$$

with corresponding Gram matrix:

$$G = \frac{1}{n-1} Y_0 Y_0^T \tag{18.15}$$

where C and G share the same singular values. Additionally and equally importantly the eigencomponents of the covariance matrix C are related to the eigencomponents of the Gram matrix G. The proof is straightforward: it entails expressing Y_0 as USV^T

and expressing the eigen decomposition of C as $\frac{1}{n-1}VS^2V^T$ and G as $\frac{1}{n-1}US^2U^T$ (Good 1969). Using this property, if either $n \ll p$ or $n \gg p$, one can compute the principal modes of variation of the original sample stored in the matrix Y_0 by considering only $\min(n,p) \times \min(n,p)$ matrices.

Clearly there are situations where the number of observed functions n is large. In these situations the most common solutions fall under the category of randomized algorithms (Papadimitriou 2003). The basic idea is that one uses a random projection matrix Ω to multiply the original large sample A to get the matrix $B = A\Omega$. Subsequently one computes the matrix Q that defines an orthonormal basis for the range of B. A caveat is that, as B is an approximation to the range of A and thus projecting A by using Q^TQ, we will get $A \approx Q^TQA$ (Witten and Candès 2015); this is natural as given that Q is approximately orthonormal then $Q^TQ \approx I$. There is extensive work on how to sample these random projections. Randomized algorithms have been shown to be promising for very large datasets (Mahoney 2011). For example, even in cases where less than a hundredth of the original dataset can be stored in the computer's RAM, efficient algorithms have been derived to obtain nearly optimal decompositions (Halko et al. 2011). Another option is simply to work with a random subsample of size $n^* < n$, for a suitably small n^* at the cost of increased variability.

18.5 Functional Variance Process

When analyzing functional data, there is sometimes valuable information in small oscillations that can be captured by the *functional variance process* (Müller et al. 2006) (FVP). FVP is a non-parametric tool to analyze stochastic time-trends in the noise variance of functional data. In contrast to parametric techniques that impose particular assumptions regarding the form of potential heteroskedastic trends, non-parametric variance function estimation techniques allow for data-driven detection of locally varying patterns of variations that may contain relevant information to quantify the variation observed in the data.

More formally, we assume that $Y_1, \ldots, Y_n$ denote n continuous smooth random functions defined over a real interval $[0, T]$. Here, we also assume that these functions are observed at a grid of dense time-points $t_j = \frac{j-1}{m-1}T$, $j = 1, \ldots, m$, with measurements:

$$Y_{ij} = Y_i(t_j) + R_{ij}, \quad i = 1, \ldots, n, \quad j = 1, \ldots, m, \tag{18.16}$$

and $R_{i,j}$ is additive noise such that $E\{R_{i,j}R_{i'k}\} = 0$ for all $i \neq i'$, $E\{R\} = 0$ and $E\{R^2\} < \infty$.

Importantly, the noise process R is assumed to generate the squared errors $R^2_{i,j}$ which themselves are assumed to be the product of two non-negative components $V(t_{i,j})$ and $W_{i,j}$ such that $R^2_{i,j} = \exp[V(t_{i,j})]\exp[W_{i,j}]$; $V(t_{i,j})$ and $W_{i,j}$ representing the underlying smooth functional variance process and a noise component, respectively. To draw conclusions about the underlying variance process of the original data Y we then focus on the examination of its two components $V(t_{i,j})$ and $W_{i,j}$.

The initial step is to work on the log-domain to transform the multiplicative relation between V and W to an additive relation. This leads to the transformed errors:

$$Z_{i,j} = \log(R_{i,j})^2 = V(t_{i,j}) + W_{i,j}. \tag{18.17}$$

We assume that:

$$E\{Z_{i,j}\} = E\{V(t_{i,j})\} = \mu_V(t_{i,j}), \tag{18.18}$$

$$\text{Cov}(Z_{i,j}, Z_{i,j'}) = \text{Cov}(V(t_{i,j}), V(t_{i,j'})). \tag{18.19}$$

Furthermore we assume that the $V(t_{i,j})$ are derived from a smooth *variance process* $V(t_{i,j})$ with mean:

$$E\{V(t)\} = \mu_V(t), \tag{18.20}$$

and smooth symmetric auto-covariance:

$$C_{VV}(s,t) = \sum_{k=1}^{\infty} \rho_k \psi_k(s) \psi_k(t) \tag{18.21}$$

where in analogy with Eq. (18.3), the ρ_k are the non-negative ordered eigenvalues $\rho_1 \geq \rho_2 \geq \ldots 0$ and ψ_k are the corresponding orthonormal eigenfunctions of V describing its principal modes of variation. The $W_{i,j}$ are assumed to be independent with $E\{W_{i,j}\} = 0$ and $\text{Var}(W_{i,j}) = \sigma_W^2$. The functional data can then be decomposed into two processes Y and V with Karhunen–Loève representations :

$$Y(t) = \mu_Y(t) + \sum_{k=1}^{\infty} \xi_k \phi_k(t) \quad \text{(Eq. (18.5))and} \tag{18.22}$$

$$V(t) = \mu_V(t) + \sum_{k=1}^{\infty} \zeta_k \psi_k(t). \tag{18.23}$$

Operationally, after we compute the FPCA of the original data $Y_{i,j}$ yielding fits $\tilde{Y}_{i,j} = \hat{y}_i(t_{i,j})$ for the underlying smooth processes, we use the logarithms of the squared differences between Y_i and $\tilde{Y}_i$ as our new sample to conduct a second FPCA step. In the notation of Eq. (18.16), the total decomposition of the observed measurements then becomes:

$$Y_{i,j} = Y_i(t_j) + \text{sign}(R_{i,j})\{\exp[V_i(t_{i,j}) + W_{i,j}]\}^{\frac{1}{2}}. \tag{18.24}$$

18.6 Functional Data Analysis for Temperature Trends

In this section we illustrate the insights one can obtain by using functional data analysis techniques on historical temperature data. We begin by showing the findings obtained by a basic FPCA approach, followed by the insights gained from the inspection of the functional cross-covariance operator and finally from the subsequent FVPA (Functional Variance Process Analysis) of our data. In related work, Horváth and Kokoszka present a detailed introductory application on the subject of detecting changes in the mean function (Horváth and Kokoszka 2012) for temperature data.

Our illustration data are daily temperature time-series for two cities in Northern California, Redwood City (37°28′N, 122°14′W) and Davis (38°33′N, 121°44′W—Station IDs: USC00047339 and USC00042294 for Redwood City and Davis respectively), downloaded from the Global Historical Climatology Network(GHCN) (Menne et al. 2012). While the two cities are relatively close ($\sim$150 km) and of similar elevation ($\sim$15 m), Davis is landlocked while Redwood City is adjacent to the south San Francisco Bay; as large bodies of water are known to act as natural "heat reservoirs" we expect that the two cities have different temperature patterns. For both locations we were able to obtain daily temperature data for approximately a whole century (86 and 108 years of observations for Redwood City and Davis, respectively). This allows us to look at changes across time. We focus our attention to two variables of interest, the daily maximum temperature and the daily minimum temperature.

In addition to the difference between coastal and continental weather patterns, we are interested to see if there are long-term trends such as a warming effect. Both stations are in urban or peri-urban locations. Therefore, an urban warming effect may be present. Our basic unit of analysis is the daily trajectory of the temperature extrema records across the span of a whole year; we make no assumptions of seasonality or stationarity.

The original data are obviously non-smooth (Fig. 18.1) in their raw form; nevertheless, it is reasonable to assume that there is an underlying smoothly

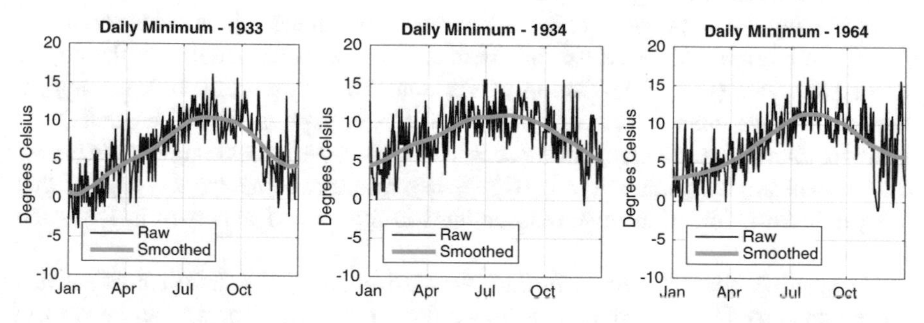

Fig. 18.1 The raw and smoothed temperature curves for the daily minimum in Redwood City in 1933, 1934 and 1964

varying process that encapsulates the dynamics behind the temperature extrema recorded. The deviations from a smooth function can be viewed as aberrations or as measurement errors. We will study the decomposition of the temperature signals into a functional process that reflects the long-term trends and a variance process that reflects the aberration patterns.

For the regular FDA part, the smoothing procedures can be either directly applied to densely sampled data for each subject (here: actual temperature profiles) or the pooled data before obtaining the mean and covariance functions. Here we choose to smooth the data directly for computational reasons. First, smoothing the raw data allows us to massively parallelize the smoothing procedure. When dealing with big data it is paramount to avoid moving data around for efficient computation. One should aim to move the computation close to the data rather than the other way around. If someone chooses to smooth the covariance, then one would deal with one (potentially massive) object that would need to be segmented, distributed across nodes, processed and then recombined. We avoid this by employing the fact that data might be already split and thus we "bring the computation" to the data.

Second, all smoothing procedures experience edge effects which is a problem that needs to be addressed. The remedy is often to avoid smoothing near the edges (e.g., Hadjipantelis et al. 2012), to truncate the smoothed curves (e.g., Petersen and Müller 2016), to enforce knot locations for splines (e.g., Ramsay and Silverman 2005), or to rebalance the weights intelligently (e.g., Hall and Müller 2003). We minimize edge effects during smoothing as well as parallelize our smoothing procedure by combining adjacent years. In Fig. 18.2, we show the basic idea where one segments data into two main data segments (Chunks A and B). After segmenting the data into chunks, the chunks can be processed independently. Importantly, a small part of the data will be duplicated (shown in a dark gray) in both chunks. This duplication serves to avoid edge effects. When smoothing the data we smooth only half of the duplicated part present in each chunk (shown in gray and black dotted lines for chunks A and B respectively). Then we avoid duplicating calculations and distribute the computation among all available resources. As a final comment we note that the actual duplicate chunk is relatively small in comparison with the overall space requirements; it needs to be only of size $2bq$, b being the bandwidth used and q being the number of segments used.

Analyzing the data and looking at the respective means, it is evident that the temperature curves of Davis exhibit more extreme weather patterns in the course of a typical year (Fig. 18.3). During winter-time the daily extrema in Davis appear marginally lower than in Redwood City (Fig. 18.3, both plots), signifying that Davis is colder than Redwood city. During summer Davis experiences significantly higher daily maximum temperatures (Fig. 18.3, RHS). These patterns strongly suggest that the yearly variation of average temperatures is higher in Davis than in Redwood City.

To quantify the differences further, we also quantify the daily spread, which corresponds to the daily maximum minus the daily minimum. As can be seen in Fig. 18.4 which shows the means as well as the covariance of the daily spreads in each region, the temperature spread in Davis is higher than in Redwood City for

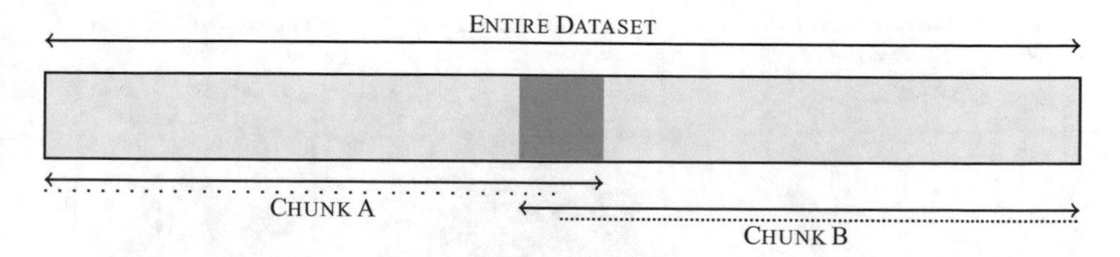

Fig. 18.2 Example where the original dataset to be smoothed is segmented into two chunks *A* and *B*. The two chunks overlap (dark gray box) so information is shared and smoothing edge effects are alleviated near the segmentation limits. The smoothing output grid-points (shown in widely and densely dotted lines for chunks *A* and *B*, respectively) do not overlap so one does not duplicate any calculations

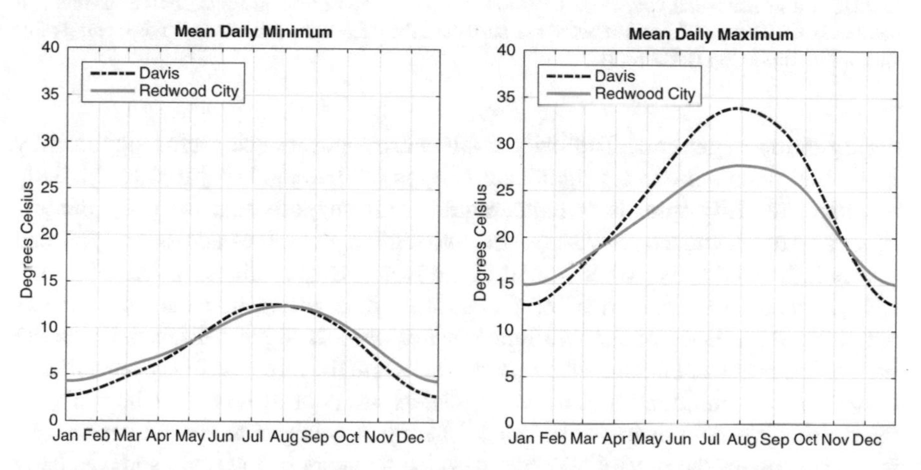

Fig. 18.3 Smoothed estimates for the mean daily temperature extrema recorded in Davis from 1908 to 2015 and in Redwood City from 1930 to 2015 using an Epanechnikov kernel and bandwidth of 49 days

approximately 10 months (Fig. 18.4, LHS). In addition it is evident that the variance of the spread across the year is of higher amplitude in Davis compared to Redwood City. Davis in late autumn (mid-October to late-November) appears to have very strong intra-day temperature changes indicating that the temperature patterns in Davis exhibit more variation that those in Redwood City, the latter being possibly modulated by the coastal location of Redwood City.

An issue of interest is the covariance/correlation of the annual temperature curves between the two locations and the detection of the periods where synergy is strongest. The cross-covariance estimator [Eq. (18.6)] is of interest here. Figure 18.5 shows the cross-covariance surfaces between the three temperature related measurements we examined. First of all it is notable that the surfaces are not symmetric across their diagonal; this is expected.

The off-diagonal elements appear to be routinely of lower absolute magnitude hinting at to little covariance between the temperatures in Davis and Redwood City

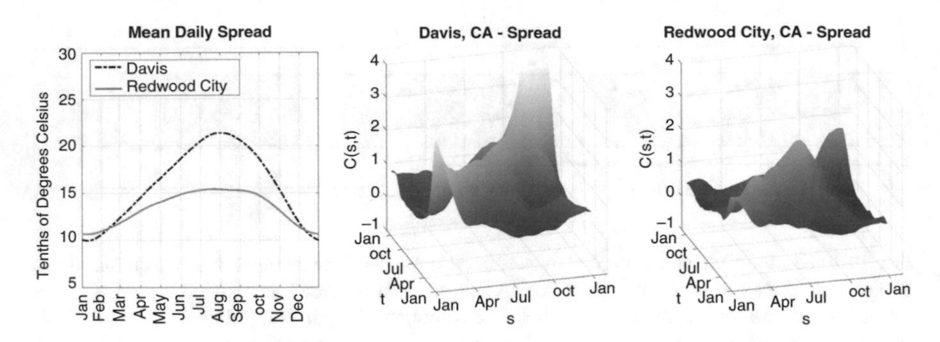

Fig. 18.4 Smoothed estimates for the mean daily temperature spread recorded in Davis from 1908 to 2015 and in Redwood City from 1930 to 2015 (LHS plot) using an Epanechnikov kernel and bandwidth of 49 days. The auto-covariance surface of the daily temperatures in Davis (central plot) and in Redwood City (RHS plot)

during different seasons. The daily minima cross-covariance seems significantly flatter and does not exhibit significant troughs (or bumps) compared to the daily maxima and daily spreads of temperatures. This suggests that the daily minima covary more homogeneously across the two regions as well as across time; a trend that is in line with the lower spread in the mean daily minima curves (Fig. 18.3, LHS). Focusing on the ridges of the surfaces it is telling that the time periods where the covariance of the two functional datasets is highest differ between the type of measurement examined. In particular, while the minima covary most during January, suggesting that cool or warm winters are cool or warm for both places, the maxima do not covary at the same time. Spring daily maximum temperatures appear to covary the most while, based on the behavior of the daily spreads, daily spreads appear covary the most in late autumn or early winter. This can also be seen by visually inspecting the cross correlation matrices (Fig. 18.6).

A question of interest is whether the data reflect climate change. It is generally accepted (Hansen et al. 2010) that the world as a whole is getting warmer. We note that in our analysis we do not incorporate any soil moisture, solar radiance or ocean-wind information which limits the interpretability of our data analysis as these factors are generally thought to be important confounders of the Northern California temperature dynamics.

We begin by examining the first estimated eigenfunction ϕ_1 (Fig. 18.7). Larger absolute values are associated with higher variation in the direction of the first eigenfunction, as the sign of the eigenfunction is arbitrary. The peaks of the various components are located at different times between different locations suggesting that the periods of maximum variation do not fully coincide.

Examining the second principal mode of variation ϕ_2 (Fig. 18.8) it appears that it encapsulates additional strongly expressed shapes in each of the respective underlying temperature processes. Something that is noteworthy is that neither of the eigenfunctions is periodic, thus reflecting the non-monotonicity of the original sample series (Fig. 18.1). We also note that one can assess the fraction of variance

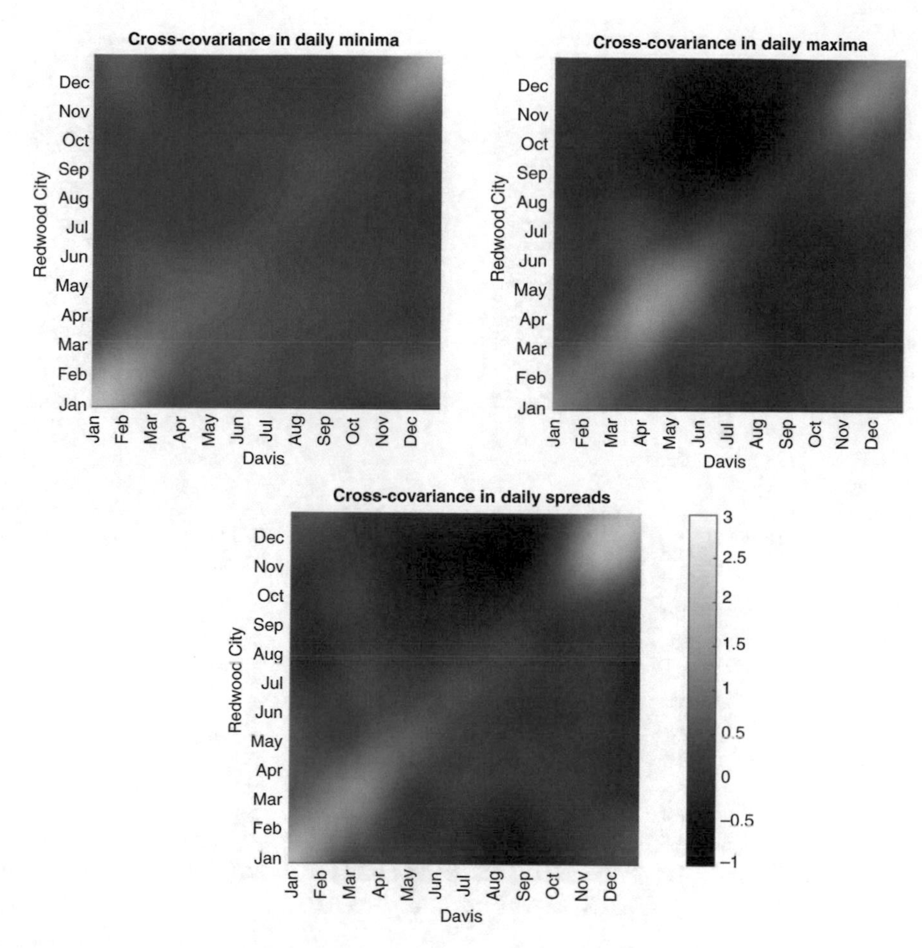

Fig. 18.5 Smoothed estimates for the cross-covariance surfaces between Davis and in Redwood City from 1930 to 2015 for daily temperature minima (LHS), maxima (central plot) and spread (RHS)

explained by each component using its respective eigenvalue ratio [Eq. (18.4)]. Using ϕ_1 and ϕ_2 we can obtain the FPC scores ξ_1 and ξ_2 by $\xi_j = \int(y - \mu)(t)\phi_j(t)dt, j = 1, 2$. A question of interest is whether there are trends in ξ_1 and ξ_2 over calendar years. Overall climate trends could be reflected in trends in some of the FPC scores. We therefore regress the scores ξ_j against calendar time t (measured t in years) for $j = \{1, 2\}$, applying either linear or non-parametric regression.

Fitting simple linear regressions, we obtain the fits in Fig. 18.9 and the t-values of Table 18.2. While the maxima estimates are relatively flat (Fig. 18.2, middle plot) the minima process has a clear upward pattern along time. This pattern suggests that nights are becoming warmer. In the temperature spread process the estimated

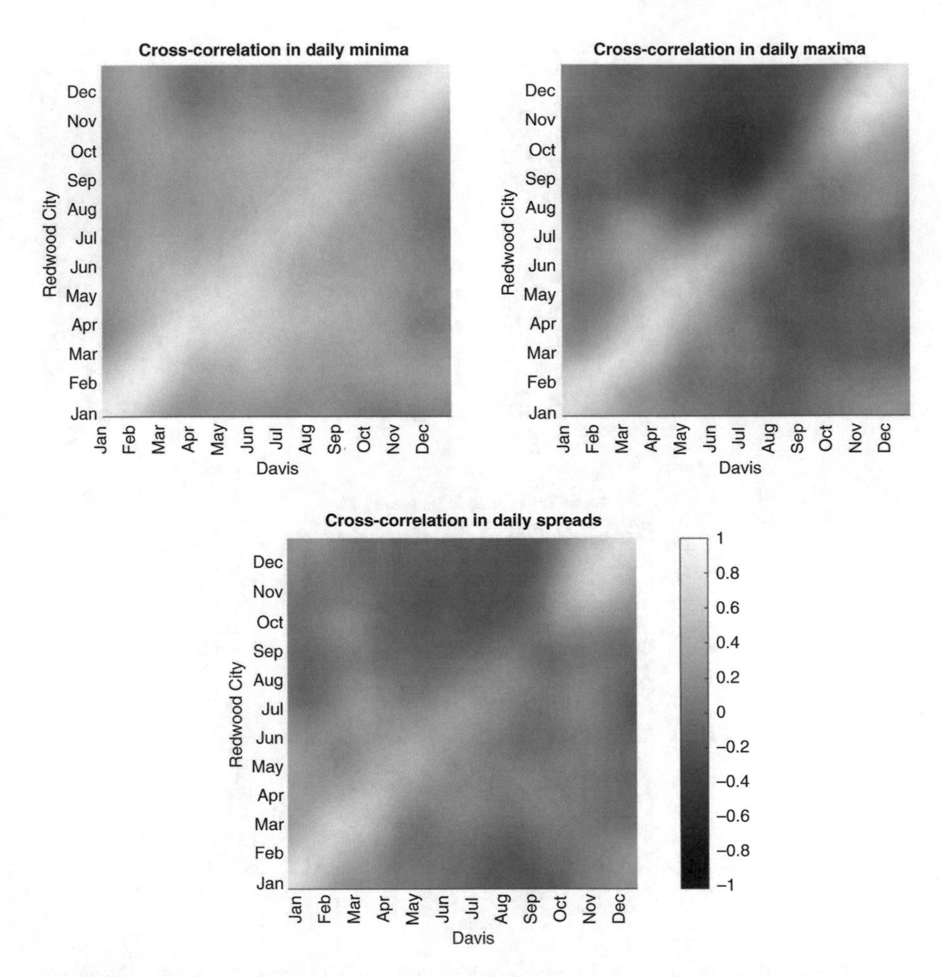

Fig. 18.6 Smoothed estimates for the cross-correlation surfaces between Davis and in Redwood City from 1930 to 2015 for daily temperature minima (LHS), maxima (central plot) and spread (RHS)

$\hat{\beta}_1$ show a clear decreasing trend. This shows that the daily spread is getting smaller in amplitude because the daily minima get closer to the daily maxima.

For the non-parametric regression approach we used a local linear smoother with Gaussian kernel to compute smooth estimates for the time-trends in this data; the bandwidth used was equal to 12 years. Figure 18.10 shows that the trends are indeed roughly linear in the case of temperature minima and temperature spread and have increasing and decreasing tendancies respectively, across the twentieth century. On the contrary, temperature maxima are out of sync between the two locations.

In a similar manner we examined the estimated slopes for the second mode of variation. While visually some slopes (Fig. 18.11, LHS) appear significant at

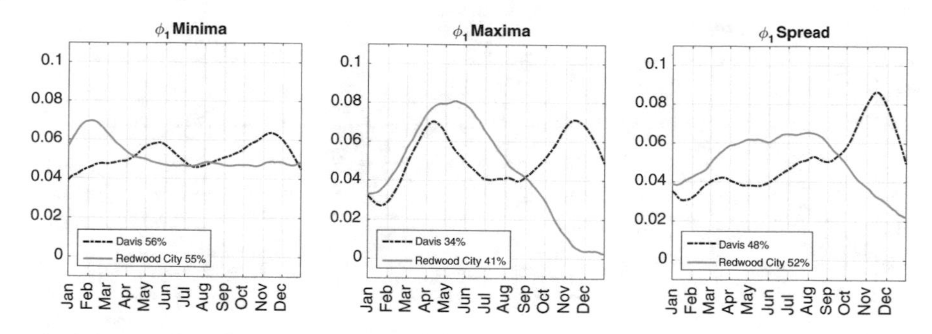

Fig. 18.7 The first estimated eigenfunction for the daily temperature minima (LHS), maxima (central plot) and spread (RHS) as a function of the day of the year. The Fraction-of-Variance-Explained for each component is shown in the legend

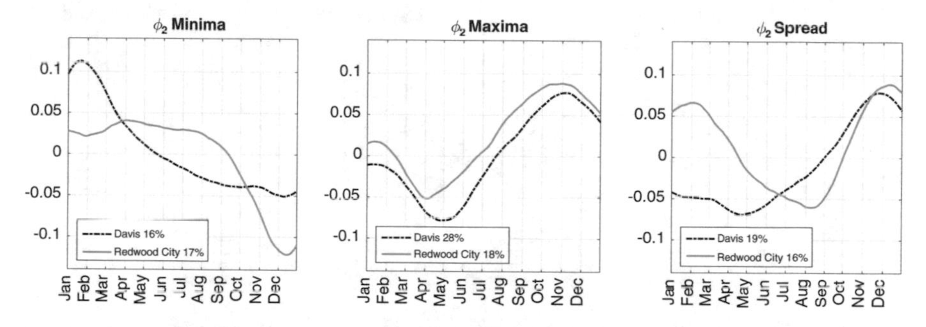

Fig. 18.8 The second estimated eigenfunction for the daily temperature minima (LHS), maxima (central plot) and spread (RHS) as a function of the day of the year. The Fraction-of-Variance-Explained for each component is shown in the legend

Table 18.2 The t-values and p-values for the slopes of the fitted simple linear regressions, assuming normality for the first principal component scores ξ_1

t-Values (p-values)	Minima	Maxima	Spread
Davis	10.480 (<1e−9)	−0.090 (0.928)	−7.311 (<1e−9)
Redwood City	5.471 (<1e−5)	0.395 (0.694)	−4.414 (<1e−4)

first glance, after correcting for multiple comparisons their effects are statistically insignificant.

Based on these findings we conclude that there are suggestions of warming effect in our current dataset. As mentioned before, several extrema influencing phenomena (e.g., urban microclimate, soil moisture, etc.) have not been actively accounted for; a full climate change assessment would definitely need to account for such confounders; for example, it is known that the increasing irrigation in California mitigates nightly cooling.

As a final step of this analysis we study patterns of the random noise in the data; for example, to assess potential issues of heteroskedasticity; to see whether the behavior of temperature patterns over shorter time spans has become more erratic

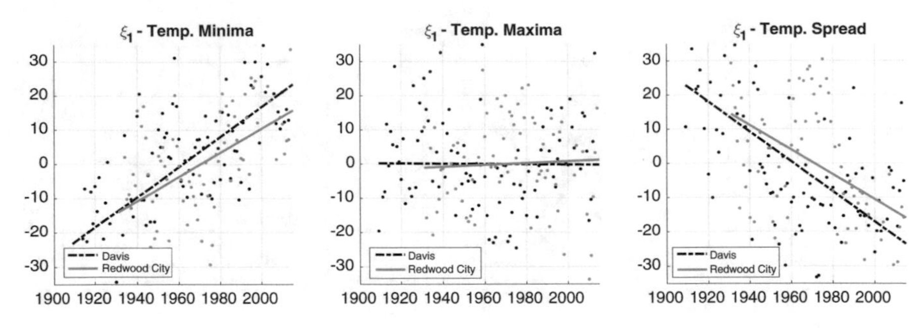

Fig. 18.9 The estimates for the first principal scores ($\hat{\xi}_1$) for daily temperature minima (LHS), maxima (central plot) and spread (RHS), plotted against calendar year, and overlaid with least squares fit of simple linear regression

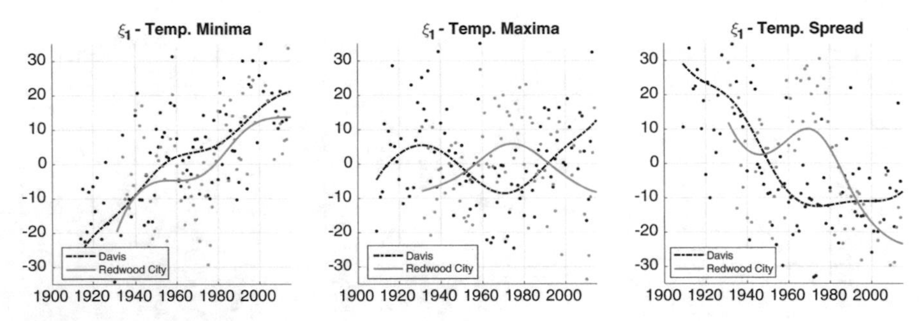

Fig. 18.10 The estimates for the first principal scores ($\hat{\xi}_1$) for daily temperature minima (LHS), maxima (central plot) and spread (RHS), plotted against calendar year, and overlaid with non-parametric fit of local linear kernel regression

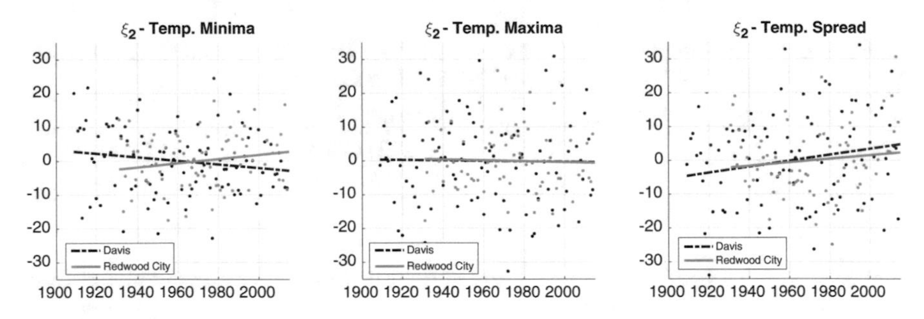

Fig. 18.11 The estimates second principal scores ($\hat{\xi}_2$) for the daily temperature minima (LHS), maxima (central plot) and spread (RHS)

in recent times. We represent each temperature series with two trajectories; one for the actual smooth temperature extrema process and a second one for the realization of the smooth variance process that reflects patterns in the deviations from smooth temperature curves. We examine the same three processes as previously: temperature daily minimum, maximum, and spread (Table 18.3).

Table 18.3 The t-values and p-values for the slopes of the fitted simple linear regressions, assuming normality for the second principal component scores ξ_2

t-Values (p-values)	Minima	Maxima	Spread
Davis	−1.642 (0.104)	−0.168 (0.867)	1.904 (0.060)
Redwood City	1.463 (0.148)	−0.318 (0.751)	1.019 (0.312)

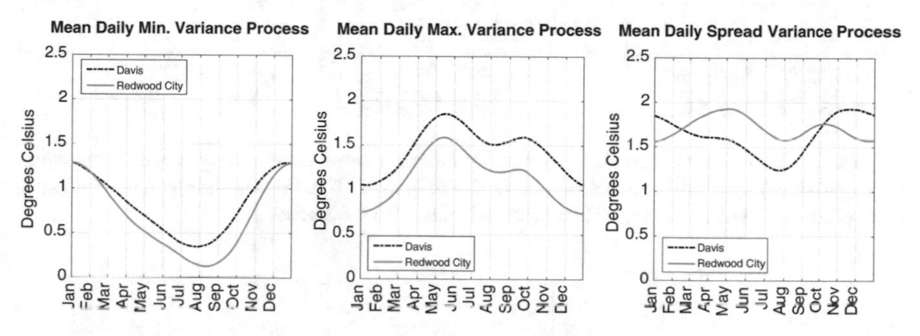

Fig. 18.12 The estimated mean functions for the functional variance processes V in Eq. (18.20) for the daily temperature minima (LHS) and maxima (RHS). The smoothing bandwidth for the Epanechnikov kernel used was 63 days

Looking first at the respective mean processes we notice that they appear very similar between the two different locations (Fig. 18.12). This suggests that the noise variance patterns are not location-specific between the two locations. This appears in line with findings about the correlation of temperature anomalies for neighboring stations (Hansen and Lebedeff 1987); as Hansen et al. note: "[correlation] typically remains above 50% to distances of about 1200 miles at most latitudes" (Hansen et al. 2010).

In Davis the overall mean variance is typically slightly higher than in Redwood City, again hinting towards the tampering effect of the nearby sea mass in Redwood City. The minima variation trend appears more variable during the winter. Similarly the variance in the maxima seems to be highest in late spring with a small maximum in early autumn.

Examining the principal modes of variation (Figs. 18.13 and 18.14) the Fraction-of-Variance-Explained (FVE) from the first estimated eigenfunction is comparable to the FVE of the second estimated eigenfunction in the case of daily minimum and daily maximum variation. This strongly suggests that two or more strong independent sources of variance are in place. In addition, particularly in the variance processes of the daily temperature maximum and temperature spread, there are clearly defined peaks of variation as reflected in the first eigenfunction. The peaks are partially aligned in the daily temperature spread. The second estimated eigenfunction for all three variances processes shows strong seasonal patterns.

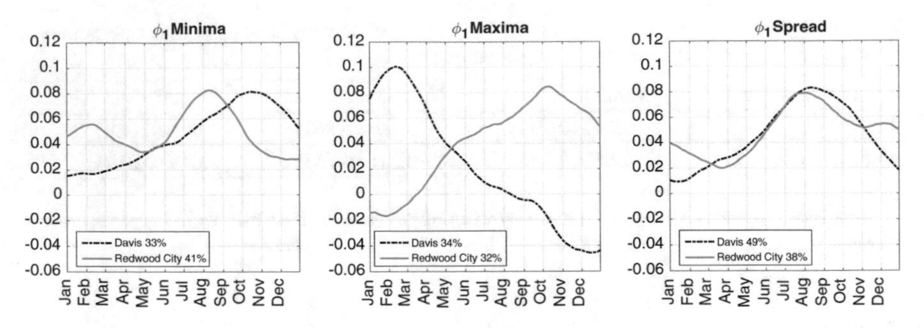

Fig. 18.13 The first estimated eigenfunction for the variance process of daily temperature minima (LHS) and maxima (RHS) as a function of the day of the year. The Fraction-of-Variance-Explained from each component in its respective process is shown in the legend

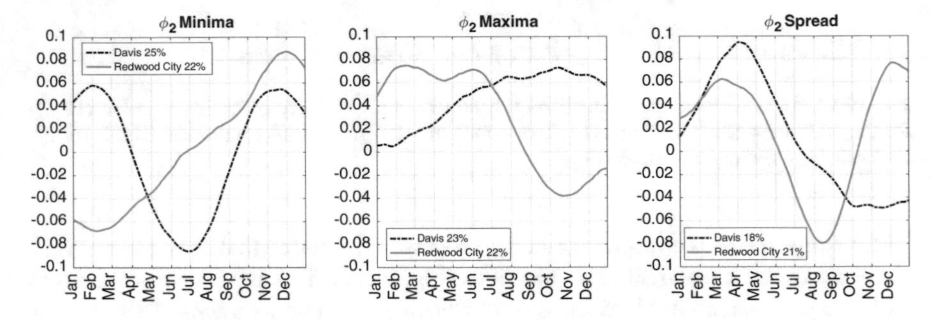

Fig. 18.14 The second estimated eigenfunction for the variance process of daily temperature minima (LHS), maxima (RHS) as a function of the day of the year. The Fraction-of-Variance-Explained from each component in its respective process is shown in the legend

In terms of time-trends, perhaps counter-intuitively the variance processes appear to have clear downward trends (Fig. 18.15, Table 18.4 for the linear regression and Fig. 18.16 for the non-parametric smoothed trends). This means that local variation from the smooth process if at all, declined over the years. Confounders of such trends can be urbanization, increased CO_2 in the air or increased irrigation, all or some of which might conspire to force temperatures closer to their overall smooth trend. Analyzing the second estimated eigenfunction similarly revealed no graphically obvious or statistically significant time trends (analysis not shown).

Overall the FVPA shows that indeed there is a clear presence of structured noise in the data. In this section we showed how functional data analysis can be used on a (potentially) massive dataset to gain insights about a system that exhibits complex time dynamics.

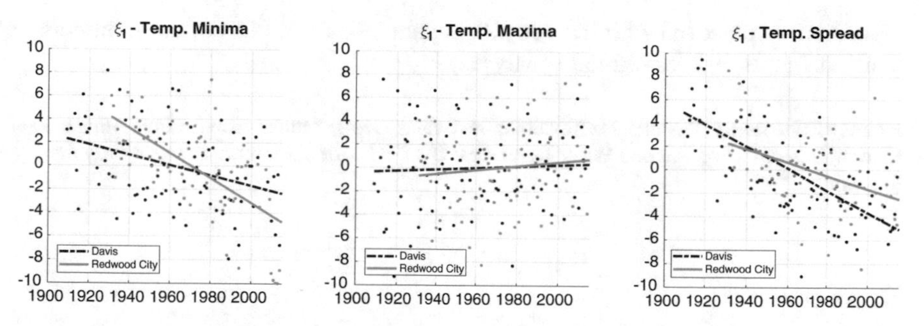

Fig. 18.15 The estimates for the first principal scores ($\hat{\zeta}_1$) for the variance process of daily temperature minima (LHS), maxima (central plot) and spread (RHS), plotted against calendar year, and overlaid with least squares fit of simple linear regression

Table 18.4 The t-values and p-values for the slopes of the fitted simple linear regressions, assuming normality for the second principal component scores ζ_1 of the functional variance process

t-Values (p-values)	Minima	Maxima	Spread
Davis	−4.216 (<1e−4)	0.552 (0.582)	−9.263 (<1e−9)
Redwood City	−6.991 (<1e−7)	0.918 (0.362)	−3.888 (<1e−3)

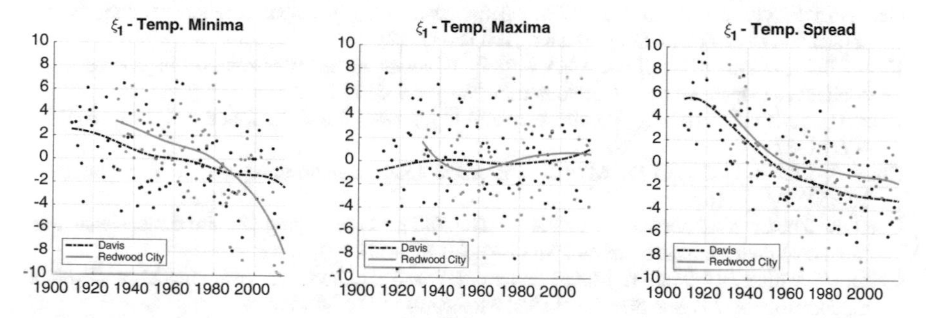

Fig. 18.16 The estimates for the first principal scores ($\hat{\zeta}_1$) for the variance process of daily temperature minima (LHS), maxima (central plot) and spread (RHS), plotted against calendar year, and overlaid with non-parametric fit of local linear kernel regression

18.7 Conclusions

While in this work we focused on temperature data, it should be noted that functional data analysis applications are routinely used in areas like genomics and medical imaging where massive datasets are typically encountered; compare (Chen et al. 2015). As the presence of sensors in mobile as well as wearable devices becomes more commonplace, even larger data collections of function-valued characteristics are poised to become part of standard analytics tasks. Meaningful and principled techniques from functional data analysis will therefore play an increasing role. Research of speeding up common FDA computations is still needed. With fast algorithms in place that will include proper parallelization, FDA offers an important analysis framework for big-data analysts.

Acknowledgements Pantelis Hadjipantelis and Hans-Georg Müller would like to thank Paul Ullrich, Dept. of Land, Air and Water Resources, UC Davis, for his comments on an early draft of this work.

References

Antoniadis A, Gregoire G, McKeague IW (1994) Wavelet methods for curve estimation. J Am Stat Assoc 89(428):1340–1353
Aston JAD, Chiou JM, Evans JP (2010) Linguistic pitch analysis using functional principal component mixed effect models. J R Stat Soc Ser C (Appl Stat) 59(2):297–317
Barber D (2012) Bayesian reasoning and machine learning, chaps 12, 16, 19 & 21. Cambridge University Press, Cambridge
Breiman L (1996) Bagging predictors. Mach Learn 24:123–140
Bruns A (2004) Fourier-, Hilbert- and wavelet-based signal analysis: are they really different approaches? J Neurosci Methods 137(2):321–332
Bühlmann P, De Geer Sara V (2011) Statistics for high-dimensional data: methods, theory and applications. Springer Science & Business Media, Berlin
Cangelosi R, Goriely A (2007) Component retention in principal component analysis with application to cDNA microarray data. Biol Direct 2:2+
Castro PE, Lawton WH, Sylvestre EA (1986) Principal modes of variation for processes with continuous sample curves. Technometrics 28(4):329–337
Chen D, Müller H-G (2012) Nonlinear manifold representations for functional data. Ann Stat 40(1):1–29
Chen K, Müller H-G (2012) Modeling repeated functional observations. J Am Stat Assoc 107(500):1599–1609
Chen K, Zhang X, Petersen A, Müller H-G (2015) Quantifying infinite-dimensional data: functional data analysis in action. Stat Biosci 9:582–604
Chiou JM, Müller HG, Wang JL (2003) Functional quasi-likelihood regression models with smooth random effects. J R Stat Soc Ser B (Stat Methodol) 65(2):405–423
Coffey N, Harrison AJ, Donoghue OA, Hayes K (2011) Common functional principal components analysis: a new approach to analyzing human movement data. Hum Mov Sci 30(6):1144–1166
Cunningham JP, Gilja V, Ryu SI, Shenoy KV (2009) Methods for estimating neural firing rates, and their application to brain-machine interfaces. Neural Netw 22(9):1235–1246
Davison AC Statistical models, Chap 4. Cambridge University Press, Cambridge
Delicado P (2007) Functional k-sample problem when data are density functions. Comput Stat 22(3):391–410
Delicado P, Giraldo R, Comas C, Mateu J (2010) Statistics for spatial functional data: some recent contributions. Environmetrics 21(3–4):224–239
Dryden IL (2005) Statistical analysis on high-dimensional spheres and shape spaces. Ann Stat 33:1643–1665
Dryden IL, Koloydenko A, Zhou D (2009) Non-euclidean statistics for covariance matrices, with applications to diffusion tensor imaging. Ann Appl Stat 3:1102–1123
Efron B, Tibshirani RJ (1994) An introduction to the bootstrap. CRC Press, Boca Raton
Eubank RL (1999) Nonparametric regression and spline smoothing. CRC Press, Boca Raton
Fan J, Gijbels I (1996) Local polynomial modelling and its applications: monographs on statistics and applied probability, vol 66. CRC Press, Boca Raton
Ferré L, Yao A-F (2003) Functional sliced inverse regression analysis. Statistics 37(6):475–488

Freedman D, Diaconis P (1981) On the histogram as a density estimator: L 2 theory. Probab Theory Relat Fields 57(4):453–476

Gartner Inc. (2011) Gartner says solving 'big data' challenge involves more than just managing volumes of data. http://www.gartner.com/newsroom/id/1731916. Accessed 1 Sept 2015

Gasser T, Müller H-G (1984) Estimating regression functions and their derivatives by the kernel method. Scand J Stat 11:171–185

Ghodsi A (2006) Dimensionality reduction a short tutorial. Technical report, Department of Statistics and Actuarial Science, University of Waterloo, Waterloo

Good IJ (1969) Some applications of the singular decomposition of a matrix. Technometrics 11(4):823–831

Grabe E, Kochanski G, Coleman J (2007) Connecting intonation labels to mathematical descriptions of fundamental frequency. Lang Speech 50(3):281–310

Graves S, Hooker G, Ramsay J (2009) Functional data analysis with R and MATLAB. Springer, New York

Guo W (2002) Functional mixed effects models. Biometrics 58:121–128

Hadjipantelis PZ, Aston JAD, Evans JP (2012) Characterizing fundamental frequency in Mandarin: a functional principal component approach utilizing mixed effect models. J Acoust Soc Am 131(6):4651–4664

Hadjipantelis PZ, Müller H-G, Aston JAD, Evans JP (2015) Unifying amplitude and phase analysis: a compositional data approach to functional multivariate mixed-effects modeling of mandarin Chinese. J Acoust Soc Am 110(510):545–559

Halko N, Martinsson P-G, Shkolnisky Y, Tygert M (2011) An algorithm for the principal component analysis of large data sets. SIAM J Sci Comput 33(5):2580–2594

Hall P, Müller H-G (2003) Order-preserving nonparametric regression, with applications to conditional distribution and quantile function estimation. J Am Stat Assoc 98:598–608

Hall P, Müller HG, Wang JL (2006) Properties of principal component methods for functional and longitudinal data analysis. Ann Stat 34(3):1493–1517

Hall P, Müller H-G, Yao F (2008) Modelling sparse generalized longitudinal observations with latent Gaussian processes. J R Stat Soc Ser B (Stat Methodol) 70(4):703–723

Hansen J, Lebedeff S (1987) Global trends of measured surface air temperature. J Geophys Res Atmos (1984–2012) 92(D11):13345–13372

Hansen J, Ruedy R, Sato M, Lo K (2010) Global surface temperature change. Rev Geophys 48(4). https://doi.org/10.1029/2010RG000345

Harezlak J, Wu MC, Wang M, Schwartzman A, Christiani DC, Lin X (2008) Biomarker discovery for arsenic exposure using functional data. Analysis and feature learning of mass spectrometry proteomic data. J Proteome Res 7(1):217–224

Horváth L, Kokoszka P (2012) Inference for functional data with applications, vol 200. Springer Science & Business Media, Berlin

Hoyle DC (2008) Automatic PCA dimension selection for high dimensional data and small sample sizes. J Mach Learn Res 9(12):2733–2759

Illian J, Benson E, Crawford J, Staines H (2006) Principal component analysis for spatial point processes — assessing the appropriateness of the approach in an ecological context. In: Case studies in spatial point process modeling. Springer, Berlin, pp 135–150

Izenman AJ (2008) Modern multivariate statistical techniques: regression, classification and manifold learning, Chap 6. Springer, New York

James GM, Hastie TJ (2001) Functional linear discriminant analysis for irregularly sampled curves. J R Stat Soc Ser B Stat Methodol 63:533–550

James GM, Hastie TJ, Sugar CA (2000) Principal component models for sparse functional data. Biometrika 87(3):587–602

Jiang C-R, Aston JAD, Wang J-L (2009) Smoothing dynamic positron emission tomography time courses using functional principal components. NeuroImage 47(1):184–193

Jiang C-R, Yu W, Wang J-L et al (2014) Inverse regression for longitudinal data. Ann Stat 42(2):563–591

Jolliffe I (2005) Principal component analysis. In: Encyclopedia of statistics in behavioral science, Chap 3. Wiley Online Library
Kenobi K, Dryden IL, Le H (2010) Shape curves and geodesic modelling. Biometrika 97(3):567–584
Kleffe J (1973) Principal components of random variables with values in a separable Hilbert space. Stat J Theor Appl Stat 4:391–406
Knight CG, Kassen R, Hebestreit H, Rainey PB (2004) Global analysis of predicted proteomes: functional adaptation of physical properties. Proc Natl Acad Sci USA 101(22):8390–8395
Kohavi R, Sahami M (1996) Error-based and entropy-based discretization of continuous features. In: KDD, pp 114–119
Kurtek S, Klassen E, Ding Z, Srivastava A (2010) A novel Riemannian framework for shape analysis of 3d objects. In: IEEE conference on computer vision and pattern recognition, CVPR 2010. IEEE, New York, pp 1625–1632
Lee DD, Seung HS (2001) Algorithms for non-negative matrix factorization. In: Advances in neural information processing systems, pp 556–562
Levitin DJ, Nuzzo RL, Vines BW, Ramsay JO (2007) Introduction to functional data analysis. Can Psychol 48(3):135
Li K-C (1991) Sliced inverse regression for dimension reduction. J Am Stat Assoc 86(414):316–327
Liu B, Müller H-G (2009) Estimating derivatives for samples of sparsely observed functions, with application to online auction dynamics. J Am Stat Assoc 104(486):704–717
Ma Y, Zhu L (2013) A review on dimension reduction. Int Stat Rev 81(1):134–150
Mahoney MW (2011) Randomized algorithms for matrices and data. Found Trends Mach Learn 3(2):123–224
Mallat SG (1989) A theory for multiresolution signal decomposition: the wavelet representation. IEEE Trans Pattern Anal Mach Intell 11(7):674–693
Menne MJ, Durre I, Vose RS, Gleason BE, Houston TG (2012) An overview of the global historical climatology network-daily database. J Atmos Oceanic Technol 29(7):897–910
Mercer J (1909) Functions of positive and negative type, and their connection with the theory of integral equations. Philos Trans R Soc Lond Ser A. Containing Pap Math Phys Character. 209:415–446
Minka TP (2001) Automatic choice of dimensionality for PCA. Adv Neural Inf Proces Syst 15:598–604
Morris JS, Carroll RJ (2006) Wavelet-based functional mixed models. J R Stat Soc Ser B 68:179–199
Müller H-G, Stadtmüller U, Yao F (2006) Functional variance processes. J Am Stat Assoc 101(475):1007–1018
Müller H-G, Wu S, Diamantidis AD, Papadopoulos NT, Carey JR (2009) Reproduction is adapted to survival characteristics across geographically isolated medfly populations. Proc R Soc Lond B Biol Sci. https://doi.org/10.1098/rspb.2009.1461
Papadimitriou CH (2003) Computational complexity, Chap 11. Wiley, New York
Petersen A, Müller H-G (2016) Functional data analysis for density functions by transformation to a Hilbert space. Ann Stat 44(1):183–218
Petrone S, Guindani M, Gelfand AE (2009) Hybrid Dirichlet mixture models for functional data. J R Stat Soc Ser B (Stat Methodol) 71(4):755–782
Pigoli D, Sangalli LM (2012) Wavelets in functional data analysis: estimation of multidimensional curves and their derivatives. Comput Stat Data Anal 56(6):1482–1498
Ramsay JO (2002) Multilevel modeling of longitudinal and functional data. In: Modeling intraindividual variability with repeated measures data: methods and applications. CRC Press, Boca Raton, pp 171–201
Ramsay JO, Silverman BW (2005) Functional data analysis, Chaps 3, 4 and 7. Springer, New York
Ramsay JO, Munhall KG, Gracco VL, Ostry DJ (1996) Functional data analyses of lip motion. J Acoust Soc Am 99(6):3718–3727

Rao CR (1965) The theory of least squares when the parameters are stochastic and its application to the analysis of growth curves. Biometrika 52(3/4):447–458
Rasmussen CE (2004) Gaussian processes in machine learning. In: Advanced lectures on machine learning. Springer, New York, pp 63–71
Roweis ST, Saul LK (2000) Nonlinear dimensionality reduction by locally linear embedding. Science 290(5500):2323–2326
Sangalli LM, Ramsay JO, Ramsay TO (2013) Spatial spline regression models. J R Stat Soc Ser B (Methodol) 75(4):1–23
Schölkopf B, Smola A, Müller K-R (1998) Nonlinear component analysis as a kernel eigenvalue problem. Neural Comput 10(5):1299–1319
Scott DW (1979) On optimal and data-based histograms. Biometrika 66(3):605–610
Silverman BW (1985) Some aspects of the spline smoothing approach to non-parametric regression curve fitting. J R Stat Soc Ser B (Methodol) 47:1–52
Sturges HA (1926) The choice of a class interval. J Am Stat Assoc 21(153):65–66
Tenenbaum JB, De Silva V, Langford JC (2000) A global geometric framework for nonlinear dimensionality reduction. Science 290(5500):2319–2323
Theis S (2005) Deriving probabilistic short-range forecasts from a deterministic high-resolution model. PhD thesis, University of Bonn - Universität Bonn
Tibshirani R (1996) Regression shrinkage and selection via the lasso. J R Stat Soc Ser B (Methodol) 58(1):267–288
Tucker LR (1958) Determination of parameters of a functional relationship by factor analysis. Psychometrika 23:19–23
Valderrama MJ (2007) An overview to modelling functional data. Comput Stat 22(3):331–334
Wang S, Jank W, Shmueli G (2008) Explaining and forecasting online auction prices and their dynamics using functional data analysis. J Bus Econ Stat 26(2):144–160
Weinberger KQ, Sha F, Saul LK (2004) Learning a kernel matrix for nonlinear dimensionality reduction. In: Proceedings of the twenty-first international conference on Machine learning. ACM, New York, p 106
Wink AM, Roerdink JBTM (2004) Denoising functional MR images: a comparison of wavelet denoising and Gaussian smoothing. IEEE Trans Med Imag 23(3):374–387
Witten R, Candès E (2015) Randomized algorithms for low-rank matrix factorizations: sharp performance bounds. Algorithmica 72(1):264–281
Worsley KJ, Liao CH, Aston J, Petre V, Duncan GH, Morales F, Evans AC (2002) A general statistical analysis for FMRI data. Neuroimage 15(1):1–15
Yang W, Müller H-G, Stadtmüller U (2011) Functional singular component analysis. J R Stat Soc Ser B (Stat Methodol) 73(3):303–324
Yao F, Müller HG, Wang J-L (2005) Functional data analysis for sparse longitudinal data. J Am Stat Assoc 100(470):577–590
Zhang X, Shao X, Hayhoe K, Wuebbles DJ (2011) Testing the structural stability of temporally dependent functional observations and application to climate projections. Electron J Stat 5:1765–1796

Chapter 19
Bayesian Spatiotemporal Modeling for Detecting Neuronal Activation via Functional Magnetic Resonance Imaging

Martin Bezener, Lynn E. Eberly, John Hughes, Galin Jones, and Donald R. Musgrove

Abstract We consider recent developments in Bayesian spatiotemporal models for detecting neuronal activation in fMRI experiment. A Bayesian approach typically results in complicated posterior distributions that can be of enormous dimension for a whole-brain analysis, thus posing a formidable computational challenge. Recently developed Bayesian approaches to detecting local activation have proved computationally efficient while requiring few modeling compromises. We review two such methods and implement them on a data set from the Human Connectome Project in order to show that, contrary to popular opinion, careful implementation of Markov chain Monte Carlo methods can be used to obtain reliable results in a matter of minutes.

Keywords Bayesian variable selection · fMRI · MCMC · Spatiotemporal · Areal model

---

Authors are listed in alphabetical order.

M. Bezener
Stat-Ease, Inc., Minneapolis, MN, USA

L. E. Eberly · D. R. Musgrove
Division of Biostatistics, University of Minnesota, Twin Cities, Minneapolis, MN, USA
e-mail: lynn@biostat.umn.edu; musgr007@umn.edu

J. Hughes
Department of Biostatistics and Informatics, Colorado School of Public Health, University of Colorado, Denver, CO, USA
e-mail: j.hughes@ucdenver.edu

G. Jones (✉)
School of Statistics, University of Minnesota, Twin Cities, Minneapolis, MN, USA
e-mail: galin@umn.edu; galin@stat.umn.edu

© Springer International Publishing AG, part of Springer Nature 2018

W. K. Härdle et al. (eds.), *Handbook of Big Data Analytics*, Springer Handbooks of Computational Statistics, https://doi.org/10.1007/978-3-319-18284-1_19

19.1 Introduction

Functional neuroimaging experiments often aim to either uncover localized regions where the brain activates during a task or describe the networks required for a particular brain function. Our focus is on functional magnetic resonance imaging (fMRI) techniques to study localized neuronal activation in response to a task. Neuronal activation occurs in milliseconds and is not observed directly in fMRI experiments. However, activation of neurons leads to an increase in metabolic activity, resulting in an increase of oxygenated blood flow to the activated regions of the brain. The magnetic properties of oxygen can then be exploited to measure the so-called blood oxygen level dependent (BOLD) signal contrast.

The BOLD signal response is not observed at the neuronal level. Instead the image space is partitioned into voxels in a rectangular three-dimensional lattice. The partition size is often between 200,000 and 500,000 voxels. The BOLD response is typically observed for each voxel at each of several hundred time points 2–3 s apart. The nature of the BOLD response is somewhat complicated. The BOLD response increases above baseline roughly 2 s after the onset of neuronal activation, peaks 5–8 s after activation, and falls below baseline for 10 or so seconds (see, e.g., Aguirre et al. 1997). While this describes the general shape of the hemodynamic response function (HRF), it is well known that the specific hemodynamic response can depend on the location of the voxel and the nature of the task (Aguirre et al. 1998). There is also a complicated spatial dependence; activation tends to occur in groups of voxels, but activation is not limited to spatially contiguous voxels since long-range spatial associations are common. Thus, even for a single subject, there can be an enormous amount of data that exhibits complicated spatiotemporal dependence.

fMRI analyses begin by preprocessing the data to adjust for motion, physiologically-based noise (e.g., cardiac and respiratory sources), and scanner drift. Preprocessing can also include segmentation, spatial co-registration, normalization, and spatial smoothing. Preprocessing is not our focus, but the reader can find much more about these topics in Friston et al. (2007), Huettel et al. (2009), Kaushik et al. (2013), Lazar (2008), Lindquist (2008), Mikl et al. (2008), and Triantafyllou et al. (2006) among many others.

Once preprocessing is complete, statistical modeling continues to play a crucial role in the analysis. There can be several goals in an fMRI experiment, including characterization of the HRF, estimation of the magnitude and volume of neuronal activation, and assessment of functional connectivity. Our focus is on detecting neuronal activation, but it has been argued that HRF estimation and activation detection are inextricable (cf. Makni et al. 2008).

Classical approaches to detecting activation are based on voxel-wise univariate statistics, often using a linear model for each voxel, which are displayed in a statistical parametric map (SPM). Of course, SPMs do not account for the inherent spatial correlation among voxels, and there is a problem of multiplicity in conducting inference. These issues are typically addressed through the use of Gaussian random field theory (Friston et al. 1994, 1995, 2007; Worsley et al.

1992). SPMs are conceptually simple and computationally efficient. Hence, they see widespread use in the neuroimaging community. However, these methods do not result in a full statistical model, and the required assumptions have often been criticized as unrealistic (see, e.g., Holmes et al. 1996).

There has been a recent explosion in the development of Bayesian models for neuroimaging applications (see Bowman 2014; Friston et al. 2007; Lazar 2008; Zhang et al. 2015, for comprehensive reviews). The most common approach to constructing Bayesian models for detecting local activation begins with a general linear model. For voxel $v = 1, \ldots, N$ and time $t = 1, \ldots, T$, let $Y_{v,t}$ be the value of the BOLD signal, and assume

$$Y_{v,t} = z_t^T a_v + x_{v,t}\beta_v + \epsilon_{v,t} \tag{19.1}$$

where $z_t^T a_v$ is the baseline drift, which is modeled as a linear combination of basis functions, and $\epsilon_{v,t}$ is the measurement error. The part of the linear model of primary interest is $x_{v,t}\beta_v$. Here $x_{v,t}$ is a fixed and known transformed input stimulus (see Hensen and Friston 2007, for a thorough introduction to this topic), and β_v is the activation amplitude. When β_v is nonzero the voxel is "active," and hence our goal is to find the voxels for which this occurs. Accounting for the spatiotemporal nature of the response can be accomplished by making distributional assumptions on the $\epsilon_{v,t}$ and using appropriate prior distributions on the parameters.

A Bayesian approach typically results in complicated posterior distributions that can be of enormous dimension for a whole-brain analysis, thus posing a formidable computational challenge. One common approach to addressing the computational difficulties is to make modeling compromises, such as accounting for spatial dependence while ignoring temporal dependence (Genovese 2000; Smith et al. 2003; Smith and Fahrmeir 2007). Even so, the required computation is typically still too intensive for the methods to become widely adopted.

Recently developed Bayesian approaches to detecting local activation have proved computationally efficient while requiring few modeling compromises. In Sect. 19.2 we discuss two novel Bayesian areal models. In Sects. 19.2.1.3 and 19.2.2.5 we implement Markov chain Monte Carlo (MCMC) algorithms which, although the posteriors are high dimensional, illustrate that MCMC methods can be implemented so that reliable results are obtained in a matter of minutes. In the rest of this section we describe the data which is analyzed in Sects. 19.2.1.3, 19.2.2.5, and 19.2.3.

19.1.1 Emotion Processing Data

The data was collected as part of the Human Connectome Project (Essen et al. 2013), and aims to evaluate emotional processing. The experiment was a modified version of the design proposed by Hariri et al. (2002), which we now summarize.

The subject laid in a scanner and completed one of two tasks arranged in a block design. In the first task, two faces were displayed in the top half of a screen. One of

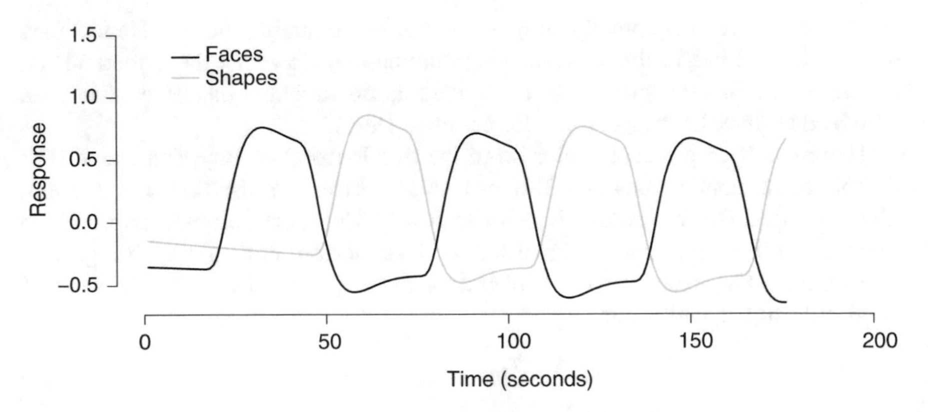

Fig. 19.1 Hemodynamic response functions corresponding to the modified Hariri task

the faces had a fearful expression, and the other had an angry expression. A third face was displayed in the bottom half of the screen. The third face had either a fearful expression or an angry expression. The subject chose which of the two faces in the top half of the screen matched the expression of the third face in the bottom half of the screen. Each set of faces was displayed for 2 s, after which there was a 1-s pause.

A second task was functionally identical to the first task, except that geometric shapes were used instead of faces, and the subject had to choose which of the two shapes in the top half of the screen matched the shape in the bottom of the screen. This task was used as a control. Each of the face and shape blocks was 18 s long, with an 8 s pause between successive task blocks. Each pair of blocks was replicated three times. The goal here is to detect which regions of the brain are involved in distinguishing emotional facial expressions.

A total of 176 scans were collected on a 3 T scanner on over 500 subjects. We will consider the data from one randomly selected subject to illustrate our methods. Before data collection, the image space was partitioned into a $91 \times 109 \times 91$ rectangular lattice comprising voxels of size 2 mm^3. After standard preprocessing and masking, a total of 225,297 voxels remained to be analyzed. Spatial smoothing was applied at 5 mm in each direction. Each of the two task stimulus functions was convolved with a gamma probability density function to produce the hemodynamic response functions shown in Fig. 19.1.

19.2 Variable Selection in Bayesian Spatiotemporal Models

Detecting activation using (19.1) is equivalent to selecting the voxels with nonzero β_v, and hence is a variable selection problem. Bezener et al. (2015), Lee et al. (2014), Musgrove et al. (2015), Smith and Fahrmeir (2007), and Smith et al. (2003)

built on the approach of George and McCulloch (1993, 1997) to variable selection. However, Smith and Fahrmeir (2007) and Smith et al. (2003) ignored temporal correlation, although they did incorporate spatial dependence in their models. Lee et al. (2014) extended the approach of Smith and Fahrmeir (2007) and Smith et al. (2003) to include both spatial and temporal dependence. All three of these papers rely on using a binary spatial Ising prior to model the spatial dependence. While appealing from a modeling perspective, the Ising prior results in substantial computational challenges that can be avoided with the approaches described below. Both approaches are based on partitioning the image into three-dimensional parcels and using a sparse spatial generalized linear mixed model (SGLMM). While there are many commonalities between the two models, there are substantial differences between the models and the required computation.

19.2.1 Bezener et al.'s (2015) Areal Model

Let $Y_v = (Y_{v,1}, \ldots, Y_{v,T_v})^T$ be the time series of BOLD signal image intensities for voxel $v = 1, \ldots, N$. Suppose there are p experimental tasks or stimuli, and let X_v be a known $T_v \times p$ design matrix and β_v be a $p \times 1$ vector. If Λ_v is a $T_v \times T_v$ positive definite matrix, assume

$$Y_v = X_v\beta_v + \epsilon_v \qquad \epsilon_v \sim \mathcal{N}_{T_v}(0, \sigma_v^2\Lambda_v) \ . \tag{19.2}$$

The regression coefficients correspond to activation amplitudes, and detecting neuronal activation is equivalent to detecting the nonzero $\beta_{v,j}$. We will address this through the introduction of latent variables. Let $\gamma_{v,j}$ be binary random variables such that $\beta_{v,j} \neq 0$ if $\gamma_{v,j} = 1$, and $\beta_{v,j} = 0$ if $\gamma_{v,j} = 0$. Let $\gamma_v = (\gamma_{v,1}, \gamma_{v,2}, \ldots, \gamma_{v,p})$, so that $\beta_v(\gamma_v)$ is the vector of nonzero coefficients from β_v, and $X_v(\gamma_v)$ is the corresponding design matrix. Model (19.2) can be expressed as

$$Y_v = X_v(\gamma_v)\beta_v(\gamma_v) + \varepsilon_v \ . \tag{19.3}$$

Consider the covariance matrix $\sigma_v^2\Lambda_v$ from (19.2). We assume that the σ_v^2 are a priori independent and that each is given the standard invariant prior. That is,

$$\pi(\sigma_v^2) \propto \frac{1}{\sigma_v^2} \ .$$

Note that temporal dependence can be modeled through the structure chosen for Λ_v. In addition to the nature of the hemodynamic response, other cyclical neuronal events and the nature of the measurement process indicate that temporal autocorrelation can be substantial in fMRI experiments. Moreover, autoregressive (such as AR(p) for $p = 1$ or $p = 2$) and autoregressive moving average (ARMA) structures are sensible starting points, and are common in neuroimaging applications (see, e.g.,

Lee et al. 2014; Lindquist 2008; Xia et al. 2009; Locascio et al. 1997; Monti 2011). We assume an AR(1) structure for Λ_v and will use an empirical Bayes approach for the prior on Λ_v by estimating it with maximum likelihood to obtain $\hat{\Lambda}_v$ in a pre-processing step. A major advantage to this approach is that it avoids a prohibitively expensive matrix inversion in the MCMC algorithm. In addition, it has been demonstrated to result in reasonable inferences (Lee et al. 2014; Bezener et al. 2015).

We will use an instance of Zellner's g-prior (Zellner 1996) for the prior on $\beta_v(\gamma_v)$. Let

$$\hat{\beta}_v(\gamma_v) = [X_v^T(\gamma_v)\hat{\Lambda}_v^{-1}X_v(\gamma_v)]^{-1}X_v^T(\gamma_v)\hat{\Lambda}_v^{-1}Y_v,$$

and assume the $\beta_v(\gamma_v)$ are conditionally independent and that

$$\beta_v(\gamma_v) \mid Y_v, \sigma_v^2, \gamma_v \sim \mathcal{N}\{\hat{\beta}_v(\gamma_v),\ T_v\sigma_v^2[X_v^T(\gamma_v)\hat{\Lambda}_v^{-1}X_v(\gamma_v)]^{-1}\}\ .$$

This is a data-dependent prior since both $\hat{\Lambda}_v$ and $\hat{\beta}_v(\gamma_v)$ depend on Y_v. Zellner's g-prior depends on a parameter denoted g, and we set $g = T_v$, yielding a unit information prior. The major advantage to this prior is that it results in simpler computation, but similar inferences, than alternative priors (Lee 2010).

Finally, we need priors for the $\gamma_{v,j}$. We choose to work with the prior probabilities of activation $\pi(\gamma_{v,j} = 1)$ since this has been shown to produce activation maps with better edge-preservation properties and classification accuracies (Smith and Fahrmeir 2007). We will assume that the spatial dependence is governed by an underlying areal model (Cressie 1993; Haran 2011; Banerjee et al. 2003), and parcellate the image into G non-overlapping regions, or *parcels*. To ensure efficient computation, we recommend using no more than $G = 500$ parcels.

Let $\gamma_{(j)} = (\gamma_{1,j}, \gamma_{2,j}, \ldots, \gamma_{N,j})$ be the vector of indicators for all voxels for task j, and let $\mathcal{R}_g$ denote the collection of all voxels in parcel g. Let the spatial random effects be denoted $S_{(j)} = (S_{1,j}, S_{2,j}, \ldots, S_{G,j})$. Given that voxel $v \in \mathcal{R}_g$, we assume that the $\gamma_{v,j}$ are independent and

$$\gamma_{v,j} \mid S_{g,j} \overset{\text{ind}}{\sim} \text{Bern}\left(\frac{1}{1+e^{-S_{g,j}}}\right). \tag{19.4}$$

Let c_i and c_k denote the centroid coordinates of parcel i, k, let $||\cdot||$ denote Euclidean distance, and let $r_j > 0$. Then the matrix Γ_j with (i, k)th element given by

$$\Gamma_j(i,k) = \exp\left(-\frac{||c_i - c_k||}{r_j}\right) \tag{19.5}$$

is a valid correlation matrix. Next assume that

$$S_{(j)} \mid \delta_j^2, r_j \overset{\text{ind}}{\sim} \mathcal{N}_G(0,\ \delta_j^2\,\Gamma_j),$$

where δ_j^2 is a smoothing parameter that controls the spatial continuity of the spatial random effects and hence the γ_{vj}.

Finally, Bezener et al. (2015) assume that the δ_j^2 and r_j are a priori independent and have priors

$$\pi(\delta_j^2) \propto \frac{1}{\delta_j^2}$$

and $r_j \sim \chi^2$.

19.2.1.1 Posterior Distribution and MCMC Algorithm

The posterior distribution is thus given by

$$\begin{aligned} q\{\beta(\gamma), \gamma, S, \delta^2, r, \sigma^2 \mid y\} &\propto p\{y \mid \beta(\gamma), \gamma, S, \delta^2, r, \sigma^2\}\pi\{\beta(\gamma), \gamma, S, \delta^2, r, \sigma^2\} \qquad (19.6) \\ &\propto p\{y \mid \beta(\gamma), \gamma, \sigma^2\}\pi\{\beta(\gamma) \mid \gamma, \sigma^2\}\pi(\sigma^2) \\ &\quad \times \pi(\gamma \mid S)\pi(S \mid \delta^2, r)\pi(\delta^2)\pi(r). \end{aligned}$$

The dimension of the posterior in (19.6) is $2p(N + 1) + N + pG$, which can be up to several millions of variables. Our main goals are to determine which tasks and stimuli result in voxel activation as well as to determine the amount of spatial dependence in the images. Thus, it is sufficient to work with the marginal posterior

$$q(\gamma, S, r \mid y) = \int q\{\beta(\gamma), \gamma, S, \delta^2, r, \sigma^2 \mid y\}\, d\beta(\gamma)\, d\sigma^2\, d\delta^2\,, \qquad (19.7)$$

which is derived explicitly by Bezener et al. (2015).

The posterior in (19.7) is still analytically intractable, and so MCMC methods are required to sample from it. Bezener et al. (2015) develop a component-wise MCMC approach based on the posterior full conditional densities. That is,

$$\begin{aligned} q(\gamma \mid S, r, y) &\propto \pi(\gamma \mid S) \prod_{v=1}^{N} (1 + T_v)^{-q_v/2} K(\gamma_v)^{-T_v/2} \\ q(S \mid \gamma, r, y) &\propto \pi(\gamma \mid S)\pi(S \mid r) \\ q(r \mid S, \gamma, y) &\propto \pi(S \mid r)\pi(r)\,. \end{aligned}$$

Schematically, one update of the MCMC algorithm looks like

$$(S, \gamma, r) \to (S', \gamma, r) \to (S', \gamma', r) \to (S', \gamma', r'),$$

where each update is a Metropolis–Hastings step based on the relevant conditional density (for full details, see Bezener et al. 2015).

19.2.1.2 Starting Values

Selection of starting values for the MCMC simulation is an especially critical issue with a high-dimensional posterior density. We suggest two strategies for choosing the MCMC starting values. The first method is straightforward:

1. Set each $\gamma_{v,j}^{(0)} = 0$.
2. Initialize each spatial random effect as $S_{g,j}^{(0)} \sim N(0, \tau^2)$, where τ^2 is small (e.g., $\tau^2 = 0.001$).
3. Set each spatial correlation parameter to its prior mean: $r_j^{(0)} = E[\pi(r_j)]$.

An alternative and more efficient way to choose starting values is to first perform a preliminary frequentist analysis (e.g., using SPM) and choose the starting values as follows:

1. Set each $\gamma_{v,j}^{(0)} = \hat{\gamma}_{v,j}^{\text{freq}}$.
2. Initialize each spatial random effect as $S_{g,j}^{(0)}$ by first computing

$$\hat{\pi}_{g,j} = \frac{1}{n_g} \sum_{v \in \mathcal{R}_g} \hat{\gamma}_{v,j}^{\text{freq}}$$

where n_g is the number of voxels in the gth parcel and then solving (19.4) to get

$$S_{g,j}^{(0)} = \log\left(\frac{\hat{\pi}_{g,j}}{1 - \hat{\pi}_{g,j}}\right).$$

3. Use a variogram with the $S_{g,j}^{(0)}$ from the previous step to determine $r_j^{(0)}$.

19.2.1.3 Emotion Processing Data

We will consider implementation of the method in Sect. 19.2.1 in the emotion processing data described in Sect. 19.1.1. The image was parcellated into 300 regions of approximately equal size. For the MCMC simulation we used the starting values based on a frequentist analysis as described in Sect. 19.2.1.2. The tuning parameters of the MCMC algorithm were chosen so that the acceptance rates for the Metropolis–Hastings steps were approximately 50%. We used standard diagnostic measures to assess convergence. For example, we checked trace plots of the spatial dependence parameters and a subset of 30 randomly selected spatial random effects under both tasks. All diagnostics indicated the Markov chain mixes well.

Table 19.1 Timing for MCMC samples

MCMC iterations	Time (min)
20k	40
100k	187
200k	372

Table 19.2 The number and percentage of active voxels as well as the estimated spatial correlation parameter is reported

	Face task					
	20k		100k		200k	
Iterations	No burn-in	Burn-in	No burn-in	Burn-in	No burn-in	Burn-in
Active (%)	6254 (2.77)	6255 (2.77)	6255 (2.77)	6255 (2.77)	6254 (2.77)	6254 (2.77)
$\hat{r}_{\text{face}}$ (MCSE)	25.08 (0.09)	25.09 (0.12)	24.97 (0.04)	24.92 (0.06)	24.93 (0.03)	24.93 (0.04)

Table 19.3 The number and percentage of active voxels as well as the estimated spatial correlation parameter is reported

	Shape task					
	20k		100k		200k	
Iterations	No burn-in	Burn-in	No burn-in	Burn-in	No burn-in	Burn-in
Active (%)	4197 (1.86)	4197 (1.86)	4200 (1.86)	4199 (1.86)	4200 (1.86)	4200 (1.86)
$\hat{r}_{\text{shape}}$ (MCSE)	22.99 (0.08)	22.92 (0.11)	22.94 (0.04)	22.94 (0.05)	22.93 (0.03)	22.93 (0.04)

We then implemented the MCMC simulation for each of 20,000, 100,000, and 200,000 iterations; Table 19.1 shows the time required for each of these implementations. We estimated the spatial dependence parameters and posterior probabilities of activation using all of the MCMC samples and after discarding the first 50% as burn-in. The batch means method was used to calculate Monte Carlo standard errors for the estimated quantities.

The results of our implementation are reported in Tables 19.2 and 19.3. The estimation is remarkably stable. Not only were the same number of voxels active, the same voxels were active. Burn-in seemed to have little impact except to increase the Monte Carlo standard errors on the estimates. This analysis indicates that we could easily use only 20,000 iterations to obtain stable results which requires only 40 min of sampling time.

19.2.2 Musgrove et al.'s (2015) Areal Model

This approach makes use of a parcellation technique that divides the image into many non-overlapping parcels. Within each parcel, a spatial Bayesian variable selection method is applied that also accounts for voxel-level temporal correlation. A sparse SGLMM prior is used to model the spatial dependence among the activation indicators. Since the parcels are treated as independent the required computation can be done in parallel. Thus parcellation and the sparse SGLMM together permit efficient sampling even though the model is fully Bayesian.

19.2.2.1 Partitioning the Image

There are two natural parcellation techniques: parcellation based on an anatomical atlas (Tzourio-Mazoyer et al. 2002) and uniform parcellation. In the analysis of the emotion processing data in Sect. 19.2.2.5 we will use the uniform parcellation. See Musgrove et al. (2015) for more information on the anatomical parcellation.

The uniform approach is used primarily in the interest of computational efficiency. First, the image is divided into H cubes, each of which has n_0 voxels on a side. For example, $n_0 = 9$ results in parcels of size 729 voxels. Since the brain is not rectangular, many of the parcels will include fewer than n_0^3 voxels. The sparse SGLMM performs best with a minimum of 500 voxels per parcel. Thus, an algorithm is implemented that iteratively identifies parcels with less than 500 voxels, combines them with adjacent parcels, and creates new parcels with a minimum of 500 voxels while ensuring that the underlying graph is connected. The final dataset for analysis comprises G parcels, with the gth parcel having n_g voxels.

19.2.2.2 Spatial Bayesian Variable Selection with Temporal Correlation

Recall the notation of Sect. 19.2.1. The approach here is also based on a Bayesian variable selection scheme (George and McCulloch 1997) for the regression coefficients of the voxel-level regression of

$$Y_v = X_v(\gamma_v)\beta_v(\gamma_v) + R_v\rho_v + \varepsilon_v, \qquad \varepsilon_v \sim \mathcal{N}(0, \sigma_v^2 I), \tag{19.8}$$

where R_v is a voxel-level lagged prediction matrix that is introduced to model temporal correlation. Each of β_v, γ_v, ρ_v, and σ_v^2 is unknown. Variable selection is carried out in part by placing a spike-and-slab mixture prior on the regression coefficients β_v such that each $\beta_{v,j}$, $j = 1, \ldots, p$, is drawn from a diffuse normal distribution (the slab) or a point mass at zero (the spike). This structure reflects the prior belief that a coefficient is nonzero or zero, respectively. To facilitate MCMC sampling, latent indicator variables $\gamma_v = (\gamma_{v,1}, \ldots, \gamma_{v,p})$ are used such that the mixture prior for each β_{vj} has the form

$$\pi(\beta_{v,j} \mid \gamma_{v,j}) = \gamma_{v,j}\mathcal{N}\left(0, \tau_j^2\right) + (1 - \gamma_{v,j})I_0, \tag{19.9}$$

where τ_j^2 is an unknown stimulus-level variance and I_0 denotes a point mass at zero. This prior specification makes the natural assumption that the regression coefficients are a priori independent conditional on the indicator variables (George and McCulloch 1997). Spatial dependence between voxels is modeled by placing a spatial prior on the indicator variables.

To account for the serial correlation present in the univariate voxel time series, an AR model of order r is easily implemented and computationally efficient. Similar to Penny et al. (2003), the matrix of lagged prediction errors, denoted R_v, is included in the regression model. The AR coefficients $\rho_{v,r} = (\rho_{v,1}, \ldots, \rho_{v,r})'$ are assumed a priori independent and normally distributed with mean zero and known variance, which is typically taken to be "large." To complete the voxel-level prior specification, the error variance and stimulus-level variance are assumed a priori independent with default priors $\pi(\sigma_v^2) \propto 1/\sigma_v^2$ and $\pi(\tau_j^2) \propto 1/\tau_j^2$, respectively. In this way, regression coefficients across voxels share a prior variance, resulting in additional smoothing beyond that induced by the spatial prior.

19.2.2.3 Sparse SGLMM Prior

The spatial prior is used to model the voxel- and task-specific binary indicator variables γ_{vj}. The chosen spatial prior is the sparse areal generalized linear mixed model (Hughes and Haran 2013) and is used to account for spatial dependence for each of $\gamma_j = (\gamma_{1,j}, \ldots, \gamma_{n_g,j})'$, $j = 1, \ldots, p$. Specifically, the γ_j are conditionally independent Bernoulli random variables with a probit link function such that

$$\begin{aligned} \gamma_{v,j} \mid \eta_{v,j} &\overset{\text{ind}}{\sim} Bern\{\Phi(a_{v,j} + \eta_{v,j})\} \\ \eta_{v,j} &= m_v' \delta_j + \epsilon_{v,j} \\ \epsilon_{v,j} &\sim \mathcal{N}(0,1), \end{aligned} \tag{19.10}$$

where $\Phi(\cdot)$ denotes the cumulative distribution function of a standard normal random variable, m_v is a vector of synthetic spatial predictors, $\delta_j = (\delta_{1,j}, \ldots, \delta_{q,j})$ is a vector of spatial random effects, $a_{v,j}$ is an offset that controls the prior probability of activation, and $\eta_{v,j}$ is an auxiliary variable that is introduced to facilitate Gibbs sampling (Holmes and Held 2006). Voxels are located at the vertices of an underlying undirected graph, the structure of which reflects spatial adjacency among voxels. For a partition of G parcels, with each parcel indexed by g, the graph is represented using its parcel-level adjacency matrix A, which is $n_g \times n_g$ with entries given by diag$(A) = 0$ and $(A)_{u,v} = I(u \sim v)$. In the context of a two-dimensional analysis, a voxel neighborhood might comprise the four nearest voxels. With three-dimensional fMRI data, a neighborhood contains the 26 nearest voxels.

The prior for the spatial random effects is

$$\pi(\delta_j \mid \kappa_j)\pi(\kappa_j) = \mathcal{N}\left\{\delta_j \mid 0, (\kappa_j M'QM)^{-1}\right\} \times \mathcal{G}amma\left(\kappa_j \mid a_\kappa, b_\kappa\right), \tag{19.11}$$

where κ_j is a smoothing parameter; M is an $n_g \times q$ matrix, the columns of which are the q principal eigenvectors of A; and $Q = D - A$ is the graph Laplacian, where D is the diagonal degree matrix. Note that m_v' is the vth row of M. The columns of M are multiresolutional spatial basis vectors that are well suited for spatial smoothing and capture both the small-scale and large-scale spatial variation typically exhibited by fMRI data (Woolrich et al. 2004). Sparsity is introduced by selecting the columns of M corresponding to eigenvalues greater than 0.05. This choice permits appropriate spatial smoothing while reducing the dimensionality considerably (typically, $q < n_g/2$). The choice of prior on the smoothing parameter κ_j follows Kelsall and Wakefield (1999) by using $a_\kappa = 1/2$ and $b_\kappa = 2000$, which does not lead to artifactual spatial structure in the posterior.

19.2.2.4 Posterior Computation and Inference

Denote the voxel-level parameters as $\theta_v = (\beta_v', \gamma_v', \rho_v', \sigma_v^2,)'$, and the parcel-level parameters as $\Theta_g = (\delta', \kappa', (\tau^2)')'$. Within the gth parcel (having n_g voxels), the prior distribution is

$$\pi(\theta_v, \Theta_g) = \prod_{v=1}^{n_g} \pi(\rho_v)\pi(\sigma_v^2)\pi(\beta_v \mid \gamma_v) \prod_{j=1}^{p} \pi(\gamma_j \mid \delta_j)\pi(\delta_j \mid \kappa_j)\pi(\kappa_j)\pi(\tau_j^2),$$

which implies that the between-voxel and between-task parameters are conditionally independent a priori. The posterior distribution is obtained in the usual way by combining priors and the likelihood.

To obtain updates for each $\gamma_{v,j}$, a voxel-level likelihood is used where $\beta_{v,j}$ has been integrated out analytically. For $W_{v,t,(j)} = Y_{v,t} - \sum_{l \neq j}^{p} X_{t,l}\beta_{v,l}$, let $W^*_{v,t,(j)} = W_{v,t,(j)} - \sum_{k=1}^{r} \rho_{v,k} W_{v,t-k,(j)}$, and let $X^*_{t,j} = X_{t,j} - \sum_{k=1}^{r} \rho_{v,k} X_{t-k,j}$. Then, conditional on $\gamma_{v,j}$, the likelihood can be written as a mixture with two components:

$$L_1 = \tau_j^{-1} \exp\left\{ -\frac{1}{2\sigma_v^2} \sum_{t=1}^{T} \left(W^*_{v,t,(j)} - X^*_{t,j}\beta_{v,j} \right)^2 - \frac{1}{2\tau_j^2}\beta_{v,j}^2 \right\}$$

and

$$L_0 = \exp\left\{ -\frac{1}{2\sigma_v^2} \sum_{t=1}^{T} W^{*2}_{v,t,(j)} \right\},$$

where L_1 is the voxel-level likelihood when $\gamma_{v,j} = 1$, and L_0 is the likelihood when $\gamma_{v,j} = 0$. Integrating $\beta_{v,j}$ out of L_1, it is straightforward to show that

$$L_1 = \tau_j^{-1} \sigma^{*-1}_{v,j} \exp\left\{ -\frac{1}{2\sigma_v^2} \sum_{t=1}^{T} W^{*2}_{v,t,(j)} + \frac{1}{2\sigma^{*2}_{v,j}} \left(\frac{1}{\sigma_v^2} \sum_{t=1}^{T} W^*_{v,t,(j)} X^*_{t,j} \right)^2 \right\},$$

where $\sigma^{*2}_{v,j} = \sigma_v^{-2} \sum_{t=1}^{T} X^{*2}_{t,j} + \tau_j^{-2}$. The posterior probability that $\gamma_{v,j} = 1$ is $q\left(\gamma_{v,j} = 1 \mid Y_v, \cdot\right) = (1 + \mathcal{P})^{-1}$, where

$$\mathcal{P} = \frac{L_0\, q\left(\gamma_{v,j} = 0 \mid \eta_j\right)}{L_1\, q\left(\gamma_{v,j} = 1 \mid \eta_j\right)}.$$

Conditional on $\gamma_{v,j} = 0$, set $\beta_{v,j} = 0$. Otherwise, $\beta_{v,j}$ is updated from its full conditional distribution. Writing $W^*_{v,t,(j)} = X^*_{t,j}\beta_{v,j} + \varepsilon_{v,t}$, and using the fact that the error term is normally distributed, each $\beta_{v,j}$ has a normal prior distribution. Conditional on $\gamma_{v,j} = 1$, the posterior distribution of $\beta_{v,j}$ is $\mathcal{N}(\hat{\beta}_{v,j},\ \hat{\tau}^2_{v,j})$, where

$$\hat{\beta}_{v,j} = \hat{\tau}^2_{v,j} \sum_{t=1}^{T} W^*_{v,t,(j)} X^*_{t,j} \quad \text{and} \quad \hat{\tau}^2_{v,j} = \left(\sum_{t=1}^{T} X^{*2}_{t,j} + \sigma^2_v / \tau^2_j \right)^{-1}.$$

Posterior sampling of each of $\eta_j = (\eta_{1,j}, \ldots, \eta_{n_g j})'$, δ_j, and κ_j uses probit regression with auxiliary variables, conditional on $\gamma_{v,j}$ only (Holmes and Held 2006). The full conditional distributions are given in Musgrove et al. (2015).

19.2.2.5 Emotion Processing Data

We consider implementation of the method in Sect. 19.2.2 in the emotion processing data described in Sect. 19.1.1. The image was parcellated into 321 regions ranging in size from 500 to 1000 voxels. The spatial dimension reduction offered by the sparse SGLMM resulted in an average reduction of 72%, i.e., for a parcel with $n_g =$ 1000 voxels, there were 280 spatial random effects. At the voxel level, we used an autoregression model of order 2. Thus, for two covariates there were approximately two million parameters to be estimated.

We implemented the MCMC simulation for each of 20,000, 100,000, and 200,000 iterations. Estimation was done both using 50% burn-in and no burn-in. Starting values of all parameters were taken to be the maximum likelihood estimates. Since the parcellation method results in assumed independent parcels, the parcels are analyzed separately and in parallel to speed computation. Thus, the computational speed is limited by the number of parcels and the availability of a computing cluster.

The results are reported in Table 19.4. The estimation is remarkably stable and the use of burn-in seemed to have little impact except to increase the Monte Carlo standard errors on the estimates. This analysis indicates that we could easily use only 20,000 iterations to obtain stable results, which required 217 s of sampling time for the largest parcel.

Table 19.4 Max MCSE is the maximum Monte Carlo standard error of all activation probabilities

	MCMC iterations					
	20k		100k		200k	
	No burn-in	Burn-in	No burn-in	Burn-in	No burn-in	Burn-in
Max MCSE	0.042	0.050	0.028	0.033	0.022	0.028
Active voxels (%)	9053 (4.02)	9031 (4.01)	9001 (4.00)	9007 (4.00)	8943 (3.7)	8933 (3.97)
Max run time	217		1195		2195	

Max run time is the time in seconds required to analyze the largest parcel (parcel 168 with 997 voxels)

19.2.3 Activation Maps for Emotion Processing Data

In this section we used 100,000 MCMC samples to implement both methods on the emotion processing data with the goal of producing activation maps. Activation regions were found by overlaying the parcellation of Tzourio-Mazoyer et al. (2002) to the activation results. The parcellation included 116 regions. Regional activation occurred if at least ten voxels within a region were estimated to be significantly active.

Results of the face task for both procedures are displayed in Fig. 19.2. Six slices were chosen to illustrate the results for the two approaches. Slices 20, 30, and 35 show activation in the occipital and temporal lobes, with significant activation in the left calcarine and right lingual regions. The method of Sect. 19.2.2 found more extensive activation in the frontal lobe. Slice 40 shows activation in the right angular and the right frontal lobe, with the method of Sect. 19.2.2 detecting more extensive

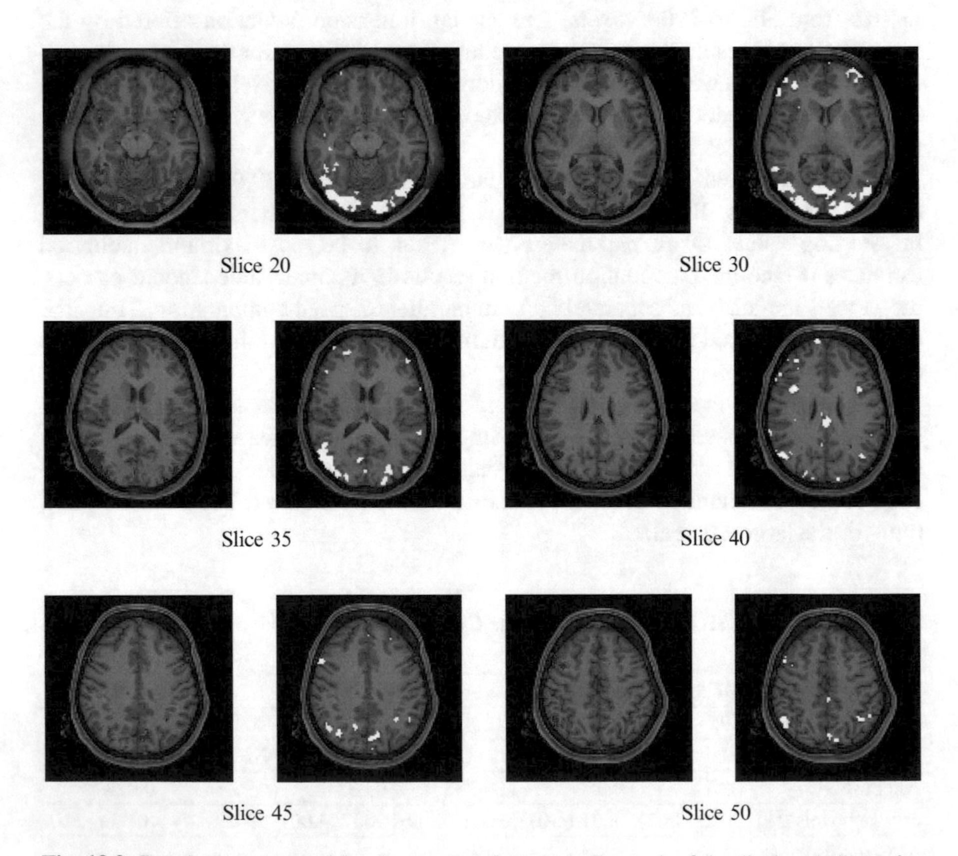

Fig. 19.2 Results are presented for the emotion-faces task. For each of the displayed slice pairs, the left slice (red blobs) displays the results for the model of Sect. 19.2.1, and the right slice (white blobs) displays the results for the model of Sect. 19.2.2

activation in the right frontal lobe. Slice 45 shows activation in the right angular and the left precuneus, with the method of Sect. 19.2.2 detecting more extensive activation in the left precuneus. Finally, slice 50 shows activation in the parietal lobes. The method of Sect. 19.2.1 found no activation in the right parietal lobe while the method of Sect. 19.2.2 found a small amount of activation.

19.3 Discussion

We considered two Bayesian areal models (and the associated MCMC algorithms) for detecting activation in fMRI experiments. Contrary to popular opinion we have demonstrated that both approaches are computationally efficient and produce stable results in a matter of minutes, rather than hours or days. Both methods are based on parcellations. In Sect. 19.2.1 the parcellations are not assumed independent, while in Sect. 19.2.2 they are. The advantage of assuming independence is that the computation may then be parallelized. The disadvantage is that it is a somewhat awkward assumption that seems to have little relevance to the underlying science. On the other hand, the computation required in Sect. 19.2.1 is not parallelizeable and hence takes longer. In the emotion processing data example we see that both approaches yield similar results, although the approach of Sect. 19.2.2 yields more active voxels than does the approach of Sect. 19.2.1.

Acknowledgements Data were provided by the Human Connectome Project, WU-Minn Consortium (Principal Investigators: David Van Essen and Kamil Ugurbil; 1U54MH091657) funded by the 16 NIH Institutes and Centers that support the NIH Blueprint for Neuroscience Research; and by the McDonnell Center for Systems Neuroscience at Washington University.

John Hughes was supported by the Simons Foundation. Galin Jones was supported by the National Institutes of Health and the National Science Foundation. Donald R. Musgrove was supported by University of Minnesota Academic Health Center Faculty Research Development Grant.

References

Aguirre GK, Zarahn E, D'Esposito M (1997) Empirical analyses of BOLD fMRI statistics. II. Spatially smoothed data collected under null-hypothesis and experimental conditions. NeuroImage 5:199–212

Aguirre, GK, Zarahan E, D'Esposito M (1998) The variability of human, BOLD hemodynamic responses. NeuroImage 8:360–369

Banerjee S, Carlin BP, Gelfand AE (2003) Hierarchical modeling and analysis for spatial data, 1st edn. Chapman and Hall/CRC, New York

Bezener M, Hughes J, Jones GL (2015) Bayesian spatiotemporal modeling using hierarchical spatial priors with applications to functional magnetic resonance imaging. Preprint

Bowman FD (2014) Brain imaging analysis. Ann Rev Stat Appl 1:61–85

Cressie NA (1993) Statistics for spatial data, revised edn. Wiley Interscience, New York

Essen DCV, Smith SM, Barch DM, Behrens TE, Yavoub E, Ugurbil K (2013) The WU-Minn Human Connectome Project: an overview. Neuroimage 80:62–79
Friston KJ, Worsley K, Frackowiak R, Mazziotta J, Evans A (1994) Assessing the significance of focal activations using their spatial extent. Hum Brain Mapp 1:210–220
Friston KJ, Holmes A, Worsley KJ, Polin JB, Frith C, Frackowik R (1995) Statistical parametric maps in functional imaging: a general linear approach. Hum Brain Mapp 2:189–210
Friston KJ, Ashburner JT, Kiebel SJ, Nichols TE, Penny WD (2007) Statistical parametric mapping: the analysis of functional brain images. Academic, London
Genovese CR (2000) A Bayesian time-course model for functional magnetic resonance imaging data. J Am Stat Assoc 95(451):691–703
George EI, McCulloch RE (1993) Variable selection via Gibbs sampling. J Am Stat Assoc 88(423):881–889
George EI, McCulloch RE (1997) Approaches for Bayesian variable selection. Stat Sin 7:339–373
Haran M (2011) Gaussian random field models for spatial data. In: Brooks SP, Gelman AE, Jones GL, Meng XL (eds) Handbook of Markov chain Monte Carlo. Chapman and Hall/CRC, London, pp 449–478
Hariri AR, Mattay VS, Tessitore A, Kolachana B, Fera F, Goldman D, Egan MF, Weinberger DR (2002) Serotonin transporter genetic variation and the response of human amygdala. Science 297:400–403
Hensen R, Friston K (2007) Convolution models for fMRI. In: Statistical parametric mapping: the analysis of functional brain images. Academic, Cambridge
Holmes CC, Held L (2006) Bayesian auxiliary variable models for binary and multinomial regression. Bayesian Anal 1(1):145–168
Holmes AP, Blair RC, Watson JD, Ford I (1996) Nonparametric analysis of statistic images from functional mapping experiments. J Cereb Blood Flow Metab 16:7–22
Huettel SA, Somng AW, McCarthy G (2009) Functional magnetic resonance imaging. Sinauer Associates, Sunderland
Hughes J, Haran M (2013) Dimension reduction and alleviation of confounding for spatial generalized linear mixed models. J R Stat Soc Ser B Stat Methodol 75(1):139–159
Kaushik K, Karesh K, Suresha D (2013) Segmentation of the white matter from the brain fMRI images. Int J Adv Res Comput Eng Technol 2:1314–1317
Kelsall J, Wakefield J (1999) Discussion of “Bayesian Models for spatially correlated disease and exposure data”, by Best et al. In: Berger J, Bernardo J, Dawid A, Smith A (eds) Bayesian statistics, vol 6. Oxford University Press, Oxford
Lazar NA (2008) The statistical analysis of fMRI data. Springer, New York
Lee K-J (2010) Computational issues in using Bayesian hierarchical methods for the spatial modeling of fMRI data. PhD thesis, University of Minnesota, School of Statistics
Lee K-J, Jones GL, Caffo BS, Bassett SS (2014) Spatial Bayesian variable selection models on functional magnetic resonance imaging time-series data. Bayesian Anal 9:699–732
Lindquist MA (2008) The statistical analysis of fMRI data. Stat Sci 23:439–464
Locascio J, Jennings PJ, Moore CI, Corkin S (1997) Time series analysis in the time domain and resampling methods for studies of functional magnetic brain imaging. Hum Brain Mapp 5: 168–193
Makni S, Idier J, Vincent T, Thirion B, Dehaene-Lambertz G, Ciuciu P (2008) A fully Bayesian approach to the parcel-based detection-estimation of brain activity in fMRI. NeuroImage 41:941–969
Mikl M, Mareček R, Hluštík P, Pavlicová M, Drastich A, Chlebus P, Brázdil M, Krupa P (2008) Effects of spatial smoothing on fMRI group inferences. Magn Reson Imaging 26:490–503
Monti MM (2011) Statistical analysis of fMRI time-series: a critical review of the GLM approach. Front Hum Neurosci 5:28
Musgrove DR, Hughes J, Eberly LE (2015) Fast, fully Bayesian spatiotemporal inference of fMRI data. Preprint
Penny W, Kiebel S, Friston K (2003) Variational Bayesian inference for fMRI time series. NeuroImage 19:727–741

Smith M, Fahrmeir L (2007) Spatial Bayesian variable selection with application to functional magnetic resonance imaging. J Am Stat Assoc 102(478):417–431
Smith M, Pütz B, Auer D, Fahrmeir L (2003) Assessing brain activity through spatial Bayesian variable selection. NeuroImage 20:802–815
Triantafyllou C, Hoge R, Wald L (2006) Effect of spatial smoothing on physiological noise in high-resolution fMRI. NeuroImage 32:551–557
Tzourio-Mazoyer N, Landeau B, Papathanassiou D, Crivello F, Etard O, Delcroix N, Mazoyer B, Joliot M (2002) Automated anatomical labeling of activations in SPM using a macroscopic anatomical parcellation of the MNI MRI single-subject brain. Neuroimage 15(1):273–289
Woolrich MW, Jenkinson M, Brady JM, Smith SM (2004) Fully Bayesian spatio-temporal modeling of fMRI data. IEEE Trans Med Imaging 23:213–231
Worsley K, Marrett S, Neelin P, Evans A (1992) A three-dimensional statistical analysis for CBF activation studies in human brain. J Cereb Blood Flow Metab 12:900–918
Xia J, Liang F, Wang YM (2009) fMRI analysis through Bayesian variable selection with a spatial prior. In: IEEE international symposium on biomedical imaging (ISBI), pp 714–717
Zellner A (1996) On assessing prior distributions and Bayesian regression analysis with *g*-prior distributions. In: Bayesian inference and decision techniques: essays in honor of Brunode Finetti. North-Holland/Elsevier, Amsterdam, pp 233–243
Zhang L, Guindani M, Vannucci M (2015) Bayesian models for functional magnetic resonance imaging data analysis. WIREs Comput Stat 7:21–41

Chapter 20
Construction of Tight Frames on Graphs and Application to Denoising

Franziska Göbel, Gilles Blanchard, and Ulrike von Luxburg

Abstract Given a neighborhood graph representation of a finite set of points $x_i \in \mathbb{R}^d, i = 1, \ldots, n$, we construct a frame (redundant dictionary) for the space of real-valued functions defined on the graph. This frame is adapted to the underlying geometrical structure of the x_i, has finitely many elements, and these elements are localized in frequency as well as in space. This construction follows the ideas of Hammond et al. (Appl Comput Harmon Anal 30:129–150, 2011), with the key point that we construct a tight (or Parseval) frame. This means we have a very simple, explicit reconstruction formula for every function f defined on the graph from the coefficients given by its scalar product with the frame elements. We use this representation in the setting of denoising where we are given noisy observations of a function f defined on the graph. By applying a thresholding method to the coefficients in the reconstruction formula, we define an estimate of f whose risk satisfies a tight oracle inequality.

Keywords Neighborhood graph · Tight frame · Dictionary learning · Denoising · Thresholding · Oracle inequality

20.1 Introduction

20.1.1 Motivation

When dealing with high-dimensional data, a general principle is that the curse of dimensionality can be efficiently fought if one assumes the data points to lie on a structure of smaller intrinsic dimensionality, typically a manifold. Some well-known

F. Göbel · G. Blanchard (✉)
Institute of Mathematics, University of Potsdam, Potsdam, Germany
e-mail: goebel@uni-potsdam.de; gilles.blanchard@math.uni-potsdam.de

U. von Luxburg
Department of Computer Science, University of Tübingen, Tübingen, Germany
e-mail: luxburg@informatik.uni-tuebingen.de

© Springer International Publishing AG, part of Springer Nature 2018

W. K. Härdle et al. (eds.), *Handbook of Big Data Analytics*, Springer Handbooks of Computational Statistics, https://doi.org/10.1007/978-3-319-18284-1_20

methods to discover such a lower dimensional structure include Isomap (Tenenbaum et al. 2000), LLE (Roweis and Saul 2000), and Laplacian Eigenmaps (Belkin and Niyogi 2003).

In this work, our main interest is not in visualizing or representing by an explicit mapping the underlying structure of the observed data points; rather, we want to represent or estimate efficiently a real-valued function on these points. More specifically, we focus on the following *denoising* problem: assuming we observe a noisy version of the function f, $y_i = f(x_i) + \varepsilon_i$ at points $(x_1, \ldots, x_n)$, we would like to recover the values of f at these points. An important step for solving this problem is to find a dictionary of functions to represent the signal f, which is adapted to the structure of the data. Ideally, we would like this dictionary to exhibit the features of a wavelet basis. In traditional signal processing on a flat space, with data points on a regular grid, orthogonal wavelet bases offer a very powerful tool to sparsely represent signals with inhomogeneous regularity (such as a signal that is very smooth everywhere except at a few singular points where it is discontinuous). Such bases are in particular well suited to the denoising task. Can this be generalized to irregularly scattered data on a manifold?

We present such a method to construct a so-called *Parseval frame* of functions exhibiting wavelet-like properties while adapting to the intrinsic geometry of the data. Furthermore, we use this dictionary for the denoising task using a simple coefficient thresholding method.

This work is organized as follows. In the coming section, we discuss the relationship to previous work on which the present chapter is built, as well as pointing out our new contributions. In Sect. 20.2, we recall important notions of frame theory as well as of neighborhood graphs needed for our construction. The construction of the frame and its properties is presented in Sect. 20.3. In Sect. 20.4, we develop a coefficient thresholding strategy for the denoising problem. In Sect. 20.5, we present numerical results and method comparison on testbed data.

20.1.2 Relation to Previous Work

Regression methods that adapt to an underlying lower dimension of the data have been considered by Bickel and Li (2007), Kpotufe and Dasgupta (2012) and Kpotufe (2011) using local polynomial estimates, random projection trees, and nearest-neighbors, respectively. However, these methods are not constructed to adapt to an inhomogeneous regularity of the target function: in these three cases, the smoothing scale (determined by the smoothing kernel bandwidth, the tree partition's average data diameter, or the number of neighbors, respectively) is fixed globally. In the experimental Sect. 20.5, for data lying on a smooth manifold but a target function exhibiting a sharp discontinuity, we demonstrate the advantage of our method over kernel smoothing.

Based on the motivations similar to ours, a method for constructing a wavelet-like basis on scattered data was proposed by Gavish et al. (2010). It is based on a hierarchical tree partition of the data, on which a Haar-like basis of 0–1 functions is constructed. However, the performance of that method is then adapted to the geometry of the *tree*, in the sense that the distance of two points is measured through tree path distance. This can strongly distort the original distance: two close points in original distance can find themselves in very separated subtrees.

The construction proposed here, based on a transform of the spectral decomposition of the graph Laplacian, follows closely the ideas of Hammond et al. (2011). Two important contributions brought forth in the present work are that we construct a Parseval (or tight) frame, rather than a general frame; and we consider an explicit thresholding method for the denoising problem. The former point is crucial to obtain sharp bounds for the thresholding method, and also eliminates the computational problem of signal reconstruction from the frame coefficients, since Parseval frames enjoy a reconstruction formula similar to that of an orthonormal basis. The choice of multiscale bandpass filter functions leading to the tight frame is inspired by the recent work of Coulhon et al. (2012), where the spectral decomposition principle is also studied, albeit in the setting of a quite general metric space.

20.2 Notation and Basics

20.2.1 Setting

We consider a sample of n points $x_i \in \mathbb{R}^d$. These points are assumed to belong to an unknown low-dimensional submanifold $\mathscr{M} \subset \mathbb{R}^d$. We denote the design by $\mathfrak{D} = \{x_1, \dots, x_n\} \subset \mathscr{M}$. Furthermore, we observe on these points the (noisy) value of a function $f : \mathfrak{D} \to \mathbb{R}$. Since $\mathfrak{D}$ is finite, we can represent the function f as vector $f = (f(x_1), \dots, f(x_n))^t \in \mathbb{R}^n$. The space of all (square-integrable) functions f defined on $\mathfrak{D}$ is denoted $L^2(\mathfrak{D})$ and endowed with the usual Euclidean inner product.

We denote by $y_i = f(x_i) + \epsilon_i$ the noisy observation of f at x_i, where ϵ_i are independent identically distributed centered random variables. The problem we consider in this work is that of denoising, that is, try to recover the underlying value of the function f at the points x_i.

While the existence of a low-dimensional supporting manifold $\mathscr{M}$ for the design points motivates the construction of the proposed method, we underline (again) that $\mathscr{M}$ is not known to the user and the method only uses the knowledge of the design points. In such a setting, a key idea to recover implicitly some information on the geometry of $\mathscr{M}$ is to construct a neighborhood graph based on the design points (see Sect. 20.2.3 for details).

20.2.2 Frames

For the construction in Sect. 20.3, we rely on the notion of a *vector frame*, for which we recall here some important properties (see, e.g., Casazza et al. 2013; Han 2007; Christensen 2008). A frame is an overcomplete dictionary with particular properties allowing it to act almost as basis.

Definition 1 Let $\mathscr{H}$ be a Hilbert space. Then a countable set $\{z_i\}_{i\in I} \subset \mathscr{H}$ is a frame with frame bounds A and B for $\mathscr{H}$ if there exists constants $0 < A \leq B < \infty$ such that

$$\forall z \in \mathscr{H}: \;\; A\,\|z\|^2 \leq \sum_{i\in I} |\langle z, z_i\rangle|^2 \leq B\,\|z\|^2 . \tag{20.1}$$

A frame is called tight if $A = B$, in particular the frame is called Parseval if $A = B = 1$.

In the remainder of this work we consider the case of a Euclidean space $\mathscr{H} = \mathbb{R}^n$, and assume that $\{z_i\}_{i\in I}$ is a frame with a finite number of elements. Two important operators associated to the frame are the *analysis* operator

$$T : \mathbb{R}^n \to \mathbb{R}^I,\; Tz := (\langle z, z_i\rangle)_{i\in I} \tag{20.2}$$

(sequence of frame coefficients), and its adjoint the *synthesis* operator:

$$T^* : \mathbb{R}^I \to \mathbb{R}^n,\; T^*a = T^*(a_i)^t_{i\in I} = \sum_{i\in I} a_i z_i . \tag{20.3}$$

Further, the *frame* operator is defined as $S = T^*T$:

$$S : \mathbb{R}^n \to \mathbb{R}^n,\; Sz = T^*Tz = \sum_{i\in I} \langle z, z_i\rangle\, z_i , \tag{20.4}$$

and finally the *Gramian* operator as $U = TT^*$,

$$U : \mathbb{R}^I \to \mathbb{R}^I,\; Ua = TT^*a = \left\{\left\langle \sum_{i\in I} a_i z_i, z_k \right\rangle\right\}_{k\in I} . \tag{20.5}$$

In matrix form, the columns of T^* are the vectors $z_i, i \in I$, T is its transpose and $U_{ij} = \langle z_i, z_j\rangle$.

The definition of a frame implies that S is invertible, and it is possible to reconstruct any z from its frame coefficients by $z = \sum_{i\in I} \langle z, z_i\rangle\, z_i^* = \sum_{i\in I} \langle z, z_i^*\rangle\, z_i$, where $z_i^* := S^{-1} z_i, i \in I$ is called the *canonical dual* frame of $(z_i)_{i\in I}$.

We recall some properties of finite Parseval frames over Euclidean spaces (see, e.g., Han 2007, chapter 3).

Theorem 1 (Properties of Parseval frames) *Let $\mathscr{H}$ be a Hilbert space with* $\dim \mathscr{H} = n < \infty$. *The following statements are equivalent:*

1. $\{z_i\}_{1 \leq i \leq k} \subset \mathscr{H}$ *is a Parseval frame.*
2. $\forall y \in \mathscr{H}: \ y = \sum_{i=1}^{k} \langle y, z_i \rangle z_i$.
3. *The frame operator S is the identity on $\mathbb{R}^n$.*
4. *The Gramian operator U is an orthogonal projector of rank n in $\mathbb{R}^k$.*

Furthermore if $\{z_i\}_{1..k} \subset \mathscr{H}$ is a Parseval frame, then

- $\|z_i\| \leq 1$ *for* $i \in \{1, \ldots, k\}$;
- $\dim \mathscr{H} = n = \sum_{i=1}^{k} \|z_i\|^2$;
- *the canonical dual frame is the frame itself.*

For the present work, the two most important points of this theory are the following: first, the reconstruction formula (point two above), where we see that a Parseval frame acts similarly to an orthonormal basis; secondly, if we construct a vector $v = T^* a = \sum_i a_i z_i$ from an arbitrary vector of coefficients (a_i), then

$$\left\| \sum_i a_i z_i \right\|^2 = \langle T^* a, T^* a \rangle = \langle a, Ua \rangle = \|Ua\|^2 \leq \|a\|^2 , \tag{20.6}$$

which follows from property 4 above.

20.2.3 Neighborhood Graphs

In order to exploit the structure and geometry of the unknown submanifold $\mathscr{M}$ on which the sample $\mathfrak{D}$ is supposed to lie, a powerful idea is to use a graph-based representation of the data $\mathfrak{D}$ through a *neighborhood graph*. The points in $\mathfrak{D}$ correspond to the vertices of the graph, and two vertices of the graph are joined by an edge when the two corresponding points are neighbors (in some appropriate sense) in $\mathbb{R}^n$. The underlying idea is that the local geometry of $\mathbb{R}^n$ is reflected in the local connectivity of the graph, while the long-range geometry of the graph reflects the geometrical properties of the manifold $\mathscr{M}$, rather than those of $\mathbb{R}^n$.

Formally, a finite graph $G = (V, E)$ is given by a finite set of vertices V and a set of edges $E \subset V \times V$. The $|V| \times |V|$ adjacency matrix A of the graph is defined by $A_{i,j} = 1$ if $(v_i, v_j) \in E$ and $A_{i,j} = 0$ otherwise. An undirected graph is such that its adjacency matrix is symmetric.

The graph is called weighted if every edge $e \in E$ has a positive weight $w(e) \in \mathbb{R}_+$. In this case the notion of adjacency matrix is extended to $A_{i,j} = w((v_i, v_j))$ if $(v_i, v_j) \in E$ and $A_{i,j} = 0$ otherwise. The degree of a vertex v_i in a (possibly weighted) graph is defined as $d_i = d(i) = \sum_{j=1}^{|V|} A_{i,j}$.

As announced, we focus on geometric graphs, which (can) approximate the structure of the unknown $\mathscr{M}$. Each point x_i is represented by a vertex, say v_i. An edge between two vertices represents a small distance, or a high similarity, of the two associated points. The weight of an edge can quantify the similarity more finely.

We use the Euclidean distance $d(x_i, x_j) = \left\| x_i - x_j \right\|$. We recall three usual ways to construct the edges of a neighborhood graph:

- (undirected) k-nearest-neighbor graph: an undirected edge connects the two vertices v_i and v_j iff x_i belongs to the k nearest neighbors of x_j, or x_j belongs to the k nearest neighbors of x_i ("the k-NN-graph").
- ϵ-graph: an undirected edge connects two vertices v_i and v_j iff $d(x_i, x_j) \leq \epsilon$.
- complete weighted neighborhood graph: for each pair of vertices there exists an undirected edge with a weight depending on the distance/similarity of the two vertices.

A k-NN graph or an ϵ-graph can be made weighted by additionally assigning weights to the edges depending on $d(x_i, x_j)$, for instance by choosing Gaussian weights $w(\{i, j\}) = \exp(-d^2(x_i, x_j)/2\lambda^2)$.

20.2.4 Spectral Graph Theory

If one considers real-valued functions $f : \mathcal{M} \rightarrow \mathbb{R}$ defined on a submanifold $\mathcal{M} \subset \mathbb{R}^d$, it is known that under some regularity assumptions on the submanifold $\mathcal{M}$, the eigenfunctions of the Laplace-Beltrami-operator give a basis of the space of squared-integrable functions on $\mathcal{M}$. Since $\mathcal{M}$ is unknown in our setting, the principle of the *Laplacian Eigenmaps* method (Belkin and Niyogi 2003) is to use a discrete analogon, namely the graph Laplace operator L on a neighborhood graph.

Given a finite weighted undirected graph with adjacency matrix A ($n \times n$) and vertex degrees $(d_i)_i$, as introduced in the previous section, we will either use the unnormalized graph Laplace operator L^u or the normalized (symmetric) graph Laplace operator L^{norm} defined by

$$\begin{aligned} L^u &= D - A \qquad (20.7) \\ L^{\text{norm}} &= \mathbf{I}_n - D^{-1/2} A D^{-1/2}, \end{aligned}$$

where $D = \operatorname{diag}(d_1, \ldots, d_n)$ is a diagonal matrix with entries d_i on the diagonal. By construction L^u and L^{norm} are symmetric matrices. The positive semidefiniteness follows from

$$f^t L^u f = 0.5 \sum_{(i,j)} A_{i,j} (f_i - f_j)^2 \text{ and } f^t L^{\text{norm}} f = 0.5 \sum_{(i,j)} A_{i,j} \Big(\frac{f_i}{\sqrt{d_i}} - \frac{f_j}{\sqrt{d_j}}\Big)^2,$$

respectively. The spectral theorem for matrices indicates that the normalized eigenvectors Φ_i of the graph Laplace operator L (L^u resp. L^{norm}) form an orthonormal basis of $\mathbb{R}^n$ and all eigenvalues are nonnegative. Furthermore the number of components of the graph is given by the number of eigenvalues equal to 0.

20.3 Construction and Properties

20.3.1 Construction of a Tight Graph Frame

As discussed earlier, the principle of Laplacian Eigenmaps is to use the basis $(\Phi_i)_{1\leq i\leq n}$ to represent and process the data. An important advantage of this basis as compared with the natural basis of $\mathbb{R}^d$ is that it will be *adapted* to the geometry of the underlying submanifold $\mathscr{M}$ supporting the data distribution. For instance, in the denoising problem, a reasonable estimator of f could be a truncated expansion of the noisy vector of observations Y in the basis $(\Phi_i)_{1\leq i\leq n}$.

On the other hand, a disadvantage of this basis is that it is not *spatially localized*. To get an intuitive view, consider the simple case of the interval [0, 1] with uniformly distributed data. In the population view, the eigenbasis of the Laplacian is the Fourier basis. While a truncated expansion in this basis is well-adapted to represent functions that are uniformly regular, it is not well-suited for functions exhibiting locally varying regularity (as an extreme example, a signal that is very smooth everywhere except at a few singular points where it is discontinuous). By contrast, wavelet bases, because they are localized both in space and frequency, allow for an efficient (i.e., sparse) representation of signals with locally varying regularity.

If we now think of data supported on a one-dimensional submanifold (curve) of $\mathbb{R}^d$, we can expect that the Laplacian eigenmaps method will discover a warped Fourier basis following the curve; and, for a more general submanifold $\mathscr{M}$, "harmonics" on $\mathscr{M}$.

In order to go from this basis to a spatially localized dictionary, following ideas of Coulhon et al. (2012) and Hammond et al. (2011), we use the principle of the *Littlewood-Paley* decomposition.

Let G be an undirected geometric neighborhood graph with adjacency matrix A constructed from $\mathfrak{D}$, and L be an associated symmetric graph Laplace operator with increasing eigenvalues $0 = \lambda_1 \leq \lambda_2 \leq \cdots \leq \lambda_n$ and normalized eigenvectors $\Phi_i \in \mathbb{R}^n, i = 1 \ldots n$.

We first define a set of vectors using a decomposition of unity and a splitting operation and we will show that this vector set is a Parseval frame.

Definition 2 Let $\{\zeta_k\}_{k\in\mathbb{N}}$ be a sequence of functions $\zeta_k : \mathbb{R}_+ \to [0, 1]$ satisfying

(DoU) $\sum_{j\geq 0} \zeta_j(x) = 1$ for all $x \geq 0$;
(FD) $\#\{\zeta_k : \zeta_k(\lambda_i) \neq 0\} < \infty$ for $i = 1, \ldots, n$.

Then we define the set of column vectors $\{\Psi_{kl} \in \mathbb{R}^n, 0 \leq k \leq Q, 1 \leq j \leq n\}$ by

$$\Psi_{kl} = \sum_{i=1}^{n} \sqrt{\zeta_k(\lambda_i)}\Phi_i(x_l)\Phi_i. \tag{20.8}$$

with $Q := \max\{k : \exists i \in \{1, \ldots, n\} \text{ with } \zeta_k(\lambda_i) > 0\}$.

Theorem 2 *$\{\Psi_{kl}\}_{k,l}$ is a Parseval frame for $\mathscr{H} = \mathbb{R}^n$, that is for all $x \in \mathbb{R}^n$:*

$$\sum_{k,l} |\langle x, \Psi_{kl}\rangle|^2 = \|x\|^2 . \tag{20.9}$$

Proof If we can show that $\sum_{(k,l)} \Psi_{kl}\Psi_{kl}^t = \mathbf{I}_n$, we get immediately

$$y = \mathbf{I}_n y = \left(\sum_{(k,l)} \Psi_{kl}\Psi_{kl}^t\right) y = \sum_{(k,l)} \langle y, \Psi_{kl}\rangle \Psi_{kl}, \tag{20.10}$$

for $y \in \mathbb{R}^n$. According to Theorem 1 this equation is equivalent to the condition (20.1) with $A = B = 1$. So we are done. It remains to show $\sum_{k,l} \Psi_{kl}\Psi_{kl}^t = \mathbf{I}_n$. We have (since we sum over a finite number of elements)

$$\begin{aligned}\sum_{(k,l)} \Psi_{kl}\Psi_{kl}^t &= \sum_{k,l,i,j} \sqrt{\zeta_k(\lambda_i)}\sqrt{\zeta_k(\lambda_j)}\Phi_i(x_l)\Phi_j(x_l)\Phi_i\Phi_j^t \\ &= \sum_{i=1}^{n}\sum_{k=0}^{Q} \zeta_k(\lambda_i)\,\Phi_i\Phi_i^t \\ &= \sum_{i=1}^{n} \Phi_i\Phi_i^t = \mathbf{I}_n.\end{aligned} \tag{20.11}$$

For the second equality, we have used that $\sum_l \Phi_i(x_l)\Phi_j(x_l) = \langle \Phi_i, \Phi_j\rangle = \mathbf{1}\{ i = j\}$, since $\{\Phi_i\}_i$ is an orthonormal basis (onb). For the third equality, we used (DoU), and for the last again the onb property. □

We now choose a special sequence of functions satisfying the decomposition of unity (DoU) condition while also ensuring (a) a spectral localization property for the frame elements and (b) a multiscale decomposition interpretation of the resulting decomposition. This construction follows Coulhon et al. (2012), and is known in the context of functional analysis as a smooth Littlewood-Paley decomposition.

Definition 3 (Multiscale Bandpass Filter) Let $g \in C^\infty(\mathbb{R}_+)$, $\operatorname{supp} g \subset [0, 1]$, $0 \le g \le 1$, $g(u) = 1$ for $u \in [0, 1/b]$ (for some constant $b > 1$). For $k \in \mathbb{N} = \{0, 1, \ldots\}$ the functions $\zeta_k : \mathbb{R}_+ \to [0, 1]$ are defined by

$$\zeta_k(x) := \begin{cases} g(x) & \text{if } k = 0 \\ g(b^{-k}x) - g(b^{-k+1}x) & \text{if } k > 0 \end{cases} \tag{20.12}$$

The sequence $\{\zeta_k\}_{k\ge 0}$ is called multiscale bandpass filter.

This definition leads to the following properties: $\zeta_k(x) = \zeta_1(b^{-k}x)$ for $k \ge 1$ (multiscale decomposition), $\zeta_k \in C^\infty(\mathbb{R}_+)$, $0 \le \zeta_k \le 1$, $\operatorname{supp} \zeta_0 \subset [0, 1]$,

$\operatorname{supp} \zeta_k \subset [b^{k-2}, b^k]$ for $k \geq 1$ (spectral localization property). Moreover, one can check readily

$$\sum_{j \geq 0} \zeta_j(x) = 1, \tag{20.13}$$

i.e., the (DoU) condition holds. In practice, we use a dyadic bandpass filter, that is, $b = 2$. The functions $\zeta_0, \ldots, \zeta_5$ with $b = 2$ are displayed in Fig. 20.1b. By construction, the parameter k in Ψ_{kl} is naturally a spectral scale parameter, while l is a spatial localization parameter: the frame element Ψ_{kl} is localized around the point x_l, as we discuss next.

20.3.2 *Spatial Localization*

By construction, the elements of the frame are band-limited, i.e. localized in the spectral scale, in the sense that for a fixed k, the frame elements Ψ_{kl} ($l = 1, \ldots, n$) are linear combinations of the eigenvectors of the graph Laplacian ("graph harmonics") corresponding to eigenvalues in the range $[b^{k-2}, b^k]$ only.

From our initial motivations, it is desirable that in contrast with the eigenfunctions of the Laplace operator, the frame elements Ψ_{kl} are spatially localized functions. In the classical Littlewood-Paley construction for the usual Laplacian on the interval $[0, 1]$, this is a well-known fact: the use of linear combination of trigonometric functions $\Psi_{kl}(y) := \sin(kl)\sin(ky)$ via smooth multiscale bandpass filters weights as described in Definition 3 gives rise to strongly localized functions (as illustrated in Fig. 20.1).

Regarding the corresponding discrete construction based on the graph Laplacian, this localization property is certainly observed in practice (as illustrated in Figs. 20.2 and 20.3, see Sect. 20.5 for the setup of the numerical experiments).

Concerning the theoretical perspective, we first review briefly the existing results of Hammond et al. (2011), denote d the shortest path distance in the graph. Theorem 5.5 of Hammond et al. (2011) gives the following localization result for graph frames:

$$\frac{\Psi_{kl}(x)}{\|\Psi_{kl}\|_2} \leq C b^{-k}, \tag{20.14}$$

for all x with $d(x, x_l) \geq K$, under the assumption that the scaling function ζ_1 is K-times differentiable with vanishing first $(K-1)$ derivatives in 0, non-vanishing K-th derivative, and the scale parameter k is big enough. This says that Ψ_{kl} is "localized" around the point x_l. Unfortunately, this result is not informative in our framework for two reasons: first, we chose a function ζ_1 (see (20.12)) vanishing in a neighborhood of zero, so that all derivatives vanish in the origin, contradicting one of the above assumptions. Secondly, and independently of this first issue, the condition "k is big

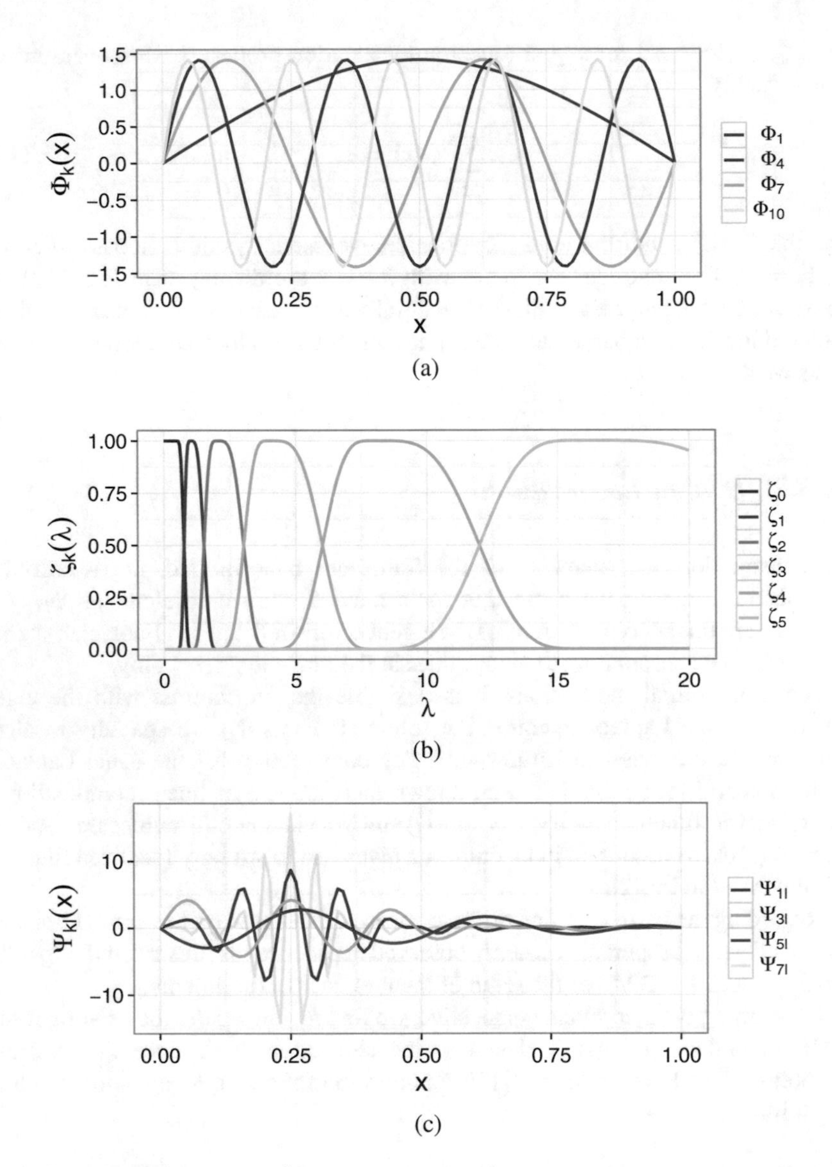

Fig. 20.1 Littlewood-Paley on $L^2(0,1)$: (**a**) eigenfunctions; (**b**) multiscale bandpass filter; (**c**) frame elements

enough," and the factor C depend on the size n of the graph and of the largest eigenvalue of the Laplacian. As a consequence it is unclear if this bound covers any interesting part of the spectrum (for k too large, the spectral support $[b^{k-2}, b^k]$ does not contain any eigenvalues, so that Ψ_{kl} is trivial). Finally, for fixed k the bound also does not give information on the behavior of $\Psi_{kl}(x)$ when the path distance of x to x_l becomes very large.

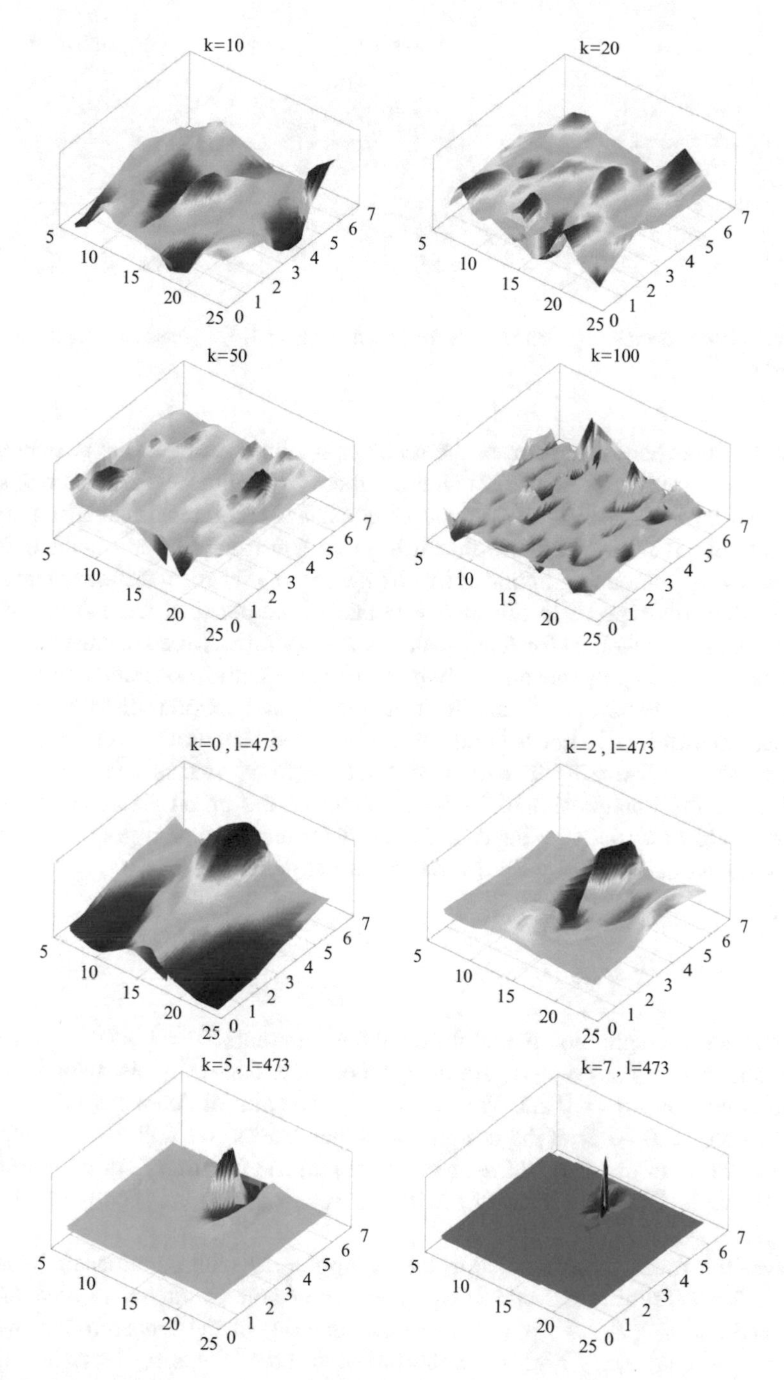

Fig. 20.2 Swiss roll data: top: eigenvectors Φ_j for $j = 10, 30, 50, 100$; bottom: frame elements Ψ_{kl} for l fixed and $k = 0, 2, 5, 7$ (construction from actual swiss roll data, then "unrolled" for clearer graphical representation)

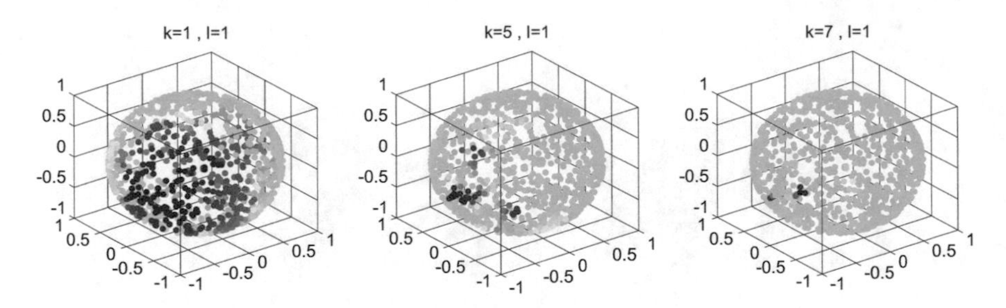

Fig. 20.3 Sphere data: frame elements Ψ_{kl} for l fixed and $k = 1, 5, 7$ (color encodes the value of the function)

On the other hand, the form of the scaling function ζ_1 used in the present work is based on Coulhon et al. (2012) where a theory of multiscale frame analysis is developed on very general metric spaces under certain geometrical assumptions. In a nutshell, it is proved there that using this construction, the obtained frame functions $\Psi_{kl}(x)$ are upper bounded by $O((d(x, x_l)/b^k)^{-\nu})$ for ν arbitrary large. We observe that this type of localization estimate is sharper than (20.14) for fixed x and growing k, as well as for fixed scale k and varying x. We conjecture that these theoretical results apply meaningfully in the discrete setting considered here, under the assumption that $x_1, \ldots, x_n$ are iid from a sufficiently regular distribution $\mathbf{P}_0$ on a regular manifold $\mathscr{M}$, but it is out of the intended scope of the present chapter to establish this formally. In particular "meaningfully" means that the constants involved in the bounds should be independent of the graph size (otherwise the bounds could potentially be devoid of interest for any particular graph, as pointed out above), a question that we are currently investigating.

20.4 Denoising

We consider the regression model for fixed design points $\mathfrak{D} = \{x_i, i = 1 \ldots n\}$ and observations $y_i = f(x_i) + \epsilon_i$ (ϵ_i are independent and identically distributed random variables with $\mathbf{E}(\epsilon_i) = 0$ and $\mathbf{Var}(\varepsilon)_i = \sigma^2$). The aim of denoising is to recover the function $f : \mathfrak{D} \to \mathbb{R}$ at the design points themselves. We will use the proposed Parseval frame in order to define an estimate $\widehat{f}$ of the function f. In what follows, since the $\mathfrak{D}$ is fixed, we identify f with the vector $(f(x_1), \ldots, f(x_n))$ and denote $y = (y_1, \ldots, y_n)$.

Given the frame $\mathscr{F}$ associated to the data points $\mathfrak{D}$ with a multiscale bandpass filter as from Definitions 2 and 3, we denote the frame coefficients $a_{kl} = \langle \Psi_{kl}, f\rangle$ for f and $b_{kl} = \langle \Psi_{kl}, y\rangle$ for y. Due to the linearity of the inner product we get $a_{kl} = b_{kl} - \langle \Psi_{kl}, \epsilon\rangle$. We estimate the unknown coefficients a_{kl} by adjusting the known coefficients b_{kl} by soft-thresholding:

$$S_s(z, c) = \text{sgn}(z)\,(|z| - c)_+. \tag{20.15}$$

In order to take into account that the frame elements Ψ_{kl} are not normalized, and generally have different norms, we use element-adapted thresholds of the form $c_{kl} = \sigma \|\Psi_{kl}\| t$ which depend on the variance of $\langle \epsilon, \Psi_{kl} \rangle$ and some global parameter t. Equivalently, this corresponds to first normalizing the observed coefficients b_{kl} by dividing by their variance, then applying a global threshold to the normalized coefficients, and finally inverting the normalization.

The estimator of f is then the plug-in estimator

$$\widehat{f}_{S_s} = \sum_{k,l} S_s(b_{kl}, c_{kl}) \Psi_{kl} = T^* S_s(b, c), \tag{20.16}$$

where $S_s(b, c)$ denotes the vector of thresholded coefficients, and T^* is the synthesis operator of the frame as introduced in Sect. 20.2.2.

To measure the performance of this estimator, we use the risk measure

$$Risk(\widehat{f}, f) = \mathbf{E}_\epsilon \left(\left\| \widehat{f} - f \right\|^2 \right), \tag{20.17}$$

that is, the expected quadratic norm at the sampled points (where $\|f\|^2 = \sum_{i=1}^n f(x_i)^2$ is the Euclidean vector norm of f on the observation points), for the performance analysis of an estimator $\widehat{f} \in \mathbb{R}^n$.

For bounding the risk of the thresholding estimator $\widehat{f}_{S_s}$, rather than assuming some specific regularity properties on the function f, it is useful to compare the performance of $\widehat{f}_{S_s}$ to that of a group of reference estimators. This is called the *oracle* approach (Candès 2006; Donoho and Johnstone 1994): can the proposed estimator have a performance (almost) as good as the best estimator (for this specific f) in a reference family (that is to say, as good as if an oracle would have given us advance knowledge of which reference estimator is the best for this function f). We review here briefly some important results.

A suitable class of simple reference estimators consists of "keep or kill" (or diagonal projection) estimators, that keep without changes the observed coefficients $b_{k,l}$ for (k, l) in some subset I, and put to zero the coefficients for indices outside of I:

$$\widehat{f}_I := \sum_{(k,l)\in I} b_{kl}\Psi_{kl} = T^* \widehat{a}^I_{kl}, \tag{20.18}$$

where $\widehat{a}^I_{kl} = b_{kl}\mathbf{1}\{(k, l) \in I\}$. Now using the frame reconstruction formula and (20.6), we obtain

$$\begin{aligned}\mathbf{E}_\epsilon \left(\left\| \widehat{f}_I - f \right\|^2 \right) &= \mathbf{E}_\epsilon \left(\left\| T^*(a - \widehat{a}^I) \right\|^2 \right) \\ &\leq \mathbf{E}_\epsilon \left(\left\| a - \widehat{a}^I \right\|^2 \right)\end{aligned}$$

$$= \sum_{(k,l)} \left(a_{kl}^2 \mathbf{1}\{(k,l) \notin I\}\right.$$

$$\left.+\sigma^2 \|\psi_{kl}\|^2 \mathbf{1}\{(k,l) \in I\}\right). \tag{20.19}$$

Therefore, the optimal (oracle) choice of the index set I^* obtained by minimizing the above upper bound is given by

$$\begin{aligned} (k,l) \in I^* \quad &\Leftrightarrow \quad \langle f, \Psi_{kl}\rangle^2 \geq \sigma^2 \|\Psi_{kl}\|^2 \quad \text{(keep)} \\ (k,l) \notin I^* \quad &\Leftrightarrow \quad \langle f, \Psi_{kl}\rangle^2 \leq \sigma^2 \|\Psi_{kl}\|^2 \quad \text{(kill)}\,. \end{aligned} \tag{20.20}$$

One deduces from this that

$$\inf_I \mathbf{E}_\epsilon \left(\left\|\widehat{f_I} - f\right\|^2\right) \leq \sum_{(k,l)\in N} \min\left(\langle f, \Psi_{kl}\rangle^2, \sigma^2 \|\Psi_{kl}\|^2\right) =: OB(f)\,. \tag{20.21}$$

The relation of soft thresholding estimators to the collection of keep-or-kill estimators on a Parseval frame is captured by the following oracle-type inequality (see Candès 2006, Section 9)[1]:

Theorem 3 *Let $\{\Psi_{kl}\}_{k,l}$ be a Parseval frame and consider the denoising observation model with Gaussian noise. Let $\widehat{f_{S_s}} = \sum_{k,l} S_s\left(\langle y, \Psi_{kl}\rangle, t_{kl}\right) \Psi_{kl}$ be the soft-threshold frame estimator from (20.16). Then with $t_{kl} = \sigma \|\Psi_{kl}\| \sqrt{2\log(n)}$ the following inequality holds:*

$$\mathbf{E}_\epsilon \left(\left\|\widehat{f_{S_s}} - f\right\|^2\right) \leq (2\log(n) + 1)\left(\sigma^2 + OB(f)\right). \tag{20.22}$$

To interpret this result, observe that if we renormalize the squared norm by $\frac{1}{n}$, so that it represents averaged squared error per point, we expect (depending on the regularity of f) the order of magnitude of $n^{-1}OB(f)$ to be typically a polynomial rate $O(n^{-\nu})$ for some $\nu < 1$. Then the term σ^2/n is negligible in comparison, and the oracle inequality states that the performance of $\widehat{f_{S_s}}$ is only worse by a logarithmic factor than the performance obtained with the optimal, f-dependent choice of I in a keep-or-kill estimator.

For this tight oracle inequality to hold, it is particularly important that a Parseval frame is used. While thresholding strategies can also be applied to the coefficients of a frame that is not Parseval, the reconstruction step is less straightforward (the canonical dual frame must be computed for reconstruction from the thresholded coefficients, see Sect. 20.2.2); furthermore, an additional factor B/A comes into the bound ($A \leq 1 \leq B$ being the frame bounds from definition (20.1)) (see, for instance, Haltmeier and Munk 2014, Prop. 3.10). Therefore, the performance of simple thresholding estimates deteriorates when used with a non-Parseval frame.

[1]Candès (2006) only hints at the proof; we provide a proof in the appendix for completeness.

20.5 Numerical Experiments

We investigate the performance of the proposed method for denoising on two testbed datasets where the ground truth is known and the design points are drawn randomly iid from a distribution on a manifold. More precisely, we will consider one example where the design points $\mathfrak{D}$ are drawn uniformly ($n = 500$) on the unit square, which is then rolled up into a "swiss roll" shape in 3D. We consider a very simple target function represented (on the original unit square) as a piecewise constant function (with values 5 and -3) on two triangles, displaying a sharp discontinuity along one diagonal of the square and very smooth regularity elsewhere. This function is observed with an additional Gaussian noise of variance $\sigma^2 = 1$. In the second example the design points $\mathfrak{D}$ are drawn uniformly ($n = 500$) on the unit sphere in $\mathbb{R}^3$. The target function remains a piecewise constant function, defined on the two parts of the sphere when intersecting it with a chosen plane. Again, this function is observed with an additional Gaussian noise of variance $\sigma^2 = 1$. For the swiss roll example as well as for the sphere example, one sample consisting of design points and noisy function values is displayed in Fig. 20.4.

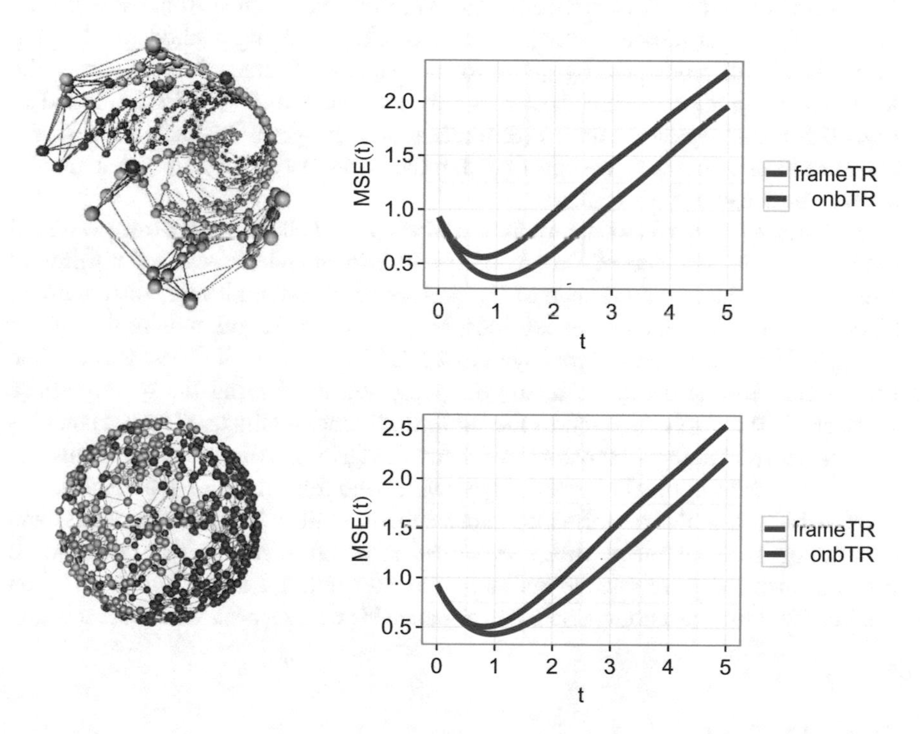

Fig. 20.4 Left: noisy function on swiss roll data (top) and sphere data (bottom), graph representation. Right: MSE for two representative settings (weighted ε-Graph and k-NN-Graph) as a function of threshold level. Red is thresholding in the original Laplacian Eigenmaps ONB, blue is thresholding of frame coefficients

In each example, we consider the different types of neighborhood graphs described in Sect. 20.2.3. Following usual heuristics, for the construction of the k-NN graph we take $k = 7 \approx \log n$; for the ε-graph, we take for ε the average distance to the $k = 7$th nearest neighbor, and for weighted graphs we take Gaussian weights, where the bandwidth λ is calibrated so that points at the distance ε defined above are given weight 0.5.

After constructing the (weighted or unweighted) graph Laplacian, we compute explicitly its eigendecomposition. For the construction of the frame via the multiscale bandpass filter, we use a $\mathscr{C}^3$ piecewise polynomial plateau function g satisfying the support constraints of Definition 3 for $b = 2$ (i.e., constant equal to 1 for $x \leq 0.5$, and zero for $x \geq 1$). While this function is not $\mathscr{C}^\infty$, it has the advantage of fast computation.

We compare the denoising performance of the following competitors: Parseval frame with soft thresholding, soft thresholding applied to the Laplacian Eigenmaps orthonormal basis, and truncated expansion in the Laplacian Eigenmaps basis (only the k coefficients corresponding to the first eigenvalues are kept, without thresholding). The latter method is in the spirit of Belkin and Niyogi (2002). It is well-known (from the regular grid case) that the "universal" theoretical threshold $\sigma\sqrt{\log n}$ is often too conservative in practice. For a fair comparison, we therefore compute the mean squared error (MSE) of both thresholding methods for varying threshold t (still modulated by $\|\Psi_{kl}\|$ for the Parseval frame). Comparison of the MSE for one sample across the t-range for two particular settings is plotted in Fig. 20.4. For all studied settings (different graph and graph Laplacian types), for the same threshold level t we observed that the frame-based method systematically shows a noticeable improvement.

In Table 20.1 we report the minimum MSEs and their standard error (averaged over $m = 50$ samples of design points and independent noise) for different methods over the possible range of the parameter (threshold level t, resp. number of coefficients for truncated expansion), both for the swissroll and for the sphere example. We observe an improvement of 20–25% across the different settings (the best overall results being obtained with weighted graphs and the unnormalized Laplacian). We also compared to the more traditional methods of kernel smoothing (Nadaraya-Watson estimator) and kernel ridge regression, using a Gaussian kernel (also with optimal choices of bandwidth and regularization parameter), and observed a comparable performance improvement. While it is not realistic to assume that the optimal parameter choice is known in practice, it is fair to compare all methods under their respective optimal parameter settings, as parameter selection methods will induce a comparable performance hit with respect to the best setting.

20.6 Outlook

Following the recently introduced idea of generalizing the Littlewood-Paley spectral decomposition, we constructed explicitly a Parseval frame of functions on a neighborhood graph formed on the data points. We established that a thresholding strategy

Table 20.1 MSE performance under optimal parameter choice

Graph	L	FrTh	LETh	LETr
Example 1: sphere, jump function, $\sigma^2 = 1, n = 500, m = 50$				
kNN	U	0.510 (0.050)	0.693 (0.061)	0.905 (0.108)
kNN	N	0.538 (0.046)	0.712 (0.055)	0.931 (0.094)
WkNN	U	0.521 (0.049)	0.652 (0.050)	0.800 (0.097)
WkNN	N	0.530 (0.049)	0.674 (0.057)	0.749 (0.091)
CGK	U	0.520 (0.055)	0.638 (0.065)	0.821 (0.107)
CGK	N	0.530 (0.052)	0.670 (0.050)	0.725 (0.081)
ϵG	U	0.505 (0.058)	0.650 (0.068)	0.865 (0.115)
ϵG	N	0.557 (0.052)	0.710 (0.059)	0.902 (0.106)
WϵG	U	0.482 (0.055)	0.622 (0.064)	0.787 (0.111)
WϵG	N	0.530 (0.049)	0.674 (0.057)	0.749 (0.091)
Smoothing Kernel Regression: min. MSE = 0.612 (0.066)				
Kernel Ridge Regression: min. MSE = 0.594 (0.051)				
Example 2: swiss roll, jump function, $\sigma^2 = 1, n = 500, m = 50$				
kNN	U	0.462 (0.043)	0.647 (0.039)	0.876 (0.079)
kNN	N	0.494 (0.043)	0.676 (0.043)	0.902 (0.071)
WkNN	U	0.443 (0.045)	0.600 (0.050)	0.790 (0.102)
WkNN	N	0.500 (0.043)	0.659 (0.045)	0.775 (0.079)
CGK	U	0.491 (0.053)	0.625 (0.057)	0.844 (0.096)
CGK	N	0.520 (0.047)	0.648 (0.049)	0.713 (0.079)
ϵG	U	0.459 (0.049)	0.610 (0.053)	0.872 (0.095)
ϵG	N	0.532 (0.045)	0.681 (0.050)	0.884 (0.089)
WϵG	U	0.441 (0.049)	0.574 (0.049)	0.793 (0.113)
WϵG	N	0.503 (0.045)	0.643 (0.051)	0.744 (0.089)
Smoothing Kernel Regression: min. MSE = 0.589 (0.082)				
Kernel Ridge Regression: min. MSE = 0.779 (0.052)				

FrTh: Frame Thresholding; LETh/LETr: Laplacian Eigenmaps Thresholding/Truncated expansion. Prefix W indicates edge weighting in the graph. CGK is the complete graph with Gaussian weights. U/N is un/normalized graph Laplacian. Standard error in brackets. Top: Sphere example. Bottom: Swiss roll example

on the frame coefficients has superior performance for the denoising problem as compared to usual, spectral or non-spectral, approaches. Future developments include extension of this methodology to the semisupervised learning setting, and a stronger theoretical basis for spatial localization.

Acknowledgements The authors acknowledge the financial support of the German DFG, under the Research Unit FOR-1735 "Structural Inference in Statistics—Adaptation and Efficiency."

Appendix

Proof of Theorem 3

Theorem 3 states a oracle-type inequality which captures the relation of soft thresholding estimators $\hat{f}_{S_s} = \sum_{k,l} S_s(\langle y, \Psi_{kl}\rangle, t_{kl})\, \Psi_{kl}$ defined in (20.16) to the collection of keep-or-kill estimators on a Parseval frame. This result is known in the literature (see Candès 2006, Section 9), but we provide a short self-contained proof for completeness, modulo a technical result from Donoho and Johnstone (1994) for soft thresholding of a single one-dimensional Gaussian variable, which is basic for the Proof of Theorem 3.

Lemma 1 *For* $0 \le \delta \le 1/2$, $t = \sqrt{2\log(\delta^{-1})}$ *and* $X \sim \mathcal{N}(\mu, 1)$

$$\mathbf{E}_X\left((S_s(X,t) - \mu)^2\right) \le (2\log(\delta^{-1}) + 1)(\delta + \min(1, \mu^2))$$
$$= (t^2 + 1)\left(\exp\left(-\frac{t^2}{2}\right) + \min(1, \mu^2)\right). \quad (20.23)$$

The proof of this lemma can be found in appendix 1 of Donoho and Johnstone (1994). Now we are able to prove Theorem 3.

Proof First note that for $y = \tau x$, $\tau > 0$, we have

$$S_s(y, u) = \tau S_s\left(x, \frac{u}{\tau}\right). \quad (20.24)$$

Secondly we remark that

$$\frac{\langle y, \Psi_{kl}\rangle}{\sigma \|\Psi_{kl}\|} \sim \mathcal{N}\left(\frac{a_{kl}}{\sigma \|\Psi_{kl}\|}, 1\right). \quad (20.25)$$

Considering now the risk of the soft thresholding estimator $\hat{f}_{S_s}$ we get

$$\mathbf{E}\left(\left\|\hat{f}_{S_s} - f\right\|^2\right) = \mathbf{E}\left(\left\|\sum_{k,l}(S_s(\langle y, \Psi_{kl}\rangle, t_{kl}) - a_{kl})\,\Psi_{kl}\right\|^2\right)$$
$$\le \mathbf{E}\left(\sum_{k,l}(S_s(\langle y, \Psi_{kl}\rangle, t_{kl}) - a_{kl})^2\right)$$
$$= \sum_{k,l}\mathbf{E}\left((S_s(\langle y, \Psi_{kl}\rangle, t_{kl}) - a_{kl})^2\right). \quad (20.26)$$

by using inequality (20.6). By applying (20.24) and then (20.23) with $t = \sqrt{2\log(n)}$ it follows that

$$\mathbf{E}\left(\left\|\hat{f}_{S_s} - f\right\|^2\right) \leq \sum_{k,l} \sigma^2 \|\Psi_{kl}\|^2 \mathbf{E}\left(\left(S_s\left(\frac{\langle y, \Psi_{kl}\rangle}{\sigma\|\Psi_{kl}\|}, \sqrt{2\log(n)}\right) - \frac{a_{kl}}{\sigma\|\Psi_{kl}\|}\right)^2\right)$$

$$\leq \sum_{k,l} \sigma^2 \|\Psi_{kl}\|^2 (2\log(n) + 1)\left(\exp\left(-\frac{2\log(n)}{2}\right) + \min\left(1, \frac{a_{kl}^2}{\sigma^2\|\Psi_{kl}\|^2}\right)\right)$$

$$= \sum_{k,l} (2\log(n) + 1)\left(\frac{1}{n}\sigma^2\|\Psi_{kl}\|^2 + \min\left(\sigma^2\|\Psi_{kl}\|^2, a_{kl}^2\right)\right)$$

$$= (2\log(n) + 1)\left(\frac{1}{n}\sum_{k,l}\sigma^2\|\Psi_{kl}\|^2 + \sum_{k,l}\min\left(\sigma^2\|\Psi_{kl}\|^2, a_{kl}^2\right)\right). \tag{20.27}$$

Recalling the Parseval frame property $\sum_{k,l}\|\Psi_{kl}\|^2 = n$, we finally obtain

$$\mathbf{E}\left(\left\|\hat{f}_{S_s} - f\right\|^2\right) \leq (2\log(n) + 1)\left(\frac{1}{n}n\sigma^2 + \sum_{k,l}\min\left(\sigma^2\|\Psi_{kl}\|^2, a_{kl}^2\right)\right)$$

$$= (2\log(n) + 1)\left(\sigma^2 + \sum_{k,l}\min\left(\sigma^2\|\Psi_{kl}\|^2, a_{kl}^2\right)\right). \tag{20.28}$$

where we recognize the upper bound $\sum_{k,l}\min\left(\sigma^2\|\Psi_{kl}\|^2, a_{kl}^2\right) = OB(f)$ for the oracle. □

References

Belkin M, Niyogi P (2002) Using manifold structure for partially labeled classification. In: NIPS, pp 929–936

Belkin M, Niyogi P (2003) Laplacian eigenmaps for dimensionality reduction and data representation. Neural Comput 15(6):1373–1396

Bickel P, Li B (2007) Local polynomial regression on unknown manifolds. In: Complex datasets and inverse problems: tomography, networks and beyond. IMS lecture notes, vol 54. Institute of Mathematical Statistics, Bethesda, pp 177–186

Candès E (2006) Modern statistical estimation via oracle inequalities. Acta Numer 15:257–325

Casazza P, Kutyniok G, Philipp F (2013) Introduction to finite frame theory. In: Casazza PG, Kutyniok G (eds) Finite frames, applied and numerical harmonic analysis. Birkhäuser, Boston, pp 1–53

Christensen O (2008) Frames and bases: an introductory course. In: Applied and numerical harmonic analysis. Birkhäuser, Boston

Coulhon T, Kerkyacharian G, Petrushev P (2012) Heat kernel generated frames in the setting of Dirichlet spaces. J Fourier Anal Appl 18(5):995–1066

Donoho DL, Johnstone IM (1994) Ideal spatial adaptation by wavelet shrinkage. Biometrika 81(3):425–455

Gavish M, Nadler B, Coifman RR (2010) Multiscale wavelets on trees, graphs and high dimensional data: theory and applications to semi supervised learning. In: Fürnkranz J, Joachims T (eds) ICML. Omnipress, Madison, pp 367–374

Haltmeier M, Munk A (2014) Extreme value analysis of empirical frame coefficients and implications for denoising by soft-thresholding. Appl Comput Harmon Anal 36(3):434–460. https://doi.org/10.1016/j.acha.2013.07.004

Hammond DK, Vandergheynst P, Gribonval R (2011) Wavelets on graphs via spectral graph theory. Appl Comput Harmon Anal 30(2):129–150

Han D (2007) Frames for undergraduates. In: Student mathematical library. American Mathematical Society, Providence

Kpotufe S (2011) k-NN regression adapts to local intrinsic dimension. In: NIPS, pp 729–737

Kpotufe S, Dasgupta S (2012) A tree-based regressor that adapts to intrinsic dimension. J Comput Syst Sci 78(5):1496–1515

Roweis S, Saul L (2000) Nonlinear dimensionality reduction by locally linear embedding. Science 290:2323–2326

Tenenbaum J, de Silva V, Langford J (2000) A global geometric framework for nonlinear dimensionality reduction. Science 290:2319–2323

Chapter 21
Beta-Boosted Ensemble for Big Credit Scoring Data

Maciej Zięba and Wolfgang Karl Härdle

Abstract In this work we present the novel ensemble model for credit scoring problem. The main idea of the approach is to incorporate separate beta binomial distributions for each of the classes to generate balanced datasets that are further used to construct base learners that constitute the final ensemble model. The sampling procedure is performed on two separate ranking lists, each for one class, where the ranking is based on probability of observing positive class. The two strategies are considered in the studies: one assumes mining easy examples and the second one force good classification of hard cases. The proposed solutions are tested on two big datasets from credit scoring domain.

Keywords Credit scoring · Ensemble model · Beta distribution · Beta boost · Big data

21.1 Introduction

The problem of constructing the decision model to distinguish good and bad consumers can be defined as dichotomous classification task, where the positive class (usually less numerous) represents "bad" applicants and the negative class stays behind "good" cases. Usually, instead of obtaining the binary classification result we aim at estimating the probability of credit repayment for each of the consumers. Basing on the probabilities the financial institution is capable to define the various profiles of the consumers. The common procedure for that kind of applications is

M. Zięba
Wroclaw University of Science and Technology, Wroclaw, Poland
e-mail: maciej.zieba@pwr.edu.pl

W. K. Härdle (✉)
Humboldt-Universität zu Berlin, Berlin, Germany

School of Business, Singapore Management University, Singapore, Singapore
e-mail: haerdle@wiwi.hu-berlin.de

© Springer International Publishing AG, part of Springer Nature 2018
W. K. Härdle et al. (eds.), *Handbook of Big Data Analytics*, Springer Handbooks of Computational Statistics, https://doi.org/10.1007/978-3-319-18284-1_21

to separate from training some group of labeled consumers and sort them according to the predictive probability using the trained model. The sorted group with the given labels is further used to distinguish the profiles. As a consequence, the higher patience is given to construct the models that are characterized by good sorting capabilities than to the typical classifiers used for binary classification. Instead of maximizing the *accuracy of prediction* the community working on the credit scoring models aims at achieving the highest value of *AUC (area under ROC curve)* criterion that stays behind sorting capabilities of the models.

Various machine learning algorithms were applied to solve credit scoring and fraud detection problems, such as: neural networks (Lee et al. 2002; Oreski et al. 2012; Zhao et al. 2015), Gaussian Processes (Huang 2011), various extensions of SVMs (Support Vector Machines) (Bellotti and Crook 2009; Chen et al. 2010, 2011; Härdle et al. 2009, 2012; Harris 2015; Zhou et al. 2009) or comprehensible models based on neural structures (Tomczak and Zięba 2015) or SVMs (Martens et al. 2007).

Ensemble methods have also gained particular attention in the field of credit scoring. The general idea of this type of models is based on constructing many component models (so-called base learners) that are joined together as one complex classifier. Usually, the base model is so-called weak learner that is characterized by poor individual performance, but strong learners are also used for particular ensemble models. In work Nanni and Lumini (2009) authors present very beneficial comparison of the standard ensemble procedures in application to credit scoring tasks. Some more up-to-date analysis of this kind of models for this particular application were presented in Abellán and Mantas (2014) and Zhu et al. (2016). The most recent models make use of various types of base learners (Koutanaei et al. 2015), joined two strategies of diversification on features and data levels (Marqués et al. 2012), switching class labels (Zięba and Świątek 2012), boosting neural networks (Tsai and Wu 2008) or using ensemble of cost-sensitive SVMs trained with active learning strategy (Zięba and Tomczak 2015). Most recent studies show the great benefit of using Extreme Boosted Trees (Zięba et al. 2016).

In this work we aim at constructing novel boosting approach that works independently on selected base model and performs well on big credit scoring datasets. The key idea of this approach is to apply sampling strategy to sample examples for each of the boosting iterations to construct the base learners. We make use of particular Beta Binomial distributions that are applied to the sorted training data according to the prediction probabilities returned by current ensemble model. In this work we distinguish two sampling strategies: the first strategy aims at sampling with the higher probability the examples that are already well located in the ranking. The other strategy is an example of so-called *hard examples* mining where the higher probabilities are given to the examples badly predicted and badly located in the ranking. Our approach was tested on the two benchmark datasets using two base models: Logistic Regression and Decision Tree classifier. The results show that the first strategy works fine with the stable models like Logistic Regression, while the second strategy improves the quality of weak learners like Decision Trees.

The chapter is organized as follows. In Sect. 21.2 we present *BetaBoost* algorithm. In Sect. 21.3 we introduce some experimental studies investigating the performance of the approach. The chapter is summarized with some conclusions in Sect. 21.4.

21.2 Method Description

The main idea of the proposed approach is to create the ensemble model that makes use of re-sampling diversification technique to increase its sorting capabilities. To achieve the goal, each of the base learners is trained using re-sampled training data. The re-sampling procedure makes use of two particular beta binomial distributions (one for each class) that are used to generate indexes of examples that are going to be taken in the next boosting iteration. The crucial step in the training procedure is sorting the training data according to predictive capabilities of the so far created ensemble model. As a consequence, the examples with higher probability value have higher indexes and are going to be selected more often in training iterations. For the sampling procedure we propose to use *Beta Binomial* distribution which is going to be characterized in the next subsection.

21.2.1 Beta Binomial Distribution

The beta binomial distribution is selected because it is capable to assign high probabilities to particular regions of the sorted data according to predictive probability values of the training examples. Practically, it means that we are capable to concentrate our model either on learning from difficult-to-distinguish credit consumers or put the higher impact on learning from the easy-to-classify client applicants.

The flexibility of beta binomial distribution is controlled by three parameters:

- Shape parameters a and b that are characteristic for beta distribution ($a, b > 0$).
- Parameter N that represents the number of trials characteristic for binomial distribution ($N \in \mathbf{N_0}$).

The probability function for beta binomial distribution ($BBin(a, b, N)$) can be presented in the following form:

$$p(k; a, b, N) = \binom{N}{k} \frac{B(k + a, N - k + b)}{B(a, b)}, \tag{21.1}$$

where $B(a, b)$ is the beta function.

The presented distribution has the important property for the particular values of shape parameters a and b. In this application we are concentrating on particular families of beta binomial distributions:

- The subset of distributions, where $a \leq 1$, $b \geq 1$ and $a \neq b$. If $k_1 > k_2$, then $p(k_1; a, b, N) < p(k_2; a, b, N)$.
- The subset of distributions, where $a \geq 1$, $b \leq 1$ and $a \neq b$. If $k_1 > k_2$, then $p(k_1; a, b, N) > p(k_2; a, b, N)$.

The selection of the particular distributions is indicated by the strategies that are going to be applied to train the ensemble model. For the first strategy we aim at putting the higher impact on selecting better located examples in the ranking so for the ranking list for negative examples (sorted according to probability of observing positive class) we apply the family of distributions that satisfies $p(k_1; a, b, N) < p(k_2; a, b, N)$, while for the ranking list for positive cases we use family of distribution that satisfies $p(k_1; a, b, N) > p(k_2; a, b, N)$. As a consequence it is more probable to select the examples properly located on the both of the lists.

For the second strategy we make use of the first family of distributions for positive ranking list and the second family for negative sorted samples. Contrary to previous strategy we aim at mining rather hard positive and negative examples and omitting well-classified examples.

In the next section we present, how the beta binomial sampling is used in constructing the boosted model.

21.2.2 Beta-Boosted Ensemble Model

In this work we aim at constructing the ensemble classifier for binary classification $y \in \{0, 1\}$, composed of T base models:

$$p_T(y|\mathbf{x}) = \sum_{t=0}^{T} p_t p(y|\mathbf{x}, t)^y \left\{1 - p(y|\mathbf{x}, t)\right\}^{(1-y)}, \qquad (21.2)$$

where $\mathbf{x}$ is the vector of input features, $p(y|\mathbf{x}, t)$ represents the t-th base learner, and p_t is the prior distribution over base learners.

For further work we assume that base learners are characterized by uniform distribution, so we can present the ensemble model given by Eq. (21.2) in the following form:

$$p_T(y = 1|\mathbf{x}) = \frac{1}{T+1} \sum_{t=0}^{T} p(y = 1|\mathbf{x}, t). \qquad (21.3)$$

We are interested in obtaining probability value for a given positive class therefore we will further operate on probability for this class, $p(y = 1|\mathbf{x})$.

For the given predictor $p(y = 1|\mathbf{x})$ and the set of examples $X_N = \{\mathbf{x}_n\}_{n=1}^N$ we can define the rank function $h(\mathbf{x}, X_N, p)$:

$$h(\mathbf{x}, X_N, p) = \sum_{n=1}^{N} \mathbb{I}\Big\{p(y = 1|\mathbf{x}) > p(y = 1|\mathbf{x_n})\Big\} \tag{21.4}$$

The procedure for creating the ensemble classifier can be described by Algorithm 1. To create the classifier we make use of training data $D_N = \{(\mathbf{x}_n, y_n)\}_{n=1}^N$, that contains N training examples: N_1 positive and N_0 negative instances. We aim at constructing the ensemble model given by Eq. (21.3).

To initialize the training procedure we distinguish positive and negative examples denoting them by $\mathbf{X}_{N_1}$ and $\mathbf{X}_{N_2}$, respectively. We also initialize the ensemble structure by training the first base learner $p(y|\mathbf{x}, 0)$ using initial training set $D_N = \{(\mathbf{x}_n, y_n)\}_{n=1}^N$. In the next step we perform constructing the committee of T base classifiers in the training loop. Before creating the base learner we perform beta binomial sampling using separate distributions for each of the classes to obtain $N/2$ samples for each class. We use distributions for each of the classes, $BBin_0(a, b, N)$ to sample negatives and $BBin_1(a, b, N)$ to sample positives. We recommend to use particular families of distributions that were characterized in Sect. 21.2.1.

The procedure of sampling the data makes use of the currently created ensemble model $p_{t-1}(y = 1|\mathbf{x})$ to determine the ranking position of the example $\mathbf{x}$ in the given set X_N using ranking function $h(\mathbf{x}, X_N, p)$ given by Eq. (21.4). The sampling procedure is performed independently for each of the classes and is described by Algorithm 2. First, we sample the integer k from $BBin(a, b, N-1)$ distribution. Second, we identify the sample that has ranking value equal to the sampled k value

---

Algorithm 1: BetaBoost

Input: Training data: $D_N = \{(\mathbf{x}_n, y_n)\}_{n=1}^N$
Output: Ensemble model: $p_T(y = 1|\mathbf{x})$ (see Eq. (21.3))
Parameters : $BBin_0(\cdot)$ parameters for negative class: a_0, b_0,
$BBin_1(\cdot)$ parameters for positive class: a_1, b_1,
number of base learners: $T + 1$.

1 Set $\mathbf{X}_{N_1} = \{\mathbf{x}_n : y_n = 1\}$ and $\mathbf{X}_{N_0} = \{\mathbf{x}_n : y_n = 0\}$;
2 Train weak learner $p(y|\mathbf{x}, 0)$ with data D_N;
3 **for** $t \leftarrow 1$ **to** T **do**
4 | Create ensemble predictor $p_{t-1}(y = 1|\mathbf{x}) = \frac{1}{t+1}\sum_{j=0}^{t} p(y = 1|\mathbf{x}, j)$;
5 | Generate $\widetilde{\mathbf{X}}_{N/2}^{(1)}$ with $sample(\mathbf{X}_{N_1}, p_{t-1}, a_1, b_1, N/2)$ (see Algorithm 2);
6 | Generate $\widetilde{\mathbf{X}}_{N/2}^{(0)}$ with $sample(\mathbf{X}_{N_0}, p_{t-1}, a_0, b_0, N/2)$ (see Algorithm 2);
7 | Create new training data $\widetilde{D}_N = (\widetilde{\mathbf{X}}_{N/2}^{(1)}, \mathbf{1}) \cup (\widetilde{\mathbf{X}}_{N/2}^{(1)}, \mathbf{0})$;
8 | Train weak learner $p(y|\mathbf{x}, k)$ with data $\widetilde{D}_N$;
9 **end**

Algorithm 2: Sampling procedure: $sample(X_N, p, a, b, N_{out})$

Input: Predictor $p(y = 1|\mathbf{x})$, the set of examples $X_N = \{\mathbf{x}_n\}_{n=1}^{N}$, number of output samples N_{out}.
Output: Set of data samples $\tilde{X}_{Nout} = \{\mathbf{x}_n\}_{n=1}^{N_{out}}$
Parameters : $BBin(\cdot)$ parameters: a, b.
1 $\tilde{X}_0 \leftarrow \emptyset$;
2 **for** $n \leftarrow 1$ **to** N_{out} **do**
3 Sample $k \sim BBin(a, b, N - 1)$;
4 $\tilde{X}_n \leftarrow \tilde{X}_{n-1} \cup \{\mathbf{x} \in X_N : h(\mathbf{x}, X_N, p) = k\}$, $h(\mathbf{x}, X_N, p)$ is given by Eq. (21.4);
5 **end**

and include it into the set of output samples $\tilde{X}_n$. The sampling procedure is repeated N_{out} times to obtain the output set of examples, $\tilde{X}_{N_{out}}$. The procedure is equivalent to sorting the given data according to the given predictions and then sampling their position with beta binomial distribution.

The sampling procedure is performed separately for the sets of positive and negative examples $\mathbf{X}_{N_1}$, $\mathbf{X}_{N_0}$ and, as a consequence, the new sets $\tilde{\mathbf{X}}_{N/2}^{(1)}$ and $\tilde{\mathbf{X}}_{N/2}^{(0)}$ are created and each of them contains $N/2$ sampled examples. The two sets are then labeled and concatenated to the new training data $\tilde{D}_N$ that is further used to train the k-th base learner $p(y|\mathbf{x}, k)$. The procedure is repeated T times to obtain ensemble model composed of $T + 1$ base learners.

21.2.3 Toy Example

Consider the toy example in which we have set of 15 examples, 5 from positive class and 10 from negative class. Assume that we have the committee of the models that sorted the training examples according to the predictive probability $p(y = 1|\mathbf{x})$ (see Fig. 21.1a). Further, we assign individual ranking position for each of the considered classes (see Fig. 21.1b). Next, we assume individual Beta binomial distribution for each of the classes:

- $BBin_0(0.8, 2, 9)$ for negative examples.
- $BBin_1(2, 0.8, 4)$ for positive examples.

The selected distributions are consistent with the first strategy described in Sect. 21.2.1, where we aim at mining easy examples from both classes. The selection of the a and b is crucial for the training procedure. If the both values are close to 0 the distribution approaches uniform distribution, while for large a and small b examples with high positions are going to be selected multiple times. To select proper parameters for the distributions model selection procedure should be applied (Fig. 21.2).

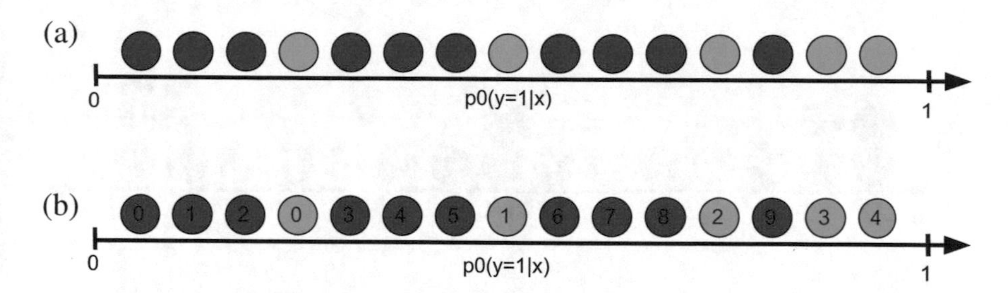

Fig. 21.1 The set of data examples sorted according to $p_0(y = 1|\mathbf{x})$. Red circles represent negative examples, and green circles stand behind positive cases. (**a**) Sorted data points according to the predictive distribution. (**b**) Sorted data points with individual rankings for each class

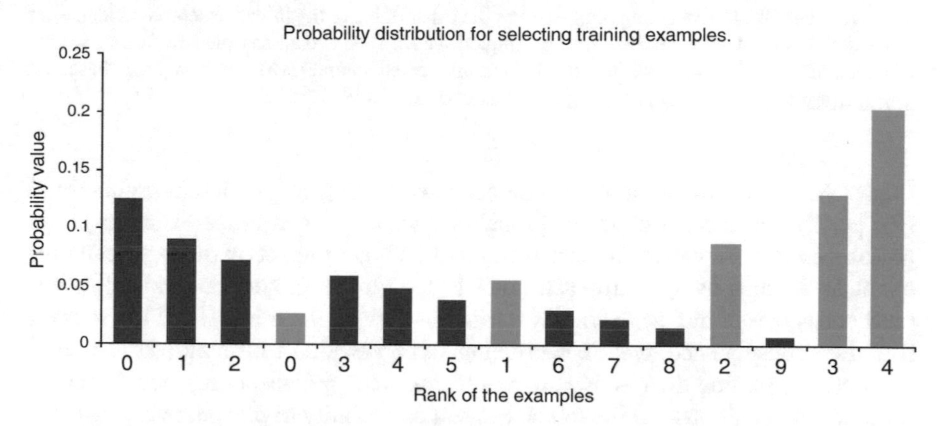

Fig. 21.2 Sampling distribution for examples presented in Fig. 21.1—$BBin_0(0.8, 2, 9)$ for negative and $BBin_1(2, 0.8, 4)$ for positive examples

If we assume equal prior probabilities for selecting examples from minority and majority class, the sampling distribution for the next boosting iteration is presented in Fig. 21.1.

If we perform sampling with replacement from the given distribution we can obtain the set of examples that should be taken into next boosting iteration that is presented in Fig. 21.3a. After learning the second base learner $p(y = 1|x, 1)$ and adding it to the ensemble model $p_1(y = 1|\mathbf{x}) = \frac{p(y=1|x,0)+p(y=1|x,1)}{2}$ we obtain the better sorting of the data (see Fig. 21.3b).

If we consider the *AUC* criterion (*area under ROC curve*) that represents the quality of the sorting capabilities for the binary classification models it increases from 0.76 to 0.92.

The idea that stays behind the proposed procedure is a proper selection of the sampling distributions the satisfy the conditions that are described in Sect. 21.2.1. In this variant we take the distribution for sampling positive examples that satisfies: $a_1 \geq 1$, $b_1 \leq 1$ and for sampling negative instances we use the distribution with parameters: $a_0 \leq 1$, $b_0 \geq 1$. Practically it means that we aim at putting the higher

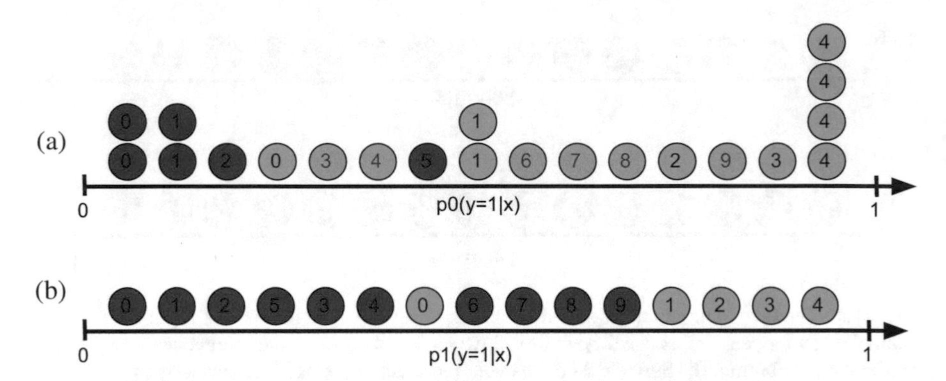

Fig. 21.3 The illustrative example presenting the capabilities of the joined ensemble model, after training the second base learner on the sampled data. (**a**) The data sampled from distribution presented in Fig. 21.2 (grey circle stays behind unselected sample). (**b**) The new order based on the classification of the ensemble model $p_1(y = 1|\mathbf{x}) = \frac{p(y=1|x,0)+p(y=1|x,1)}{2}$

impact on the examples that are characterized by higher predictive probabilities (for positive examples) or lower probability values (for negative examples). Our philosophy for this particular case is to put the higher impact on distinguishing the examples located away from each other in the global ranking determined by the predictions comparing to examples located in the weighted middle of the ranking list. As a consequence, we are sacrificing some portion of difficult to distinguish examples by putting them to unsure region, but we avoid observing them in low or high ranking positions. So the model has some capability to prevent overfitting that can be caused by discursive (or even noise) examples in training data. We also aim at dealing with imbalanced data phenomenon by sampling equal number of positive and negative examples.

Quite opposite strategy is observed for the following sampling distribution:

- $BBin_0(2, 0.8, 9)$ for negative examples.
- $BBin_1(0.8, 2, 4)$ for positive examples.

In this case, the sampling distribution for the next boosting iteration is presented in Fig. 21.4. Following this strategy we aim at correct classification of the improperly ranked examples, assuming that they are rather hard examples that we manage to classify by the ensemble model.

The two presented strategies aim at different cases. In the first case we trust our base model, but we do not trust our data assuming that there are some portion of the examples that are impossible to be distinguished. Therefore, we are leaving some portion of examples in controversial area on the ranking, cleaning low and high ranking regions with improperly located samples. For the second strategy, we use rather untrusted weak learner as a base model, but we aim at creating the complex model that will properly classify hard instances if their impact is going to be decreased.

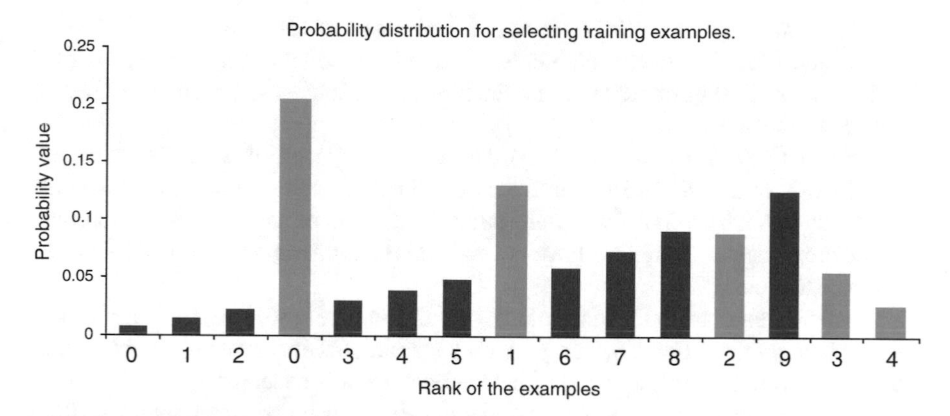

Fig. 21.4 Sampling distribution for examples presented in Fig. 21.1—$BBin_0(2, 0.8, 9)$ for negative and $BBin_1(0.8, 2, 4)$ for positive examples

21.2.4 Relation to Existing Solutions

The presented work is inspired by existing *RankBoost* (Freund et al. 2003) (for which the equivalence to well-known *AdaBoost* was described in Rudin and Schapire 2009) method and couple of other approaches. In contrast to the *RankBoost* we define two separate ranking functions for positive and negative examples. First of all, the *Rankboost* approach is very sensitive to the noisy examples located in training data. *BetaBoost* model presented in this chapter deals well with insecure and noisy data because the distribution is not updated in iterations and does not depend on global ranking. Moreover, it is also more beneficial to use more flexible sampling distribution that is characterized by two parameters (a and b) contrary to the specific exponential-based distribution used in typical boosting approaches. The proposed solution is also inheriting self-paced philosophy (Kumar et al. 2010) if the strategy with the increasing probabilities for positive and with decreasing probabilities for negative examples is applied.

As the procedure is independent on global ranking it is crucial to apply proper model selection procedure that will fit proper sampling curves for each of the classes.

21.3 Experiments

We are going to evaluate our approach on two large datasets from credit scoring domain that are available in *Kaggle* repository:

- *Give me Some Credit* (Give Me Some Credit 2011).
- *Lending Club Loan Data* (Lending Club 2016).

Give me Some Credit (GMSC) dataset is composed of 150,000 examples, 10,026 positive and 139,974 negative elements. Each of the credit consumers is represented by the vector of 10 numeric features. Each of the attributes was normalized before using it for training.

Lending Club Loan Data (LCLD) dataset is composed of 887,379 examples, 67,429 positive and 819,955 negative cases. Each of the examples was described by 12 features, where 6 of them were numeric, and the remaining 6 were nominal. On the preprocessing stage we have normalized the numeric features and binarized nominal attributes.

We divide each of the initial datasets to: training set (80% examples) and test set (20% examples). From training set we separate 10% instances for validation to monitor the training progress and select the best set of base learners.

For the evaluation we use *AUC (area under ROC curve)* criterion, which is often for evaluating credit scoring models and measures well the sorting capabilities of learners. For each of the scenarios we apply model selection of the sampling parameters (a_1, b_1, a_0, b_0) from the set of candidates and select the parameters with the highest *AUC* obtained on the validation set.

We consider the two scenarios that were described in this work. In the first of the scenarios we aim at putting higher weights to the "secure" examples, assuming that controversial examples are hard to classify.

Therefore we propose to use *Logistic Regression* as a stable base learner:

$$p(y = 1|x, k) = \sigma(\mathbf{w}_k^{\mathrm{T}}\mathbf{x}) = \frac{1}{1 - \exp\{-\mathbf{w}_k^{\mathrm{T}}\mathbf{x}\}} \tag{21.5}$$

At first we analyze the training capabilities of the *BetaBoost* model trained using the following beta parameters: $a_0 = 0.8$, $b_0 = 2$, $a_1 = 2$, and $b_1 = 0.8$. We compare the proposed approach with the so-called *Balanced Bagging* that performs sampling with replacement from uniform distribution to obtain $N/2$ samples from each class. The results of the comparison are presented in Figs. 21.5 and 21.6.

It can be observed that *Logistic Regression* is a very stable model characterized by small variance of the performance. Practically, it means that small changes in data caused by uniform sampling do not affect the overall performance of the model. If we apply sampling for the procedure characteristic for *BetaBoost* model, we would obtain the improvement of *AUC* measure as it is observed in Figs. 21.5 and 21.6. As a consequence of increasing probabilities for positive examples ($a_1 > 1$ and $b_1 < 1$) and decreasing probabilities for negative cases ($a_0 < 1$ and $b_0 > 1$) we aim at good quality prediction of the positive examples that are located on higher ranking positions and negative examples that are located on low positions. To obtain the goal we sacrifice the "difficult" examples that are suspected to be "noisy" instances, that are located in the discussion area. As a consequence, the improvement of *AUC* is observed for both of the considered datasets.

(a)

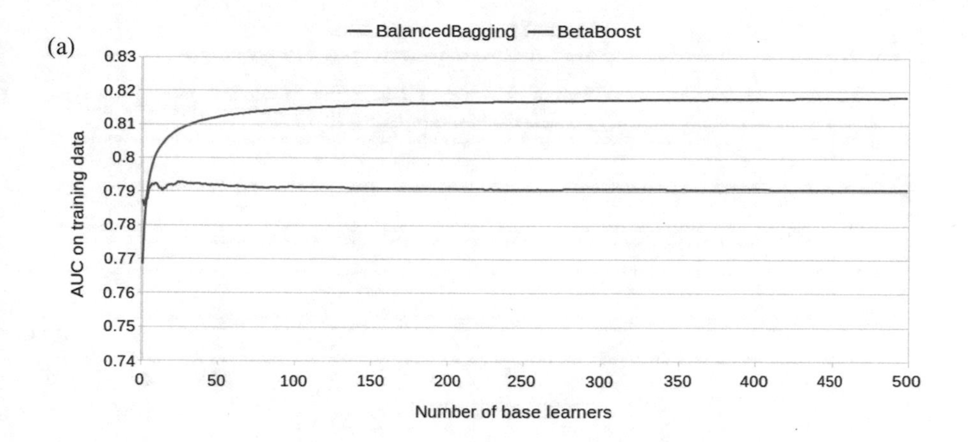

(b)

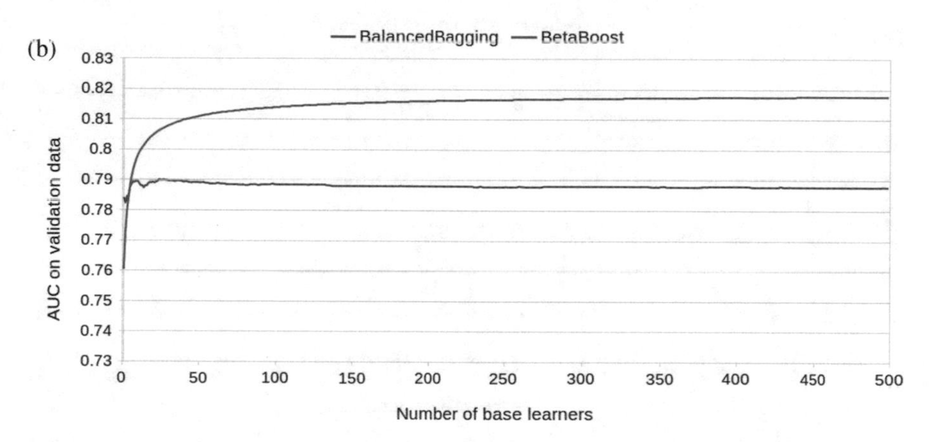

Fig. 21.5 A comparison analysis of *BetaBoost* ($a_0 = 0.8$, $b_0 = 2$, $a_1 = 2$ and $b_1 = 0.8$) and *Balanced Bagging* for the growing number of base learners on *GMSC* dataset. We consider *Logistic Regression* as base learner. *AUC* is taken as quality criterion. (**a**) Training set. (**b**) Validation set

As a second base model we propose to use *Decision Trees*. Due to the fact that this model is recognized as so-called *weak learner*, we propose the following sampling parameters to train the *BetaBoost* models:

- $a_0 = 1.5$, $b_0 = 0.8$, $a_1 = 0.8$ and $b_1 = 1.5$ for *GMSC* dataset,
- $a_0 = 1.2$, $b_0 = 0.8$, $a_1 = 0.8$ and $b_1 = 1.2$ for *LCLD* dataset.

The results are presented in Figs. 21.6 and 21.7. We can see that sampling with replacement using the second strategy ($a_0 \geq 1$, $b_0 \leq 1$ and $a_1 \leq 1$, $b_1 \geq 1$) makes significant improvement of *AUC* criterion comparing to *BetaBoost* strategy, that also uses decision tree as a base learner. We also consider in the analysis the *AdaBoost* classifier that learns the component base model using the similar strategy that increases the impact of "hard examples," decreasing the significance of well-predicted instances. The *AdaBoost* model needs more iterations

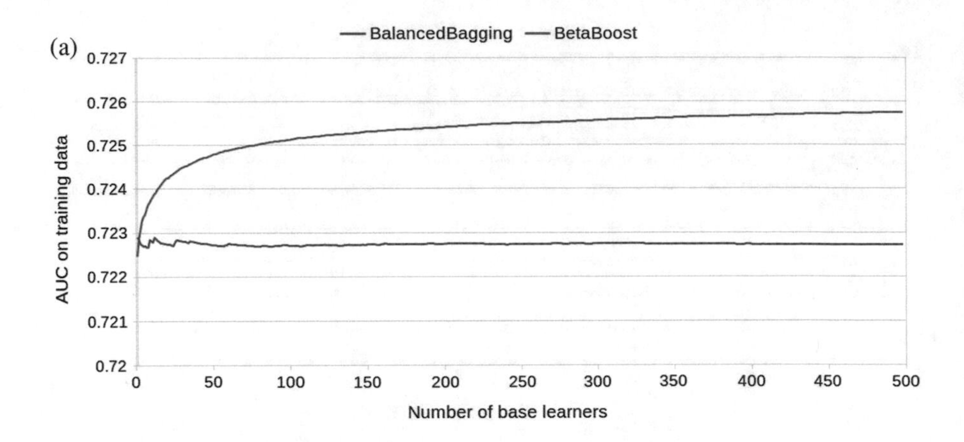

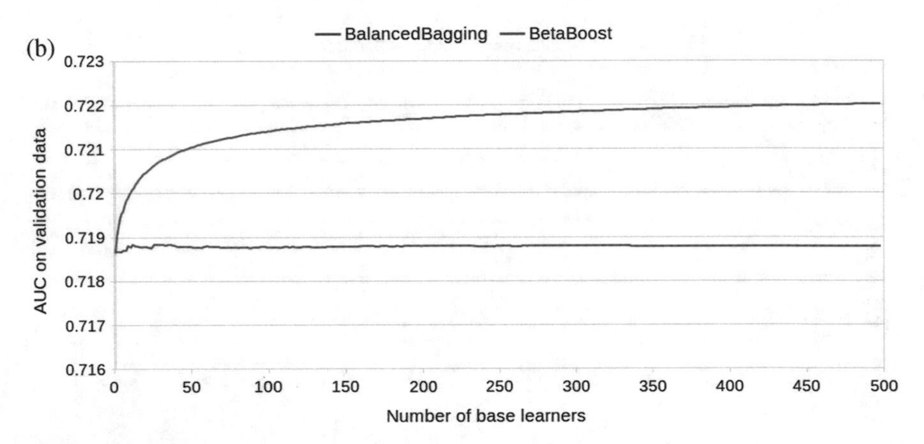

Fig. 21.6 A comparison analysis of *BetaBoost* ($a_0 = 0.8$, $b_0 = 2$, $a_1 = 2$ and $b_1 = 0.8$) and *Balanced Bagging* for the growing number of base learners on *LCLD* dataset. We consider *Logistic Regression* as base learner. *AUC* is taken as quality criterion. (**a**) Training set. (**b**) Validation set

to achieve acceptable *AUC* level because both datasets are spoiled by imbalanced data phenomenon. The performance of *AdaBoost* is similar to *BetaBoost* on *GMSC* dataset, but on *LCLD* dataset it gives significantly worse results. We also present the results on validation data to show that overfitting problem is not observed for the considered models (Fig. 21.8).

We presented the final results obtained by the considered models in Fig. 21.9 (*GMSC* dataset) and Fig. 21.10 *LCLD* dataset. The considered models are as follows:

- **BetaBoostL**. *BetaBoost* with *Logistic Regression* as base learner trained with the first strategy ($a_0 \leq 1$, $b_0 \geq 1$ and $a_1 \geq 1$, $b_1 \leq 1$).
- **BalBagL**. *Balanced Bagging* with *Logistic Regression* as a base learner.

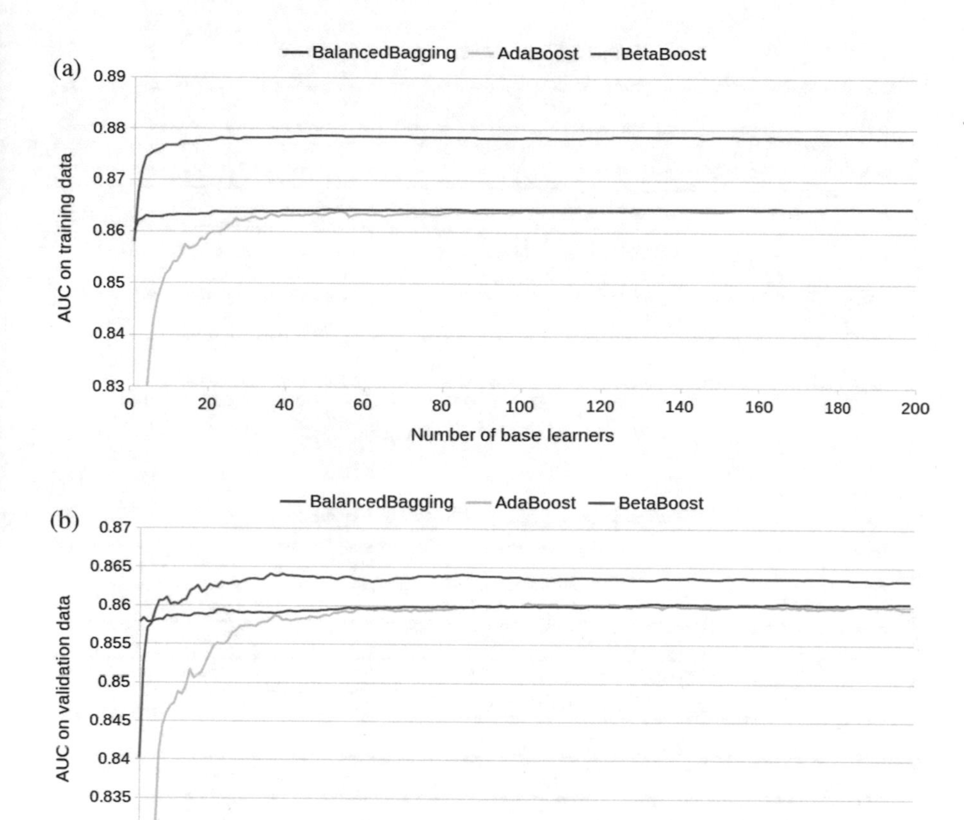

Fig. 21.7 A comparison analysis of *BetaBoost* ($a_0 = 1.5, b_0 = 0.8, a_1 = 0.8$ and $b_1 = 1.5$) and *Balanced Bagging* for the growing number of base learners on *GMSC* dataset. We consider *Logistic Regression* as base learner. *AUC* is taken as quality criterion. (**a**) Training set. (**b**) Validation set

- **BetaBoostDT**. *BetaBoost* with a decision tree as a base learner trained with the second strategy ($a_0 \geq 1, b_0 \leq 1$ and $a_1 \leq 1, b_1 \geq 1$).
- **BalBagDT**. *Balanced Bagging* with a decision tree as a base learner.
- **AdaBoost**. *AdaBoost* classifier with a decision tree as a base learner.

It can be observed that the *BetaBoost* with decision tree as a base learner train with the second strategy performed better than the reference approaches considered in the experiments. On (*GMSC* dataset) we observed only slight increase in quality of *BetaBoost* comparing to *Balanced Bagging* from 0.8652 to 0.8673. However, we operate on *big data*, so the slight improvements in quality criterion may have great impact on financial benefit. The improvement observed on the *LCLD* dataset is indisputable.

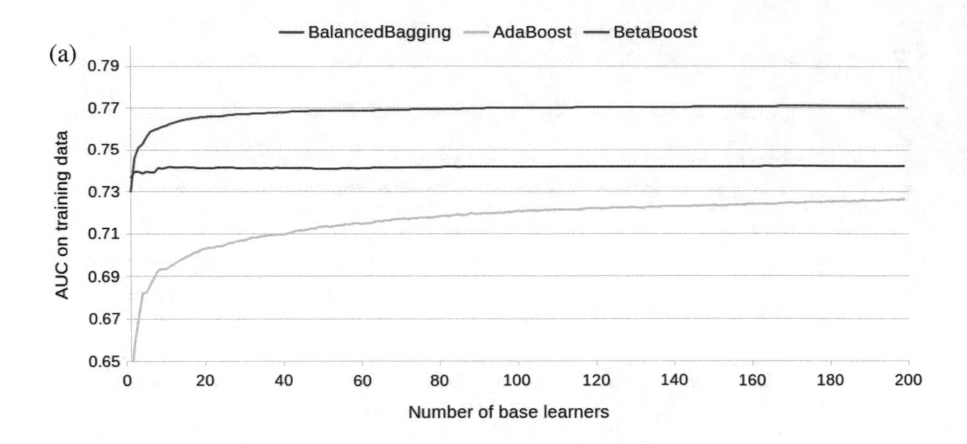

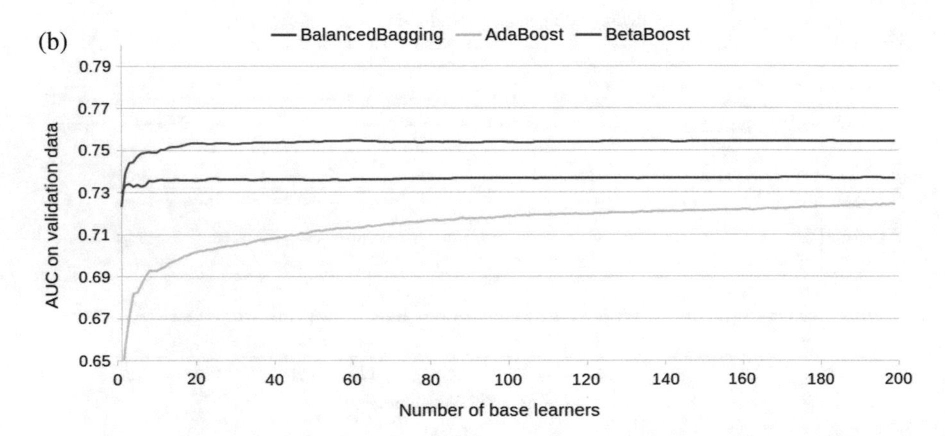

Fig. 21.8 A comparison analysis of *BetaBoost* ($a_0 = 1.2$, $b_0 = 0.8$, $a_1 = 0.8$, and $b_1 = 1.2$), *Balanced Bagging* and *AdaBoost* for the growing number of base learners on *LCLD* dataset. We consider *Decision Tree* as base learner. *AUC* is taken as quality criterion. (**a**) Training set. (**b**) Validation set

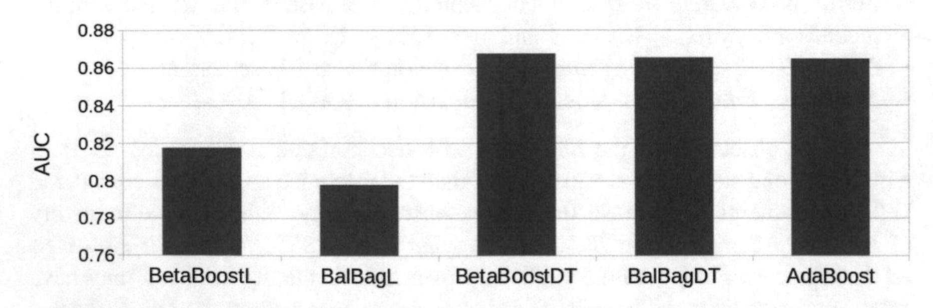

Fig. 21.9 Final results for considered models—*GMSC* dataset (test data)

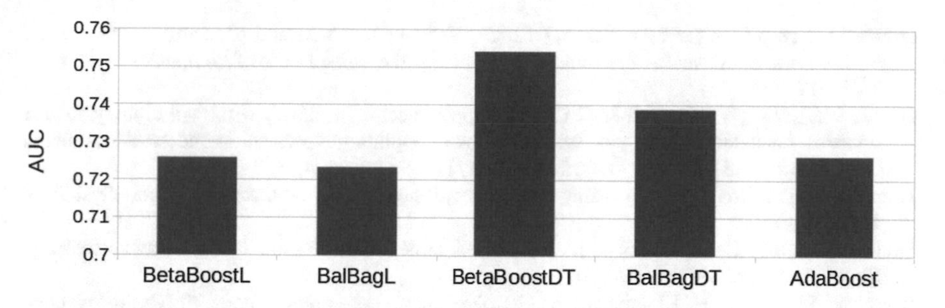

Fig. 21.10 Final results for considered models—*LCLD* dataset (test data)

21.4 Conclusion and Future Work

In this work we propose alternative ensemble based strategy that makes use of beta binomial sampling to create the base models. Two strategies can be distinguished while taking the sampling distribution. In the first strategy we aim at putting higher impact on "easy examples," we bestow trust on the base model and do not trust in data quality. In the second strategy we take rather weak and unstable base model and we put the higher impact on training "hard examples."

Contrary to existing approaches like *AdaBoost*, we update the sampling distribution basing only on individual ranking for each of the classes. As a consequence, the impact of noisy examples in training data is not high.

The crucial step for the proposed *BetaBoost* model is to find proper parameters for sampling distributions. It can be performed by grid search, but this approach is ineffective for large data sets. In the future works we plan to propose the smart model selection approach to solve that issue. Additionally, we are going to perform more formal discussion of the properties of the proposed model. Moreover, the weighted variant of ensemble model is going to be proposed.

References

Abellán J, Mantas CJ (2014) Improving experimental studies about ensembles of classifiers for bankruptcy prediction and credit scoring. Expert Syst Appl 41(8):3825–3830

Bellotti T, Crook J (2009) Support vector machines for credit scoring and discovery of significant features. Expert Syst Appl 36(2):3302–3308

Chen S, Härdle WK, Jeong K (2010) Forecasting volatility with support vector machine-based GARCH model. J Forecast 29(4):406–433

Chen S, Härdle W, Moro R (2011) Modeling default risk with support vector machines. Quant Finan 11(1):135–154

Freund Y, Iyer R, Schapire RE, Singer Y (2003) An efficient boosting algorithm for combining preferences. J Mach Learn Res 4(Nov):933–969

Give Me Some Credit (2011) Give me some credit. https://www.kaggle.com/c/GiveMeSomeCredit

Härdle W, Lee YJ, Schäfer D, Yeh YR (2009) Variable selection and oversampling in the use of smooth support vector machines for predicting the default risk of companies. J Forecast 28(6):512–534

Härdle WK, Prastyo DD, Hafner C (2012) Support vector machines with evolutionary feature selection for default prediction. In: Handbook of applied nonparametric and semi-parametric econometrics and statistics. Oxford University Press, Oxford, pp 346–373

Harris T (2015) Credit scoring using the clustered support vector machine. Expert Syst Appl 42(2):741–750

Huang SC (2011) Using Gaussian process based kernel classifiers for credit rating forecasting. Expert Syst Appl 38(7):8607–8611

Koutanaei FN, Sajedi H, Khanbabaei M (2015) A hybrid data mining model of feature selection algorithms and ensemble learning classifiers for credit scoring. J Retail Consum Serv 27:11–23

Kumar MP, Packer B, Koller D (2010) Self-paced learning for latent variable models. In: Advances in neural information processing systems. MIT Press, Cambridge, pp 1189–1197

Lee TS, Chiu CC, Lu CJ, Chen IF (2002) Credit scoring using the hybrid neural discriminant technique. Expert Syst Appl 23(3):245–254

Lending Club (2016) Lending club loan data. https://www.kaggle.com/wendykan/lending-club-loan-data

Marqués A, García V, Sánchez JS (2012) Two-level classifier ensembles for credit risk assessment. Expert Syst Appl 39(12):10916–10922

Martens D, Baesens B, Van Gestel T, Vanthienen J (2007) Comprehensible credit scoring models using rule extraction from support vector machines. Eur J Oper Res 183(3):1466–1476

Nanni L, Lumini A (2009) An experimental comparison of ensemble of classifiers for bankruptcy prediction and credit scoring. Expert Syst Appl 36(2):3028–3033

Oreski S, Oreski D, Oreski G (2012) Hybrid system with genetic algorithm and artificial neural networks and its application to retail credit risk assessment. Expert Syst Appl 39(16):12605–12617

Rudin C, Schapire RE (2009) Margin-based ranking and an equivalence between AdaBoost and RankBoost. J Mach Learn Res 10(Oct):2193–2232

Tomczak JM, Zięba M (2015) Classification restricted Boltzmann machine for comprehensible credit scoring model. Expert Syst Appl 42(4):1789–1796

Tsai CF, Wu JW (2008) Using neural network ensembles for bankruptcy prediction and credit scoring. Expert Syst Appl 34(4):2639–2649

Zhao Z, Xu S, Kang BH, Kabir MMJ, Liu Y, Wasinger R (2015) Investigation and improvement of multi-layer perceptron neural networks for credit scoring. Expert Syst Appl 42(7):3508–3516

Zhou L, Lai KK, Yen J (2009) Credit scoring models with AUC maximization based on weighted SVM. Int J Inf Technol Decis Mak 8(04):677–696

Zhu Y, Xie C, Wang GJ, Yan XG (2016) Comparison of individual, ensemble and integrated ensemble machine learning methods to predict China's SME credit risk in supply chain finance. Neural Comput Appl 28:1–10

Zięba M, Świątek J (2012) Ensemble classifier for solving credit scoring problems. In: Doctoral conference on computing, electrical and industrial systems. Springer, Berlin, pp 59–66

Zięba M, Tomczak JM (2015) Boosted SVM with active learning strategy for imbalanced data. Soft Comput 19(12):3357–3368

Zięba M, Tomczak SK, Tomczak JM (2016) Ensemble boosted trees with synthetic features generation in application to bankruptcy prediction. Expert Syst Appl 58:93–101

Printed in the United States
By Bookmasters